The Garnaut Climate Change Review

Final Report

Ross Garnaut is one of Australia's most distinguished and well-known economists. He is an Officer of the Order of Australia for services to education and international relations, a Fellow of the Australian Academy of Social Sciences, and Honorary Professor of the Chinese Academy of Social Science. Based in the Research School of Pacific and Asian Studies at the Australian National University from 1972 to 2008, he was Professor of Economics from 1989 to 2008, and head of the economics department for much of that time. Professor Garnaut is now Vice-Chancellor's Fellow and Professorial Fellow in Economics at Melbourne University and Distinguished Professor at the Australian National University. He is Chairman of the International Food Policy Research Institute of Washington DC, Chairman of Lihir Gold Limited, Chairman of Papua New Guinea Sustainable Development Limited, Chairman of the editorial boards of the academic journals *Asian Pacific Economic Literature* and the *Bulletin of Indonesian Economic Studies*, and a Director of the Lowy Institute for International Policy.

Professor Garnaut has had periods of secondment to senior positions in government. He was head of financial and economic policy in Papua New Guinea's Department of Finance in the years straddling independence in the mid-1970s, the principal economic adviser to Australian Prime Minister R.J.L. Hawke in the 1980s, and Australian Ambassador to China from 1985 to 1988.

The many books of which he has been author, co-author or editor include *The Taxation of Mineral Rent; Australian Protectionism*; *Australia and the Northeast Asian Ascendancy*; *The Third Revolution in the Chinese Countryside*; *Social Democracy in Australia's Asian Future;* and *China's Ownership Transformation.*

The Garnaut Climate Change Review

Final Report

Ross Garnaut

University Printing House, Cambridge CB2 8BS, United Kingdom

One Liberty Plaza, 20th Floor, New York, NY 10006, USA

477 Williamstown Road, Port Melbourne, VIC 3207, Australia

314-321, 3rd Floor, Plot 3, Splendor Forum, Jasola District Centre, New Delhi - 110025, India

79 Anson Road, #06-04/06, Singapore 079906

Cambridge University Press is part of the University of Cambridge.

It furthers the University's mission by disseminating knowledge in the pursuit of education, learning and research at the highest international levels of excellence.

www.cambridge.org
Information on this title: www.cambridge.org/9780521744447

First published 2008

Editorial and artwork by WHH Publishing
Index by Trevor Matthews

A catalogue record for this publication is available from the British Library

National Library of Australia Cataloging-in-Publication entry

Author:	Garnaut, Ross.
Title:	The Garnaut climate change review/Ross Garnaut.
ISBN:	9780521744447 (pbk.)
Notes:	Includes index.
Subjects:	Climatic changes—Economic aspects—Australia Power resources—Economic aspects—Australia Greenhouse effect, Atmospheric—Australia—Economic aspects Carbon taxes—Australia Greenhouse gas mitigation—Australia Sustainable development—Australia.
Dewey number:	363.738740994
Front cover image:	Sidney Nolan, *Rainbow over Pilbara*, 1982 Enamel spray on canvas 121 x 152 cm Private collection © The Trustees of the Sidney Nolan Trust/The Bridgeman Art Library
Back cover image:	Sidney Nolan, *Storm over Pilbara*, 1982 Enamel spray on canvas 122 x 152.5 cm Private collection © The Trustees of the Sidney Nolan Trust/The Bridgeman Art Library

ISBN 978-0-521-74444-7 Paperback

Contents

Preface xiii

Acknowledgments xiv

Terms of reference xvi

Introduction xvii

Synopsis of key points xxxv

1 A decision-making framework 1

1.1 The costs of mitigation 3

1.2 Risk and uncertainty 7

1.3 Four types of benefits from mitigation 9

1.4 How effective adaptation reduces the costs of climate change 13

1.5 Measuring the benefits of mitigation against the costs 14

1.6 A graphical representation of the benefits and costs 15

1.7 Valuing the future relative to the present 18

2 Understanding climate science 23

2.1 The earth's atmosphere 24

2.2 Understanding climate change 27

2.3 Linking emissions and climate change 30

2.4 The task of global mitigation 42

3 Emissions in the Platinum Age 53

3.1 Greenhouse gas emissions by source and country 53

3.2 Recent trends in carbon dioxide emissions from fossil fuels 55

3.3 Existing emissions projections 58

3.4 The Review's no-mitigation projections: methodology and assumptions 59

3.5 Results from the Review's projections and comparisons with existing projections 62

3.6 The impact of high energy prices 67

3.7 Resource limits 69

4 Projecting global climate change 75

4.1 How has the climate changed? 75

4.2 Understanding climate change projections 83

4.3 Projected climate change for the three emissions cases 87

4.4 Assessing the climate risk 96

5 Projecting Australian climate change 105

5.1 Attributing climate change to humans 106

5.2 How has the climate changed in Australia? 106

5.3 Projected climate change in Australia 113

6 Climate change impacts on Australia 121

6.1 Understanding Australia's vulnerability to climate change 124

6.2 Australia without global mitigation 125

6.3 Direct impacts of climate change on Australia 128

6.4 Indirect impacts of climate change on Australia 145

7 Australia's emissions in a global context 153

7.1 Australia's emissions profile and international comparisons 153

7.2 Emissions profiles of Australian industries 165

8 Assessing the international response 173

8.1 The evolving international framework for addressing climate change 174

8.2 National commitments and policies to mitigate climate change 177

8.3 Assessment of progress under the Kyoto Protocol 180

8.4	Projections given the current trajectory of mitigation effort	183
8.5	Accelerating progress	184
9	Towards global agreement	191
9.1	Agreeing on a global goal	192
9.2	What form should national commitments take?	195
9.3	A graduated approach to national commitments	198
9.4	Principles for allocating emissions entitlements across countries	200
9.5	Modelling a per capita approach to the allocation of emissions entitlements	205
9.6	Reaching agreement on 550 or 450: is it possible?	212
10	Deepening global collaboration	217
10.1	International public funding for mitigation	218
10.2	International public funding for adaptation	223
10.3	Promoting collaborative research to assist developing countries	226
10.4	International trade in emissions rights	227
10.5	Price-based sectoral agreements for the trade-exposed, emissions-intensive sectors	230
10.6	Climate change and trade policy	232
10.7	International aviation and shipping	234
10.8	Land-use change and forestry	235
10.9	Enforcement mechanisms	239
11	Costing climate change and its avoidance	245
11.1	The three global scenarios	246
11.2	Comparing the costs of climate change and mitigation	247
11.3	Modelling mitigation	250
11.4	The decision to mitigate	252
11.5	How much mitigation?	268

12 Targets and trajectories 277

12.1 Determining our conditional and unconditional targets 278

12.2 The benefits of global cooperation 285

12.3 Solving a diabolical problem in stages 287

12.4 Hastening progress towards greater emissions reductions 289

12.5 Moving from a 550 to a 450 goal 290

12.6 Does Australia matter for global mitigation? 291

12.7 Interim targets 294

12.8 Implications for an Australian emissions trading scheme 298

13 An Australian policy framework 299

13.1 Confronting uncertainty: the policy challenges of climate change 300

13.2 Avoiding the greatest market failure ever seen 307

13.3 Bungling Australia's emissions trading scheme 314

14 An Australian emissions trading scheme 321

14.1 The framework to guide efficient scheme design 322

14.2 Elemental design features 324

14.3 Releasing permits into the market 330

14.4 Lowering the costs of meeting targets 334

14.5 Addressing the distortion faced by trade-exposed, emissions-intensive industries 341

14.6 Transition period: Australia's emissions trading scheme to the end of 2012 350

14.7 Governance: institutional arrangements 351

14.8 Addressing the relationships between an emissions trading scheme and other policies 353

14.9 Summary of design features of an Australian emissions trading scheme 357

15 Adaptation and mitigation measures for Australia 363
15.1 Information and understanding 365
15.2 The role of markets and market-based policies 370
15.3 Scaling the challenges: five examples 376

16 Sharing the burden in Australia 385
16.1 Effects of mitigation policy in the short term 386
16.2 A framework for government intervention 393
16.3 Long-term impacts and structural change 400

17 Information barriers to known technologies 403
17.1 The impact of information and agency barriers 404
17.2 Information barriers 406
17.3 Principal–agent problems 413
17.4 Minimum performance standards 415

18 The innovation challenge 423
18.1 What is innovation? 424
18.2 Ensuring optimal levels of early research 428
18.3 Rewarding early movers 433
18.4 Overcoming barriers from technological lock-in 441

19 Network infrastructure 445
19.1 The transmission of electricity 446
19.2 The distribution of electricity 451
19.3 Gas transmission infrastructure 453
19.4 The transportation of carbon dioxide 453
19.5 The transport of passengers and freight 455
19.6 Water supply infrastructure 458
19.7 The planning of urban settlements 460

20 Transforming energy 467
20.1 The energy sector today 468
20.2 Drivers of the transformation 472
20.3 The transformation 478
20.4 Modelling results for the energy sector 482
20.5 Major economic impacts 490
20.6 Risks to the transformation 499

21 Transforming transport 503
21.1 The role of transport and its current structure 504
21.2 Causes of the transformation 505
21.3 Economic modelling results: a possible future? 511
21.4 The path to transformation: a picture of future transport 517
21.5 Fostering the transformation 526

22 Transforming rural land use 531
22.1 Drivers of a transformation towards lower emissions 532
22.2 Economic modelling results: a possible future? 537
22.3 An alternative future 542
22.4 Barriers and limits to a low-emissions future 558

23 Towards a low-emissions economy 565
23.1 The dynamics of economic adjustment with global mitigation 566
23.2 The economy to and at 550 ppm 570
23.3 The difference between 550 and 450 575
23.4 Australia in the low-emissions world energy economy 576
23.5 The downside risks 579
23.6 The upside in technology assumptions 580

23.7 The importance of flexible global and national markets 584

23.8 The importance of education and training 586

23.9 Global mitigation and ongoing prosperity 587

23.10 Australia in a successful world of change 588

24 Fateful decisions 591

List of figures and tables 599

List of shortened forms 606

Glossary 608

Index 617

Preface

The Garnaut Climate Change Review was initiated in April 2007 by the then Leader of the Opposition, Kevin Rudd, and by the Premiers of the six states and the Chief Ministers of the two territories of Australia. It was commissioned by the First Ministers on 30 April 2007. The Commonwealth Government joined the Review in January 2008 after Mr Rudd became Prime Minister of Australia.

The Review was required to examine the impacts of climate change on the Australian economy, and to recommend medium- to long-term policies and policy frameworks to improve the prospects of sustainable prosperity.

The Review's secretariat was established in June 2007. Based within the Victorian Department of Premier and Cabinet, it included members from the public services of Queensland, Western Australia and South Australia. A secretariat office within the federal Department of Climate Change was set up in January 2008.

As part of its research and analysis, the Review consulted with a wide range of experts and stakeholders in Australia and overseas: academics, officials, government departments and public bodies, business leaders and representatives, and non-government organisations.

The Review produced four major documents as a basis for public discussion. An interim report was presented to First Ministers and released in February 2008; a discussion paper on the proposed emissions trading scheme in March, a draft report on 4 July and a supplementary draft report on 5 September.

The Review commissioned a number of papers on the impacts of climate change on Australia, which represent major contributions to the growing body of knowledge about these impacts. The papers are available on the Review's website at <www.garnautreview.org.au>.

The methodology applied in, and the results of, the Review's modelling have generated large volumes of analysis and information, of which this final report presents only a small proportion. A technical appendix to the report on the modelling is available on the website.

The Review has benefited substantially from interactions with other organisations and the community more generally at specialist forums, and public forums and lectures held around the country between August 2007 and September 2008. More than 10 000 people participated in these events over the course of the Review.

A formal submission process was also conducted, which attracted almost 4000 submissions. Interested stakeholders were encouraged to respond to a series of five issues papers, the discussion paper on the emissions trading scheme and the interim report, all of which stimulated considerable public discussion and debate on climate change mitigation and adaptation in Australia.

This final report is the last stage of a wide-ranging process that has transparently examined how Australia, as a single country, is likely to be affected by climate change, and how we can best contribute to climate change mitigation and start to adapt.

Acknowledgments

The Review's work, with its scale and complexity, could only be completed through the generous contributions of many people and organisations in Australia and across the world.

I had the good fortune of being supported by a management team led by Ron Ben-David as head of the secretariat, and Ian de Cruz, Kevin Keeffe and Stephen Howes. I am grateful that Ron took on the job of helping me get it all together on the day that I was given my presumptuous task. We have worked together on it every day since. Ian led the states-based team in Melbourne from the beginning in late June 2007 and Kevin the Commonwealth component of the secretariat from the time of its establishment in January 2008. Stephen held together the international work from his base at the Australian National University. The end product has been the beneficiary of their exceptional skills, capacities and professionalism.

Thank you to the team members who have been with me from the outset: Jonathan Chew, Elizabeth Edye, Ana Markulev, Nina Rogers and Karlie Tucker; and to our team of expert modellers from the Queensland Treasury, led by Greg Watts and Matt Clark.

It has been a long and hard road, but we have reached the place of which we spoke and maybe dreamed more than a year ago.

Tony Wood added valuable knowledge of technologies and much else when he joined me as a private adviser. Frank Jotzo introduced me to areas of economics in which he was an old hand and I was not.

From earlier this year, the team was strengthened with great professionalism by Dominic Burke, Alison Carrington, Clare Chick, Peter Corcoran, Charles Edlington, Kylie Meakins, Helen Morrow, Rob Murray-Leach and Claire Ruedin. During this time Daniel Adams, James Allen, Jacqueline Boreham, Conrad Buffier, Karen Dempsey, Darren Gladman, Lana Kelly and Kirsten Mann have all made valuable contributions to the Review. Anna Freeman managed to keep us in touch with a high proportion of the extraordinary number of Australians and others who were interested in the Review's work. Veronica Webster held together the threads of my absurd program of commitments against all the odds.

Many of the individuals mentioned above, and some others, made exceptional and particular intellectual and other contributions to the work. It would be invidious of me to single out those contributions here, but I will find ways of acknowledging these exceptional contributions in more personal ways.

My thanks go to several Commonwealth and state government agencies. The secretariat began its work in the Victorian Department of Premier and Cabinet, where the former Secretary, Terry Moran, made sure that things started strongly, and the current Secretary, Helen Silver, continued the excellent support. The Review became a joint Commonwealth–states project from early 2008, with the Commonwealth Department of Climate Change and its Secretary, Martin Parkinson, as its central point. The Australian Bureau of Agricultural and Resource

Economics provided strong professional support for the work from the beginning in 2007, with special contributions from the modelling team led by Don Gunasekera, ably supported by Helal Ahammad. The Review's joint work on modelling with the Australian Treasury team, led by David Gruen and Meghan Quinn, has broken fertile ground that will be cultivated by Australians for many years. Thanks also to the Office of Economic and Statistical Research in the Queensland Treasury for its ongoing support through the dedication of staff to the Review. The Australian Academy of Science and Academy of Technological Science and Engineering made significant contributions.

I am grateful for the assistance, services and information throughout the Review provided by the staff at the Australian Bureau of Statistics (particularly Steve Gelsi and Sandra Waters); Roger Jones and his many colleagues at the Commonwealth Scientific and Industrial Research Organisation; the Bureau of Meteorology; the Productivity Commission; Philip Adams at the Centre of Policy Studies (Monash University), who made large contributions directly and through the joint modelling with Treasury; the Melbourne Institute of Applied Economic and Social Research at Melbourne University; and WHH Publishing (especially but not only Virginia Wilton and Larissa Joseph).

Over the last 16 months, my team and I have held thousands of conversations with leaders in their fields in Australia and internationally, and received countless emails and letters offering advice, support and criticism. It is simply not possible to acknowledge all of these contributions. But many have been instrumental in informing the ideas and proposals presented in this Report, and some have been immensely important. Many of you will recognise your influence.

To Jayne, and to the partners and families of all members of my team, I thank you for your patience over many months, and your willingness to bear the demands of this project.

Ross Garnaut
Canberra and Melbourne
30 September 2008

Terms of reference

30 April 2007

To report to the Governments of the eight States and Territories of Australia, and if invited to do so, to the Prime Minister of Australia, on:

1. The likely effect of human induced climate change on Australia's economy, environment, and water resources in the absence of effective national and international efforts to substantially cut greenhouse gas emissions;
2. The possible ameliorating effects of international policy reform on climate change, and the costs and benefits of various international and Australian policy interventions on Australian economic activity;
3. The role that Australia can play in the development and implementation of effective international policies on climate change; and
4. In the light of 1 to 3, recommend medium to long-term policy options for Australia, and the time path for their implementation which, taking the costs and benefits of domestic and international policies on climate change into account, will produce the best possible outcomes for Australia.

In making these recommendations, the Review will consider policies that: mitigate climate change, reduce the costs of adjustment to climate change (including through the acceleration of technological change in supply and use of energy), and reduce any adverse effects of climate change and mitigating policy responses on Australian incomes.

This Review should take into account the following core factors:

- The regional, sectoral and distributional implications of climate change and policies to mitigate climate change;
- The economic and strategic opportunities for Australia from playing a leading role in our region's shift to a more carbon-efficient economy, including the potential for Australia to become a regional hub for the technologies and industries associated with global movement to low carbon emissions; and
- The costs and benefits of Australia taking significant action to mitigate climate change ahead of competitor nations; and
- The weight of scientific opinion that developed countries need to reduce their greenhouse gas emissions by 60 percent by 2050 against 2000 emission levels, if global greenhouse gas concentrations in the atmosphere are to be stabilised to between 450 and 550 ppm by mid century.

Consult with key stakeholders to understand views and inform analysis. A draft Report is to be distributed for comment by June 30 2008. The final Report is to be completed and published by September 30 2008. Interim draft reports on particular issues may be released before that time for public discussion. The Report will embody the independent judgments of its author.

INTRODUCTION

The weight of scientific evidence tells us that Australians are facing risks of damaging climate change.

The risk can be substantially reduced by strong, effective and early action by all major economies. Australia will need to play its full proportionate part in global action. As one of the developed countries, its full part will be relatively large, and involve major early changes to established economic structure.

The work of the Review shows that the costs of Australia playing its proportionate part in an effective global effort, while considerable, are manageable. There is a path to Australia being a low-emissions economy by the middle of the 21st century, consistently with continuing strong growth in material living standards (chapters 11 and 23). By the end of the 21st century, and beyond, more so with each passing decade material living standards would be higher with than without mitigation of climate change.

Scientific opinion and dissent

There is no doubt about the position of most reputed specialists in climate science, in Australia and abroad, on the risks of climate change (Chapter 2). There is no doubt about the position of the leaders of the relevant science academies in all of the major countries.[1] The outsider to climate science has no rational choice but to accept that, on a balance of probabilities, the mainstream science is right in pointing to high risks from unmitigated climate change.

There are nevertheless large uncertainties in the science. There is debate and recognition of limits to knowledge about the times and ways in which the risk will manifest itself. Every climate scientist has views on some issues that differ from the mainstream in detail.

There are prominent dissenters on this matter, gathered under the rubric of 'sceptic'. For the most part 'sceptic' is a misnomer for their position, because these dissenters hold strongly to the belief that the mainstream science is wrong.

In a different category are a small number of climate scientists of professional repute who maintain that the mainstream science embodies misjudgments about quantities. These scientists, who accept the theory of the warming effects of higher concentrations of greenhouse gases, hold the view that these warming effects are relatively or even trivially small in comparison with many other causes of climate variations that are beyond the control of humans.

The dissent took a curious turn in Australia in 2008, with much prominence being given to assertions that the warming trend had ended over the last decade. This is a question that is amenable to statistical analysis, and we asked econometricians with expertise in analysis of time series to examine it. Their response—that the temperatures recorded in most of the last decade lie above the confidence level produced by any model that does not allow for a warming trend—is reported in Chapter 4 (Box 4.1).

The prisoner's dilemma in international collective action

Effective international action is necessary if the risks of dangerous climate change are to be held to acceptable levels, but deeply problematic. International cooperation is essential for a solution to a global problem. However, such a solution requires the resolution of a genuine prisoner's dilemma: each country benefits from a national point of view if it does less of the mitigation itself, and others do more. If all countries act on this basis, without forethought, communication and cooperation, there will be no resolution of the dilemma. Future generations will judge the outcome to have been insufficient and unsatisfactory.

Resolution of the prisoner's dilemma requires communication, to find a division of costs and benefits of cooperation that is acceptable to all essential participants in a solution. The eventual solution can involve a range of cooperative arrangements, and not only matters related to the division of the mitigation task.

But resolution of the international prisoner's dilemma will take time—possibly more time than we have. The world squandered the time that it did have in the 1990s to experiment with various approaches to mitigation.

A diabolical problem and a saving grace

Climate change is a diabolical policy problem. It is harder than any other issue of high importance that has come before our polity in living memory. Climate change presents a new kind of challenge. It is uncertain in its form and extent, rather than drawn in clear lines. It is insidious rather than (as yet) directly confrontational. It is long term rather than immediate, in both its impacts and its remedies. Any effective remedies lie beyond any act of national will, requiring international cooperation of unprecedented dimension and complexity. While an effective response to the challenge would play out over many decades, it must take shape and be put in place over the next few years. Without such action, if the mainstream science is broadly right, the Review's assessment of likely growth in global greenhouse gas emissions in the absence of effective mitigation tells us that the risks of dangerous climate change, already significant, will soon have risen to dangerously high levels.

Observation of daily debate and media discussion in Australia and elsewhere suggests that this issue might be too hard for rational policy making. It is too complex. The special interests are too numerous, powerful and intense. The time frames within which effects become evident are too long, and the time frames within which action must be effected too short.

But there is a saving grace that may make all the difference. This is an issue in which a high proportion of Australians are deeply interested. A high proportion of Australians say that they are prepared to pay for mitigation in higher goods and services prices. Most of them say that they are prepared to pay even if Australia is acting independently of other countries. There is a much stronger base of support for reform and change on this issue than on any other big question of structural change in recent decades, including trade, tax and public business ownership reform. People in other countries, to varying degrees, seem to share Australians' interest in and preparedness to take action on global warming.

Public attitudes in Australia and in other countries create the possibility of major reform on emissions reductions, despite the inherent difficulty of the policy problem.

This report aims to nurture the chance that Australia and the world will manage to develop a position that strikes a good balance between the costs of dangerous climate change and the costs of mitigation. It does this by examining approaches to mitigation in one country within a framework which, if followed elsewhere, would add up to a solution. The Review recognises that other approaches may also add up to a solution. If others were also to develop proposals that add up to a solution to the problem, that would provide the basis for the type of realistic discussion across the international community that will be essential if a basis is to be found for effective global action.

The Australian economy and the challenge of climate change

Australia has a larger interest in a strong mitigation outcome than other developed countries. We are already a hot and dry country; small variations in climate are more damaging to us than to other developed countries. We live in a region of developing countries, which are in weaker positions to adapt to climate change than wealthy countries with robust political and economic institutions. The problems of our neighbours would inevitably become our problems. And the structure of our economy means that our terms of trade would be damaged more by the effects of climate change than would those of any other developed country (see chapters 11 and 23).

At the same time, Australia carries some major assets into this challenge. Australians are facing this new kind of challenge in the best of times. These are the times that earlier generations of Australians had hoped for their country. Australia is fortunate that humanity is enjoying the harvest of modern economic development in Asia and beyond. More people are emerging from poverty more quickly than ever before in human history. Australia's geographic location and economic structure make it a large beneficiary of these historic developments.

Australia is enjoying a double harvest. The internationally oriented market reforms in Australia from the 1980s were put in place just in time to take advantage of the new opportunities in Asia, and more broadly in the developing world. We are now riding the extension of the beneficent processes of modern economic growth into the heartlands of the populous countries of Asia.

In the early years of our federation Australians took pride in having the highest living standards in the world. On the eve of World War I, Australia's output per person was a bit above that of the United States, then and still the benchmark for economic modernity. Then, for seven decades, we turned in on ourselves, and paid the price. For seven decades, we fell further and further behind the global frontiers of productivity and incomes. The value of our output per person fell to less than two-thirds that of the United States.

Then, a quarter of a century ago, we caught that tide which taken at the flood leads on to fortune. On such a full sea we are now afloat. In recent times, the value of output per person in Australia has again been comparable to that in the United States when both are measured in the national accounts and converted into a common currency at today's exchange rates.

So we have much to contribute and much to lose as we face the diabolical policy challenge of climate change. We would surrender to this challenge if we left climate change unmitigated. We would also surrender if we bungled the attempt to mitigate climate change, which would bring back into the centre of Australian national policy all of the self-interested pressure groups and arbitrary interventions that retarded our progress for so long. It would encumber the international polity with another layer of barriers to and complications of international exchange.

Australians' recent relative economic prosperity has had two direct causes. The first is our decisive rejection and reversal of the mistakes of the early decades after federation at the beginning of the 20th century: our protectionism, xenophobia and bureaucratic trammelling of the market.

The second cause is the Asian economic boom. Australia's resources and human capacities are more closely complementary to those of the densely populated countries of Asia than are those of any other economies on earth. For other developed and many developing countries, the strong growth in industrial production and demand for raw materials and food that accompanies economic growth in China, India, Indonesia and other Asian countries is seen as a competitive and inflationary threat. For Australia, it is an unbridled opportunity. Strong Chinese and other Asian economic growth has been the main factor behind the lift in Australia's terms of trade by about two-thirds over the past six years. This has lifted the average value of Australian output and incomes by more than one-eighth from the effects of increased export prices alone.

The combination of internationally oriented economic reform and opportunities provided by strong growth in the developing countries has so far set Australia apart as the 2008 financial crisis threatens to bring recession to many developed countries.

The increase in concentrations of greenhouse gases in the atmosphere over the last two centuries has generated the climate change that we have experienced to date and will experience over the next couple of decades and beyond. This is the result of economic activity in the countries that are now rich.

The Asian economic boom, half the cause of our prosperity, is also the source of the sharper immediacy of the climate change problem. The rapid increase in concentrations expected over the next several decades is primarily the result of activities that are expected in the developing countries that are becoming rich. The rapid increase in developing country emissions is what makes action to avert dangerous climate change urgent.

The links between Australia's own prosperity and the increase in greenhouse gas emissions in Asian developing countries are rather more direct than the general terms of trade effects would suggest. Fossil fuels have been a major component of increased Australian exports through the Asian boom of the early 21st century. The contribution to the value of Australian exports of the increase in price alone of just one fossil fuel commodity—coal—in 2008–09 is projected to equal about 2 per cent of Australian GDP.

It is neither desirable, nor remotely feasible, to seek to lower the climate change risk by substantially slowing the rise in living standards anywhere, least of all in developing countries. If such an approach were thought to be desirable in some expression of distant and idiosyncratic values, neither Australians, nor people in the developing countries, would accept it. Nor would it be in Australia's interests for Asia's developing countries to accept a dampening of their people's hopes for rising living standards in the interests of climate change mitigation. Their prosperity or its end is translated quickly into our own.

The solutions to the climate change challenge must be found in removing the links between economic activity and greenhouse gas emissions. For Australia, the commitment to the mitigation of climate change can be seen as the reinvestment of a part of the immense gains that have come from accelerated Asian economic growth, in contributing to reduction of an adverse side effect of that growth. In this, we are in a privileged position. We are different from most other countries, and certainly from all other developed countries except Norway.

Vested interests in the public policy process

These realities need to be kept in mind if we are to retain perspective in the domestic debate about mitigation and the introduction of an emissions trading scheme. Some elements of the Australian resources sector have expressed concern about the threat that a price on carbon poses to their competitiveness and to Australian prosperity. Our trade-exposed, emissions-intensive industries have valid concerns. The Review has acknowledged these from the beginning, and sought to accommodate them in its proposals for emissions trading scheme design. The Review proposes arrangements that deal with the valid concerns, without getting in the way of Australia's efficient transition to a low-emissions economy (Chapter 14). Along with some of our farm industries, metals processing would be the most affected, and have the first claims for assistance.

Every element of costs matters, and no increase in costs should be imposed on business without good reason. But when assessments of the reasonableness of arrangements for trade-exposed industries are made, we should be mindful of the wider context. The highest possible obligations under an emissions trading scheme, at the top end of the range of possibilities for permit prices for the foreseeable future, would represent a small fraction of the resource sector's increased revenue from higher export prices in recent years.

It is only to be expected that each firm, industry and sector will argue its own case in its own interests. Senior corporate executives are paid to do exactly that.

But in taking these arguments into the national debate, we must make sure that there is also a strong and independent voice for the public interest in the policy-making process that can keep sectoral claims in perspective.

Public policy in the national interest

The public interest must rely on the clarity of the analysis of the issues, on the dissemination of sound information, and in the end on the judgment of a public that is interested enough in the issue to make the effort to use effectively the information available to it.

Balance, reason and understanding of the premises, information and logic leading to policy conclusions are the keys to Australia and the world using well its last chance to get this difficult policy problem right. The Review's first aim is to lay out the issues for policy choice in a transparent way. The Review will have done its job if Australian governments and the community make their choices in full knowledge of the consequences of their decisions. It is hoped that the approach and analysis communicated in this final report will be helpful to developing transparent approaches to decisions in other countries as well.

No answers to questions as complex and difficult as those introduced here and discussed at length through the report, would seem right, or palatable, to everyone. Perhaps no answers, with their many parts, would seem right or palatable to anyone. Many will disagree with elements or the whole of the conclusions of the Review. Many will disagree with the policy proposals that flow from the conclusions. They would prefer cheaper, more certain, later and less disruptive approaches to reductions in emissions, or higher levels and urgency of mitigation ambition.

The Review would prefer cheaper, more certain, later and less disruptive ways forward—if any were available that were not associated with large risks of damage from climate change. Tempting though it is to do so, it is neither rational nor helpful to reject conclusions because we do not like them. The conclusions will only be wrong if the premises, information or logic leading to them are wrong. The Review has sought to be clear in its premises, information and methodology, so that they can be contested transparently. If the subsequent public policy debate follows these lines, we will improve the chances of Australian and other governments taking good decisions in the year ahead on a sound basis and with widespread community support, and therefore with prospects of policy continuity.

On no issue will this be more important than on the targets and trajectories for Australian mitigation. There has been considerable comment, from people concerned to solve the climate change problem, since the Review's proposals on targets and trajectories were released for discussion in the supplementary draft report on 5 September 2008. The material in this report allows discussion to move beyond expressions of like or dislike for conclusions, into identification of premises, information or logic that may require debate and, if flaws are identified, modification.

Measuring the costs and benefits of mitigation

The Review examines analytically whether and how much mitigation is justified. We do this by comparing the costs of mitigation with the benefits of climate change avoided by mitigation.

The assessment of whether and how much mitigation is worthwhile takes the reader on a long conceptual and quantitative journey. The methodology is new, and can be applied to other countries. Many of the numbers and other facts are specific to Australia.

Mitigation at a given level is justified if the benefits exceed the costs. If the benefits exceed the costs for more than one level of mitigation, the appropriate level is the one that generates the largest excess of benefits over costs. The conceptual framework for assessing whether various levels of mitigation are justified is set out in Chapter 1.

The costs of mitigation are felt through standard economic processes, and can be calculated through computable general equilibrium models.

The benefits of mitigation are the avoided costs of climate change. The assessment of the benefits of mitigation begins with analysis of the costs of climate change with no mitigation and at various levels of mitigation. The benefit of any extent of mitigation is the difference between the costs of climate change with no mitigation, and at the specified extent of mitigation.

There are four types of costs of climate change—and therefore four types of benefits from avoided climate change. Two of these take the form of standard economic costs, felt through markets, from mid-points in the probability distributions of climate change impacts that are derived from mainstream science. These two can, in principle, be assessed quantitatively through general equilibrium models. However, data in the precise form necessary for quantitative analysis through general equilibrium modelling are available for only one of these, Type 1. Type 2 costs of climate change, comprising standard economic impacts for which data are not available in a form that is sufficiently precise for modelling, have to be estimated. Type 3 costs of climate change (and benefits of mitigation) comprise the special and additional costs (value of avoiding) of extreme outcomes, and can be considered as insurance value. Type 4 benefits of mitigation comprise the conservation value related to all of the non-market benefits that would be lost through climate change in the absence of mitigation.

That is not the end of the complications in calculating the costs and benefits of climate change mitigation. The models used for assessing the costs of mitigation and climate change depend critically on the assumptions that are fed into them about structural relationships in the economy. The further we go into the future, and the more we introduce large structural pressures on the models that we use, whether from an ever-rising carbon price or from increasing climate change impacts, the more speculative are the assumptions that make up the model. By the time we get to the end of the 21st century, we have stretched the capacity of the models to the limits of usefulness.

This is not a problem for assessing the costs of mitigation, which will stabilise over time. But it cuts off the costs of climate change and therefore the benefits of mitigation just as they are starting to become large, and long before their peaks.

So what we derive from the quantitative analysis is the net costs of mitigation up to the end of the 21st century, taking into account only the Type 1 and Type 2 benefits of mitigation. The decision that has to be made is whether anticipated Type 3 and Type 4 benefits in the 21st century, and net benefits of all kinds after 2100, justify the calculated net costs of mitigation up to the end of the 21st century.

That is not the end of the conceptual complexity. The costs of mitigation come much earlier than the benefits of avoided climate change. We therefore need to apply a discount rate in comparison of costs and benefits accruing at different points of time.

Perhaps most complex of all is the assessment of the connection between the global mitigation effort, and a single country's contribution to it. The benefits of mitigation come from a global effort, and not from any single country's contribution to that effort. The Review has based its assessments of the costs and benefits of various levels of mitigation on the premise that Australia will make its full proportionate contribution to any global effort. This allows the benefits of a specified global mitigation effort to be associated with the costs of a corresponding Australian mitigation effort.

How can we calculate each country's, in this case Australia's, proportionate share of any global mitigation effort? We have to articulate a set of principles that has a chance of being seen as fair across the international community. Any specified mitigation objective—for example, to hold emissions concentrations to 550 ppm CO_2-e, or 450 ppm CO_2-e with overshooting—will be associated with a global trajectory for emissions over the period leading to the realisation of the objective. How can this emissions trajectory be allocated among countries in a way that can be the subject of an agreement among countries, because it is seen as being fair and practical?

Perceptions and realities of fairness can be influenced by transfers related to technology or support for adaptation, from developed to developing countries. The shape of an agreement could be influenced by the prospect of penalties, for example related to emissions-related restrictions on trade (chapters 8 to 10).

It is unlikely that any allocation of a global trajectory for emissions entitlements will be seen as being fair if it is not based on the idea that, sooner or later, there will be equal per capita rights to use the atmosphere's limited capacity to absorb more greenhouse gases. To be seen as being practical, it will need to allow some time to move from the currently highly unequal assumption of emissions rights across countries, to equal per capita rights. The basis thought to be most likely to be successful is what has become known as 'contraction and convergence', modified to allow faster growth in emissions from fast-growing developing countries for a transition period.

For purposes of analysis, the Review assumed that per capita entitlements would converge to equal entitlements in 2050. The timing of convergence would be a substantive issue in international negotiations on dividing a global emissions entitlements budget among countries.

Note that we are talking about entitlements and not actual emissions. Countries that are able to hold actual emissions below their entitlements will be able to sell their surplus entitlements on international markets. It is not essential for success of the international mitigation effort that every country choose to engage in international trade in permits, rather than to live within its national trajectory. Costs can be reduced by international trade in permits, because trade allows reductions in emissions to occur where they are cheapest. However, the choice to make use of this opportunity to lower costs can be left to individual countries. We can be reasonably confident that enough high-income countries will want to take advantage of opportunities for trade for developing countries to be able to sell surplus permits, and therefore have additional incentives to join the international mitigation effort.

Australia's emissions entitlements within 550 and 450 mitigation objectives were derived from the global emissions trajectories associated with each of them, and from the modified contraction and convergence framework (Chapter 10).

These were the building blocks for the Review's assessment of whether and how much mitigation was worthwhile from an Australian perspective.

Mitigation on the basis of 550 objectives was judged to generate benefits that exceeded the costs. Mitigation on the basis of 450 was thought to generate larger net benefits than 550.

Emissions growth in the Platinum Age

The reassessment of business-as-usual emissions, described in Chapter 3, is a feature of the Review's work. It has large implications for climate change analysis and policy.

The Review has replaced outdated scenarios on emissions growth, embodied in the reports of the Intergovernmental Panel on Climate Change and in earlier studies, by assessments of business-as-usual emissions that are based realistically on growth trends in the early 21st century. These recent years have seen stronger rates of economic growth in the large developing countries than in any earlier period. We call this period of accelerated economic growth the 'Platinum Age', because for most of the world's people it involves stronger growth than the decades after World War II, which economic historians have called the 'Golden Age'. The strong growth is concentrated in countries, first of all China, but also India and Indonesia and others, that, because of their levels of development and economic structure, are experiencing energy-dependent growth. These happen to be countries in which coal is the lowest cost option for increasing energy supply, and coal happens to be the major energy source that is associated with the highest levels of emissions.

None of these realities underlying exceptionally high growth in emissions is going to change at an early date except in the context of climate change policy decisions by governments.

It is a consequence of the reworking of emissions scenarios that the costs of both climate change, and of mitigation to achieve specified concentrations objectives, are higher than had been anticipated by earlier studies. These reworked projections bring forward the critical points for high risks of damaging climate change.

It follows that, if specified mitigation goals are to be reached, it is at once more difficult and more urgent to put in place an effective global agreement. Mitigation efforts that were once thought reasonable now appear to be inadequate.

At the time of presentation of this report to the Prime Minister, state premiers and territory chief ministers of Australia on 30 September 2008, global financial markets are experiencing major instability. Some analysts are suggesting that this will be seriously destabilising for economic growth throughout the world.

Will the positive view of global and in particular of developing country growth, and the associated negative view of greenhouse gas emissions under business as usual (Chapter 3), remain valid in these circumstances?

It is likely that aggregate global growth in 2008 and 2009 will be significantly lower than in the preceding five years or so. However, the acceleration of economic growth in the developing world in the early 21st century has firm foundations. It is unlikely that the current turmoil on financial markets will derail long-term global growth in developing countries permanently from its new and stronger course.

Adaptation

The international community is too late with effective mitigation to avoid significant damage from climate change. So in the best of circumstances, Australians and people everywhere will be adapting to substantial climate change impacts through the 21st century.

The international community may yet fail to put in place effective global mitigation, in which case the challenge of adaptation to climate change will be more daunting. Sound policy on adaptation involves costs, but in many circumstances can later reduce the costs of climate change impacts. Chapters 13 and 15 describe the Review's approach to adaptation policy.

Adaptation to some of the possible consequences of climate change would test humans and their values and preferences in profound ways.

Contemplating the adaptation challenges of people in future times helps to focus our minds on the more difficult dimensions of mitigation choices. We are led to think about how we value future against current generations. We are forced to decide what we would be prepared to pay in terms of consumption of goods and services forgone, to avoid uncertain prospects of possibly immensely unhappy outcomes. We are forced to decide what current and early material consumption we would be prepared to forgo to avoid loss of things that we value, but are not accustomed to valuing in monetary terms.

In making their choices, Australians will have to decide whether and how much they value many aspects of the natural order and its social manifestations that have been part of their idea of their country. In the discussion of the costs of climate change, much is made of damage to natural wonders—to the Great Barrier and Ningaloo reefs, the wetlands of Kakadu, the karri forests of the south-west. We know that we value them highly, and now we will need to think about whether we are prepared to pay for their preservation.

As a changed future approaches, Australians will find themselves thinking about how much they care about other dimensions of our national life that have always been taken for granted. As we will see, with unmitigated climate change, the risks are high that there will be change beyond recognition in the heartlands of old, rural Australia, in Victoria, Western Australia, South Australia, and in the Murray-Darling Basin, which features prominently in our analysis of the possible impacts of climate change. The loss of these heartlands of old Australian identity would be mourned.

Main policy themes

Five general themes that are connected to the Review's policy recommendations run through the report.

Domestic policy must be integrated into global

The first theme is that domestic policy must be deeply integrated into global discussions and agreements. Only a global agreement has any prospect of reducing risks of dangerous climate change to acceptable levels. The costs of achieving any target or holding any trajectory for reducing Australian greenhouse gas emissions will be much lower within the framework of an international agreement. The continuation for long periods of strong Australian mitigation outside a global agreement is likely to corrode the integrity of the Australian market economy. The continuation for long of strong national mitigation in a number of countries without an international framework is likely to corrode the global trading system. It is therefore important to see any period in which an Australian mitigation effort is in place prior to an effective global arrangement as short, transitional and contributing to the achievement of a sound global agreement.

The international dimension of policy is relevant in almost every sphere: in the establishment of targets and trajectories for reduction in emissions; in all dimensions of an increased research and development effort, from climate science to low-emissions technologies; in adaptation to the impacts of climate change; in the importance of equitable distribution of the burden of climate change mitigation.

Strong mitigation must be consistent with prosperity

The second general theme is that global and national mitigation is only going to be successful if reductions in emissions can be made and demonstrated to be consistent with continued economic growth and rising living standards. For Australia, our prime asset in meeting the climate change challenge is the prosperous, flexible, market-oriented economy that has emerged from difficult reforms over the past quarter century. This gives Australia the resources to join other developed countries in sharing the global leadership responsibility for mitigation and adaptation. It provides a basis for market-oriented domestic approaches to mitigation and adaptation that can reduce their costs. It suggests the primacy of preservation of the integrity of market institutions in designing the approach to mitigation and adaptation.

It is a corollary of the second theme that an effective market-based system must be as broadly based as possible, with any exclusions driven by practical necessity and not by short-term political considerations. This will allow abatement to occur in the enterprises, households, industries and regions in which it can be achieved at lowest cost. We do not know now what those firms and industries and regions will be, or how households will respond. Application of similar incentive structures over as much of the economy as possible allows market processes to guide the emergence of favourable outcomes.

Policies must be practical

The third theme is the importance of practicality. The climate change policy discussion has been bogged in delusion, in Australia and elsewhere. Mitigation targets are defined, and sometimes agreed internationality, without the difficult work being done, to make sure that the separate numbers add up to desired solutions, and to make sure that there are realistic paths to where we commit ourselves to go.

The most inappropriate response to the climate change challenge is to take measures and to reach international agreements that create an appearance of action, but which fail to solve or to move substantially towards a solution to the problem. Such an approach risks the integrity of our market economy and political processes to no good effect. It also weakens the political base for later efforts.

It is delusion for one country to develop its own views on the amount of mitigation that it is prepared to undertake without analysing whether that contribution fits into a global outcome that solves the problem.

It is delusion for people in one or many countries to think that they can commit to reductions in emissions in order to solve the climate change problem, without having in mind steps that can actually be taken to implement that commitment.

Some of the past delusion has arisen out of difficulties of working out how to respond to uncertainty.[2]

It is an error to think that uncertainty provides good reason for delaying decisions to start with effective mitigation. Uncertainty surrounding the climate change issue is a reason for disciplined analysis and decision, not for delaying decisions. Under uncertainty, knowledge has high value, and this makes the case for increased investment in applied climate science. Rigorous decision-making under uncertainty recognises that options have value, and that option values decay with time. The rate of decay of good options is faster than was thought by the proponents of strong mitigation only a few years ago, because of the 21st century acceleration of growth in greenhouse gas emissions under business as usual.

The acceleration of emissions growth in recent years—itself the other side of the coin to a beneficent acceleration of growth in many developing countries—has underlined the significance of another delusion: the delusion that this problem can be solved without developing countries playing a major part in the process from an early date.

Policies must be equitable

The fourth theme is that to be practical, any policies on national or international mitigation will need to be and to be seen to be equitable. While there will be no satisfactory solution to the global warming problem without active participation of developing countries from an early date, equity requires developed countries to accept a major part of the costs in the initial years. This was recognised in early international meetings, at Rio de Janeiro and Kyoto, but the recognition so far has been honoured mainly in the breach.

Domestically, in developed and developing countries alike, there is a likelihood that, in the absence of deliberate policies aimed at equitable distribution of the costs of adjustment to a low-emissions economy, the burden would be carried disproportionately by people on low incomes. This reality has the potential for generating resistance to mitigation. As in the international sphere, concerns for equity merge into concerns for practicality.

Good governance is critical

The fifth theme is that there will be no success in mitigation, at a national or international level, without good governance in relation to climate change policies. Proposals that can work on climate change are complex, and cut across strong vested interests of many kinds. These are circumstances in which it is easy, indeed natural, for vested interests to capture policy, and for the national or international interest, and the ultimate reasons for policy, to be forgotten. The only antidote to these tendencies is good governance: the articulation of clear and soundly based principles as a foundation for policy, and the establishment of strong, effective and well-resourced institutions to implement the principles. This is important in both the international and national spheres.

Main policy recommendations for Australia

The main policy recommendations to Australian governments fall within four clusters. One cluster relates to Australia's contributions to the emergence of an effective global agreement. A second relates to efficient implementation of mitigation policies within Australia, and in particular to design of an emissions trading scheme. A third relates to research, and in some spheres development and commercialisation of the products of research. The fourth cluster relates to equitable distribution of the burden. There is obvious overlap across the clusters, but it is useful to introduce them separately.

Australia's commitments in global context

The first is at the centre of the others. The only effective mitigation is global. It is unlikely that the sum of each country's mitigation efforts could add up to effective global mitigation except in the context of an agreement among countries. There will be no international agreement unless each country, and in the first instance each developed country, contributes positively to it. So the first policy issue relates to the contribution that each country, and in this case Australia, makes to an effective global agreement.

Strong mitigation, with Australia playing its proportionate part, is in Australia's interests. In preparation for Copenhagen, Australia should support the objective of reaching international agreement around an objective of holding concentrations to 450 ppm CO_2-e—inevitably with overshooting. It should express its willingness to reduce its own entitlements to emissions from 2000 levels by 25 per cent by 2020 and by 90 per cent by 2050 in the context of an international agreement, so long as the components of that agreement add up to the concentrations objective.

While desirable for Australia and the world, such an agreement will not be easy to reach in one step. It would place constraints on emissions from both developed and developing countries that go beyond what is being contemplated in any but a few countries.

The chances of achieving an effective, soundly based agreement that adds up to 550 are much stronger (chapters 9 and 12). An effective and realistic agreement around a 550 ppm objective would be a major step forward in its own terms. It would also support the beginning of effective international cooperation in emissions reduction and the development and transfer of low-emissions technologies, which would build confidence that ambitious mitigation is consistent with continued economic growth in developed and developing countries. It could therefore be a path towards a subsequent agreement with a more ambitious mitigation objective.

While maintaining its support for the 450 objective, the Commonwealth Government should make it clear that it is prepared to play its full proportionate part in an effective international agreement to hold greenhouse gas concentrations to 550 ppm CO_2-e. This would involve reducing emissions entitlements by 10 per cent by 2000 levels by 2020, and by 80 per cent by 2050.

Consistently with the anticipated content of international agreements at Copenhagen, the offers would relate to binding commitments for 2020, and indicative commitments for 2050. Targets for 2020 are best expressed in relation to the Kyoto Protocol as offers over a base that assumes compliance with the Protocol in the first commitment period (2008–12). In these terms, Australia's offer would be a 17 per cent reduction under a 550 ppm agreement and a 32 per cent reduction under a 450 ppm agreement.

If there were no comprehensive global agreement at Copenhagen, Australia, in the context of an agreement among developed countries only, should commit to a reduction in emissions entitlements by 5 per cent from 2000 levels by 2020 (25 per cent per capita) or 13 per cent from Kyoto compliance in 2008–12. This would be Australia's unconditional offer.

The international agreement would need to go beyond allocation of emissions entitlements across countries. It would need to cover commitments from developed countries to provide support for research, development and commercialisation of low-emissions technologies. Australia should express its willingness to play its full proportionate part in a commitment of a total of US$100 billion per annum by developed countries, with commitments calibrated to income. Australia's share in current circumstances would be in the order of $2.7 billion per annum. Details of the proposed International Low-Emissions Technology Commitment are set out in Chapter 10.

Australia should express its willingness to play its proportionate part in a commitment by high-income countries to support adaptation to climate change in developing countries (Chapter 10).

Within its commitment to support the development of low-emissions technologies, Australia should play a leading role in the management and funding of an expanded international effort to develop and to commercialise carbon capture and storage technologies for carbon dioxide.

It is in Australia's interests to work with other countries towards international sectoral agreements to create a level playing field for major trade-exposed, emissions-intensive industries, including metals, international shipping and aviation. A World Trade Organization agreement would support international mitigation efforts by establishing rules for trade measures to be taken against countries doing too little on climate change.

Australia can also take the lead in continuing what it has already begun, in building productive cooperation on climate change issues with its developing country neighbours, first of all Papua New Guinea and Indonesia. An example of successful cooperation between developed and developing countries that was advantageous for development, covering trading in emissions entitlements, transfer of technology, technical assistance on mitigation in the forestry sector, and cooperation on adaptation could be influential beyond Southeast Asia and the South Pacific.

Australia, alongside others who are willing to play this role, could promote the idea that heads of governments with commitments to strong outcomes at Copenhagen could appoint representatives to a group that is given the task of developing detailed proposals that add up to a range of different concentrations objectives. This would increase the chances that discussion at Copenhagen is constructed around practical alternatives.

Above all, Australia can take a lead by putting in place domestic institutional arrangements and policies that are capable of delivering its share of an agreed mitigation objective at the lowest possible cost. An efficient emissions trading scheme and the introduction of supporting arrangements that reduce the costs of adjusting to the carbon constraint will demonstrate that Australia's emissions reduction commitments are credible.

Design of an emissions trading scheme

The second cluster of policy recommendations relate to the Australian system for reduction in greenhouse gas emissions.

An emissions trading scheme will not be the best instrument of greenhouse gas emissions reduction in every country. It is, if designed and implemented well, the best approach for Australia. Chapter 14 describes a simple emissions trading scheme with broad coverage that can be closely integrated into international markets.

Chapter 14 stresses the importance of assisting trade-exposed, emissions-intensive industries to the extent that other countries do not have comparable carbon pricing. It cautions against compensating them for the effects of the introduction of carbon pricing in Australia. This approach supports adjustment towards the structures that are required in a low-emissions global economy.

Chapter 14 also stresses the importance of the administration and long-term stability of the scheme being placed in the hands of an independent authority—an independent carbon bank—working within clear principles established in law.

Trajectories for reductions in emissions over time should be set consistently with Australia's international commitments. All emissions permits should be sold competitively by the independent carbon bank, periodically to deliver the required emissions reduction trajectories. The trajectories would be changed only with five years' notice, and following certification by the federal government that specified international conditions for change had been met.

Australia's emissions trading system should be established at the earliest possible date, in 2010. During the remainder of the Kyoto compliance period, to the end of 2012, permits should be sold at a fixed price, rising over time, as discussed in Chapter 14.

The sale of permits would generate large amounts of revenue. Some of the revenue would be preempted by the issue of credits for trade-exposed industries. It is judged that the application of the principles outlined in Chapter 14 would entail less than 30 per cent of the permit value being preempted by issue to trade-exposed industries, falling over time as other countries adopted comprehensive or sectoral carbon pricing. Actual revenue from permit sales would be allocated either as payments to households (about 50 per cent), as support for research, development and commercialisation of new technologies (about 20 per cent, contributing a major part of Australia's obligations under the International Low-Emissions Technology Commitment), or to business as credits for trade-exposed activities or as cuts in taxes (about 30 per cent).

There is large scope for biosequestration in Australia, and to a lesser extent in other countries. Full realisation of this potential requires comprehensive carbon accounting in relation to land use, and a determined program, policy and research effort.

No useful purpose is served by other policies that have as their rationale the reduction of emissions from sectors covered by the trading scheme. The Mandatory Renewable Energy Target should be phased out (Chapter 14).

There are structural reasons to expect market failure in response to carbon pricing in relation to the information required for optimal use of known technologies; to research, development and commercialisation of new technologies; and to network infrastructure. Policies to correct these failures are discussed in chapters 17, 18 and 19.

Research and application of new knowledge

The third cluster of recommendations relate to research and development. These matters have already been discussed in relation to the International Low-Emissions Technology Commitment. A stronger Australian climate science research effort is required. It is required as a basis for continued readjustment of Australia's contribution to identifying appropriate mitigation targets and trajectories and it is required to provide a stronger Australian basis for adaptation responses (chapters 13 and 15).

There is an important gap in Australia's research capacity, relating to extending and bringing together research related to climate policy. It is recommended that Australia establish a climate change policy research institute, with disciplinary strengths in the physical and biological sciences, economics and other relevant social sciences.

Sharing the burden of mitigation

Finally, the fourth cluster of policies relate to income distribution. Equitable distribution of the burden of mitigation internationally and in each country is at the heart of the practicality of mitigation. This has already been discussed in its international dimension.

Within Australia, the maintenance of full employment and an effective social safety net are the most important requirements for equity in the process of emissions reduction and adaptation to climate change.

The use of revenue generated by the competitive sale of permits is an important instrument of equity. For the most part, the value will have been created by the recoupment from households of the scarcity rents of permits, and passing back proceeds from sales can reduce the effect of mitigation-induced prices on household living standards.

The Review recommends that half the permit revenue be paid to households, with a focus on the lower half of the income distribution. Before and in the early years of the emissions trading scheme, a 'green credit' arrangement should facilitate energy-saving adjustment for low-income households. Other payments to households should be made through the taxation and social security systems (Chapter 16).

There have been demands for compensation of Australian business for loss of income or wealth associated with the introduction of an emissions trading scheme. The most vocal have been coal-based electricity generators. The Review's analysis indicates that such claims must be assessed against other equity claims by people who have been adversely affected by the scheme. The case for business compensation is weak alongside the claims of low-income Australian households. There is, however, a case for structural adjustment assistance during the transition to low-emissions technologies in coal-based electricity-generating regions, to prevent the emergence of disadvantaged regions in the transition to a low-emissions economy.

Notes

1 Issued in a statement by the national academies of science of Brazil, Canada, China, France, Germany, India, Japan, Mexico, Russia, South Africa, the United Kingdom and the United States in 2008 (Joint Science Academies 2008).

2 This is not uncommon in the early stages of coming to grips with new problems. The pity is that the analysis of decision making under uncertainty has been taken so far in other contexts that we did not need to learn it all again in a new sphere (Hacking 1990).

References

Hacking, I. 1990, *The Taming of Chance*, Cambridge University Press, Cambridge.

Joint Science Academies 2008, *Joint Science Academies' Statement: Climate change adaptation and the transition to a low carbon society*, joint statement released by the National Academies of Science of the G8 Countries plus Brazil, China, India, Mexico and South Africa.

SYNOPSIS OF KEY POINTS

Chapter 1 A decision-making framework

The central policy issue facing the Review can be simply stated: what extent of global mitigation, with Australia playing its proportionate part, provides the greatest excess of gains from reduced risks of climate change over costs of mitigation?

The mitigation costs are experienced through conventional economic processes and can be measured through formal economic modelling.

Only some of the benefits of mitigation are experienced through conventional market processes (Types 1 and 2) and only one is amenable to modelling (Type 1). Others take the form of insurance against severe and potentially catastrophic outcomes (Type 3), and still others the avoidance of environmental and social costs, which are not amenable to conventional measurement (Type 4).

The challenge is to make sure that important, immeasurable effects are brought to account.

The long time frames involved create a special challenge, requiring us to measure how we value the welfare of future generations relative to our own.

Chapter 2 Understanding climate science

The Review takes as its starting point, on the balance of probabilities and not as a matter of belief, the majority opinion of the Australian and international scientific communities that human activities resulted in substantial global warming from the mid-20th century, and that continued growth in greenhouse gas concentrations caused by human-induced emissions would generate high risks of dangerous climate change.

A natural carbon cycle converts the sun's energy and atmospheric carbon into organic matter through plants and algae, and stores it in the earth's crust and oceans. Stabilisation of carbon dioxide concentrations in the atmosphere requires the rate of greenhouse gas emissions to fall to the rate of natural sequestration.

There are many uncertainties around the mean expectations from the science, with the possibility of outcomes that are either more benign—or catastrophic.

Chapter 3 Emissions in the Platinum Age

Greenhouse gas emissions have grown rapidly in the early 21st century. In the absence of effective mitigation, strong growth is expected to continue for the next two decades and at only somewhat moderated rates beyond.

So far, the biggest deviations from earlier expectations are in China. Economic growth, the energy intensity of that growth, and the emissions intensity of energy use are all above projections embodied in earlier expectations. China has recently overtaken the United States as the world's largest emitter and, in an unmitigated future, would account for about 35 per cent of global emissions in 2030.

Other developing countries are also becoming major contributors to global emissions growth, and will take over from China as the main growing sources a few decades from now. Without mitigation, developing countries would account for about 90 per cent of emissions growth over the next two decades, and beyond.

High petroleum prices will not necessarily slow emissions growth for many decades because of the ample availability of large resources of high-emissions fossil fuel alternatives, notably coal.

Chapter 4 Projecting global climate change

As a result of past actions, the world is already committed to a level of warming that could lead to damaging climate change.

Extreme climate responses are not always considered in the assessment of climate change impacts due to the high level of uncertainty and a lack of understanding of how they work. However, the potentially catastrophic consequences of such events mean it is important that current knowledge about such outcomes is incorporated into the decision-making process.

Continued high emissions growth with no mitigation action carries high risks. Strong global mitigation would reduce the risks considerably, but some systems may still suffer critical damage.

There are advantages in aiming for an ambitious global mitigation target in order to avoid some of the high-consequence impacts of climate change.

Chapter 5 Projecting Australian climate change

Australia's dry and variable climate has been a challenge for the continent's inhabitants since human settlement.

Temperatures in Australia rose slightly more than the global average in the second half of the 20th century. Streamflow has fallen significantly in the water catchment areas of the southern regions of Australia. Some of these changes are attributed by the mainstream science to human-induced global warming.

Effects of future warming on rainfall patterns are difficult to predict because of interactions with complex regional climate systems. Best-estimate projections show considerable drying in southern Australia, with risk of much greater drying. The mainstream Australian science estimates that there may be a 10 per cent chance of a small increase in average rainfall, accompanied by much higher temperatures and greater variability in weather patterns.

Chapter 6 Climate change impacts on Australia

The Review has conducted detailed studies of impacts of climate change on Australia. These studies are available in full on the Review's website.

Growth in emissions is expected to have a severe and costly impact on agriculture, infrastructure, biodiversity and ecosystems in Australia.

There will also be flow-on effects from the adverse impact of climate change on Australia's neighbours in the Pacific and Asia.

These impacts would be significantly reduced with ambitious global mitigation.

The hot and dry ends of the probability distributions, with a 10 per cent chance of realisation, would be profoundly disruptive.

Chapter 7 Australia's emissions in a global context

Australia's per capita emissions are the highest in the OECD and among the highest in the world. Emissions from the energy sector would be the main component of an expected quadrupling of emissions by 2100 without mitigation.

Australia's energy sector emissions grew rapidly between 1990 and 2005. Total emissions growth was moderated, and kept more or less within our Kyoto Protocol target, by a one-off reduction in land clearing.

Relative to other OECD countries, Australia's high emissions are mainly the result of the high emissions intensity of energy use, rather than the high energy intensity of the economy or exceptionally high per capita income. Transport emissions are not dissimilar to those of other developed countries. Australia's per capita agricultural emissions are among the highest in the world, especially because of the large numbers of sheep and cattle.

The high emissions intensity of energy use in Australia is mainly the result of our reliance on coal for electricity. The difference between Australia and other countries is a recent phenomenon: the average emissions intensity of primary energy supply for Australia and the OECD was similar in 1971.

Chapter 8 Assessing the international response

Climate change is a global problem that requires a global solution.

Mitigation effort is increasing around the world, but too slowly to avoid high risks of dangerous climate change. The recent and projected growth in emissions means that effective mitigation by all major economies will need to be stronger and earlier than previously considered necessary.

The existing international framework is inadequate, but a better architecture will only come from building on, rather than overturning, established efforts.

Domestic, bilateral and regional efforts can all help to accelerate progress towards an effective international agreement.

The United Nations meeting in Copenhagen in December 2009 is an important focal point in the attempt to find a basis for global agreement. Australia must be prepared to play its full proportionate part as a developed country.

Chapter 9 Towards global agreement

Only a comprehensive international agreement can provide the wide country coverage and motivate the coordinated deep action that effective abatement requires.

The only realistic chance of achieving the depth, speed and breadth of action now required from all major emitters is allocation of internationally tradable emissions rights across countries. For practical reasons, allocations across countries will need to move gradually towards a population basis.

An initial agreement on a global emissions path towards stabilisation of the concentration of greenhouse gases at 550 CO_2-e is feasible. 450 CO_2-e is a desirable next step. Agreement on, and the beginnings of implementation of, such an agreement, would build confidence for the achievement of more ambitious stabilisation objectives.

All developed and high-income countries, and China, need to be subject to binding emissions limits from the beginning of the new commitment period in 2013.

Other developing countries—but not the least developed—should be required to accept one-sided targets below business as usual.

Chapter 10 Deepening global collaboration

International trade in permits lowers the global cost of abatement, and provides incentives for developing countries to accept commitments.

Trade in emissions rights is greatly to be preferred to trade in offset credits, which should be restricted.

A global agreement on minimum commitments to investment in low-emissions new technologies is required to ensure an adequate level of funding of research, development and commercialisation. Australia's commitment to support of research, development and commercialisation of low-emissions technology would be about $2.8 billion.

An International Adaptation Assistance Commitment would provide new adaptation assistance to developing countries that join the mitigation effort.

Early sectoral agreements would seek to ensure that the main trade-exposed, emissions-intensive industries face comparable carbon prices across the world, including metals and international civil aviation and shipping.

A WTO agreement is required to support international mitigation agreements and to establish rules for trade measures against countries thought to be doing too little on mitigation.

Chapter 11 Costing climate change and its avoidance

Type 1 (modelled median outcomes) plus Type 2 (estimates of other median outcomes) costs of climate change in the 21st century are much higher than earlier studies suggested. The Platinum Age emissions grow much faster than earlier studies contemplated.

The modelling of the 550 mitigation case shows mitigation cutting the growth rate over the next half century, and lifting it somewhat in the last decades of the century.

GNP is higher with 550 mitigation than without by the end of the century. The loss of present value of median climate change GNP through the century will be outweighed by Type 3 (insurance value) and Type 4 (non-market values) benefits this century, and much larger benefits of all kinds in later years.

Mitigation for 450 costs almost a percentage point more than 550 mitigation of the present value of GNP through the 21st century. The stronger mitigation is justified by Type 3 (insurance value) and Type 4 (non-market values) benefits in the 21st century and much larger benefits beyond. In this context, the costs of action are less than the costs of inaction.

Chapter 12 Targets and trajectories

Australia should indicate at an early date its preparedness to play its full, proportionate part in an effective global agreement that 'adds up' to either a 450 or a 550 emissions concentrations scenario, or to a corresponding point between.

Australia's full part for 2020 in a 450 scenario would be a reduction of 25 per cent in emissions entitlements from 2000 levels, or one-third from Kyoto compliance levels over 2008–12, or 40 per cent per capita from 2000 levels. For 2050, reductions would be 90 per cent from 2000 levels (95 per cent per capita).

Australia's full part for 2020 in a 550 scenario would be a reduction in entitlements of 10 per cent from 2000 levels, or 17 per cent from Kyoto compliance levels over 2008–12, or 30 per cent per capita from 2000. For 2050, reductions would be 80 per cent per capita from 2000 levels or 90 per cent per capita.

If there is no comprehensive global agreement at Copenhagen in 2009, Australia, in the context of an agreement amongst developed countries only, should commit to reduce its emissions by 5 per cent (25 per cent per capita) from 2000 levels by 2020, or 13 per cent from the Kyoto compliance 2008–12 period.

Chapter 13 An Australian policy framework

Australia's mitigation effort is our contribution to keeping alive the possibility of an effective global agreement on mitigation.

Any effort prior to an effective, comprehensive global agreement should be short, transitional and directed at achievement of a global agreement.

A well-designed emissions trading scheme has important advantages over other forms of policy intervention. However, a carbon tax would be better than a heavily compromised emissions trading scheme.

The role of complementary measures to the emissions trading scheme is to lower the cost of meeting emissions reduction trajectories, as well as adapting to the impacts of climate change by correcting market failures.

Once a fully operational emissions trading scheme is in place, the Mandatory Renewable Energy Target will not address any additional market failures. Its potentially distorting effects can be phased out.

Governments at all levels will inform the community's adaptation response. More direct forms of intervention may be warranted when events unfold suddenly or when communities lack sufficient options or capacity for dealing with the impacts of climate change.

Chapter 14 An Australian emissions trading scheme

A principled approach to the design of the Australian emissions trading scheme is essential if the scheme is to avoid imposing unnecessary costs on Australians.

The integrity, efficiency and effectiveness of the scheme will require:

- establishment of an independent carbon bank with all the necessary powers to oversee the long-term stability of the scheme
- implementation of a transition period from 2010 to the conclusion of the Kyoto period (end 2012) involving fixed price permits
- credits to trade-exposed, emissions-intensive industries to address the failure of our trading partners to adopt similar policies
- no permits to be freely allocated

- no ceilings or floors on the price of permits (beyond the transition period)
- intertemporal use of permits with 'hoarding' and 'lending' from 2013
- a judicious and calibrated approach to linking with international schemes
- scheme coverage that is as broad as possible, within practical constraints

Seemingly small compromises will quickly erode the benefits that a well-designed emissions trading scheme can provide.

The existing, non-indexed shortfall penalty in the Mandatory Renewable Energy Target needs to remain unchanged in the expanded scheme.

Chapter 15 Adaptation and mitigation measures for Australia

Every Australian will have to adapt to climate change within a few decades. Households and businesses will take the primary responsibility for the maintenance of their livelihoods and the things that they value.

Information about climate change and its likely impacts is the first requirement of good adaptation and mitigation policies. This requires strengthening of the climate-related research effort in Australia. The Australian Climate Change Science Program should be provided with the financial resources to succeed as a world-class contributor to the global climate science effort from the southern hemisphere.

A new Australian climate change policy research institute should be established to raise the quality of policy-related research.

Flexible markets using the best available information are the second essential component for successful adaptation and mitigation policies. It will be important to strengthen markets for insurance, water and food.

Government regulatory intervention and provision of services will be required in relation to emergency management services and preservation of ecosystems and biodiversity.

Chapter 16 Sharing the burden in Australia

Low-income households spend much higher proportions of their incomes than other households on emissions-intensive products. The effects of the emissions trading scheme will fall heavily on low-income households, so the credibility, stability and efficiency of the scheme require the correction of these regressive effects by other measures.

At least half the proceeds from the sale of all permits could be allocated to households, focusing on the bottom half of the income distribution. The bulk could be passed through the tax and social security systems, with energy efficiency commitments to low-income households in the early years.

To assist in early adjustment of low-income households, a system of 'green credits' should be introduced to help with funding of investments in energy efficiency in housing, household appliances and transport.

It is possible but not certain that regional employment issues could arise in coal regions. They would not emerge in the early years of an emissions trading scheme. Up to $1 billion in total should be made available for matched funding for investment in reducing emissions in coal power generation, as a form of preemptive structural adjustment assistance.

Chapter 17 Information barriers to known technologies

There are potentially large and early gains from better utilisation of known technologies, goods and services, including energy efficiency and low-emissions transport options.

Externalities in the provision of information and principal–agent issues inhibit the use of distributed generation and energy-saving opportunities in appliances, buildings and vehicles.

A combination of information, regulation and restructuring of contractual relationships can reduce the costs flowing from many of the market failures blocking optimal utilisation of proven technologies and practices.

Chapter 18 The innovation challenge

Basic research and development of low-emissions technologies is an international public good, requiring high levels of expenditure by developed countries.

Australia should make a proportionate contribution alongside other developed countries in its areas of national interest and comparative research advantage. This would require a large increase in Australian commitments to research, development and commercialisation of low-emissions technologies, to more than $3 billion per annum by 2013.

A new research council should be charged with elevating, coordinating and targeting Australia's effort in low-emissions research.

There are externalities associated with private investment in commercialising new, low-emissions technologies.

To achieve an effective commercialisation effort on a sufficiently early time scale, an Australian system of matching funding should be available automatically where there are externalities associated with private enterprise investment in low emissions innovation.

Research in adaptation technologies is required. Existing arrangements are well placed to meet immediate priorities.

Chapter 19 Network infrastructure

There is a risk that network infrastructure market failures relating to electricity grids, carbon dioxide transport systems, passenger and freight transport systems, water delivery systems and urban planning could increase the costs of adjustment to climate change and mitigation.

The proposed national electricity transmission planner's role should be expanded to include a long-term economic approach to transmission planning and funding. The Building Australia Fund should be extended to cover energy infrastructure. A similarly planned approach is necessary to facilitate timely deployment of large-scale carbon capture and storage.

There is a limited case for carefully calculated rates for feed-in tariffs for household electricity generation and co-generation.

The need to reduce the costs of mitigation reinforces other and stronger reasons for giving higher priority to increasing capacity and improving services in public transport, and for planning for greater urban density.

Chapter 20 Transforming energy

Australians have become accustomed to low and stable energy prices. This is being challenged by rapidly rising capital costs and large price increases for natural gas and black coal. These cost effects will be joined by pressures from rising carbon prices, and will be larger than the impact of the emissions trading scheme for some years.

Australia is exceptionally well endowed with energy options, across the range of fossil fuel and low-emissions technologies.

The interaction of the emissions trading scheme with support for research, development and commercialisation and for network infrastructure will lead to successful transition to a near-zero emissions energy sector by mid-century.

The future for coal-based electricity generation, for coal exports and for mitigation in developing Asia depends on carbon capture and storage becoming commercially effective. Australia should lead a major international effort towards the testing and deployment of this technology.

Chapter 21 Transforming transport

Transport systems in Australia will change dramatically this century, independently of climate change mitigation. High oil prices and population growth will change technologies, urban forms and roles of different modes of transport.

An emissions trading scheme will guide this transformation to lower-emissions transport options.

Higher oil prices and a rising emissions price will change vehicle technologies and fuels. The prospects for low-emissions vehicles are promising. It is likely that zero-emissions road vehicles will become economically attractive and be the most important source of decarbonisation from the transport sector.

Governments have a major role to play in lowering the economic costs of adjustment to higher oil prices, an emissions price and population growth, through planning for more compact urban forms and rail and public transport. Mode shift may account for a quarter of emissions reductions in urban passenger transport, lowering the cost of transition and delivering multiple benefits to the community.

Chapter 22 Transforming rural land use

Rural Australia faces pressures for structural change from both climate change and its mitigation.

Effective mitigation would greatly improve the prospects for Australian agriculture, at a time when international demand growth in the Platinum Age is expanding opportunities.

Choices for landowners will include production of conventional commodities, soil carbon, bioenergy, second-generation biofuels, wood or carbon plantations, and conservation forests.

There is considerable potential for biosequestration in rural Australia. The realisation of this potential requires comprehensive emissions accounting.

The realisation of a substantial part of the biosequestration potential of rural Australia would greatly reduce the costs of mitigation in Australia. It would favourably transform the economic prospects of large parts of remote rural Australia.

Full utilisation of biosequestration could play a significant role in the global mitigation effort. This is an area where Australia has much to contribute to the international system.

Chapter 23 Towards a low-emissions economy

Australian material living standards are likely to grow strongly through the 21st century, with or without mitigation, and whether 450 or 550 ppm is the mitigation goal. Botched domestic and international mitigation policies are a risk.

Substantial decarbonisation by 2050 to meet either the 450 or 550 obligation is feasible. It will go fastest in the electricity sector, then transport, with agriculture being difficult unless, as is possible, there are transformative developments in biosequestration.

There is considerable technological upside. This could leave Australian energy costs relatively low, so that it remains a competitive location for metals processing.

Australia's human resource strengths in engineering, finance and management related to the resources sector are important assets in the transition to a low-emissions economy. They will need to be nurtured by high levels of well-focused investment in education and training.

The introductory impact of the Australian emissions trading scheme will not be inflationary if permit revenue is used judiciously to compensate households.

Chapter 24 Fateful decisions

There are times in the history of humanity when fateful decisions are made. The decision this year and next on whether to enter a comprehensive global agreement for strong action is one of them.

Australia's actions will make a difference to the outcome, in several ways.

The chances of success at Copenhagen would be greater if heads of government favouring a strong outcome set up an experts group to come up with a practical approach to global mitigation that adds up to various environmental objectives.

On a balance of probabilities, the failure of our generation on climate change mitigation would lead to consequences that would haunt humanity until the end of time.

A DECISION-MAKING FRAMEWORK

1

Key points

The central policy issue facing the Review can be simply stated: what extent of global mitigation, with Australia playing its proportionate part, provides the greatest excess of gains from reduced risks of climate change over costs of mitigation?

The mitigation costs are experienced through conventional economic processes and can be measured through formal economic modelling.

Only some of the benefits of mitigation are experienced through conventional market processes (Types 1 and 2) and only one is amenable to modelling (Type 1). Others take the form of insurance against severe and potentially catastrophic outcomes (Type 3), and still others the avoidance of environmental and social costs, which are not amenable to conventional measurement (Type 4).

The challenge is to make sure that important, immeasurable effects are brought to account.

The long time frames involved create a special challenge, requiring us to measure how we value the welfare of future generations relative to our own.

This chapter puts forward a framework for looking at these issues.

How do we assess whether Australian mitigation action is justified? Would the substantial costs of mitigation be exceeded by avoided costs of climate change? What degree of mitigation would lead to the largest net benefits?

These turn out to be immensely complex questions. The answers depend on our judgments about the prospects for effective international mitigation. They depend on the efficiency of measures to achieve reductions in greenhouse gas emissions, including supporting measures that affect the market response to the mitigation regime, and therefore the costs of achieving various levels of abatement. They depend on the efficiency of supporting measures to share the costs of mitigation across the Australian community, and on the international distribution of the mitigation burden. They depend on the options for and costs of adaptation. These decisions need to be taken under conditions of uncertainty and risk.

The answers also depend on our ability to measure accurately the conventional economic effects of climate change, and the likely reduction in those effects due to mitigation. Not all of the effects on output and consumption through market processes are amenable to precise quantification. Our conclusions depend on our ability to form sound judgments about the magnitude of any changes that are excluded from attempts at formal measurement because adequate information is not available at this time. The answers depend fundamentally on the approach taken to decision making under conditions of risk and uncertainty, and in particular, on the insurance value that is placed on avoiding the possibility of large negative outcomes.

The answers depend also on the value we place on outcomes not related to consumption of goods and services, but on Australians' valuation of environmental amenity in many dimensions. These assessments are affected by how we view the inter-relationship between these and other non-material values with conventional consumption in determining welfare.

The answers are affected by the relative value that is placed on the welfare of people living in the future relative to the welfare of those living at present.

This chapter introduces an approach to decision making to openly deal with these immensely complex and difficult issues. This allows people who are uneasy or unhappy about the conclusions to understand or take issue with the underlying premises and logic.

We are seeking to assist community choice on the extent of mitigation that provides the greatest excess of gains from reduced climate change over costs of mitigation. The complexity of the influences on that choice makes simplicity especially challenging and particularly important. Here, even more than in other areas of public policy choice, focus on the central underlying issues is essential if we are to reach conclusions through a transparent process, open to challenge, as a basis for long-term community support, policy continuity and stability.

Climate change mitigation decisions in 2008, and for the foreseeable future, are made under conditions of great uncertainty. There is great uncertainty about the climatic outcomes of varying concentrations of greenhouse gases; about the impact of various climate outcomes; and about the costs and effectiveness of adapting to climate change. There is uncertainty about the costs of various degrees of mitigation in Australia; about the extent to which the international community will make effective commitments to mitigation; and about the relationship of global to Australian mitigation efforts.

Under conditions of such uncertainty, it is sensible to ask whether it would be better to delay decisions while information is gathered and analysed. However, it is as much a decision to do nothing, or to delay action, as it is to decide to take early action. The issue is whether delay would be a good decision.

When global warming first became a major international public policy issue nearly two decades ago, it may have been good policy to take modest and low-cost steps on mitigation, while investing heavily in improving the information base for later decisions.

In 2008, the costs of delay—in the probabilistic terms that frame a good decision under conditions of uncertainty—are high. The work of the Review has contributed to changing international perceptions on the rate at which emissions will grow over the next several decades under business as usual. Australia and the world are running towards high risks of dangerous climate change at a more rapid rate than was previously understood. The opportunity costs of delaying decisions are high.

Australia and its partners in the international community will, for good reasons, make historic and fateful decisions about their approaches to climate change mitigation in the three years ahead. They will do this on the basis of currently available information and analysis, however sound or weak that may be.

The sceptical economist—and the Review counts itself within this tradition—insists on equally rigorous evaluation for a decision to delay as for a decision to take action now.

The Review's approach to the important questions about mitigation policy starts with scientific assessment of the costs of climate change to Australia and Australians. We have to be able to compare the costs of climate change without mitigation, and with varying degrees of effective mitigation and adaptation effort. These costs include indirect costs through effects on other countries, to the extent that these feed back into impacts on Australia, or in themselves are valued by Australians. The scientific assessments are highly uncertain, and their impacts on human activity and welfare even more so. We have no alternative to making decisions on complex issues of valuation under conditions of great uncertainty.

1.1 The costs of mitigation

The increase in greenhouse gas emissions is a product of the advances in science, technology and economic organisation that have transformed humanity as well as its natural context over the last two centuries. In the history of life on earth, and even of human life, we are talking about an almost infinitesimally short period of extraordinary dynamism.

A modern acceleration in rates of human-induced greenhouse gas emissions is the source of contemporary concerns about climate change.

Economic development over the past two centuries has taken most of humanity—but certainly not all—from lives that were insecure, ignorant and short, to personal health and security, material comfort and knowledge unknown to the elites of the wealthiest and most powerful societies in earlier times.

In the first millennium after the life of Jesus Christ, global economic output increased hardly at all—by only one sixth. All of the small increase was contributed by population growth, and none by increased production per person. By contrast, output increased 300-fold in the second millennium, with population increasing 22 times and per capita production 13 times. Most of the extraordinary expansion took place towards the end of the period. From 1820 until the end of the 20th century, per capita output increased more than eight times and population more than five times (Maddison 2001).

In most of its first two centuries, the cornucopia of modern economic growth was located in a small number of countries, in Western Europe and its overseas offshoots in North America and Oceania, and in Japan. In the third quarter of the 20th century it extended into a number of relatively small economies in East Asia.

A new era began in the fourth quarter of the last century, with the rapid extension of the beneficent processes of modern economic development into the heartland of the populous countries of Asia, including China, India and Indonesia. From this has emerged what can be described as the Platinum Age of global economic growth in the early 21st century (Garnaut & Huang 2007).[1] Incomes are growing rapidly in a large proportion of the developing world. In the absence of a major dislocation of established trends, this is likely to continue for a considerable period. There will be a greater absolute increase in annual human output and consumption in the first two decades of the 21st century than was generated in the whole previous history of our species. Similarly strong growth in output can be expected in the next following decade to 2030.

Increasingly through the 21st century, the expansion of production will be associated with rising output per person, rather than increase in population. In all of the economically successful countries, higher incomes, the increased survival rates of children and the expansion of education and choice for women are leading to declining rates of population increase. Before the end of the 21st century, a continuation of these processes should lead to stabilisation (by about 2080), and then, at least for a while, a gradual decline in global human population. By that time, nearly three billion will have been added to the global population.

The era of modern economic growth has been intimately linked to rapid expansion in the use of fossil fuels. This is returning to the atmosphere a part of the carbon that was sequestered naturally over billions of years, through a process that created the conditions necessary for the emergence of human life on earth. While the share of carbon returned to the atmosphere is small relative to the stock, it is large enough to throw the equilibrium of heat trapping in the atmosphere out of balance.

The amount of fossil fuel in the earth's crust, in the forms of petroleum, natural gas, coal, tar sands and shale, is finite. However, the amount is so large that its limits are of no practical importance for climate change policies.

However, there is a much tighter engineering limit to the availability for human use of fossil fuels: the point at which the energy used to extract the resources would be greater than their energy content.

Tighter still is the economic limit: the availability of fossil fuels in forms and locations that can be extracted for human use at costs below the prices of oil, gas and coal in global markets. There is debate on whether the economic limits will constrain global economic growth in the period immediately ahead or in the foreseeable future. The limit will be reached much earlier for liquid petroleum than for natural gas, and for gas much earlier than for coal.

It was once common for economists to see constraints on the availability of natural resources and in particular fossil fuels as placing limits on modern economic growth (Malthus 1798; Jevons 1865). The success of technological improvement and economic processes in easing supposed constraints in the first centuries of modern economic growth established confidence that these constraints could be overcome in ways that allowed global economic growth to continue.

Rapid growth from the early 1950s to the early 1970s, and extraordinary Japanese growth at the end of that period, rekindled old concerns about resource constraints on growth.

Fossil fuel resource availability was one element in the cautions of the Club of Rome, and their prophecy about limits to growth in the early 1970s (Club of Rome 1972). The extraordinary growth in demand for fossil fuels in the early years of the Platinum Age—and the immense and unexpected increases in prices that accompanied it—have rekindled interest in the issue. Will the supply of fossil fuels slow down the growth in greenhouse gas emissions enough to do the mitigation task?

It is clear from the present state of knowledge—as it was not to earlier generations—that it would be possible for humanity to break the link between economic growth and combustion of fossil fuels. This would make it possible for the world economy to adjust to the approach of economically relevant limits to fossil fuel availability, without bringing the increase in human consumption of goods and services to an end.

For the time being, the pervasive and rapidly growing use of fossil hydrocarbons in economic activity is a matter of economic optimisation and not of technological necessity. If the human species avoids some catastrophic truncation of the triumphs of modern economic development, it will need to pursue a transition out of reliance on fossil fuels—and it will succeed in doing so.

The constraints on the economic availability of fossil fuels will aid the climate change mitigation process. But the Review's analysis suggests that in the time available, the reduction in use of fossil fuels, associated with scarcity and high prices, will be nowhere near enough to avoid high risks of dangerous climate change.

To the extent that mitigation is effective, reduced demand for petroleum and other fossil fuels associated with effective mitigation would reduce the global price of these resources, improve the terms of trade of importing countries, and probably have favourable effects on global economic growth. This would be an offset for some countries against the cost of mitigation.

The beneficiaries of lower fossil fuel prices would not include Australia, whose terms of trade rise with high global energy prices. Lower export prices for resources hurt producers in resource-based industries, and the beneficiaries of government revenue generated from these industries. But they also tend to lower interest rates and the exchange rate, and increase incomes, in some rural manufacturing and service industries, and for many households.

Adjusting to limits on the use of fossil fuels required to mitigate climate change would be less costly than adjusting to naturally imposed economic constraints on the availability of fossil fuels. This is because sequestration through physical processes (geosequestration) or biological processes (biosequestration) can ease the mitigation task but cannot ease natural constraints on fossil fuel supply.

However, mitigation needs to be imposed through political processes. Such decisions in single countries are hard enough. Achieving mitigation outcomes through cooperation of many sovereign entities, each with an incentive to shift the cost of adjustment to other countries, is more challenging.

A dramatic transformation in humanity's use of fossil-fuel-based energy would be necessary sooner or later to sustain and to extend modern standards of living. It will be required sooner if the world is to hold the risks of climate change to acceptable levels. The costs incurred in making an early adjustment will bring forward, and reduce for future times, the costs of the inevitable adjustment away from fossil fuels. How much sooner and at what extra cost are central questions before the Review.

The costs of mitigation depend on the extent to which, and the time over which, reductions in emissions are achieved. Costs depend on the efficiency of the chosen policy instruments. There are cost advantages in having a single price on emissions as the main instrument of policy, supported by measures to correct market failures in utilisation of the commercial opportunities created by the price on emissions.

If mitigation is approached through an efficient set of policies, its costs are determined by the extent and the rate of emissions reductions to be achieved. These, in turn, are determined by the ambitions of a global effort to which Australia has subscribed, and by what Australia is prepared to do in the context of global action.

The costs of mitigation can be calculated for various levels and rates of reductions in emissions. Each level and rate of Australian mitigation can be related to a global mitigation outcome. The global mitigation outcome will define a benefit to Australia in terms of reduced risks of climate change. The benefits of reduced risks of climate change to Australia can be identified. The costs and benefits of mitigation can then be compared. The policy task in setting Australian mitigation objectives, therefore, begins with identification of the costs and benefits (in reduced risks of loss from climate change) for various mitigation ambitions.

The higher the market prices of petroleum, coal and natural gas, the lower the costs of mitigation will be. The costs of business as usual, compared with the costs of using alternative, low-emissions technologies, will be higher. The historically high fossil fuel prices make this of current interest.

The more ambitious the extent and speed of reductions in emissions, the higher the costs of mitigation will be. The costs of mitigation will be lower the more efficient the instruments chosen to give effect to policy.

An economically efficient approach to mitigation would generate a rising carbon price over time, imposing increasingly strong pressure for adjustment out of high-emissions technologies, and increasingly strong incentives for sequestration. For a given abatement task, emissions costs will be lowest if the emissions price rises at the interest rate, which will lead to optimal timing in investment in the mitigation effort.

The challenge is to allocate efficiently over time access to a limited global capacity to absorb additional greenhouse gases without unacceptably high risks of dangerous climate change. The allocation problem is familiar as one of optimal depletion of a finite resource. This frames the economics of the timing of the mitigation effort, and suggests the relevance of the 'Hotelling curve' to the price curve for the right to emit (Hotelling 1931).

The annual costs of mitigation are likely to rise for some time, as a rising carbon price forces deeper abatement. While the price would be expected to continue to rise over time at the interest rate, the cost to the economy would not rise at that rate. At some point, the tendency for costs to rise would be moderated and eventually reversed by improvements in low-carbon technologies.

At some time in the future—no later and perhaps much earlier than the time when economic constraints on the use of fossil fuels would be forcing structural change comparable with what had been achieved for mitigation purposes—the incremental costs of mitigation will become negative. The sunk costs of technological improvement and structural change associated with mitigation will avoid the need for investments to accommodate the constraints on availability of fossil fuels.

Above all else, the cost of mitigation in Australia, and not only the benefits in avoided climate change, will be shaped by the nature of the global mitigation effort. An effective global effort would open a wide range of opportunities for trade in mitigation responsibilities, assigning greater reductions in emissions to countries in which it can be achieved at lowest cost. A global effort would increase and distribute more efficiently and equitably the world's investment in new technologies to develop lower-emissions paths to consumption and production. It would obviate the need for special policy measures to avoid carbon leakage—the shift of emissions-intensive industries from high-mitigation to low-mitigation countries—a policy requirement that is likely to distort both domestic economic efficiency and political integrity.

1.2 Risk and uncertainty

Climate change policy requires us to come to grips with both risk and uncertainty. Keynes (1921) and Knight (1921) drew a distinction between the two that is still useful today.

Risk relates to an event that can be placed on a known probability distribution. When we toss a coin, we do not know whether or not we will see a head. If we toss the coin enough times, it will fall as a head about half of the time.

In many spheres of human life, an activity has similarities with others that have been repeated many times, so that participants have a reasonable idea of the odds. A piece of surgery with some risk of death and short-term investments in financial markets have some similar properties to the toss of a coin. No new piece of surgery, and no new investment, is exactly the same as any other. But there have been enough similar events for players to feel that they can form judgments with some confidence about the probabilities.

There is uncertainty when an event is of a kind that has no close precedents, or too few for a probability distribution of outcomes to be defined, or where an event is too far from understood events for related experience to be helpful in foreseeing possible outcomes. Humans are often required to form judgments about events that are unique, or so unusual that analysis based on secure knowledge and experience is an absent or weak guide. Columbus sailing west in search of China, or Oxley heading west along the rivers of Australia in search of an inland sea, are historically important examples (Figure 1.1).

Figure 1.1 The risk–uncertainty spectrum

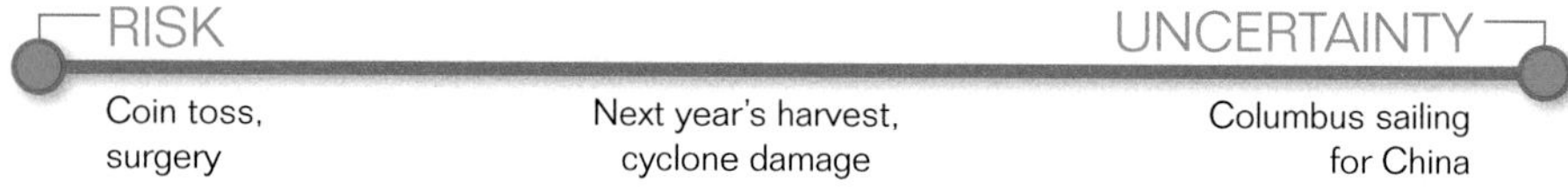

The 18th century British philosopher Bayes has given his name to a well-developed approach to decisions under uncertainty. Bayesian decision theory encourages us to treat decisions under uncertainty as if we were taking a risk (Raiffa 1968; Raiffa & Schlaifer 1961). We will make the best possible decisions under uncertainty if we force those who are best placed to know to define subjective probabilities that they would place on various outcomes, and work through the implications of those assessments as if they were probability distributions based on experience (Figure 1.2). These subjective probability distributions can then be updated on the basis of experience.

Figure 1.2 A probability distribution

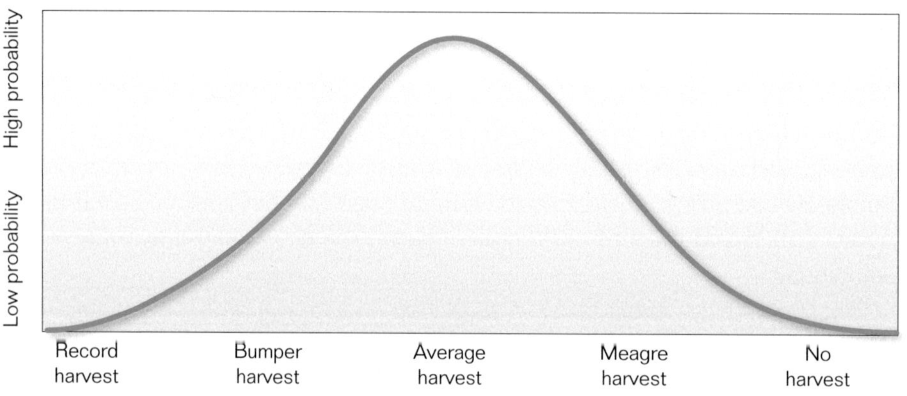

While the distinction between risk and uncertainty is analytically helpful, it does not distinguish discrete and separate phenomena. Rather, risk and uncertainty are the extreme ends of a single spectrum. Next year's harvest can be assessed as a risk on the basis of past experience but carries an element of uncertainty, because it is affected by various climatic parameters that are not at all predictable from experience or with current knowledge. The risk of a cyclone hitting a tropical city can be assessed using data on past occurrence of cyclones, although aspects of the potential damage are uncertain.

If it is correct to treat a subjectively formed assessment of a probability distribution as if it were drawn from a distribution based on repeated experience, what is the difference between risk and uncertainty? Perceptions of the probability distribution formed under conditions of uncertainty are more likely to change materially with a small number of new observations or amount of experience or further analysis.

The Review's work on climate change has made some contact with risk, more with uncertainty, and most of all with the wide territory between them. The mainstream science, embodied in the work of the Intergovernmental Panel on Climate Change (IPCC), sometimes discusses possible outcomes in terms of fairly precise probability distributions, yet describes its assessments in terms of 'uncertainties'. This suggests that they are applying Bayesian approaches to decisions under uncertainty. The decision framework is rarely made explicit, and sometimes is not clear.

The climate models on which the assessments are based are themselves diverse. They provide numerous observations on possibilities out of their diversity; in addition, each generates numerous results from repeated experiments. These are the senses in which the IPCC science draws from probability distributions. There are many points at which judgment rather than experience informs the model relationships. The resulting conclusions are therefore located somewhere on the uncertainty side of the middle of the risk–uncertainty spectrum.

1.3 Four types of benefits from mitigation

Three types of benefit from avoided climate change—that is, mitigation—can be measured in monetary values, as a change in the value of output or consumption. The fourth type of benefit of mitigation requires a different measurement unit.

The four types of climate change impacts, which in part can be mitigated, are illustrated in Figure 1.3.

1.3.1 Type 1: currently measurable market impacts

The first type of benefit from mitigation comprises currently measurable market impacts of climate change, which are avoided by mitigation. The measurement can be brought together through a computable general equilibrium economic model. The starting point for assessment is the estimation of climate impacts based on the means of the relevant probability distributions for these outcomes. These effects are typically measured as an impact on GDP or consumption, with monetary values as the unit of measurement.

Figure 1.3 The four types of climate change impacts

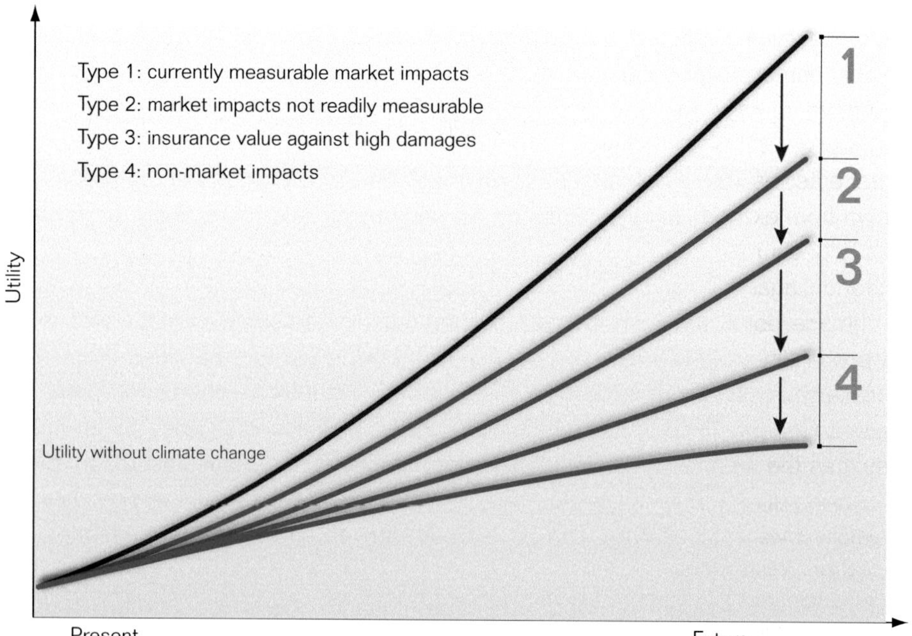

1.3.2 Type 2: market impacts not readily measurable

The second type of benefit of mitigation comprises market impacts similar in nature to the first, but not amenable to measurement in the current state of knowledge. For the Review, these impacts were not defined precisely enough in time for the modelling, but are, in principle, amenable to quantitative analysis. We seek to use what we know of these effects roughly to compare their possible size with the impacts that have been subject to formal modelling. As with the effects that are subject to modelling, we focus on the medians of the probability distributions of possible outcomes. We are drawing these judgments from views of the impacts that are closer to the uncertainty than the risk end of the risk–uncertainty spectrum. There is no reason to expect our estimates of these impacts to be too low rather than too high, but they are more likely than the estimates of the first type of benefit to be subject to large adjustments, in one direction or another, with the advance of knowledge. Examples from the Review include the impact of climate change on the tourism industry. As with the first type of benefit, the estimation of these effects would be in monetary values of GDP or consumption.

1.3.3 Type 3: insurance value against high damages

The third type of benefit of mitigation is the insurance value that it provides. On many impacts, there is large asymmetry between human evaluation of outcomes that are much more benign or much more damaging than the median. Humans tend to be risk averse when the outcomes include the possibility of large loss. Some of the possible outcomes at the bad end of the probability distribution would be thought by many people to be catastrophic. In such cases, mitigation has additional insurance value. What would we be prepared to pay to avoid a small probability of a highly damaging or possibly catastrophic outcome? It is probably more than we would be prepared to pay to avoid the certain prospect of the catastrophic event's contribution to the median outcome. Uncertainty strongly plays into this category of benefits, as the probability of extreme or catastrophic climate impacts is not known from experience, and must instead be based on expert judgment.

It is not a new idea for governments to make large financial commitments for insurance against low-probability, high-impact events. Defence absorbs several percentage points of GDP per annum, most of it on insurance against genuinely low-probability developments.

The possibility of outcomes that most people would consider to be catastrophic makes this a particularly important element of the assessment. Weitzman (2008a) sees it as the main element:

> [T]he burden of proof in the economics of climate change is presumptively upon whoever wants to model or conceptualize the expected present discounted utility of feasible trajectories under greenhouse warming without considering that structural uncertainty might matter more than discounting or pure risk per se. Such a middle-of-the-distribution modeler should be prepared to explain why the bad fat tail of the ... distribution is not empirically relevant and does not play a very significant role in the analysis.

1.3.4 Type 4: non-market impacts

The fourth type of benefit is more difficult to conceptualise, and quantify. Let us evoke an old tradition in economics, and talk about units of utility. (We could just as well call it welfare, if we removed ourselves from a modern interpretation in terms of social security for disadvantaged people.) That unit, 'utility', is used in economics to represent welfare in a demonetised fashion.

The focus of Australian policy making, as in other countries, is on maximising the welfare—or utility—of Australians. We can think of a utility function as rising with Australian consumption of goods and services, and also with a number of non-monetary services, such as environmental amenity (which itself may have a number of components), longevity, health, and welfare of people in other countries. If the comparisons of costs and benefits of the first three types of benefits from mitigation suggest a particular outcome, and it is clear from inspection that inclusion of the fourth might lead elsewhere, it is necessary explicitly to compare the monetary with the non-monetary effects on welfare of a particular position. This could in principle be done by forcing a monetary value onto particular non-monetary outcomes. An alternative is to leave the comparison of the monetary and non-monetary outcomes until after the possible conflict between them is known.

Examples of such non-market impacts include Australians' valuation of environmental amenity. They include the value that Australians place on the integrity of the Great Barrier Reef, Ningaloo, Kakadu and other features of the

Australian and international landscapes, on known shorelines, on genetic diversity and on the survival of species. They include the value that Australians place on long-established communities and social structures built around particular patterns of climate, or the use of green urban gardens and playing fields for recreation. To include such elements in an Australian utility function is not to place intrinsic value on environmental conservation, as some argue that we should. It is only necessary to accept that many Australians value such things, including as options for their offspring and future generations, and would be prepared to sacrifice some consumption of goods and services to retain them. Australians also value the avoidance of poverty and trauma in other countries, as demonstrated in their continued support of public and private international development assistance and disaster relief.

Non-market elements in a utility function, and the level of utility when the function includes market and non-market elements, are in their nature difficult to measure. Any politically derived mitigation policy decision will implicitly value them alongside changes that can be measured more easily.

Traditional welfare economics contains a few important insights into the roles that non-market factors, such as environmental amenity and concern for the welfare of others, might play in determining utility in a world of climate change.

The non-market values are likely to be 'superior goods', in that the relative value that people assign to them rises with incomes. In the late 21st century, when the average purchasing power of incomes over material goods and services is likely to be several times the present level, much higher relative value will be placed on any truncation of the natural estate that has occurred in the intervening years.

It is likely that at higher incomes, the price elasticity of substitution between conventional consumption and access to such non-market values as environmental amenity and concern for others' welfare will be low, and much lower than today. Near subsistence levels of consumption, few people would willingly sacrifice goods and services for greater environmental amenity, or for improved development prospects of others at home or abroad. In the likely material affluence of the late 21st century, many more people are likely to trade substantial amounts of access to material consumption for small amounts of improved values of services not available through market processes.

An extremely low rate of substitution between non-market services and conventional consumption of goods and services at high incomes, in the presence of large impacts from climate change, would challenge the proposition that continuing economic growth would necessarily lead to higher average utility in the distant future.

One implication is that the utility of Australians under policies that allocate priority to such non-market values as the services provided by the natural environment, and development in poor countries, is likely to be much higher than the application of today's preference systems at today's material consumption levels would suggest.

1.4 How effective adaptation reduces the costs of climate change

Some of the costs of climate change can be diminished by the adaptive behaviour of individuals and firms, and by policies that support productive adaptation.

Effective adaptation requires a strong applied science base; good markets for reallocation of resources, goods and services; and capital for investment in defensive structures and new productive capacity that is more suitable to the new environment.

All of these capacities are more abundant in developed than low-income developing countries. For the latter, the impact of climate change is likely to be undiluted and more severe. Australia's location in an immediate region of vulnerable developing countries will make some of its neighbours' challenges our own. Investment in adaptive responses in the arc of island countries and regions from Timor-Leste through eastern Indonesia, Papua New Guinea and the South Pacific is likely to become an important component of Australia's own cost-reducing adaptation to climate change.

The modifying impact of adaptation is illustrated by Australian agriculture. Better and earlier knowledge will allow farmers to make timely decisions on whether new money should continue to be invested in locations that seem to be severely damaged by climate change, or whether it is better to find new livelihoods in less challenging locations. Investment in plant and animal genetics may be able to diminish the loss of productivity associated with higher temperatures and changing rainfall patterns. Investment in water retention or storage will sometimes be an economically sensible response to more variable rainfall.

Hardest of all, the most effective adaptive responses in agriculture to climate change will sometimes require fundamental changes in attitudes, policies and institutions. For example, the loss in irrigated agricultural value under moderate warming and drying scenarios could be greatly reduced by shifting from established to free market allocation patterns of water allocation, so that limited water resources are directed without qualification to their most productive uses. Livestock industries in these same circumstances would suffer less if established patterns of quarantine on feed imports were to be relaxed. We can presume that change of such a fundamental kind would not be achieved without rancour and disputation over policy, and would require public policy management of exceptional dexterity and quality.

In assessing the costs and benefits of mitigation, the costs of adaptation need to be subtracted from business as usual and mitigated output and consumption. The benefits of adaptation through reduced climate change damage need to be subtracted from the gains from mitigation. The Review takes this partly into account by presuming a substantial adaptive response in assessing the costs of climate change at various levels of mitigation. For example, the presumed wheat yields are based on the expectation that planting times and new seed varieties will be developed rapidly for changing conditions.

The costs of adaptive responses will generally come early, and the benefits from reduced costs of climate change later. On the whole, the Review has only been partially able to take account of the costs of adaptation; and the assessment of reduced costs of climate change on output and consumption is incomplete—a task for future analysts.

Some of the most important adaptive responses to climate change, and the most difficult to bring to account in an analysis of optimal levels of mitigation, involve changes in attitudes and values. The city dwellers of densely populated regions of Northeast Asia have long been accustomed to life that is almost entirely separated from the natural environment. If climate change separates more and more humans from natural environmental conditions, will they simply change in their values and preferences, and learn to accept without a feeling of loss what they have never known, or knew only in distant memory? Will it matter to Australians of the future if their children do not enjoy the grassed playing fields that were once formative in what we imagined as our community culture? Could Australians learn to love living in a country and a world without many of the natural features that are now enjoyed with comfortable familiarity?

Humans adapt to changed and difficult environments when they must. There will still be joys of learning and life even in the most unhappy scenarios of future climate change. But it must be doubted that humans will have changed so much that they fail to regret what they still had in the Australian natural environment in the early 21st century. For want of reason to do otherwise, the Review will assess the value that Australians place on environmental amenity in terms of today's perspectives and preferences, manifested in a future world of greatly increased consumption of conventional goods and services.

1.5 Measuring the benefits of mitigation against the costs

To a sceptical economist, the case for action is not made simply by comparing the cost of unmitigated climate change with the cost of mitigation.

The relevant comparator is the reduction in the cost of climate change that is achieved as a result of the mitigation action. If we are evaluating Australian mitigation action, the reduction in costs of climate change that is relevant is that associated with the total global mitigation action that it enables—either by Australia, or by the set of countries that are undertaking joint action.

The benefit from mitigation is the costs of climate change avoided, after the costs and ameliorating effects of adaptation had been taken into account. Do the benefits of mitigation exceed the costs for Australians?

The costs of mitigation come earlier and are more certain. The benefits come later and are less certain. How do we compare later with earlier benefits? How do we compare more with less certain outcomes?

The costs and benefits of mitigation, in Australia and in other countries, fall on and accrue to different groups in the community. They are also felt and valued in various ways by different people. How do we weigh the relative effects on welfare of different people? In particular, what relative weight do we give to costs and benefits to the rich and to the poor? It may be that an overall assessment of whether mitigation is worthwhile will depend on the distribution of costs and benefits across the community.

The Stern Review (2007) addressed the question of whether mitigation action was justified for the world as a whole. This turns out to be an easier question than whether mitigation action is justified from the point of view of an individual country. An assessment of whether mitigation action is justified for an individual country must deal with all of the complexities that Stern addressed for the world as a whole, plus one. And that additional source of complexity is perhaps the most difficult of all.

The relevant mitigation is global. A single country's action is relevant only in its direct and indirect contribution to global mitigation. The costs of various levels of mitigation for a single country depend mainly on the extent of its own mitigation—although these costs are substantially reduced in a global agreement within which at least major economies apply similar emissions pricing regimes. The benefits depend overwhelmingly on what other countries are doing. Each country's evaluation of whether some mitigation action of its own is justified depends on its assessment of the interaction between its own decisions and those of others. Thus its own decision framework must depend on its assessment of the dynamics of complex games, among many countries. The games are framed within an awful reality, that each country has a narrow national interest in doing as little as it can, whatever others do, so long as its own action does not diminish the mitigation action that others actually take.

The global mitigation effort is the sum of the separate but inter-related mitigation decisions of individual sovereign countries. It is the sum of implicit or explicit decision processes in all countries, of the kind that we are attempting for Australia. The sum of the decision processes in many countries—democratic and authoritarian, soft and hard states, rich and poor—will determine the global mitigation effort.

1.6 A graphical representation of the benefits and costs

Let us plot our expectations of the level of national utility or welfare over time in the absence of mitigation, national or global (Figure 1.4). National utility will generally rise over time, in line with the Australian experience through its history. On the same graph, now plot expectations of welfare over time at a given level of national mitigation, which is associated with a defined degree of global mitigation.

Figure 1.4 Utility with and without mitigation

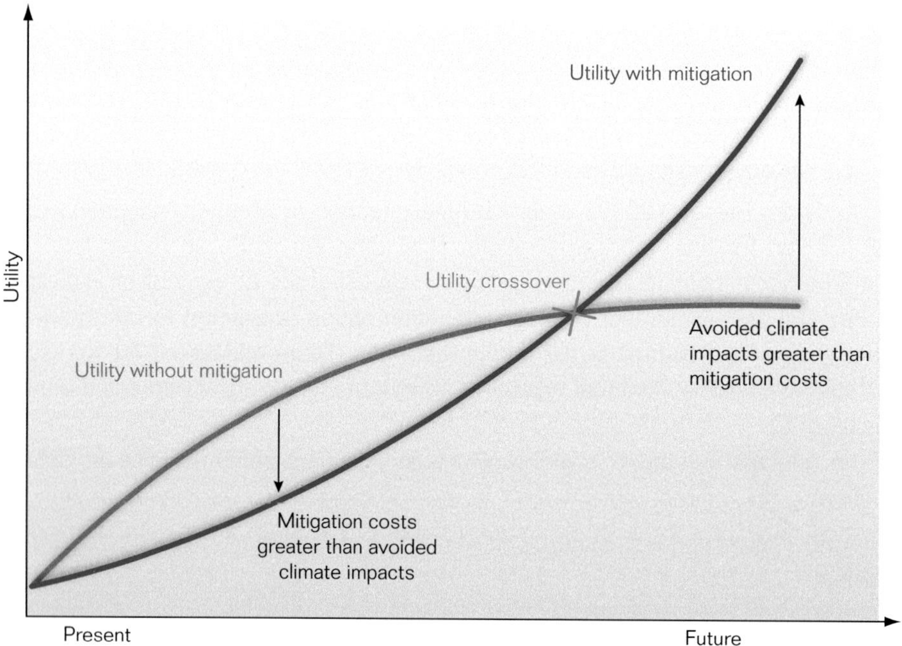

Figure 1.5 Utility under a more ambitious level of mitigation

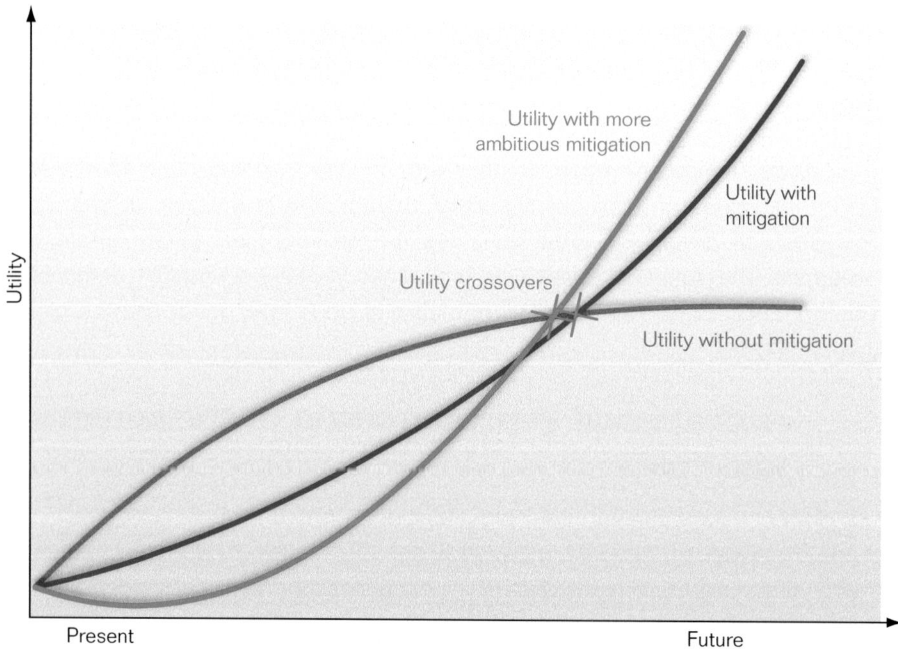

Figure 1.6 Utility with more climate change impacts taken into account

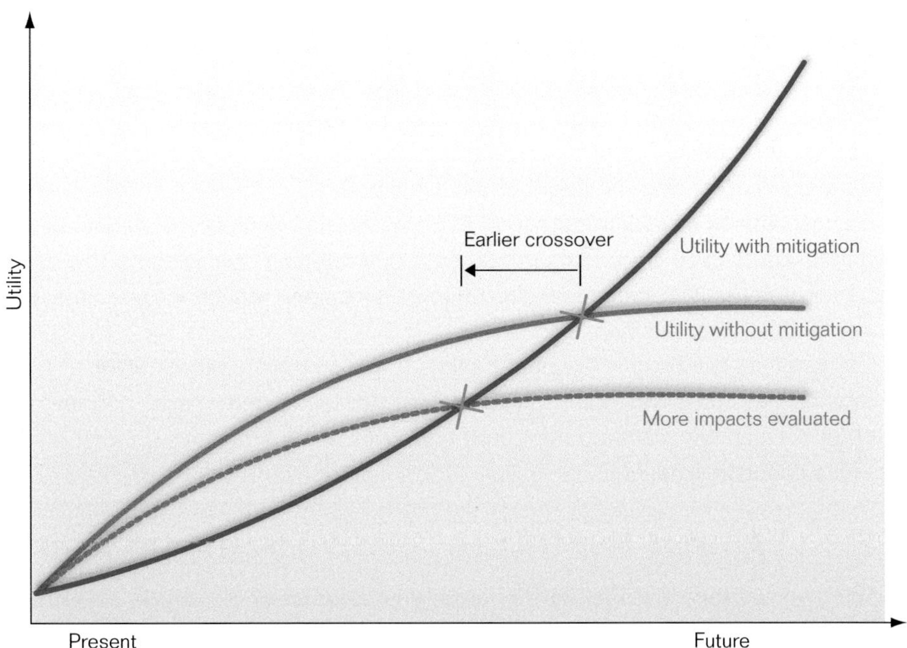

As shown in Figure 1.4, the utility curve without mitigation is above the utility curve with mitigation in the early years. Mitigation has a cost. However, in cases in which national mitigation is associated with substantial global mitigation, at some point the mitigation utility curve may rise above the utility curve in the absence of mitigation. We can call the point at which utility associated with mitigation exceeds utility in the absence of mitigation for the first time, the utility crossover point.

The two curves together describe the shape of a fish. The body of the fish covers years in which the net benefits of mitigation are negative. The area of the body of the fish represents the excess costs of mitigation in the years to the crossover point. The tail of the fish covers years in which the net benefits of mitigation are positive. The tail of the fish grows in depth and total area with time.

Figure 1.5 shows the utility curves with and without mitigation for a more ambitious level of mitigation. The fish will tend to have a fatter body and tail, as both the costs of mitigation and the benefits are increased. The relative sizes of body and tail, and locus of the crossover point, are matters for empirical analysis and judgment. Figure 1.6 shows the utility curves with a greater degree of climate change impacts being taken into account. As more costs of climate change are quantified, the utility curve for the unmitigated case shifts downward.[2] The crossover point comes earlier and the tail becomes larger relative to the body of the fish—implying higher net benefits from mitigation. These are also matters for empirical analysis. The policy question is whether the area of the body of the fish exceeds that of the tail of the fish. Future utility has to be valued at present values, so the difference in the annual levels of utility between the two curves defining the fish have to be discounted to the present at an appropriate rate. The choice of discount rate will have a major influence on the size of the body relative to the size of the tail. We turn to the discount rate after looking more closely at other influences on the sizes of the body and tail of the fish.

The cost of mitigation (augmented somewhat, and more in the low-mitigation cases, by the greater cost of investment in adaptation in the low-mitigation scenarios) will determine the depth of the fish's body.

For any chosen mitigation outcome, there will be an optimum distribution of the mitigation effort over time. For purposes of presentation, we assume that ideal mitigation policies, including ideal allocation of emissions reduction over time, are chosen for implementation.

The length of the body is the time it takes to get to the crossover point.

Beyond the crossover point, the length of the tail is determined only by any limit that society may place on the future time over which it remains concerned for the utility of Australians.

1.7 Valuing the future relative to the present

Should society place any limit on the future time over which it remains concerned for the utility of Australians? It is not obvious why it should do so.

The value of avoided irreversible effects of climate change extends forward to the point at which the life of the human species may have been extinguished by some separate influence. There is some chance of extinction at any time, at least in the contemporary human state of knowledge about weapons of mass destruction and capacity to control their use, and the low level of knowledge and capacity relevant to avoidance of the risks of the earth colliding with extra-terrestial bodies. The probability of extinction is not high in any year and perhaps in any century, but it is above zero.

So if we are to include the welfare of all future generations in our assessment of utility, how should we value the future relative to the present?

How much value should be attached to climate change impacts that occur beyond the lifetimes of most of those alive today? In comparing utility across generations, we need to determine the discount rate. In the Review's quantitative assessments, only the market costs and benefits of median impacts (types 1 and 2) of climate change mitigation are assessed, and only to the year 2100. Nevertheless, any view formed on discounting is important also in an assessment of qualitative climate change impacts in the longer term, and their implication for mitigation policy today.

There are two reasons why society may place less value on income and consumption in the future than on the same income and consumption today. The first element in the discount rate is the rate of pure time preference, that is, the rate at which future utility is discounted simply because it is in the future. Many of the philosopher kings of economics, from Ramsey (1928) to Sen (1961), have argued for a pure rate of time preference that is close to zero, thus placing no discount on the utility of people in the future just because today these people are young or not yet born. For example, as DeLong (2006) pointed out, a rate of pure time preference of 3 per cent means that somebody born in 1995 'counts for twice as much as someone born in 2020'. This flies in the face of the utilitarian principle underlying most economic analysis: that equal weight should be placed on each person. Dasgupta (2006), in his critique of the Stern Review's approach to discounting, also supports a rate of pure time preference close to zero. By contrast, some economists, including Nordhaus (1994, 2008), a pioneer of economic modelling of climate change, use a pure rate of time preference of 1.5 per cent or higher, calibrated to yield an overall discount rate that matches the overall cost of capital to the economy (see below).

The Review judges that a near-zero pure rate of time preference is appropriate. The only reason for a positive rate of pure time preference is the risk of human extinction in any one year. This should be a low number, and the Review uses a rate of pure time preference of 0.05 per cent. This is similar to the parameter value used in the Stern Review (2007).

The second element in the discount rate is the marginal elasticity of utility with respect to consumption. This is a measure of society's concern for equity in income distribution. We accept that a dollar of incremental income means less to the utility of the rich than of the poor. The people of tomorrow will have higher material incomes and wealth than people today, although this is likely to be offset by reduced environmental amenity in assessment of utility. It is reasonable to value future income at a lower rate than current income, insofar as future income is higher. How much less? Higher and lower values have been suggested, but no one contests that income has diminishing marginal utility with increased income. There are compelling theoretical reasons for using an elasticity of 1, and Quiggin (2008) argues that this is the most common choice in the literature (see also Jensen & Webster 2007). By contrast, Dasgupta (2007) argued that an elasticity of 1 implies that 'distribution of well-being among people doesn't matter much', and that higher values more adequately reflect distributional concerns and observed savings rates. These savings rates, however, vary greatly between countries and through time. Stern used a parameter value of 1, Nordhaus of 2.

The Review uses two alternative parameter values for the marginal elasticity of utility, 1 and 2, a range that accommodates strongly diverging views on how much should be spent now to benefit future, presumably richer, generations. Under an elasticity of 1, future income is discounted at the same rate as the increase in per capita income (plus the rate of pure time preference), while at an elasticity of 2 it is discounted at twice that rate. The average annual growth rate in Australian per capita income from 2013 to 2100 in the base case is 1.3 per cent. Thus the two real discount rates used by the Review for assessment of discounted net costs of mitigation of climate change in Australia are 1.35 per cent and 2.65 per cent.

The argument for being careful about the sacrifice of current utility through expenditure on mitigation in pursuit of future income is a powerful one. But there is one important qualification of this case for caution about strong mitigation on intergenerational income distribution grounds. The rate of substitution between conventional consumption and non-market services is likely to be low when incomes and material consumption are much higher than they are today. Climate change may greatly diminish the availability of non-market services for future generations. As a result, one cannot be sure that, despite much higher material consumption, the average utility of people in future will be greater than the average utility today. Hence, linking the marginal elasticity of utility to the growth in per capita income may lead to higher than intended discount rates. Furthermore, if considerable weight is given to the bad end of the probability distribution of outcomes from climate change, there is a possibility that utility may be lower for many people in future than at present.

There is another view, that market discount rates reflect the time preferences that are revealed in actual decisions on savings and investment, which are the vehicles for arbitrage between future and current economic activity. This raises two questions. What is the appropriate market discount rate? The other and more fundamental question is whether discounting is a normative or a positive issue, and so whether it is appropriate to use a market rate (Baker et al. 2007).

On the first question, a case can be made for using the market rate for sovereign debt in countries like Australia. The mid-point of the range of normative discount rates discussed above roughly coincides with the inflation-adjusted long-term market rate of return of government bonds in Australia and the United States, which stands at 2.2 and 2.1 per cent respectively. These would seem to be more appropriate than equity market rates, which are much higher, reflecting perceptions of firm-specific and other risks that are not relevant to the current analysis.

Is there a contradiction between using a discount rate in the range 1.25 to 2.65 per cent in summing utility of income over the generations, and applying a higher rate, 4 per cent, in pricing emissions permits and allocating capital over the 21st century in the Garnaut–Treasury modelling that is discussed later in this report? No. The Garnaut–Treasury rate is that at which investors choose to allocate capital between permits and other financial investments over time. It is our assessment of the 'Hotelling rate' that the market would come to apply in decisions that determine the optimal rate of depletion of the atmosphere's limited capacity to absorb greenhouse gases. It is advisedly a rate embodying all of the commercial risks involved in holding permits and investing in emissions-related activities. Its nearest analogue in currently operating markets is the gold contango, where the implicit real discount rate is typically lower than the 4 per cent applied in the Garnaut–Treasury modelling for the permit market contango. This reflects the view agreed between the Review and Treasury, that there would be additional risk and therefore a higher risk premium in the permit market. The Review uses a discount rate of 4 per cent in the modelling analysis of permit price trajectories, embodying a risk-free real interest rate of 2 per cent and a risk premium in markets for permits of 2 per cent (see Chapter 11).

On the second question, strong arguments can be made against any approach using observed market rates in the case of climate change. They include that a market portfolio approach averaging over high-yielding and underperforming projects is not applicable to climate change where winners and losers live in different generations (Beckerman & Hepburn 2007), and that the genuine uncertainties over which interest rate should be used in the long term favour lower discount rates (Weitzman 2007).

The Review judges that a normative approach is warranted on an issue that affects society as a whole over long time frames and on fundamental issues. Yet the justification of the discount rates used does not rely on using a normative approach. Rates that the Review derived from analysis, presented above, straddle the market rate that is judged to be most appropriate. In this case at this time, there is no conflict between normative and positive approaches.

The analysis for the Review has been calibrated with percentage points of GNP or consumption. The use of a discount rate that is higher than the rate of growth of GNP will cause the present value of a percentage point of current GNP to be greater than that of a percentage point of future GNP. The use of a discount rate that is lower than the rate of growth of GNP causes the present value of a percentage point of future GNP to exceed that of a percentage point of current GNP.

In Australia's case, unlike for most developed countries, the modelling points to the expected rate of GNP growth (2.1 per cent over the remainder of the 21st century) falling within the middle or higher than the middle of the range of discount rates thought to be relevant (1.35 to 2.65 per cent). It follows that at the lower discount rate, the present value of a percentage point of GDP in the early 22nd century will exceed that of a percentage point of GDP now.

Notes

1 The 'Platinum Age' is so named because global economic growth in this period has been and is expected to continue to be more expensive and stronger than in the 'Golden Age' of the 1950s and 1960s.

2 For ease of exposition, Figure 1.6 assumes that utility with mitigation already accounts for the lesser economic consequences of the additional impacts. If this were not the case, the utility-with-mitigation curve would shift downwards but the downward shift would be less than the downward shift of the utility-without-mitigation curve. The crossover point would still be earlier.

References

Baker, R., Barker, A., Johnston, A. & Kohlhaas, M. 2007, *The Stern Review: An Assessment of its Methodology, Productivity Commission Staff Working Paper*, Australian Government, January.

Beckerman, W. & Hepburn, C. 2007, 'Ethics of the Discount Rate in the Stern Review on the Economics of Climate Change', *World Economics*, vol. 8, no. 1, pp. 187–210.

Club of Rome 1972, *The Limits to Growth: A report for the Club of Rome's project on the predicament of mankind*, D.H. Meadows, D.L. Meadows, J. Randers & W.W. Behrens (eds), Universe Books, New York.

Dasgupta, P. 2007, 'Commentary: the Stern Review's economics of climate change', *National Institute Economic Review* 199: 4–7.

DeLong, B. 2006, *Partha Dasgupta Makes a Mistake in His Critique of the Stern Review*, November 30, <http://delong.typepad.com/sdj/2006/11/partha_dasgupta.html>.

Garnaut, R. & Huang, Y. 2007, 'Mature Chinese growth leads the global Platinum Age', in R. Garnaut & Y. Huang (eds), *China: Linking Markets for Growth*, Asia Pacific Press, Australian National University, Canberra.

Hotelling, H. 1931, 'The economics of exhaustible resources', *Journal of Political Economy* 39(2): 137–75.

Jensen, P. & Webster, E. 2007, *Inter-generational Justice, Discount Rates and Climate Change*, report commissioned for the Garnaut Climate Change Review, Melbourne Institute of Applied Economic and Social Research, Melbourne.

Jevons, W.S. 1865, *The Coal Question: An inquiry concerning the progress of the nation, and the probable exhaustion of our coal-mines*, Macmillan, London.

Keynes, J.M. 1921, *A Treatise on Probability*, Macmillan, London.

Knight, F. 1921, *Risk, Uncertainty and Profit*, Houghton Mifflin, Boston.

Little, I.M.D. & Mirrlees, J.A. 1968, *Manual of Industrial Project Analysis in Developing Countries*, OECD, Paris.

Maddison, A. 2001, *The World Economy: A millennial report*, OECD, Paris.

Malthus, T. 1798, *An Essay on the Principle of Population*, 1st edn, J. Johnson, London.

Nordhaus, W. 1994, *Managing the Global Commons: The economics of climate change*, MIT Press, Boston.

Nordhaus, W. 2008, *The Challenge of Global Warming: Economic models and environmental policy*, Yale University, New Haven, Connecticut.

Quiggin, J. 2008, 'Stern and his critics on discounting and climate change: an editorial essay', *Climatic Change*, doi:10.1007/s10584-008-9434-9.

Raiffa, H. 1968, *Decision Analysis: Introductory lectures on choices under uncertainty*, Addison-Wesley, Reading, Massachusetts.

Raiffa, H. & Schlaifer, R.O. 1961, *Applied Statistical Decision Theory*, Harvard Business School, Boston.

Ramsey, F.P. 1928, 'A mathematical theory of saving', *Economic Journal* 38: 543–59.

Sen, A.R. 1961, 'On optimising the rate of saving', *Economic Journal* 71: 479–96.

Stern, N. 2007, *The Economics of Climate Change: The Stern Review*, Cambridge University Press, Cambridge.

Weitzman, M.L. 2008a, 'On modeling and interpreting the economics of catastrophic climate change', Harvard University, Cambridge, Massachusetts, available at <www.economics.harvard.edu/faculty/weitzman/files/modeling.pdf>.

Weitzman, M.L., 2008b. 'A review of the Stern Review on the economics of climate change', *Journal of Economic Literature* 45(3): 703–24.

2 UNDERSTANDING CLIMATE SCIENCE

Key points

The Review takes as its starting point, on the balance of probabilities and not as a matter of belief, the majority opinion of the Australian and international scientific communities that human activities resulted in substantial global warming from the mid-20th century, and that continued growth in greenhouse gas concentrations caused by human-induced emissions would generate high risks of dangerous climate change.

A natural carbon cycle converts the sun's energy and atmospheric carbon into organic matter through plants and algae, and stores it in the earth's crust and oceans. Stabilisation of carbon dioxide concentrations in the atmosphere requires the rate of greenhouse gas emissions to fall to the rate of natural sequestration.

There are many uncertainties around the mean expectations from the science, with the possibility of outcomes that are either more benign—or catastrophic.

Climate change policy must begin with the science. When people who have no background in climate science seek to apply scientific perspectives to policy, they are struck by the qualified and contested nature of the material with which they have to work. Part of the uncertainty derives from the complexity of the scientific issues. In the public discussion of the science, additional complexity derives from the enormity of the possible consequences, which encourages a millennial perspective. Part derives from the large effects of possible policy responses on levels and distributions of income, inviting intense and focused involvement in the discussion by those with vested interests.

The Review is not in a position to independently evaluate the considerable body of scientific knowledge; it takes as a starting point the majority opinion of the Australian and international scientific communities that human-induced climate change is happening, will intensify if greenhouse gas emissions continue to increase, and could impose large costs on human civilisation.

This chapter draws extensively on the Fourth Assessment Report of the Intergovernmental Panel on Climate Change, and on detailed reports prepared by Australian scientists, research published since the IPCC Fourth Assessment Report and work commissioned specifically for the Review. It aims to build an understanding of the way humans can influence the climate and the limitations in our current understanding of the climate system, and introduces key terminology and concepts relevant to people who are interested in climate change policy.

In drawing on the work of the IPCC, and the large majority of Australian scientists who are comfortable working within that tradition, we are still faced with immense uncertainties, which have informed (and at times bedevilled) the Review's analysis. These perspectives may cease to be the mainstream as the development of climate science proceeds and the uncertainties narrow. At this time, the Review believes it is appropriate to give the main weight to them.

2.1 The earth's atmosphere

2.1.1 A changing atmosphere

The earth is surrounded by an atmosphere that protects it from high-energy radiation and absorbs heat to provide a moderate climate that supports life.

The earth's atmosphere has not always been the same as it is today. Billions of years ago, the atmosphere was composed mainly of ammonia, water vapour and methane, but over time release of gases from within the planet through volcanic eruptions and discharge of gases from ocean vents changed conditions so that carbon monoxide, carbon dioxide and nitrogen became dominant.

Around 3.5 billion years ago, algae-like organisms first began to use the energy from the sun to convert carbon dioxide from the air into carbohydrates, with oxygen as a by-product. Over time, oxygen accumulated and reached the current levels of 21 per cent of atmospheric volume.

2.1.2 The natural greenhouse effect

The earth's atmosphere acts like the roof of a greenhouse, allowing short-wavelength (visible) solar radiation from the sun to reach the surface, but absorbing the long-wavelength heat that is emitted back. This process is referred to as 'the greenhouse effect', and the gases that absorb the emitted heat are known as greenhouse gases. The main naturally occurring greenhouse gases are water vapour, carbon dioxide, methane, nitrous oxide and ozone (see section 2.3).

Compared to nitrogen and oxygen, which together comprise 99 per cent of the volume of the atmosphere, greenhouse gases occur only at trace levels, making up just 0.1 per cent of the atmosphere by volume (IPCC 2001a). Despite this, their presence means that the earth has an average global surface temperature of about 14°C—about 33°C warmer than if there were no greenhouse gases at all (IPCC 2007a: 946).

2.1.3 Changes in greenhouse gases and temperature over time

Records of carbon dioxide concentrations taken from proxy measures such as fossil plants and ice core data are available for the last 400 million years. These records indicate that atmospheric concentrations of carbon dioxide have fluctuated between about 180 ppm (parts per million by volume), levels similar to pre-industrial concentrations of 280 ppm, and levels higher than 4000 ppm (Royer 2006; IPCC 2007a: 444).

There is a high degree of uncertainty in historical measurements of temperature before modern times, which must be estimated from a range of indirect sources such as annual growth rings in trees and corals, and small fossils in ocean and lake sediments. During the last 2.5 million years, climate records document a saw-tooth pattern of changes in temperature and ice volume. In the last 600 000 years the fluctuations show a periodicity of around 100 000 years (Ruddiman 2008). The periods when polar ice caps were greatly expanded, which resulted in ice sheets covering large parts of the northern continents, are known as glacial periods or ice ages. Those without extended polar ice caps are known as interglacials. The last glacial maximum occurred 21 000 years ago. For the last 10 000 years, the earth has been in an interglacial period (IPCC 2007a: 447).

Glacial periods occur when solar radiation is reduced, and interglacials when solar radiation is more intense. Consistent changes in the intensity of solar radiation from regular variations in the shape of the earth's orbit and the tilt of its axis, as well as the sunspot cycle, are seen as drivers of these cyclical climate changes (Ruddiman 2008).

Fluctuations in carbon dioxide and methane concentrations have occurred in the 600 000-year period before the present, but the role of greenhouse gases in contributing or responding to glacial–interglacial fluctuations is complex.

2.1.4 How are the recent changes different?

Why are we so concerned about the current changes in climate and greenhouse gas concentrations if they have fluctuated so much over the earth's history?

Apart from the earliest identified hominids, the history of our species has been within the period of relatively low carbon dioxide concentrations. Modern forms of our species first appeared only around 200 000 years ago. The development of agriculture, large-scale social organisation, writing, cities and the behaviours we associate with modern civilisation has occurred only in the last 10 000 years. The period in which human civilisation has developed, located within an interglacial period known as the Holocene, has been one of equable and reasonably stable temperatures.

Concentrations of carbon dioxide now exceed the natural range of the last two million years by 25 per cent, of methane by 120 per cent and of nitrous oxide by 9 per cent (IPCC 2007a: 447). The anthropogenically driven rise in carbon dioxide since the beginning of the industrial revolution (around 100 ppm) is about double the normal 'operating range' of carbon dioxide during glacial–interglacial cycling

(180–280 ppm) (Steffen et al. 2004). Trends in the atmospheric concentration of carbon dioxide, methane and nitrous oxide for the last 250 years are shown in Figure 2.1—it is not just the *magnitude* of the post-industrial increase in greenhouse gas concentrations that is unusual, but also the *rate* at which it has occurred.

There is high natural variability in global temperatures in recent millennia. The current high temperatures, though unusual over the last 1000 years, are not unusual on longer time frames. However, the rapid rate of the current warming is highly unusual in the context of the past millennium (CASPI 2007).

Figure 2.1 Trends in atmospheric concentrations of carbon dioxide, methane and nitrous oxide since 1750

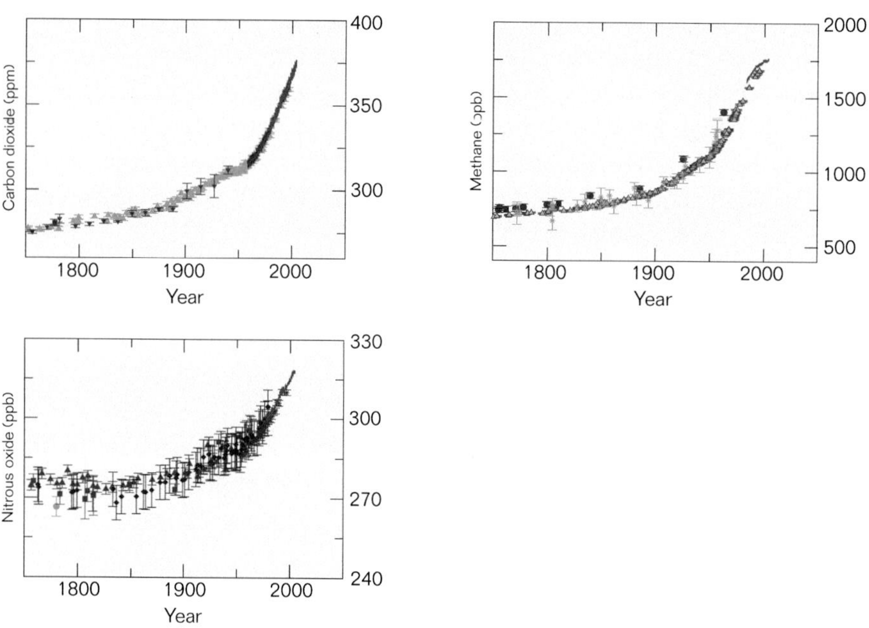

Note: Measurements are shown from ice cores (symbols with different colours for different studies) and atmospheric samples (red lines).

Source: IPCC (2007a: 3), formatted for this publication.

2.1.5 Is human activity causing the earth to warm?

The development of climate-related research and modelling has allowed increasingly more definitive assessments of the human impacts on climate. The IPCC Fourth Assessment Report noted an improvement in the scientific understanding of the influence of human activity on climate change. The report concluded that the warming of the climate system is 'unequivocal' (IPCC 2007a: 5), that there is a greater than 90 per cent chance that 'the global average net effect of human activities since 1750 has been one of warming' (IPCC 2007a: 3). Confidence in the influence of humans on other elements of climate change, such as droughts and severe weather events, is not as high, but is increasing as modelling techniques and observation databases improve.

2.2 Understanding climate change

2.2.1 Definitions of climate change

The IPCC (2007a: 943) defines climate change as 'a change in the state of the climate that can be identified (for example, by using statistical tests) by changes in the mean and/or variability of its properties, and that persists for an extended period, typically decades or longer'. Climate change may be due to natural internal processes or external influences, or to persistent anthropogenic changes in the composition of the atmosphere or land use.

By contrast, the United Nations Framework Convention on Climate Change defines climate change as 'change of climate which is attributed directly or indirectly to human activity that alters the composition of the global atmosphere and which is in addition to natural climate variability observed over comparable time periods' (UN 1992).

This report uses the IPCC definition, so that the discussion of climate change includes changes to the climate caused by natural phenomena such as volcanic eruptions.

2.2.2 The climate system

In a narrow sense, climate can be defined as the 'average weather' and described in terms of the mean and range of variability of natural factors such as temperature, rainfall and wind speed.

More broadly, the climate is a system involving highly complex interactions between the atmosphere, the oceans, the water cycle, ice, snow and frozen ground, the land surface and living organisms. This system changes over time in response to internal dynamics and variations in external influences such as volcanic eruptions and solar radiation.

The atmospheric component is the most unstable and rapidly changing part of the climate system. The atmosphere is divided into five layers with different temperature characteristics. The lower two—the troposphere and the stratosphere—have the most influence on the climate system.

The troposphere extends from the surface of the earth to an altitude of between 10 and 16 km. Clouds and weather phenomena occur in the troposphere, and greenhouse gases absorb heat radiated from the earth.

The stratosphere, which extends from the boundary of the troposphere to an altitude of around 50 km, is the second layer of the atmosphere. The stratosphere holds a natural layer of high ozone concentrations, which absorb ultraviolet radiation from the sun (Figure 2.2). The balance of energy between the layers of the atmosphere is a major driver of atmospheric and ocean circulation, which leads to weather and climate patterns (IPCC 2007a: 610).

Figure 2.2 A stylised model of the natural greenhouse effect and other influences on the energy balance of the climate system

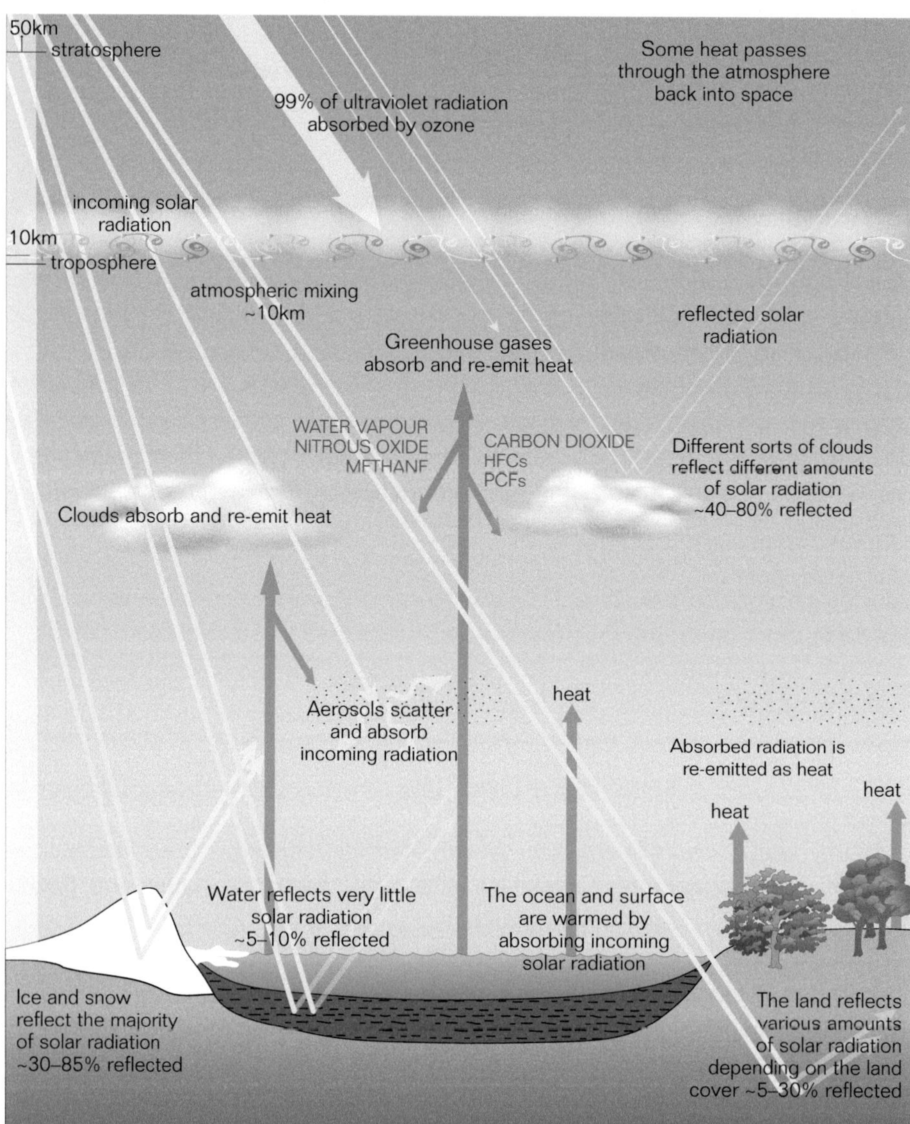

2.2.3 The energy balance of the climate system

Energy enters the climate system as visible and ultraviolet radiation from the sun, and is absorbed, scattered or reflected, re-emitted as heat, transferred between different elements of the system, used by organisms and emitted back into space.

The balance between the energy entering and leaving the system is what determines whether the earth gets warmer or cooler, or stays the same. Changes in the strength of radiation from the sun change the energy that enters the system, while the flow of energy within the system and the amount that is released can be modified by a wide range of factors.

The composition of gases in the atmosphere plays a big part in the amount of heat that is retained in the climate system, but there are many other influences. Aerosols (Box 2.1) can scatter the incoming radiation so that it never reaches the surface, and can also cause changes in cloud cover. More clouds will reflect more sunlight from their upper surfaces, but also absorb more heat radiating from the earth. Variations in land cover affect the amount of sunlight that is reflected from the surface (the 'albedo effect'), and how much is absorbed and re-emitted as heat.

Human activity can affect the energy balance of the climate system in a number of ways. Examples include changes in land use; emissions of aerosols and other pollutants; emissions of greenhouse gases from activities such as agriculture, energy production, industry and land clearing; emissions of other pollutants that react in the atmosphere to form greenhouse gases; and influences on cloud formation through aviation.

Box 2.1 Aerosols

Aerosols are tiny particles or droplets in the atmosphere, including sulphates, ash, soot, dust and sea salt, that can be natural or anthropogenic in origin. A major natural source of aerosols is volcanic eruptions. The major anthropogenic source is fossil fuel combustion.

Aerosols generally create a cooling effect, but there is great uncertainty about the magnitude of this effect. Black carbon, or soot, has a warming effect because it absorbs solar radiation. Recent research suggests this effect may be considerably higher than previously estimated (Ramanathan & Carmichael 2008). Because the lifetime of aerosols in the atmosphere is much shorter than that of greenhouse gases, the effects are more likely to be felt in the region in which the aerosol is produced.

2.2.4 Factors leading to warming of the climate system

The warming of the climate system evident in the last half century is a result of the cumulative effect of all the natural and human drivers that influence the amount of warming or cooling in the system. The contribution of different factors leading to an overall warming of the atmosphere since 1750 is shown in Figure 2.3.

The dominant influence since 1750 has been an increase in concentrations of carbon dioxide. Aerosols have had a net cooling influence, although this effect is poorly understood. Natural variability in solar radiation has had a small warming influence, but there is a high level of uncertainty in the magnitude of the effect (IPCC 2007a: 192).

Even if there were no further human-induced increases in aerosols and greenhouse gases, the long-lived greenhouse gases would remain for hundreds and even thousands of years, leading to continued warming. Aerosols are removed from the atmosphere over much shorter periods, so their cooling effect would no longer be present. Therefore, in the long term, the major influence of humans on the climate will be through activities that lead to increased concentrations of greenhouse gases in the atmosphere.

Figure 2.3 Contribution of human and natural factors to warming since 1750

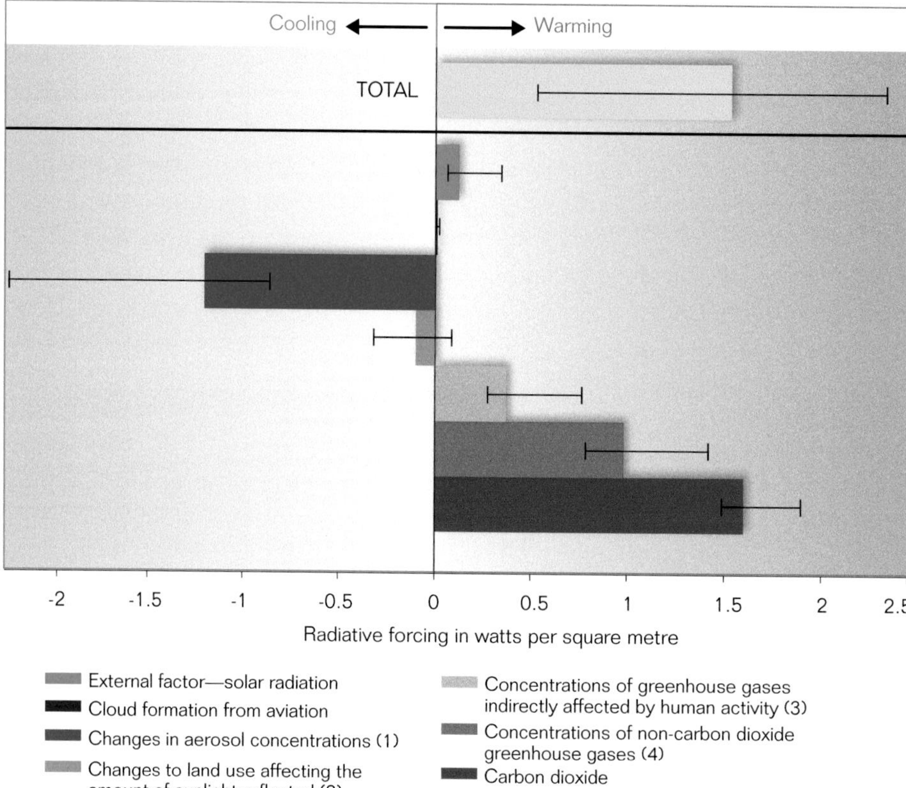

Note: 'Radiative forcing' is a measure of the induced change to the energy balance of the atmosphere. Warming and cooling influences are indicated by positive and negative values respectively. When elements are grouped, uncertainty bands are approximated from the highest uncertainty in an individual element.

(1) Includes both the direct effect and the cloud albedo effect.
(2) Includes the cooling effect of changes to land use and the warming effect of black carbon on snow.
(3) Includes tropospheric and stratospheric ozone and stratospheric water vapour.
(4) Includes methane, nitrous oxide, HFCs, PFCs and sulphur hexafluoride.

Source: Based on IPCC (2007a: 204).

2.3 Linking emissions and climate change

The high natural variability and complex internal interactions create uncertainty in the way the climate will respond to increased emissions. Figure 2.4 illustrates the relationship between emissions from human activities and climate change as a causal chain. This causal chain does not explicitly include the feedbacks and non-linearities in the climate system that are important in its response to human forcings.

Figure 2.4 Steps in the causal chain of greenhouse gas emissions leading to climate change

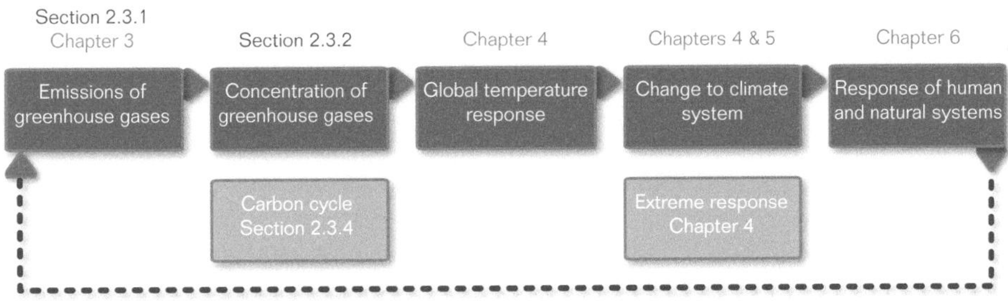

2.3.1 Emissions of greenhouse gases from human activities

The greenhouse gases with the greatest influence on warming of the atmosphere are water vapour (H_2O), carbon dioxide (CO_2), nitrous oxide (N_2O), methane (CH_4) and ozone (O_3). In addition, there is a range of human-made halocarbons (such as perfluorocarbons (PFCs), hydrofluorocarbons (HFCs), chlorofluorocarbons (CFCs)) and sulphur hexafluoride (SF_6) that are present in small amounts but are potent and contribute significantly to the total warming. Table 2.1 summarises the natural and anthropogenic sources of these gases.

Not all of these gases are directly emitted through human activities. Humans have less direct control over gases such as water vapour and ozone, although concentrations of these gases can be affected by human emissions of other reactive gases.

Table 2.1 Sources of greenhouse gases

Gas	Natural sources	Dominant anthropogenic sources
Carbon dioxide	Respiration from living organisms Volcanic eruptions Forest fires Decomposition of dead animals and plants Outgassing from the ocean	Combustion of fossil fuels and cement manufacture (more than 75 per cent of the increase in concentration since pre-industrial times) (IPCC 2007a: 512) Land-use changes (deforestation and changing agricultural practices)
Methane	Oceans Termites Natural wetlands Hydrates	Fossil fuel mining Vegetation burning Waste treatment Rice cultivation Ruminant livestock Landfill
Nitrous oxide	Processes in soils and oceans Oxidation of ammonia in the atmosphere	Nitrogenous fertiliser use Biomass burning Management of livestock manure Fossil fuel combustion Industrial activities such as nylon manufacture

Table 2.1 Sources of greenhouse gases *(continued)*

Gas	Natural sources	Dominant anthropogenic sources
Fluorinated gases		
HFCs	Some PFCs and all HFCs have no detected natural sources Other PFCs, and SF_6, are present in small amounts in the Earth's crust and released into the atmosphere through volcanic activity (Harnisch et al. 2000)	Refrigeration Air conditioning Solvents Fire retardants Foam manufacture Aerosol propellants
PFCs		Aluminium production
Sulphur hexafluoride		Electricity supply industry (switches and high-voltage systems)
Water vapour	The amount of water vapour in the atmosphere is a function of temperature and tends to fluctuate regionally and on short timescales	Irrigation, artificial dams and lakes, and fossil fuel production—less than 1 per cent of emissions from natural sources (IPCC 2007a: 28)
CFCs and HCFCs	No known natural sources	Propellants in aerosol cans Refrigerants in refrigerators and air conditioners Manufacture of foam packaging
Tropospheric ozone	Chemical reaction between other gases ('precursor species'), including carbon monoxide, methane and nitrogen oxides	Humans have limited direct influence, but influence concentrations through the emission of precursor species such as methane, nitrogen oxides and organic compounds from industry, power generation and transport

2.3.2 Accumulation of greenhouse gases in the atmosphere

The accumulation of greenhouse gases in the atmosphere that leads to warming is a function of both the rate of emissions and the rate of natural removal from the atmosphere.

Each greenhouse gas has specific characteristics that affect how long it stays in the atmosphere. The 'lifetime' of a gas in the atmosphere is a general term used for various timescales that characterise the rate of processes affecting the destruction or removal of trace gases.

With the exception of carbon dioxide, physical and chemical processes generally remove a specific fraction of the amount of a gas in the atmosphere each year. In some cases, the removal rate may vary with atmospheric properties such as temperature or background chemical conditions (IPCC 2007a: 23).

The lifetime of a gas affects the speed at which concentrations of that gas change following changes in the level of emissions. Gases with short lifetimes such as methane respond quickly to changes in emissions. Other gases with longer lifetimes respond very slowly. Over the course of a century, half of the carbon dioxide emitted in any one year will be removed, but around 20 per cent will remain in the atmosphere for millennia (IPCC 2007a: 824).

Carbon dioxide

After water vapour, carbon dioxide is the most abundant greenhouse gas in the atmosphere. Most gases are either removed from the atmosphere by chemical reaction or destroyed by ultraviolet radiation. Carbon dioxide, however, is very stable in the atmosphere. When it enters the atmosphere, carbon dioxide exchanges rapidly with plants and the surface ocean, and is then redistributed over hundreds to thousands of years through various forms of carbon storage, or 'sinks', as part of the carbon cycle (see section 2.3.4).

The concentration of carbon dioxide in the atmosphere increased from about 280 ppm in 1750 to 383 ppm in 2007. Over the last 10 years, carbon dioxide in the atmosphere has increased at an average rate of 2 ppm per year (Tans 2008).

Methane

Methane is a reactive compound of carbon and hydrogen that is short lived in the atmosphere, where it reacts to form water vapour and carbon dioxide. Considerable methane is stored in frozen soils and as methane hydrates in ocean sediments.

The methane concentration in the atmosphere has more than doubled since pre-industrial times to about 1774 ppb (parts per billion) after a slow fluctuation between around 580 ppb and 730 ppb over the last 10 000 years. Since the early 1990s growth rates have declined, and virtually no growth occurred between 1999 and 2005 (IPCC 2007a: 27), but more recent data for 2007 show concentrations rising for the first time since 1998 (NOAA 2008).

Nitrous oxide

Nitrous oxide is relatively stable in the atmosphere, but is eventually destroyed in the stratosphere when it reacts with ultraviolet light and charged oxygen molecules.

Nitrous oxide concentrations have increased by about 18 per cent to 319 ppb since pre-industrial times.

Fluorinated gases

Generally, fluorinated gases (HFCs, PFCs, sulphur hexafluoride) are non-flammable and non-toxic and have a low boiling point, which makes them useful for a number of manufacturing processes. HFCs and PFCs have been developed as replacements for ozone-depleting gases being phased out under the Montreal Protocol on Substances that Deplete the Ozone Layer, which entered into force in 1989.

Fluorinated gases are predominantly human in origin. They are destroyed and removed from the atmosphere mainly through reaction with ultraviolet light and other agents in the atmosphere such as chlorine; uptake in ocean surface waters; and chemical and biological degradation processes. Concentrations of these gases are relatively small but increasing rapidly (IPCC 2007a: 28).

Water vapour

Water vapour is the most abundant and important greenhouse gas in the atmosphere, accounting for around 60 per cent of the natural greenhouse effect for clear skies (IPCC 2007a: 271). Humans have a limited ability to directly influence its concentration (IPCC 2007a: 135).

Stratospheric water vapour has shown significant long-term variability. It is produced by the oxidation of methane, and the rate of reaction increases as methane concentrations rise. However, this effect is estimated to be equivalent to only about 1 to 4 per cent of the total change caused by long-lived greenhouse gases (IPCC 2007a: 28).

Human activities that contribute to warming indirectly affect the amount of water vapour in the atmosphere, because a warmer atmosphere can hold more water vapour than a cooler one. Increases in global temperature as a result of climate change are likely to affect water vapour concentrations substantially (IPCC 2007a: 135).

The lack of understanding of the way water vapour will respond to climate change, specifically its role in cloud formation, is a key factor in the uncertainty surrounding the response of the climate to increased temperatures.

Chlorofluorocarbons and hydrochlorofluorocarbons

Chlorofluorocarbons and hydrochlorofluorocarbons (CFCs and HCFCs) are human-made gases that are odourless, non-toxic, non-flammable and non-reactive. These gases are at low concentrations but have a strong warming effect: as a group they contribute about 12 per cent of the warming from long-lived greenhouse gases.

The low reactivity of CFCs and HCFCs allows them to remain in the atmosphere for long periods and cycle up to the stratosphere. Through a series of chemical reactions, CFCs destroy ozone and other similar compounds in the stratosphere.

The Montreal Protocol has caused a substantial reduction in emissions of these gases. Emissions of CFC-11 and CFC-13 decreased substantially between 1990 and 2002 (IPCC 2007a: 513), but with lifetimes of 45 and 85 years, respectively, their concentrations in the atmosphere are falling much more slowly.

The Montreal Protocol, which aims to phase out emissions of CFCs and HCFCs by 2030, has removed the need to discuss these gases in detail in the context of greenhouse gas mitigation policy.

Tropospheric ozone and precursor species

Tropospheric ozone is produced by chemical reaction between gases rather than being directly emitted. Increases in tropospheric ozone have accounted for 10–15 per cent of the positive change in the earth's energy balance since pre-industrial times (IPCC 2007a: 204). Tropospheric ozone is short-lived, and concentrations are likely to be localised around the main sources of precursor emissions.

2.3.3 How is the warming from different gases compared?

Global warming potential is an index that compares the radiative forcing from a given mass of greenhouse gas emissions to the radiative forcing caused by the same mass of carbon dioxide (CASPI 2007). Actual emissions of different gases are multiplied by their global warming potential to give a value for the mass of emissions in carbon dioxide equivalent (CO_2-e). Global warming potential depends both on the intrinsic capability of a molecule to absorb heat, and the lifetime of the gas in the atmosphere.

Global warming potential values take into account the lifetime, the concentration at the start of the time period and the warming potential of the gas. Sulphur hexafluoride has the highest global warming potential of all gases at 22 800 times that of carbon dioxide, but has a low impact on overall warming due to its low concentrations.

Global warming potential is used under the Kyoto Protocol to compare the magnitude of emissions and removals of different greenhouse gases from the atmosphere. It is also the concept used in the design and implementation of multi-gas emissions trading schemes for calculating the value of a trade between the reductions in emissions of different greenhouse gases.

2.3.4 The carbon cycle

The 'carbon cycle' refers to the transfer of carbon, in various forms, through the atmosphere, oceans, plants, animals, soils and sediments. As part of the carbon cycle, plants and algae convert carbon dioxide and water into biomass using energy from the sun (photosynthesis). Living organisms return carbon to the atmosphere when they respire, decompose or burn. Methane is released through the decomposition of plants, animals and other hydrocarbon material (fossil fuels and waste) when no oxygen is present.

Carbon sinks

The parts of the carbon cycle that store carbon in various forms are referred to as 'carbon sinks'. The majority of carbon that was present in the early atmosphere is now stored in sedimentary rocks and marine sediments. Other carbon sinks are the atmosphere, oceans, fossil fuels such as coal, petroleum and natural gas, living plants and organic matter in the soil.

Carbon dioxide dissolves in the ocean and is returned to the atmosphere through dissolution in a continuous exchange. Dissolved carbon dioxide is carried deep into the oceans through the sinking of colder water and waste and debris from dead organisms, where it is either buried or redissolves. The transfer of carbon to the deep ocean is slow. Water at intermediate depths mixes with the surface water over decades or centuries, but deep waters mix only on millennial timescales and thus provide a long-term carbon sink.

Table 2.2 provides estimates of the amount of carbon stored in different sinks in 1750 and how they changed up to the end of the 20th century.

Table 2.2 Estimates of the amount of carbon stored in different sinks in 1750 and how they have changed

Carbon sink	Gigatonnes carbon stored in 1750	Percentage of total cycling carbon	Net change in sink between 1750 and 1994	Percentage change
Atmosphere	597	1.3	165	27.6
Vegetation, soil and detritus	2 300	5.1	-39	-1.7
Fossil fuels	3 700	8.3	-244	-6.6
Surface ocean	900	2.0	18	2.0
Marine biota	3	0.0	–	–
Deeper ocean	37 100	82.9	100	0.3
Surface sediments	150	0.3	–	–
Sedimentary rocks	>66 000 000	n/a	–	–

Note: Due to the very slow exchange with other carbon sinks, percentages of total carbon do not include storage in sedimentary rocks.

Sources: For sedimentary rocks, UNEP & GRID–Arendal (2005); for all others, IPCC (2007a: Figure 7.3).

The deep and surface ocean accounts for more than 85 per cent of the carbon being cycled more actively. Although the atmosphere accounts for just over 1 per cent of carbon storage, it shows the largest percentage increase since pre-industrial times. Vegetation and soil have had a net decrease in carbon stored—a considerable loss from land-use change has been partially offset by carbon uptake by living organisms.

The major change to the carbon cycle from human activity is increased emissions of carbon dioxide to the atmosphere from the burning of fossil fuels. The rate of exchange between the ocean and the atmosphere has increased in both directions, but with a net movement to the ocean. Terrestrial ecosystems are also a significant carbon sink. However, absorption by both the ocean and land is less than emissions from fossil fuels, and almost 45 per cent of human emissions since 1750 have remained in the atmosphere. It is generally accepted that future climate change will reduce the absorptive capacity of the carbon cycle so that a larger fraction of emissions remain in the atmosphere compared to current levels (IPCC 2007a: 750).

Carbon–climate feedbacks

Carbon–climate feedbacks occur when changes in climate affect the rate of absorption or release of carbon dioxide from land and ocean sinks. Examples of climate–carbon feedbacks include the decrease in the ability of the oceans to remove carbon dioxide from the atmosphere with increasing water temperature, reduced circulation and increased acidity (IPCC 2007a: 531); and the weakening of the uptake of carbon in terrestrial sinks due to vegetation dieback and reduced growth from reduced water availability, increased soil respiration at higher temperatures and increased fire occurrence (IPCC 2007a: 527; Canadell et al. 2007).

Large positive climate–carbon feedbacks could result from the release of carbon from long-term sinks such as methane stored deep in ocean sediments and in frozen soils as temperatures increase (IPCC 2007a: 642). There is a high level of uncertainty about how the carbon cycle will respond to climate change.

2.3.5 Greenhouse gases and temperature rise

How do different greenhouse gases change the energy balance?

The radiative forcing of a greenhouse gas represents the change in the effect of that gas on the energy balance of the atmosphere. It takes into account the greenhouse gas's concentration in the atmosphere at the start of a period (in this report, pre-industrial), the amount the concentration has changed due to human activities, and the way a molecule of that gas absorbs heat. Because radiative forcing measures change between two specific points in time, the lifetime of a gas in the atmosphere has a considerable influence on future forcing.

For most gases, removal from the atmosphere is minimally influenced by changes in the climate system. However, for the calculation of carbon dioxide concentrations, the complexity of the carbon cycle and the uncertainty over how it will change over time must be taken into account.

Carbon dioxide molecules absorb heat in a particular range of wavelengths, and as concentrations increase the additional heat of those wavelengths gets absorbed. If concentrations keep growing, carbon dioxide added later will cause proportionately less warming than carbon dioxide added now. The same amount of warming will occur from a doubling from 280 ppm (pre-industrial levels) to 560 ppm as from another doubling from 560 ppm to 1120 ppm.

The forcing due to carbon dioxide, methane, nitrous oxide and halocarbons is relatively well understood. However, the contributions of ozone at different levels in the atmosphere, aerosols and linear clouds from aviation are poorly understood (CASPI 2007).

A measure commonly used in the literature and policy discussions is the concept of carbon dioxide equivalent of a gas *concentration*, measured in parts per million. This is a different but related measure to carbon dioxide equivalent *emissions* calculated using the global warming potential index. The former is the concentration of carbon dioxide that would cause the same amount of radiative forcing as a particular concentration of a greenhouse gas. This term is often used in discussions of global stabilisation or concentration targets.

The total radiative forcing of the long-lived greenhouse gases is 2.63 (± 0.26). In terms of carbon dioxide equivalence, this equates to a concentration of around 455 ppm CO_2-e (range: 433–477 ppm CO_2-e) (IPCC 2007b: 102).

However, the warming that would result from this is offset by the cooling effects of aerosols and land-use changes, which reduce the concentration to a range of 311 to 435 ppm CO_2-e, with a central estimate of about 375 ppm CO_2-e (IPCC 2007b: 102).

Climate sensitivity

'Climate sensitivity' is the measure of the climate system's response to sustained radiative forcing. More precisely, it is the global average surface warming (measured in degrees Celsius) that will occur when the climate reaches equilibrium following a doubling of carbon dioxide concentrations above the pre-industrial value. Such doubling of carbon dioxide levels is approximately equivalent to reaching 560 ppm carbon dioxide, which is twice pre-industrial levels of 280 ppm. Climate models predict a wide range of climate sensitivities due to differing assumptions about the magnitude of feedbacks in the climate system. Feedbacks include the response of water vapour to increased temperatures, changes in cloud formation, and the implications of the melting of ice and snow for the amount of heat absorbed by the surface.

Climate sensitivity relates to the equilibrium temperature reached when all elements of the climate system have responded to induced changes. Due to the long timescale of response, this may not occur for thousands of years. The IPCC estimates that it is likely (that is, a greater than 66 per cent chance) that a doubling of carbon dioxide will lead to a long-term temperature increase of between 2°C and 4.5°C (IPCC 2007a: 12). It is considered unlikely that climate sensitivity will be less than 1.5°C. Values substantially higher than 4.5°C cannot be excluded, but these higher outcomes are less well supported (IPCC 2007a: 799). The best estimate of the IPCC is about 3°C (IPCC 2007a: 12).

When considering the impacts of climate change on human society in the coming century, and how to respond to those impacts, temperature change over shorter time frames is more relevant. The effective climate sensitivity reflects the warming occurring in the short term, and takes into account climate feedbacks at a particular time. Assumptions on the rate of warming of the oceans in different models have a considerable effect on the short-term temperature outcomes.

Climate sensitivity is the largest of the uncertainties affecting the amount of warming when a single future pathway of greenhouse gases is selected (IPCC 2007a: 629).

Rate of change in global temperatures

Much of the analysis of projected climate change impacts focuses on the temperature increase under a certain emissions pathway by a point in time, often 2100. However, it is not just the magnitude of temperature rise, but also the rate at which it occurs that determines climate change impacts, as a higher rate of change in temperature affects the adaptive capacity of natural and human systems (Warren 2006; Ambrosi 2007; IPCC 2007a: 774).

2.3.6 Changes to the climate system

Elements of the climate system respond in different ways to changes in the energy balance. The climate response can vary considerably over time and space.

Many considerations of climate response focus on the changes to mean climate at a particular point in time. These considerations do not address the inherent variability of the climate system or the long time frames over which the system will continue to respond even if the energy balance ceases to change.

Climate variability

Climate variability refers to the natural variations in climate from the average state, and occurs over both space and time.

Natural climate variability occurs as a result of variations in atmospheric and ocean circulations, larger modes of variability such as the El Niño – Southern Oscillation (see Box 2.2), and events such as volcanic eruptions and changes in incoming solar radiation.

Box 2.2 Large-scale patterns of climate variability

Analysis of variability in global climate over time has shown that a significant component can be described in terms of a relatively small number of large-scale patterns of variability in atmospheric and oceanic circulation (IPCC 2007a: 39).

A key example is the El Niño – Southern Oscillation, which is a coupled fluctuation in the atmosphere and the equatorial Pacific Ocean. The El Niño – Southern Oscillation leads to changed conditions in surface temperature across the central equatorial Pacific Ocean and are thought to be once every three to seven years, which in turn leads to changes in rainfall, floods and droughts on both sides of the Pacific. It is characterised by large exchanges of heat between the ocean and atmosphere, which affect global mean temperatures but also have a profound effect on the variability of the climate in Australia.

Another example relevant to Australia is the Southern Annular Mode, which is the year-round fluctuation of a pattern with low sea surface pressure in the Antarctic and strong mid-latitude westerlies (IPCC 2007a: 39). The Southern Annular Mode has a considerable influence over climate in much of the southern hemisphere, affecting surface temperatures in the Southern Ocean and the distribution of sea ice. It influences rainfall variability and storm tracks in southern Australia.

The extent of natural climate variability differs from place to place. In Australia, a notable feature of the climate is the high variability in rainfall from year to year, influenced by the El Niño – Southern Oscillation. This affects the ability of scientists to identify long-term trends in the climate system and to establish whether the changes result from human activities.

Sustained changes to the energy balance of the climate system will cause a change in the long-term means of elements of the climate system such as temperature and rainfall, but may also lead to a change in the pattern of variability about a given mean. In Australia, observed changes in the climate suggest that the frequency of extremes in rainfall events is increasing at a faster rate than the mean (CSIRO & BoM 2007).

Severe weather events

Natural variability in the weather is reflected in the occurrence of 'severe weather events', defined in this report as an event of an intensity that is rare at a particular place and time of year. Examples of severe weather events include:

- hot days and nights (including heatwaves)
- cold days and nights (including frosts)
- heavy rainfall events
- droughts
- floods
- hail and thunderstorms
- tropical cyclones
- bushfires
- extreme winds.

The characteristics of what is called severe weather may vary from place to place in an absolute sense—for example, the temperature required to define a heatwave in Hobart would be lower than in Darwin. Weather events may also be considered severe if they cause extensive damage due to timing or location, even if they are not considered rare in terms of their likelihood.

Climate change may result in changes to the frequency, intensity and distribution of weather events that are considered 'severe' today. Figure 2.5 demonstrates the effect of a change in mean temperature on the lower probability temperature events—in some cases, an increase in the frequency at one end of the probability (more heatwaves) will be associated with a decrease in the opposite end (fewer frosts). However, if the variability (or variance) were to change rather than the mean, hot days and frosts would both increase. If both the variability and the mean were to change, there would be fewer frost events but a more significant increase in the number of hot days.

Figure 2.5 Effect on extremes of temperature from an increase in mean temperature, an increase in variance, and an increase in both mean temperature and variance

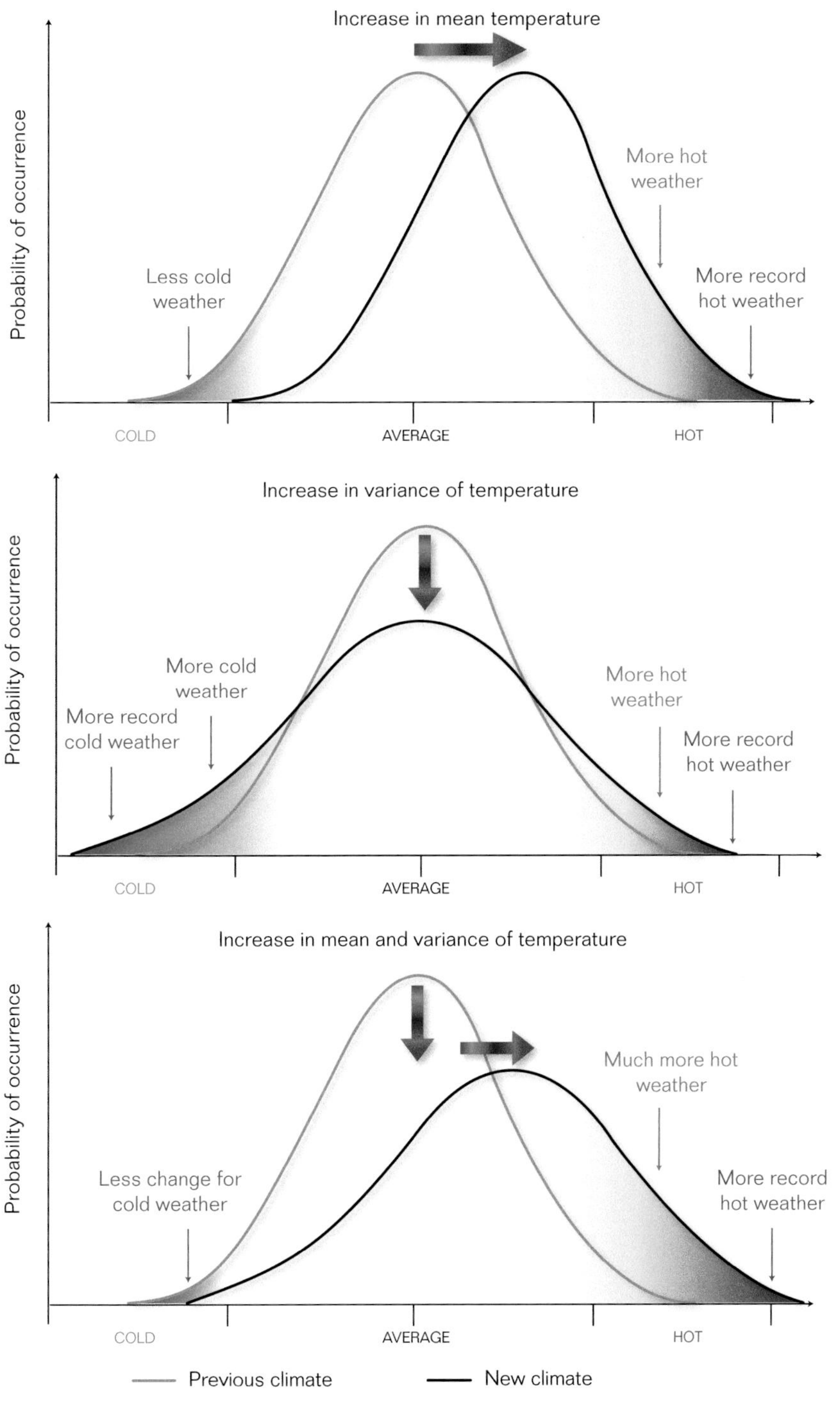

Source: IPCC (2001b: Figure 4.1), reformatted for this publication.

The slow response of the climate system

Figure 2.6 shows estimates of the time it takes for different parts of the climate system to respond to a situation where emissions are reduced to equal the rate of natural removal. While greenhouse gas concentrations stabilise in around a hundred years, the temperature and sea-level rise due to thermal expansion of the oceans takes much longer to stabilise. The melting of ice sheets is still increasing the sea level even after a thousand years.

Figure 2.6 Inertia in the climate system

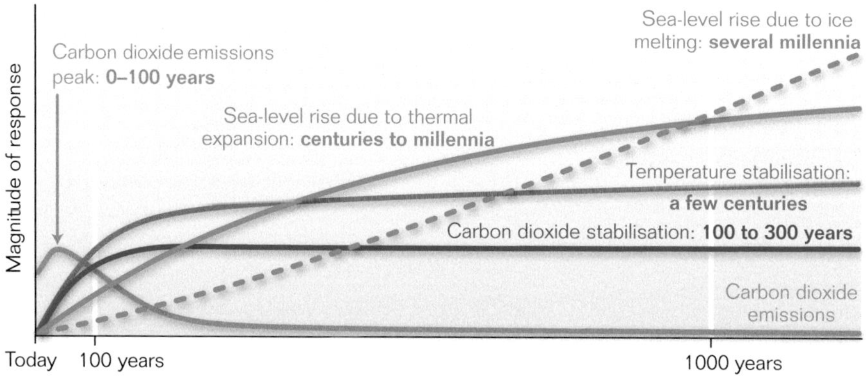

Source: IPCC (2001b: Figure 5.2), reformatted for this publication.

2.4 The task of global mitigation

Goals for global mitigation have typically been cast in terms of stabilisation of greenhouse gas concentrations in the atmosphere. Article 2 of the United Nations Framework Convention on Climate Change (UNFCCC) states as its ultimate objective:

> [s]tabilisation of greenhouse gas concentrations in the atmosphere at a level that would prevent dangerous anthropogenic interference with the climate system. Such a level should be achieved within a time frame sufficient to allow ecosystems to adapt naturally to climate change, to ensure that food production is not threatened and to enable economic development to proceed in a sustainable manner (UN 1992).

However, the UNFCCC does not define the point at which 'dangerous anthropogenic interference with the climate system' or 'dangerous climate change' might occur. Even if the climate change resulting from a given pathway of future emissions were known with certainty, there would be different approaches to defining 'danger', and interpretation of the UNFCCC goal will not be defined only by the science. Ethical, economic and political judgments will also be required (IPCC 2007b: 99).

The Review's terms of reference required it to analyse two specific stabilisation goals: one at which greenhouse gases are stabilised at 550 ppm CO_2-e and one at which they are stabilised at 450 ppm CO_2-e. A stabilisation target of 450 ppm

CO_2-e gives about a 50 per cent chance of limiting the global mean temperature increase to 2°C above pre-industrial levels (Meinshausen 2006), a goal endorsed by the European Union (Council of the European Union 2007) among others. Stabilisation at 500 ppm or 550 ppm CO_2-e would be less costly than a more ambitious target, but is associated with higher risks of dangerous climate change.

Based on a best-estimate climate sensitivity of 3°C, stabilisation at 550 ppm CO_2-e is likely to lead to an equilibrium global mean temperature increase of 3°C above pre-industrial levels (IPCC 2007b; Meinshausen 2006).

2.4.1 What is stabilisation?

Stabilisation of a greenhouse gas is achieved when its atmospheric concentration is constant. For a group of greenhouse gases, stabilisation is achieved when the combined warming effect (radiative forcing) of the gases is maintained at a constant level.

Stabilisation of long-lived greenhouse gases does not mean the climate will stop changing—temperature and sea-level changes, for example, will continue for hundreds of years after stabilisation is achieved (see Figure 2.6).

How does the lifetime of a gas influence stabilisation?

For all greenhouse gases, if emissions continue to increase over time their atmospheric concentration will also increase. However, the way in which the concentration of a gas will change in response to a decrease in emissions is dependent on the lifetime of the gas (IPCC 2007a: 824). Stabilisation of greenhouse gas emissions is therefore not the same as stabilisation of greenhouse gas concentrations in the atmosphere.

Carbon dioxide is naturally removed slowly from the atmosphere through exchange with other parts of the carbon cycle. The current rate of emissions is well above the natural rate of removal. This has caused the accumulation of carbon dioxide in the atmosphere. If carbon dioxide emissions were stabilised at current levels, concentrations would continue to increase over this century and beyond. To achieve stabilisation of carbon dioxide concentrations, emissions must be brought down to the rate of natural removal.

The rate of absorption of carbon by sinks depends on the carbon imbalance between the atmosphere, the oceans and the land, and the amount already contained in these sinks. Once stabilisation in the atmosphere is reached, the rate of uptake will decline (Figure 2.7). Long-term maintenance of a stable carbon dioxide concentration will then involve the complete elimination of carbon dioxide emissions as the net movement of carbon dioxide to the oceans gradually declines (IPCC 2007a: 824; CASPI 2008).

The response of other greenhouse gases to decreases in emissions is more straightforward: the level at which concentrations are stabilised is proportional to the level at which emissions are stabilised. For gases with a lifetime of less than a century (such as methane) or around a century (such as nitrous oxide), keeping emissions constant at current levels would lead to the stabilisation of concentrations at slightly higher levels than today within decades or centuries, respectively (IPCC 2007a: 824). If anthropogenic emissions of these gases were to cease completely, their concentration levels would eventually return to pre-industrial levels.

For greenhouse gases other than carbon dioxide with lifetimes of thousands of years (such as sulphur hexafluoride), stabilisation would only occur many thousands of years after emissions stopped increasing. In the policy context they are treated in the same way as carbon dioxide, with the long-term aim of bringing emissions to zero in order to stabilise their warming effect.

Figure 2.7 Response of different carbon sinks to the rate of emissions over time

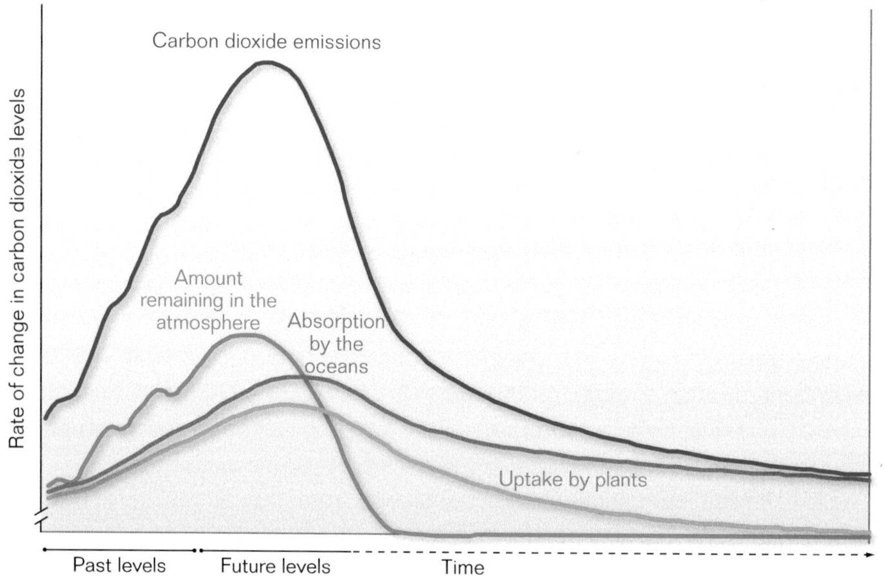

Source: Based on CASPI (2008).

How can stabilisation be achieved?

Any number of emissions pathways could lead to stabilisation of a gas at a given concentration. For carbon dioxide, these pathways generally involve a trade-off between the level at which emissions peak and the maximum rate of reductions required in the future. Figure 2.8 shows some of the possible emissions pathways to achieve the same stabilisation target. These curves are stylised—in the real world, annual emissions would fluctuate. The pathways that have a higher peak in emissions have a much greater rate of reduction at a later point in time, shown by the steepness of the curve.

The timing of emissions reductions influences the efficiency of uptake of carbon dioxide by sinks, the rate of temperature increase and potentially the timing of climate–carbon feedback effects. Delays in the realisation of a specified stabilisation target through more rapid mitigation would, if they were large, give greater environmental benefits (O'Neill & Oppenheimer 2004).

Is a target of 450 ppm CO_2-e or below achievable?

The concentration of long-lived greenhouse gases in the atmosphere for 2005 is equivalent to the warming effect of 455 ppm of carbon dioxide (see section 2.3.5). However, when the cooling influence of aerosols is included, the equivalent carbon dioxide concentration is estimated at 375 ppm CO_2-e. The concentration of carbon dioxide in 2007 was 383 ppm (Tans 2008).

Figure 2.8 Different pathways of emissions reductions over time to achieve the same concentration target

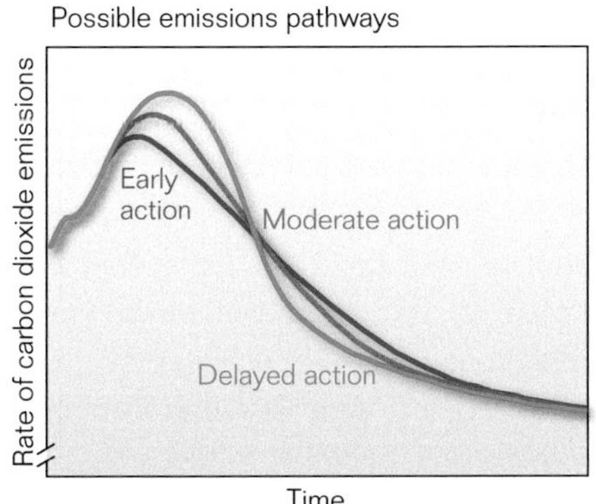

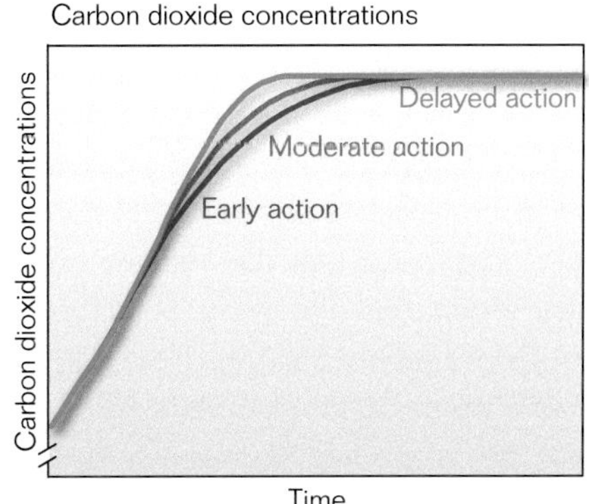

Source: Based on CASPI (2008).

Due to the short lifetime of aerosols in the atmosphere, it is not appropriate to include their influence in a long-term target. The presence of aerosols is expected to lessen through a reduction in the burning of fossil fuels as a result of climate change policies as well as through separate efforts to reduce air pollution.

The 2005 long-lived greenhouse gas concentration of 455 ppm CO_2-e includes the warming influence of gases such as methane, which can be reduced in a relatively short period of time. If the target were set for some point in the future (such as 2050 or 2100), it would be scientifically feasible to bring CO_2-e emissions down to a target level of 450 ppm CO_2-e if immediate and deep cuts were made in emissions of most greenhouse gases.

However, for a target of 450 ppm CO_2-e to be achieved, global emissions would have to peak and then fall almost immediately at a high rate. In the context of current emissions trends, it would take a major change in many countries' positions for early international agreement on such dramatic changes in emissions to be feasible.

Overshooting

There is increasing recognition in both science and policy communities that stabilising at low levels of CO_2-e (around or below 450 ppm) requires 'overshooting' the concentration target (den Elzen et al. 2007; Meinshausen 2006; IPCC 2007a: 827).

The climate change impacts of the higher levels of greenhouse gas concentrations reached in an overshoot profile are dependent on the length of time the concentrations stay above the desired target, and how far carbon dioxide overshoots.

Figure 2.9 shows the different temperature outcomes for a range of cases of overshooting. All three cases show stabilisation at the same level in a similar time frame, but with varying amounts of overshooting. The temperature output demonstrates that while the 'small overshooting' case remains under the target temperature, the other cases do not. Hence, due to inertia in the climate system, a large and lengthy overshooting will influence the transient temperature response, while a small, short one will not (den Elzen & van Vuuren 2007).

Increasing attention is being paid in the environmental and scientific communities to low stabilisation scenarios. In particular, a number of organisations in Australia have suggested that the Review should focus as well on a 400 ppm objective. They argue that the risks of immense damage to the Australian environment, including the Great Barrier Reef and Kakadu National Park, are unacceptably high at 450 ppm. Some scientists have also expressed the view that stabilisation at 450 ppm is too high (Hansen et al. 2008). For any such scenarios to be feasible, there would need to be a considerable period of overshooting.

Figure 2.9 Temperature outcomes of varying levels of overshooting

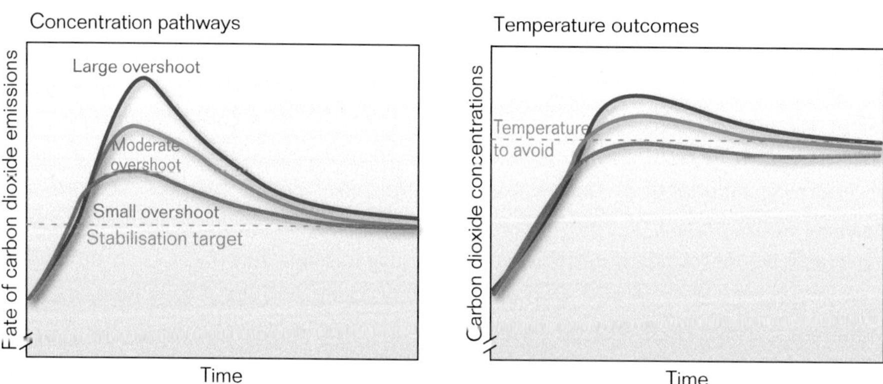

Source: Concentration and temperature pathways developed using SIMCAP (Meinshausen et al. 2006).

What is a peaking profile?

An overshooting profile requires a period in which emissions are below the natural level of sequestration before they are stabilised. Another mitigation option is to follow a 'peaking profile'.

Under a peaking profile, the goal is to cap concentrations at a particular level (the peak) and then to start reducing them indefinitely, without aiming for any explicit stabilisation level. Stabilisation is therefore not conceived as a policy objective for the foreseeable future.

The key benefit of a peaking profile is that it allows concentrations to increase to or above the level associated with a given long-term temperature outcome, but reduces the likelihood of reaching or exceeding that temperature outcome. The higher level of peak concentrations means that current trends in emissions growth do not need to be reversed as quickly to achieve any given temperature goal. This decreases the costs of meeting a given temperature target (den Elzen & van Vuuren 2007).

Following a peaking profile could be a disadvantage if the climate is found to be more sensitive to increases in greenhouse gases than anticipated. Due to the higher concentrations reached under a peaking profile, there is less flexibility to adjust to a lower concentration target at a later point in time, so there is greater risk that a threshold may be crossed.

Is overshooting feasible?

Designing a mitigation pathway—whether an overshooting or a peaking profile—that requires a decrease in the concentration of greenhouse gases assumes that emissions can be brought below the natural level of sequestration. Figure 2.10 shows the emissions pathways required to achieve a low concentration target following an overshoot. A lower concentration target following an initial overshoot will require negative emissions net of natural sequestration for a longer period.

Figure 2.10 Emissions pathways required to achieve a low concentration target following an overshoot

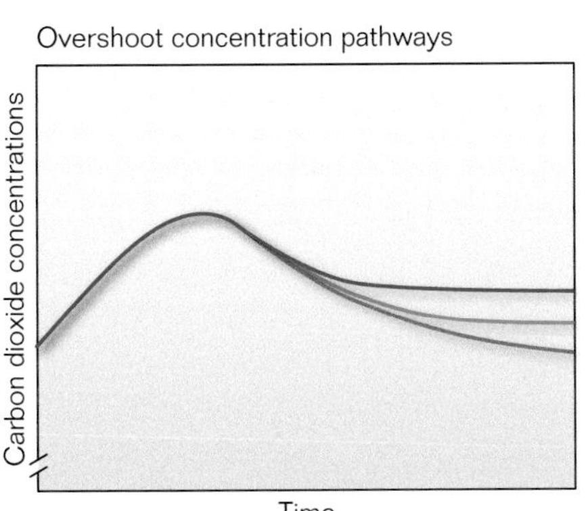

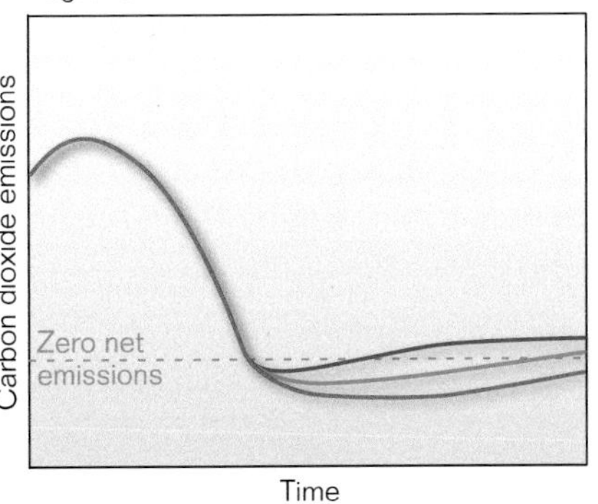

Source: Based on CASPI (2008).

The costs of reducing emissions below natural sequestration levels would be lower if controls on gross emissions were supported by cost-effective means of removing carbon dioxide from the atmosphere. Bringing emissions below the natural rate of sequestration would require rigorous reduction of emissions from all sources, but might also require extraction of carbon dioxide from the air. Possible methods include:

- increasing absorption and storage in terrestrial ecosystems by reforestation and conservation and carbon-sensitive soil management
- the harvest and burial of terrestrial biomass in locations such as deep ocean sediments where carbon cycling is slow (Metzger et al. 2002)
- capture and storage of carbon dioxide from the air or from biomass used for fuel
- the production of biochar from agricultural and forestry residues and waste (Hansen et al. 2008).

The simplest way to remove carbon dioxide from the air is to use the natural process of photosynthesis in plants and algae. Over the last few centuries, clearing of vegetation by humans is estimated to have led to an increase in carbon dioxide concentration in the atmosphere of 60 ± 30 ppm, with around 20 ppm still remaining in the atmosphere (Hansen et al. 2008). This suggests that there is considerable capacity to increase the level of absorption of carbon dioxide through afforestation activities. The natural sequestration capacities of algae were crucial to the decarbonisation of the atmosphere that created the conditions for human life on earth, and offer promising avenues for research and development.

Technologies for capture and storage of carbon from the combustion of fossils fuels currently exist, and the same process could be applied to the burning of biomass.

As yet, there are no large-scale commercial technologies that capture carbon from the air. Some argue, however, that it will be possible to develop air capture technologies at costs and on timescales relevant to climate policy (Keith et al. 2006). Research and development in Australia on the use of algae is of global importance. Captured carbon dioxide could be stored underground or used as an input in biofuel production.

Under a carbon price applying broadly across all opportunities for carbon dioxide reduction and removal, and with strong research and development support, there will be more rapid commercial development of both existing and new technologies to achieve negative emissions on a large scale.

2.4.2 Geo-engineering

'Geo-engineering' is a term used to describe 'technological efforts to stabilize the climate system by direct intervention in the energy balance of the earth' (IPCC 2007b: 815).

A range of geo-engineering proposals have been put forward, including:

- the release of aerosols into the stratosphere to scatter incoming sunlight (Crutzen 2006)
- cloud seeding through the artificial generation of micro-meter sized seawater droplets (Bower et al. 2006)
- fertilisation of the ocean with iron and nitrogen to increase carbon sequestration (Buesseler & Boyd 2003)
- changes in land use to increase the albedo (reflectivity) of the earth's surface (Hamwey 2005).

Geo-engineering proposals appear to have several advantages. First, they may be cheap in comparison to reductions in greenhouse gas emissions. They can be implemented by one or a small number of countries and thus do not require the widespread global action that stabilisation of greenhouse gases will require (Barrett 2008). They may act quickly, with a lag from implementation to impact of months rather than decades. Geo-engineering techniques could potentially be deployed to avoid reaching a tipping point related to temperature increase.

However, such proposals also have disadvantages.

- Those that focus on reducing solar radiation will do nothing to prevent the acidification of the ocean as a result of increased atmospheric concentrations of carbon dioxide, and therefore leave untouched part of the wider environmental problem.
- Geo-engineering techniques are generally untried. Some studies have been undertaken including through small-scale experiments on ocean fertilisation (Buesseler & Boyd 2003), investigation of similar natural phenomenon such as the release of aerosols from Mount Pinatubo in 1991, and computer simulations (Wigley 2006; Govindasamy & Caldeira 2000). However, there will always be the risk of unanticipated consequences, which need to be considered analytically.
- The fact that these solutions can be implemented unilaterally may also give rise to risks of conflict.

So far, the disadvantages of geo-engineering approaches have tended to outweigh the advantages in most minds that have turned to the issue. However, in recent years such proposals have received more support from a number of prominent scientists and economists, with calls for more research into their feasibility, costs, side effects and frameworks for implementation (IPCC 2007c: 79; Crutzen 2006; Cicerone 2006; Barrett 2008).

References

Ambrosi, P. 2007, 'Mind the rate! Why rate of global climate change matters, and how much', in R. Guesnerie & H. Tulkens (eds), *Conference Volume of the 6th CESIfo Venice Summer Institute—David Bradford Memorial Conference on the Design of Climate Policy*, MIT Press, Cambridge, Massachusetts.

Barrett, S. 2008, 'The incredible economics of geoengineering', *Environmental and Resource Economics* 39(1): 45–54.

Bower, K., Choularton, T., Latham, J., Sahraei, J. & Salter, S. 2006, 'Computational assessment of a proposed technique for global warming mitigation via albedo-enhancement of marine stratocumulus clouds', *Atmospheric Research* 82: 328–36.

Buesseler, K. & Boyd, P. 2003, 'Will ocean fertilisation work?' *Science* 300: 67–8.

Canadell, J., Le Quéré, C., Raupach, M., Field, C., Buitehuis, E., Ciais, P., Conway, T., Gillett, N., Houghton, R. & Marland, G. 2007, 'Contributions to accelerating atmospheric CO_2 growth from economic activity, carbon intensity and efficiency of natural sinks', *Proceedings of the National Academy of Sciences of the USA* 104(47): 18866–70.

CASPI (Climate Adaptation – Science and Policy Initiative) 2007, *The Global Science of Climate Change*, report prepared for the Garnaut Climate Change Review, University of Melbourne.

CASPI 2008, *The Science of Stabilising Greenhouse Gas Concentrations*, report prepared for the Garnaut Climate Change Review, University of Melbourne.

Cicerone, R. 2006, 'Geoengineering: encouraging research and overseeing implementation', *Climatic Change* 77(3–4): 221–6.

Council of the European Union 2007, *Presidency Conclusions*, 8/9 March 2007, 7224/1/07 Rev 1, Brussels.

Crutzen, P. 2006, 'Albedo enhancement by stratospheric sulphur injections: a contribution to resolve a policy dilemma?' *Climatic Change* 77(3–4): 211–19.

CSIRO (Commonwealth Scientific and Industrial Research Organisation) & BoM (Bureau of Meteorology) 2007, *Climate Change in Australia: Technical report 2007*, CSIRO, Melbourne.

den Elzen, M., Meinshausen, M. & van Vuuren, D. 2007, 'Multi-gas emission envelopes to meet greenhouse gas concentration targets: costs versus certainty of limiting temperature increase', *Global Environmental Change* 17(2007): 260–80.

den Elzen, M. & van Vuuren, D. 2007, 'Peaking profiles for achieving long-term temperature targets with more likelihood at lower costs', *Proceedings of the National Academy of Sciences of the USA* 104(46): 17931–6.

Govindasamy, B. & Caldeira, K. 2000, 'Geoengineering earth's radiation balance to mitigate CO_2-induced climate change', *Geophysical Research Letters* 27(14): 2141–4.

Hamwey, R. 2005, 'Active amplification of the terrestrial albedo to mitigate climate change: an exploratory study', *Mitigation and Adaptation Strategies for Global Change* 12: 419–39.

Hansen, J. 2007, 'Scientific reticence and sea level rise', *Environment Research Letters* 2: 1–6.

Hansen, J., Sata, M., Kharecha, P., Beerling, D., Masson-Delmotte, V., Pagani, M., Raymo, M., Royer, D. & Zachos, J. 2008, 'Target atmospheric CO_2: where should humanity aim?', <www.columbia.edu/~jeh1/2008/TargetCO2_20080407.pdf>.

Harnisch, J., Frische, M., Borchers, R., Eisenhauer, A. & Jordan, A. 2000, 'Natural fluorinated organics in fluorite and rocks', *Geophysical Research Letters* 27(13): 1883–6.

IPCC (Intergovernmental Panel on Climate Change) 2001a, *Climate Change 2001: The scientific basis. Contribution of Working Group I to the Third Assessment Report of the Intergovernmental Panel on Climate Change*, J.T. Houghton, Y. Ding, D.J. Griggs, M. Noguer, P.J. van der Linden, X. Dai, K. Maskell & C.A. Johnson (eds), Cambridge University Press, Cambridge and New York.

IPCC 2001b, *Climate Change 2001: Synthesis report. A Contribution of Working Groups I, II, and III to the Third Assessment Report of the Intergovernmental Panel on Climate Change*, R. Watson and the Core Writing Team (eds), Cambridge University Press, Cambridge and New York.

IPCC 2007a, *Climate Change 2007: The physical science basis. Contribution of Working Group I to the Fourth Assessment Report of the Intergovernmental Panel on Climate Change*, S. Solomon, D. Qin, M. Manning, Z. Chen, M. Marquis, K.B. Averyt, M. Tignor & H.L. Miller (eds), Cambridge University Press, Cambridge and New York.

IPCC 2007b, *Climate Change 2007: Mitigation of climate change. Contribution of Working Group III to the Fourth Assessment Report of the Intergovernmental Panel on Climate Change*, B. Metz, O.R. Davidson, P.R. Bosch, R. Dave & L.A. Meyer (eds), Cambridge University Press, Cambridge.

Keith, D., Ha-Duong, M. & Stolaroff, J. 2006, 'Climate strategy with CO_2 capture from the air', *Climatic Change* 74(1–3): 17–45.

Meinshausen, M. 2006, 'What does a 2°C target mean for greenhouse gas concentrations? A brief analysis based on multi-gas emission pathways and several climate sensitivity uncertainty estimates', in H.J. Schellnhuber, C. Cramer, N. Nakicenovic, T. Wigley & G. Yohe (eds), *Avoiding Dangerous Climate Change*, Cambridge University Press, Cambridge, pp. 265–80.

Meinshausen, M., Hare, B., Wigley, T.M.L, van Vuuren, D., den Elzen, M. & Swart, R. 2006, 'Multi-gas emissions pathways to meet climate targets', *Climatic Change* 75: 151–94.

Metzger, R., Benford, G. & Hoffert, M. 2002, 'To bury or to burn: optimum use of crop residues to reduce atmospheric CO_2', *Climatic Change* 54(3): 369–74.

NOAA (National Oceanic and Atmospheric Administration) 2008, 'Carbon dioxide, methane rise sharply in 2007', April 23, <www.noaanews.noaa.gov/stories2008/20080423_methane.html>.

O'Neill, B. & Oppenheimer, M. 2004, 'Climate change impacts sensitive to concentration stabilization path', *Proceedings of the National Academy of Sciences of the USA* 101(47): 16411–16.

Ramanathan, V. & Carmichael, G. 2008, 'Global and regional changes due to black carbon', *Nature Geoscience* 1: 221–7.

Royer, D. 2006, 'CO_2-forced climate thresholds during the Phanerozoic', *Geochimica et Cosmochimica Acta* 70(2006): 5665–75.

Ruddiman, W.F. 2008, *Earth's Climate: Past and future*, 2nd edn, W.H. Freeman and Company, New York.

Steffen, W., Sanderson, A., Tyson, P., Jäger, J., Matson, P., Moore, B. III, Oldfield, F., Richardson, K., Schellnhuber, H.-J., Turner, B.L. II & Wasson, R. 2004, *Global Change and the Earth System: A planet under pressure*, IGBP Global Change Series, Springer-Verlag, Berlin.

Tans, P. 2008, 'Trends in atmospheric carbon dioxide—global', National Oceanic and Atmospheric Administration, Earth System Research Laboratory, Global Monitoring Division, <www.esrl.noaa.gov/gmd/ccgg/trends/>.

UN (United Nations) 1992, *United Nations Framework Convention on Climate Change*, available at <unfccc.int/essential_background/convention/background/items/1349.php>.

UNEP (United Nations Environment Programme) & GRID–Arendal 2005, 'Vital climate change graphics: February 2005', Vital Climate Change Graphics update, <www.vitalgraphics.net/climate2.cfm>.

Warren, R. 2006, 'Impacts of global climate change at different annual mean global temperature increases', in H.J. Schellnhuber, C. Cramer, N. Nakicenovic, T. Wigley & G. Yohe (eds), *Avoiding Dangerous Climate Change*, Cambridge University Press, Cambridge, pp. 94–131.

Wigley, T.M.L. 2006, 'A combined mitigation/geoengineering approach to climate stabilization', *Science* 314: 452–54.

EMISSIONS IN THE PLATINUM AGE

3

Key points

Greenhouse gas emissions have grown rapidly in the early 21st century. In the absence of effective mitigation, strong growth is expected to continue for the next two decades and at only somewhat moderated rates beyond.

So far, the biggest deviations from earlier expectations are in China. Economic growth, the energy intensity of that growth, and the emissions intensity of energy use are all above projections embodied in earlier expectations. China has recently overtaken the United States as the world's largest emitter and, in an unmitigated future, would account for about 35 per cent of global emissions in 2030.

Other developing countries are also becoming major contributors to global emissions growth, and will take over from China as the main growing sources a few decades from now. Without mitigation, developing countries would account for about 90 per cent of emissions growth over the next two decades, and beyond.

High petroleum prices will not necessarily slow emissions growth for many decades because of the ample availability of large resources of high-emissions fossil fuel alternatives, notably coal.

3.1 Greenhouse gas emissions by source and country

The Fourth Assessment Report of the Intergovernmental Panel on Climate Change (IPCC 2007) estimates that in 2004 greenhouse gas emissions from human activity were about 50 Gt CO_2-e.

Almost 60 per cent of this total was emissions of carbon dioxide from fossil fuel combustion and other carbon dioxide-emitting industrial processes (such as cement production and natural gas flaring).

Other greenhouse gas emissions are measured less accurately. The IPCC (2007) reports that carbon dioxide emissions from land-use change and forestry make up 17 per cent of total emissions. Slightly less than one-quarter of emissions are other gases (which are converted to CO_2-e using their global warming potential—see Chapter 2). Methane is responsible for 14 per cent of the total; nitrous oxide for 7 per cent; and a range of industrial gases for the remaining 1 per cent.

The bulk of greenhouse gas emissions arise from the countries at the centre of global economic activity. The largest emitters are China, the United States and the European Union, which between them are responsible for more than 40 per cent of global emissions. The 20 largest emitters (including emissions from land-use change and forestry) are responsible for more than 80 per cent of global emissions. (See Figure 3.1.)

Richer countries tend to have much higher per capita emissions than poorer countries. The exceptions are poorer countries with high emissions from land-use change and forestry. (See Figure 3.2.)

Developed and transition[1] countries produce about half of current global emissions. However, the growth of emissions is much faster in developing countries, and their share of global emissions will grow over time.

Figure 3.1 The 20 largest greenhouse gas emitters: total emissions and cumulative share (%) of global emissions, c. 2004

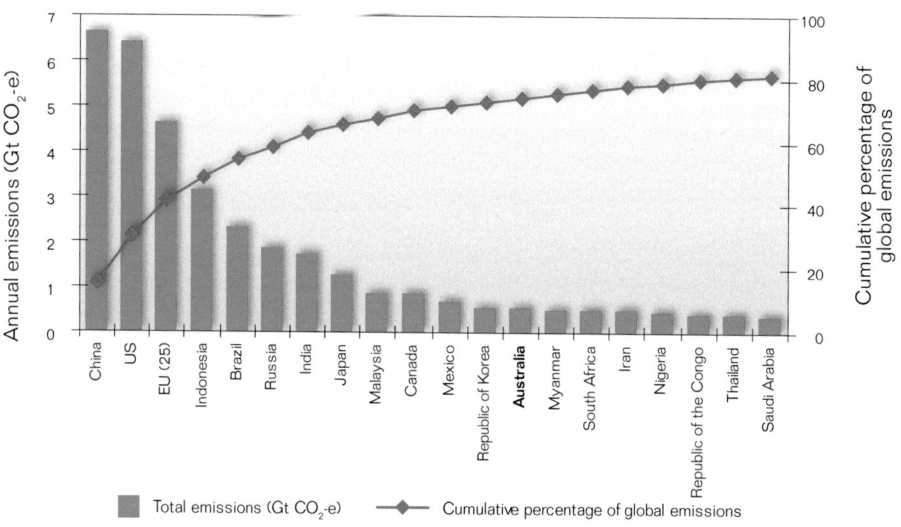

Source: UNFCCC (2007) 2004 data for US, EU (25), Russia, Japan and Canada; Department of Climate Change (2008) 2004 data for Australia (using UNFCCC accounting); and World Resources Institute (2008) for other countries (2000 data except for CO_2 emissions from fossil fuels, which are for 2004).

Figure 3.2 The 20 largest greenhouse gas emitters: per capita emissions including and excluding emissions from land-use change and forestry, c. 2004

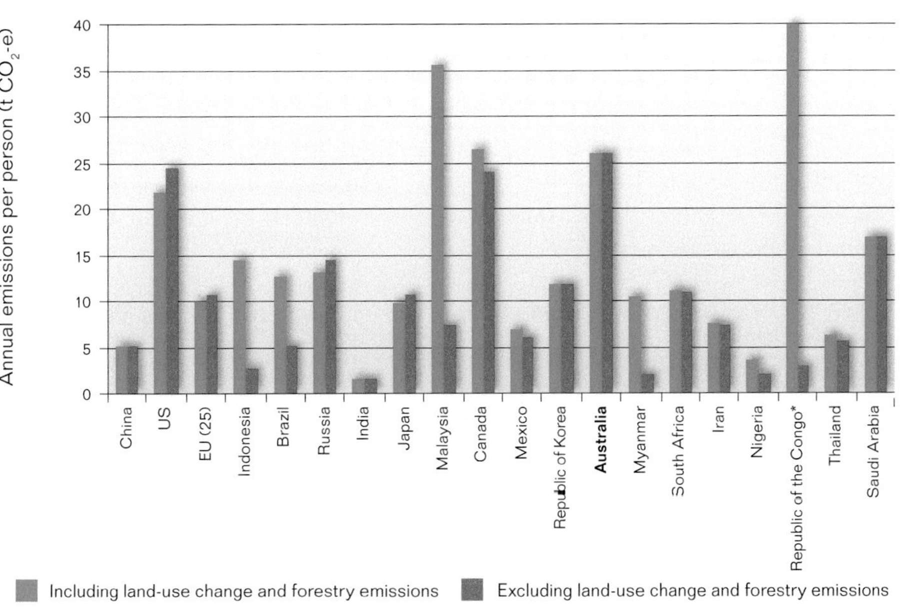

Note: Estimates of forestry-related emissions are subject to large uncertainties in many of the main emitting countries.

* Dark blue bar is truncated; per capita emissions including land-use change and forestry emissions are 105 t.

Sources: UNFCCC (2007) 2004 data for US, EU (25), Russia, Japan and Canada; Department of Climate Change (2008) 2004 data for Australia (using UNFCCC accounting); and World Resources Institute (2008) for other countries (2000 data except for CO_2 emissions from fossil fuels, which are for 2004) and for population (2004).

3.2 Recent trends in carbon dioxide emissions from fossil fuels[2]

Carbon dioxide emissions from fossil fuels are the largest and fastest-growing source of greenhouse gases.

Carbon dioxide emissions from fossil fuel combustion increased by about 2 per cent a year in the 1970s and 1980s, and by only 1 per cent a year in the 1990s. They have expanded by 3 per cent a year in the early 21st century (Table 3.1). While comprehensive national data are only available up to 2005, global estimates show a continuation of rapid emissions growth. The Netherlands Environmental Agency (2008) reports that global emissions of carbon dioxide from fossil fuel use and cement production increased by 3.5 per cent in 2006 and 3.1 per cent in 2007.

A comparison of OECD (developed) and non-OECD (developing including transition) countries shows that the latter group is driving global trends (Table 3.1). In the early 1970s, non-OECD countries were responsible for roughly one-third of global emissions, energy use and output. Since 2000, non-OECD emissions have been growing almost eight times faster than OECD emissions, accounting for 85 per cent of the growth in global emissions. In 2005 non-OECD countries were responsible for just over half of global energy use and emissions, and 45 per cent of global output.

The OECD countries show a slowdown in growth in emissions, GDP and energy in this decade compared to the last. In the non-OECD countries, the rate of growth in all three has increased significantly in this decade. The high rate of global economic growth seen this decade, at times even above that seen in the 'Golden Age' of the 1950s and 1960s, defines the new 'Platinum Age' that the world has entered (Garnaut & Huang 2007).

Table 3.1 Growth in CO_2 emissions from fuel combustion, GDP and energy

	Average annual growth rates (%)		
	1971–90	1990–2000	2000–05
World			
Emissions growth	2.1	1.1	2.9
GDP growth	3.4	3.2	3.8
Energy growth	2.4	1.4	2.7
OECD			
Emissions growth	0.9	1.2	0.7
GDP growth	3.2	2.7	2.1
Energy growth	1.5	1.6	0.8
Non-OECD			
Emissions growth	4.2	0.9	5.5
GDP growth	3.8	4.0	6.2
Energy growth	3.8	1.0	4.6

Notes: Emissions growth is CO_2 from fossil fuel combustion. Energy growth is total primary energy supply measured in millions of tonnes of oil equivalent. GDP growth is measured using 2000 US$ purchasing power parities.

Source: IEA (2007a).

There have also been significant changes among non-OECD countries in the energy intensity of economic activity (the energy to GDP ratio) and the carbon intensity of energy use (the emissions to energy ratio). The 1990s saw a rapid decline in energy intensity in the non-OECD group. Energy use grew at only a quarter of the rate of GDP, and emissions at below the rate of energy. This decade has seen the resumption of energy-intensive and carbon-intensive growth in the developing and transition world: energy use has grown at three-quarters the rate of GDP, and carbon emissions 20 per cent faster than energy use.

Figure 3.3 shows just how differently energy intensity has behaved in OECD and non-OECD countries. In the developed world, the energy/GDP curve has declined smoothly and continuously. In the developing world, energy intensity fell only slowly over the 1970s and 1980s, plunged in the 1990s, and has now flattened out, at around 70 per cent of its 1971 level.

Figure 3.3 CO_2 emissions/GDP, energy/GDP and CO_2 emissions/energy for the world, OECD and non-OECD, 1971–2005 (1971 = 1)

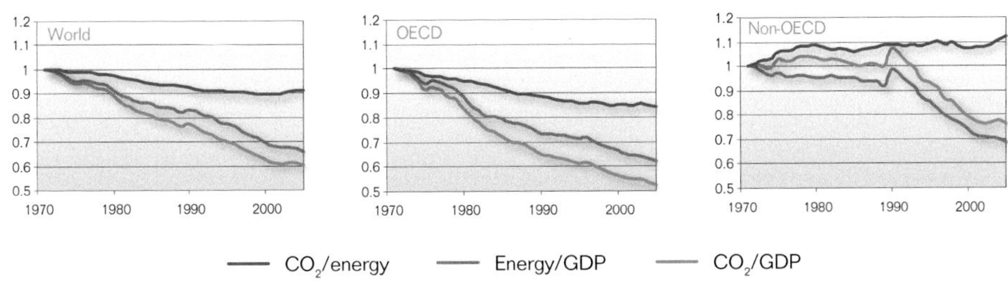

Notes: Emissions are CO_2 from fossil fuels. Energy is total primary energy supply measured in millions of tonnes of oil equivalent. GDP is measured using 2000 US$ purchasing power parities.

Source: IEA (2007a).

Figure 3.4 shows energy intensity separately for China and other developing countries. Energy intensities are remarkably steady for developing countries once China is excluded. China started out with an enormously high energy intensity in the 1970s. The ratio declined through the 1980s and 1990s, due to a shift away from subsidised prices and central planning. It flattened only at the turn of the century.

Figure 3.4 Energy intensities of GDP for China and other developing countries, 1970–2005

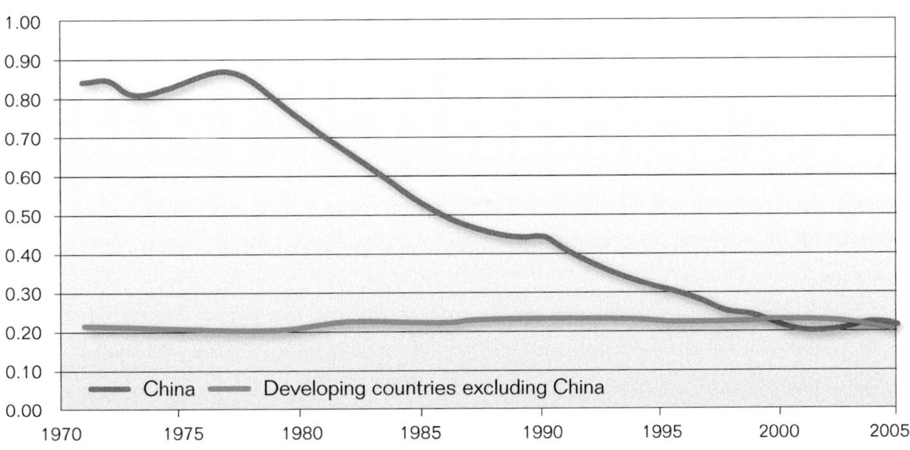

Note: The ratio is of energy (total primary energy supply measured in millions of tonnes of oil equivalent) over GDP (in 2000 US$ purchasing power parities).

Source: IEA (2007a).

Globally, increasing reliance on coal, which is more carbon intensive than oil and natural gas,[3] has raised the carbon intensity of energy above what it would otherwise have been, keeping it roughly constant, with a slight upward trend in recent years. In the 1980s and 1990s, a reduction in the share of oil in total energy demand was associated with an increase in the share of gas. Since 2000, the share of gas has remained constant, and the share of coal has increased.

For coal, the same trends are evident in both developed and developing regions, though in much more dramatic terms in the latter. Between 2000 and 2005, coal use increased in developing countries on average by 9.5 per cent annually, and by 11.7 per cent in China.[4] In 2005, 61 per cent of the world's coal was consumed in developing countries, up from 51 per cent just five years earlier. In 2005, coal provided 63 per cent of China's energy, 39 per cent of India's, and only 17 per cent of the rest of the world's (IEA 2007b).

In summary, the acceleration of emissions growth this decade has been caused by three factors: the rapid acceleration of growth in developing countries; the ending of the period of rapid decline in energy intensity in China, which lasted from the 1970s to the 1990s; and the end to the decarbonisation of energy supply in both the developed and (especially) the developing world.

3.3 Existing emissions projections

The most influential projections used in climate change analysis are still those set out in the *Special Report on Emissions Scenarios* (SRES) of the IPCC (2000). These provide a range of future emissions paths out to 2100 under four different 'storylines' about growth and technology.

The SRES authors did not assign likelihoods to particular scenarios, but rather argued that they were all equally plausible. In practice, most attention has been given to low- and mid-range emissions growth scenarios, with high-end scenarios often dismissed as extreme or unrealistic. Other analyses give all SRES scenarios equal weight, rather than asking which ones are more soundly based. Reliance on only the more pessimistic SRES scenarios is seen as unbalanced. One of the criticisms of the Stern Review has been that the SRES scenario it relied on showed 'high range greenhouse gas emissions' (Baker et al. 2008: xi). Stern himself, however, in his Ely lecture (2008), following interaction with the Garnaut Climate Change Review, noted that his review underestimated the likely growth of emissions.

Post-SRES scenarios do not show very different results to those of SRES. GDP growth, total energy use and carbon dioxide emissions are all lower in the median post-SRES no-mitigation scenario than in the median pre-SRES/SRES scenario (IPCC 2007). Energy forecasting agencies have not significantly adjusted emissions forecasts upwards despite the acceleration of growth seen so far this decade. The US Energy Information Administration reference scenarios for emissions and energy consumption growth over the period 2000 to 2020 were no higher in 2007 than in 2000, despite the higher levels in both variables seen so far this decade.[5]

Results from a range of existing projections are shown in figures 3.8, 3.9 and 3.10 along with the results from the updated projections carried out for the Review.

3.4 The Review's no-mitigation projections: methodology and assumptions

Two new sets of projections were developed for the Review. Both are constant policy scenarios where no further policies are put in place to mitigate climate change, and no additional impacts of climate change are felt.

The Review's 'Platinum Age' projections cover the period out to 2030, and are based on work by Garnaut et al. (2008). These projections use the most recent International Energy Agency projections (IEA 2007b), which make use of extensive information on energy systems in a partial equilibrium framework. Using an emissions growth decomposition framework, adjustments are made to selected macroeconomic assumptions. The strength of this approach is that it builds on the specialist knowledge of the IEA, and identifies the assumptions that need rethinking. Its limitation is that it does not capture the general equilibrium effects that would derive from the changes in assumptions.

The Review's reference case runs to 2100, and was developed by the Australian Treasury and the Garnaut Review in consultation with other experts. This scenario was implemented in the Global Trade and Environment Model (GTEM), a computable general equilibrium model of the world economy used in the joint modelling exercise by the Review and the Treasury. The top–down modelling of GTEM is complemented by a series of bottom–up models of electricity generation, transport, and land-use change and forestry.

Key GDP and population assumptions up to 2030 are broadly consistent for the two projections.[6] Population projections (figures 3.5 and 3.6) are the United Nations 'medium variant' population projections to 2050, and UN long-term

Figure 3.5 The reference case: global population, GDP and GDP per capita, 2001 to 2100

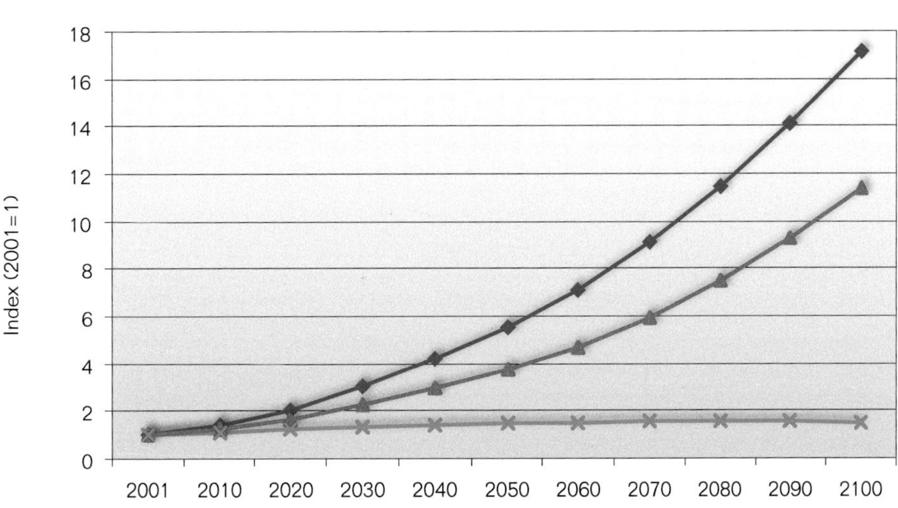

population projections to 2100. Population projections for Australia, however, are based on Australian Treasury (2007) projections, revised in the light of Treasury's most recent analysis of immigration trends. These are higher than UN projections, significantly so in the second half of the century.

By the end of the century the world's population is over 40 per cent larger than at the start. Population growth is above 1 per cent per year in the current decade, then steadily falls to zero annual growth around 2080, when global population peaks. After 2080, population falls by 0.1 per cent a year on average, with nearly all regions showing zero or negative growth. Many developing countries including India gain in population share over the century; Australia, Canada and the United States hold a broadly constant share as a result of immigration; and the shares of China, Europe, Russia and Japan drop.

Assumptions on nearer-term GDP per capita growth rates are based on growth accounting and judgments informed by recent experience, both of which suggest the continuation of high growth, albeit falling over time, in the developing world. Longer term, GDP per capita is assumed to converge over time towards that of the United States, which is assumed in the long term to grow at 1.5 per cent a year. Growth slows in developing countries as the income gap with the United States diminishes. Countries are assumed not to close the gap completely by the end of the century, with average world per capita incomes around half US levels at 2100. The global annual per capita GDP growth peaks at just over 3 per cent in the middle of the 2020s, then falls to 2 per cent by the end of the century. Global annual GDP growth peaks around 4 per cent in the early 2020s, then falls to just below 2 per cent by the end of the century (Figure 3.6).

Figure 3.6 The reference case: global population, GDP, GDP per capita, and CO_2-e emissions, 2000 to 2100—average growth rates by decade

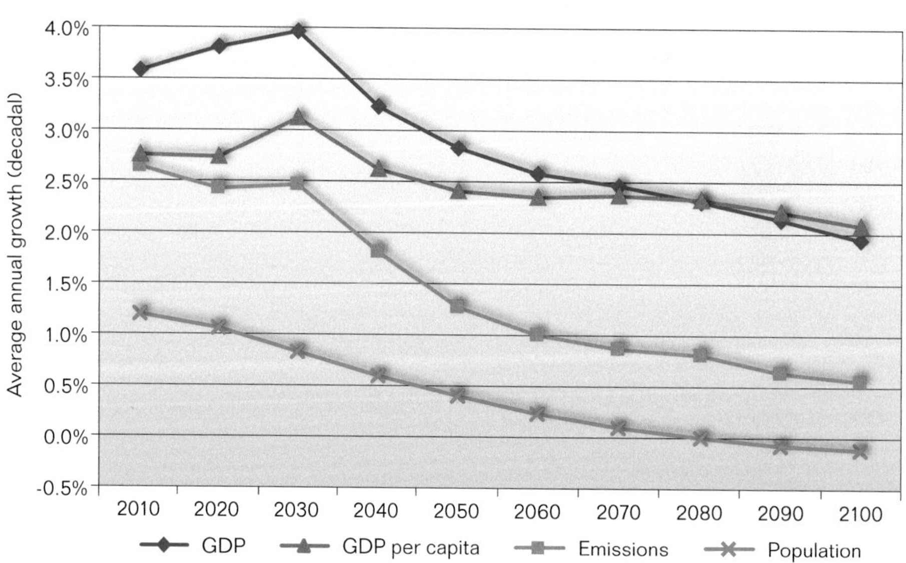

By 2100, total global GDP is 17 times its 2000 level and average per capita GDP increases by 11 times over the century. The distribution of global output is also very different from that seen today (Figure 3.7). The share of China in global output rises sharply until the 2030s, but then declines. The share of India continues to rise and by about 2080 overtakes that of China. The shares of the European Union, North America and the rest of Asia decline. The share of the rest of the world rises over the period, reflecting higher population growth in many low-income developing countries.

Growth in greenhouse gas emissions in the reference case are a function of changes in production and consumption structures in different countries, changes in relative prices including for different sources of energy, and improvements in energy efficiency and the efficiency of intermediate input use. The reference projections also include emissions of methane, nitrous oxide and various industrial gases, as well as a subset of forestry-related emissions and sequestration.

Carbon dioxide emissions from fossil fuel are modelled in the Platinum Age projections as a function of the carbon intensity of energy, and the energy intensity of GDP. Carbon intensity is assumed to remain roughly constant over time, in linc with IEA projections: the share of oil decreases, with substitution towards coal as well as low-emissions energy sources. Energy intensity is assumed to decline in both developed and developing countries. This is in contrast to the historical experience in developing countries, where energy intensities have been constant (see Figure 3.4) and is assumed to represent the effect of high energy prices. In particular, in contrast to the experience of the past few years, energy intensity is assumed to fall in China, but not as rapidly as projected by the IEA (see Box 3.1). Platinum Age projections for methane and nitrous oxide update US Environmental Protection Agency projections for higher forecast global growth. Platinum Age

Figure 3.7 Shares in global output of various countries and regions, 2001 to 2100, under the reference case

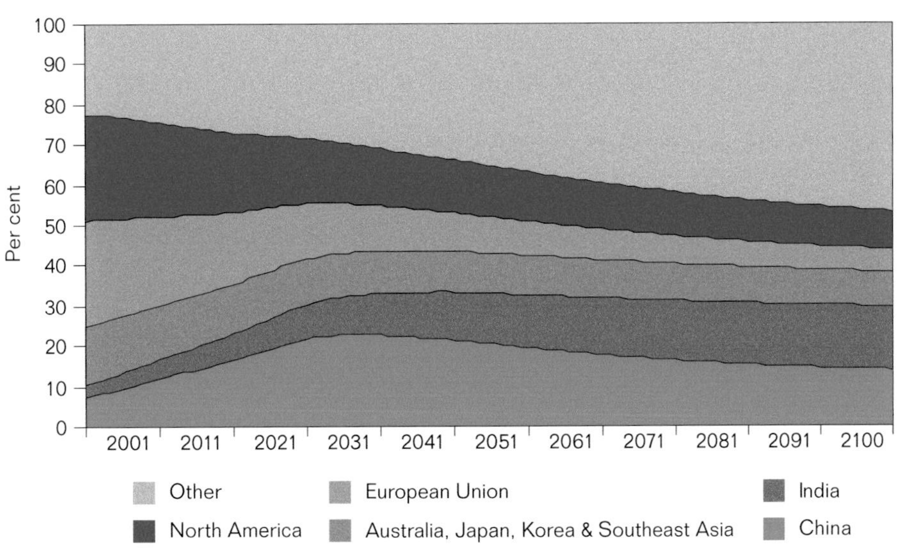

projections for forestry-related carbon dioxide emissions are based on IPCC baselines, and assume a constant value in these emissions out to 2030. Detailed assumptions are available in Garnaut et al. (2008).

3.5 Results from the Review's projections and comparisons with existing projections

Figure 3.8 compares the average growth rates for carbon dioxide emissions from fossil fuels for the Platinum Age and Garnaut–Treasury reference case with a number of SRES (in green) and post-SRES (white) scenarios for the period circa 2005–30. It also shows (in red) average emissions growth in the 1970s and 1980s, the previous decade and so far in this decade (2000 to 2005).

Most carbon dioxide emissions projections for growth out to 2030 forecast annual average growth significantly below the 2.9 per cent annual average growth seen between 2000 and 2005. Even A1FI, the SRES scenario that shows the most rapid emissions growth over the century, often regarded as extreme,

Figure 3.8 Global CO_2 emissions growth rates from fossil fuels and industrial processes to 2030: a comparison of Garnaut Review no-mitigation projections with SRES and post-SRES scenarios and historical data

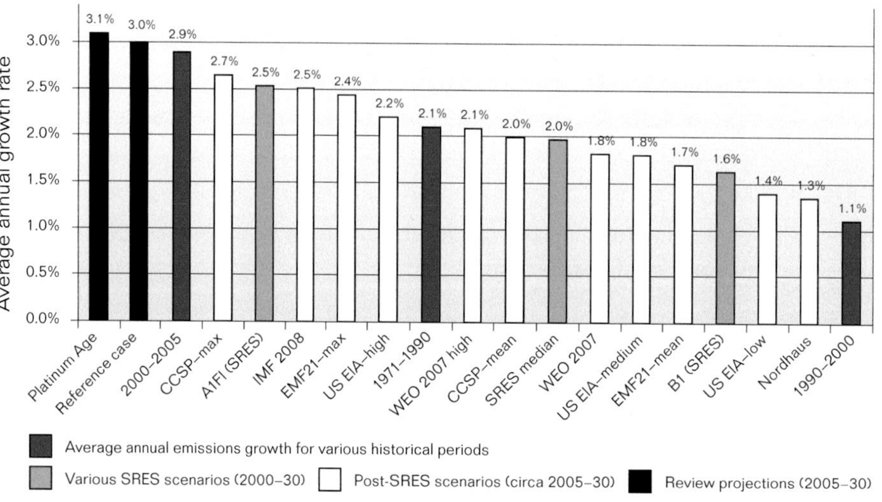

Sources: This figure is modified from Garnaut et al. (2008). Historical data are from Table 1. The SRES scenarios (IPCC 2000) used are A1FI (AIG MINICAM), which shows the most rapid emissions growth, both to 2030 and to 2100; B1 (BI IMAGE), which is at the lower end of the range; and the median SRES scenario (which is defined as the median for each variable and each decade of the four SRES marker scenarios). The SRES scenarios give projections for every 10 years from 1990 to 2100; we report here projections for 2000 to 2030. Post-SRES scenarios included are the mean and maximum emission baselines from the EMF-21 project (Weyant et al. 2006), which included 18 different emissions projection models for 2000 to 2025; the mean and maximum projections from the US Climate Change Science Program (Clarke et al. 2007), which used three models; the base case from the Nordhaus (2007) model for 2005 to 2035; projections for 2005 to 2030 from IEA (2007b) (both the base case and a rapid-growth scenario with higher growth projected for China and India); the high, medium and low projections from the US Energy Information Administration (2007) for 2004 to 2030; and the IMF World Economic Outlook baseline for 2002 to 2030 (IMF 2008).

projects carbon dioxide emissions growth of only 2.5 per cent out to 2030. The SRES median scenario shows growth of 2.0 per cent, and the moderate B1 SRES scenario shows growth of only 1.6 per cent. The post-SRES scenarios lie in a similar range.

The Platinum Age projections and the reference case, however, project growth in carbon dioxide emissions of 3.1 per cent and 3.0 per cent out to 2030. They suggest that the existing range of scenarios underestimates the future growth of emissions in the early 21st century. In the absence of unexpected dislocations in the global economy and in the absence of effective mitigation, emissions growth is unlikely to ease significantly over the next two decades.

Figures 3.9 and 3.10 project greenhouse gas emissions from human activity for the Review's no-mitigation projections and other projection exercises. The Platinum Age projections and the reference case both give annual average growth in greenhouse gas emissions of 2.5 per cent over the period 2005–30, at the top end of existing projections and comparable with the growth rate in emissions seen in the first years of this decade (Figure 3.9).

Emissions levels at 2030 are significantly higher in the Platinum Age than in existing rapid-growth scenarios because of the higher forestry-related emissions built into the base. The Platinum Age projects emissions of 83 Gt CO_2-e by 2030, almost double their 2005 level, 11 per cent higher than the A1FI scenario, and a level of emissions reached only in 2050 in the business-as-usual scenario used by the Stern Review (Stern 2007: 202). Reference case emissions are lower in 2030 at 72 Gt CO_2-e, largely due to lower base-year estimates of forestry-related and non-carbon dioxide emissions.

Figure 3.9 Global greenhouse gas emissions growth rates to 2030: a comparison of Garnaut Review no-mitigation projections, SRES and post-SRES scenarios, and historical data

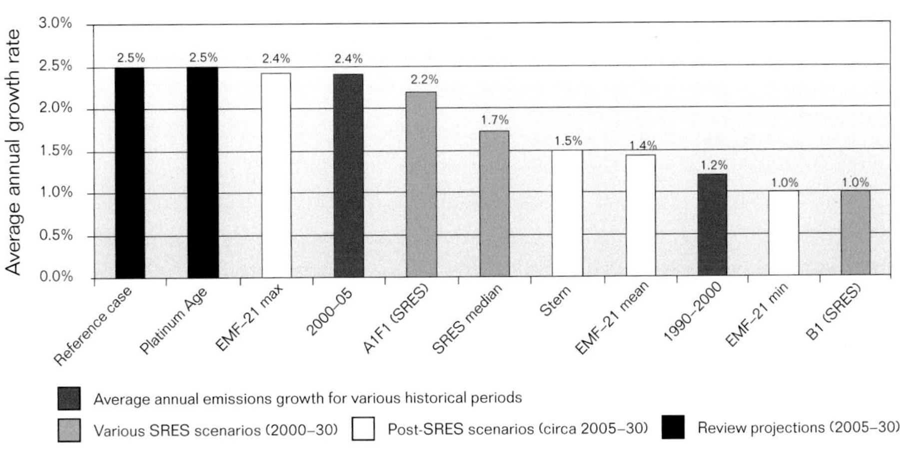

Sources: This figure is modified from Garnaut et al. (2008), which includes detailed notes. See also the notes to Figure 3.8.

Figure 3.10 Global greenhouse gas emissions to 2100: a comparison of Garnaut Review no-mitigation projections and various SRES scenarios

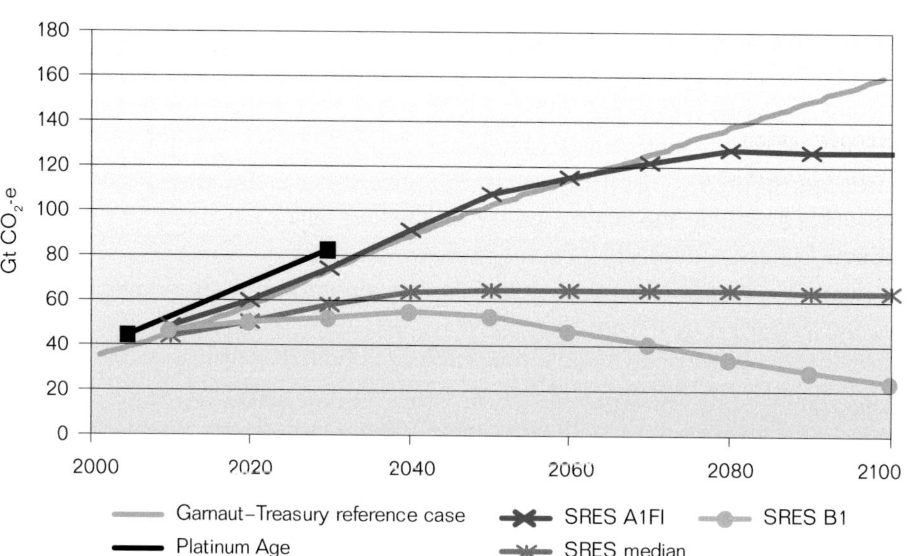

Sources: See notes to Figure 3.8.

Emissions under the reference case continue to rise post-2030 though more slowly (Figure 3.6). Figure 3.10 compares emissions over the century under the two Garnaut Review business-as-usual projections with the three SRES scenarios highlighted in figures 3.8 and 3.9. Although the rise in emissions in the reference case is much slower in the latter decades of the century (the annual growth rate drops to 0.7 per cent at the end of the century), emissions in the reference case are more than twice those in the SRES median scenario by the end of the century. Over the course of the century, emissions in the reference case are comparable with or above those of the 'extreme' A1FI scenario.

Table 3.2 shows the projected composition of emissions across countries in the reference case using the regional breakdown which GTEM deploys. The share of most developing countries is growing rapidly. More than 90 per cent of growth in emissions over the century occurs in developing countries.

China emerges as the most important country determining emissions, especially up to 2030. By 2030, its emissions easily exceed those of all developed and transition countries combined. The factors behind the explosive growth in emissions in China are explored in Box 3.1.

Table 3.2 Shares of total greenhouse gas emissions by country/region in the Garnaut–Treasury reference case

	Share of emissions (%)		
	2005	2030	2100
United States	18.3	11.1	5.1
European Union	12.6	7.1	3.6
Former Soviet Union	8.4	6.9	4.5
Japan	3.5	1.8	0.7
Canada	2.0	1.3	1.0
Australia	1.5	1.1	1.0
China	18.3	33.0	21.2
India	4.6	8.0	16.8
Indonesia	2.0	1.8	2.4
Other Southeast Asia and Korea	4.3	2.9	3.3
South Africa	1.3	1.3	1.4
OPEC	4.7	5.4	7.8
Rest of world	18.5	18.2	31.3
World	100	100	100

Box 3.1 Determinants of China's emissions trajectory to 2030

China has the world's largest population, and the highest economic growth rate of any major country. Its energy is unusually carbon intensive: out of 51 countries with a population greater than 20 million, China has the fifth most carbon-intensive energy mix (IEA 2007a). China's energy supply is carbon intensive because coal is the only domestic energy source in which China is even moderately well-endowed per capita on a global basis.

In the coming decades, China will have more impact on global emissions than any other country. Assumptions about future economic growth and energy patterns in China are therefore of critical importance to emissions projections.

China's influence out to 2030 is particularly pronounced. The Garnaut–Treasury reference case projects China's share of global emissions to rise from 18 per cent in 2005 to 33 per cent in 2030. This follows from average annual growth in emissions of 5 per cent. After 2030, China's growth is projected to slow, as its population starts to fall and per capita incomes reach relatively high levels. Emissions growth slows to just 1.3 per cent for 2030 to 2050.

What will drive China's rapid growth in emissions to 2030? First, China's economy will continue to grow rapidly. Using growth accounting, Garnaut et al. (2008) project GDP growth of China of 9.0 per cent from 2005 to 2015 and 6.8 per cent from 2015 to 2025. This rapid GDP growth will take place on the back of continued very high investment levels and total factor productivity growth. This projection is higher than most literature forecasts, but below performance seen in recent years.

Box 3.1 Determinants of China's emissions trajectory to 2030 *(continued)*

Second, China's economic expansion will continue to be energy intensive. Figure 3.11 shows the acceleration in China's energy consumption in recent years. This reflects rapid growth in China in heavy energy-intensive industry: between 2000 and 2006, crude steel production grew in China by an annual average of 22 per cent, pig iron by 21 per cent, and cement by 13 per cent (National Bureau of Statistics of China 2007a). Analysts differ on the extent to which energy efficiency will improve in China, even without a concern for climate change. The Review expects that, in the absence of deliberate mitigation, and given China's high level of investment and the rapid growth of its heavy industry, the energy intensity of output in China will decline no more rapidly than in other developing countries. The Review's Platinum Age projections adjust the IEA's more rapid energy efficiency improvements downwards to be broadly consistent with the analysis of Sheehan and Sun (2007), who predict elasticities of energy with respect to GDP of 0.8 declining to 0.7.

Third, in the absence of a price on carbon, China is unlikely to move away from its heavy reliance on coal, or make any movement towards sequestration of emissions from coal combustion—the only factors that would reduce the high carbon intensity of its energy system.

While the Review's emissions growth projections for China are substantially higher than those by the IEA, they are supported by other recent studies. Auffhammer and Carson (2008), using Chinese provincial 1985 to 2004 data, project 11–12 per cent annual growth in CO_2 emissions from fossil fuel combustion for the period 2000 to 2010.

Figure 3.11 China total energy consumption, levels and growth, 1978 to 2006

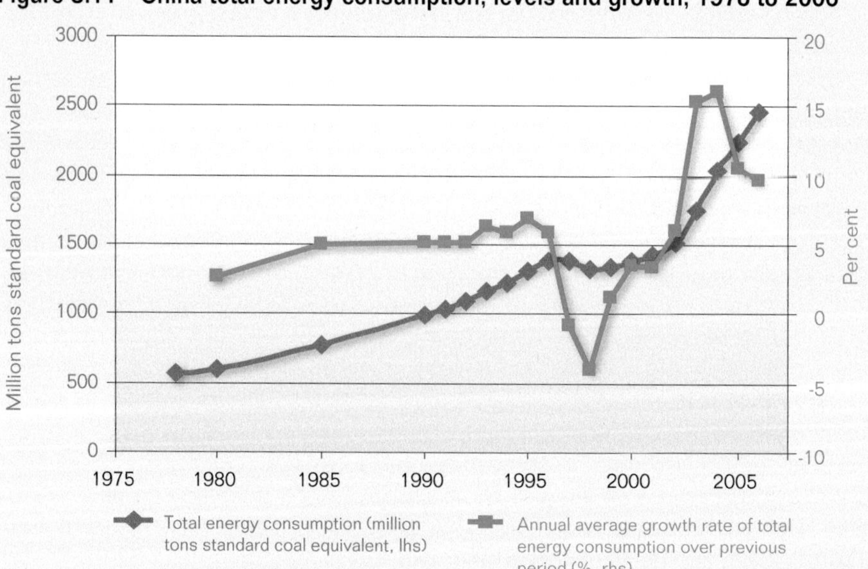

Source: National Bureau of Statistics of China (2007a).

3.6 The impact of high energy prices

Global energy prices have risen dramatically over the last few years. The oil price is now at a historic high, having reached a level in real terms not seen since the early 1980s (Figure 3.12). Natural gas prices are following suit. Most recently, coal prices have also risen sharply. The price rises are driven by increasing demand and limitations on expansion of production. In the case of oil and gas, there is a resource constraint, whereas for coal the supply constraint is purely in terms of mining and transport capacity.

Continued high fossil energy prices, if across the board, will cause reductions in energy consumption and a substitution towards non-fossil-fuel energy sources. These effects by themselves would dampen growth in carbon dioxide emissions. However, substitution away from oil and gas towards coal and synthetic liquid hydrocarbons (derived from coal, tar sands or natural gas) will increase growth in emissions. Making liquid fuels from coal can be cheaper than petroleum at oil prices reached in 2008, and for many countries is attractive as a more secure supply. In the medium term, coal prices are expected to fall as supply capacity is increased in response to excess demand. This in turn will reduce incentives to shift into renewable energy sources and nuclear power, and to reduce energy use. The share of high-carbon fuels in the energy mix, and with it the carbon intensity of energy, will not necessarily fall as a result of high oil prices.

Figure 3.12 Oil, gas and coal prices, 1970 to 2008

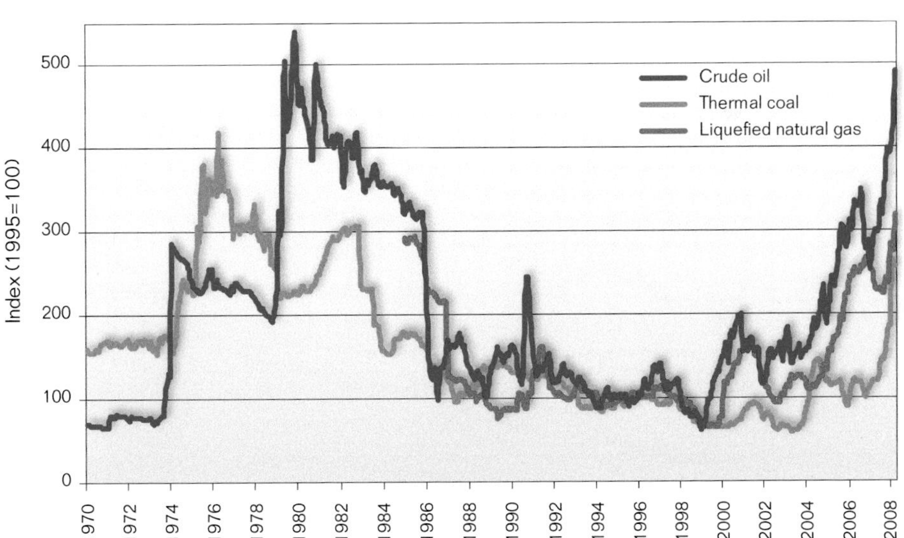

Note: Nominal prices converted to SDRs and deflated by the G7 CPI. Indexed to 1995. Prices are as at January for 1970–2007 and as at April for 2008.

Source: Table compiled by the Centre for International Economics based on IMF IFS Statistics, OECD Main Economic Indicators, *Financial Times*, and CIE estimates.

Recent data suggest that the increase in oil prices is not resulting in lower global emissions (Figure 3.13). Since 2005, growth in global oil use has slowed to around 1 per cent annually, but total energy use has grown by almost 3 per cent annually. Gas use has grown roughly in line with total energy use, while coal consumption has grown at 5 per cent, and use of other energy sources (principally renewables and nuclear) at only around 2 per cent (BP 2008). Energy-related carbon dioxide emissions have grown slightly faster than total energy use. There is strong momentum in growth of liquid fuel production from Canadian tar sands. Looking ahead, investment in coal-fired electricity generation remains strong, particularly in Asia but also other parts of the world. China is investing in coal-to-liquid plants and is expected to start operating the largest such facility outside South Africa later in 2008 (Nakanishi & Shuping 2008). Coal liquefaction is also being considered in the United States.

It is instructive to examine the oil price shocks in the 1970s and especially the 1980s (Figure 3.12). In both episodes coal prices rose later than oil prices, and fell back to or below earlier prices more quickly than oil prices. In both cases, the drop in global oil consumption was more pronounced than for other fuels (Figure 3.13). Electricity generation from renewables and nuclear power in particular grew in the aftermath of the oil price shocks, but by less than energy from coal in absolute terms. The carbon intensity of global energy supply fell markedly in the first half of the 1980s, then stagnated. It fell in the 1990s primarily because of restructuring in the former Soviet Union, and in this century has been on the rise again.

Figure 3.13 Global energy use and CO_2 emissions, 1970 to 2007

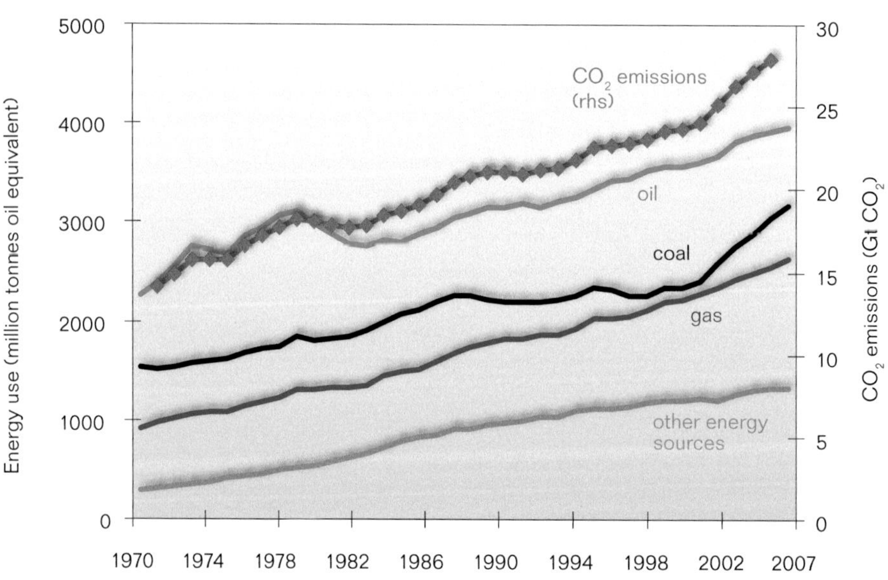

Sources: Energy use from BP (2008); CO_2 emissions from IEA (2007b) and Carbon Dioxide Information and Analysis Centre (2008).

The upshot is that high petroleum prices do not necessarily mean lower greenhouse gas emissions, and may actually lead to higher emissions. On the one hand, high prices accelerate improvements in energy efficiency and promote non-fossil-fuel energy. On the other, high oil prices increase demand for coal. Coal prices will remain high only if investment in new capacity cannot keep up with growth in demand. If coal prices remain high, renewable energy will become more attractive at the margin, but only in a context of rapid growth in demand for coal. Although it is impossible to know which influence will dominate, recent experience suggests that high petroleum prices are as consistent with an acceleration as with a deceleration of emissions growth.

3.7 Resource limits

By 2100, under the Garnaut–Treasury reference case projections, global output will be 17 times its current level. Australia's real per capita income would be US$137 000, compared to US$36 000 in 2005.

By 2100, today's developing countries would achieve higher levels of per capita expenditure than current levels in today's rich countries. India, which only has 5 per cent of Australia's per capita income today, has in the reference case in 2100 a per capita income 2.2 times Australia's current level. Does the world have the resources to support consumption based on today's preferences at these levels of income?

Concerns about natural resource limits to growth were raised by the Club of Rome and others in the period of high resource prices associated with the latter stages of Japanese industrialisation and rapid growth in the early 1970s (Club of Rome 1972). This group included eminent Japanese economist Saburo Okita, Director of Japan's Economic Planning Agency at the high tide of rapid Japanese post-war growth and author of the Ikeda administration's income-doubling plan of 1960. The Club of Rome was not from the fringe of modern development policy. Its central prediction was not for an imminent collapse in economic activity, as is often reported, but for a collapse in the middle of this century due to resource constraints and pollution (Turner 2008). Similar pessimistic expectations about the availability of natural resources to support rising human living standards had been raised by eminent economists from time to time in the first century of modern economic development (Malthus 1798; Jevons 1865).

The last quarter of a century of low resource prices has led economists generally to dismiss such concerns and forecasts on the grounds that they underestimate human ingenuity and the capacity for markets to support far-reaching structural change. The failures immunised the economics profession against acknowledgment of the possibility of resource supplies being a fundamentally important constraint on growth. However, the possibility that natural resource constraints might force fundamental changes in consumption patterns has been seeping recently into the professional consciousness, as real commodity prices across a wide front have now been sustained at exceptionally high levels for longer than ever before. High commodity prices across the board, despite the US economy teetering on the edge of recession, are concentrating many minds.

The prospects of much higher levels of income for high proportions of the world's people later in the 21st century focuses minds even more keenly. Will resource constraints prevent total global output from increasing by 17 times from the levels of the early 21st century that are already stretching supplies of many natural resource–based commodities?

There are many potential limits that could conceivably constrain output, from fuel to food to water. Pressures on global agricultural resources could be particularly problematic, especially if climate change diminishes productivity. They could seriously undermine political stability in some developing countries.

Could limits on minerals and fossil fuels also constrain growth? Table 3.3 looks at the number of years that reserves and the known reserve base can sustain production at current and assumed 2050 levels for several mineral resources and fossil fuels. By 2050, global output is projected in the reference case to be almost five times its current level. For this illustrative exercise, it is assumed that the production of metals and minerals is at three times current levels in 2050. Predicted production of oil and coal in 2050 is based on US Energy Information Administration projections. These rates of production are compared to estimates of reserves and the reserve base. Reserves are that portion of the reserve base which can be economically extracted. For fossil fuels, the reserve base is the total global 'ultimately recoverable' resource base, including an estimate of undiscovered resources. For metals and minerals, the reserve base is that portion of the global resource which has been identified, whether or not it is economic.

Current reserves for several metals will not last long, especially at 2050 levels of production. High prices will push more of the reserve base into reserves. The current reserve base will support current production levels for all the minerals reported for at least 30 years, but will support production at 2050 levels for 20 years or less. This suggests that extreme pressures on supplies of minerals and metals, if they arise, will not do so until towards or after the middle of the century. Whether they do arise then depends on the gap between the reserve base and the total resource base. This can be large. For example, the US Geological Survey estimates the global reserve base for copper to be 940 million tons, but world resources (including deep-sea nodules) to be 3.7 billion tons. Note that any tendency towards exhaustion of reserves would raise prices, which would convert resources into reserves. It would also stimulate exploration, leading to expansion of reserves and the reserve base.

In relation to fossil fuels, Table 3.3 suggests that conventional oil reserves may well come under pressure over the next several decades, but that there are ample supplies of coal. Some argue that conventional oil reserves are exaggerated (Campbell & Laherrère 1998). But unconventional sources, including oil sands in Canada, extra-heavy oil in Venezuela and shale oil in the United States, Australia and several other countries, not included in the table below, are thought to amount to at least 1 trillion barrels, or almost 50 per cent of ultimately recoverable conventional oil resources (IEA 2006).

Table 3.3 Time to exhaustion of current estimates of reserves and reserve base for various metals and minerals, and fossil fuels

	Number of years after which exhaustion will be reached of			
	Reserves		Reserve base	
	At the production rates of			
	2007	2050	2007	2050
Metals and minerals				
Nickel	40	13	90	30
Zinc	17	6	46	15
Copper	31	10	60	20
Bauxite	132	44	168	56
Platinum group metals	154	51	173	58
Lead	22	7	48	16
Tin	20	7	37	12
Tungsten	32	11	70	23
Iron ore	79	26	179	60
Fossil fuels				
Coal	139	66	214	101
Gas	60	32	110	58
Oil	40	23	70	40

Note: For metals and minerals, current production, reserves and reserve base are the latest estimates from the US Geological Survey (http://minerals.usgs.gov/minerals/pubs/commodity). Reserves are defined by the US Geological Survey to be 'that part of the reserve base which could be economically extracted or produced at the time of determination'. The reserve base is 'resources whose location, grade, quality, and quantity are known or estimated from specific geologic evidence'. Production rates for 2050 are simply assumed to be three times current levels.

Fossil fuel figures are the latest estimates from the US Energy Information Administration, the International Energy Agency and the World Energy Council. Coal includes both black and brown coal. Fossil fuel reserves are recoverable reserves: those quantities which geological and engineering information indicates with reasonable certainty can be extracted in the future under existing economic and operating conditions. (US EIA 2007: Table 8, Chapter 5). The reserve base for fossil fuels is the global resource base: all 'ultimately recoverable resources', including an estimate of 'undiscovered conventional resources that are expected to be economically recoverable' (IEA 2006: 91).

Source: Table compiled by the Centre for International Economics.

This analysis suggests that mineral and fossil fuel shortages will not be a constraint on growth in the first half of this century. By that time, if the world were still on a business-as-usual path, the environmental damage would have already been done, as dangerous levels of temperature increase would already have been locked in, if not already realised. Shortages of minerals and fossil fuels will not solve the world's emissions problems.

The recent and projected continued rapid growth in emissions has major implications for the global approach to climate change mitigation. Earlier and more ambitious action than previously thought will be required by all major emitters, if the world is to hold the risks of dangerous climate change to acceptable levels.

Notes

1 Transition countries are countries in central and eastern Europe and the former Soviet Union defined in the United Nations Framework Convention on Climate Change and Kyoto Protocol as 'undergoing the process of transition to a market economy'.

2 This section draws on Garnaut et al. (2008). See also Raupach et al. (2007).

3 The US Energy Information Administration (1998) reports that on average oil emits 40 per cent more carbon dioxide than gas, and coal 27 per cent more than oil per unit of energy input.

4 In 2006, China's coal consumption grew by 11.9 per cent and in 2007, according to preliminary estimates, by 7.8 per cent (see National Bureau of Statistics of China 2007a, 2007b).

5 See annual US Energy Information Administration *International Energy Outlooks.* See IMF (2008) and Sheehan et al. (2008) for two recent projections with more rapid rates of emissions growth.

6 Note that GDP is measured using purchasing power parities rather than market exchange rates in both sets of projections.

References

Auffhammer, M. & Carson, R.T. 2008, 'Forecasting the path of China's CO_2 emissions using province-level information', *Journal of Environmental Economics and Management* 55(3): 229–47.

Australian Treasury 2007, *Intergenerational Report 2007*, available at <www.treasury.gov.au/igr/IGR2007.asp>.

Baker, R., Barker, A., Johnson, A. & Kohlhaas, M. 2008, *The Stern Review: An assessment of its methodology*, Productivity Commission Staff Working Paper, Commonwealth of Australia, Melbourne.

BP 2008, *BP Statistical Review of World Energy June 2008*, BP, London.

Campbell, C.J. & Laherrère, J.H. 1998, 'The end of cheap oil', *Scientific American* 278(3): 78–83.

Carbon Dioxide Information and Analysis Centre 2008, <http://cdiac.ornl.gov>.

Clarke, L., Edmonds, J., Jacoby, H., Pitcher, H., Reilly, J. & Richels, R. 2007, *Scenarios of Greenhouse Gas Emissions and Atmospheric Concentrations*, Sub-report 2.1A of Synthesis and Assessment Product 2.1 by the US Climate Change Science Program and the Subcommittee on Global Change Research, Department of Energy, Office of Biological & Environmental Research, Washington DC.

Club of Rome 1972, *The Limits to Growth: A report for the Club of Rome's project on the predicament of mankind*, D.H. Meadows, D.L. Meadows, J. Randers & W.W. Behrens (eds), Universe Books, New York.

Department of Climate Change 2008, *Australia's National Greenhouse Accounts*, Australian Greenhouse Emissions Information System, <www.ageis.greenhouse.gov.au>.

Garnaut, R., Howes, S., Jotzo, F. & Sheehan, P. 2008, 'Emissions in the Platinum Age: the implications of rapid development for climate change mitigation', *Oxford Review of Economics and Policy* 24(2): 1–25.

Garnaut, R. & Huang, Y. 2007, 'Mature Chinese growth leads the global Platinum Age', in R. Garnaut & Y. Huang (eds), *China: Linking Markets for Growth*, Asia Pacific Press, Australian National University, Canberra.

IEA (International Energy Agency) 2006, *World Economic Outlook 2006*, IEA, Paris.

IEA 2007a, *CO_2 Emissions from Fuel Combustion: 1971–2005*, IEA, Paris.

IEA 2007b, *World Energy Outlook 2007: China and India insights*, IEA, Paris.

IMF (International Monetary Fund) various years to 2008, *World Economic Outlook*, IMF, Washington DC.

IPCC (Intergovernmental Panel on Climate Change) 2000, *Special Report on Emissions Scenarios. A Special Report of Working Group III of the Intergovernmental Panel on Climate Change*, N. Nakicenovic & R. Swart (eds), Cambridge University Press, Cambridge.

IPCC 2007, *Climate Change 2007: Mitigation of climate change. Contribution of Working Group III to the Fourth Assessment Report of the Intergovernmental Panel on Climate Change*, B. Metz, O.R. Davidson, P.R. Bosch, R. Dave & L.A. Meyer (eds), Cambridge University Press, Cambridge and New York.

Jevons, W.S. 1865, *The Coal Question: An inquiry concerning the progress of the nation, and the probable exhaustion of our coal-mines*, Macmillan, London.

Malthus, T. 1798, *An Essay on the Principle of Population*, 1st edn, J. Johnson, London.

Nakanishi, N. & Shuping, N. 2008, 'China builds plant to turn coal into barrels of oil', Reuters, 4 June.

National Bureau of Statistics of China 2007a, *2007 China Statistical Yearbook*, Beijing.

National Bureau of Statistics of China 2007b, *Communiqué on National Energy Consumption for Unit GDP in the First Half of 2007*, 31 July, <www.stats.gov.cn/english/newsandcomingevents/t20070731_402422194.htm>.

Netherlands Environmental Agency 2008, 'Global CO2 emissions: increase continued in 2007', 13 June, <www.mnp.nl/en/publications/2008/GlobalCO2emissionsthrough2007.html>.

Nordhaus, W. 2007, *The Challenge of Global Warming: Economic models and environmental policy*, Yale University, New Haven, Connecticut, <http://nordhaus.econ.yale.edu/dice_mss_072407_all.pdf>.

Raupach, M.R., Marland, G., Ciais, P., Le Quéré, C., Canadell, J.G., Klepper, G. & Field, C.B. 2007, 'Global and regional drivers of accelerating CO_2 emissions', *Proceedings of the National Academy of Sciences of the USA*, <www.pnas.org/cgi/content/abstract/0700609104v1>.

Sheehan, P., Jones, R., Jolley, A., Preston, B., Clarke, M., Durack, P., Islam, S.M.N. & Whetton, P. 2008, 'Climate change and the new world economy: implications for the nature and timing of policy responses', *Global Environmental Change*, doi:10.1016/j.gloenvcha.2008.04.008

Sheehan, P. & Sun, F. 2007, *Energy Use and CO_2 Emissions in China: Interpreting changing trends and future directions*, CSES Climate Change Working Paper No. 13. Centre for Strategic Economic Studies, Victoria University, Melbourne.

Stern, N. 2007, *The Economics of Climate Change: The Stern Review*, Cambridge University Press, Cambridge.

Stern, N. 2008, 'The economics of climate change', Richard T. Ely Lecture, *American Economic Review* 98(2): 1–37.

Turner, G. 2008, 'A comparison of the limits to growth with thirty years of reality', *Socio-Economics and the Environment in Discussion,* CSIRO Working Paper Series 2008–09.

UNFCCC (United Nations Framework Convention on Climate Change) 2007, 'National greenhouse gas inventory data for the period 1990–2005: Note by the Secretariat', FCCC/SBI/2007/30, 24 October, UNFCCC, Bonn.

US EIA (Energy Information Administration) 1998, *Natural Gas 1998: Issues and trends*, <www.eia.doe.gov>.

US EIA 2007, *International Energy Outlook 2007*, <www.eia.doe.gov>.

Weyant, J.P., De la Chesnaye, F.C. & Blanford, G. 2006, 'Overview of EMF–21: multigas mitigation and climate policy', *The Energy Journal*, Multi-Greenhouse Gas Mitigation and Climate Policy, Special Issue #3.

World Resources Institute 2008, *Climate Analysis Indicators Tool (CAIT) 2008*, <http://cait.wri.org>.

PROJECTING GLOBAL CLIMATE CHANGE

4

Key points

As a result of past actions, the world is already committed to a level of warming that could lead to damaging climate change.

Extreme climate responses are not always considered in the assessment of climate change impacts due to the high level of uncertainty and a lack of understanding of how they work. However, the potentially catastrophic consequences of such events mean it is important that current knowledge about such outcomes is incorporated into the decision-making process.

Continued high emissions growth with no mitigation action carries high risks. Strong global mitigation would reduce the risks considerably, but some systems may still suffer critical damage.

There are advantages in aiming for an ambitious global mitigation target in order to avoid some of the high-consequence impacts of climate change.

Some climate change has been observed already, and the mainstream science anticipates much more, even with effective global mitigation. The possible outcomes extend into the catastrophic. For any degree of mitigation, or its absence, there are most likely (median) outcomes, and considerable uncertainty around them.

In exploring likely and possible climate change outcomes, the Review has drawn on the IPCC Fourth Assessment Report and, where appropriate, research undertaken since the Fourth Assessment Report was compiled.

4.1 How has the climate changed?

The IPCC states that 'warming of the climate system is unequivocal', and that this is evident in the measured increase in global average air and surface temperatures, and also in the widespread melting of snow and ice and the rising global sea level (IPCC 2007a: 5). The climate system varies considerably on a local and regional basis, so that consideration of global averages can mask large regional variations (see Figure 4.1).

Figure 4.1 Selected regional climate change observations

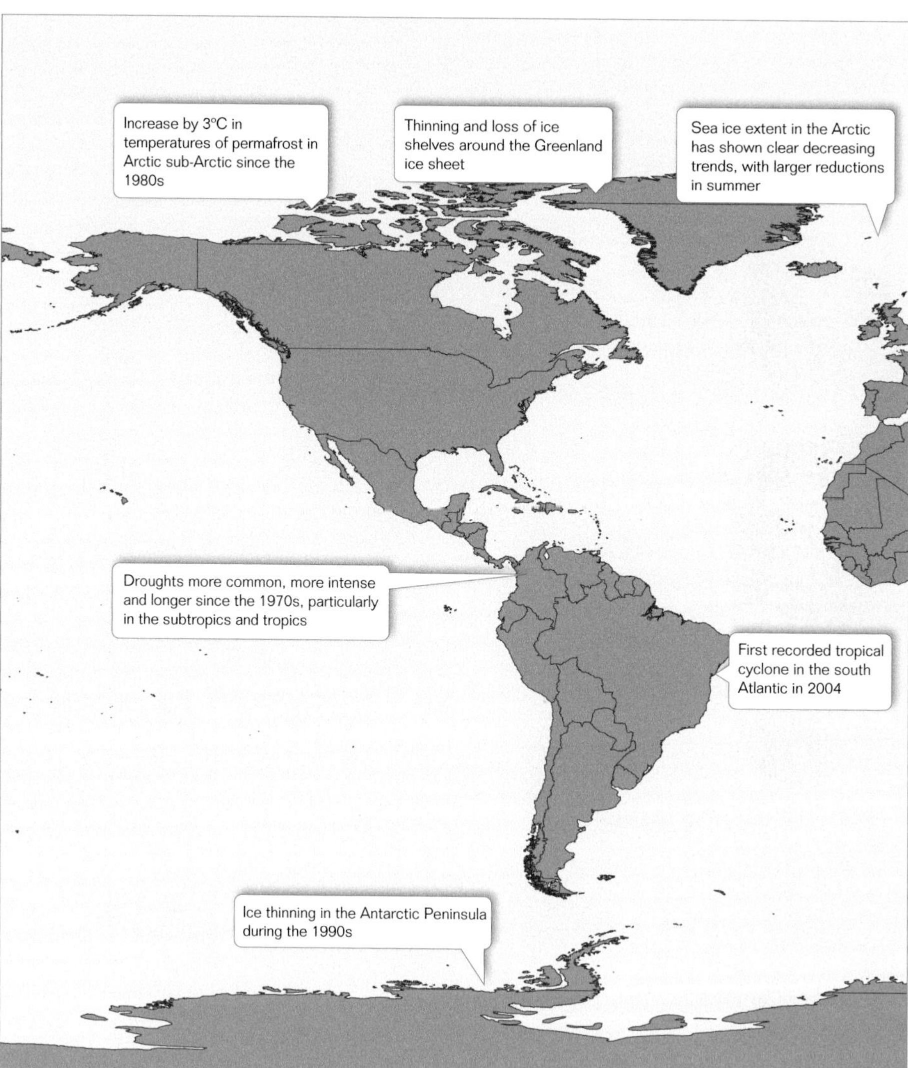

Source: IPCC (2007a); Church et al. (2008); CSIRO & BoM (2007).

Figure 4.1 Selected regional climate change observations *(continued)*

Increase in average arctic temperatures almost twice the global rate in the past 100 years
Decreased snow cover in northern hemisphere for every month except November and December
Increased precipitation in northern and central Asia
Permafrost warming observed on Tibetan plateau
Drying observed in southern Asia
Increase in number and proportion of tropical cyclones reaching categories 4 and 5 in intensity since 1970, particularly in the north Pacific, Indian and south-west Pacific oceans
Rapid reduction of tropical glaciers such as those on Mt Kilimanjaro
Substantial declines in rainfall in southern Australia since 1950
Extreme sea-level events off east and west coasts of Australia occurred three times more often in the second half of the 20th century than in the first half
Moderate to strong increases in annual precipitation in north-west Australia

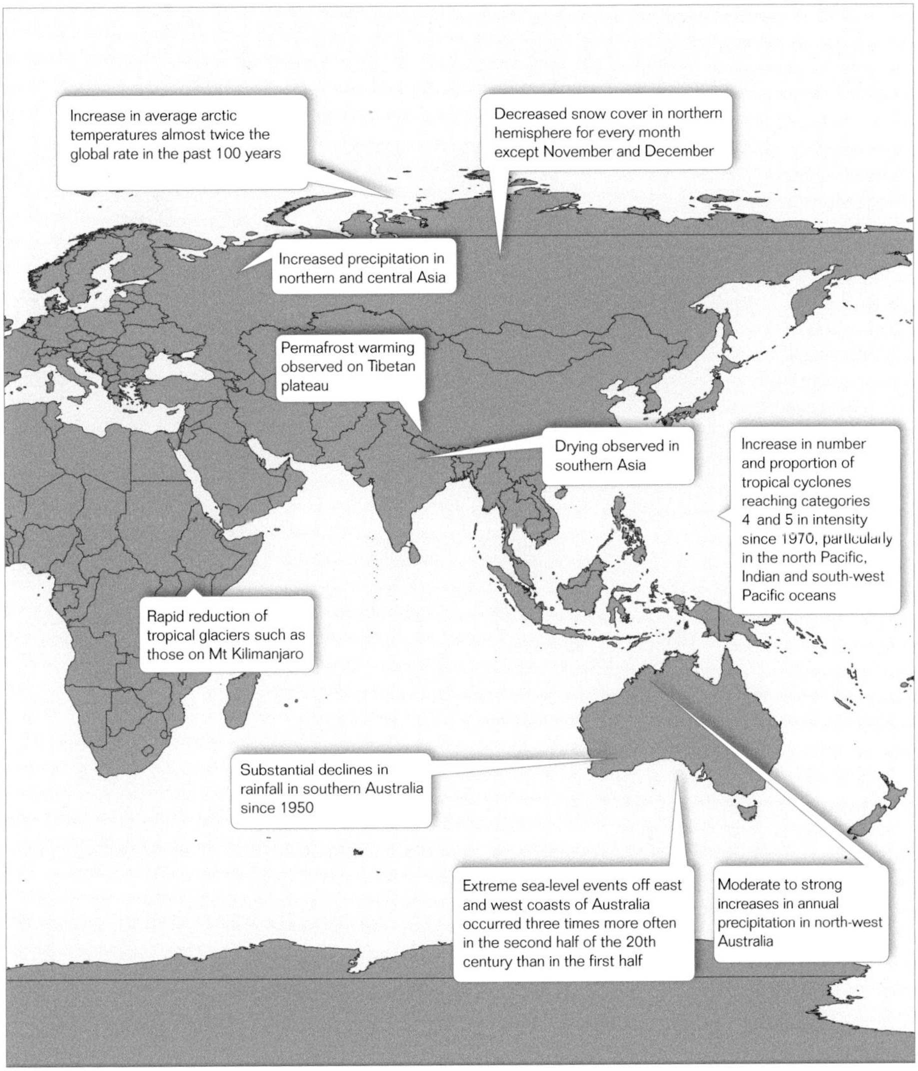

4.1.1 Changes in temperatures

Global average temperatures have risen considerably since measurements began in the mid-1800s (Figure 4.2). Since early industrial times (1850–99) the total global surface temperature increase has been estimated at 0.76°C ± 0.19°C.

Figure 4.2 Average global air temperature anomalies, 1850–2005

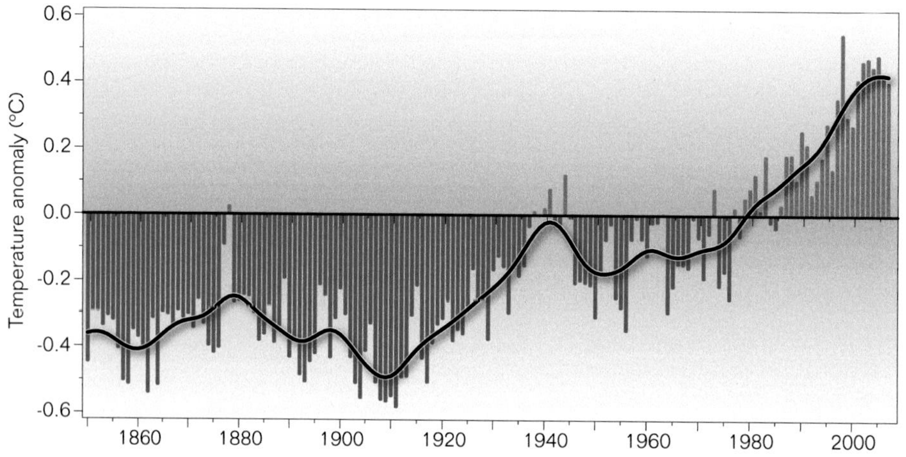

Note: The data show temperature difference from the 1961–1990 mean. The black line shows the annual values after smoothing with a 21-point binomial filter.

Source: Brohan et al. (2006, updated 2008).

Since 1979, the rate of warming has been about twice as fast over the land as over the ocean. During the last century, the Arctic has warmed at almost twice the global average rate (IPCC 2007a: 237).

The warming of the ocean since 1955 has accounted for more than 80 per cent of the increased energy in the earth's climate system (IPCC 2007a: 47). Warming in the top 700 m is widespread, with deeper warming occurring in the Atlantic Ocean.

The rate of warming in the lower atmosphere (the troposphere) has exceeded surface warming since 1958, while substantial cooling has occurred in the lower stratosphere. The pattern of tropospheric warming and stratospheric cooling is most likely due to changes in stratospheric ozone concentrations and greenhouse gas concentrations in the troposphere (IPCC 2007a: 10). Both the troposphere and the stratosphere have reacted strongly to events that have suddenly increased the volumes of aerosols in the atmosphere (IPCC 2007a: 270).

Box 4.1 Is there a warming trend in global temperature data?

Observations show that global temperatures have increased over the last 150 years (Figure 4.2). The data also suggest that the warming has been relatively steep over the last 30–50 years. A comparison of three datasets shows that they differ slightly on the highest recorded temperatures—data from the Hadley Centre in the United Kingdom show 1998 as the highest year, while data from the National Aeronautics and Space Administration and the National Climatic Data Centre in the United States show 2005 as the highest year.[1] All three datasets show that seven of the hottest 10 years on record have been in the last nine years, between 1999 and 2007. There has been considerable debate in mid-2008 in Australia on the interpretation of global temperatures over the past decade. Questions have been raised about whether the warming trend ended in about 1998.

To throw light on this question, the Review sought assistance from two econometricians from the Australian National University. Trevor Breusch and Farshid Vahid have specific expertise in the statistical analysis of time series—a specialty that is well developed in econometrics. They were asked two questions:

- Is there a warming trend in global temperature data in the past century?
- Is there any indication that there is a break in any trend present in the late 1990s, or at any other point?

They concluded:

> It is difficult to be certain about trends when there is so much variation in the data and very high correlation from year to year. We investigate the question using statistical time series methods. Our analysis shows that the upward movement over the last 130–160 years is persistent and not explained by the high correlation, so it is best described as a trend. The warming trend becomes steeper after the mid-1970s, but there is no significant evidence for a break in trend in the late 1990s.Viewed from the perspective of 30 or 50 years ago, the temperatures recorded in most of the last decade lie above the confidence band produced by any model that does not allow for a warming trend (Breusch & Vahid 2008).

4.1.2 Changes in the oceans and sea level

The ocean has the ability to store a thousand times more heat than the atmosphere. The heat absorbed by the upper layers of the ocean plays a crucial role in short-term climatic variations such as the El Niño – Southern Oscillation (IPCC 2007a: 46).

Sea level has varied extensively throughout history during the glacial and interglacial cycles as ice sheets formed and decayed. As oceans warm they expand, causing the volume of the ocean to increase and global mean sea level to rise. Sea level also rises when mass is added through the melting of grounded ice sheets and glaciers.

The total sea-level rise for the 20th century, including contributions from thermal expansion and land ice-melt, was 170 mm (Figure 4.3). The average rate of sea-level rise in the period 1961–2003 was almost 1.8 ± 0.5 mm per year. For 1993–2003 it was 3.1 ± 0.7 mm per year (IPCC 2007a: 387). Measurements show that widespread decreases in non-polar glaciers and ice caps have contributed to sea-level rise. The Greenland and Antarctic ice sheets are also thought to have contributed, but the proportions resulting from ice melt and the instability of the large polar ice sheets have yet to be fully understood (IPCC 2007a: 49).

Sea level varies spatially due to ocean circulation, local temperature differences, land movements and the salt content of the water. Regional changes in ocean salinity levels have occurred due to changes in precipitation that affect the inflow of freshwater. Changes in temperature and salinity have the potential to modify ocean currents and atmospheric circulation at the global scale. On an interannual to decadal basis, regional sea level fluctuates due to influences such as the El Niño – Southern Oscillation. Regional changes can lead to rates of sea-level change that exceed the annual increases in global average sea level (Cazenave & Nerem 2004).

Ocean acidity has increased globally as a result of uptake of carbon dioxide, with the largest increase in the higher latitudes where the water is cooler (IPCC 2007a: 405). The oceans are now more acidic than at any time in the last 420 000 years (Hoegh-Guldberg et al. 2007).

Figure 4.3 Global average sea-level rise, 1870–2005

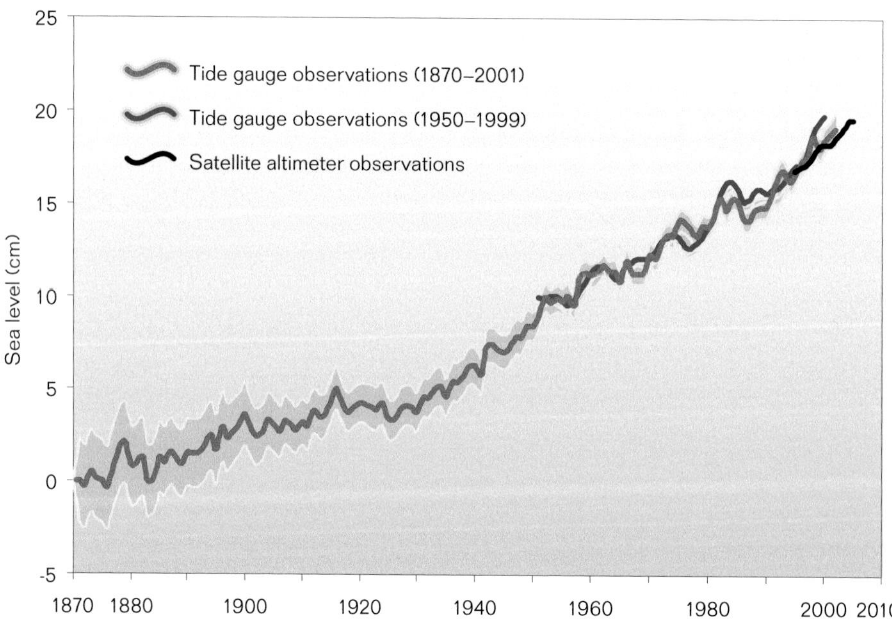

Note: Observed global average sea-level rise inferred from tide-gauge data (with 95 per cent confidence limits shown as blue shading) and satellite altimeter data.

Source: Church & White (2006); Holgate & Woodworth (2004); Leuliette at al. (2004).

4.1.3 Changes in water and ice

Precipitation

Increases in temperature affect the amount of water vapour the air can hold and lead to increased evaporation of water from the earth's surface. Together these effects alter the water cycle and influence the amount, frequency, intensity, duration and type of precipitation.

Over oceans and areas where water is abundant the added heat acts to moisten the air rather than to warm it. This can reduce the increase in air temperature and lead to more precipitation.

Where the surface is too dry to exchange much water with the atmosphere, increased evaporation can accelerate surface drying without leading to more rainfall. Cloudiness will also fall in the warmer and drier atmosphere, leading to further temperature increases from the higher amount of sunlight reaching the surface (IPCC 2007a: 505). These effects can cause an increase in the occurrence and intensity of droughts (IPCC 2007a: 262). Local and regional changes in precipitation are highly dependent on climate phenomena such as the El Niño – Southern Oscillation, changes in atmospheric circulation and other large-scale patterns of variability (IPCC 2007a: 262).

There is high variability in precipitation over time and space, and some pronounced long-term trends in regional precipitation have been observed. Between 1900 and 2005, annual precipitation increased in central and eastern North America, northern Europe, northern and central Asia and south-eastern South America (IPCC 2007a: 258). Decreases in annual precipitation have been observed in parts of Africa, southern Asia and southern Australia (IPCC 2007a: 256).

In addition to changes in mean precipitation, studies of certain regions show an increase in heavy rainfall events over the last 50 years, and some increases in flooding, even in areas that have experienced an overall decrease in precipitation (IPCC 2007a: 316).

Ice caps, ice sheets, glaciers and frozen ground

About 75 per cent of the fresh water on Earth is stored in ice caps, ice sheets, glaciers and frozen ground, collectively known as the cryosphere. At a regional scale, variations in snowfall, snowmelt and glaciers play a crucial role in the availability of fresh water.

Ice and snow have a significant influence on local air temperature because they reflect about 90 per cent of the sunlight that reaches them, while oceans and forested lands reflect about 10 per cent (IPCC 2007a: 43).

Frozen ground is the single largest component of the cryosphere by area, and is present on a seasonal and permanent basis at both high altitudes and high latitudes (around the poles). Thawing of permanently frozen ground can lead to changes in the stability of the soil and in water supply, with subsequent impacts on ecosystems and infrastructure (IPCC 2007a: 369).

Extensive changes to ice and frozen ground have been observed in the last 50 years, some at a rate that is dramatic and unexpected. Arctic sea ice coverage has shown a consistent decline since 1978. The average sea ice extent for September 2007 was 23 per cent lower than the previous record set in 2005 (NSIDC 2007).

Data for the 2008 northern summer show that the rate of daily sea ice loss in August was the fastest that scientists have ever observed during a single August (NSIDC 2008). The rate of ice loss is partly due to a reduction in sea ice thickness over large areas, which means that a small increase in energy can lead to a fast reduction in ice cover (Haas et al. 2008). At the time of printing, sea ice extent in 2008 had not reached the record low set in 2007, but still shows as the second lowest August sea ice extent on record (NSIDC 2008).

There has been a general reduction in the mass of glaciers and ice caps, except in parts of Greenland and Antarctica (IPCC 2007a: 44).

4.1.4 Changes to severe weather and climate events

Changes in the intensity and frequency of certain severe weather events have been observed throughout the world. Observed changes in temperature extremes have been consistent with the general warming trend—cold days, cold nights and frost have been occurring less frequently in the last 50 years, and hot days, hot nights and heatwaves have been occurring more frequently (IPCC 2007a: 308).

The area affected by droughts has increased in certain regions, largely due to the influence of sea surface temperatures and changes in atmospheric circulation and precipitation (IPCC 2007a: 317). Full assessments of changes in droughts are limited by difficulties in the measurement of rainfall and poor data on soil moisture and streamflow (IPCC 2007a: 82).

Tropical storm and hurricane frequency, lifetime and intensity vary from year to year and are influenced by the El Niño – Southern Oscillation, which can mask trends associated with general warming. There is a global trend towards storms of longer duration and greater intensity, factors that are associated with tropical sea surface temperatures (IPCC 2007a: 308).

4.1.5 Human attribution of observed changes

The climate system varies naturally due to external factors such as the sun's output and volcanic eruptions, and internal dynamics such as the El Niño – Southern Oscillation. Over the longer term, changes in climate are associated with changes in the earth's orbit and the tilt of its axis (Ruddiman 2008). To establish whether human activities are causing the observed changes in climate, it must be established that the changes cannot be explained by these natural factors.

When only natural factors are included in the modelling of 20th century temperature change, the resulting models cannot account for the observed changes in temperature. However, when human influences are included, the models produce results that are similar to the observed temperature changes

(IPCC 2007a: 703). Using this technique, the influence of human activities on regional temperatures can be established for every continent except Antarctica, for which limited observed data are available.

Modelling has been used to attribute a range of observed changes in the climate to human activity, including the low rainfall in the south-west of Western Australia since the 1970s (CSIRO & BoM 2007) and the reduction in extent of Arctic sea ice (CASPI 2007).

Apart from modelling exercises, there are other gauges of observed change that suggest a human influence on the climate. These include measurements of higher rates of warming over land than over sea, which are not associated with the El Niño – Southern Oscillation, and differential warming in the troposphere and stratosphere, which can be explained by increases in greenhouse gases in the troposphere and the depletion of the ozone layer in the stratosphere (CASPI 2007).

4.2 Understanding climate change projections

Future changes in the climate will depend on a range of natural changes—or forcings—as well as human activity, and on the way in which the climate responds to these changes. These forcings are difficult to predict, can occur randomly and may interact in a way that amplifies or reduces the effects of particular elements. The inherent variability and complexity of the climate system is complicated further by the possibility of a non-linear and unpredictable response to levels of greenhouse gases that are well outside the range experienced in recent history.

Box 4.2 What is a climate outcome?

In this report, the term climate *outcome* (or response) is used to refer to a future climate that is one of a range of possible outcomes for a given aspect of the climate, such as rainfall or temperature. The occurrence of a particular climate outcome excludes the occurrence of a different one—it is not possible for more than one estimate of equilibrium climate sensitivity to be correct. There is uncertainty surrounding which outcome will occur, and often about when it will happen and the magnitude of its impact.

Modelling techniques can be used to assess the potential range of climate outcomes based on current understanding. This information can be presented in the form of a probability density function, where the median has the highest probability of occurring, but where there is also a chance of outcomes that are either much more benign or positive, or much more damaging.

A weather or climate *event*—such as a big storm or a drought—can occur repeatedly as a function of climate variability. The occurrence of an event of a given intensity does not exclude the occurrence of a related event at a different intensity. The likelihood of weather and climate events occurring is often referred to in language that reflects the magnitude of the outcome as well as the likelihood—as in the phrase 'a one-in-a-hundred-year storm'.

The most important direct human forcings are greenhouse gas emissions, a process that humans can control through policy and management. Identifying specific pathways of human-induced greenhouse gas emissions—the dominant mechanism for human influence on the climate—simplifies the projection of climate change. But, of course, future changes in climate will be influenced by natural factors as well.

4.2.1 Confidence in the projection of climate change

Climate models provide a wide range of estimates of temperature response and changes in climate variables such as rainfall. It is important to understand the uncertainty surrounding the model outputs, and how these are reflected in climate change projections.

Confidence in climate models

The ability of climate models to accurately simulate responses in the climate system depends on the level of understanding of the processes that govern the climate system, the availability of observed data for various scales of climate response, and the computing power of the model. All of these have improved considerably in recent years (CSIRO & BoM 2007).

Confidence in models stems from their ability to represent patterns in the current climate and past climates, and is generally higher at global and continental scales. For some elements of the climate system, such as surface temperature, there is broad agreement on the pattern of future climate changes. Other elements, such as rainfall, are related to more complex aspects of the climate system, including the movement of moisture, and are not represented with the same confidence in models.

The likelihood of a particular outcome can be assessed through the use of a range of models. However, outcomes at the high or low end of a range of model results may also be plausible.

There is significant uncertainty around how the atmosphere will respond to a given change in carbon dioxide concentration.

The extent to which the climate warms in response to changes in greenhouse gas concentrations is centrally important to expectations of projected climate change. Many aspects of projected climate change relate closely to global mean temperature (IPCC 2007a: 630). For example, the extent of melting of glaciers and permafrost is related to the magnitude of temperature increase. However, changes in other dimensions of climate, such as regional variation in rainfall, are not so closely correlated with temperature change. Changes in the magnitude, spatial pattern and seasonality of rainfall cannot be directly inferred from temperature change.

Confidence in other elements of climate change modelling

To assess the risk of climate change it is necessary to understand the different elements that are included in or excluded from the model outcomes. It is also necessary to understand the uncertainty associated with these elements. To gain a comprehensive understanding of potential climate change for a given temperature outcome, the elements should be considered together.

- **Well-constrained climate outcomes**—Some elements of the climate system have a well-established response to increased temperatures or other parameters, and models are reliable in reflecting the possible outcomes. Examples include the pattern of regional temperature response, sea-level rise from thermal expansion, melting of permafrost and damage to reefs.
- **Partially constrained climate outcomes**—Some elements of the climate system have a relatively well-constrained pattern or direction of change in response to temperature rise, but are known to be poorly represented in models. Some uncertainty arises from differences between models. The extent and direction of rainfall change is an example of the latter, although this is becoming more reliable for some parts of Australia.
- **Poorly constrained climate outcomes**—Some elements of the climate system have an unknown response to changes in global temperature. An example is the response of the El Niño – Southern Oscillation, where scientists are unsure whether the fluctuations will increase in frequency or intensity, or perhaps decline in influence.

4.2.2 Responding to climate change now

The human-induced warming and the associated changes in climate that will occur over the next few decades will largely be the result of our past actions and be fairly insensitive to our current actions.

An exception is if large changes in emissions of sulphate aerosols occur, where a localised warming (decrease) and cooling (increase) can occur within months of a change in emissions.

Models show that the warming out to 2030 is little influenced by greenhouse gas emissions growth from the present (IPCC 2007a: 68), because there are lags in the climate system. By 2050, however, different trends in emissions have a clear influence on the climate outcomes. By 2100 the potential differences are substantial.

To estimate the magnitude of climate change in the future it is necessary to make assumptions about the future level of global emissions of greenhouse and other relevant gases.

While the SRES scenarios (IPCC 2000) show a range of possible outcomes for a world with no mitigation, the Review looks at three possible futures based on different levels of mitigation.

The three emissions cases

Figure 4.4 shows the concentration pathways for the three emissions cases considered by the Review:

- **No-mitigation case**—A global emissions case in which there is no action to mitigate climate change—the Garnaut–Treasury reference case—was developed as part of the Review. This emissions case recognises recent high trends in the emissions of carbon dioxide and other greenhouse gases. Emissions continue to increase throughout the 21st century, leading to an accelerating rate of increase in atmospheric concentrations. By the end of the century, the concentration of long-lived greenhouse gases is 1565 ppm CO_2-e, and carbon dioxide concentrations are over 1000 ppm—more than 3.5 times higher than pre-industrial concentrations.
- **550 mitigation case**—Emissions peak and decline steadily, so that atmospheric concentrations stop rising in 2060 and stabilise at around 550 ppm CO_2-e—one-third of the level reached under the no-mitigation case.
- **450 mitigation case**—Emissions are reduced immediately and decline more sharply than in the 550 case. Atmospheric concentrations overshoot to 530 ppm CO_2-e in mid-century and decline towards stabilisation at 450 ppm CO_2-e early in the 22nd century.

Figure 4.4 Concentrations of greenhouse gases in the atmosphere for the three emissions cases, 1990–2100

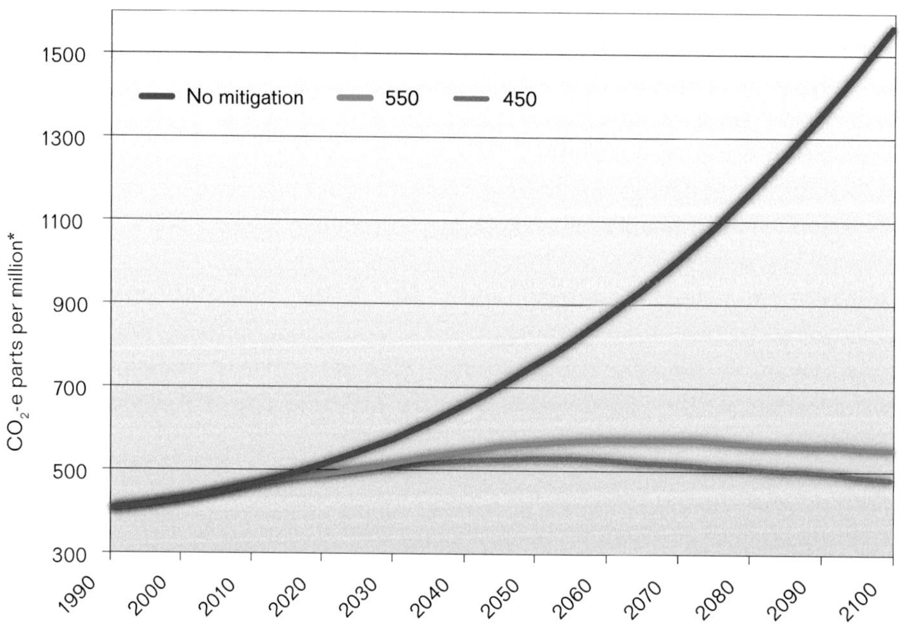

* Kyoto gases and CFCs only.

4.2.3 Limitations in the Review's assessment

The uncertainty in projected climate change and the range of model outputs for a given emissions pathway means that there is value in investigating the climate outcomes from a large number of models. The SRES emissions datasets have been publicly available since 1998, and they have been extensively used to provide a consistent basis for climate change projections. The availability of multiple models allows for a better assessment of the level and sources of uncertainty in climate change projections. This provides more robust results and reduces the influence of individual model bias (IPCC 2007a: 754).

The summary of projected climate change for the three emissions cases considered by the Review is based on a range of data from the outcomes of the climate models used in the Review's modelling exercise and interpretation of the 2007 IPCC summary of projected climate change (IPCC 2007a), based on the SRES scenarios.

4.3 Projected climate change for the three emissions cases

Quantitative climate change projections for well-constrained elements of the climate system are provided for each emissions case. Best-estimate outcomes are presented along with an indication of the magnitude and possible range of lower-probability outcomes. Severe weather events and changes in variability are considered at a general level.

4.3.1 Aspects of temperature change

Many of the changes to the climate system are related to changes in global average temperature. They tend to increase in magnitude and intensity as temperature levels rise. As a result, temperature change over time and space is a key indicator of the extent of climate change. This section explores a range of aspects of temperature change under the three emissions cases.

Box 4.3 Temperature reference points

Various reports and studies on mitigation may use different points of comparison for temperature increases. Temperature rise may be framed in terms of the increase from pre-industrial times, or from a given year.

Unless otherwise specified, temperature changes discussed by the Review are expressed as the difference from the period 1980–99, usually expressed as '1990 levels' in the IPCC Fourth Assessment Report.

Following the same convention, temperatures over the period 1850 to 1899 are often averaged to represent 'pre-industrial levels'. To compare temperature increases from 1990 levels to changes relative to pre-industrial levels, 0.5°C should be added.

Projected changes to the end of the 21st century are generally calculated from the average of 2090–99 levels, but are often expressed as '2100'.

Committed warming

'Committed warming' refers to the future change in global mean temperature from past emissions, even if concentrations are held constant.

The IPCC estimated that the warming resulting from atmospheric concentrations of greenhouse gases being kept constant at 2000 levels would produce an increase of 0.6°C by 2100. The increase for each of the next two decades would be 0.1°C from past emissions alone (IPCC 2007a: 79).

As a result of committed warming, the temperature outcomes for the next few decades are minimally affected by mitigation in the intervening years. The average warming for the period 2011 to 2030 for the middle to lower SRES scenarios is 0.64–0.69°C above 1990 levels, with high agreement between models (IPCC 2007a: 749).

Global mean temperatures post-2030

Projections of global mean surface air temperature for the 21st century show the increases continuing for all emissions cases. Figure 4.5 shows the projected temperature increases for the three emissions cases for the best-estimate climate

Figure 4.5 Global average temperature outcomes for three emissions cases, 1990–2100

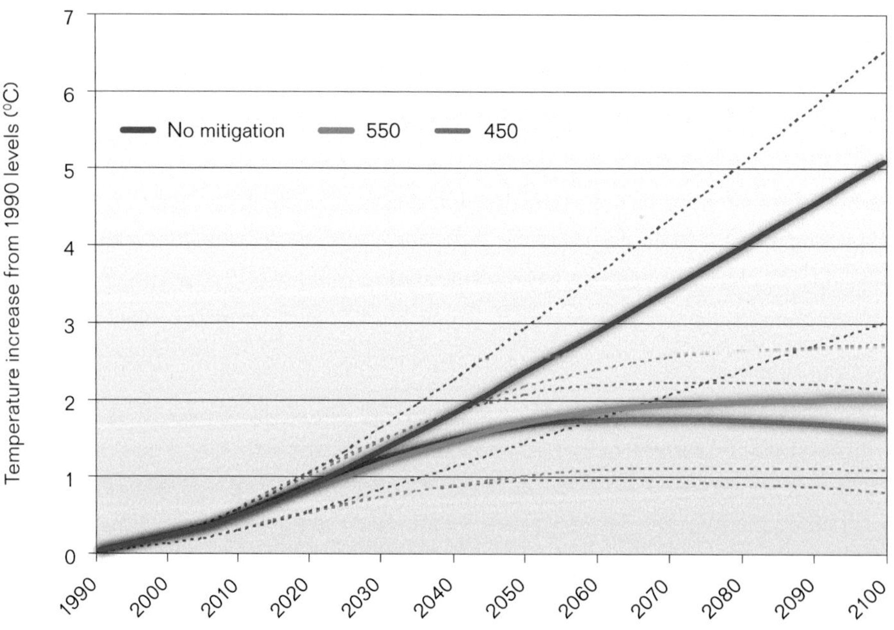

Note: Temperature increases from 1990 levels are from the MAGICC climate model (Wigley 2003). The solid lines show the temperature outcome for the best-estimate climate sensitivity of 3°C. The dashed lines show the outcomes for climate sensitivities of 1.5°C and 4.5°C for the lower and upper temperatures respectively. The IPCC considers that climate sensitivities under 1.5°C are considered unlikely (less than 33 per cent probability), and that 4.5°C is at the upper end of the range considered likely (greater than 66 per cent probability).

sensitivity of 3°C, with dashed lines indicating outcomes for climate sensitivities of 1.5°C and 4.5°C. Temperatures are projected to be slightly higher between 2020 and 2030 under the 450 case than under the 550 case, as rapid declines in aerosol emissions are associated with reductions in fossil fuel emissions, and the cooling influence decreases.

By the end of the century the global average temperature increase under the no-mitigation case is 5.1°C, and still increasing at a high rate. The 550 and 450 cases reach 2.0°C and 1.6°C respectively, with the temperatures changing only minimally by 2100 in both cases.

Spatial variation in temperature

Figure 4.6 shows the spatial variation in the simulated temperature changes at the end of the 21st century under the three emissions cases. The greatest warming occurs at the poles and over large land masses, with less warming over the oceans. A small region of cooling lies over the north Atlantic Ocean.

The strong warming in the polar regions arises from feedbacks caused by changes in the reflection of solar radiation from the loss of ice and snow. Melting ice and snow expose the darker ocean or land surface beneath. This then absorbs a greater fraction of incoming solar radiation, leading to further warming. In the no-mitigation case, by the end of the century temperatures in parts of the Arctic are more than 10°C above 1990 levels. In the 550 and 450 mitigation cases, the feedback effects of melting ice are restricted and the temperature response in the Arctic is more subdued.

The north Atlantic region is significantly influenced by oceanic circulation. The Gulf Stream transports warm surface waters northward, warming this region relative to others at the same latitude. Climate change is projected to lead to a weakening of the Gulf Stream, resulting in cooler temperatures in the north Atlantic than would otherwise be the case.

Extremes in global mean temperature response

The discussion of climate sensitivity in Chapter 2 outlined the IPCC assessment of a best-estimate climate sensitivity of 3°C, with a likely (>66 per cent probability) outcome between 2°C and 4.5°C (IPCC 2007a: 12). The lower end of the range of possible outcomes is much better constrained than the upper end—most studies investigating climate sensitivity find a lower 5 per cent probability limit of between 1°C and 2.2°C, while the upper 95 per cent probability limit for the same studies ranges from 5°C to greater than 10°C (IPCC 2007a: 721).

For the no-mitigation case, the chance of avoiding the 2°C pre-industrial warming threshold, which the European Union has announced as a mitigation goal (1.5°C above 1990 levels), is virtually zero. A temperature rise of 6.5°C is still within the range considered likely by the end of the century. The chance of avoiding the 2°C threshold by 2100 is around 25 per cent and 50 per cent for the 550 and 450 cases respectively (Jones 2008).

Figure 4.6 Spatial variation in temperature change in 2100 for the three emissions cases

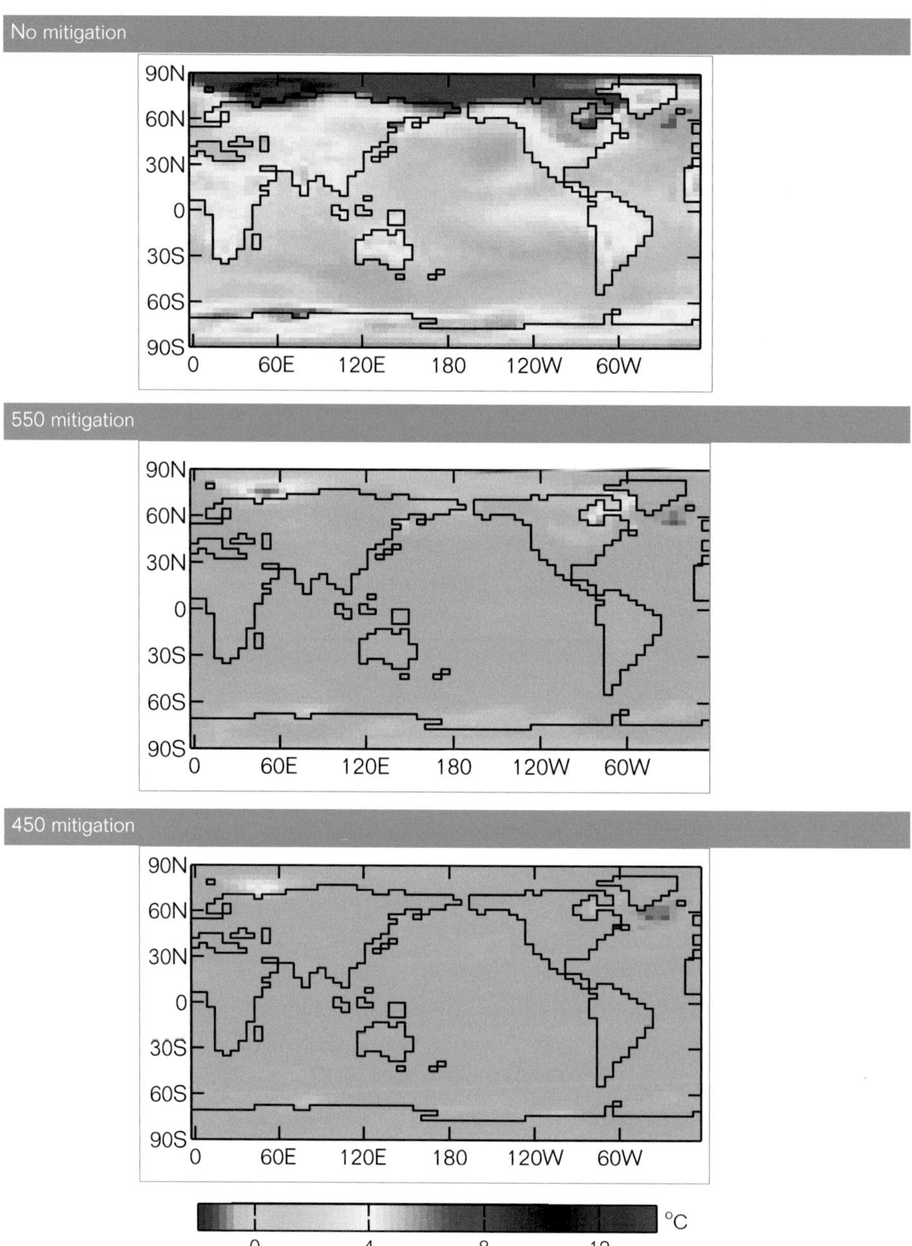

Note: Temperature outcomes are from the CSIRO Mk3L model (Phipps 2006), which demonstrates a lower global average temperature response than the MAGICC model used to calculate the global average temperatures in Figure 4.5.

If a higher temperature response were to occur, the large temperature increases evident in the polar regions in the no-mitigation case would be even more extreme. In the mitigation cases, a higher temperature response would lead to positive feedbacks and amplified temperatures in the polar regions.

Temperature outcomes for other mitigation pathways

The 550 and 450 global mitigation cases represent a future where the world responds in a coordinated way to a specific target for stabilisation of greenhouse gases in the atmosphere. There are many possibilities for the choice of a global mitigation target, as well as for the pathway to get there.

Due to the slow response of the ocean to changes in the energy balance of the climate system, it may be hundreds or even thousands of years before the equilibrium temperature is reached. Policy makers considering the impacts of climate change in the future should also consider the transient, or short-term, temperature response. Different models can show a range of short-term temperature outcomes for the same emissions pathway.

The pathway to a given stabilisation target will affect the short-term temperature response. To achieve some of the lower concentration targets, an overshoot in concentration is required, leading to higher temperature responses for a period.

The short-term temperature response can also be influenced by the mix of greenhouse gases emitted. Under an emissions trading scheme, higher reductions in some gases could be traded off against lower reductions in other gases based on what was most economically or technically viable at the time. Methane has a much greater warming effect than carbon dioxide, but this effect will be relatively short-lived due to its shorter residence time in the atmosphere. Reduction of methane emissions could therefore have a proportionally larger effect than reduction of carbon dioxide emissions on global temperature. A focus on methane reduction could potentially offset the warming caused by reductions in aerosol emissions, substitute for reducing other greenhouse gas emissions or be used to avoid key tipping points if a rapid reduction in the rate of temperature increase was required. However, if priority is given to methane reductions in the short term and carbon dioxide concentrations continue to rise, any benefits would be temporary because of the longer residence time of carbon dioxide in the atmosphere. This short-term temperature response could lead to greater impacts and an increased risk of reaching key temperature thresholds.

Post-2100 temperatures

Climate change comes with long lags. Greenhouse gases emitted today will have warming effects in the atmosphere for decades and centuries, and many of the effects of warming on earth systems, will play themselves out over even longer time frames. The melting of the Greenland ice sheet is an example.

Beyond 2100, modelling of projected climate change becomes even more uncertain; many emissions pathways and scenarios do not extend into the 22nd century. The economic modelling that underpins the three emissions cases

Figure 4.7 Temperature increases above 1990 levels for the three emissions cases

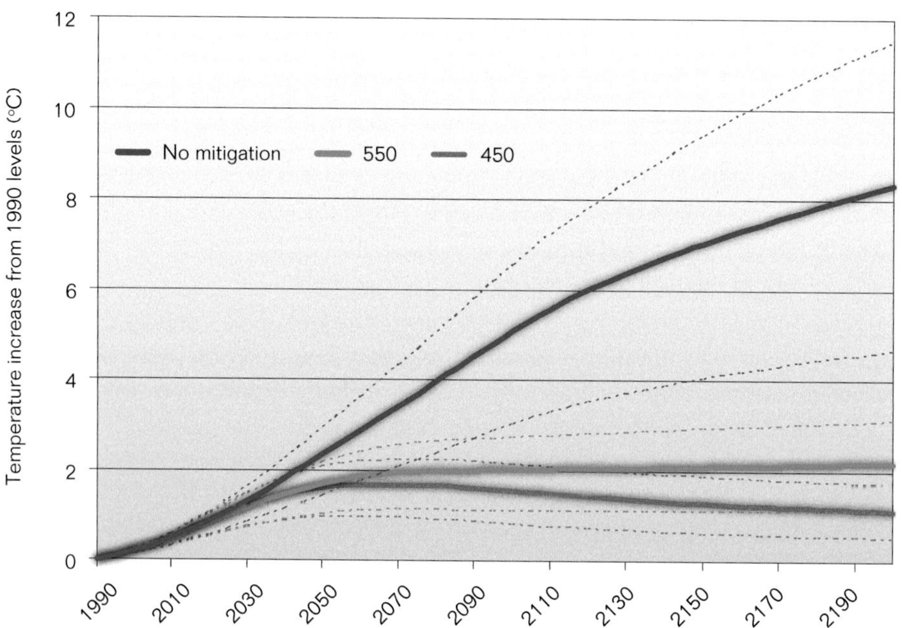

Note: Best-estimate figures, shown in solid lines, are calculated with a climate sensitivity of 3°C. The dashed lines show the outcomes for climate sensitivities of 1.5°C and 4.5°C for the lower and upper temperatures respectively. Temperature increases from 1990 levels are derived from the MAGICC climate model (Wigley 2003). Temperature outcomes beyond 2100 are calculated under the simplified assumption that emissions levels reached in each scenario in the year 2100 continue unchanged. They do not reflect an extension of the economic analysis underlying these scenarios out to 2100, and are illustrative only. It is unlikely that emissions in the no-mitigation case will stabilise abruptly in 2101 with no policies in place, and hence the temperatures shown underestimate the likely warming outcomes if continued growth in emissions is assumed.

modelled by the Review only goes to 2100. However, the climate will continue to change beyond 2100, particularly in the no-mitigation case where emissions can be expected to continue at high levels. Figure 4.7 shows the temperature outcomes for an illustrative modelling exercise looking at the temperature in the 22nd century for the three emissions cases, where emissions are held constant at 2100 levels. By 2200, temperatures in the no-mitigation case rise by 8.3°C under the best-estimate climate sensitivity, with a rise of 11.5 °C still within the likely range. Under the other two mitigation scenarios, temperatures are maintained at 2.2°C above 1990 levels for the 550 case, and drop to 1.1°C above 1990 levels in the 450 case due to the low levels of emissions in 2100.

4.3.2 Precipitation

Climate model simulations show that, as temperatures increase, there will be increased precipitation in the tropics and at high latitudes, and decreased precipitation in the subtropical and temperate regions (IPCC 2007a: 750).

Rainfall is set to increase at high latitudes and over equatorial oceans, where the atmosphere can be rapidly resupplied with moisture after rainfall. In the subtropics, air tends to descend as a result of atmospheric circulation patterns, and humidity is relatively low. This process is intensified as the climate warms. Changes in the circulation patterns also push the weather systems that bring rain further towards the poles, causing a decrease in rainfall over many of the subtropical and temperate regions. Major regions with substantial decreases in rainfall include Australia, the Mediterranean, Mexico, and north-west and south-west Africa (Watterson 2008). Rainfall is expected to increase in the Asian monsoon (although variability is expected to increase season to season) and the southern part of the West African monsoon.

The interactions among the processes that control rainfall in a particular region are complicated, and rainfall changes can vary considerably at the local level. For example, while most models suggest the likelihood of drying of southern Australia with increases in global average temperature, others suggest some possibility of an increase in rainfall.

Climate change is also likely to affect the seasonal and daily patterns of rainfall intensity, which may not be reflected in annual average data. There is expected to be an increased risk of drought in the mid-latitudes. There is also an increase in the risk of flooding, as precipitation will be concentrated into more intense events (IPCC 2007a: 782).

4.3.3 Sea-level rise

Sea-level rise will come from two main sources—thermal expansion and the melting of land-based ice sheets and glaciers. The greater the temperature increase, the greater the sea-level rise from thermal expansion and the faster the loss of glacial mass (IPCC 2007a: 830). Current IPCC estimates predict that thermal expansion of the oceans will contribute 70 per cent to 75 per cent of the projected rise to 2100. The level of understanding about the magnitude and timing of contributions to sea-level rise from ice melt is low. Melting of some ice sheets on Greenland and the west Antarctic has accelerated in recent decades (IPCC 2007a: 44), and observed data suggest that sea-level rise has been at the high end of previous IPCC projections (Rahmstorf et al. 2007). The accelerated response in Greenland may be the result of meltwater from the surface lubricating the movement at the base of the ice sheet and increasing the dynamic flow of solid ice into the sea. The west Antarctic ice sheet is grounded below sea level, which allows warming ocean water to melt the base of the ice sheet, making it more unstable (Oppenheimer & Alley 2005).

The IPCC estimated sea-level rise in 2100 for a scenario similar to the no-mitigation case (SRES scenario A1FI) at 26–59 cm. This figure does not include the potential for rapid dynamic changes in ice flow, which could add 10 and 20 cm to the upper bound of sea-level rise predicted for the 21st century. A key conclusion of the IPCC sea-level rise projections was that larger values above the upper estimate of 79 cm by 2100 could not be excluded (IPCC 2007a:14).

A significant change between the IPCC's Third and Fourth Assessment Reports was the revision of the lower estimate of sea-level rise upwards. However, the lower end of the range is still only slightly higher than the sea-level 'committed' rise that would occur if greenhouse gas emissions ceased, and leaves little room for contributions for additional ocean warming and land-ice melt, making such a low outcome unlikely (Rahmstorf 2007; Pew Center 2007).

Ongoing melt of the Greenland and west Antarctic ice sheets

If a sufficiently warm climate were sustained, the Greenland and west Antarctic ice sheets would be largely eliminated over a long period. If the Greenland and west Antarctic ice sheets were to melt completely, they would add an estimated 7 m and 6 m to global sea level respectively (IPCC 2007a: 752; Oppenheimer & Alley 2005).

Current models suggest that once a certain temperature is exceeded, major reduction of the Greenland ice sheet would be irreversible. Even if temperatures were to fall later, the climate of an ice-free Greenland might be too warm for the accumulation of ice (IPCC 2007a: 752, 776). Sufficient global temperature rise to initiate ongoing melt of the Greenland ice sheet lies in the range of 1.2–3.9°C relative to 1990 (IPCC 2007a: 752). A simple reading of the scientific literature suggests a high probability that, under business as usual, the point of irreversible commitment to the melting of the Greenland ice sheet will be reached during this century.

Accelerated sea-level rise

Some of the recent scientific work suggests that future sea-level rise could be much worse than the sea-level rise outcomes projected in the IPCC's Fourth Assessment Report.

Climate records show that at the end of the last interglacial, when the northern hemisphere ice sheets disintegrated, sea-level rise peaked at a rate of 4 m per century. This indicates that substantially higher rates of sea-level rise than those predicted by the IPCC have occurred historically (Church et al. 2008).

Estimating the likelihood of rapid disintegration of ice sheets remains very difficult and poorly represented in current models (Oppenheimer & Alley 2005). Recent evidence and comparison with historical rates suggest it is more likely than previously thought (Hansen 2005; Lenton et al. 2008). The IPCC estimates that the complete melting of the Greenland ice sheet would take longer than 1000 years (IPCC 2007a: 794). However, Lenton et al. (2008) suggest that a lower limit of 300 years is conceivable if the rapid disappearance of continental ice at the end of the last ice age were to be repeated.

It has been suggested that if the relationship between temperature and sea-level rise observed during the 20th century were to continue, under strong warming scenarios a rise of up to 1.4 m by 2100 could occur (Rahmstorf 2007).

4.3.4 Other climate outcomes

Current limitations in scientific understanding and current levels of uncertainty mean that quantifying outcomes under various scenarios is difficult. Box 4.4 summarises trends in projected climate change under the three emissions cases.

Some climate variables are likely to respond to temperature increase in a nearly linear manner. Change in other variables could occur much faster than the rate of temperature increase. An example of a non-linear relationship is sea-level rise due to the dynamic flow of ice sheets and glaciers, which occurs at an increasing rate as the mass of ice is reduced (IPCC 2007a: 70).

Box 4.4 Summary of trends in projected climate change

Temperature extremes

Heatwaves will become more intense, more frequent and last longer as the climate warms. Cold episodes and frost days will decrease (IPCC 2007a: 750).

Future increases in temperature extremes generally follow increases in the mean (IPCC 2007a: 785).

Snow and ice

As the climate warms, snow cover and sea ice extent will decrease. Glaciers and ice caps will lose mass, as summer melting will dominate increased winter precipitation (IPCC 2007a: 750).

Snow cover and permafrost area is strongly correlated to temperature, so decreases are expected as warming occurs. In some northern regions where precipitation is expected to increase, snow cover may also increase. This is also true for much of Antarctica.

Some models project that summer sea ice in the Arctic could disappear entirely by 2100 under high-emissions scenarios like the no-mitigation case (IPCC 2007a: 750). Comparisons with observed loss in Arctic sea ice suggest that current models may underestimate the rate of decline.

The extent of sea-ice melting is related to the magnitude of temperature rise, which is amplified in the polar regions due to the warming feedback of melting ice. The small change in global average temperature between the 550 and 450 mitigation pathways could have a relatively large impact on sea-ice extent.

Carbon cycle

As temperatures increase, the capacity of the land and ocean to absorb carbon dioxide will decrease so that a larger fraction of emissions will remain in the atmosphere and cause greater warming (IPCC 2007a: 750). The extent of the carbon–climate feedback is dependent on the level of emissions or stabilisation. The higher the temperature increase, the larger the impact on the carbon cycle (IPCC 2007a: 750).

Box 4.4 Summary of trends in projected climate change *(continued)*

Tropical cyclones and storms

Tropical cyclones are likely to increase in intensity and to generate greater precipitation in their vicinity. Most models suggest a decrease in the total number of storms. There will be a poleward shift of storm tracks, particularly in the southern hemisphere.

Increased wind speeds will cause more extreme wave heights in these regions (IPCC 2007a: 783).

Ocean acidification

As carbon dioxide concentrations increase in the atmosphere, a greater amount is absorbed by the ocean where it reacts with water to create carbonic acid. This increases the acidity of the ocean, which will increase the rate at which carbonate sediments in shallow water dissolve. This may affect marine calcifying organisms.

The level of ocean acidification will depend on the carbon dioxide concentration in the atmosphere. Under the no-mitigation case, carbon dioxide concentrations are more than 1000 ppm in 2100—more than three-and-a-half times pre-industrial concentrations.

4.4 Assessing the climate risk

Understanding the potential risk from climate change requires consideration of both the likelihood and the consequence of an event or outcome occurring. This section identifies climate change outcomes with high consequences due to the extent or speed of the change itself, as well as those with high consequences for human populations.

4.4.1 Thresholds and tipping points

Many of the processes within the climate and other earth systems (such as the carbon cycle) are well buffered and appear to be unresponsive to changes until a threshold is crossed. Once the threshold has been crossed, the response can be sudden and severe and lead to a change of state or equilibrium in the system. This is often referred to as rapid or abrupt climate change (see Figure 4.8). Under abrupt climate change, a small change can have large, long-term consequences for a system (Lenton et al. 2008).

The threshold at which a system is pushed into irreversible or abrupt climate change occurs is often referred to as the 'tipping point'.

In other cases, the crossing of a threshold may lead to a gradual response that is difficult or even impossible to reverse. An example is the melting of the Greenland ice sheet, which may occur over a period of hundreds of years, but once started would be difficult to reverse.

Figure 4.8 Abrupt or rapid climate change showing the lack of response until a threshold is reached

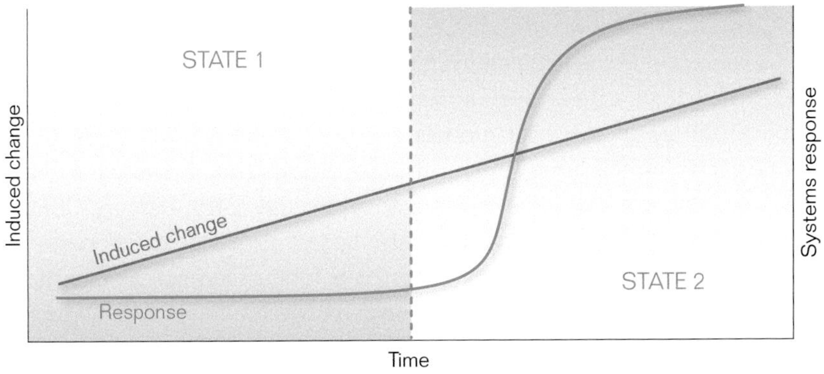

Source: Based on Steffen et al. (2004).

For most elements of the climate system there is considerable uncertainty as to when the tipping point might occur, but with each increment of temperature rise, the likelihood of such an event or outcome increases.

4.4.2 Extreme climate outcomes

Due to the high level of uncertainty and the range of possible climate responses, there is a tendency in the policy community to focus on the median, or best-estimate, outcomes of climate change.

For all climate outcomes, there are levels of change that are at the bounds of the probability distribution—sometimes referred to as the 'extremes'. While they are by definition less likely to occur based on current knowledge, an understanding of the potential for more damaging climate outcomes is vital to assessing risk.

The potential for high levels of climate sensitivity and abrupt sea level rise are examples of extreme climate outcomes that have already been discussed. This section identifies some other extreme climate outcomes that could have considerable impact at the subcontinental scale.

Changes to the El Niño – Southern Oscillation

The El Niño – Southern Oscillation is a large-scale pattern of climate variability that leads to climatic effects in both the Pacific region and some other parts of the world (IPCC 2007a: 945). Its fluctuations involve large changes in the transfer of heat between the ocean and atmosphere, which has a considerable influence on year-on-year changes in global mean temperature and rainfall patterns.

Palaeoclimatic data suggest that the nature of the El Niño – Southern Oscillation, and the way it affects the global climate, has changed over time. In the 20th century, El Niño events occurred every three to seven years, but were more intense in the second half of the century (IPCC 2007a: 288).

Based on knowledge of the mechanisms behind the El Niño – Southern Oscillation, as well as historical evidence, it would be expected that changes in ocean temperature and density would change the current pattern of the El Niño – Southern Oscillation (Lenton et al. 2008). Model outcomes suggest that such events will continue, but some simulations have shown an increase in its variability, while others exhibit no change or even a decrease.

Based on a survey of available models, the IPCC states that 'there is no consistent indication at this time of discernable changes in projected El Niño – Southern Oscillation amplitude or frequency in the 21st century' (IPCC 2007a: 751). Lenton et al. (2008) disagree, and based on the available evidence consider there to be 'a significant probability of a future increase in El Niño – Southern Oscillation amplitude', with the potential for the threshold warming to be reached this century. However, the existence and location of any 'threshold' that may result in such a change is highly uncertain.

Climate–carbon feedbacks

There is agreement across climate models that climate change in the future will reduce the efficiency of the oceans and the terrestrial biosphere in absorbing carbon dioxide from the atmosphere. There is large uncertainty regarding the sensitivity of the response. When the impacts of climate change are taken into account in the calculation of carbon dioxide concentrations over time, an additional climate warming of 0.2–1.5°C occurs by 2100 (Friedlingstein et al. 2006).

There are a number of climate–carbon feedback effects that are not well understood, but which could have considerable influence on the temperature response to an increase in carbon dioxide emissions. These include:

- **Release of methane from methane hydrates in the ocean**—Methane hydrates are stored in the seabed along the continental margins in stable form due to the low temperatures and high pressure environment deep in the ocean. Warming may cause the hydrates to become unstable, leading to the release of methane into the atmosphere.
- **Release of methane from melting permafrost**—Methane trapped in frozen soils may be released as temperatures increase and melting occurs.
- **Abrupt changes in the uptake and storage of carbon by terrestrial systems**—Terrestrial systems could change from a sink to a source of carbon. Potential sources could be the increased rate of oxidation of soil carbon and dieback of the Amazon rainforest and high-latitude forests in the northern hemisphere.
- **Reduced uptake of carbon dioxide by oceans**—The uptake of carbon dioxide by oceans may be reduced due to decreased solubility of carbon dioxide at higher temperatures and changes in ocean circulation.

As the climate warms, the likelihood of the system crossing the thresholds of temperature or emissions that might trigger such outcomes increases (IPCC 2007a: 642).

4.4.3 High-consequence climate outcomes

Some climate outcomes may be highly damaging due to the vulnerability or adaptive capacity of the human or natural systems affected, rather than the severity of the outcome itself.

Severe weather events, or best-estimate climate change outcomes, could be considered of high consequence, or catastrophic, due to a range of factors, including:

- the magnitude, timing, persistence or irreversibility of the changes to the climate system or impacts on natural and human systems
- the importance of the systems at risk
- the potential for adaptation
- the distributional aspects of impacts and vulnerabilities (IPCC 2007b: 781).

Melting of the Himalayan glaciers

After the polar regions, the Himalayas are home to the largest glacial areas. Together, the Himalayan glaciers feed seven of the most important rivers in Asia—the Ganga, Indus, Brahmaputra, Salween, Mekong, Yangtze and Huang.

These glaciers are receding faster than any other glaciers around the world, and some estimates project that they may disappear altogether by 2035 (WWF Nepal Program 2005).

Rivers fed from glaciers are projected to experience increased streamflows over the next few decades as a result of glacial melt, followed by a subsequent decline and greater instability of inflows as glaciers begin to disappear altogether, leaving only seasonal precipitation to feed rivers (WWF Nepal Program 2005).

Glacial retreat can also result in catastrophic discharges of water from meltwater lakes, known as glacial lake outburst floods, which can cause considerable destruction and flooding downstream.

Failure of the Indian monsoon

The Indian monsoon has been remarkably stable for the last hundred years. The monsoon is central to south Asia's economy and social structure (Challinor et al. 2006). Any change in the timing or intensity of the monsoon is likely to have significant consequences for the region.

There is limited scientific understanding of the processes underpinning the development of the Indian monsoon (Challinor et al. 2006). The monsoon is the result of the complex interactions among the ocean, atmosphere, land surface, terrestrial biosphere and mountains.

Central to projections of how the Indian monsoon may change with a changing climate is the uncertain response of the El Niño – Southern Oscillation. Some projections indicate a reduction in the frequency of rainfall events associated with the monsoon but an increase in their intensity. The monsoon also exhibits variations within seasons that lead to severe weather events with potentially large consequences. The ability of current climate models to predict these seasonal cycles is limited, but changes in the intensity, duration and frequency of these cycles may constitute the most profound effects of climate change on the monsoon system (Challinor et al. 2006).

Destruction of coral reefs

Coral reefs are highly sensitive to changes in the temperature and acidity of the ocean. As carbon dioxide concentrations increase, a greater amount is absorbed by the ocean where it reacts with the water to create carbonic acid. Higher ocean acidity reduces the availability of calcium carbonate for reef-building corals to create their hard skeletons. The concentration of calcium carbonate in the ocean is a key factor in the current distribution of reef ecosystems.

Long-term records show that sea temperature and acidity are higher than at any other time in the last 420 000 years (Hoegh-Guldberg et al. 2007).

Reef-building corals have already been pushed to their thermal limits by increases in temperature in tropical and subtropical waters over the past 50 years. Many species have a limited capacity to adapt quickly to environmental change, so the rate at which these changes occur is critical to the level of impact. In combination with other stressors such as excessive fishing and declining coastal water quality, increases in acidity and temperature can push reefs from a coral- to algae-dominated state. If the reef ecosystem is pushed far enough, a tipping point is likely to be reached (Hoegh-Guldberg et al. 2007).

At a carbon dioxide concentration of 450 ppm, the diversity of corals on reefs will decline under the combined affects of elevated temperature and ocean acidity. Atmospheric carbon dioxide concentrations as low as 500 ppm will result in coral communities that no longer produce sufficient calcium carbonate to be able to maintain coral reef structures.

Risk of species extinction

The patterns of temperature and precipitation in the current climate are important determinants in the core habitat of a species. These affect the abundance and distribution of species.

Recent research has shown that significant changes in ecosystems are occurring on all continents and in most oceans, and that anthropogenic climate change is having a significant impact on these systems globally and on some continents (Rosenzweig et al. 2008). Projected climate change under high-emissions scenarios is expected to exacerbate the effects of existing stressors and lead to even further loss of biodiversity (Steffen et al. in press). The projected rate of climate change will be beyond the capability of many organisms to adapt. A further loss of species is likely, particularly among those that are already threatened or endangered (Sheehan et al. 2008). Changes in biodiversity would be an irreversible consequence of projected climate change.

Recently there has been increased recognition of the so-called ecosystem services that biodiversity provides. Industries such as forestry, agriculture and tourism that rely directly on ecosystem services are most exposed to risks linked to declines in biodiversity as a result of climate change (UNEP FI 2008). Ultimately all human beings, even those in highly urbanised areas, are completely dependent on a wide range of ecosystem services for their well-being and even their existence (Millennium Ecosystem Assessment 2005).

4.4.4 Evaluating the likelihood of extreme climate outcomes

Table 4.1 summarises the outcomes of a range of studies (Sheehan et al. 2008; Warren 2006; Lenton et al. 2008), considering the potential for a selection of high-consequence and extreme climate outcomes for each of the three emissions cases studied (shown for a range of temperature outcomes). The consideration of higher climate sensitivities in section 4.3.1 suggests that much higher temperature outcomes are possible by the end of the century, even if they have a lower probability of occurring. Under the higher climate sensitivities the likelihood of these outcomes occurring before 2100 becomes much higher.

The climate outcomes discussed in this section are sometimes referred to in the literature as 'high-consequence, low-probability outcomes' under human-induced climate change. This assessment shows that these events are of high consequence, but not always of low probability.

Table 4.1 Summary of extreme climate responses, high-consequence outcomes and ranges for tipping points for the three emissions cases by 2100

Extreme climate response or impact	450	550	No mitigation
Temperature outcomes	1.5 (0.8–2.1) °C	2 (1.1–2.7) °C	5.1 (3–6.6) °C
(a) Species at risk of extinction	7 (3–13)%	12 (4–25)%	88 (33–98)%
(b) Likelihood of initiating large-scale melt of the Greenland ice sheet	10 (1–31)%	26 (3–59)%	100 (71–100)%
(c) Area of reefs above critical limits for coral bleaching	34 (0–68)%	65 (0–81)%	99 (85–100)%
Estimated lower threshold exceeded by 2100			
(d) Threshold for initiating accelerated disintegration of the west Antarctic ice sheet	No	No	Yes
(e) Threshold for changes to the variability of the El Niño – Southern Oscillation	No	No	Yes
(f) Threshold where terrestrial sinks could become carbon sources	Possibly	Possibly	Yes

Notes:

The temperatures shown are increases from 1990 levels. The central number is based on a best-estimate climate sensitivity of 3°C, while the numbers in brackets relate to the temperature in 2100 for climate sensitivities of 1.5°C and 4.5°C for the lower and upper temperatures respectively. The approach is different to that used in Table 5.1 of the Review's draft report, which showed the range of median outcomes from three separate climate sensitivity studies.

(a) The percentage of all species 'committed to extinction' due to shifts in habitat caused by temperature and climate changes, from sample regions covering 20 per cent of the earth's land surface. The upper limit (>3.5°C) is based on less comprehensive datasets and is therefore more uncertain (Sheehan et al. 2008).

(b) Cumulative probability based on four estimates from the literature. The percentage represents the likelihood of triggering the commencement of partial or complete deglaciation. This is considered virtually certain under the best-estimate temperature outcomes in the no-mitigation case (Sheehan et al. 2008).

(c) Percentage of reef area in which there is widespread mortality in slow-growing, tolerant reef species on a frequency of less than 25 years, based on a range of studies from the literature (Sheehan et al. 2008).

(d) A range in which the threshold for initiating accelerated disintegration of the west Antarctic ice sheet is expected to occur. The outcomes combine a literature review and expert judgment (Lenton et al. 2008).

(e) A range in which the threshold for changes to the variation of El Niño – Southern Oscillation is expected to occur. The outcomes combine a literature review and expert judgment (Lenton et al. 2008).

(f) A range in which the threshold where terrestrial sinks could be damaged to the extent that they become carbon sources is expected to occur. This includes a combination of outcomes from Lenton et al. (2008), relating to the threshold for extensive damage to the Amazon rainforest boreal forest systems, and Warren (2006), relating to desertification leading to widespread loss of forests and grasslands.

Note

1 The three datasets used in this analysis were (1) Hadley Centre HadCRUT3 (Brohan et al. 2006), <www.cru.uea.ac.uk/cru/data/temperature/hadcrut3gl.txt>; (2) the Goddard Institute for Space Studies, NASA, <http://data.giss.nasa.gov/gistemp/tabledata/GLB.Ts.txt>; and (3) the National Climatic Data Center, US Department of Commerce, <ftp://ftp.ncdc.noaa.gov/pub/data/anomalies/annual.land_and_ocean.90S.90N.df_1901-2000mean.dat>.

References

Breusch, T. & Vahid, F. 2008, *Global Temperature Trends*, report prepared for the Garnaut Climate Change Review, Australian National University.

Brohan, P., Kennedy, J.J., Harris, I., Tett, S.F.B. & Jones, P.D. 2006, 'Uncertainty estimates in regional and global observed temperature changes: a new dataset from 1850', *Journal of Geophysical Research* 111, D12106, updated from <www.cru.uea.ac.uk/cru/info/warming>.

CASPI (Climate Adaptation – Science and Policy Initiative) 2007, *The Global Science of Climate Change*, report prepared for the Garnaut Climate Change Review, University of Melbourne.

Cazenave, A. & Nerem, R.S. 2004, 'Present-day sea level change: observations and causes', *Reviews of Geophysics* 42, RG3001, doi:10.1029/2003RG000139.

Challinor, A., Slingo, J., Turner, A. & Wheeler, T. 2006, *Indian Monsoon: Contribution to the Stern Review*, University of Reading, United Kingdom.

Church, J. & White, N.J. 2006, 'A 20th century acceleration in global sea-level rise', *Geophysical Research Letters* 33, L01602, doi:10.1029/2005GL024826.

Church, J., White, N., Aarup, T., Stanley Wilson, W., Woodworth, P., Domingues, C., Hunter, J. & Lambeck, K. 2008, 'Understanding global sea levels: past, present and future', *Sustainability Science* 3(1): 9–22.

CSIRO (Commonwealth Scientific and Industrial Research Organisation) & BoM (Bureau of Meteorology) 2007, *Climate Change in Australia: Technical report 2007*, CSIRO, Melbourne.

Friedlingstein, P., Cox, P., Betts, R., Bopp, L., von Bloh, W., Brovkin, V., Cadule, P., Doney, S., Eby, M., Fung, I., Bala, G., John, J., Jones, C., Joos, F., Kato, T., Kawamiya, M., Knorr, W., Lindsay, K., Matthews, H.D., Raddatz, T., Rayner, P., Reick, C., Roeckner, E., Schnitzler, K.-G., Schnur, R., Strassmann, K., Weaver, A.J., Yoshikawa, C. & Zeng, N. 2006, 'Climate-carbon cycle feedback analysis: results from the C4MIP model intercomparison', *Journal of Climate* 19(14): 3337–53.

Hansen, J. 2005, 'A slippery slope: how much global warming constitutes "dangerous anthropogenic interference"?' *Climatic Change* 68: 269–79.

Haas, C., Pfaffling, A., Hendricks, S., Rabenstein, L., Etienne, J-L., & Rigor, I., 2008, 'Reduced ice thickness in Arctic Transpolar Drift favors rapid ice retreat', *Geophysical Research Letters*, 35, L17501, doi:10.1029/2008GL034457.

Hoegh-Guldberg, O., Mumby, P., Hooten, A., Steneck, R., Greenfield, P., Gomez, E., Harvell, C., Sale, P., Edwards, A., Caldeira, K., Knowlton, N., Eakin, C., Iglesias-Prieto, R., Muthiga, N., Bradbury, R., Dubi, A. & Hatziolos, M. 2007, 'Coral reefs under rapid climate change and ocean acidification', *Science* 318(5857): 1737–42.

Holgate, S. & Woodworth, P.L. 2004, 'Evidence for enhanced coastal sea level rise during the 1990s', *Geophysical Research Letters* 31, L07305, doi:10.1029/2004GL019626.

IPCC (Intergovernmental Panel on Climate Change) 2000, *Special Report on Emissions Scenarios. A Special Report of Working Group III of the Intergovernmental Panel on Climate Change*, N. Nakicenovic & R. Swart (eds), Cambridge University Press, Cambridge.

IPCC 2007a, *Climate Change 2007: The physical science basis. Contribution of Working Group I to the Fourth Assessment Report of the Intergovernmental Panel on Climate Change*, S. Solomon, D. Qin, M. Manning, Z. Chen, M. Marquis, K.B. Averyt, M. Tignor & H.L. Miller (eds), Cambridge University Press, Cambridge and New York.

IPCC 2007b, *Climate Change 2007: Impacts, adaptation and vulnerability. Contribution of Working Group II to the Fourth Assessment Report of the Intergovernmental Panel on Climate Change*, M.L. Parry, O.F. Canziani, J.P. Palutikof, P.J. van der Linden & C.E. Hanson (eds), Cambridge University Press, Cambridge and New York.

Jones, R. 2008, *Warming Probabilities for the Garnaut Review Emissions Cases*, data prepared for the Garnaut Review, CSIRO, Aspendale, Victoria.

Lenton, T.M., Held, H., Kriegler, E., Hall, J.W., Lucht, W., Rahmstorf, S. & Schellnhuber, H.J. 2008, 'Tipping elements in the Earth's climate system', *Proceedings of the National Academy of Sciences of the USA* 105(6): 1786–93.

Leuliette, E.W., Nerem, R.S. & Mitchum, G.T. 2004, 'Calibration of TOPEX/Poseidon and Jason altimeter data to construct a continuous record of mean sea level change', *Marine Geodesy* 27(1–2): 79–94.

Millennium Ecosystem Assessment 2005, *Ecosystems and Human Well-being: Synthesis report*, Island Press, Washington DC.

NSIDC (National Snow and Ice Data Centre) 2007, 'Arctic sea ice shatters all previous record lows', press release, 1 October, <http://nsidc.org/news/press/2007_seaiceminimum/20071001_pressrelease.html>.

NSIDC 2008, 'Record sea ice loss in August', *Arctic Sea Ice News and Analysis*, 4 September, <http://nsidc.org/arcticseaicenews>.

Oppenheimer, M. & Alley, R. 2005, 'Ice sheets, global warming, and Article 2 of the UNFCCC', *Climatic Change* 68: 257–67.

Pew Center on Global Climate Change 2007, 'Sea level rise: the state of the science', February 2, <http://www.pewclimate.org/global-warming-basics/slr.cfm>.

Phipps, S.J. 2006, *The CSIRO Mk3L Climate System Model, Technical Report 3*, Antarctic Climate & Ecosystems Cooperative Research Centre, Hobart.

Rahmstorf, S. 2007, 'A semi-empirical approach to projecting future sea-level rise', *Science* 315, <www.sciencemag.org/cgi/content/full/1135456/DC1>.

Rahmstorf, S., Cazenave, A., Church, J., Hansen, J., Keeling, R., Parker, D. & Somerville, R. 2007, 'Recent climate observations compared to projections' *Science* 316(5825): 709.

Rosenzweig, C., Karoly, D., Vicarelli, M., Neofotis, P., Wu, Q., Casassa, G., Menzel, A., Root, T.L., Estrella, N., Seguin, B., Tryjanowski, P., Liu, C., Rawlins, S. & Imeson, A. 2008, 'Attributing physical and biological impacts to anthropogenic climate change', *Nature* 453: 353–7.

Ruddiman, W.F. 2008, *Earth's Climate: Past and future*, 2nd edn, W.H. Freeman and Company, New York.

Sheehan, P., Jones, R., Jolley, A., Preston, B., Clarke, M., Durack, P., Islam, S., Whetton, P. 2008, 'Climate change and the new world economy: implications for the nature and timing of policy responses', *Global Environmental Change,* doi:10.1016/j.gloenvcha.2008.04.008.

Steffen, W., Burbidge, A., Hughes, L., Kitching, R., Lindenmayer, D., Mummery, J., Musgrove, W., Stafford Smith, M. & Werner, P. in press, *Conserving Our Biotic Heritage in a Rapidly Changing World*, CSIRO Publishing.

Steffen, W., Sanderson, A., Tyson, P., Jäger, J., Matson, P., Moore, B. III, Oldfield, F., Richardson, K., Schellnhuber, H.-J., Turner, B.L. II & Wasson, R. 2004, *Global Change and the Earth System: A planet under pressure*, IGBP Global Change Series, Springer-Verlag, Berlin.

UNEP FI (United Nations Environment Programme Finance Initiative) 2008, *Biodiversity and Ecosystem Services: Bloom or Bust? A Document of the UNEP FI Biodiversity & Ecosystem Services Work Stream (BESW)*, Geneva.

Warren, R. 2006, 'Impacts of global climate change at different annual mean global temperature increases', in H.J. Schellnhuber, C. Cramer, N. Nakicenovic, T. Wigley & G. Yohe (eds), *Avoiding Dangerous Climate Change*, Cambridge University Press, Cambridge, pp. 94–131.

Watterson, I. 2008, *Global Climate Change Projections Results of Interest to the Garnaut Review*, paper prepared for the Garnaut Review, CSIRO, Aspendale, Victoria.

Wigley, T.M.L. 2003, *MAGICC/SCENGEN 4.1: Technical Manual*, National Center for Atmospheric Research, Colorado.

WWF Nepal Program, 2005, *An Overview of Glaciers, Glacier Retreat, and Subsequent Impacts in Nepal, India and China: WWF Nepal Program March, 2005*, S. Chamling Rai (ed.).

PROJECTING AUSTRALIAN CLIMATE CHANGE

5

Key points

Australia's dry and variable climate has been a challenge for the continent's inhabitants since human settlement.

Temperatures in Australia rose slightly more than the global average in the second half of the 20th century. Streamflow has fallen significantly in the water catchment areas of the southern regions of Australia. Some of these changes are attributed by the mainstream science to human-induced global warming.

Effects of future warming on rainfall patterns are difficult to predict because of interactions with complex regional climate systems. Best-estimate projections show considerable drying in southern Australia, with risk of much greater drying. The mainstream Australian science estimates that there may be a 10 per cent chance of a small increase in average rainfall, accompanied by much higher temperatures and greater variability in weather patterns.

Australia is a vast continent characterised by high regional and seasonal climate variability. It has a wide range of ecosystems within its borders, from tropical to alpine and Mediterranean to arid desert. There are multiple influences on Australia's climate, ranging from the global, to the regional, such as the El Niño – Southern Oscillation, and the local, such as the Great Dividing Range.

The Commonwealth Scientific and Industrial Research Organisation (CSIRO) and the Bureau of Meteorology (2007) have undertaken major work projecting Australia's future climate. The Review commissioned the CSIRO to extend its projections for a number of variables out to 2100 on the basis of the Review's projections of emissions, as utilised in Chapter 6,[1] under a no-mitigation and specified global mitigation cases.

5.1 Attributing climate change to humans

For climate variables that exhibit only a limited range of natural variability, small deviations can be significant. The role of human influence can be attributed with some confidence. Global average temperature is one such factor.

Changes in climate variables such as rainfall, which can exhibit high interannual and interseasonal variability, are much harder to attribute in this way. For these it is difficult to distinguish the human-induced element from natural variability. Changes in climate variables that manifest themselves over longer timescales, such as decades or centuries, are harder still to attribute to human activity (CSIRO & BoM 2007).

Single events, such as an intense tropical cyclone or a long-lived heatwave, cannot be directly attributed to climate change. Climate change may, however, affect the factors that lead to such events. It may make certain events, like the heatwave that occurred in Adelaide during the summer of 2007–08, more likely (CSIRO & BoM 2007).

Some changes in the Australian climate system have been attributed to human-induced climate change. Examples include the increase in average temperatures since the middle of the 20th century. Up to 50 per cent of the reduction in rainfall in the south-west of Western Australia, and of the decline in snow cover in the south-east Australian alpine region, has been attributed to human-induced climate change (CSIRO & BoM 2007; Cai & Cowan 2006).

Rainfall decline in other parts of the country, such as south-east Australia, has not been definitively attributed to human-induced climate change. By contrast, the higher temperatures that have accompanied and exacerbated the effects of drought conditions have been so attributed.

5.2 How has the climate changed in Australia?

Over the last century, and the last 50 years in particular, marked changes have been observed in a number of key climate variables.

5.2.1 Temperature

Annual average temperature in Australia has increased by 0.9°C since 1910 (CSIRO & BoM 2007). Figure 5.1 shows Australian annual average temperature anomalies from 1900 to 2007.

The warming tendency since the middle of the 20th century has not been uniform across the country. The greatest warming has occurred in central Australia (Murphy & Timbal 2008). In south-eastern Australia, average maximum temperatures have increased. As a result, droughts have become hotter, with effects on rainfall, evaporation and runoff, and, more generally, water availability for human use (Nicholls 2004).

Figure 5.1 Australian annual average temperature anomalies, 1910–2007

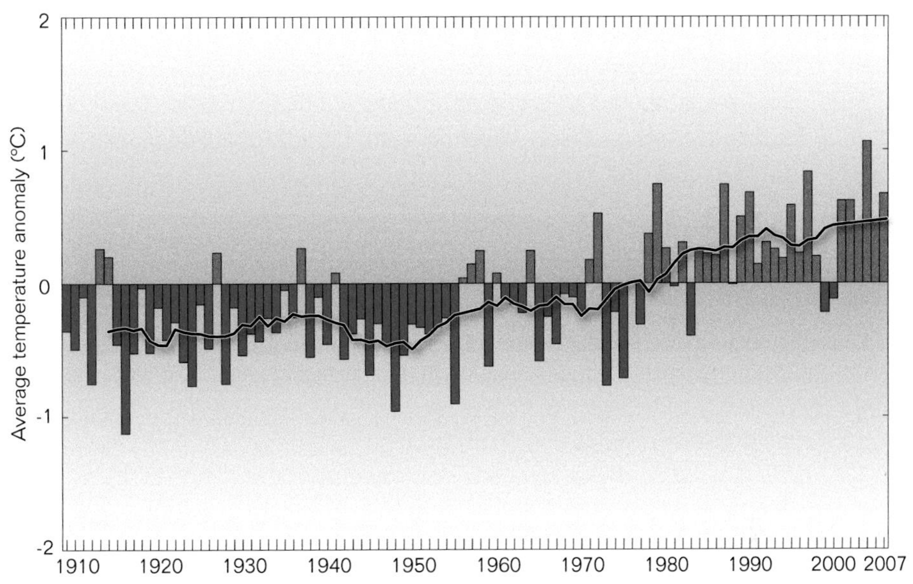

Note: The data show temperature difference from the 1961–90 average. The black line shows the eleven-year running averages.

Source: BoM (2008).

Ocean temperature has changed at a slower pace, due to the ocean's large heat content and enhanced evaporative cooling. Nevertheless, substantial warming has occurred in the three oceans surrounding Australia. The Indian Ocean is warming faster than all other oceans, with significant warming off the coast of Western Australia (CSIRO & BoM 2007).

5.2.2 Rainfall

There has been a major change in rainfall patterns since the 1950s, with large geographic variation. North-west Australia has seen a significant increase in annual rainfall, whereas most of the eastern seaboard and south-west Australia have seen a significant decrease (CSIRO & BoM 2007). Rainfall changes over the longer period from 1900 to 2007 are generally positive and are largest in the north-west. Drying tendencies over this period are evident in south-west Australia, some other parts of southern Australia, including much of Tasmania, and over much of north-east Australia.

In Australia, the rate of change in the frequency and intensity of rainfall extremes is greater than the rate of change for average rainfall (Alexander et al. 2007).

The rainfall decline observed in the 1990s in south-east Australia shares many characteristics with the decline in the south-west, but has only recently become the subject of extensive research (CSIRO & BoM 2007). The factors affecting rainfall decline in the south-east appear more complex. This region is affected to a greater extent by major systems such as the El Niño – Southern Oscillation and the subtropical ridge (Murphy & Timbal 2008) (see Box 5.1).

Rainfall can be affected by anthropogenic emissions other than greenhouse gases. It has been suggested that the increased rainfall in the north-west of Australia could be attributed to aerosols drifting south from Asia. Aerosols can create a localised cooling effect, which in turn affects convection and rainfall patterns (Rotstayn et al. 2007).

Box 5.1 Drought in Australia

Drought can be defined in many ways. The main contributing factors for all definitions, however, are rainfall, temperature and evaporation. Due to the strong connection between anthropogenic emissions and warming in Australia, the CSIRO and the Bureau of Meteorology (2007) have concluded that the drought in many parts of the country is linked to, or at least exacerbated by, global warming.

Drought can have many causes. For example, the drought in south west Western Australia has been attributed to a combination of natural variability, an increase in greenhouse gas concentrations, and land-use change (Timbal et al. 2006).

In south-east Queensland, the drought appears to have been caused by changes in two key climate variables: the El Niño – Southern Oscillation and tropical cyclones (Department of Natural Resources and Water 2007). Since 1950, there has been a strong decline in rainfall across eastern Australia, and since 1977 there has been an increase in the frequency of El Niño events (Power & Smith 2007). This shift in the system's El Niño behaviour has been related to a reduction in the number of tropical cyclones in the region. Over time, such cyclones contribute a large amount to total rainfall.

Since 1997, south-east Australia has recorded a number of changes in average rainfall (Murphy & Timbal 2008; B. Timbal 2008, pers. comm.):

- From 1997 to 2007 only one year had rainfall above the 1961–90 annual average.
- Only one autumn from 1990 had rainfall above the 1961–90 autumn average (a critical rainfall period for the establishment of crops).
- Melbourne streamflows have been below the long-term average every year since 1996 (Timbal & Jones 2008).

A similar decline in annual average rainfall in south-east Australia has occurred once before in historical records (Murphy & Timbal 2008; B. Timbal 2008, pers. comm.). This was for the period 1936–45. However, the current drought is marked by reduced interannual variability and an increase in average maximum temperatures, affecting evaporation and further reducing streamflows.[2]

Box 5.1 Drought in Australia *(continued)*

The decline in autumn rainfall in south-east Australia has strong qualitative similarities with the decline observed in the 1970s in Perth (Murphy & Timbal 2008). Unlike south-west Western Australia, however, the south-east of Australia is affected by several climate systems including the El Niño – Southern Oscillation and the Southern Annular Mode. There is currently no consensus on the magnitude of the influence of these systems, or on how they will respond to global warming (Murphy & Timbal 2008; Timbal & Murphy 2007; Cai & Cowan 2008a; Hendon et al. 2007). One estimate is that the El Niño – Southern Oscillation and the Southern Annular Mode each account for around 15 per cent of the observed rainfall variability in south-west Australia during winter and in south-east Australia during winter, spring and summer (Hendon et al. 2007).

It has recently been suggested that increased temperature has a large impact on streamflow. After accounting for interdependencies, such as the effect of rainfall and clouds on minimum temperatures, Cai and Cowan (2008b) conclude that a 1°C increase in maximum temperature results in a 15 per cent decrease in streamflow in the Murray-Darling Basin.

Streamflows

A reduction in rainfall results in a proportionately larger fall in streamflows. Generally, a decrease in rainfall can result in a two- to threefold decrease in streamflow (Chiew 2006). In the Murray-Darling Basin, a 10 per cent change in rainfall has been found to result in a 35 per cent change in streamflow (Jones et al. 2001).

Low streamflows have been recorded in the rivers supplying most major urban water storage systems over the last decade (Water Services Association of Australia 2007). For Melbourne, Sydney, Brisbane, Adelaide and Canberra, average streamflows over the period 1997–2007 are notably below the long-term average.[3] The period of the long-term average is between 84 and 108 years, depending when measurements began. Recent streamflows supplying Canberra are 43 per cent of the long-term average, in Melbourne 65 per cent, Adelaide 62 per cent, Sydney 40 per cent and Brisbane 42 per cent.

The greatest, and earliest, decline in streamflows of rivers supplying major urban water storages has been observed in Perth (Figure 5.2). There has been a marked decline since the 1970s, which has continued and appears to have intensified over the last decade. Annual streamflows over recent years (2001–07) are only 25 per cent of the long-term average before this observed decline. The decline in rainfall in the region, which occurred at approximately the same time, has been partly attributed to human-induced climate change (Cai & Cowan 2006).

Figure 5.2 Annual streamflows into Perth's dams (excluding Stirling and Samson dams)

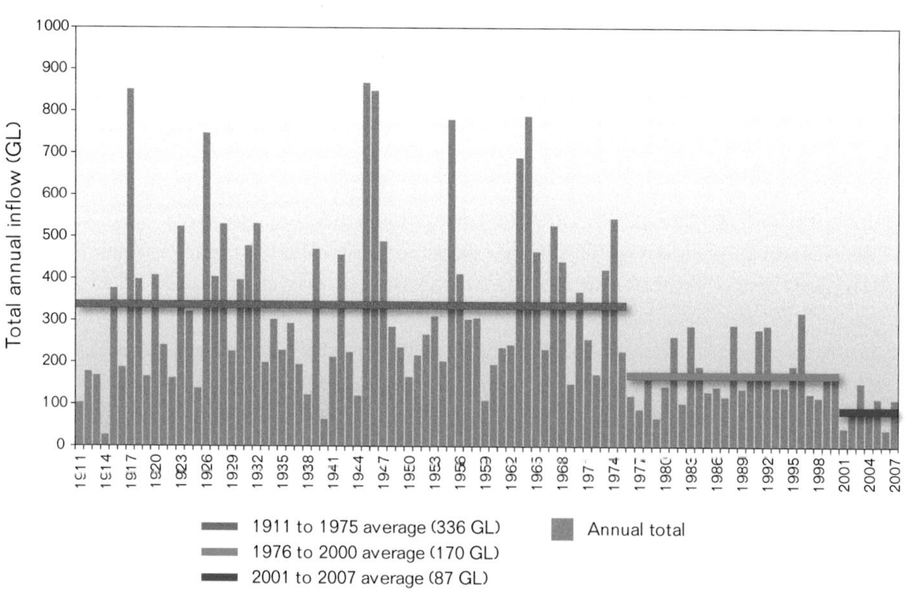

Note: Values represent totals for May–April.

Source: Western Australia Water Corporation (2008).

5.2.3 El Niño – Southern Oscillation and the Southern Annular Mode

El Niño – Southern Oscillation

The El Niño – Southern Oscillation is a naturally occurring phenomenon that temporarily disrupts climate patterns in many countries in the Pacific and Asia, including Australia (Power et al. 2006; Ropelewski & Halpert 1984; Nicholls 1992; Allan et al. 1996). It is often monitored using the Southern Oscillation Index, which is a measure of the difference in mean sea-level pressure across the Pacific Ocean, between Darwin and Tahiti. In Australia, a positive sustained index is associated with stronger trade winds, increased rainfall and flooding, and decreased temperatures across eastern Australia (Power et al. 1998). This is known as a La Niña episode. A negative index is associated with warmer than normal sea surface temperatures in the eastern Pacific, weaker trade winds, and a reduction in rainfall in eastern and northern Australia. This is known as an El Niño episode (BoM 2005; Power & Smith 2007).

Many major Australian droughts are associated with an El Niño event, though not all El Niño events trigger a drought. The effect of the El Niño – Southern Oscillation on climate varies across the country. South-west Western Australia, the west coast of Tasmania and coastal New South Wales are affected less than inland eastern Australia (BoM 2005).

The impact of the El Niño – Southern Oscillation on Australia varies substantially from decade to decade and generation to generation. There was a record high number of El Niño events in the period 1977–2006 and a corresponding record low number of rain-bearing La Niña events. This is the most El Niño-dominated period on record (Power & Smith 2007).

The extent to which the decline in the Southern Oscillation Index is influenced by global warming is unknown. Some climate models anticipate weakened trade winds in response to global warming (Vecchi et al. 2006), so global warming may be contributing to some of the observed changes evident in and around the tropical Pacific (Power & Smith 2007).

Southern Annular Mode

Another dominant driver of variability at high southern latitudes is the Southern Annular Mode. When this mode is in its positive phase, there will be weaker westerly winds in southern Australia and stronger westerly winds at high latitudes (CSIRO & BoM 2007).

The Southern Annular Mode and the El Niño – Southern Oscillation are estimated to drive approximately 15 per cent each of the variability in winter rainfall in south-west Australia and in winter, spring and summer rainfall in south-east Australia (Hendon et al. 2007).

Over recent decades, the Southern Annular Mode has spent more time in its positive phase (CSIRO & BoM 2007). The positive phase is associated with significantly reduced winter rainfall in southern Australia and significant rainfall increases in the Murray-Darling Basin in summer (Hendon et al. 2007).

This shift has led to a decrease in the potential for storm formation over southern Australia. In south-west Australia, a decrease in the number of rain-bearing synoptic systems and a deflection southward of some storms is associated with a reduction in winter rainfall (CSIRO & BoM 2007).

5.2.4 Other climate variables

Cyclones and storms

Tropical cyclones in Australia are subject to multidecadal variability in frequency and intensity (CSIRO & BoM 2007). When this variability is combined with the varying quality of historical records it becomes difficult to draw definitive conclusions on whether observed changes in tropical cyclones can be attributed to climate change (B. Buckley 2008, pers. comm.).

Similarly, hailstorms are highly sensitive to small-scale variations in meteorological and oceanographic conditions, as well as to other factors. Data collection methods have not yet evolved to allow rigorous climate change analyses and attribution.

Limited observations suggest that there was a substantial increase in tropical cyclone numbers on the east coast in the 1950s, followed by a reduction in the 1970s. This reduction appears to be linked to an increasing number of El Niño events. In general, there are fewer tropical cyclones in Australia during El Niño than during La Niña (CSIRO & BoM 2007).

On the west coast, there appears to have been an increase in the proportion of severe (category 3 and 4) cyclones (CSIRO & BoM 2007). During the period 1974–88, severe cyclones accounted for 29 per cent of the total. In the period 1989–98 they accounted for 41 per cent.

Bushfires

During the period 1973–2007, there was a general increase in the Forest Fire Danger Index across the east and south-east of the country. A recent review of 23 measuring locations over this period analysed the three years with the highest index (Lucas et al. 2007). Fifty out of 69 of the selected years were after the year 2000, with the increasing trend statistically significant above the 95 per cent probability level for most inland locations.

Box 5.2 Heatwaves

The number of hot days and warm nights per year has increased since 1955 (CSIRO & BoM 2007) and heatwaves have become increasingly common (Lynch et al. 2008).

In February 2004, for example, maximum temperatures were 5–6°C above average throughout large areas, reaching 7°C above average in parts of New South Wales (National Climate Centre 2004). Adelaide had 17 successive days over 30°C (the previous record was 14 days). Sydney had 10 successive nights over 22°C (the previous record was six). Around two-thirds of the continent recorded maximum temperatures over 39°C and temperatures peaked at 48.5°C in western New South Wales (Lynch et al. 2008).

In March 2008, Adelaide had 15 consecutive days of 35°C or above and 13 consecutive days of 37.8°C or above, surpassing the previous record of eight and seven days respectively. Hobart matched its previous record high temperature of 37.3°C and Melbourne recorded a record high overnight minimum of 26.9°C (National Climate Centre 2008).

5.3 Projected climate change in Australia

In assessing projections of climate change in Australia, the Review used a combination of the IPCC SRES scenarios, as used by the CSIRO and the Bureau of Meteorology (2007), and global mitigation cases based on stabilisation of greenhouse gas concentrations at 450 and 550 parts per million carbon dioxide equivalent. Emissions scenarios are based on global models that are necessarily simplified. Given the significant uncertainty in the timing of temperature increases, actual outcomes are unlikely to exactly match a given scenario or case.

The future climate is a function of both human-induced climate change and natural climate variability. In some decades natural variability will reinforce the climate change signal. In other decades, it will offset the signal to some degree.

Projections of global average temperature across different emissions cases show little variation up to the decade beginning in 2030. Australian average temperature responds in a similar fashion. After this point, projections of climate variables are increasingly dependent on emissions pathways.

There is no consensus among models as to how climate change will affect the El Niño – Southern Oscillation (see IPCC 2007: 779–80; CSIRO & BoM 2007; Lenton et al. 2008). In some models it intensifies, while in others it weakens.

5.3.1 Temperature

Annual average temperatures in Australia are expected to rise in parallel with rises in global average temperature. Significant regional variation, however, is projected across Australia. In general, the north-west is expected to warm more quickly than the rest of the country.

By 2030, annual average temperature over Australia will be around 1°C above 1990 levels (CSIRO & BoM 2007).[4] The range of uncertainty (10th to 90th percentiles) produces a national increase of between 0.4°C and 1.8°C for 2030. Coastal areas will experience slightly less warming (in the range 0.7–0.9°C), whereas inland Australia will experience greater warming (in the range 1.0–1.2°C).

From 2030 to the end of the century there are marked differences between emissions cases. This is shown in Figure 5.3, which considers the best-estimate 50th percentile.

By 2100, at both the 10th and 90th percentile outcomes, there are noticeable temperature increases across the country in a no-mitigation case. The 90th percentile of outcomes includes an increase of more than 7°C in some areas. The 10th percentile of outcomes shows an increase of more than 3°C for most of the country, increasing to 4.9°C over an extensive area in north-west Australia.

Figure 5.3 Best estimate (50th percentile) of Australian annual temperature change at 2030, 2070 and 2100 under three emissions cases

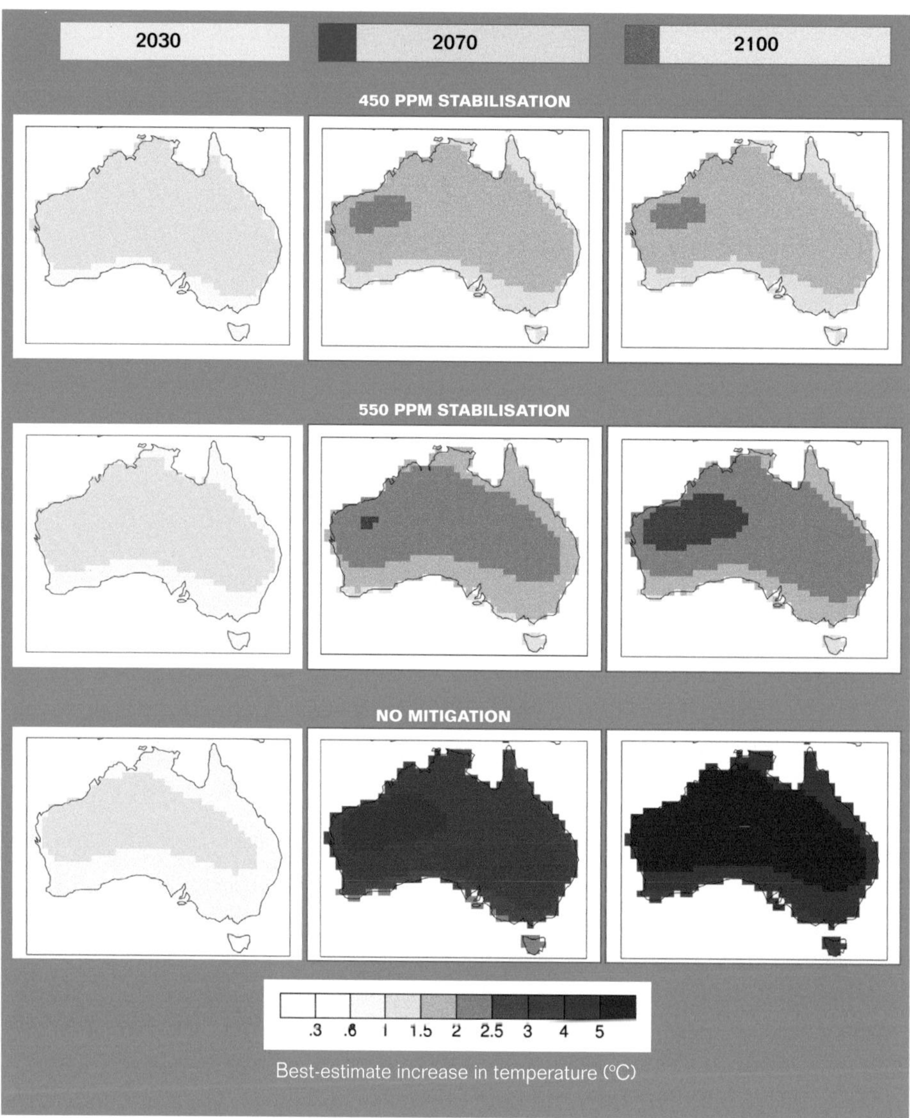

Note: The no-mitigation case is based on the SRES A1FI scenario. Values greater than or equal to 5 are represented with the same colour.

Source: CSIRO (2008c).

5.3.2 Rainfall

The relationship between local precipitation and atmospheric temperature is complex. Local rainfall patterns are highly sensitive to the amount of water available for evaporation, the local topography and land cover, and atmospheric and ocean circulation patterns. At the global level, a decrease in precipitation is indicated as the best-estimate outcome for Australia. However, because of the localised nature of influences on precipitation, there is considerable regional variation in precipitation change within Australia. Some areas are expected to experience an increase in rainfall. The complexities also lead to disagreement among climate models about the potential extent, and even direction, of the change.

Best-estimate annual average change in rainfall

Table 5.1 shows the median or best-estimate (50th percentile) annual average rainfall outcomes for Australia in a no-mitigation case in 2030, 2070 and 2100.[5] The best-estimate outcomes of change in annual average rainfall in 2030 are minimally different between the different climate cases due to climate change commitments (see section 4.3.1), but later in the century the rainfall outcomes are more dependent on the level of mitigation. The changes under the 550 and 450 global mitigation cases follow the same patterns, but the reductions are considerably more subdued. The extent of rainfall change under the 450 mitigation case in 2100 is less than under the no-mitigation case in 2070.

Table 5.1 Projected changes to statewide annual average rainfall, best-estimate outcome in a no-mitigation case (per cent change relative to 1990)

	NSW	Vic.	Qld	SA	WA	Tas.	NT	ACT
2030	-2.5	-3.5	-2.4	-4.2	-4.1	-1.4	-2.5	-2.8
2070	-9.3	-12.9	-8.6	-15.5	-14.9	-5.1	-9.0	-10.3
2100	-13.7	-19.0	-12.7	-22.8	-21.9	-7.6	-13.3	-15.2

Source: CSIRO (2008b).

The 'dry' and 'wet' ends of precipitation projections

The best-estimate outcomes do not reflect the extent of the uncertainty in potential rainfall outcomes for Australia under climate change. Rainfall projections are highly sensitive to small changes in model assumptions and inputs. The range of precipitation outcomes predicted by various climate models for Australia is large.

Table 5.2 shows the average annual changes for rainfall in Australia for the no-mitigation case for the 'dry' (10th percentile) and 'wet' (90th percentile) ends of projections in 2030, 2070 and 2100.

Table 5.2 Projected changes to statewide average rainfall, dry and wet outcomes in a no-mitigation case (per cent change relative to 1990)

Dry outcome (10th percentile)

	NSW	Vic.	Qld	SA	WA	Tas.	NT	ACT
2030	-10.1	-8.3	-11.5	-13.1	-12.7	-5.2	-11.4	-8.2
2070	-37.0	-30.3	-42.0	-48.0	-46.5	-19.2	-41.8	-30.1
2100	-54.6	-44.7	-61.8	-70.8	-68.5	-28.3	-61.6	-44.4

Wet outcome (90th percentile)

	NSW	Vic.	Qld	SA	WA	Tas.	NT	ACT
2030	4.2	0.9	6.0	4.0	4.2	2.6	6.0	2.0
2070	15.5	3.4	22.0	14.8	15.5	9.5	22.0	7.4
2100	22.8	5.1	32.5	21.9	22.8	14.0	32.4	10.9

Source: CSIRO (2008b). The methodology for the preparation of these distributions is described in CSIRO & BoM (2007).

Temporal variation in rainfall

Changes in annual rainfall often mask significant interseasonal variation. Similarly, average annual rainfall may mask changes in rainfall patterns, rainfall intensity and the number of rain events. For example, it is possible for average annual rainfall to remain the same while rain intensity increases and the number of rain days decreases (see Pitman & Perkins 2008).

In summer and autumn, rainfall decreases are more limited and some areas experience slight increases. Larger decreases in rainfall are experienced in winter and spring (CSIRO & BoM 2007).

As well as changes to annual average rainfall, the character of daily rainfall may change. There is expected to be an increase in the intensity of rainfall events in some areas, and the number of days without rainfall is also expected to increase. This suggests that the future precipitation regime may have longer dry spells broken by heavier rainfall events (CSIRO & BoM 2007; Pitman & Perkins 2008).

Consideration of a progressive change in annual average rainfall does not reflect the considerable interannual and decadal variation in Australian rainfall. In terms of precipitation, observed decadal variability in the 20th century and models run with a no-climate-change assumption show natural variability in rainfall of between 10 and 20 per cent (CSIRO & BoM 2007). Natural variability may therefore mask, or exaggerate, changes that are due to high concentrations of greenhouse gases.

Variation in the intensity and temporal pattern of daily rainfall, differences in seasonal change, and the influence of natural decadal variability could have considerable impacts on sectors such as agriculture and infrastructure.

5.3.3 Other climate variables

Cyclones and storms

Tropical cyclone frequency and intensity display high variability across seasonal, annual, decadal and longer timescales (CSIRO & BoM 2007). The El Niño – Southern Oscillation has a strong effect on tropical cyclone numbers (Abbs et al. 2006). Since it is as yet unknown what effect climate change will have on the El Niño – Southern Oscillation, it is difficult to project changes in the frequency and intensity of tropical cyclones.

Studies suggest that the frequency of east coast cyclones will either remain the same or decrease by up to 44 per cent (see CSIRO & BoM 2007). Abbs et al. (2006) estimate that category 3–5 storms will increase in intensity by 60 per cent by 2030 and 140 per cent by 2070.

Projections also indicate that the regions of east Australian cyclone genesis could shift southward by two degrees latitude (approximately 200 km) by 2050 (Leslie et al. 2007), while the average decay location could be up to 300 km south of the current location (Abbs et al. in CSIRO & BoM 2007). Models estimate that the number of strong cyclones reaching the Australian coastline will increase, and 'super cyclones', with an intensity hitherto unrecorded on the Australian east coast, may develop over the next 50 years (Leslie et al. 2007).

Projections indicate an increase in the intensity and frequency of hailstorms in the Sydney basin region with only a 1–2°C rise in temperature (Leslie et al. 2008).

Heatwaves

A strong increase in the frequency of hot days and warm nights is projected. The number of days per year above 35°C for 2030, 2070, and 2100 in all capital cities is shown in Table 5.3.

Most notable is the marked increase of hot days in Darwin. Under a no-mitigation case, there are 221 days over 35°C in 2070. By 2100, this increases to 312 days—or fewer than eight weeks in the year with days under 35°C.

Table 5.3 Projected increases in days over 35°C for all capital cities under a no-mitigation case

	Current	2030	2070	2100
Melbourne	9	12	21	27
Sydney	3.3	4.4	9	14
Brisbane	0.9	1.7	8	21
Adelaide	17	22	34	44
Perth	27	35	56	72
Canberra	5	8	21	32
Darwin	9	36	221	312
Hobart	1.4	1.7	2.5	3.4

Source: CSIRO (2008a).

Bushfires

Recent projections of fire weather (Lucas et al. 2007) suggest that fire seasons will start earlier, end slightly later, and generally be more intense. This effect increases over time, but should be directly observable by 2020.

Table 5.4 shows projections of the percentage increase in the number of days with very high and extreme fire weather.[6]

Table 5.4 Projected per cent increases in the number of days with very high and extreme fire weather for selected years

	Approximate year		
	2013	2034	2067
Very high	+2–13	+10–30	+20–100
Extreme	+5–25	+15–65	+100–300

Note: This study was based on scenarios producing 0.4°C, 1.0°C and 2.9°C temperature increases, which equate to the years in this table under a no-mitigation case.

Source: Lucas et al. (2007).

The Lucas et al. study defined two new categories of fire weather: 'very extreme' and 'catastrophic'.[7] Of the 26 sites used in the study, only 12 have recorded catastrophic fire weather days since 1973. At a 0.4°C increase in temperature there is little or no change in the number of catastrophic fire weather days. At a 1.0°C increase, catastrophic days occur at 20 sites. For half of these sites, the return period is around 16 years or less. At a 2.9°C increase, 22 sites record catastrophic days. Nineteen of these have a return period of around eight years or less. Seven sites have return periods of three years or less.

Notes

1 The emissions pathways that underpin the Australian climate projections in this chapter and Chapter 6 differ slightly from the pathways used in Chapter 4.

2 The standard deviation for annual rainfall for the period 1936–45 was 105.5 mm, compared to 85.8 mm for the current drought (Murphy & Timbal 2008; B. Timbal 2008, pers. comm.).

3 Hobart has not been included since normally only 40 per cent of supply is taken from storage facilities. The remainder is extracted from the Derwent River, whose catchment area is approximately 20 per cent of the area of the state. During scarce periods, larger quantities are drawn from the Derwent River, though on average this represents less than 1 per cent of the total streamflow. Darwin has not been included as streamflows in its catchment areas have been largely unaffected over the last decade (R. Young 2008, pers. comm.).

4 All temperature increases are from a 1990 baseline.

5 Changes in precipitation are reported as percentage changes from a 1990 baseline.

6 'Very high' fire weather has a Forest Fire Danger Index (FFDI) of 25–50, and 'extreme' fire weather has an FFDI of 50+. Suppression of fires during 'extreme' fire weather is 'virtually impossible on any part of the fire line due to the potential for extreme and sudden changes in fire behaviour. Any suppression actions such as burning out will only increase fire behaviour and the area burnt' (Vercoe in Lucas et al. 2007).

7 'Very extreme' fire weather has an FFDI of 75–100 and 'catastrophic' fire weather has an FFDI of over 100 (Lucas et al. 2007).

References

Abbs, D., Aryal, S., Campbell, E., McGregor, J., Nguyen, K., Palmer, M., Rafter, T., Watterson, I. & Bates, B. 2006, *Projections of Extreme Rainfall and Cyclones*, report to the Australian Greenhouse Office, Canberra.

Alexander, L.V., Hope, P., Collins, D., Trewin, B., Lynch, A. & Nicholls, N. 2007, 'Trends in Australia's climate means and extremes: a global context', *Australian Meteorological Magazine* 56: 1–18.

Allan, R.J., Lindesay, J. & Parker, D. 1996, *El Niño Southern Oscillation and Climatic Variability*, CSIRO, Collingwood, Victoria.

BoM (Bureau of Meteorology) 2005, *El Niño, La Niña and Australia's Climate*, BoM, Melbourne.

BoM 2008, 'Timeseries: Australian Climate Variability and Change', <www.bom.gov.au/cgi-bin/silo/reg/cli_chg/timeseries.cgi?variable=tmean®ion=aus&season=0112>.

Cai, W. & Cowan, T. 2006, 'SAM and regional rainfall in IPCC AR4 models: can anthropogenic forcing account for southwest Western Australian winter rainfall reduction?' *Geophysical Research Letters* 33, L24708, doi:10.1029/2006GL028037.

Cai, W. & Cowan, T. 2008a, 'Dynamics of late autumn rainfall reduction over southeastern Australia', *Geophysical Research Letters* 35, L09708, doi:10.1029/2008GL033727.

Cai, W. & Cowan, T. 2008b, 'Evidence of impacts from rising temperature on inflows to the Murray-Darling Basin', *Geophysical Research Letters* 35, L07701, doi:10.1029/2008GL033390.

Chiew, F.H.S. 2006, 'An overview of methods for estimating climate change impact on runoff', report for the 30th Hydrology and Water Resources Symposium, 4–7 December, Launceston, Tasmania.

CSIRO (Commonwealth Scientific and Industrial Research Organisation) 2008a, 'Projections of days over 35°C to 2100 for all capital cities under a no-mitigation case', data prepared for the Garnaut Climate Change Review, CSIRO, Aspendale, Victoria.

CSIRO 2008b, 'Regional rainfall projections in Australia to 2100 for three climate cases', data prepared for the Garnaut Climate Change Review, CSIRO, Aspendale, Victoria.

CSIRO 2008c, 'Regional temperature projections in Australia to 2100 for three climate cases', data prepared for the Garnaut Climate Change Review, CSIRO, Aspendale, Victoria.

CSIRO & BoM 2007, *Climate Change in Australia: Technical report 2007*, CSIRO, Melbourne.

Department of Natural Resources and Water 2007, *The South East Queensland Drought to 2007*, Government of Queensland, Brisbane.

Hendon, H.H., Thompson, D.W.J. & Wheeler, M.C. 2007, 'Australian rainfall and surface temperature variations associated with the Southern Hemisphere annular mode', *Journal of Climate* 20: 2452–67.

IPCC 2007, *Climate Change 2007: Impacts, adaptation and vulnerability. Contribution of Working Group II to the Fourth Assessment Report of the Intergovernmental Panel on Climate Change*, M. Parry, O. Canziani, J. Palutikof, P. van der Linden & C. Hanson (eds), Cambridge University Press, Cambridge.

Jones, R., Whetton, P., Walsh, K. & Page, C. 2001, *Future Impact of Climate Variability, Climate Change and Land Use Change on Water Resources in the Murray-Darling Basin. Overview and Draft Program of Research*, CSIRO.

Lenton, T.M., Held, H., Kriegler, E., Hall, J.W., Lucht, W., Rahmstorf, S. & Schellnhuber, H.J. 2008, 'Tipping elements in the Earth's climate system', *Proceedings of the National Academy of Sciences of the USA* 105(6): 1786–93.

Leslie, L.M., Karoly, D.J., Leplastrier, M. & Buckley, B.W. 2007, 'Variability of tropical cyclones over the southwest Pacific Ocean using a high resolution climate model', *Meteorology and Atmospheric Physics* 97: 171–80.

Leslie, L.M., Leplastrier, M. & Buckley, B.W. 2008, 'Estimating future trends in severe hailstorms over the Sydney Basin: a climate modelling study', *Atmospheric Research* 87(1): 37–51.

Lucas, C., Hennessey, K., Mills, G. & Bathols, J. 2007, *Bushfire Weather in Southeast Australia: Recent trends and projected climate change impacts*, consultancy report prepared for the Climate Institute of Australia.

Lynch, A., Nicholls, N., Alexander, L. & Griggs, D. 2008, *Defining the Impacts of Climate Change on Extreme Events*, report prepared for the Garnaut Climate Change Review.

Murphy, B.F. & Timbal, B. 2008, 'A review of recent climate variability and climate change in southeastern Australia', *International Journal of Climatology* 28(7): 859–80.

National Climate Centre 2004, 'Eastern Australia experiences record February heatwave', *Bulletin of the Australian Meteorological and Oceanographic Society* 17: 27–9.

National Climate Centre 2008, 'An exceptional and prolonged heatwave in Southern Australia', Special Climate Statement 15, Bureau of Meteorology, Melbourne.

Nicholls, N. 1992, 'Historical El Niño/Southern Oscillation variability in the Australian region', in H.F. Diaz & V. Markgraf (eds), *El Niño: Historical and paleoclimatic aspects of the Southern Oscillation*, Cambridge University Press, Cambridge, pp. 151–74.

Nicholls, N. 2004, 'The changing nature of Australian droughts', *Climatic Change* 63: 323–36.

Pitman, A.J. & Perkins, S.E. 2008, 'Regional projections of future seasonal and annual changes in rainfall and temperature over Australia based on skill-selected AR4 models', *Earth Interactions* 12: 1–50.

Power, S., Haylock, M., Colman, R. & Wang, X. 2006, 'The predictability of interdecadal changes in ENSO and ENSO teleconnections', *Journal of Climate* 8: 2161–80.

Power, S., Tseitkin, F., Torok, S., Lavery, B. & McAvaney, B. 1998, 'Australian temperature, Australian rainfall, and the Southern Oscillation, 1910–1996: coherent variability and recent changes', *Australian Meteorological Magazine* 47: 85–101.

Power, S.B. & Smith, I.N. 2007, 'Weakening of the Walker Circulation and apparent dominance of El Niño both reach record levels, but has ENSO really changed?' *Geophysical Research Letters* 34, L18702, doi:10.1029/2007GL030854.

Ropelewski, C.F. & Halpert, M.S. 1984, 'Global and regional scale precipitation patterns associated with ENSO', *Monthly Weather Review* 15: 1606–26.

Rotstayn, L.D., Cai, W., Dix, M.R., Farquhar, G.D., Feng, Y., Ginoux, P., Herzog, M., Ito, A., Penner, J.E., Roderick, M.L. & Wang, M. 2007, 'Have Australian rainfall and cloudiness increased due to the remote effects of Asian anthropogenic aerosols?' *Journal of Geophysical Research* 112, doi:10.1029/2006JD007712.

Timbal, B., Arblaster, J. & Power, S. 2006, 'Attribution of the late 20th century rainfall decline in Southwest Australia', *Journal of Climate* 19(10): 2046–62.

Timbal, B. & Jones, D.A. 2008, 'Future projections of winter rainfall in southeast Australia using a statistical downscaling technique', *Climatic Change* 86: 165–87.

Timbal, B. & Murphy, B. 2007, 'Observed climate change in the south-east of Australia and its relation to large-scale modes of variability', *Bureau of Meteorology Research Centre Research Letters* 6: 6–11.

Vecchi, G.A., Soden, B.J., Wittenberg, A.T., Held, I.A., Leetma, A. & Harrison, M.J. 2006, 'Weakening of tropical Pacific atmospheric circulation due to anthropogenic forcing', *Nature* 441(7089): 73–6.

Water Services Association of Australia 2007, *The WSAA Report Card 2006/07: Performance of the Australian urban water industry and projections for the future*, WSAA, Melbourne.

Western Australia Water Corporation 2008, 'Annual streamflows into Perth's dams', data supplied by the Western Australia Water Corporation to the Garnaut Climate Change Review.

6 CLIMATE CHANGE IMPACTS ON AUSTRALIA

Key points

This chapter provides a sample of conclusions from detailed studies of Australian impacts. These studies are available in full on the Review's website.

Growth in emissions is expected to have a severe and costly impact on agriculture, infrastructure, biodiversity and ecosystems in Australia.

There will also be flow-on effects from the adverse impact of climate change on Australia's neighbours in the Pacific and Asia.

These impacts would be significantly reduced with ambitious global mitigation

The hot and dry ends of the probability distributions, with a 10 per cent chance of realisation, would be profoundly disruptive.

This chapter focuses on Australia's exposure and sensitivity to climate change and considers the impacts of climate change on Australia in six key sectors and areas, chosen either because they make a large economic contribution to Australia, or because the impacts on market or non-market values are expected to be pronounced. These areas, sectors and subsectors are presented in Table 6.1.

The Review considers both the direct (section 6.3) and indirect (section 6.4) impacts of climate change on Australia. 'Direct' refers to those impacts that are experienced within Australia's land and maritime boundaries. 'Indirect' refers to impacts experienced in other countries with consequences for Australia. The first focus is on the medians of the probability distributions of the impacts identified by the main climate models. For some sectors the middle-of-the-road assessment is supplemented with analysis of the higher ends of the probability distribution of impacts. The standard IPCC projections, and those based on them, provide cases that correspond most closely to those we expect from no mitigation or from effective global mitigation policies (Table 6.2). The more serious implications of climatic tipping points are not examined, due to the rapidly developing nature of the relevant science and the limited time available to the Review.

Two time periods are discussed. The first is up to 2030. Impacts over the next two decades can be considered to be locked in because of past and present greenhouse gas emissions. The magnitude of these impacts can only be tempered by our adaptation effort. The second is the period from 2030 to the close of the century. The magnitude of impacts on Australia in 2100 will be determined by international greenhouse gas mitigation and also by Australia's continued adaptation effort.

This chapter offers an illustrative selection rather than a complete assessment of the impacts that are likely to be experienced across Australia. It reflects the insights provided in a series of papers commissioned by the Review, which are available on the Review's website. In addition to the sectors and impacts discussed in this chapter, the papers cover livestock, horticulture, viticulture, forestry, Australia's World Heritage properties, tourism in the south-west of Western Australia, the Ross River and dengue viruses, ports, roads and telecommunications. These studies are an important part of the base from which the modelling of economic impacts on Australia has been developed.

The Review encourages readers to examine the commissioned studies. It has drawn on Australian experts in 30 fields of inquiry (listed in Table 6.1) to provide a wide-ranging collection of analyses of impacts.

Further details on potential impacts can be found in various synthesis reports (CSIRO & BoM 2007; IPCC 2007; PMSEIC 2007; Pittock 2003; Preston & Jones 2006).

Table 6.1 Sectors and areas considered in this chapter

Sector or area	Discussed in this chapter	Modelled by the Review
Resource-based industries and communities		
Subsector or area		
Dryland cropping	Yes – wheat	Yes
Irrigated cropping	Yes – in the Murray-Darling Basin	Yes – nationally
Livestock carrying capacity	No	Yes
Fisheries and aquaculture	No	No
Forestry	No	No
Mining	No	No
Horticulture	No	No
Viticulture and the wine industry	No	No
Australia's World Heritage properties	No	No
Alpine zone of south-east Australia	Yes	No
South-west Western Australia	No	No

Table 6.1 Sectors and areas considered in this chapter *(continued)*

Sector or area	Discussed in this chapter	Modelled by the Review
Great Barrier Reef	Yes	No
Critical infrastructure		
Subsector or area		
Buildings in coastal settlements	Yes	Yes
Urban water supply	Yes	Yes
Electricity transmission and distribution network	No	Yes
Port operations	No	Yes
Roads and bridges	No	No
Telecommunications	No	No
Cyclone impacts on dwellings	No	Yes
Human health		
Subsector or area		
Temperature-related death and serious illness	Yes – death	Yes
Ross River virus	No	No
Dengue virus	No	Yes
Bacterial gastroenteritis	No	Yes
Health of remote northern Australian Indigenous communities	No	No
Rural mental health	No	No
Ecosystems and biodiversity		
Subsector or area		
A range of ecosystems and impacts on plants and animals	Yes	No
Changes in demand and terms of trade		
	Yes	Yes
Geopolitical stability		
Subsector or area		
Geopolitical instability in the Asia–Pacific region and the subsequent aid and national security response from Australia	Yes	No
Catastrophic events as affect Australia		
	Yes	No
Severe weather events in Australia		
	Yes	No

To illustrate the impacts of climate change out to 2100, the Review considered a set of physically plausible climate outcomes for Australia, as shown in Table 6.2.

Table 6.2 Climate cases considered by the Review

Case	Emissions	Climate sensitivity	Rainfall and relative humidity (surface)	Temperature (surface)	Mean global warming in 2100
Unmitigated 1 Hot, dry	A1FI path	3°C	10th percentile	90th percentile	~4.5°C
Unmitigated 2 Best estimate (median)			50th percentile	50th percentile	
Unmitigated 3 Warm, wet			90th percentile		
550 mitigation Dry	CO_2-e stabilised at 550 ppm by 2100 (CO_2 500 ppm)		10th percentile	90th percentile	~2°C
550 mitigation 2 Best estimate (median)			50th percentile	50th percentile	
550 mitigation 3 Wet			90th percentile		
450 mitigation Best estimate (median)	CO_2-e stabilised at 450 ppm by 2100 (CO_2 420 ppm)		50th percentile		~1.5°C

Note: For each of the above cases global mean temperature is presented from a 1990 baseline. To convert to a pre-industrial baseline add 0.5°C.

6.1 Understanding Australia's vulnerability to climate change

The effect of climate change on the Australian population and natural assets will depend on *exposure* to changes in the climate system, *sensitivity* to those exposures and the *capacity to adapt* to the changes to which we are sensitive. These components of *vulnerability* to climate change are illustrated in Figure 6.1.

Australia's level of exposure and sensitivity to the impacts of climate change is high. The extent to which these impacts are realised will depend on the success and timing of global greenhouse gas mitigation and on national adaptation efforts.

As a nation, Australia has a high level of capacity to plan for and respond to the impacts of climate change—that is, its adaptation potential is high.

The consideration of impacts in this chapter assumes some adaptation at the level of an individual or firm.

Figure 6.1 Vulnerability and its components

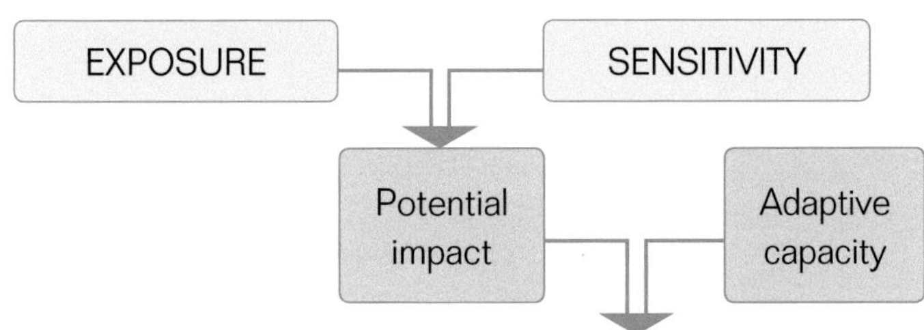

6.2 Australia without global mitigation

If global development continues without effective mitigation, the mainstream science tells us that the impacts of climate change on Australia are likely to be severe.

For the next two decades or so, the major impacts of climate change are likely to include stressed urban water supply and the effects of changes in temperature and water availability on agriculture. All major cities and many regional centres are already feeling the strain of declining rainfall and runoff into streams. Most major cities are beginning to develop high-cost infrastructure for new water sources. In the absence of effective global mitigation, continued investment in expensive new sources of water is likely to be a necessity.

By mid-century, there would be major declines in agricultural production across much of the country. Irrigated agriculture in the Murray-Darling Basin would be likely to lose half of its annual output. This would lead to changes in our capacity to export food and a growing reliance on food imports, with associated shifts from export parity to import parity pricing.

A no-mitigation case is likely also to see, by mid-century, the effective destruction of the Great Barrier Reef and other reef systems such as Ningaloo. The three-dimensional coral of the reefs is likely to disappear. This will have serious ramifications for marine biodiversity and the tourism and associated service industries reliant on the reefs.

By the close of the century, the impacts of a no-mitigation case can be expected to be profound (see Figure 6.2). The increased frequency of drought, combined with decreased median rainfall and a nearly complete absence of runoff in the Murray-Darling Basin, is likely to have ended irrigated agriculture for this region. Depopulation will be under way.

Much coastal infrastructure along the early 21st century lines of settlement is likely to be at high risk of damage from storms and flooding.

Key Australian export markets are projected to have significantly lower economic activity as a result of climate change. This is likely to feed back into significantly lower Australian export prices and terms of trade. As some states

Figure 6.2 State and territory impacts of climate change by 2100 under the no-mitigation case

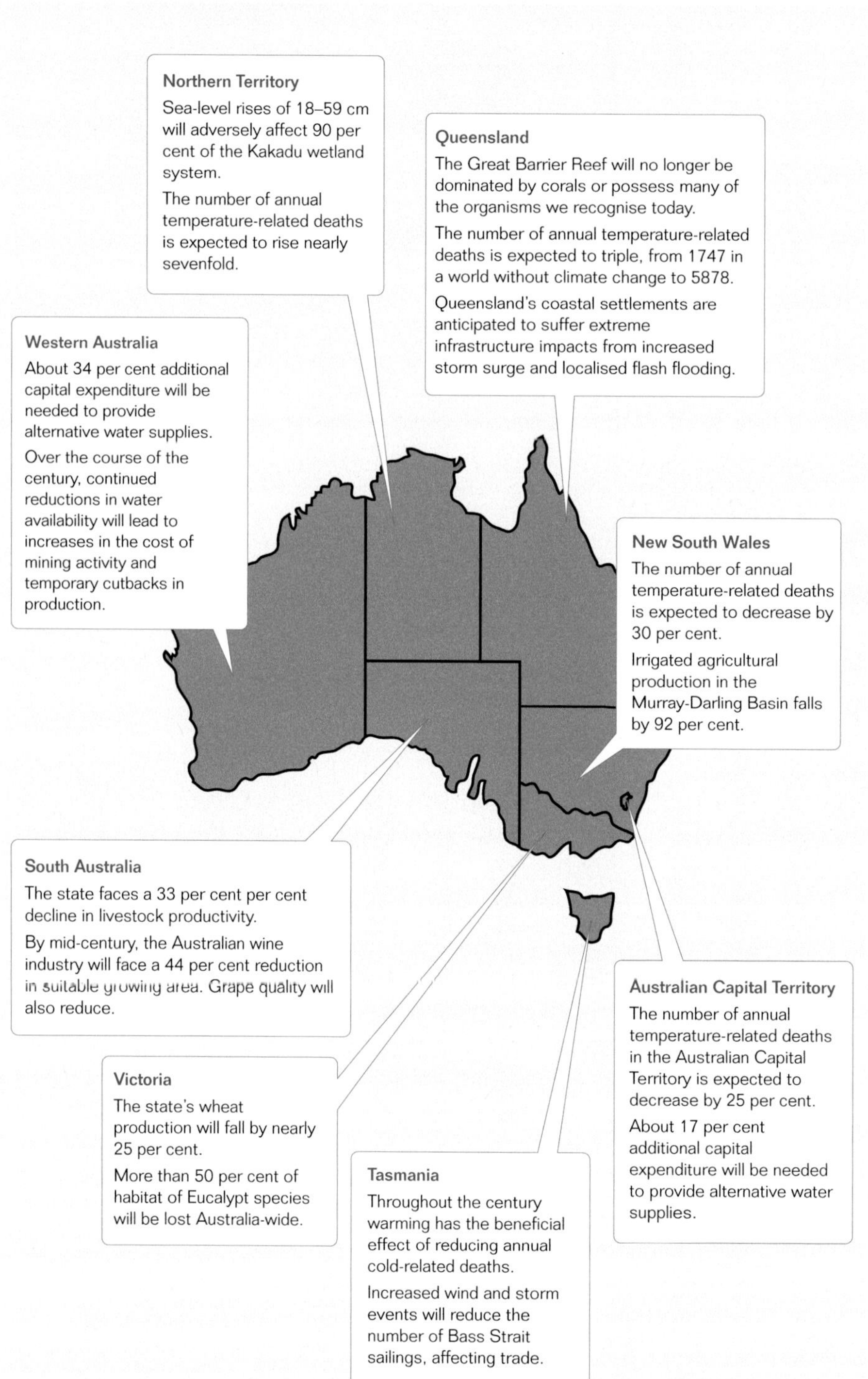
Northern Territory
Sea-level rises of 18–59 cm will adversely affect 90 per cent of the Kakadu wetland system.
The number of annual temperature-related deaths is expected to rise nearly sevenfold.
Queensland
The Great Barrier Reef will no longer be dominated by corals or possess many of the organisms we recognise today.
The number of annual temperature-related deaths is expected to triple, from 1747 in a world without climate change to 5878.
Queensland's coastal settlements are anticipated to suffer extreme infrastructure impacts from increased storm surge and localised flash flooding.
Western Australia
About 34 per cent additional capital expenditure will be needed to provide alternative water supplies.
Over the course of the century, continued reductions in water availability will lead to increases in the cost of mining activity and temporary cutbacks in production.
New South Wales
The number of annual temperature-related deaths is expected to decrease by 30 per cent.
Irrigated agricultural production in the Murray-Darling Basin falls by 92 per cent.
South Australia
The state faces a 33 per cent per cent decline in livestock productivity.
By mid-century, the Australian wine industry will face a 44 per cent reduction in suitable growing area. Grape quality will also reduce.
Victoria
The state's wheat production will fall by nearly 25 per cent.
More than 50 per cent of habitat of Eucalypt species will be lost Australia-wide.
Tasmania
Throughout the century warming has the beneficial effect of reducing annual cold-related deaths.
Increased wind and storm events will reduce the number of Bass Strait sailings, affecting trade.
Australian Capital Territory
The number of annual temperature-related deaths in the Australian Capital Territory is expected to decrease by 25 per cent.
About 17 per cent additional capital expenditure will be needed to provide alternative water supplies.

in our Asia–Pacific neighbourhood are further weakened by the effects of climate change, we can expect Australian defence personnel and police to be more heavily committed in support of peacekeeping operations.

Australians will be substantially wealthier in 2100 in terms of goods and services, despite setbacks from climate change. They are likely to be substantially poorer in terms of environmental amenity of various kinds. Australians over a century of change will have demonstrated the capacity to adapt in various ways. In some regions, retreat will have been the only viable strategy.

If the world were to have agreed and implemented global mitigation so that greenhouse gas concentrations were stabilised at 450 ppm or even 550 ppm CO_2-e, the impacts on Australia could be radically different. The differences are summarised in Table 6.3, again in terms of the median of the probability distributions emerging from the assessments of contemporary mainstream science. The difference between the median of the distributions at 550 ppm to 450 ppm CO_2-e is generally of material importance.

Table 6.3 Differences between probable unmitigated and mitigated futures at 2100—median of probability distributions

Sector	No mitigation	Mitigation	
		550 ppm CO_2-e	450 ppm CO_2-e
Irrigated agriculture in the Murray-Darling Basin	92% decline in irrigated agricultural production in the Basin, affecting dairy, fruit, vegetables, grains.	20% decline in irrigated agricultural production in the Basin.	6% decline in irrigated agricultural production in the Basin.
Natural resource–based tourism (Great Barrier Reef and Alpine areas)	Catastrophic destruction of the Great Barrier Reef. Reef no longer dominated by corals.	Disappearance of reef as we know it, with high impact to reef-based tourism. Three-dimensional structure of the corals largely gone and system dominated by fleshy seaweed and soft corals.	Mass bleaching of the coral reef twice as common as today.
	Snow-based tourism in Australia is likely to have disappeared. Alpine flora and fauna highly vulnerable because of retreat of snowline.	Moderate increase in artificial snowmaking.	

Table 6.3 Differences between probable unmitigated and mitigated futures at 2100—median of probability distributions *(continued)*

Sector	No mitigation	Mitigation 550 ppm CO_2-e	450 ppm CO_2-e
Water supply infrastructure	Up to 34% increase in the cost of supplying urban water, due largely to extensive supplementation of urban water systems with alternative water sources.	Up to 5% increase in the cost of supplying urban water. Low-level supplementation with alternative water sources.	Up to 4% increase in the cost of supplying urban water. Low-level supplementation with alternative water sources.
Buildings in coastal settlements	Significant risk to coastal buildings from storm events and sea-level rise, leading to localised coastal and flash flooding and extreme wind damage.	Significantly less storm energy in the climate system and in turn reduced risk to coastal buildings from storm damage.	Substantially less storm energy in the climate system and in turn greatly reduced risk to coastal buildings from storm damage.
Temperature-related death	Over 4000 additional heat-related deaths in Queensland each year. A 'bad-end story' (10% chance) would lead to more than 9500 additional heat-related deaths in Queensland each year.	Fewer than 80 additional heat-related deaths in Queensland each year.	Fewer deaths in Queensland than at present because of slight warming leading to decline in cold-related deaths.
Geopolitical stability in the Asia–Pacific region	Sea-level rise beginning to cause major dislocation in coastal megacities of south Asia, south-east Asia and China and displacement of people in islands adjacent to Australia.	Substantially lower sea-level rise anticipated and in turn greatly reduced risk to low-lying populations. Displacement of people in small island countries of South Pacific.	

Note: The assessment of impacts in this table does not build in centrally coordinated adaptation. The median of the probability distribution is used for the scenarios considered.

6.3 Direct impacts of climate change on Australia

6.3.1 Resource-based industries and communities

Climate variability has long posed a challenge to Australian communities and industries that rely on access to or use of natural resources. This challenge is now compounded by risks of human-induced climate change. Australia's forestry, agriculture, mining, horticulture and natural resource–based tourism sectors are exploring the implications of climate change for their operations.

Agriculture

Climate change is likely to affect agricultural production through changes in water availability, water quality and temperatures. Crop production is likely to be affected directly by changes in average rainfall and temperatures, in distribution of rainfall during the year, and in rainfall variability. The productivity of livestock industries will be influenced by the changes in the quantity and quality of available pasture, as well as by the effects of temperature increases on livestock (Adams et al. 1999).

Some agricultural impacts are positive. Increases in carbon dioxide concentration will increase the rate of photosynthesis in some plants where there is adequate moisture to support it (Steffen & Canadell 2005). The positive impacts of carbon fertilisation are likely to be restricted by higher temperatures and lower rainfall, which are both expected to become more important through the 21st century. A 10 per cent reduction in rainfall would be likely to remove the carbon dioxide fertilisation benefit (Howden et al. 1999; Crimp et al. 2002). Higher concentrations of carbon dioxide could reduce crop quality, however, by lowering the content of protein and trace elements (European Environment Agency 2004).

Severe weather events, including bushfires and flooding, are likely to reduce agricultural production, through effects on crop yields and stock losses (Ecofys BV 2006). Changes in temperatures are also projected to alter the incidence and occurrence of pests and diseases.

Irrigated agriculture in the Murray-Darling Basin

Note: This section draws heavily on a paper commissioned by the Review on the impacts of climate change to irrigated agriculture in the Murray-Darling Basin. See Quiggin et al. (2008), available at <www.garnautreview.org.au>.

The Murray-Darling Basin covers over one million km^2 of south-eastern Australia. Water flows from inside the Great Dividing Range from Queensland, New South Wales, the Australian Capital Territory and Victoria, eventually draining into the Southern Ocean off South Australia.

The Basin produces more than 40 per cent of Australia's total gross value of agricultural production, uses over 75 per cent of the total irrigated land in Australia, and consumes 70 per cent of Australia's irrigation water (ABS 2007).

The Review considered the impacts of climate change on several irrigated production groups: beef and sheep products, dairy, other livestock, broad-acre (cotton, rice and other grains), and other agriculture (grapes, stone fruit and vegetables).

A crucial feature of the analysis is that inflows to river systems vary much more than precipitation, and particularly rainfall. Inflows are a residual variable, consisting of water flows that are not lost to evapo-transpiration or absorbed by the soil. Quite modest changes in precipitation and evaporation reduce inflows substantially. For the Review's modelling, the reductions in runoff were capped at 84 per cent based on advice from R. Jones (2008, pers. comm.), reflecting the likelihood that there would continue to be occasional opportunities for irrigated cropping in the north of the Basin.

In aggregating our findings across all production groups, the Review found big differences between the implications of a no-mitigation case and one of global mitigation. The differences between runoff levels and consequently economic activity within the Basin have large implications for the viability of many aspects of life in the Murray-Darling. The change in economic value of production in the Murray-Darling Basin from a world with no human-induced climate change through to 2100 is presented in Table 6.4.

Table 6.4 Decline in value of irrigated agricultural production in the Murray-Darling Basin out to 2100 from a world with no human-induced climate change

	No-mitigation case	Global mitigation with CO_2-e stabilisation at 550 ppm by 2100	Global mitigation with CO_2-e stabilisation at 450 ppm by 2100	Hot, dry extreme case (the 'bad-end story')
Year	Decline in economic value of production (%)			
2030	12	3	3	44
2050	49	6	6	72
2100	92	20	6	97

Note: Moving from left to right, the first three cases are best-estimate cases and use the 50th percentile rainfall and relative humidity, and 50th percentile temperature for Australia (see Table 6.2 for a description of each case). The fourth case is an illustrative 'bad-end story' that uses the 10th percentile rainfall and relative humidity and 90th percentile temperature for Australia (a hot, dry extreme).

In an unmitigated case, irrigation will continue in the Basin for some time. Later in the century, decreasing runoff and increased variation in runoff are likely to limit the Basin's ability to recharge storages. By 2030, economic production falls by 12 per cent. By 2050 this loss increases to 49 per cent and, by 2100, 92 per cent has been lost due to climate change. Beyond 2050, fundamental restructuring of the irrigated agriculture industry will be required.

If the world were to achieve greenhouse gas concentrations of 450 ppm CO_2-e by 2100, it is likely that producers would be able to adjust their production systems with technological improvement to adapt with little cost to overall economic output from the Basin (such adaptation efforts have not been modelled). By 2030, economic production falls by 3 per cent. By 2050, this loss increases to 6 per cent. By 2100, 20 per cent has been lost.

While the differences between economic output in the 450 and 550 ppm CO_2-e mitigation cases are not substantial until the end of the century, the additional considerations of environmental flows and water quality in the Basin create a presumption that there is greater value in higher degrees of global mitigation.

In the 10th percentile hot, dry case, in the absence of mitigation, by 2050 the rivers in the Basin would be barely flowing. This would be well outside the range of natural variation observed in the historical record. By 2070 all except one catchment would be operating on the maximum possible reduction on which the model has been allowed to run (84 per cent decline in runoff from baseline). By 2030 economic production falls by 44 per cent. By 2050, 72 per cent of production has been lost. By 2100, 97 per cent has been lost. Only opportunistic upstream production might persist in 2100.

Box 6.1 Is there potential for a positive irrigation story? The possibility of a wetter Murray-Darling Basin

Mainstream science says that there is a 10 per cent chance of Australia becoming wetter under a no-mitigation case. This would be associated with a significant increase in rainfall in the northern part of the Murray-Darling Basin by 2050 (that is, a 20–30 per cent increase). The increased water supply flowing south would support higher levels of economic activity.

Under a warm, wet no-mitigation case, the average value of irrigated agricultural production in the Murray-Darling Basin would be less than 1 per cent greater than with no human-induced climate change.

Dryland cropping: wheat

Note: This section draws heavily on a paper commissioned by the Review on the impacts of climate change to Australia's wheat industry. See Crimp et al. (2008), available at <www.garnautreview.org.au>.

Wheat is the major crop in Australia in terms of value ($5.2 billion in 2005–06), volume (25 Mt in 2005–06) and area (12.5 Mha in 2005–06) (ABARE 2008). On average, over the 10 years to 2005–06, about 80 per cent of the Australian wheat harvest has been exported, worth on average about $3.2 billion a year (ABARE 2008). Yields are generally low due to low rainfall, high evaporative demand and low soil fertility. Thus the Australian wheat industry is highly sensitive to climatic influences. Average crop yields can vary by as much as 60 per cent in response to climate variability (Howden & Crimp 2005).

Both quantity and quality (protein content) determine the wheat crop's value. A range of studies indicate that grain protein contents are likely to fall in response to combined climate and carbon dioxide changes. There could be substantial protein losses (Howden et al. 2001), which would lower prices unless increased fertiliser application or more frequent pasture rotations were incorporated to reduce the effect. Increases in heat shock also may reduce grain quality by affecting dough-making qualities (Crimp et al. 2008).

The Review considered 10 study sites to understand the difference in magnitude of impacts on wheat yield between a no-mitigation case and one of global mitigation. As Table 6.5 shows, there are markedly different yield impacts between regions and also between the no-mitigation case and the 550 and 450 global mitigation cases.

Table 6.5 Percentage cumulative yield change from 1990 for Australian wheat under four climate cases

	No-mitigation case		Global mitigation with CO_2-e stabilisation at 550 ppm by 2100		Global mitigation with CO_2-e stabilisation at 450 ppm by 2100		Hot, dry extreme case (the 'bad-end story')	
	Cumulative yield change (%)							
	2030	2100	2030	2100	2030	2100	2030	2100
Dalby, Qld	8.2	-18.5	4.8	-1.0	1.6	-3.7	-6.6	-100.0
Emerald, Qld	7.2	-10.1	4.4	0.0	1.8	-2.5	-7.6	-100.0
Coolamon, NSW	11.6	1.9	9.9	12.3	8.2	7.4	1.2	-100.0
Dubbo, NSW	8.1	-5.9	6.1	6.7	4.0	2.3	-2.4	-100.0
Geraldton, WA	12.5	22.4	9.7	5.6	6.9	2.6	9.5	-16.9
Birchip, Vic.	14.8	-24.1	10.7	1.5	6.8	-0.3	-0.7	100.0
Katanning, WA	15.6	16.8	14.8	18.9	13.9	14.6	-15.7	-18.7
Minnipa, SA	0.8	-23.9	-3.4	-15.3	-7.4	-15.7	-13.8	-82.0
Moree, NSW	20.6	10.9	17.7	14.1	14.8	10.8	6.4	-79.2
Wongan Hills, WA	16.1	-21.8	13.0	5.5	10.0	4.4	5.5	-100.0

Note: Moving from left to right, the first three cases are best-estimate cases and use the 50th percentile rainfall and relative humidity, and 50th percentile temperature for Australia (see Table 6.2 for a description of each case). The fourth case is an illustrative 'bad-end story' that uses the 10th percentile rainfall and relative humidity and 90th percentile temperature for Australia (a hot, dry extreme).

Under the no-mitigation case and through adaptive management, much of Australia could experience an increase in wheat production by 2030. This would involve moving planting times in response to warming and selection of optimal production cultivars. Increases would also result from higher carbon dioxide concentrations. Over time, even with adaptive management, a number of regions would experience substantial declines in wheat yield.

Later in this century, the benefits of carbon dioxide fertilisation and adaptive management are likely to have been negated by increasing temperatures and declining available water.

In some sites in Western Australia (Geraldton and Katanning), the rainfall changes with no mitigation would improve yields. This unexpected result arises because subsoil constraints to growth, for example salinity, respond to declines in rainfall. This beneficial impact is only associated with modest rainfall declines (that is, less than 30 per cent of the long-term annual mean). Yields are negatively affected with larger declines.

Under the global mitigation cases, the carbon dioxide fertilisation effect is less marked and early yield increases are lower than for the no-mitigation case. However, for those regions that under the no-mitigation case were facing large declines in yield, this impact is reduced substantially with global mitigation by 2100.

The hot, dry extreme case has devastating consequences for the Australian wheat industry, leading to complete abandonment of production for most regions.

The cases above use *average change* in temperature and rainfall, and therefore available runoff, as the key variables. The approach implicitly assumes that there is *no change* in either the frequency or intensity of El Niño and La Niña events with climate change. There is concern among mainstream scientists, however, that the frequency of El Niño events may increase, thus changing the proportion of good and bad years. This would cause the net impact on wheat to be different from that of the average change in rainfall (Crimp et al. 2008). A world of climate change would be associated with less predictability and greater variability of rainfall, generating large problems for management of farm systems to make optimal use of available water resources.

Natural resource-based tourism

Note: This section draws heavily on a paper commissioned by the Review on the impacts of climate change to the Great Barrier Reef and associated tourism industries. See Hoegh-Guldberg and Hoegh-Guldberg (2008), available at <www.garnautreview.org.au>.

The Australian tourism industry generated value added of $37.6 billion per annum or 3.7 per cent of GDP for 2006–07 (ABS 2008a). International tourism generated $21 billion of export income in 2005–06, or 10.5 per cent of total exports (ABS 2008b).

In 2006–07, 482 000 people were employed in the tourism industry, or 4.7 per cent of total employment (ABS 2008a). Tourism is often the major non-agricultural source of livelihoods in rural and regional areas, and is the major industry in some regions.

Australia's natural landscapes are important to the Australian tourism industry. The Great Barrier Reef and rainforests of tropical north Queensland, Kakadu, the deserts of Central Australia, the Ningaloo Reef and coastal environs of south-west Western Australia, and the alpine regions of New South Wales, Victoria and Tasmania (see Box 6.2) are leading examples of tourist attractions defined by features of the natural environment. Many are World Heritage areas, which Australia has international legal obligations to protect.

Each of these attractions, and many that are less well known, would be significantly affected in a future of unmitigated climate change. Climate change would lead to loss of attractions; loss of quality of attractions; costs of adaptation; increased cost for repair, maintenance and replacement of tourism infrastructure; and increased cost for developing alternative attractions (Sustainable Tourism Cooperative Research Centre 2007).

Some tourist activities may benefit from drier and warmer conditions—beach-based activities, viewing of wildlife, trekking, camping, climbing and fishing outside the hottest times of the year. However, even in these cases, greater risks to tourism are likely from increases in hazards such as flooding, storm surges, heatwaves, cyclones, fires and droughts.

In a study by Hoegh-Guldberg (2008), 77 Australian tourism regions were assessed for prospective risk of climate change. The following three were identified as the most threatened:

- **Tropical north Queensland**, the hub of Great Barrier Reef tourism, contains the coral reef complex, severely threatened rainforest areas, beaches in danger of inundation and increasing storm damage. There are threats to tourism from increased incidence of bushfires and increased ultraviolet radiation. The threats to the region are exacerbated by a high reliance on international holiday tourism, which could be relatively easily diverted elsewhere.
- **South-west Western Australia** is the scene of Australia's only internationally recognised biodiversity hotspot, one of 34 in the world. It has high risk ratings based on the greatest diversity of vulnerable native flora, a vulnerable wine industry and, together with the Murray-Darling Basin, the greatest salinity problem in the country. It attracts a large number of holiday tourists, but few are international visitors.
- In the **Top End of the Northern Territory**, national parks and wetlands are at risk, and tourism is threatened by increased ultraviolet radiation and increased exposure to disease.

Australia is likely to be greatly diminished as an international tourist destination by climate change.

Domestically, the loss of tourism income from one region, such as the Great Barrier Reef, does not necessarily equate with overall loss of tourism income for Australia. Some of the tourism expenditure will be diverted to other Australian regions.

Box 6.2 Alpine tourism in south-east Australia

The alpine resorts in Australia are located in areas of great environmental sensitivity and are at severe risk from climate change. The total alpine environment in Australia is small: approximately 0.2 per cent of the total land mass, with alpine areas restricted to New South Wales, Victoria and Tasmania.

The alpine resorts generate 2 per cent of total Australian tourist activity (National Institute of Economic and Industry Research 2006). The industry is characterised by many small businesses, a large proportion of which only operate during the snow sports season, a period of around four months.

In the alpine regions of south-east Australia, natural snow conditions over the past 35 years have been in slow but steady decline, with increased maximum and minimum temperatures across many locations (Hennessy et al. 2003). This has created greater reliance by the ski industry on the production of artificial snow to service tourism demand (for snow depths and season length). There have also been implications for sensitive alpine flora and fauna due to changing snow conditions.

Box 6.2 Alpine tourism in south-east Australia *(continued)*

The no-mitigation case would see the average snow season contract by between 85 to 96 per cent by 2050 (Hennessy et al. 2003) and disappear before the end of the century. Conversely, if the international community were to limit concentrations to 450 ppm CO_2-e by 2100, snow depths and coverage would fall only marginally. In this latter case, it is likely that alpine resorts could continue their current operations with minimal technological adaptation. Stabilisation at 550 ppm CO_2-e by 2100 would be likely to result in maintenance of snow depth and coverage at higher elevations. However, the alpine areas located at lower elevations would experience a loss of snow coverage as the snowline moved to higher ground.

As many as a third of all visitors to the alpine region visit outside the traditional snow season, to enjoy the unique flora and fauna, as well as recreational activities such as hiking, camping and fishing. However, summer recreational activities are also at increasing risk from bushfires and storm and wind events.

6.3.2 Critical infrastructure

Climate change will have wide-ranging and significant impacts on the infrastructure critical to the operation of settlements and industry across Australia. This will occur through changes in the average climate and changes in the frequency and intensity of severe weather events.

Buildings and infrastructure being constructed will have functional lives of many decades. An understanding of the anticipated impacts from climate change over the course of the century is helpful to inform construction decisions being made now, and to avoid increased operation and maintenance costs in future, or the early retirement of infrastructure.

This section presents the impacts of climate change on two key forms of infrastructure:

- water supply infrastructure in major cities
- buildings in coastal settlements.

The Review offers a broad commentary on the magnitude of impacts in a no-mitigation case compared to a future with global mitigation.

Water supply infrastructure in major cities

Note: This section draws heavily on a paper commissioned by the Review on the impacts of climate change on urban water supply in major cities. See Maunsell (2008), available at <www.garnautreview.org.au>.

Nearly all major Australian cities are already experiencing the effects of reductions in rainfall on water supplies. All capital cities except Darwin and Hobart are now relying on severe restrictions on water use. Some regional cities are facing sharply diminished supply and extreme restrictions (Marsden Jacob Associates 2006).

Under a no-mitigation case, with outcomes near the median of the probability distributions generated by mainstream Australian science, most major population centres across the country will be required to supplement their water supply system with substantial new water sources through the 21st century. As shown in Table 6.6, there will be differing impacts across the states and territories, with Western Australia and South Australia the most severely affected by climate change. The development of new water sources for Perth and Adelaide is required now.

In a case of global mitigation, the reduced level of temperature increase relative to business as usual lessens the changes in rainfall and evaporative demand, and therefore places less stress on water supply. However, a low level of supplementation would still be required across most major centres.

Table 6.6 Magnitude of impacts to water supply infrastructure in major cities under four climate cases

	No-mitigation case		Global mitigation with CO_2-e stabilisation at 550 ppm by 2100		Global mitigation with CO_2-e stabilisation at 450 ppm by 2100		Hot, dry extreme case (the 'bad-end story')	
Region	2030	2100	2030	2100	2030	2100	2030	2100
ACT	M	E	M	L	M	L	H	E
NSW	H	E	H	L	H	L	H	E
NT	L	H	L	N	N	N	L	E
Qld	H	E	H	L	H	L	H	E
SA	E	E	E	M	E	M	E	E
Tas.	N	M	N	N	N	N	N	E
Vic.	H	E	H	M	H	L	H	E
WA	E	E	E	M	E	M	E	E

Magnitude of net impact									
N	Neutral	L	Low	M	Moderate	H	High	E	Extreme

Note: Moving from left to right, the first three cases are best-estimate cases and use the 50th percentile rainfall and relative humidity, and 50th percentile temperature for Australia (see Table 6.2 for description of each case). The fourth case is an illustrative 'bad-end story' that uses the 10th percentile rainfall and relative humidity and 90th percentile temperature for Australia (a hot, dry extreme).

A description of each level of impact is provided in Box 6.3.

Box 6.3 Infrastructure impacts criteria

The Review's assessment of impacts of climate change on infrastructure is based on determination of net impact to capital expenditure, operational expenditure and productivity. The criteria for this assessment are presented in Table 6.7.

Table 6.7 Infrastructure impacts criteria

	Magnitude of net impact	Description of impact
N	Neutral	No change in capital expenditure, operational expenditure or productivity.
L	Low	Minor increase in capital expenditure and operational costs but no significant change to cost structure of industry. Minor loss in productivity. For example, minor loss in port productivity due to increased downtime of port operations.
M	Moderate	Moderate increase in capital and operational expenditure with a minor change to cost structure of industry. Moderate loss in productivity. For example, moderate increase in capital and operational expenditure for electricity transmission and distribution due to increased design standards, maintenance regimes and damage from severe weather events.
H	High	Major increase in capital and operational expenditure with a significant change to cost structure of industry. Major loss of productivity. For example, major increase in capital expenditure for increased building design standards for new and existing residential and commercial buildings.
E	Extreme	Extreme increase and major change to cost structure of industry with an extreme increase in operational and maintenance expenditure. Extreme loss of productivity. For example, extreme increase in capital expenditure from significant investment in new water supply infrastructure.

Buildings in coastal settlements

Note: This section draws heavily on a paper commissioned by the Review on the impacts of climate change to buildings in coastal settlements. See Maunsell (2008), available at <www.garnautreview.org.au>.

More than 80 per cent of the Australian population lives within 50 km of the coastline. For most of the past century, coastal regions have experienced significant growth and are projected to continue to show the most rapid population growth (IPCC 2007). Without mitigation, the impacts of climate change on these regions are likely to be substantial (Table 6.8).

The increased magnitude of storm events and sea-level rise under a no-mitigation case is likely to exert significant pressure on coastal infrastructure through storm damage and localised flash flooding. This would cause immediate damage to assets, particularly building contents, and accelerate the degradation of buildings.

Changes in temperature, and extreme rainfall and wind, may also accelerate degradation of materials, structures and foundations of buildings, thereby reducing the life expectancy of buildings and increasing their maintenance costs. Low soil moisture before severe rainfall events would increase the impact and magnitude of flooding. In between flooding episodes, the low levels of soil moisture would lead to increased ground movement and cause building foundations to degrade.

In the medium term (2030 to 2070) the cost of climate change for coastal settlements would mainly arise from repair and increased maintenance, clean-up and emergency response. Later in the century, costs for preventive activity are likely to be higher. There will be large costs associated with altered building design, sea-wall protection and higher capital expenditure for improved drainage.

Table 6.8 Magnitude of impacts on buildings in coastal settlements under four climate cases

	No-mitigation case		Global mitigation with CO_2-e stabilisation at 550 ppm by 2100		Global mitigation with CO_2-e stabilisation at 450 ppm by 2100		Hot, dry extreme case (the 'bad-end story')	
Region	2030	2100	2030	2100	2030	2100	2030	2100
NSW	M	H	M	M	M	M	M	E
NT	L	M	L	M	L	L	L	H
Qld	M	E	M	M	M	M	M	E
SA	L	H	L	M	L	L	L	H
Tas.	L	M	L	M	L	N	L	M
Vic.	M	H	M	M	M	L	M	H
WA	L	M	L	M	L	L	L	H

Magnitude of net impact				
N Neutral	L Low	M Moderate	H High	E Extreme

Note: Moving from left to right, the first three cases are best-estimate cases and use the 50th percentile rainfall and relative humidity, and 50th percentile temperature for Australia (see Table 6.2 for description of each case). The fourth case is an illustrative 'bad-end story' that uses the 10th percentile rainfall and relative humidity and 90th percentile temperature for Australia (a hot, dry extreme).

A description of each level of impact is provided in Box 6.3.

Changes to building design are expected to improve the resilience of buildings in the latter part of the century as stock is renewed or replaced. However, even with improved design and use of materials, the magnitude of climate change leading up to 2100 under a no-mitigation case is expected to generate high impacts.

In a future with global mitigation, the reduced level of temperature increase would lessen the magnitude of temperature-driven storm energy in the Australian climate system. This would greatly reduce impacts from storm surge, severe rainfall and flash flooding. As shown in Table 6.8, overall impacts to buildings in coastal settlements would be substantially lower under the global mitigation cases.

6.3.3 Human health

Note: This section draws heavily a paper commissioned by the Review on the impacts of climate change to human health. See Bambrick et al. (2008), available at <www.garnautreview.org.au>.

Climate change is likely to affect the health of Australians over this century in many ways. Some impacts, such as heatwaves, would operate directly. Others would occur indirectly through disturbances of natural ecological systems, such as mosquito population range and activity.

Most health impacts will impinge unevenly across regions, communities and demographic subgroups, reflecting differences in location, socio-economic circumstances, preparedness, infrastructure and institutional resources, and local preventive (or adaptive) strategies. The adverse health impacts of climate change will be greatest among people on lower incomes, the elderly and the sick. People who lack access to good and well-equipped housing will be at a disadvantage.

The main health risks in Australia include:

- impacts of severe weather events (floods, storms, cyclones, bushfires)
- impacts of temperature extremes, including heatwaves
- vector-borne infectious diseases (for example, dengue virus and Ross River virus)
- food-borne infectious diseases (including those due to *Salmonella* and *Campylobacter*)
- water-borne infectious diseases and health risks from poor water quality
- diminished food production and higher prices, with nutritional consequences
- increases in air pollution (for example, from bushfire smoke)
- changes in production of aeroallergens (spores, pollens), potentially exacerbating asthma and other allergic respiratory diseases
- mental health consequences and the emotional cost of social, economic and demographic dislocation (for example, in parts of rural Australia, and through disruptions to traditional ways of living in remote Indigenous communities).

Temperature-related death

Exposure to prolonged ambient heat promotes various physiological changes, including cramping, heart attack and stroke. People most likely to be affected are those with chronic disease (such as cardiovascular disease or type 2 diabetes). These tend to be older people.

The effects of climate change on temperature-related mortality and morbidity are highly variable over place and time. Temperature-related deaths and hospitalisations may fall at some places and times (due to fewer cold-related deaths) in some parts of Australia, but increase in others. Table 6.9 illustrates the change in the number of temperature-related deaths in Australia over time under four different climate change cases. In Australia as a whole and across all cases, small declines in total annual temperature-related deaths are expected in the first half of the century due to decreased cold-related sickness and death. The winter peak in deaths is likely to be overtaken by heat-related deaths in nearly all cities by mid-century (McMichael et al. 2003).

Table 6.9 Change in likely temperature-related deaths due to climate change

	Baseline – a world with no human-induced climate change		No-mitigation case		Global mitigation with CO_2-e stabilisation at 550 ppm by 2100		Global mitigation with CO_2-e stabilisation at 450 ppm by 2100		Hot, dry extreme case (the 'bad-end story')	
	Number of temperature-related deaths									
Region	2030	2100	2030	2100	2030	2100	2030	2100	2030	2100
ACT	300	333	280	250	278	285	276	295	275	262
NSW	2 552	2 754	2 316	1 906	2 290	2 224	2 268	2 334	2 255	2 040
NT	63	61	63	407	63	93	64	76	64	768
Qld	1 399	1 747	1 276	5 878	1 274	1 825	1 278	1 664	1 286	11 322
SA	806	811	770	704	766	735	762	750	758	740
Tas.	390	375	360	240	357	313	354	327	352	211
Vic.	1 788	1 966	1 632	1 164	1 614	1 586	1 599	1 673	1 589	1 012
WA	419	515	418	685	419	529	419	519	420	835
Australia	7 717	8 562	7 155	11 234	7 061	7 590	7 020	7 638	6 999	17 190

Note: Moving from left to right, in the baseline case any increase in number of deaths shown is due to the expanding and ageing of the population. The next three cases are best-estimate cases and use the 50th percentile rainfall and relative humidity, and 50th percentile temperature for Australia (see Table 6.2 for a description of each case). The final case (right-hand side) is an illustrative 'bad-end story' that uses the 10th percentile rainfall and relative humidity and 90th percentile temperature for Australia (a hot, dry extreme).

For the no-mitigation case there is a large national increase in temperature-related deaths in the second half of the century. Much of the increase is attributable to expected deaths in Queensland and the Northern Territory. The large increases in deaths between 2030 and 2100 are avoided under the global mitigation cases.

The hot, dry extreme case would lead to twice as many temperature-related deaths annually when compared with no climate change. In Victoria, Tasmania and New South Wales, even under the hot, dry extreme case, temperature-related deaths are reduced relative to no climate change because those populations are more susceptible to cold than to heat (K. Dear 2008, pers. comm.).

6.3.4 Ecosystems and biodiversity

Note: This section draws heavily on a paper commissioned by the Review on the impacts of climate change to ecosystems and biodiversity. See Australian Centre for Biodiversity (2008), available at <www.garnautreview.org.au>.

Natural biological systems in Australia have been dramatically altered by human actions. The added stressors from climate change would exacerbate existing environmental problems, such as widespread loss of native vegetation, over-harvesting of water and reduction of water quality, isolation of habitats and ecosystems, and the influence of introduced plant and animal pests.

Some species can tolerate the changes where they are or adapt to change. Other species will move to more suitable habitat if possible. Some species may dwindle in numbers in situ, threatening their viability as a species and ultimately leading to extinction.

For biological systems, climate change will affect:

- physiology (individual organisms)
- timing of life cycles (phenology)
- population processes, such as birth and death rates
- shifts and changes in distribution (dispersal and shifts in geographic range)
- potential for adaptation (rapid evolutionary change).

These effects on individual organisms and populations cascade into changes in interactions among species. Changes in interactions further heighten extinction rates and shifts in geographic range. The ultimate outcomes are expected to be declines in biodiversity favouring weed and pest species (a few native, most introduced) at the expense of the rich variety that has developed naturally across Australia.

Many plant and animal species depend on the wide dispersal of individuals for both demographic processes and interchange of genes to avoid inbreeding effects. Over large areas and long periods, many species will respond (and have already responded) to climate change by moving, resulting in geographic range shifts. However, some species will not be able to migrate or adapt to climate change because they lack a suitable habitat into which to move, have limited or impeded mobility or do not possess sufficient and necessary genetic diversity to adapt. For these species, their geographic ranges would contract, heightening the risk of extinction.

Australia's high-altitude species are at risk. These species are already at their range limits due to the low relief of Australia's mountains, and lack suitable habitat to which to migrate. For example, a 1°C temperature rise, anticipated in about 2030 for south-eastern Australia under all cases, will eliminate 100 per cent of the habitat of the mountain pygmy possum (*Burramys parvus*). This species cannot move to higher mountains because there are no such mountains, and will not be able to stay where it is because it does not have the capacity to adapt to warmer temperatures. The potential for extinction is high.

The wet tropics of far north Queensland are also likely to face high levels of extinction. It is estimated that a 1°C rise in temperature, anticipated before 2030 under all four cases, could result in a 50 per cent decrease in the area of highland rainforests (Hilbert et al. 2001). A 2°C rise in average temperatures (anticipated by about 2050 for the no-mitigation case, 2070 for 550 ppm CO_2-e, and after 2100 for 450 ppm CO_2-e) would force all endemic Australian tropical rainforest vertebrates to extinction (Australian Centre for Biodiversity 2008).

Sea-level rise would have implications for coastal freshwater wetlands that may become inundated and saline. A well-documented example is the World Heritage and Ramsar Convention–recognised wetlands of Kakadu National Park in the Northern Territory. The wetland system at Kakadu depends on a finely balanced interaction between freshwater and marine environments. In places, the natural levees that act as a barrier between Kakadu's freshwater and saltwater systems are only 20 cm high. Sea-level rises of another 59 cm (see Chapter 4) by 2100 would adversely affect 90 per cent of the Kakadu wetland system. The area supports more than 60 species of water birds, which congregate around freshwater pools in the wetlands. The coastal wetlands are important nursery areas for barramundi, prawns and mud crabs, and are important breeding habitats for crocodiles, turtles, crayfish, water snakes and frogs. Fundamental changes in the ecological function of the national park will place severe pressure on many species of plants and animals.

Increased warming of Australia's oceans has pushed coral reefs above their thermal tolerance. This has resulted in episodes of mass coral bleaching (see Box 6.4).

Box 6.4 Climate change and the Great Barrier Reef

The Great Barrier Reef is the world's most spectacular coral reef ecosystem. Lining almost 2100 kilometres of the Australian coastline, the Reef is the largest continuous coral reef ecosystem in the world. It is home to a wide variety of marine organisms including six species of marine turtles, 24 species of seabirds, more than 30 species of marine mammals, 350 coral species, 4000 species of molluscs and 1500 fish species. The total number of species is in the hundreds of thousands. New species are described each year, and some estimates suggest that we may be familiar with less than 50 per cent of the total number of species that live within this ecosystem.

In addition to housing a significant part of the ocean's biodiversity, coral reefs provide a barrier that protects mangrove and sea grass ecosystems, which in turn provide habitat for a large number of fish species. This protection is also important to the human infrastructure that lines the coast.

The Great Barrier Reef is threatened by increased nutrients and sediments from land-based agriculture, coastal degradation, pollution and fishing pressure. Climate change is an additional and significant stressor.

The IPCC recognises coral reefs globally as highly threatened by rapid human-induced climate change (IPCC 2007). The Great Barrier Reef waters are 0.4°C warmer than they were 30 years ago (Lough 2007). Increasing atmospheric carbon dioxide has also resulted in 0.1 pH decrease (that is, the ocean has become more acidic).

These changes have already had major impacts. Short periods of warm sea temperature have pushed corals and the organisms that support their development above their thermal tolerance. This has resulted in episodes of mass coral bleaching that have increased in frequency and intensity since they were first reported in the scientific literature in 1979 (see Brown 1997; Hoegh-Guldberg 1999; Hoegh-Guldberg et al. 2007).

The Great Barrier Reef has been affected by coral bleaching as a result of heat stress six times over the past 25 years. Recent episodes have been the most intense and widespread. In the most severe episode to date, in 2002, more than 60 per cent of the reefs within the Great Barrier Reef Marine Park were affected by coral bleaching, with 5–10 per cent of the affected corals dying.

Consideration has recently been given to how reef systems will change in response to changes in atmospheric greenhouse gas composition. If atmospheric carbon dioxide levels stabilise at 420 ppm and the sea temperatures of the Great Barrier Reef increase by 0.55°C, mass bleaching events will be twice as common as they are at present.

Box 6.4 Climate change and the Great Barrier Reef *(continued)*

If atmospheric carbon dioxide concentrations increase beyond 450 ppm, together with a global temperature rise of 1°C, a major decline in reef-building corals is expected. Under these conditions, reef-building corals would be unable to keep pace with the rate of physical and biological erosion, and coral reefs would slowly shift towards non-carbonate reef ecosystems. Reef ecosystems at this point would resemble a mixed assemblage of fleshy seaweed, soft corals and other non-calcifying organisms, with reef-building corals being much less abundant, even rare. As a result, the three-dimensional structure of coral reefs would slowly crumble and disappear.

Depending on the influence of other factors such as the intensity of storms, this process may happen either slowly or rapidly. Significantly, this has happened relatively quickly (over an estimated 30 to 50 years) on some inshore Great Barrier Reef sites.

A carbon dioxide concentration of 500 ppm or beyond, and likely associated temperature change, would be catastrophic for the majority of coral reefs across the planet. Under these conditions the three-dimensional structure of the Great Barrier Reef would be expected to deteriorate and would no longer be dominated by corals or many of the organisms that we recognise today. This would have serious ramifications for marine biodiversity and ecological function, coastal protection and the tourism and associated service industries reliant on the reefs.

(Hoegh-Guldberg & Hoegh-Guldberg 2008)

The disruption of ecosystems, species populations and assemblages will also affect ecosystem services—the transformation of a set of natural assets (soil, plants, animals, air and water) into things that we value. These include clean air, clean water and fertile soil, all of which contribute directly to human health and wellbeing. The productivity of some of our natural resource–based industries, including agriculture and tourism, depends on them.

The vast majority of ecosystem services are far too complex to produce through engineering, even with the most advanced technologies. Their benefits are poorly understood but seem to be large. Human-induced environmental change has already disrupted ecosystem processes. Climate change will further degrade the services provided. The complex biotic machinery that provides ecosystem services is being disrupted and degraded. The consequences are impossible to predict accurately.

6.4 Indirect impacts of climate change on Australia

Australia will be affected indirectly by climate change as experienced by other countries.

6.4.1 International trade impacts for Australia

Climate change is likely to affect economic activity in other countries. It will therefore affect the supply of imports to Australia and demand for Australian exports and consequently Australia's terms of trade (the ratio of Australian export to import prices). The Review's modelling indicates that Australia's terms of trade are affected much more adversely than any other developed country by climate change.

The Review's modelling indicates that China, India, Indonesia and other Asian economies will be by far Australia's major export markets long before the end of the 21st century. These countries are expected to be relatively badly affected by climate change.

Climate change would be associated with a decline in international demand for Australia's mineral and energy resources and agricultural products.

The decline in Australia's terms of trade as a result of unmitigated climate change will be driven primarily by falls in the prices received for coal and other minerals. These and other commodities are projected to account for more than 60 per cent of the value of Australia's exports in 2100 in the absence of climate change impacts.[1]

6.4.2 Geopolitical stability in Asia and the Pacific region

Note: This section draws heavily on a paper commissioned by the Review on the impacts of climate change to Australia's security. See Dupont (2008), available at <www.garnautreview.org.au>.

Weather extremes and large fluctuations in rainfall and temperatures have the capacity to refashion Asia's productive landscape and exacerbate food, water and energy scarcities in Asia and the south-west Pacific. Australia's immediate neighbours are vulnerable developing countries with limited capacity to adapt to climate change.

Climate change outcomes such as displacement of human settlements by sea-level rise, reduced food production, water scarcity and increased disease, while immensely important in themselves, also have the potential to destabilise domestic and international political systems in parts of Asia and the south-west Pacific.

Should climate change coincide with other transnational challenges to security, such as terrorism or pandemic diseases, or add to pre-existing ethnic and social tensions, the impact will be magnified.

The problems of its neighbours can quickly become Australia's, as recent history attests. Over the past decade, Australia has intervened at large cost in Bougainville, Solomon Islands and Timor-Leste in response to political and humanitarian crises. Responding to the regional impacts of climate change will require cooperative regional solutions and Australian participation.

Food security

Climate change is likely to affect food production in the Asia–Pacific region for five main reasons:

- Increased temperatures could reduce crop yields by shortening growing seasons and accelerating grain sterility in crops.
- Marine ecosystems could experience major migratory changes in fish stocks and mortality events in response to rising temperatures. Fish is the primary source of protein for more than one billion people in Asia.
- Shifts in rainfall patterns could disrupt flows in rivers used for irrigation, accelerate erosion and desertification and reduce crop and livestock yields.
- Rising sea levels could inundate and make unusable fertile coastal land.
- An increase in the intensity or frequency of severe weather events could disrupt agriculture.

There is particular vulnerability associated with any disruption to the South Asian monsoon, drying of the northern China plain, and disruption to the flows of Asia's great rivers arising out of deglaciation of the Himalayas and the Tibetan plateau (see Box 6.5).

The Consultative Group on International Agricultural Research (2002) has predicted that food production in Asia will decrease by as much as 20 per cent due to climate change. These forecasts are in line with IPCC projections showing significant reductions in crop yield (5–30 per cent compared with 1990) affecting more than one billion people in Asia by 2050 (Parry et al. 2004, cited in IPCC 2007).

Poorer countries with predominantly rural economies and low levels of agricultural diversification will be at most risk. They have little flexibility to buffer shifts in food production. Higher worldwide food prices associated with climate change, its mitigation and other factors will diminish the opportunity to seek food security from international trade—compounding biophysical constraints on production and negatively affecting both rural and urban poor (Consultative Group on International Agricultural Research 2002).

In these circumstances, in the absence of international food trade liberalisation, it is likely that price volatility on world markets will increase, especially at times of pressure on global food supplies. Freer and more deeply integrated international markets for agricultural products would be a helpful adaptive response.

Box 6.5 The security challenge created by the melting of the Himalayan and Tibetan plateau glaciers

The melting of the Himalayan and Tibetan plateau glaciers illustrates the complex nexus of climate change, economic security and geopolitics.

Well over a billion people are dependent on the flow of the area's rivers for much of their food and water needs, as well as transportation and energy from hydroelectricity. Initially, flows may increase, as glacial runoff accelerates, causing extensive flooding. Within a few decades, however, water levels are expected to decline, jeopardising food production and causing widespread water and power shortages.

As water availability in China has decreased because of rising demand and diminishing freshwater reserves, China has increased its efforts to redirect waters from the Yangtze to water-deficient areas of northern China. Questions are now arising about the reliability of the Yangtze flows for these purposes. A wider problem is that rivers like the Mekong, Ganges, Brahmaputra and Salween flow through several states. China's efforts to rectify its own emerging water and energy problems indirectly threaten the livelihoods of millions of people in downstream, riparian states. Chinese dams on the Mekong are already reducing flows to Myanmar, Thailand, Laos, Cambodia and Vietnam. India is concerned about Chinese plans to channel the waters of the Brahmaputra to the over-used Yellow River. Should China go ahead with this ambitious plan, tensions with India and Bangladesh would almost certainly increase (Chellaney 2007).

Any disruption of flows in the Indus would be highly disruptive to Punjabi agriculture on both sides of the India–Pakistan border. It would raise difficult issues in India–Pakistan relations.

Any consequent conflicts between China and India, or India and Pakistan, or between other water-deficient regional states, could have serious implications for Australia, disrupting trade, displacing people and increasing strategic competition in Asia.
(Dupont 2008)

Infectious disease

Climate change can generate security risks through infectious disease. Temperature is the key factor in the spread of some infectious diseases, especially where mosquitoes are a vector as with Ross River virus, malaria and dengue virus. With warming, mosquitoes will move into previously inhospitable areas and higher altitudes, and disease transmission seasons may last longer. A study by the World Health Organization (2002) estimated that 154 000 deaths annually were already attributable to the ancillary effects of climate change due mainly to malaria and malnutrition. The study suggests that this number could nearly double by 2020.

Severe weather events

Severe weather events such as cyclones, intense storms and storm surges pose a significant security challenge for the Asia–Pacific region, because of the death and destruction that results and the political, economic and social stresses these events place on even the most developed states. The densely populated river deltas of south and Southeast Asia and south China are particularly vulnerable. Severe events may call into question the legitimacy or competence of a national government and feed into existing ethnic or inter-communal conflicts.

As an example, the 1998 monsoon season brought with it the worst flood in living memory to Bangladesh, inundating some 65 per cent of the country, devastating its infrastructure and agricultural base and raising fears about Bangladesh's long-term future in a world of higher sea levels and more intense cyclones. The 2008 Myanmar cyclone severely affected an estimated 2.4 million people (OCHA 2008).

Severe weather events have the potential to generate an increasing number of humanitarian disasters requiring national and international relief. Because it has the resources and skilled personnel to respond quickly and effectively, Australia will be called upon to shoulder a substantial part of any increase in emergency and humanitarian operations in its immediate neighbourhood, and the major part in the south-west Pacific, Timor-Leste and eastern Indonesia.

Australian defence personnel and police may also be more heavily committed in support of peacekeeping and peace enforcement operations, particularly in the south-west Pacific, should already unstable states be further weakened by the effects of climate change. This will have significant cost and human resource implications. Since 1999, Australian Defence Force regional interventions have cost the federal budget on average over half a billion dollars per annum, a figure that could rise significantly in the longer term with climate change (M. Thomson 2008, pers. comm.).

Sea-level rise

In Asia and the Pacific, millions of people are exposed to relatively high levels of risk from flooding because of the density of urban populations and industrial economic activity and the prevalence of high-value agriculture in coastal regions. The vulnerability of coasts varies dramatically for a given amount of sea-level rise. Small rises in mean sea level, when associated with storm surges and major coastal populations, can be devastating.

It is estimated that 105 million people in Asia would be at risk of their homes becoming inundated by a 1 m rise in sea level (Anthoff et al. 2006). Most of Asia's densest aggregations of people and productive lands are on coastal deltas, including the cities of Shanghai, Tianjin, Guanzhou, Tokyo, Jakarta, Manila, Bangkok, Ningbo, Mumbai, Kolkata and Dhaka. Much of Hong Kong and Singapore are on low-lying land, much of it recently reclaimed from the sea. The areas under greatest threat are the Yellow and Yangtze river deltas in China; Manila Bay in the Philippines; the low-lying coastal fringes of Sumatra, Kalimantan and Java in Indonesia; the Mekong (Vietnam), Chao Phraya (Thailand) and Irrawaddy (Myanmar) deltas (Handley 1992; Morgan 1993); and the delta cities of south Asia.

Sea-level rise would have proportionately the most severe consequences for low-lying atoll countries in the Pacific such as Kiribati (population 78 000), the Marshall Islands (population 58 000), Tokelau (population 2000) and Tuvalu (population 9000). Human habitation may not be possible on these islands even with moderate climate change. If temperature and sea-level rises are at the high end of expectations, then either the sea will eventually submerge the coral atolls or groundwater will become so contaminated by saltwater intrusion that agricultural activities will cease (IPCC 2007). Their small populations make them relatively easy to absorb into larger countries, and the international community and the islanders themselves would expect Australia and New Zealand to be the main countries of resettlement.

The numbers of people exposed to small increases in sea level are much larger in Papua New Guinea, in coastal and low-lying river areas of West Papua, and in other island areas of eastern Indonesia.

Elementary mapping of the vulnerability of people in these areas to sea-level rise has hardly begun. The tendency for settlement to proceed to the high-tide levels in coastal and river delta areas has meant that small rises in sea level have already been associated with saline intrusion into gardens and household water supplies. Village communities have been displaced by destruction of food and water supplies by unexpectedly high king tides. In addition, as Bourke (2008: 53) notes: 'There are about 100 000 people in PNG living on what have been defined as "Small Islands in Peril." These are about 140 islands smaller than 100 km^2 in size and with population densities greater than 100 persons/km^2. It is these people [who are] likely to suffer the most severe consequences of rising sea levels'.

For these reasons, climate change has risen to the top of the political agenda in the Pacific and will require an Australian response.

Climate refugees

Ecological stress in the form of naturally occurring droughts, floods and pestilence has been a significant factor in forcing people to migrate since the beginning of recorded history. So has war-related environmental destruction. In the future, however, climate refugees could constitute the fastest-growing proportion of refugees globally, with serious consequences for international security (Dupont 2008).

Climate-induced migration is set to play out in three distinct ways. First, people will move in response to a deteriorating environment, creating new or repetitive patterns of migration, especially in developing states. Second, there will be increasing short-term population dislocations due to particular climate stimuli such as severe cyclones or major flooding. Third, larger-scale population movements are possible. These may build more slowly but will gain momentum as adverse shifts in climate interact with other migration drivers such as political disturbances, military conflict, ecological stress and socio-economic change.

Australia will not be immune from the consequences of climate-induced migration in Asia and the Pacific. Although abrupt climate change that triggers a massive exodus of environmental refugees is unlikely, significant population displacement caused by sea-level rise, declining agricultural production, flooding, severe weather and step changes in the climate system are all distinct possibilities.

In developing countries with effective governments, early adaptive action can reduce the eventual impact of climate change. In these cases the security consequences may be small. Even in these countries, climate change is set to stretch the limits of adaptability and resilience. Elsewhere, climate change may overwhelm the carrying capacity of the land, disrupt traditional land management systems and make migration an attractive option to preserve quality of life (Edwards 1999). Poorer states could well be overwhelmed by the task confronting them, in which case Australia is likely to experience the ripple effects of climate-induced political disturbances and even violent conflict in the region.

Note

1. From GIAM modelling for the Garnaut Review (see Chapter 11).

References

ABARE (Australian Bureau of Agricultural and Resource Economics) 2008, *Australian Commodities March Quarter*, vol. 8, no. 1, ABARE, Canberra.

ABS (Australian Bureau of Statistics) 2007, *Year Book Australia, 2007*, cat. no. 1301.0, ABS, Canberra.

ABS 2008a, *Tourism Satellite Account: Australian National Accounts, 2006–07*, cat. no. 5249.0, ABS, Canberra.

ABS 2008b, *Year Book Australia, 2008*, cat. no 1301.0, ABS, Canberra.

Adams, R.M., Hurd, B.H. & Reilly, J. 1999, *Agriculture and Climate Change: A review of the impacts on US agricultural resources*, Pew Center on Global Climate Change, Arlington, Virginia.

Anthoff, D., Nicholls, R., Tol, R. & Vafeidis, A. 2006, *Global and Regional Exposure to Large Rises in Sea-level: A sensitivity analysis*, Tyndall Centre for Climate Change Research, United Kingdom.

Australian Centre for Biodiversity 2008, *Biodiversity and Climate Change*, report commissioned by the Garnaut Climate Change Review, Australian Centre for Biodiversity, Monash University, Victoria.

Bambrick, H., Dear, K., Woodruff, R., Hanigan, H. & McMichael, A. 2008, *The Impacts of Climate Change on Three Health Outcomes: Temperature-related mortality and hospitalisations, salmonellosis and other bacterial gastroenteritis, and population at risk from dengue*, report commissioned by the Garnaut Climate Change Review.

Bourke, R.M. 2008, 'Climate change', in R.M. Bourke & T. Harwood (eds), *Kaikai na Mani: Agriculture in Papua New Guinea*, Australian National University, Canberra.

Brown, B.E. 1997, 'Coral bleaching: causes and consequences', *Coral Reefs* 16: S129–38.

Chellaney, B. 2007, 'Climate change: a new factor in international security?', *Global Forces 2007: Proceedings of the ASPI Conference: Day 1*, Australian Strategic Policy Institute, Barton, ACT, pp. 20–36.

Consultative Group on International Agricultural Research Inter-Center Working Group on Climate Change 2002, *The Challenge of Climate Change: Research to overcome its impact on food security, poverty, and natural resource degradation in the developing world*, <www.cgiar.org>.

Crimp, S., Howden, M., Power, B., Wang, E. & De Voil, P. 2008, *Global Climate Change Impacts on Australia's Wheat Crops*, report prepared for the Garnaut Climate Change Review, CSIRO, Canberra.

Crimp, S.J., Flood, N.R., Carter, J.O., Conroy, J.P. & McKeon, G.M. 2002, *Evaluation of the Potential Impacts of Climate Change on Native Pasture Production: Implications for livestock carrying capacity*, report for the Australian Greenhouse Office.

CSIRO (Commonwealth Scientific and Industrial Research Organisation) & BoM (Bureau of Meteorology) 2007, *Climate Change in Australia: Technical report 2007*, CSIRO, Melbourne.

Dupont, A. 2008, *Climate Change and Security: Managing the risk*, report commissioned by the Garnaut Climate Change Review.

Ecofys BV 2006, *Agriculture and Forestry Sectoral Report*, report prepared for the European Climate Change Programme Working Group II Impacts and Adaptation, Brussels, <http://ec.europa.eu/environment/climat/pdf/eccp/impactsadaptation/agriforestry.pdf>.

Edwards, M. 1999, 'Security implications of a worst-case scenario of climate change in the South-West Pacific', *Australian Geographer* 30(3): 311–30.

European Environment Agency 2004, *Agriculture*, European Environment Agency, Copenhagen, <http://epaedia.eea.europa.eu/page.php?pid=346#galleryhere>.

Handley, P. 1992, 'Before the flood', *Far Eastern Economic Review*, 16 April, p. 65.

Hennessy, K., Whetton, P., Smith, I., Bathols, J., Hutchinson, M. & Sharples, J. 2003, *The Impact of Climate Change on Snow Conditions in Mainland Australia*, CSIRO, Aspendale, Victoria.

Hilbert, D.W., Ostendorf, B. & Hopkins, M. 2001, 'Sensitivity of tropical forests to climate change in the humid tropics of North Queensland', *Austral Ecology* 26: 590–603.

Hoegh-Guldberg, H. 2008, *Australian Tourism and Climate Change*, <http://economicstrategies.files.wordpress.com/2008/02/background-tourism-paper-updated.pdf>.

Hoegh-Guldberg, O. 1999, 'Climate change, coral bleaching and the future of the world's coral reefs', *Marine Freshwater Research* 50: 839–66.

Hoegh-Guldberg, O. & Hoegh-Guldberg, H. 2008, *The Impact of Climate Change and Ocean Acidification on the Great Barrier Reef and Its Tourist Industry*, report commissioned by the Garnaut Climate Change Review.

Hoegh-Guldberg, O., Mumby, P.J., Hooten, A.J., Steneck, R.S., Greenfield, P., Gomez, E., Harvell, D.R., Sale, P.F., Edwards, A.J., Caldeira, K., Knowlton, N., Eakin, C.M., Iglesias-Prieto, R., Muthiga, N., Bradbury, R.H., Dubi, A. & Hatziolos, M.E. 2007, 'Coral reefs under rapid climate change and ocean acidification', *Science* 318: 1737–42.

Howden, S.M. & Crimp, S. 2005, 'Assessing dangerous climate change impacts on Australia's wheat industry', in A. Zerger & R.M. Argent (eds), *International Congress on Modelling and Simulation: MODSIM 2005*, Modelling and Simulation Society of Australia and New Zealand, pp. 170–6.

Howden, S.M., McKeon, G.M., Meinke, H., Entel, M. & Flood, N. 2001, 'Impacts of climate change and climate variability on the competitiveness of wheat and beef cattle production in Emerald, north-east Australia', *Environment International* 27: 155–60.

Howden, S.M., Reyenga, P.J. & Meinke, H. 1999, *Global Change Impacts on Australian Wheat Cropping: CSIRO Wildlife & Ecology Working Paper 99/04*, report to the Australian Greenhouse Office, Canberra.

IPCC 2007, *Climate Change 2007: Impacts, adaptation and vulnerability. Contribution of Working Group II to the Fourth Assessment Report of the Intergovernmental Panel on Climate Change*, M. Parry, O. Canziani, J. Palutikof, P. van der Linden & C. Hanson (eds), Cambridge University Press, Cambridge.

Lough, J. 2007, 'Climate and climate change on the Great Barrier Reef', in J. Johnson & P. Marshall (eds), *Climate Change and the Great Barrier Reef: A vulnerability assessment*, Great Barrier Reef Marine Park Authority & Australian Greenhouse Office.

McMichael, A., Woodruff, R., Whetton, P., Hennessy, K., Nicholls, N., Hales, S., Woodward, A. & Kjellstrom, T. 2003, *Human Health and Climate Change in Oceania: A risk assessment 2002. Climate Change Health Risk Assessment*, Department of Health and Ageing, Canberra.

Marsden Jacob Associates, 2006, *Securing Australia's Urban Water Supplies: Opportunities and impediments*, discussion paper prepared for the Department of Prime Minister and Cabinet.

Maunsell Australia Pty Ltd, in association with CSIRO Sustainable Ecosystems 2008, *Impact of Climate Change on Infrastructure in Australia and CGE Model Inputs*, report commissioned by the Garnaut Climate Change Review.

Morgan, J. 1993, 'Natural and human hazards', in H. Brookfield & Y. Byron (eds), *Southeast Asia's Environmental Future: The search for sustainability*, Oxford University Press, Kuala Lumpur.

National Institute of Economic and Industry Research 2006, *The Economic Significance of the Australian Alpine Resorts*, National Institute of Economic and Industry Research, Victoria.

OCHA (United Nations Office for the Coordination of Humanitarian Affairs) 2008, *Myanmar Cyclone Nargis*, <http://ochaonline.un.org/roap/CycloneNargis/tabid/4579/Default.aspx>.

Pittock, B. (ed.) 2003, *Climate Change: An Australian guide to the science and potential impacts*, Australian Greenhouse Office, Canberra.

PMSEIC (Prime Minister's Science, Engineering and Innovation Council) 2007, *Climate Change in Australia: Regional impacts and adaptation—managing the risk for Australia*, Independent Working Group, Canberra.

Preston, B.L. & Jones, R.N. 2006, *Climate Change Impacts on Australia and the Benefits of Early Action to Reduce Global Greenhouse Gas Emissions*, CSIRO, Canberra.

Quiggin, J., Adamson, D., Schrobback, P. & Chambers, S. 2008, *The Implications for Irrigation in the Murray-Darling Basin*, report commissioned by the Garnaut Climate Change Review.

Steffen, W. & Canadell, P. 2005, *Carbon Dioxide Fertilisation and Climate Change Policy*, Australian Greenhouse Office, Australian Government Department of the Environment and Heritage, Canberra.

Sustainable Tourism Cooperative Research Centre 2007, *Climate Change and Australian Tourism*, Sustainable Tourism CRC, Southport, Queensland.

World Health Organization 2002, *The World Health Report 2002: Reducing risks, promoting healthy life*, World Health Organization, Geneva.

7 AUSTRALIA'S EMISSIONS IN A GLOBAL CONTEXT

Key points

Australia's per capita emissions are the highest in the OECD and among the highest in the world. Emissions from the energy sector would be the main component of an expected quadrupling of emissions by 2100 without mitigation.

Australia's energy sector emissions grew rapidly between 1990 and 2005. Total emissions growth was moderated, and kept more or less within our Kyoto Protocol target, by a one-off reduction in land clearing.

Relative to other OECD countries, Australia's high emissions are mainly the result of the high emissions intensity of energy use, rather than the high energy intensity of the economy or exceptionally high per capita income. Transport emissions are not dissimilar to those of other developed countries. Australia's per capita agricultural emissions are among the highest in the world, especially because of the large numbers of sheep and cattle.

The high emissions intensity of energy use in Australia is mainly the result of our reliance on coal for electricity. The difference between Australia and other countries is a recent phenomenon: the average emissions intensity of primary energy supply for Australia and the OECD was similar in 1971.

7.1 Australia's emissions profile and international comparisons

Australia's per capita greenhouse gas emissions are the highest of any OECD country and are among the highest in the world. In 2006 our per capita emissions (including emissions from land use, land-use change and forestry) were 28.1 tonnes carbon dioxide equivalent (CO_2-e) per person (DCC 2008d). Only five countries in the world rank higher—Bahrain, Bolivia, Brunei, Kuwait and Qatar. Australia's per capita emissions are nearly twice the OECD average and more than four times the world average (see Figure 7.1).

Figure 7.1 Per capita greenhouse gas emissions

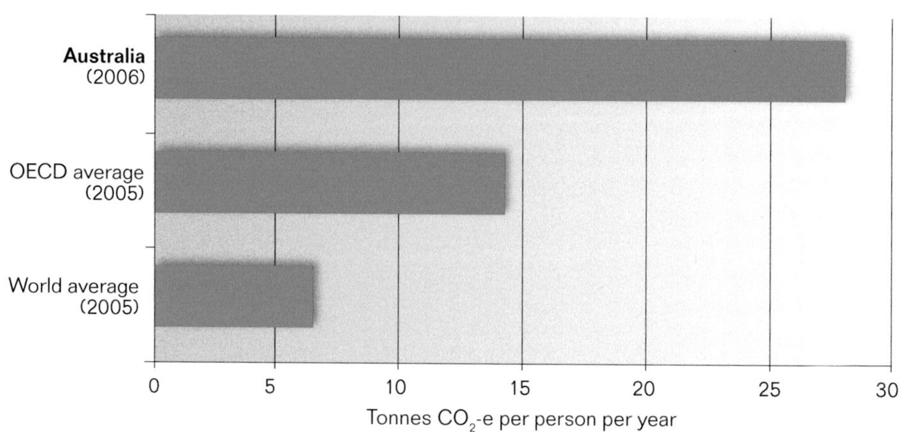

Sources: DCC (2008c); IEA (2007a).

For the calculation of per capita greenhouse gas emissions illustrated in Figure 7.1, the data source used for Australia was the Department of Climate Change, while the International Energy Agency (IEA) was the source used for all other countries. There are other data sources for developed countries, such as the United Nations Framework Convention on Climate Change (UNFCCC) and relevant national agencies. There is some variation between emissions estimates by source (see Table 7.1 for a comparison of IEA and UNFCCC estimates for the highest per capita OECD emitters).

Table 7.1 Comparison of the highest per capita emissions among OECD countries (tonnes per person per year)

	2005, excluding land use, land-use change and forestry[a]	2006, excluding land use, land-use change and forestry[b]	2006, including land use, land-use change and forestry[b]
Australia	30.3	26.0	26.7
United States	24.5	23.5	20.6
Luxembourg	24.0	26.6	26.1
New Zealand	22.6	19.0	13.4
Canada	22.5	22.1	23.1
Ireland	15.6	16.6	16.5
Czech Republic	14.3	14.4	14.1

Sources: a. IEA (2007a); b. For emissions data, UNFCCC (2008); for population data, Population Reference Bureau (2008).

7.1.1 Recent growth trends in Australia's emissions

In 2006 Australia's net greenhouse gas emissions were 576 Mt CO_2-e using Kyoto Protocol accounting provisions (DCC 2008c). From 1990 to 2006, Australia's net emissions increased by 4.2 per cent (23.4 Mt).

Emissions for 1990 and 2006 by sector are illustrated in Figure 7.2. Energy sector emissions increased by about 40 per cent between 1990 and 2006. Over the same period there was a substantial reduction (about 71 per cent) in emissions from land use, land-use change and forestry.

Stationary energy sector emissions are from energy industries (for example, electricity generation and petroleum refining), fuel combustion in the manufacturing and construction industries, and fuel combustion in other sectors such as commercial, residential, agriculture, forestry and fishing. Fugitive emissions arise during the extraction, transportation and handling of fossil fuels. Industrial process emissions are from production of metals, minerals, chemicals, pulp and paper, and food and beverages.

Figure 7.2 Greenhouse gas emissions by sector, 1990 and 2006

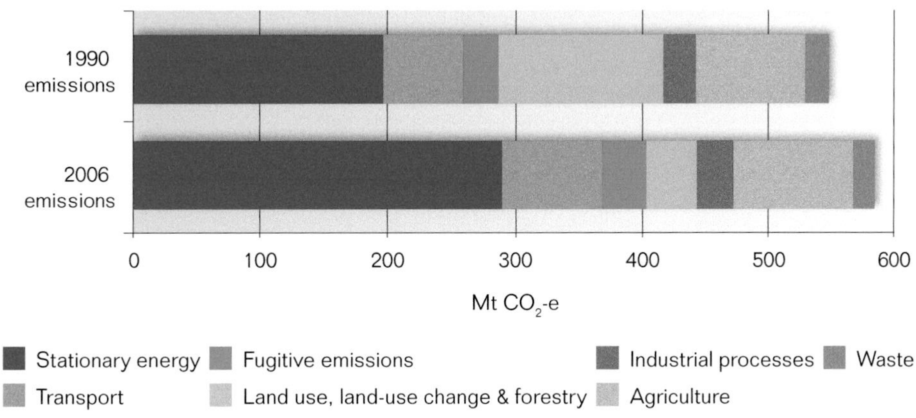

Source: DCC (2008b).

7.1.2 Future emissions growth in Australia

Figure 7.3 presents expectations of future emissions under the Garnaut–Treasury reference case (see Chapter 11). In the absence of mitigation measures, energy-related emissions are expected to grow rapidly and to increase their share of the total.

In the reference case, stationary energy sector emissions are projected to increase by about 85 per cent by 2050 and about 260 per cent by 2100. Growth in emissions from the stationary energy sector is largely driven by the structure and growth of Australia's economy, the fuel mix in electricity generation, and energy efficiency across the economy.

Figure 7.3 Greenhouse gas emissions by sector: 1990, 2006 and reference case scenarios

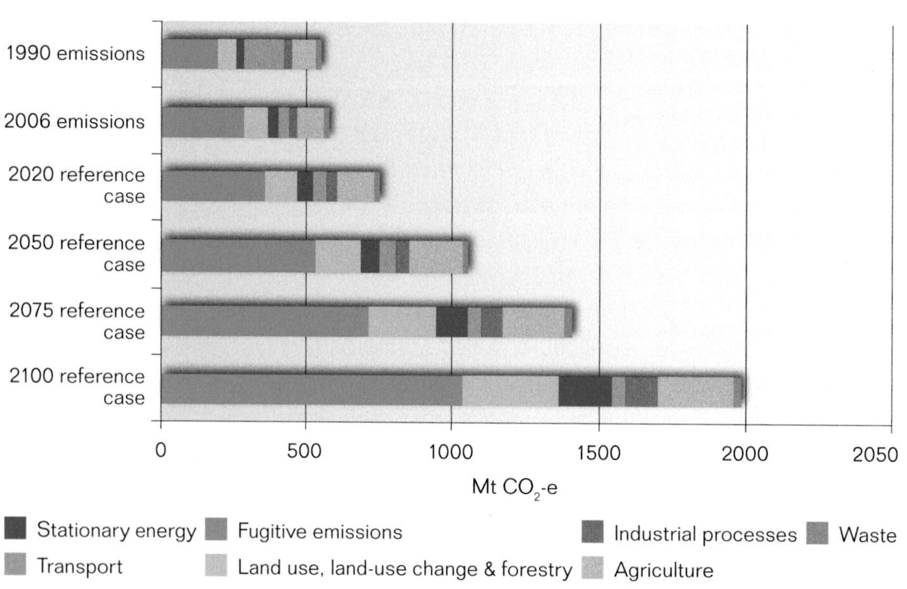

Sources: 1990 and 2006 emissions are from the most recent National Greenhouse Gas Inventory (DCC 2008b); 2020–2100 projections are from the Monash Multi Regional Forecasting model.

Transport-related emissions are also projected to grow rapidly. Transport-related emissions would almost double by 2050 and more than quadruple by 2100.

Emissions arising from land-use change depend on the area of forest cover removal, the method of forest conversion and land development. The estimates rely on assumptions about the amount of carbon sequestered in biomass and soils, which varies by vegetation type, geography and climate (DCC 2008d). Reductions in the rate of forest cover removal since 1990 have been the main source of the reduction in land-use-related emissions.

7.1.3 Australia's energy emissions relative to other countries'

In 2006 Australia's greenhouse gas emissions due to energy were 401 Mt CO_2-e (DCC 2008b), which represented about 70 per cent of Australia's total. From 1990 to 2006, Australia's net emissions from energy increased by about 40 per cent (114 Mt).

Australia's per capita greenhouse gas emissions due to energy are the third highest of any OECD country and the seventh highest in the world, after Luxembourg, the United States, Qatar, Kuwait, Bahrain, the United Arab Emirates and Netherlands Antilles. Australia's per capita greenhouse gas emissions due to energy in 2005 were about 67 per cent higher than the OECD average and more than four times the world average (see Figure 7.4).

Figure 7.4 Per capita emissions due to energy use, 2005

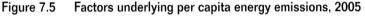

Source: IEA (2007a).

The energy intensity of an economy is a measure of the amount of energy used per unit of economic activity generated. The emissions intensity of energy is a measure of the amount of greenhouse gases emitted per unit of energy used.

Energy-associated per capita emissions are the product of per capita GDP, energy intensity (of the economy) and emissions intensity (of energy) as follows:

$$CO_2 \text{ per capita} = \text{GDP per capita} \times \text{Energy/GDP} \times CO_2\text{/Energy}$$

Figure 7.5 illustrates the factors underlying a country's per capita emissions and compares those factors for Australia, the OECD average and the world average.

Figure 7.5 Factors underlying per capita energy emissions, 2005

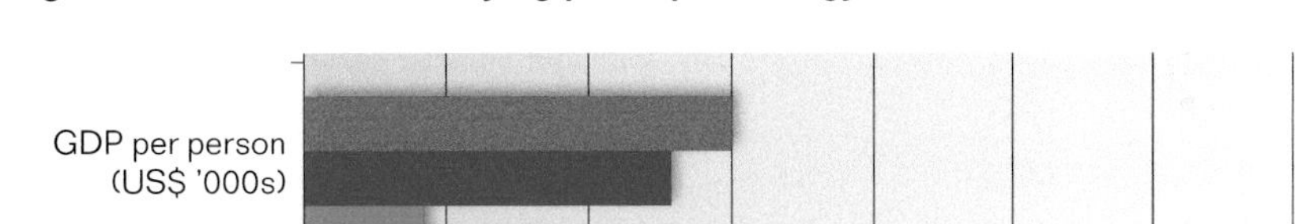

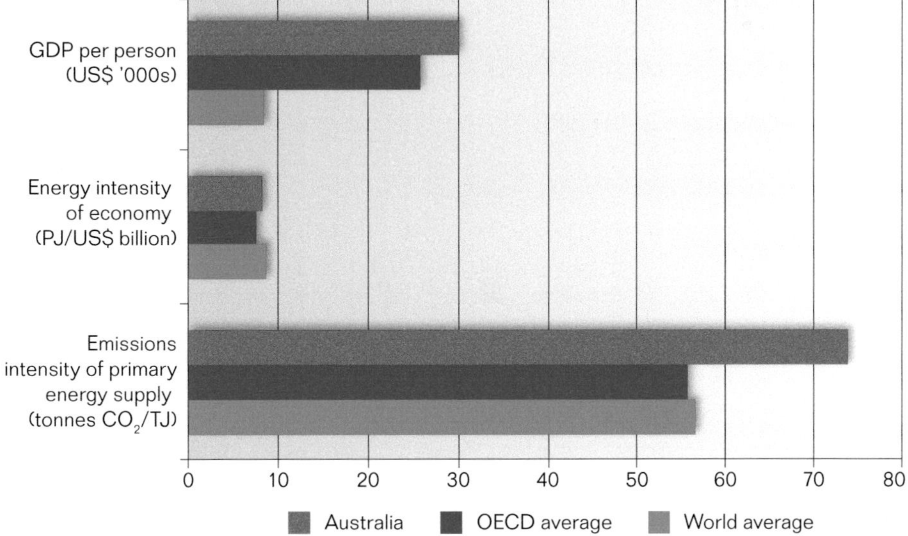

Note: All financial values are measured in 2000 US$, using purchasing power parities.

Source: IEA (2007a).

Australia's GDP per capita in 2005 was about 16 per cent higher than the OECD average (IEA 2007a). While this contributes to Australia's comparatively high per capita greenhouse gas emissions due to energy use, it does not account for why they are about two-thirds higher than the OECD average.

Australia's economy is the eighth most energy intensive among OECD countries. It is about 5 per cent less energy intensive than the world average and about 8 per cent more energy intensive than the OECD average.

The aggregate energy intensity of the Australian economy, measured as total primary energy consumption per dollar of GDP, remained broadly stable over the 1970s and 1980s, and then fell by an average of 1.1 per cent a year during the 1990s (Syed et al. 2007).

The energy intensity of Australia's economy does not account for our extremely high per capita greenhouse gas emissions.

The emissions intensity of Australia's primary energy supply is the second highest among OECD countries. It is more than 30 per cent higher than both the OECD average and the world average. There are only five countries in the world with a more emissions-intensive energy supply than Australia's—Bosnia Herzegovina, the Democratic People's Republic of Korea, Estonia, Mongolia and Poland.

Fossil fuels play a dominant role in Australia's primary energy consumption. More than 40 per cent of Australia's total primary energy supply is derived from coal. This is a much higher proportion than in other OECD countries, as illustrated in Figure 7.6.

Figure 7.6 Fuel mix contributing to total primary energy supply, 2005

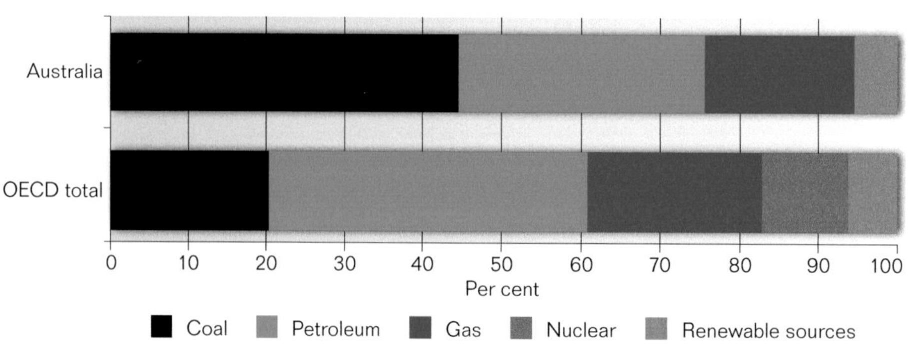

Source: IEA (2007b).

The exceptional emissions intensity of Australia's primary energy supply has only emerged in recent decades. Figure 7.7 shows the trends in Australia's average emissions intensity of primary energy supply compared with those in all OECD countries. The Australian average was similar to that of the OECD in 1971.

Figure 7.7 Trends in average emissions intensity of primary energy supply, Australia and OECD, 1971–2005

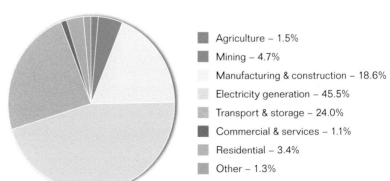

Source: IEA (2007a).

The increasing emissions intensity of Australia's primary energy supply is largely due to its increasing reliance on coal for electricity generation, at a time when other developed countries have shifted significantly to lower-emissions sources.

Figure 7.8 shows primary energy consumption in Australia by sector. The electricity generation, transport and manufacturing sectors account for nearly 90 per cent of primary energy consumption. Coal comprises about 79 per cent of primary energy supply in the electricity generation sector. The transport sector is dominated by oil, which contributes more than 97 per cent of primary energy consumed. In the manufacturing sector primary energy supply is composed of coal (27 per cent), oil (32 per cent), natural gas (34 per cent) and biomass (7 per cent).

Figure 7.8 Primary energy consumption in Australia, by sector, 2005–06

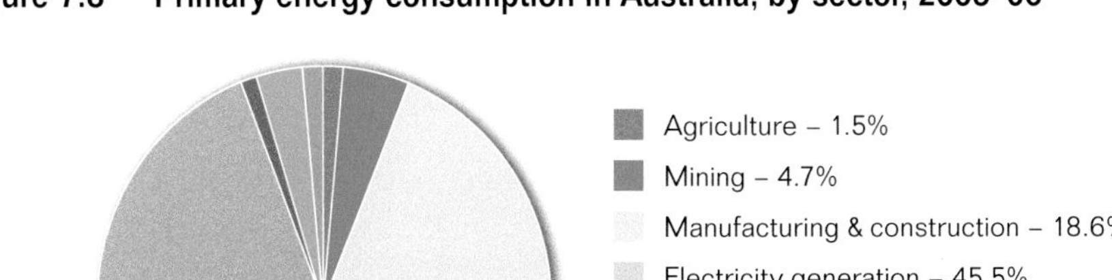

Source: Syed et al. (2007).

7.1.4 Australia's electricity emissions relative to other countries'

Australia's per capita electricity consumption is about 22 per cent above the OECD average, while our per capita emissions due to electricity generation are more than three times the OECD average (see Figure 7.9). The difference is due to the high emissions intensity of electricity generated in Australia.

Figure 7.9 Per capita emissions due to electricity, 2005

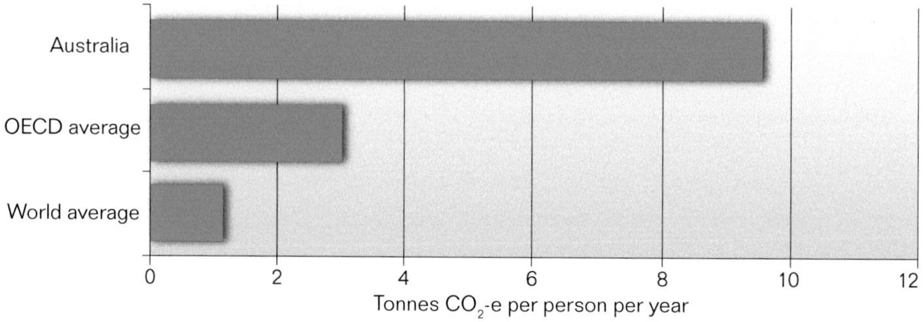

Australia
OECD average
World average
0 2 4 6 8 10 12
Tonnes CO_2-e per person per year

Source: IEA (2007a); DCC (2008b).

The emissions intensity of Australia's electricity supply is the highest of any OECD country. It is 98 per cent higher than the OECD average, and 74 per cent higher than the world average (see Figure 7.10). There are only eight countries in the world with an electricity system that is more emissions intensive than Australia's—Bahrain, Botswana, Cambodia, Cuba, India, Kazakhstan, Libya and Malta.

Figure 7.10 Factors underlying per capita electricity emissions, 2005

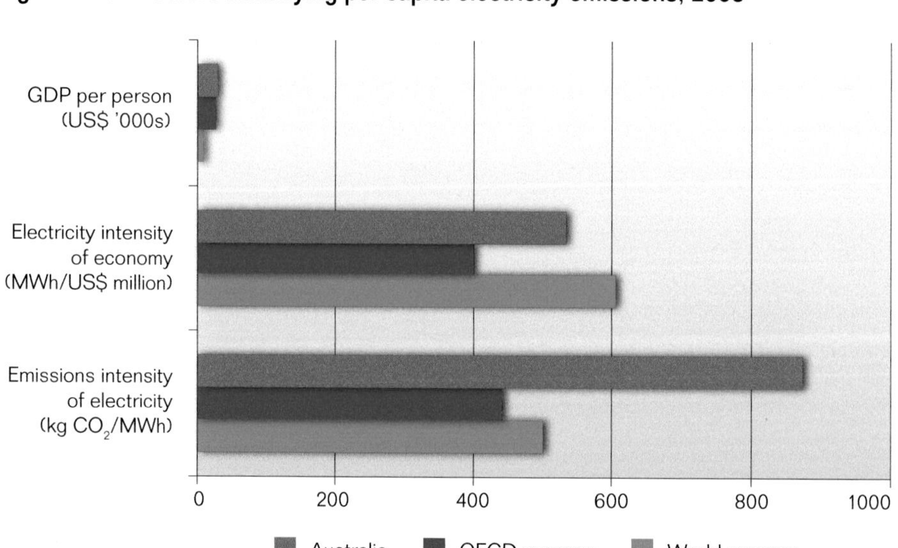

Source: IEA (2007a).

7.1.5 Australia's transport emissions relative to other countries'

In 2006 Australia's greenhouse gas emissions due to transport were 79.1 Mt CO_2-e using Kyoto accounting provisions. They do not include emissions arising from the generation of electricity used by public transport (trams and electric trains) or from fuel sold to ships or aircraft engaged in international transport (DCC 2008b).

Transport emissions represent about 14 per cent of Australia's total greenhouse gas emissions. From 1990 to 2006, Australia's net emissions from transport increased by about 27 per cent (17.0 Mt). The residential sector accounts for about 57 per cent of Australia's transport emissions. The remaining 43 per cent is attributable to Australian business.

Australia's per capita greenhouse gas emissions due to transport are the fourth highest of any OECD country and the seventh highest in the world, after Canada, Luxembourg, Netherlands Antilles, Qatar, the United Arab Emirates and the United States. Australia's per capita greenhouse gas emissions due to transport in 2005 were about 30 per cent higher than the OECD average and nearly four times the world average (see Figure 7.11).

Figure 7.11 Per capita emissions due to transport, 2005

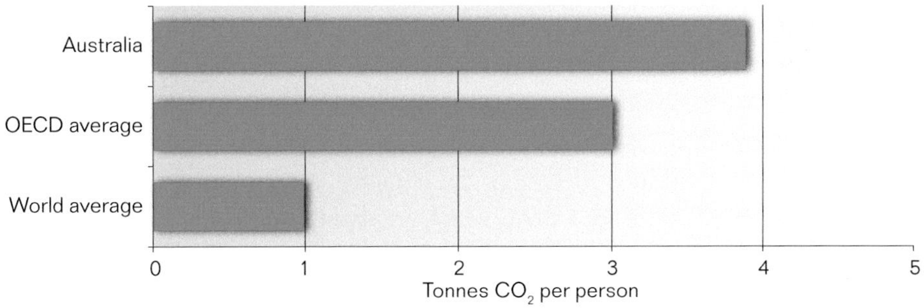

Source: IEA (2007a).

Transport-associated per capita emissions are the product of per capita GDP, the transport energy intensity of the economy and the emissions intensity of transport energy. The transport energy intensity of an economy is a measure of the amount of transport energy used per unit of economic activity generated. The emissions intensity of transport energy is a measure of the amount of greenhouse gases emitted per unit of transport energy used. Figure 7.12 illustrates the factors underlying a country's per capita transport emissions and compares those factors for Australia, the OECD average and the world average.

Figure 7.12 Factors underlying per capita transport emissions

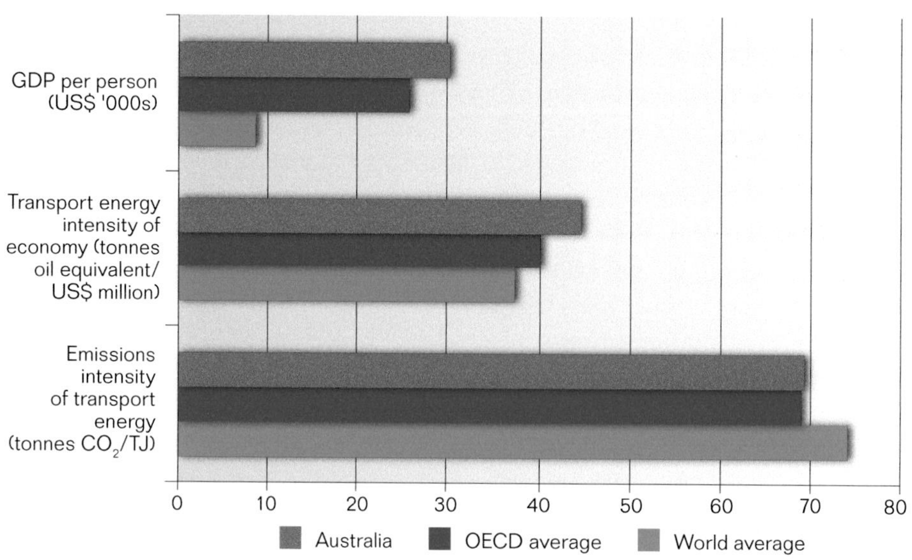

Source: IEA (2007a).

The emissions intensity of transport energy is determined by the proportions of travel undertaken using public and private transport modes and, within modes, the combination of the engine type and fuel used.

The emissions intensity of Australia's transport energy is within a fraction of 1 per cent of the OECD average. Among OECD countries there is less than 10 per cent variation between the country whose transport energy is most emissions intensive and that which is least emissions intensive. This statistical outcome is affected by the exclusion from the data of electricity used for transport.

The lack of variation in the emissions intensity of transport energy among OECD countries is a reflection of the fact that, in all OECD countries, transport emissions are dominated by the same engine type (internal combustion) and the same fuel (petroleum).

The transport energy intensity of Australia's economy is the sixth highest among OECD countries and is about 10 per cent higher than the OECD average. The five OECD countries that consume more transport energy per unit of GDP generated are Canada, Luxembourg, Mexico, New Zealand and the United States.

7.1.6 Australia's agricultural emissions relative to other countries'

Australia's per capita emissions arising from agriculture are more than six times the world average, more than four times the OECD average and third highest in the OECD (see Figure 7.13).

Figure 7.13 Per capita emissions due to agricultural production

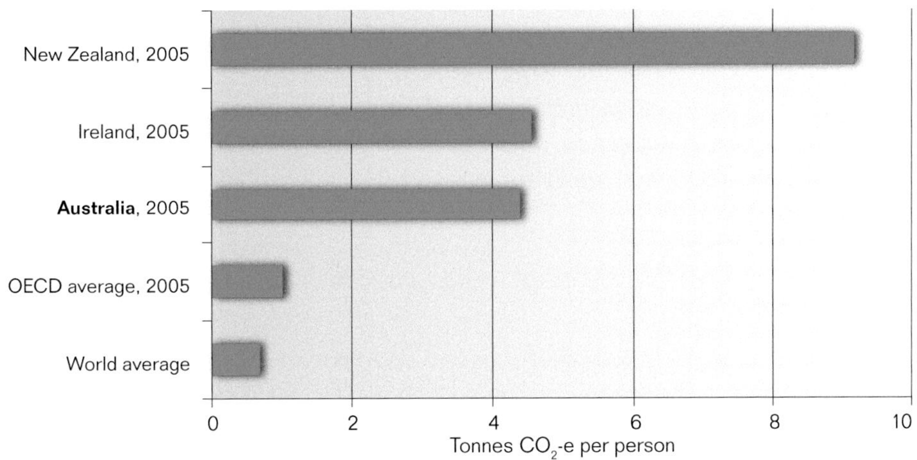

Note: Data on emissions from agricultural production in developing countries are of poor quality and a consistent dataset for recent years is not available. The world average estimate is derived from UNFCCC data on agricultural emissions of developing countries, some of which are as old as 1990.

Source: UNFCCC (2005, 2008).

Under Article 3.1 of the Kyoto Protocol and in accordance with the 1996 IPCC guidelines (IPCC 1996), agricultural emissions are reported according to the following physical processes that give rise to them:

- **enteric fermentation in livestock**—emissions associated with microbial fermentation during digestion of feed by ruminant (mostly cattle and sheep) and some non-ruminant domestic livestock
- **manure management**—emissions associated with the decomposition of animal wastes while held in manure management systems
- **rice cultivation**—methane emissions from anaerobic decay of plant and other organic material when rice fields are flooded
- **agricultural soils**—emissions associated with the application of fertilisers, crop residues and animal wastes to agricultural lands and the use of biological nitrogen-fixing crops and pastures
- **prescribed burning of savannas**—emissions associated with the burning of tropical savanna and temperate grasslands for pasture management, fuel reduction, and prevention of wildfires
- **field burning of agricultural residues**—emissions from field burning of cereal, sugar cane and crop stubble.

Livestock emissions from enteric fermentation play a large role in the emissions profile of the agriculture sector. About 34 per cent of OECD countries' agricultural emissions are due to livestock emissions. This figure is even higher in the countries with the highest per capita emissions from agricultural production—64 per cent for New Zealand, 49 per cent for Ireland and 66 per cent for Australia.

The countries with the highest per capita agricultural emissions are those with the largest numbers of cattle and sheep relative to population.

New Zealand, Ireland and Australia all produce more than 100 kilograms of beef per person per year. The world average is less than 9 kilograms and the OECD average about 22 kilograms. New Zealand, Ireland and Australia produce about 132, 18 and 29 kilograms respectively of sheep meat per person per year, compared with OECD and world averages of about 2 kilograms and 1 kilogram respectively.

7.1.7 Australia's biosequestration potential relative to other countries'

There is significant global potential for emissions removal (or carbon sequestration) through revegetation of previously cleared land and increasing the stock of carbon in forests, wooded land and soils. Management of existing forests for ecosystem services, rather than simply for fibre production, would significantly reduce emissions from degradation and deforestation.

The potential for carbon removal through existing forests and revegetation of cleared land is proportional to the land area available for those purposes. Australia has relatively large areas of forested land suitable for carbon removal and deforested land suitable for revegetation. There are about 28.8 hectares of forest and wooded land for every person in Australia (FAO 2008). This is the largest area of forest and wooded land per person in the OECD and the second largest globally behind Suriname. In OECD countries there are on average 1.4 hectares of forest and woodland per person, and across the world there are on average 0.8 hectares (see Figure 7.14).

Figure 7.14 Per capita area of forested and wooded land, 2005

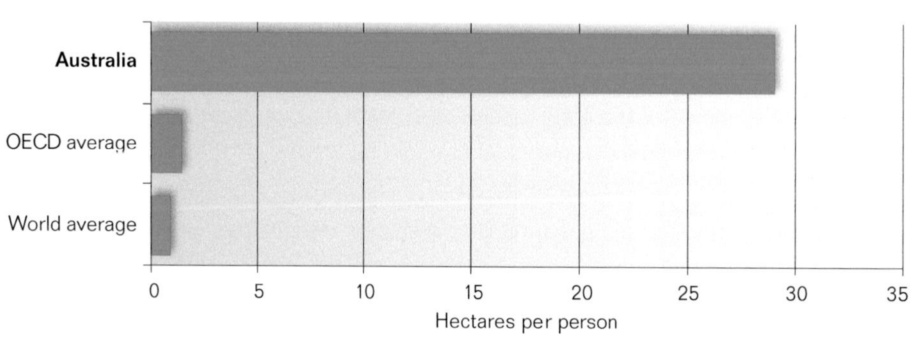

Source: FAO (2008).

In addition to land area, the amount of additional carbon that can be sequestered by existing forests and woodlands and through revegetation of cleared lands is also determined by the local climate; the fertility of the substrate; the characteristics of the plant species including growth rates, wood density and their suitability to local environmental conditions; and the impact of land use history in reducing carbon stocks below the land's carbon-carrying capacity.

Australia has an estimated 163.7 million hectares of forest and 421.6 million hectares of other wooded land (FAO 2008). The IPCC default values for temperate forests are a carbon stock of 217 tonnes carbon per hectare, 96 tonnes biomass carbon per hectare and net primary productivity of 7 tonnes carbon per hectare per year. These estimates are probably conservative for intact (unlogged) natural forests. Mackey et al. (2008) have shown that the stock of carbon for intact natural eucalypt forests in south-eastern Australia is about 640 tonnes carbon per hectare (biomass plus soil, with a standard deviation of 383), with 360 tonnes biomass carbon per hectare. The average net primary productivity of these forests is 12 tonnes carbon per hectare per year (with a standard deviation of 1.8). Mackey et al. estimate that the eucalypt forests of south-eastern Australian could remove about 136 Mt CO_2-e per year (on average) for the next 100 years. This estimate is premised on several key assumptions, including cessation of logging and controlled burning over the 14.5 million hectare study area.

In 2006 the net removal by pre-1990 plantations and native forests was estimated to be about 24.1 Mt CO_2-e (DCC 2008b). Under Article 3.3 of the Kyoto Protocol this sequestration does not contribute to Australia meeting its Kyoto target.

It is important for efficient global mitigation that the international community move to comprehensive carbon accounting related to agriculture and forestry. This is particularly important for Australia.

Comparative carbon accounting, among much else, would bring to account all carbon sequestered by and emitted from managed lands. This would provide significant revenue opportunities for landowners. It would also bring risks, especially, as would be required in logic, if all emissions arising from fires and the effects of drought are covered.

7.2 Emissions profiles of Australian industries

7.2.1 How do Australian industries contribute to emissions and GDP?

Figure 7.15 shows the total emissions attributable to each Australian industry, derived by summing a sector's direct emissions and the indirect emissions attributable to its electricity consumption. Emissions due to transport have not been attributed in the same way, due to lack of suitable data. Industry accounts for about 82 per cent of Australia's total emissions, with the remainder attributable to the residential sector. The agriculture, mining and manufacturing industries are responsible for large amounts of greenhouse gas emissions relative to their shares of GDP.

Figure 7.15 Emissions attributable to Australian industry by sector, 2006

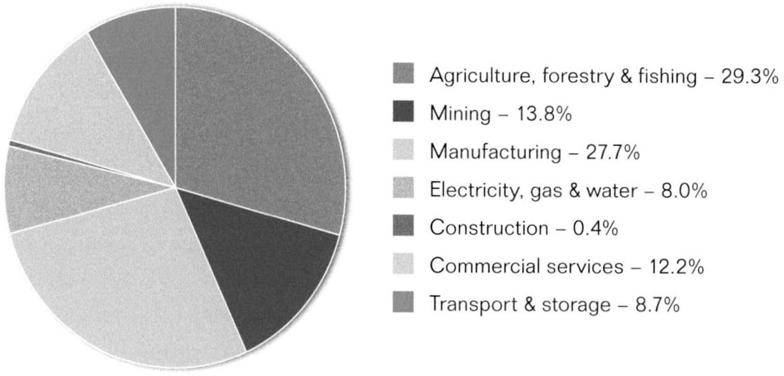

Sources: DCC (2008b); ABS (2007).

7.2.2 Which industries would be most affected by a price on emissions?

The industries whose competitiveness is most likely to be adversely affected by a price on greenhouse gas emissions are those that are exposed to international trade and that have either a high degree of energy intensity or a high level of direct greenhouse gas emissions.

The Review sought to identify the trade-exposed industries that might be most affected by a price on greenhouse gas emissions. It considered data on trade, direct emissions, and indirect emissions attributable to electricity consumption. It used the 1993 Australian and New Zealand Standard Industrial Classification and the Australian National Accounts (ABS 2008) as guides to industry classification, and selected the following for analysis of the impact of an emissions price on international competitiveness:

Manufacturing

- food, beverage and tobacco
- textile, clothing, footwear and leather
- wood, paper and printing
- petroleum refining
- petroleum and coal products
- basic chemicals
- cement, lime, plaster and concrete
- iron and steel
- non-ferrous metals and products (including aluminium production)
- machinery and equipment manufacturing

Agriculture

- dairy cattle
- beef cattle
- sheep
- grains
- pigs
- poultry
- other

Mining

- coal mining
- oil and gas extraction (including liquefied natural gas production)
- other mining (non-energy)

Mining and manufacturing industries

In 2006 the mining industry's direct emissions were 52.1 Mt CO_2-e and its indirect emissions due to purchase of electricity were 12.9 Mt CO_2-e. In the same year the manufacturing industry's direct and indirect emissions were 69.3 Mt and 61.0 Mt CO_2-e respectively.

Due to changes in methodology adopted for Australia's 2006 National Greenhouse Gas Inventory, the manufacturing industry's indirect emissions arising from electricity consumption are no longer disaggregated by sector. This is why figures 7.16 and 7.17 use 2005 data to show industry contributions to emissions from mining and manufacturing.

Figure 7.16 Emissions attributable to the Australian mining and manufacturing industries, disaggregated by sector, 2005

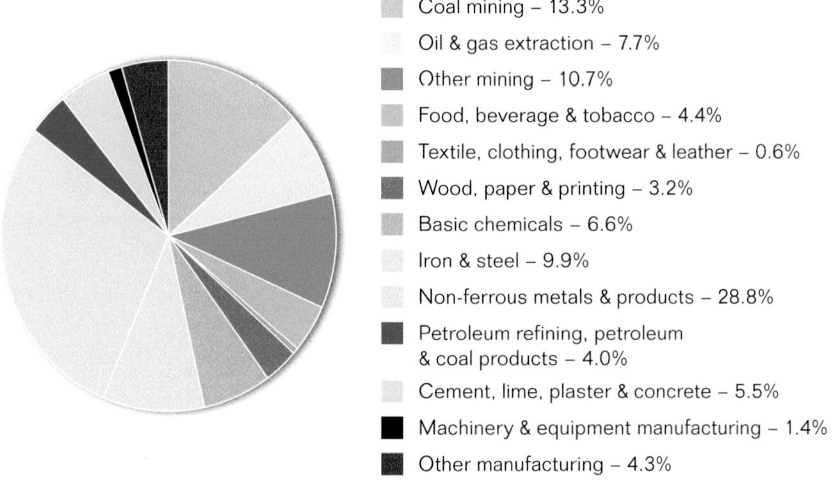

Source: DCC (2008a); ABS (2007).

Agriculture

In 2006 the agriculture industry produced an estimated 90.1 Mt CO_2-e of direct agricultural emissions and an additional 1.5 Mt CO_2-e of indirect emissions arising from the purchase of electricity. In addition, land clearing for agriculture was responsible for emissions of 74.1 Mt CO_2-e.

The guidelines of the UNFCCC require countries to report agricultural, land use and forestry emissions according to the physical processes that give rise to them. The Review calculated 2005 direct emissions and indirect emissions attributable to the purchase of electricity by commodity within the agriculture and forestry industries using the Australian National Accounts as a guide to classification (see Table 7.2).

Table 7.2 Agricultural emissions and land use, land-use change and forestry emissions, by commodity and economic sector, 2005

Commodity/ sector	Agricultural emissions (Mt)	Land use, land-use change and forestry emissions (Mt)	Total direct emissions (Mt)	Emissions attributable to purchase of electricity (Mt)	Total attributable emissions (Mt)
Sheep	19.6	–	19.6	0.1	19.7
Grains	2.2	2.7	4.9	0.2	5.1
Beef cattle	51.9	71.4	123.3	0.4	123.7
Dairy cattle	10.3	–	10.3	0.3	10.7
Pigs	1.6	–	1.6	–	1.6
Poultry	0.8	–	0.8	0.2	1.0
Other agriculture	2.9	–	2.9	0.4	3.2
Forestry	–	-21.8	-21.8	–	-21.7
Total	89.3	52.4	141.7	1.6	143.3

Sources: DCC (2008b); ABS (2008).

In order to gauge the potential impact of a price being placed on greenhouse gas emissions, the Review examined the effect of a permit price of \$10, \$20 and \$40 per tonne of CO_2-e and assumed that there would be 100 per cent pass-through of emissions costs to energy consumers. The latter is a worst-case scenario from the perspective of energy-intensive industries. If this were too high, the analysis that follows would overestimate the additional costs accruing to energy-intensive industries as a result of a price being placed on emissions. Under these assumptions, the ratio of greenhouse gas emission costs to the value of production is as shown in Figure 7.17. Note that Figure 7.17 does not include emissions due to deforestation by the grain and beef cattle industries.

Recent and projected increases in commodity prices reduce the ratio of greenhouse gas emissions to the value of production. For example, it is estimated that projected 2008–09 increases in the price for coal exports would reduce the ratio of greenhouse gas emissions costs to the value of production for coal mining to about one-third of that shown in Figure 7.17.

7.2.3 Comparison of Australia's agriculture industry with other OECD countries'

To get some indication of the likely impact of a price on emissions upon the competitiveness of the agriculture industry, a direct comparison was made of the emissions intensity of this industry across a number of countries (see Figure 7.18). Lack of data restricted the comparison to a subset of OECD countries.

Figure 7.18 includes direct emissions only. The exceptional emissions intensity of Australian electricity generation means that inclusion of these indirect emissions would set back Australia's performance in comparison with its OECD competitors.

Figure 7.17 Ratio of permit costs to value of production, 2005

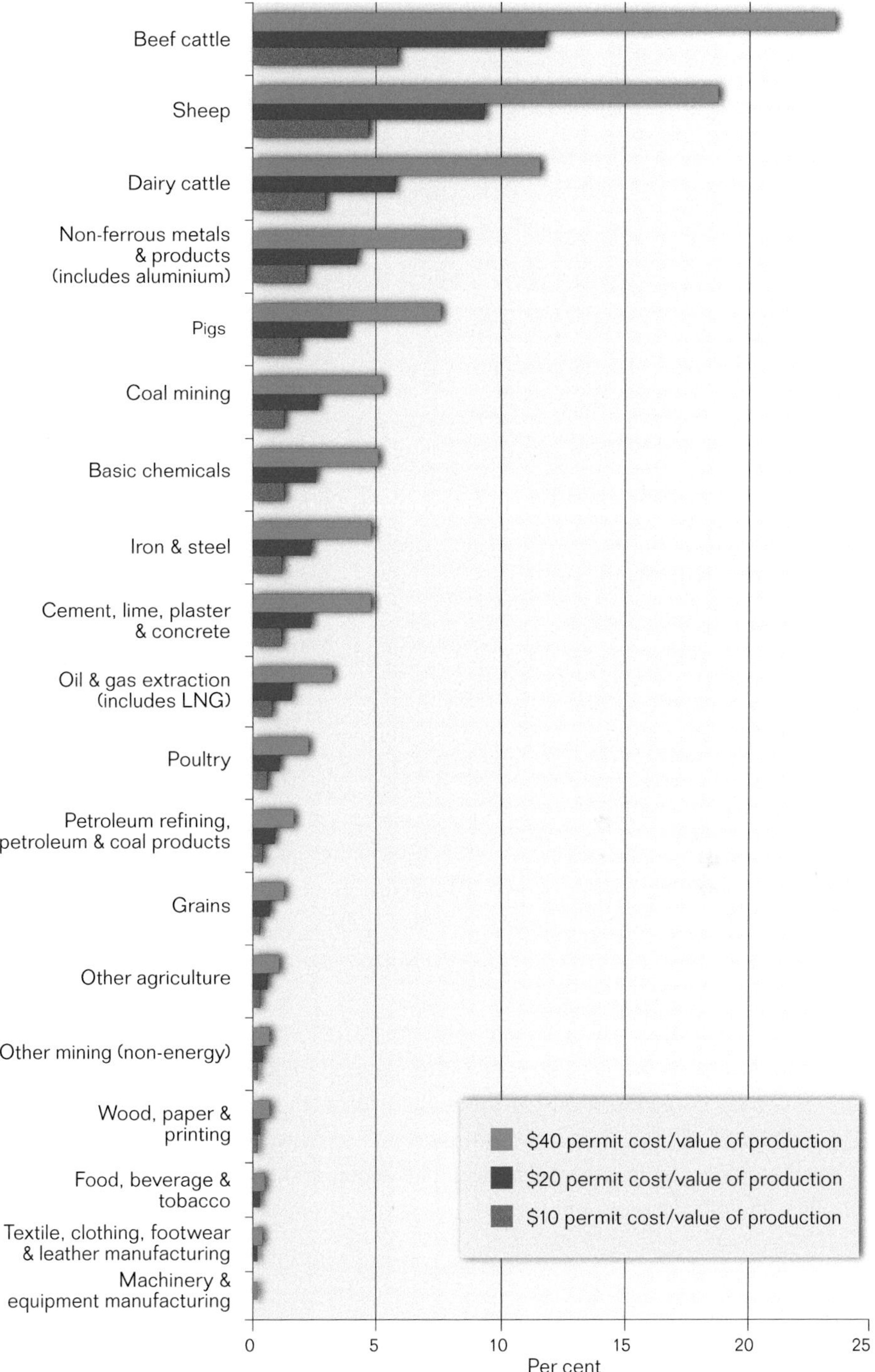

Note: Production is largely composed of sales revenue but also includes production for own final use.

Sources: DCC (2008a); ABS (2008).

Figure 7.18 Direct emissions intensity of Australia's agriculture industry compared with selected OECD countries, 2006

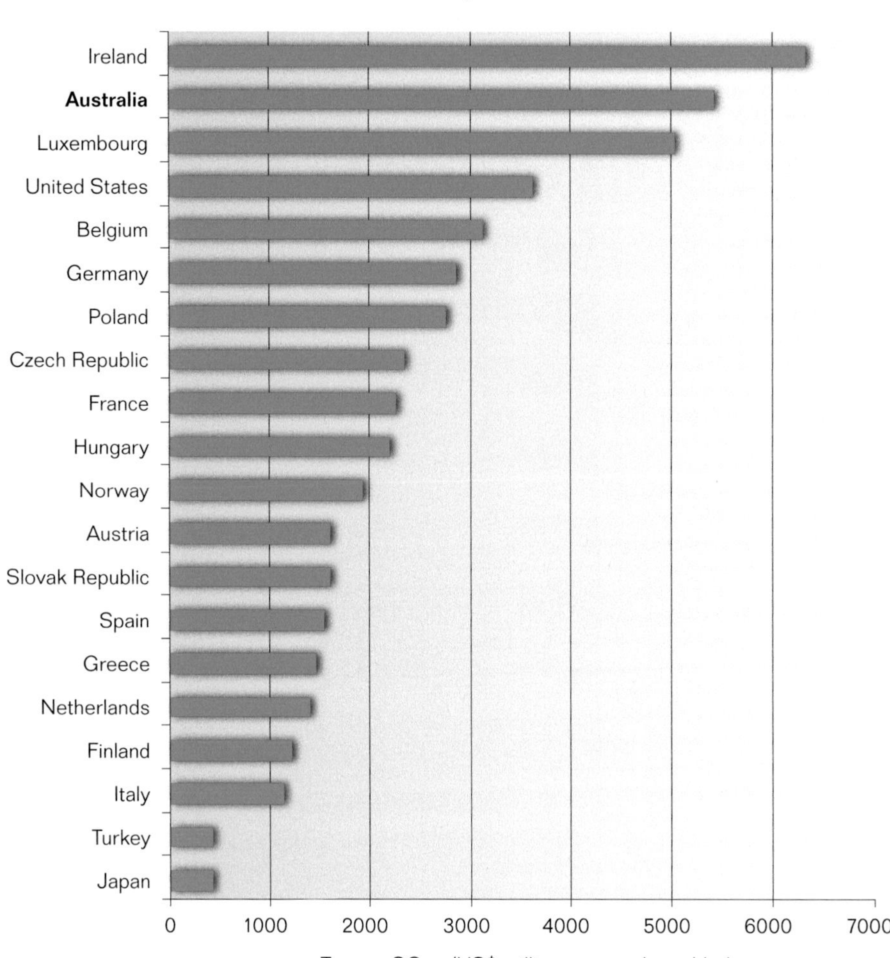

Sources: UNFCCC (2008); OECD (2008).

Australia's agriculture industry has the second highest emissions intensity of the countries considered and is more than twice the average. New Zealand's would be higher still, but data were not available to allow inclusion in Figure 7.18. This result reflects the large contribution that sheep and cattle production make to Australia's agricultural production.

References

ABS (Australian Bureau of Statistics) 2007, *Australian System of National Accounts, 2006–07*, cat. no. 5204.0, ABS, Canberra.

ABS 2008, *Australian National Accounts: Input–Output, 2004–05 (preliminary)*, cat. no. 5209.0.55.001, ABS, Canberra.

DCC (Department of Climate Change) 2008a, Australia's National Greenhouse Accounts, *Australian Greenhouse Emissions Information System*, <www.ageis.greenhouse.gov.au>, data as at 26 May 2008.

DCC 2008b, Australia's National Greenhouse Accounts, *Australian Greenhouse Emissions Information System*, <www.ageis.greenhouse.gov.au>, data as at 14 July 2008.

DCC 2008c, *National Greenhouse Gas Inventory 2006*, DCC, Canberra.

DCC 2008d, *Tracking to the Kyoto Target 2007: Australia's greenhouse emissions trends 1990 to 2008–2012 and 2020*, DCC, Canberra.

FAO (Food and Agriculture Organization of the United Nations) 2008, FORIS database, <www.fao.org/forestry/32185/en/>.

IEA (International Energy Agency) 2007a, *CO_2 Emissions from Fuel Combustion: 1971–2005*, IEA, Paris.

IEA 2007b, *Energy Balances of OECD Countries: 2004–2005*, IEA, Paris.

IPCC (Intergovernmental Panel on Climate Change) 1996, *Revised 1996 IPCC Guidelines for National Greenhouse Gas Inventories*, <www.ipcc-nggip.iges.or.jp/public/gl/invs1.html>.

Mackey, B., Keith, H., Berry, S. & Lindenmayer, D. 2008, *Green Carbon: The role of natural forests in carbon storage. Part 1. A green carbon account of Australia's south-eastern eucalypt forests, and policy implications*, Australian National University, Canberra.

OECD (Organisation for Economic Co-operation and Development) 2008, *National Accounts of OECD Countries, Volume IIa, 1995–2006, Detailed Tables*, OECD, Paris.

Population Reference Bureau 2008, *2006 World Population Data Sheet*, <www.prb.org>.

Syed, A., Wilson, R., Sandu, S., Cuevas-Cubria, C. & Clarke, A. 2007, *Australian Energy: National and state projections to 2029–30*, ABARE research report 07.24, prepared for the Australian Government Department of Resources, Energy and Tourism, Canberra.

UNFCCC (United Nations Framework Convention on Climate Change) 2005, *Key GHG Data*, UNFCCC, Bonn.

UNFCCC 2008, *2008 Annex I Inventory Submissions*, <http://unfccc.int/national_reports>.

ASSESSING THE INTERNATIONAL RESPONSE 8

Key points

Climate change is a global problem that requires a global solution.

Mitigation effort is increasing around the world, but too slowly to avoid high risks of dangerous climate change. The recent and projected growth in emissions means that effective mitigation by all major economies will need to be stronger and earlier than previously considered necessary.

The existing international framework is inadequate, but a better architecture will only come from building on, rather than overturning, established efforts.

Domestic, bilateral and regional efforts can all help to accelerate progress towards an effective international agreement.

The United Nations meeting in Copenhagen in December 2009 is an important focal point in the attempt to find a basis for global agreement. Australia must be prepared to play its full proportionate part as a developed country.

Greenhouse gas emissions are a global public 'bad'. One country's emissions affect all countries. Global warming therefore requires a global solution. Individual countries will not on their own undertake adequate mitigation, since each country is better off—from a narrow, national point of view—the more it can free ride on the efforts of others. As a country that is especially vulnerable to climate change, Australia has a strong interest in an effective international response to climate change.

An effective international response to climate change needs to cover both mitigation and adaptation. The main focus of the Review's discussion of the international response is mitigation, since adaptive responses are largely national and regional. However, an international element is required in the adaptation response.

This chapter assesses the global mitigation effort to date, and concludes that progress on the current trajectory is too slow and limited to constitute an effective global response to the risk of climate change. Chapters 9 and 10 outline a more effective response to international climate change.

8.1 The evolving international framework for addressing climate change

8.1.1 United Nations Framework Convention on Climate Change

The United Nations Framework Convention on Climate Change (UNFCCC) provides the foundation for the international collaborative effort to mitigate and adapt to climate change. The Convention was established in 1992, entered into force on 21 March 1994, and has been ratified by 192 parties to date, including Australia and the United States (both in 1992). It articulates a global goal of 'stabilization of greenhouse gas concentrations in the atmosphere at a level that would prevent dangerous anthropogenic interference with the climate system' (Article 2).

The Convention gives important guidance on the allocation of mitigation effort among countries, dividing parties into different groups according to their commitments. Annex I parties include the industrialised countries that were members of the OECD in 1992, plus countries with economies in transition.[1]

Apart from reporting duties, all countries commit to 'formulate, implement, publish and regularly update national and, where appropriate, regional programmes containing measures to mitigate climate change by addressing anthropogenic emissions' (Article 4.1(b)). Annex I countries are called on to do more. In particular, on the 'basis of equity and in accordance with their common but differentiated responsibilities and respective capabilities', developed countries 'should take the lead in combating climate change and the adverse effects thereof' (Article 3.1). Annex I countries are also called on to bear the cost of the financing 'needed by the developing country Parties to meet the agreed full incremental costs of implementing measures' to take actions to mitigate and adapt to climate change (Article 4.3).

8.1.2 Kyoto Protocol

The Kyoto Protocol was adopted by the UNFCCC parties on 11 December 1997, and entered into force on 16 February 2005. The Protocol commits developed and transition economies (essentially the Annex I countries of the UNFCCC) to limit or reduce their greenhouse gas emissions to specified levels during the commitment period from 2008 to 2012, with the aim of reducing their collective emissions by at least 5 per cent from 1990 levels (Article 3.1).[2]

The use of a five-year budget (2008 to 2012) is sometimes referred to as a 'flexibility when' provision, as it allows countries to average their emissions over time (Frankel 2007). 'Flexibility what' is also allowed under the Protocol, which includes fixed conversion factors for different greenhouse gases. Finally, the Protocol includes three 'flexibility where' mechanisms to assist countries to achieve their targets: international emissions permit trading, the Clean Development Mechanism, and Joint Implementation.

In international emissions permit trading, if a country with a target commitment reduces its emissions below its Kyoto target it can sell surplus reductions to another country. The other two flexibility mechanisms enable credits from emissions-reducing projects in one country to be used to meet the Kyoto target of another country. Under Joint Implementation, projects are hosted in countries with target commitments. Under the Clean Development Mechanism (see Box 8.2), projects are hosted in countries without target commitments (developing countries). While the supplementarity principle of the Protocol states that countries should primarily achieve their emissions reduction goals through domestic efforts, the Protocol does not place any quantitative limits on the use of flexibility mechanisms.

The Protocol also sets out specific rules regarding the accounting of emissions and removals from the land use, land-use change and forestry sector, establishes detailed accounting and reporting systems and creates a Compliance Committee.

8.1.3 The Bali Roadmap

The United Nations Climate Change Conference held in Bali, Indonesia, in December 2007 resulted in two negotiation tracks—the Convention track and the Protocol track, together known as the Bali Roadmap—aimed at achieving agreement on an arrangement to succeed the first Kyoto commitment period. While the exact shape of a future architecture is still unclear, both tracks are proceeding in parallel and have the same anticipated end date of December 2009, at which point parties will come together in Copenhagen with a view to agreeing on the way forward post-2012.

The Convention track negotiations will work towards a 'shared vision for long-term cooperative action', likely to be framed as a long-term global goal for emissions reductions. Developed countries have agreed to consider 'nationally appropriate mitigation commitments or actions, including quantified emission limitation and reduction objectives', while developing countries have agreed to consider 'measurable, reportable and verifiable' mitigation actions 'supported and enabled by technology, financing and capacity-building' (UNFCCC 2007a: 3). Underlying these undertakings is a commitment to put in place 'positive incentives for developing country parties for the enhanced implementation of national mitigation strategies and adaptation action' (UNFCCC 2007a: 5).

The purpose of the Protocol track is to agree on second commitment period (post-2012) emissions reduction commitments for UNFCCC Annex I parties. This track will need to result in quantified emissions reduction targets and agreement on the time frame of the second commitment period.

8.1.4 Other international initiatives

The UNFCCC is the focus of international climate negotiations, but is no longer the sole home of international discussions on climate change.

Major Economies Meeting on Energy Security and Climate Change

The Major Economies Meeting process on Energy Security and Climate Change was launched by the United States in September 2007 with the purpose of bringing together the largest emitters of greenhouse gases to discuss a global response to climate change.[3] US President George W. Bush nominated agreement in 2008 on a long-term global goal for emissions reduction as a key goal for the process.

Group of Eight (G8)

In 2005, climate change dominated the Gleneagles Leaders' Summit agenda, resulting in the establishment of the Gleneagles Dialogue on Climate Change, Clean Energy and Sustainable Development. Bringing together the G8 nations[4] as well as key developing countries and other major emitters, the Gleneagles Dialogue focused on technology and finance. It reported to the 2008 G8 Summit in Toyako, Japan (7–9 July 2008).

Asia–Pacific Economic Cooperation (APEC)

At the 2007 APEC Leaders Meeting in Sydney, Australia, the leaders of the 21 member economies[5] reaffirmed their commitment to the UNFCCC and agreed on an Action Agenda, which included APEC-wide aspirational goals of reducing energy intensity (the amount of energy used by unit of output) by at least 25 per cent by 2030 from 2005 and increasing forest cover in the region by at least 20 million hectares by 2020. Other agreements were to establish an Asia–Pacific Network for Energy Technology and an Asia–Pacific Network for Sustainable Forest Management and Rehabilitation.

Asia–Pacific Partnership on Clean Development and Climate

The Asia–Pacific Partnership on Clean Development and Climate is based on a model of cooperation and collaboration between partner governments,[6] business and researchers. Joint government–business task forces in eight sectors (cleaner fossil energy, aluminium, coal mining, steel, cement, buildings and appliances, power generation and transmission, and renewable energy and distributed generation) agree on projects that are then financed or provided with in-kind support by both government and industry participants. Progress to date has been limited by funding commitments.

Other international bodies

Work on climate change mitigation and/or adaptation is taking place in many other international bodies. These include UN agencies, the World Bank and regional development banks, the International Monetary Fund, the Organisation for Economic Co-operation and Development, the International Energy Agency, and others. The UN Secretary-General has made climate change a priority issue and the UN General Assembly holds regular thematic debates on the issue. Heads of state and government made declarations on the urgent need to address climate change at the Commonwealth Heads of Government Meeting and the East Asia Summit (both held in November 2007).

8.2 National commitments and policies to mitigate climate change

8.2.1 Developed countries

Some countries have proposed national emissions reduction goals beyond the end of the first Kyoto Protocol commitment period:

- **Australia**—The Australian Government has committed to an emissions reduction target of 60 per cent below 2000 levels by 2050.
- **European Union**—The European Union has put forward dual emissions reduction goals—an 'independent commitment' for a 20 per cent reduction over 1990 levels by 2020, and a conditional offer for a 30 per cent reduction over 1990 levels by 2020. The trigger announced for moving to the conditional offer is 'a satisfactory global agreement to combat climate change post-2012' (European Commission 2008b), which implies as prerequisites that 'other developed countries commit themselves to comparable emission reductions and economically more advanced developing countries commit themselves to contributing adequately according to their responsibilities and capabilities' (European Commission 2008a). The European Parliament and Environment Ministers have also proposed 2050 targets of a 60–80 per cent reduction relative to 1990 levels.
- **Individual European countries (EU member and non-member states)**—Some European countries have made separate national commitments, showing greater ambition than the EU approach. For example, the United Kingdom has committed itself to reducing emissions by 20 per cent on 1990 levels by 2010 and 60 per cent by 2050 (with scope for greater reductions if needed). Germany has committed to a 40 per cent reduction on 1990 levels by 2020. Norway is noteworthy—30 per cent reductions on 1990 levels by 2020 and carbon neutral by 2050.
- **Canada**—In April 2007, the Canadian Government announced new targets to reduce Canada's greenhouse gas emissions to 20 per cent below the 2006 level by 2020, and to 60–70 per cent below the 2006 level by 2050.
- **Japan**—In June 2008, the Japanese Government announced a target of a 60–80 per cent cut in emissions by 2050 from current levels, as well as plans for emissions trading, renewable energy targets, and low-emissions automobile targets.
- **Korea**, which is not bound by quantitative commitments under the Kyoto Protocol, has recently announced that in 2009 it will propose a 2020 emissions target below business-as-usual levels.
- **New Zealand** is in the process of introducing an emissions trading scheme. Its targets will be guided by international negotations.

- **United States**—Under the Bush administration, the United States declined to ratify the Kyoto Protocol or to take a strong stance on domestic emissions reductions. However, the signs from presidential candidates, Congress, various states and even the judiciary indicate that major changes in the US position can be expected (Box 8.1).

Box 8.1 Recent developments in US climate change policy

Active participation by the United States will be a crucial element of an effective global climate change framework.

Under the Bush administration, the United States declined to ratify the Kyoto Protocol and has taken a back seat in international negotiations. In April 2008, President Bush announced a new national goal to stop the growth in US greenhouse gas emissions by 2025.

In contrast, both presidential candidates have committed to reducing emissions to 1990 levels by 2020. The Democrats have promised an 80 per cent reduction and the Republicans 60 per cent, both from 1990 levels by 2050. Both candidates support taking on a more active international role and introducing a nationwide emissions trading scheme. This suggests that, whoever wins the November presidential election, the array of legislative cap and trade proposals introduced during the 110th Congress might be considered with a more open mind by the White House in future. The Lieberman-Warner Climate Security Act is the proposal that has had the most congressional support, though it too has so far been unable to command majority support, with Democrat legislators hesitating tactically in the lead-up to the November elections. Its provisions aim to reduce overall US greenhouse gas emissions from 2005 levels by roughly 63 per cent by 2050 (Pew Center 2007).

Meanwhile, some states have moved ahead. Multistate regional initiatives include the Regional Greenhouse Gas Initiative, involving northeastern states, the Western Climate Initiative, with California at its centre, and the Midwestern Regional Greenhouse Gas Accord. All have a cap and trade scheme at their core, although with different levels of ambition and design. California has passed legislation requiring emissions to fall to 80 per cent of their 1990 level by 2050.

Existing federal legislation, such as the Clean Air Act, is also being used to tackle climate change. The Bush administration is opposed to this course of action, but in 2007 the US Supreme Court decided that the Act gave authority to the US Environmental Protection Agency to regulate greenhouse gases and that the Agency would need to make a strong case if it decided not to exercise that discretion.

While major changes in policy can be expected after the November 2008 election, there is still uncertainty and the prospect of delay. Even with majority support in the Congress and a supportive president, US legislative processes, combined with the delays in establishing any new administration, mean that the timely passage of climate change legislation is far from guaranteed.

Many developed countries have policies in place to reduce emissions. These include emissions trading schemes, renewable energy targets, and fuel efficiency targets. In addition to its emissions trading scheme, the European Union has a goal of sourcing 20 per cent of its energy (specifically electricity, transport, and heating and cooling) from renewables by 2020. It has also legislated a suite of measures on building, appliance and vehicle standards. Japan has various renewable energy and performance standards in place for its industry. Canada aims to meet its targets by establishing a carbon trading scheme, requiring industry to improve its emissions performance, and introducing measures such as new fuel consumption standards for cars and energy efficiency standards for buildings. Many other developed countries are pursuing similar policies and measures, though most are struggling to meet their Kyoto targets (section 8.3).

The United States and European countries have introduced mandatory requirements and subsidies for the use of biofuels. These have put strong upward pressure on global food prices, with negligible environmental benefits.

8.2.2 Developing countries

All developing countries continue to reject containment of their emissions growth through the adoption of mandatory targets. Nonetheless, some developing countries have already made important domestic commitments or are on the way to doing so.

- As the largest developing country, and now the world's largest emitter, **China** is particularly important. As part of its 11th Five-Year Plan (2006–10), China has committed to reducing the energy intensity of its economic activity by 20 per cent below 2005 levels by 2010. In June 2007, China released its first National Climate Change Program, which confirmed the energy intensity target and also renewable energy and forest coverage targets. Under the program, the renewables goal is set at 10 per cent of the energy mix by 2020 (this has since been revised by the National Development and Reform Commission to 15 per cent by 2020), and an increase of carbon sinks by 50 million tons over 2005 levels by 2010. China has also announced its intention to halve its energy intensity by 2020 over 2008 (DCC 2005). These are ambitious targets that will not be easy to realise.
- **India** released its National Action Plan on Climate Change in June 2008. The plan identifies a national target area for forest and tree cover of 33 per cent (against a current area of 23 per cent) as well as a number of energy efficiency measures which will complement existing measures already expected to result in a saving of 10 000 MW by the end of 2012. The plan has a strong focus on the development and use of new technologies. It also includes a long-term commitment that India's per capita emissions will not exceed those of the developed countries (Government of India 2008).

- In 2007, **Brazil** released a white paper on its contribution to preventing climate change, focusing on energy and avoided deforestation. Specific initiatives referenced in the paper include the Program for Incentive of Alternative Electric Energy Sources, launched in 2002, which sets an overall goal of 10 per cent of annual energy consumption to come from renewables by 2022; and the National Ethanol Program, implementation of which has led to ethanol accounting for about 40 per cent of vehicle fuel use in Brazil (WRI 2008).
- The prime minister of **Papua New Guinea** has asked his country's newly established Climate Change Office to prepare an analysis of ambitious mitigation targets: a reduction in emissions of 50 per cent by 2020, and carbon neutrality by 2050 (Somare 2008). Papua New Guinea has large opportunities to reduce net emissions in the forestry sector (The National Online 2008).
- **South Africa** has launched a Long-Term Mitigation Scenarios process, designed to lay the foundations for a more comprehensive national climate change policy and eventually to 'inform a legislative, regulatory and fiscal package that will give effect to our policy at a mandatory level' (Department of Environmental Affairs and Tourism, South Africa, 2008). The South African government has not set specific targets, but has indicated that national emissions must peak by 2020–25 at the latest, and then stabilise and decline.

8.3 Assessment of progress under the Kyoto Protocol

The decisions not to ratify Kyoto by the United States and Australia after the election of the Bush administration seven years ago were of historic importance in disrupting an international approach. Australia's return to the international fold following the election of the Rudd Labor Government is an important corrective measure.

The performance of developed countries against their Kyoto Protocol targets varies (Figure 8.1).

- **Ahead of target**—Countries that were moving out of centrally planned economic systems, including Russia, Poland and Ukraine, were required to make similar reductions in emissions from 1990 levels to those of OECD countries. They currently have emissions at levels far below their targets due to the large fall in economic activity and emissions that occurred in the 1990s with the collapse of central planning. Since these emissions reductions were not the result of any mitigation effort but rather were achieved before the Kyoto Protocol was signed, the gap between emissions and the targets is often referred to as 'hot air'.
- **On target without use of flexibility mechanisms**—Australia is one of the few countries that currently have national emissions at or close to the level required by the Protocol over the period 2008–2012 (in Australia's case due to one-off reductions in land clearing).

- **On target if flexibility mechanisms are used**—The domestic emissions of most countries are above their Kyoto targets. This is true for the European Union as a whole, and for Japan and New Zealand. These countries could be in compliance with Kyoto if they were to purchase sufficient Clean Development Mechanism or Joint Implementation credits, or buy permits from those countries that are ahead of target (that is, the 'hot air' countries).
- **Off target**—In Canada, against a required 6 per cent cut, emissions had increased by 27 per cent as of 2005 compared to the 1990 base. In the United States emissions had grown by 16 per cent over the same period against a required 7 per cent reduction. The United States has not ratified the Kyoto Protocol. While Canada did ratify the Protocol, the current government has declared it will not be able to meet its target.

Figure 8.1 Kyoto targets and 2005 emissions, relative to 1990

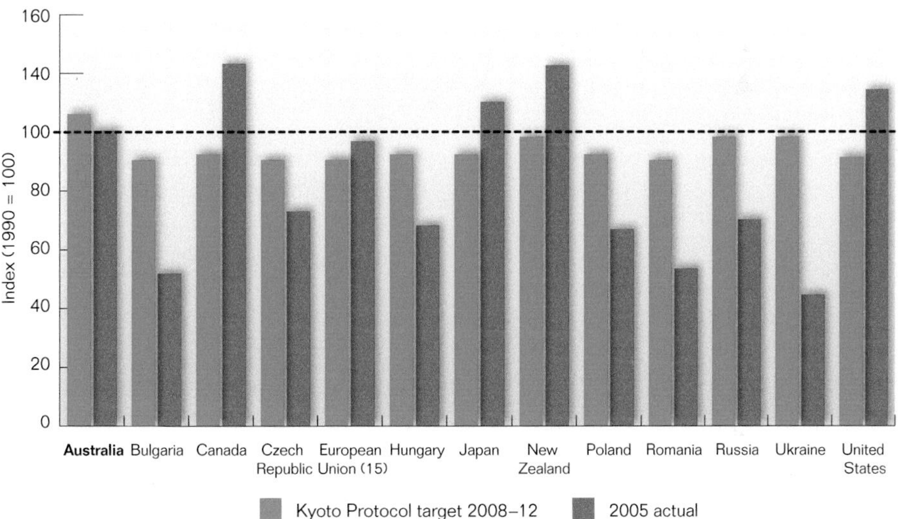

Notes: Only parties with emissions of 100 million tonnes of CO_2-e or more are included, except for New Zealand. The United States has signed but not ratified the Kyoto Protocol, and is not a party to it. The 2008–12 target is simply the Kyoto target over the 1990 baseline. Growth in greenhouse gases from 1990 to 2005 for countries other than Australia excludes land-use change and forestry. For countries other than Australia there may be discrepancies between greenhouse gas emissions as reported to the UNFCCC and as calculated in relation to Kyoto Protocol commitments. These are expected to be minor. For countries with base years other than 1990, the following years are used: Bulgaria—1988, Hungary—average of 1985–87, Poland—1988, and Romania—1989.

Sources: UNFCCC (2007b, 2008c); Australian Greenhouse Office (2007).

The Kyoto Protocol is more than a set of targets for Annex I countries. It engages developing countries through the Clean Development Mechanism, which promotes abatement projects. The Clean Development Mechanism has grown rapidly, but is flawed in a number of respects (Box 8.2).

In summary, the fact that most developed countries are in a position to achieve their Kyoto targets is positive. It is desirable for developed countries to be meeting part of their required emissions reductions through financing the mitigation efforts of developing countries, as this provides international financing for these efforts. However, the virtual repudiation of the Protocol by Canada and the failure of the United States to ratify are serious threats to its credibility. It is only in the last year or so that developed countries have started to pay more attention to, and put targeted financing into, research and development and mitigation financing in developing countries.

Box 8.2 The Clean Development Mechanism: is it flawed?

The Clean Development Mechanism (CDM) is a market-based offset mechanism. Tradable credits are awarded for emissions reductions on a project-by-project basis and the resulting credits are purchased by firms or governments under an obligation to reduce emissions. As of May 2008, there were around 3400 CDM projects under way or in preparation, covering 2.5 billion tonnes of carbon dioxide emissions equivalent until 2012 (UNEP Risoe Centre 2008). During 2007 the CDM had primary transactions worth US$7.4 billion, with demand coming mainly from private sector entities in the European Union, but also from EU governments and Japan. The World Bank (2008) estimates that in 2007, the CDM leveraged US$33 billion in additional investment for clean energy, which exceeded the cumulative amount over the previous five years.

The CDM's geographic coverage is concentrated. UNFCCC figures (2008a, 2008b) show that 65 per cent of CDM projects registered to date are in Asia and the Pacific (mainly India (31 per cent) and China (23 per cent)), 32 per cent in Latin America and the Caribbean, and only 2 per cent in Africa.

To many, the CDM is a 'win–win' solution for all countries—it provides developed countries with low-cost abatement opportunities and a way of engaging developing countries in mitigation efforts, and it provides developing countries with a source of funding for lower-emissions technologies and practices. However, it is becoming increasingly clear that the CDM is a flawed device, from both an environmental and a market perspective.

First, under CDM rules, a project must be proved to be additional, that is, it would not have been undertaken had it not been for the CDM. However, additionality is difficult to prove or disprove (Wara & Victor 2008).

Second, the project basis of CDM is problematic. It leads to high transaction costs and a patchy price signal for emissions reductions. There are moves under way to expand the CDM to cover programs of activities, but this may heighten concerns about additionality.

Third, an offset mechanism does not in itself lead to any global reduction in emissions. Rather, CDM credits are used by developed country parties wishing to emit more domestically. A CDM credit simply offsets domestic reductions in countries with targets.

Fourth, large-scale sales of CDM credits may stand in the way of developing countries taking on more comprehensive commitments. Recent signs that the European Union intends to restrict acceptance of CDM credits can be seen in this light (European Commission 2008b).

8.4 Projections given the current trajectory of mitigation effort

With emissions growing rapidly in recent years and projected to continue to grow, the current trajectory of mitigation effort is inadequate for achieving the UNFCCC goal of holding the risk of dangerous climate change to moderate levels.

The first challenge facing the world is for developed countries to commit to and implement deep reductions in emissions. The Review's modelling, reported in Chapter 11, suggests that reductions by 2020 in the order of 15 to 30 per cent of emissions over a 2000 base will be required. Given the limited progress to date, this in itself will be a major challenge.

For developing countries, the current direction of negotiations cannot be expected to deliver any reduction in global emissions beyond that credited to developed countries. If the Clean Development Mechanism continues to be the main vehicle for engaging developing countries in the international mitigation effort, then, even if it is expanded, all abatement in developing countries will continue to be on an offset basis financed by developed country payments in lieu of their own reductions in emissions. Developing country reductions could then be modelled as zero, since any actual emissions reduction in developing countries would simply lead to a correspondingly smaller reduction in emissions in developed countries (see Box 8.2).

Developing country emissions under business as usual will exceed by 2027 the global emissions limit modelled for eventual stabilisation of the concentration of greenhouse gases at 550 ppm CO_2-e, and by 2030 will exceed the global limit by 20 per cent. If the goal is eventual stabilisation at 450 ppm, then developing country emissions without any mitigation will on their own exceed the modelled global limit by 2024, and by 2030 will exceed that limit by 60 per cent.

Exceeding emissions stabilisation paths over the next decade and beyond would increase climate change risk. Offsetting the earlier overshooting would require deeper cuts in emissions in later years, possibly greatly increasing overall mitigation costs.

Clearly the current trajectory of effort traced out from the Kyoto Protocol to the Bali Roadmap and beyond will not enable the world to hold the risks posed by climate change to moderate levels. One of the reasons the current trajectory of mitigation effort is inadequate is that it has not responded to the acceleration in the growth of emissions seen so far this century, and projected to continue. Earlier scenarios forecast much slower emissions growth even in the absence of concern about climate change. This earlier outlook is captured by the 'SRES median scenario', which is representative of the various long-term scenarios developed by the IPCC in the 1990s (see Figure 3.8). Annual global emissions levels reached by the SRES median scenario in 2030 are realised by the Review's no-mitigation, or business-as-usual, scenario 10 years earlier, in 2020. As Chapter 3 showed, the SRES median scenario can no longer be regarded as a reasonable guide to future emissions growth. Other emissions trajectories that show much more rapid growth, once considered extreme, now appear moderate or even cautious. The world has changed, but climate change negotiations have not yet adjusted.

8.5 Accelerating progress

Without strong action by developed countries and firm commitments from major developing countries between now and 2020, it will be impossible to avoid high risks of dangerous climate change. Climate change negotiations have long been on a path that unhelpfully divides the world into two large groups.

Any multilateral negotiations concerning global public goods will face difficulties. The incentives facing individual delegations in a single, large multilateral negotiation are not conducive to reaching sound agreement. Each country will try to secure a better deal than others, with individual countries' perceptions of equity concerns figuring large, and with incentives for free-riding working against cooperative outcomes. Countries' circumstances and interests in the negotiations will differ widely, and geopolitical considerations will interfere. The dominant outcome is a low common denominator. This is evident from the experience with the Kyoto Protocol.

The world is dealing with a genuine international 'prisoner's dilemma', in which the cooperative outcome is the superior one, but in which countries have an incentive not to cooperate.[7] In the case of global warming, all countries are better off if they all reduce greenhouse gas emissions—but each country has an incentive to benefit from other countries' reductions in emissions without incurring any mitigation costs itself. A prisoner's dilemma can be reduced through communication and undertakings on side payments. But effective communications, and the development of understandings on the sharing of the gains from cooperation, take time.

There are four possible saving graces in international cooperation on climate change. One is the exceptional level of community interest and support for action in many countries, including Australia. The second is the issue's high international profile: the attention the issue is receiving across global forums and the growing number of countries, developed and developing, announcing emissions reduction targets and policies. The third is that a start has been made on international cooperation, with some countries taking steps towards emissions reduction. The fourth is that international climate change policy is not played out just once, but rather through interactions over time, allowing individual countries' policies to influence those of other countries (Axelrod 1984), and allowing agreements to evolve that are individually and collectively rational—and considered fair (Barrett 2003). The global success at combating ozone depletion (Esty 2007)—albeit at a much smaller scale and for a less challenging problem—shows that effective international action on environmental issues is possible.

How can the world build greater ambition into current international efforts to mitigate climate change? The two chapters following are based on four key principles for accelerating progress.

8.5.1 Building on existing architecture

While the Kyoto Protocol is inadequate and has been only partially implemented, it is a starting point. It would be counterproductive to attempt to start again with a new international architecture, based on a different set of principles, such as price rather than quantity targets. The chances of international agreement are better if existing frameworks are used to broaden participation and deepen ambition. The basic principles embodied in the Protocol are sound: the abatement burden should be distributed explicitly and equitably; and developed countries should support mitigation efforts in developing countries. Proposals to move forward should build on these principles.

8.5.2 Developed country leadership

No significant progress in the multilateral sphere will be possible until the United States shows that it is serious about addressing climate change by, among other things, adopting a credible long-term target. Legislative initiatives under way in the United States are encouraging in this regard, and a new administration is widely expected to take a positive role in international climate policy.

All developed countries need to be subject to, and meet, emissions reduction goals. It is important that developed countries show credible domestic abatement effort to demonstrate to developing countries their seriousness, and that it is possible to reduce emissions without sacrificing prosperity.[8]

A dual approach is needed. First, accelerating progress requires that developed countries show leadership and good faith by accepting binding reductions immediately and unconditionally. A number of developed countries, including Australia, have now indicated long-term reduction goals. Others need to follow suit. Second, steeper cuts can be offered if developing countries also agree to restrict emissions.

Emissions reduction goals need to be complemented by more generous offers of assistance and collaboration by developed countries through both trading and public funding.

Developed countries can exercise leadership by encouraging developing countries to come on board with regional initiatives.

Countries committed to effective international action on climate change will also need to provide negative as well as positive incentives for other countries to participate (section 10.6).

8.5.3 Developing country participation

Waiting until 2020 for any developing countries to commit to significant emissions containment policies (potentially the starting time for an agreement to follow the one currently being negotiated) would be to risk the prospect of achieving climate stabilisation at moderate levels. Reductions in developing countries' emissions below business-as-usual levels are needed in addition to developed country reductions, and not only as cheaper substitutes for them, as has been the case so far.

The differentiation between developing and developed countries, more recently reiterated in the Bali Roadmap, will continue to be important. However, interpretation of the phrase 'common but differentiated responsibilities'[9] as meaning that only one group of countries is responsible for containing emissions is no longer viable. All countries need to be jointly responsible, although poorer countries should have more flexible targets, reasonable room for growth in emissions entitlements, and the financial and technical support required to help them live within their emissions budgets.

For progress to be made, developing countries should not be seen as comprising a single category. Relevant differences in circumstances will need to be acknowledged. In particular, more can and should be expected of major emitters and of fast-growing, middle-income developing countries than of low-income countries. China, as the main source of global economic dynamism, a superpower, and already the world's largest emitter, is critical to the outcome.

Why would developing countries participate more actively in the international abatement effort? First, as they focus on the realities of prospective emissions growth and the risks associated with it, they will increasingly come to see an effective global agreement as being in their interest. China, South Africa and Brazil have already advanced a considerable way down that path. Second, major developing countries need to be offered financial incentives. The combination of transfer of public funds and technology, and the availability of funds from trading, would provide powerful incentives.

8.5.4 Action by individual countries and groups of countries

Given the limitations inherent in any multilateral process of negotiations, countries will also need to act unilaterally and in regional groupings to move from the status quo and increase the chance of a successful multilateral outcome. Early unilateral and regional efforts will help secure a more ambitious post-Kyoto framework.

Agreement on difficult political and economic issues can be much easier to achieve among small groups of countries than in large multilateral negotiations. In negotiations among small groups of countries it is easier to establish trust and take account of individual countries' circumstances and preferences. Furthermore, self-selected groups are much less subject to being held hostage by the least willing.

Formations of groups of countries that are prepared to agree on emissions reduction and technology transfer goals can accelerate global action by demonstrating that ambitious cooperative action is possible. In particular, groupings that bring developed and developing countries together into regional trading and technology transfer systems have the potential to show that developing countries can live within, and indeed benefit from, national emissions budgets. Agreements reached between major developed and developing emitters have the potential to break multilateral deadlocks and give negotiations fresh impetus. They allow for direct high-level political input, without which negotiations will languish, if not stall.

The hurdle for developing countries to take on emissions reduction commitments could be much lower in such a situation, as any commitments could be fashioned around the capabilities, needs and aspirations of each individual country. Similarly, it would make it easier for developed countries to enter into arrangements that include large-scale resource transfers to developing countries for climate change mitigation.

Unilateral, regional and multilateral efforts occurring in parallel might make for a messy process, but it is one that increases the chance of success in the short time available. The more individual countries and groups of countries undertake unilateral and regional efforts to mitigate climate change, and the sooner they do so, the greater the prospects for a comprehensive and ambitious future global framework.

To ensure compatibility, unilateral and regional schemes would need to be based around common guiding principles. Early movers on regional agreements would need to base their actions on explicit principles for allocating a global emissions budget that they consider to have good prospects for wider international acceptability. Early action on the basis of such principles would then play a role in the encouragement of international discussion of principles and in movement towards international agreement.

Notes

1 Countries with economies in transition under the UNFCCC are Belarus, Bulgaria, Croatia, Czech Republic, Estonia, Hungary, Latvia, Lithuania, Poland, Romania, Russian Federation, Slovak Republic, Slovenia and Ukraine.

2 Countries with target commitments are listed in Annex B to the Protocol, which largely coincides with Annex I to the UNFCCC.

3 Participants are the United States plus Australia, Brazil, Canada, China, the European Union (current President and European Commission representative), France, Germany, Indonesia, India, Italy, Japan, Mexico, Russia, South Africa, South Korea, the United Kingdom, and the United Nations.

4 The G8 nations are Canada, France, Germany, Italy, Japan, Russia, the United Kingdom and the United States. The European Commission is also represented at all meetings.

5 APEC's 21 member economies are Australia; Brunei Darussalam; Canada; Chile; People's Republic of China; Hong Kong, China; Indonesia; Japan; Republic of Korea; Malaysia; Mexico; New Zealand; Papua New Guinea; Peru; Philippines; Russia; Singapore; Chinese Taipei; Thailand; United States; and Vietnam.

6 Asia–Pacific Partnership on Clean Development and Climate partner governments are Australia, Canada, China, India, Japan, Republic of Korea and the United States.

7 The prisoner's dilemma is named after the situation in which two suspects would receive short sentences if neither informs on the other, and long sentences if both inform on the other. If only one suspect informs on the other, the informant will go free. The best solution for the suspects is the cooperative one (neither informs on the other), but each has an incentive not to cooperate (to inform). The prisoner's dilemma can be resolved through communication, and an agreement to shore the benefits of cooperation.

8 As Morgenstern (2007: 218) comments: 'The prospects for international progress would certainly be enhanced if one could point to genuine success in the United States or other large nation... Even though international negotiations on climate change have been under way for almost two decades, to date no major nation has yet demonstrated a viable domestic architecture suitable for achieving large-scale emission reductions and none, except for special cases like the United Kingdom, which experienced large changes in its resource base, or Germany, which benefited from economic restructuring, has made substantial progress in actually reducing emissions.'

9 The phrase 'common but differentiated responsibilities' appears in both the Rio Declaration and the UNFCCC.

References

Australian Greenhouse Office 2007, *National Greenhouse Inventory 2005: Accounting for the 108% target*, Australian Government, Canberra.

Axelrod, R. 1984, *The Evolution of Cooperation*, Basic Books, New York.

Barrett, S. 2003, *Environment and Statecraft*, Oxford University Press, Oxford.

DCC (Development Research Center of the State Council) 2005, *China's National Energy Strategy and Policy 2020*, DCC, Beijing.

Department of Environmental Affairs and Tourism, South Africa, 2008, 'SA: Van Schalkwyk: Environmental Affairs and Tourism Dept Budget Vote 2008/09', 20 May, <www.polity.org.za/article.php?a_id=133931>.

Esty, D.C. 2007, 'Beyond Kyoto: learning from the Montreal Protocol', in J.E. Aldy & R.N. Stavins (eds), *Architectures for Agreement: Addressing global climate change in the post-Kyoto world*, Cambridge University Press, New York, pp. 260–69.

European Commission 2008a, 'Proposal for a Decision of the European Parliament and of the Council on the effort of Member States to reduce their greenhouse gas emissions to meet the Community's greenhouse gas emission reduction commitments up to 2020', COM(2008) 17 final, 2008/0014 (COD), Brussels, 1 January.

European Commission 2008b, 'Questions and answers on the Commission's proposal to revise the EU Emissions Trading System', MEMO/08/35, Brussels, 23 January.

Frankel, J. 2007, 'Formulas for quantitative emissions targets', in J.E. Aldy & R.N. Stavins (eds), *Architectures for Agreement: Addressing global climate change in the post-Kyoto world*, Cambridge University Press, New York, pp. 31–56.

Government of India 2008, 'National Action Plan on Climate Change', <http://pmindia.nic.in/Pg01-52.pdf>

Morgenstern, R.D. 2007, 'The case for greater flexibility in an international climate change agreement', in J.E. Aldy & R.N. Stavins (eds), *Architectures for Agreement: Addressing global climate change in the post-Kyoto world*, Cambridge University Press, New York, pp. 209–19.

The National Online 2008, 'PM emphasises climate issues at Austrian forum', <www.thenational.com.pg/070808/nation2.php>.

Pew Center on Global Climate Change 2007, 'Lieberman-Warner Climate Security Act—S.2191—Summary of version passed by Senate Environment and Public Works Committee on December 5, 2007', <www.pewclimate.org/federal/analysis/congress/110/lieberman-warner>.

Somare, M.T. 2008, 'Climate change—policy change', speech to 13th Europa Forum, Wachau, Austria, <www.europaforum.at/en/news/show/id/5>.

UNEP Risoe Centre 2008, 'CDM/JI Pipeline Analysis and Database', 1 May.

UNFCCC (United Nations Framework Convention on Climate Change) 2007a, 'Report of the Conference of the Parties on its thirteenth session, held in Bali from 3 to 15 December 2007, Addendum, Part Two: Action taken by the Conference of the Parties at its thirteenth session', FCCC/CP/2007/6/Add.1.

UNFCCC 2007b, 'National greenhouse gas inventory data for the period 1990–2005', <http://unfccc.int/documentation/documents/advanced_search/items/3594.php?rec=j&priref=600004364#beg>.

UNFCCC 2008a, 'CDM: Registered projects by host party', <http://cdm.unfcc.int/Statistics/Registration/NumOfRegisteredProjByHostPartiesPieChart.html>.

UNFCCC 2008b, 'CDM: Registered projects by region', <http://cdm.unfccc.int/Statistics/Registration/RegisteredProjByRegionPieChart.html>.

UNFCCC 2008c, 'Kyoto Protocol base year data', <http://unfccc.int/ghg_data/kp_data_unfccc/base_year_data/items/4354.php>.

Wara, M.W. & Victor, D.G. 2008, 'A realistic policy on international carbon offsets', Program on Energy and Sustainable Development Working Paper #74, Stanford University, Stanford, California.

World Bank 2008, 'State and Trends of the Carbon Market 2008', <http://siteresources.worldbank.org/NEWS/Resources/State&Trendsformatted 06May10pm.pdf>.

WRI (World Resources Institute) 2008, National Alcohol Program (PROALCOOL), <http://projects.wri.org/sd-pams-database/brazil/national-alcohol-program-proalcool>.

9 TOWARDS GLOBAL AGREEMENT

Key points

Only a comprehensive international agreement can provide the wide country coverage and motivate the coordinated deep action that effective abatement requires.

The only realistic chance of achieving the depth, speed and breadth of action now required from all major emitters is allocation of internationally tradable emissions rights across countries. For practical reasons, allocations across countries will need to move gradually towards a population basis.

An initial agreement on a global emissions path towards stabilisation of the concentration of greenhouse gases at 550 CO_2-e is feasible. 450 CO_2-e is a desirable next step. Agreement on, and the beginnings of implementation of, such an agreement, would build confidence for the achievement of more ambitious stabilisation objectives.

All developed and high-income countries, and China, need to be subject to binding emissions limits from the beginning of the new commitment period in 2013.

Other developing countries—but not the least developed—should be required to accept one-sided targets below business as usual.

The international response to climate change is too slow and patchy to be effective. The discussion is conducted at an abstract level, and outside any requirements that numbers being discussed 'add up' to a global solution.

How can we build on existing international frameworks and negotiations to deliver an international agreement that is sufficiently ambitious to avert high risks of dangerous climate change?

A satisfactory international agreement will be difficult to reach. The prospects depend on the level of global community interest in mitigation. They depend on close communication across countries, over the years ahead, directed at developing a set of requirements which add up, and which, taken together, are widely seen as being fair.

With increasing international knowledge of the urgency of the risks, the political possibilities in the period ahead will widen.

An effective international global agreement to limit the risks of climate change will need to cover two main areas. First, the quantum of mitigation effort needs to be agreed. By how much will emissions be reduced, worldwide, and in each country? Second, while each country will be responsible for achieving its climate change mitigation goals, mechanisms for international collaboration will need to be in place to support national action. The most important of these will be international trading of emissions entitlements and public funding for technological development and adaptation.

These two areas are covered by the Kyoto Protocol, which takes as its starting point the global stabilisation goal of the United Nations Framework Convention on Climate Change (UNFCCC) and allocates emissions limits to most developed and transition countries. The Kyoto Protocol also introduces mechanisms for international collaboration. As argued in Chapter 8, while the Kyoto Protocol is not an adequate global response to climate change, any future, more effective response will have to build on it. There is no time to start again.

This chapter covers the first of the two areas: reaching agreement on global and national climate change mitigation goals. Chapter 10 discusses mechanisms for international collaboration.

9.1 Agreeing on a global goal

Determining limits over time on global emissions involves striking a balance between the benefits associated with smaller and slower climate change and the costs associated with greater and faster mitigation. The appropriate extent of mitigation is defined by the point at which the additional gains from mitigation are similar to the additional costs. In the end, judgment is required on the level of climate change that corresponds to this balancing point.

Targeted limits on climate change can be defined at three levels. At the highest level they can be defined in terms of impact or global temperature increase. At the next level they can be defined in terms of the profile for concentration of greenhouse gases in the atmosphere, which drives temperature increases. And at the third level, they can be defined in terms of emissions of greenhouse gases, which drive atmospheric concentrations.

9.1.1 Impact goals

Targets for global mean temperature compress the multiplicity of possible impacts (ranging from glacial melting to increased weather-related calamities) into a single variable. The European Union, for example, has argued that global mean warming should not be allowed to exceed 2°C from pre-industrial levels (Council of the European Union 2007).

Endorsement of a temperature threshold (and therefore of any target derived from it, for example, greenhouse gas concentration) cannot imply indifference to other factors. There may be tipping points associated with particular temperature thresholds, but the thresholds are not known with certainty.

Figure 9.1 **Different concentration goals: stabilisation, overshooting and peaking**

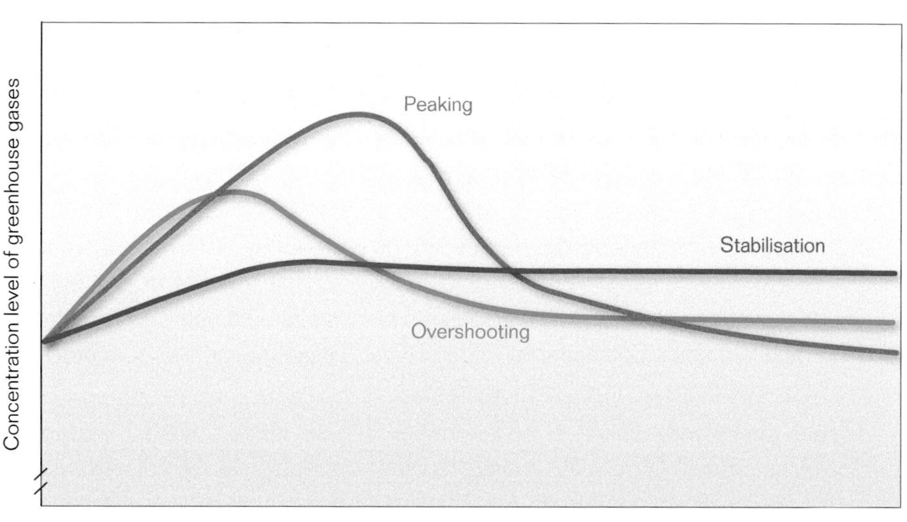

9.1.2 Concentration goals

Global warming increases temperature with a long lag. It might take more than a century after stabilisation of greenhouse gas concentrations for a new equilibrium temperature to be reached. Any goals in terms of temperature need to be translated into goals for the atmospheric concentration of greenhouse gases.

Chapter 2 introduced various types of concentration goals: stabilisation, peaking and overshooting (Figure 9.1). Most attention has focused on stabilisation scenarios, and the UNFCCC goal of the 'stabilization of greenhouse gas concentrations in the atmosphere' (Article 2).

However, as discussed in Chapter 2, special challenges are introduced by the need to reduce greenhouse gas concentrations to low levels. There is great difficulty in moving directly to that outcome from where the world is now. Whether the ultimate aim is stabilisation or prolonged decline after an initial rise (peaking), there is a good chance that the optimal response to climate change will need to involve a period (of uncertain duration) during which concentrations fall. This assumes that emissions can be brought below the natural level of sequestration. Reducing emissions below this level would probably require the development and deployment of technologies for carbon capture, such as new approaches to biosequestration.

The Review modelled two global mitigation scenarios (Chapter 4). The 550 scenario is a stabilisation scenario at which the concentration of greenhouse gases in the atmosphere approaches 550 ppm carbon dioxide equivalent (CO_2-e) and stabilises at around that level thereafter. The 450 scenario is an overshoot scenario under which concentrations peak at around 500 ppm CO_2-e and then stabilise at around 450 ppm CO_2-e. Any lower stabilisation objective, for example at 400 ppm CO_2-e, would need to involve a longer period of overshooting.

9.1.3 Emissions goals

Any concentration profile has an associated emissions trajectory. (An emissions trajectory defines the flow of greenhouse gases that converts, through various physical and chemical processes, into a stock of greenhouse gases in the atmosphere.) [1]

There are different ways in which goals for emissions can be expressed:[2]

- **End-period emissions**—This is the most common way of announcing targets (for example, that emissions will be reduced by 50 per cent by 2050). The advantage of this approach is simplicity. The disadvantage is that a target at one point of time says nothing about the rate at which emissions should approach that target level, and so does not constrain cumulative emissions or the concentration profile at that point of time (see Figure 9.2).
- **Annual emissions**—Since a concentration profile implies annual values for emissions, annual targets for emissions can be articulated. The disadvantages of this approach are complexity and inflexibility. There may be little difference in the environmental impact of two trajectories that end with similar concentrations but that have different annual emissions levels. However, the two paths could have quite different costs.
- **Cumulative emissions**—This is the budget approach, by which the total emissions determined by a target concentration profile over a number of years are summed up into a single target budget. In this approach, year-to-year variation from the target profile is allowed; what matters is to the total emissions over a number of years.

The benefit of the budget approach is its flexibility: it allows inter-temporal trade-offs and smoothing. Variations in timing would have to be large to have material environmental impacts. Variations within five-year periods as proposed in Chapter 14 would not have material effects.

The Review makes extensive use of emissions trajectories (see, for example, Chapter 12) to express emissions goals, and budgets to provide intertemporal flexibility.

Figure 9.2 Different cumulative emissions from the same end-year target (y)

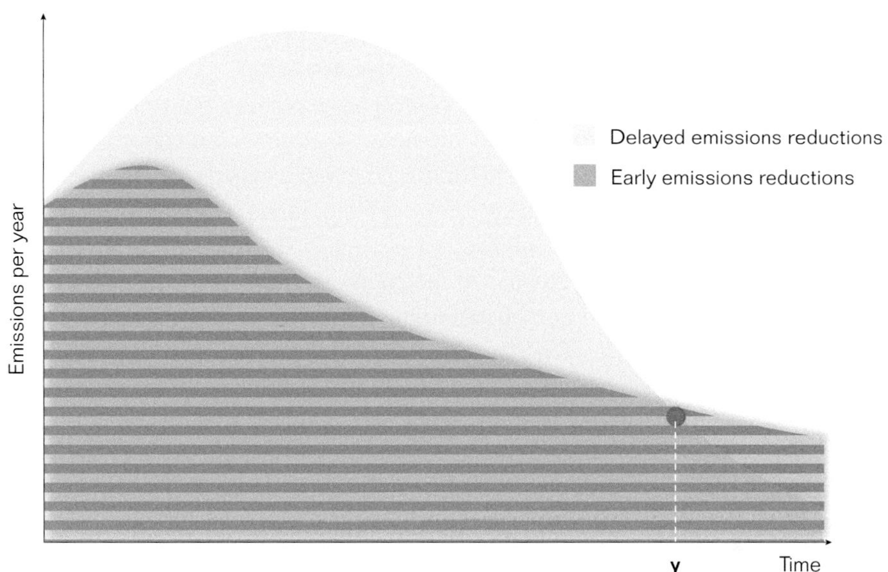

9.2 What form should national commitments take?

Once a global goal has been agreed, responsibility for its achievement needs to be allocated among countries. Unless all major economies agree to limit their emissions, it will be impossible to ensure that action at the global level adds up to an effective mitigation effort.

While any global agreement will emerge from negotiations—especially between the major emitters, and in particular between the two largest emitters, China and the United States—it is useful to spell out basic principles that could provide a framework for reaching agreement.

Proponents of price-based emissions control have argued for the adoption of national carbon taxes (Cooper 2000; Nordhaus 2008), or a common global carbon tax (Stiglitz 2006). Hybrid policies combining quantity and price controls have also been proposed, principally through emissions cap and trade schemes, but with a government-backed price cap as well (Roberts & Spence 1976; McKibbin & Wilcoxen 2002, 2008; Pizer 2002; Murray et al. 2008). A variant of the hybrid scheme has the price cap agreed internationally.[3]

At the heart of the economic argument for price control is uncertainty about abatement costs. The theory of prices versus quantities for pollution control (Weitzman 1974) shows that such uncertainty will invariably lead the policy to under- or overshoot the optimum. Imposing a quantitative target will lead to higher or lower marginal abatement costs than expected, while a given tax rate will lead to a greater or lesser abatement effort than expected. The resulting efficiency costs are thought to be lower under a price-based instrument for stock pollutants such as greenhouse gases, so getting the price wrong under a tax imposes smaller welfare losses than getting the quantity wrong under a quantity target.

Proponents of price-based emissions control have pointed out that a common global carbon tax or an agreement on an internationally harmonised price to apply in domestic permit trading schemes would avoid both questions of distribution between countries inherent in a cap and trade system, and the potentially destabilising effects of large-scale international financial flows.

While the introduction of a tax-based mitigation system would take the world significantly forward, only an international agreement that explicitly distributes the abatement burden across countries by allocating internationally tradable emissions entitlements has any chance of achieving the depth, speed and breadth of action that is now required in all major emitters, including developing countries.

There would undoubtedly be some advantages in relying instead on a carbon tax to provide the foundation for an international agreement. International trade in permit entitlements and a global carbon tax applied in all countries at a common rate would both give rise to a common international permit price. The carbon tax would avoid contentious international discussion of the allocation of entitlements. The political economy pressures for distortion are more easily avoided with a carbon tax. The simpler carbon tax would have lower transaction costs. Further, the certainty of the price under the tax would be seen as having advantages for business, although uncertainty about emissions pricing within a cap and trade scheme must be viewed in the context of the manifold other demand and supply shocks, especially the natural price volatility in the energy and resource sectors. In particular, within a cap and trade system, demand for and price of permits can be expected to fall in response to any large increase in the price of fossil fuels. This would be to some extent stabilising, unlike the rigid application of a fixed carbon tax. As a fixed carbon price cannot be expected to generate any particular abatement outcome, the tax rate would need to be adjusted from time to time. This would introduce a source of policy uncertainty that need not be present in quantity-based systems.

There are several reasons, however, why a quantity-based international agreement—where countries take on quantitative commitments to limit and reduce emissions, differentiated according to broadly accepted principles, with trade in emissions rights between countries (cap and trade)—is more likely to succeed than a tax-based one.

First, the tradable emissions entitlements approach builds on current international architecture and national practice. Quantitative targets have been the dominant form of greenhouse gas commitments so far. As in the Kyoto Protocol, quantitative targets frame the various existing and emerging national and regional climate goals and emissions trading systems, as well as the negotiations about national target commitments for the post-2012 period. The urgency of the situation means that current efforts need to be built on, not overturned. While different architectures could in theory be designed that might be superior, time has run out for new approaches and periods of trial and error.

Second, a cap and trade scheme provides incentives for developing country participation. Crucially for the goal of international cooperation, targets can be differentiated between countries without sacrificing economic efficiency. Under a price-based regime, commitments could be differentiated among countries by agreeing on lower emissions penalties for developing countries, but this would compromise the efficiency of the global mitigation effort and do less to provide a level playing field for emissions-intensive industries.

International trading in emission entitlements allows financial flows between countries. Such financial flows could offset abatement costs in developing countries, drawing them into an international policy framework.

Third, setting quantitative targets can control emissions levels more directly than setting emissions prices. This allows the extent of commitments to be more easily communicated. This could become more important as climate change risks become more urgent and the possibility of catastrophic damage from climate change gains recognition.

Fourth, trajectories and budgets can be implemented with flexibility over time and between countries to prevent cost blow-outs of the sort feared by advocates of carbon taxes. Flexibility can be provided by defining emissions budgets over a number of years, allowing inter-temporal flexibility across commitment periods, by allowing substitution between different greenhouse gases, and by allowing international trading of emissions rights.

Fifth, the adoption of national limits gives countries freedom to apply their own preferred mix of policies. A quantitative commitment under an international agreement does not mean that *domestic* policies need to be framed in quantity terms. A country could choose to introduce a tax on domestic emissions, regulation aimed at reducing emissions, a domestic emissions trading scheme, or a combination of these instruments. Some countries would undoubtedly choose to achieve their mitigation objectives through carbon taxes and regulation (the choice of domestic instruments is explored with respect to Australia in Chapter 13). International supervision of emissions commitments would be limited to monitoring emissions. By contrast, adoption of a carbon tax would require more intrusive international oversight (Frankel 2007). It would be necessary, for example, to ensure that countries did not offset a carbon tax by an increase in fossil fuel subsidies. Given the different tax treatment of fossil fuels around the world, it would be difficult, if not impossible, to ensure that national carbon taxes were both additional and comparable.

Carbon taxes could play a useful role in international commitments in some areas (Chapter 10). In the foreseeable future, it is not realistic to expect every country to be subject to quantity limits. It would be reasonable in such a situation for countries not subject to quantity limits to be under international pressure to introduce an offsetting carbon tax on the main trade-exposed, emissions-intensive industries. The revenue raised by such a tax would be retained by the government that imposed it. Such sectoral approaches would also be viable for emissions control in international aviation and shipping.

9.3 A graduated approach to national commitments

The Kyoto Protocol allocates internationally tradable emissions rights to countries that belonged to the OECD in 1992 and transition economies. This group excludes a number of high-income countries including Singapore, the Republic of Korea, Mexico and Saudi Arabia. Note that the Republic of Korea and Mexico are now members of the OECD. Many more countries will join the ranks of high-income countries in the years to come.

The principle that all high-income countries should adopt binding commitments to limit their emissions would receive widespread support. There is also broad agreement that developing countries need to take on greater obligations, although no political resolution of this issue has been in sight. So far, developing countries have resisted taking on emissions targets. The 2007 Bali Roadmap calls on developing countries only to take *actions* to reduce emissions, in contrast to the *commitments* to be taken on by developed countries. How can a way be found through this conundrum?

Clearly, differentiation is needed within the group of developing countries. The poorest, least developed economies are not ready for a national approach. They could be involved in the mitigation effort through offset mechanisms such as a strengthened Clean Development Mechanism, and international sectoral agreements where applicable (see Chapter 10).

But middle-income countries such as South Africa and Brazil and many others need to do much more. Some argue for a highly flexible approach, which would allow 'different countries to assume different types of international commitments—not only absolute targets, but also indexed targets, taxes, efficiency standards, and so forth' (Bodansky 2007: 65). Too many options, however, would make comparative assessment impossible, and therefore invite dilution of effort. Lack of a common framework would also obstruct international trading, which is the most likely route for developing countries to receive large-scale financing in support of their mitigation efforts.

Most developing countries cannot initially be expected to sign on to targets that would require them to buy emissions rights from other countries if they exceeded their emissions budgets. One-sided targets—also referred to as opt-out or non-binding targets (Philibert 2000)—could be a helpful expedient for a transitional period. With a one-sided target, countries could benefit from taking on a commitment by going further than their target required and selling emissions rights, without obligation to buy if they missed the target.

Allowing countries to adopt one-sided targets has a cost. It increases uncertainty about whether countries will follow through with their target commitments. To achieve similar global abatement as with binding targets for all countries, the countries with binding targets would need to take on more stringent commitments in order to reduce any shortfall from countries that opted out (Jotzo & Pezzey 2006). The existence of an opt-out option might weaken the resolve of national governments to follow through with mitigation policies, particularly where there are vested interests to be tackled or politically difficult decisions to be made, such as the removal of subsidies on petroleum products.

While recognising the drawbacks of one-sided commitments, the Review also recognises that most developing countries, given their low income per capita, would simply not be prepared or, in many cases, able to purchase emissions permits internationally. The risk associated with such an obligation would prevent many from accepting a binding target in the first place. The Review therefore supports the use of one-sided targets for most developing countries, to facilitate immediate uptake of target commitments and as a transitional measure in place until perhaps 2020. After that, these countries would be expected to accept binding targets.

Some argue that developing countries should be given targets set, at least initially, at their business-as-usual levels. Under this approach, promoted by Stern (2008), the Commission on Growth and Development (2008) and Frankel (2007), developing countries would only reduce emissions below business-as-usual levels if developed countries paid them to do so. Essentially, this approach amounts to an expansion of the Clean Development Mechanism to an economy-wide level.

The flaw with this business-as-usual approach is that it would put the entire burden of emission reductions on developed countries, and constraints on developed countries alone cannot reduce emissions enough to avoid high risks of dangerous climate change. Since developed countries account for a falling share of global emissions (see Chapter 4), it is unrealistic to hope to achieve substantial cuts in global emissions in this way. Developing country targets, albeit one-sided, need to be below business-as-usual levels.

The Review's proposal—for middle-income developing countries to adopt one-sided emissions targets below business as usual—goes further than most, if not all, current proposals for developing country commitments. Given the rapid growth in emissions, any less ambitious international agreement would be an inadequate response to the urgency of the problem. In the Review's framework, developing countries will have incentives to agree to such an approach: the prospect of financial gain through international selling of permits and access to international public funding in support of both mitigation and adaptation.

The Review's proposal thus requires identification of three groups of countries based on level of commitment. At the top of the income range, countries are subject to binding emissions commitments. At the bottom, countries are subject to minimal commitments. In the middle, countries are subject to one-sided commitments below business as usual. How should countries be assigned to these three groups?

It is in the global interest for as many countries as possible to be in the group with binding targets. This group should at a minimum consist of all countries currently in Annex I of the UNFCCC plus all other high-income countries. Where the high-income threshold is drawn would be a matter for negotiations.

China is a special case. Because of the country's size, current and prospective economic growth, geopolitical importance and emergence as the world's largest emitter, no global agreement would be effective unless China took on binding targets. China's fiscal, economic and technological position would allow it to do so. Of course, because of its lower income status, China's targets would not be as stringent during a transition period as those of developed countries.

The first group, if it did include China, existing Annex I members and other high-income countries (using, for this purpose, the World Bank per capita income threshold of US$11 000), would account for approximately three-quarters of global emissions of carbon dioxide from fossil fuel combustion, the main source of greenhouse gases.

The second group, expected to take on one-sided targets, would comprise most of the developing countries. This would include all members of the US-led Major Economies Meeting process not in the first group. As discussed in the next section, countries' emissions limits would be set using per capita principles. This group would account for almost all of the remaining quarter of present-day emissions from fossil fuels.

The third group would comprise countries classified as 'least developed' by the United Nations and any other developing countries that, on an objective assessment, do not yet have the necessary preconditions for a national approach—for example, those experiencing conflict or lacking the prerequisites for reliable emissions accounting. Countries in this group would be welcome but not required to take on one-sided targets. They would be able to host Clean Development Mechanism–type activities and sell offset credits, and would be expected to place a carbon tax on emissions-intensive industries producing in large amounts tradable goods that were the subject of global sectoral agreements.

It is worth reiterating that the proposed arrangements are intended only as a short transitional stage directed at achievement of a sound long-term international approach. At an early future point, desirably 2020, countries in the third group would be expected to take on one-sided targets, and countries in the second group binding targets. Countries would graduate from group to group over time.

9.4 Principles for allocating emissions entitlements across countries

In the approach outlined in the previous section, all except the least developed countries would have national emissions limits, albeit of differing types. This leaves the crucial and contentious question of how emissions rights are to be allocated across countries.

This is the question upon which the prospects of effective international agreement over the next two years will stand or fall. There are as many different possible international allocations as there are human minds to contemplate them. All can be dismissed if they do not 'add up' to a global total that meets the requirement of avoiding unacceptable risks of dangerous climate change. The proposals put forward here add up. They are based on principles that are thought to have a chance of global acceptance. Others, abroad and perhaps in Australia, can develop other proposals that also add up. These can be compared with the Review's proposal, with a view to arriving at one proposal that adds up and has wide support from heads of governments of major economies in advance of the Copenhagen meeting in December 2009.

9.4.1 Towards agreement on principles

Under the Kyoto Protocol, emissions budgets for Annex I countries for 2008–12 were defined as percentages of 1990 emissions, ranging within a relatively narrow band from 92 per cent to 110 per cent of base year emissions around the average allocation of 95 per cent, with further differentiation within the European Union. Differentiation between countries was negotiated on an ad hoc basis, with little reference to underlying principles for allocation across countries, although on average richer countries signed up to larger reductions.

In future negotiations, involving a greater number and more diverse array of countries, simply requiring somewhat differentiated reductions from a historical base, as under the Kyoto Protocol, will not underpin international agreement. The stark differences in per capita emissions levels across countries would need to be factored in. Emissions entitlements for the lower-emissions countries, which typically are also at a relatively low income level, would need to continue to grow for some time, but at a slower pace than currently anticipated under business as usual. Emissions entitlements in the richer countries would need to fall.

Leaving emissions reductions to politics, negotiations and arm-twisting, without explicit criteria, would prove deeply problematic. While politics and special circumstances will inevitably have some role, agreement on basic principles for allocation will be critical if the pace of coordinated international mitigation action is to quicken. An allocation framework based on simple principles, if it received widespread international support, could facilitate international negotiations, and in the meantime guide individual countries' commitments ahead of a new international agreement.

To be effective, a future international policy regime will require the mitigation effort to be distributed using principles that are widely accepted as being fair and practical. To be widely accepted, principles to guide the allocation of a global emissions budget across countries will need to be simple, transparent and readily applicable. To be considered fair, they will need to give much weight to population. To be considered practical, they will need to allow long periods for adjustment towards positions that give weight to population.

Various principles have been suggested. The UNFCCC emphasises *capacity*, with its call for greater and earlier mitigation effort by developed countries (those with more capacity). Graduation of a country to a more stringent level or type of commitment once it reaches some income threshold is a common feature of many proposals. Examples are the Pew Center Pocantico Dialogue (Pew Center on Climate Change 2005), the South–North Dialogue's proposal in Ott et al. (2004), and the São Paulo proposal (BASIC Project 2006). Section 9.3 argued that countries should take on more stringent types of commitments as they move from low to middle to high income status.

Some countries emphasise *responsibility*, and argue that future emissions rights should take account of how much each country has drawn historically on the atmosphere's total capacity to absorb emissions. Current industrialised countries have contributed a disproportionate share of past cumulative emissions. Historical responsibility was formally introduced to the UNFCCC by the government of Brazil (UNFCCC 1997), which called for mitigation to be shared on the basis of the contribution to climate change of countries' past emissions.

It has also been argued that emissions rights should be based on the *effort* required to meet the limits imposed. Effort could be measured in terms of the impact of mitigation action on national GDP. However, this approach takes no account of differential starting points, and would require comparing the future state of the world to the counterfactual of what would have prevailed in the absence of the scheme.

Underlying all these approaches is a concern with international *equity* made explicit in many allocative proposals. For example, the recent Greenhouse Development Rights framework (Baer et al. 2007) would apply equity considerations comprehensively to include adaptation costs and domestic income distribution. It is difficult, however, to see how broad agreement on what is equitable could be achieved in anything other than a very simple framework.

9.4.2 A per capita approach

While all of these approaches have strengths and weaknesses, the approach that seems to have the most potential to combine the desired levels of acceptability, perceived fairness and practicality is one based on gradual movement towards entitlements to equal *per capita emissions*. An approach that gives increasing weight over time to population in determining national allocations both acknowledges high emitters' positions in starting from the status quo and recognises developing countries' claims to equitable allocation of rights to the atmosphere.

Any allocative formula that does not base long-term emissions rights on population has no chance of being accepted by most developing countries. Indeed many developing countries would argue that a per capita approach does not go far enough as it does not address the issue of historical responsibility. The International Low-Emissions Technology Commitment and the International Adaptation Assistance Commitment proposed by the Review (Chapter 10) are both intended to provide additional support to developing countries and so to address the issue of historical responsibility (along the lines suggested by Bhagwati (2006)), thereby making it possible to defend a per capita approach to emissions allocation.

The per capita approach is also broadly consistent with the emerging long-term emissions-reduction goals of several developed countries. Per capita emissions of developed countries are today well above the global average of about six tonnes of CO_2-e. Per capita emissions in, for example, the United Kingdom, Japan and the United States are (as of 2000) 11.5, 10.6 and 21.6 tonnes respectively. Under the long-term emissions-reduction goals announced by or anticipated in these countries, these levels would fall by 2050 to 3.9 tonnes (United Kingdom), 4.0 tonnes (Japan) and 2.7–5.5 tonnes (United States, using the commitments made by the two presidential candidates). These levels are all below today's global per capita average, and close to the 2–3 tonnes per capita average that stabilisation scenarios summarised by the IPCC (2007), together with UN population projections, suggest will be required for stabilisation at 450 to 550 ppm CO_2-e.[4]

Indeed, it is inevitable that if global per capita emissions fall to as low as 2–4 tonnes per person by 2050, then (though variation in national emissions levels will still be possible through the trading of emissions rights) the current stark divergences in national per capita emissions entitlements will diminish over time.

The per capita approach has the virtue of simplicity. Equal per capita emissions is a natural focal point, and contestable computations based on economic variables do not need to enter the allocation formula.

9.4.3 Contraction and convergence

A precise version of the per capita approach, often referred to as 'contraction and convergence' (Global Commons Institute 2000), has figured in the international debate for some time. It has been promoted by India and has been discussed favourably in Germany and the United Kingdom (German Advisory Council on Global Change 2003; UK Royal Commission on Environmental Pollution 2000). Recent reports have shown increasing support for variations on this general approach—see, for example, Stern (2008) and the Commission on Growth and Development (2008).[5]

Under contraction and convergence, each country would start out with emissions entitlements equal to its current emissions levels, and then over time converge to equal per capita entitlements, while the overall global budget contracts to accommodate the emissions reduction objective. This means that emissions entitlements per capita would decrease for countries above the global average, and increase (albeit typically at a slower rate than unconstrained emissions growth) in countries below the global average per capita level. Emissions entitlements would be tradable between countries, allowing actual emissions to differ from the contraction and convergence trajectory.

The contraction and convergence approach addresses the central international equity issue simply and transparently. Slower convergence (a later date at which per capita emissions entitlements are equalised) favours emitters that are above the global per capita average at the starting point. Faster convergence gives more emissions rights to low per capita emitters. The convergence date is the main equity lever in such a scheme.

The group of rapidly growing middle-income countries, such as China, would have practical difficulty with a straight convergence towards equal per capita emissions. They are already around the global per capita average for greenhouse gas emissions, and would find it difficult to stop the rapid per capita growth in their emissions immediately. To account for this, the per capita approach could be modified to provide 'headroom' to allow these countries to make a more gradual adjustment, without immediately needing to buy large amounts of emissions entitlements from other countries. (See section 9.5 for more detail.)

Some argue that relying on just one criterion is simplistic. The UNFCCC itself states that developed countries' national policies to limit emissions should take into account 'differences in these Parties' starting points and approaches, economic structures and resource bases' (Article 4.2(a)). Submissions to the Review raised similar points about Australia's circumstances and resource endowments.

Contraction and convergence does take differences in starting points as the main consideration in the early years, gradually shifting the weight towards population. Moreover, country differences are handled within the per capita approach by allowing those with emissions-intensive economies to buy emissions entitlements from those with economies of lower emissions intensity. This maintains the competitiveness in emissions-intensive industries of countries with tight allocations relative to existing emissions, and with competitive advantage in emissions-intensive industries after taking carbon externalities into account—with one condition. All substantial economies must be subject to constraints that generate similar carbon prices, or, more generally, costs associated with operating within a carbon constraint. For the domestic producer of emissions-intensive goods, the higher international price for the product compensates in an economically efficient way for the need to buy permits.

Would a population-based allocation encourage environmentally damaging global population growth? This is unlikely, as population growth is decided by far more fundamental economic and social determinants. The argument is not relevant to countries—mostly developed countries and first of all Australia, the United States and Canada—where population is growing through immigration.

Another argument sometimes raised against per capita approaches is that emissions entitlement trajectories for some low-growth developing countries could be above their underlying emissions growth trajectory, allowing them to benefit from the sale of excess permits while making minimal mitigation efforts themselves. However, the opportunity to sell surplus permits is a part of the incentive for developing countries to participate in the global regime. In any case, the potential transfers, while large in some cases, are not large in comparison with other recent changes in international payments and transfers—for example, associated with fluctuations in commodity prices.

Some submissions to the Review argued that a per capita approach would be against Australia's interests because of our current high per capita emissions. This is mistaken for several reasons.

First, Australia's biggest national interest is in effective international action, and an emphasis on population is going to be required in any practicable allocation rule. While Australia would gain from an international agreement that recognised only our own special circumstances, all countries' special circumstances would then need to be recognised. Striving for such a system would be against Australia's national interest because it would create large difficulties for international agreement, and thus delay global mitigation action. Moreover, such an agreement would have its environmental benefits diluted by special pleading. Everyone would find a reason not to do very much.

Second, Australia's ongoing strong immigration and population growth means that it will be easier to cut emissions in per capita rather than absolute terms. Australia's population growth rate is above the world average. The Garnaut–Treasury reference case suggests that Australia's population will increase proportionately by almost three times global population through this century. If emissions entitlements and targets are framed in per capita terms, countries with growing populations will receive greater absolute allocations. Population growth considerations are centrally important to equitable distribution of the adjustment burden among Australia and other developed countries.

Third, reducing over time Australia's per capita emissions entitlements to the global average would not mean the end of Australia's emissions-intensive export industries. If the adjustments occur within an effective global agreement—towards which the allocation principle suggested here is directed—their continued expansion would be possible through permit purchases. Where Australia produces emissions-intensive goods for export, it is logical to cover the emissions from that production with purchases of emissions rights from international markets.

9.5 Modelling a per capita approach to the allocation of emissions entitlements

What would national emissions allocations look like under a per capita approach to the allocation of emissions as part of a global cooperative effort to mitigate climate change? The Review addressed this question in relation to the two global mitigation scenarios it modelled, the 450 ppm and 550 ppm scenarios (section 9.1.2).

Global emissions trajectories consistent with the cumulative emissions modelled to achieve the 550 stabilisation and 450 overshooting objectives are shown in Figure 9.3. They illustrate what global emissions trajectories could look like in a world of early and comprehensive mitigation. Both scenarios would represent a daunting short-term challenge, as illustrated in Table 9.1. The 550 trajectory peaks at 2021 at a level only 5 per cent above 2012 levels, and the 450 smoothed trajectory by 2020 is 3 per cent below 2012 levels. This is against a backdrop of global emissions in recent years increasing by about 2.5 per cent a year.

Table 9.1 2020, 2050 and 2100 global emissions changes for the two global mitigation scenarios, relative to 2001 (per cent)

	Change in global emissions over 2001		
	By 2020	By 2050	By 2100
550	40	-13	-60
450	29	-50	-98

Figure 9.3 Emissions trajectories for the no-mitigation, 550 and 450 scenarios, 2000–2100

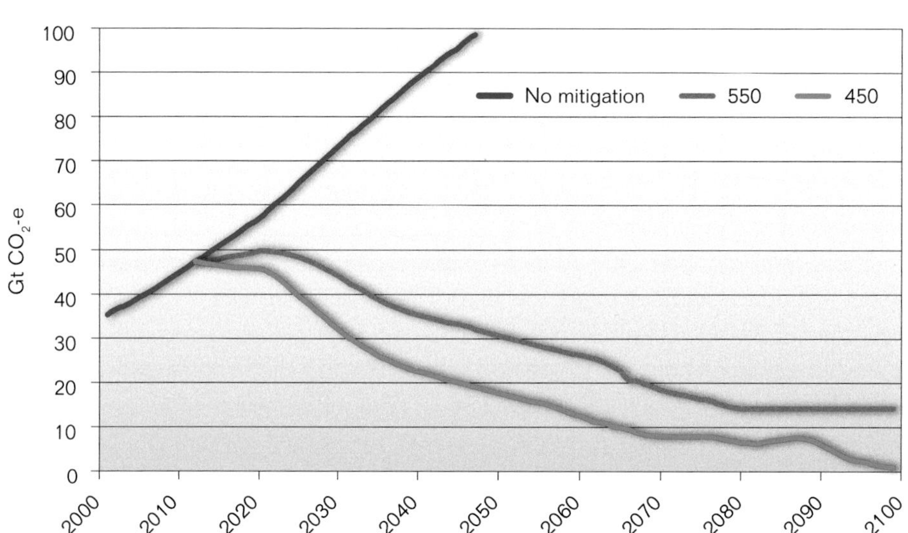

The trajectory for the 450 overshooting scenario at 2050 is close to the 50 per cent reduction in emissions relative to 2000 agreed by the G8 in Japan in July 2008. This level is at one end of the range defined by the IPCC (2007) for the most stringent stabilisation scenario, which is -50 to -85 per cent (on the 15th and 85th percentile of studies). The 550 reduction target for 2050 lies close to the middle of the relevant IPCC range (-30 per cent to 5 per cent).

This global emissions trajectory needs to be allocated between countries in the form of tradable emissions entitlements. Box 9.1 explains the assumptions used to determine national allocations based on the approach outlined in the previous section.

Box 9.1 Allocating the global emissions limit between countries using modified contraction and convergence

The main principle used by the Review to allocate emissions between countries is 'modified contraction and convergence': the idea that over time the entitlements of countries to emit should increasingly be linked to their population. A gradual shift to equal per capita allocations is a practicable principle for the allocation of emissions between countries. To give effect to this basic idea, three questions need to be answered.

First, what is the starting level of emissions from which countries converge? Convergence begins in 2013. For Annex I countries that ratified the Kyoto Protocol, the starting point is their Kyoto compliance levels, so that countries do not gain an advantage from not complying with pre-existing commitments. The one exception to this is successor states to the former Soviet Union, whose Kyoto targets are well above their business-as-usual levels.

Box 9.1 Allocating the global emissions limit between countries using modified contraction and convergence *(continued)*

There is a clear case for allowing the excess permits from the Kyoto period to remain legitimate and bankable, but not for the Kyoto special deal to be perpetuated. The former Soviet Union, the United States and all non-Annex I countries converge from their no-mitigation levels in 2012. Japan and Canada, both of which ratified the Kyoto Protocol but show domestic emissions well above Kyoto compliance levels in 2008–12, are required to make up the deficit in subsequent years. The modelling results therefore show large emissions entitlement reduction requirements for these two countries, especially Canada, by 2020 relative to 2012.

Second, what is the convergence date by which all countries have equal per capita emission allocations? The convergence date selected is 2050. This provides a substantial adjustment period, and, given the prominence of 2050 in the international debate, it is a natural focal point.

Third, how do countries move from their starting points to equal per capita emissions entitlements at the convergence date? It can be argued that an equitable solution would require that all countries move quickly to the convergence level. This is not practical, however, as time for adjustment is required to avoid unnecessary increases in costs. The basic rule applied is that countries' allocations converge in a linear manner, faster if possible or necessary, and with an initial transitional period for developing countries.

The transitional period is designed to limit the adjustment that developing countries might have to face in the initial years. This would increase the probability of their participation in a post-2012 agreement. It takes the form of allowing developing countries growth in emissions allocations at half the rate of their GDP, if this is greater than the growth in allocations under the convergence rule. The 'headroom' provided through the use of an intensity target in this way (Baumert et al. 1999) applies until 2020 or until such countries reach the developed country average, whichever occurs first. A growth of emissions at half the rate of GDP or less is implied by China's announced goals for reductions in energy intensity and its commitment to increase the proportionate role of low-emissions sources of energy. This would be an important factor in making the approach work for the world's largest emitter.

This provision of headroom is a modification to the standard contraction and convergence approach. It recognises that some developing countries will need a transitional period before they will adhere to a linear convergence line. This will be the case for rapidly growing developing countries, and for those with per capita emissions that are already relatively high, in particular (but not only) China.

Deforestation emissions are treated separately. Allocations for deforestation emissions are linearly reduced from starting levels to zero over a 30-year period.

The results for this method—the per capita approach—are shown in Figure 9.4 for the 550 scenario and in Figure 9.5 for the 450 scenario. The much greater stringency involved in the 450 scenario is evident.

Figure 9.4 Per capita emissions entitlements for the 550 scenario, 2012–2050

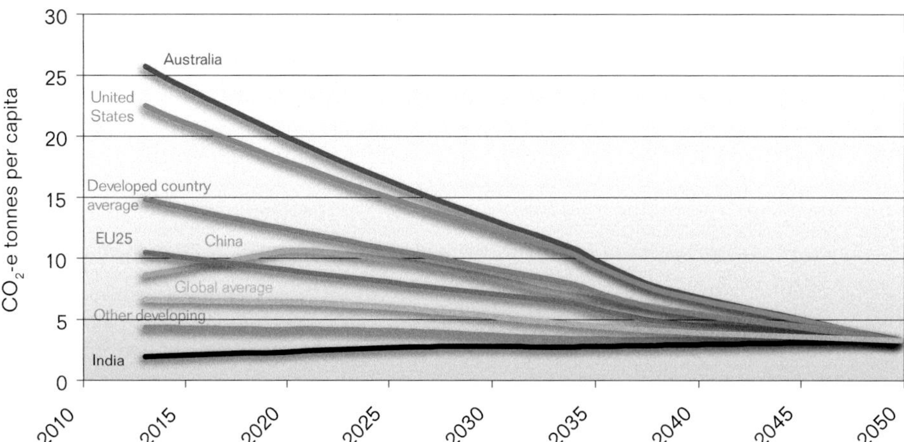

Note: The graph starts in 2012. Australia's 2012 starting value assumes Kyoto compliance, as do those for the EU25. Other countries start at their emissions level given by the reference case (the no-mitigation scenario) in 2012.

Figure 9.5 Per capita emissions entitlements for the 450 scenario, 2012–2050

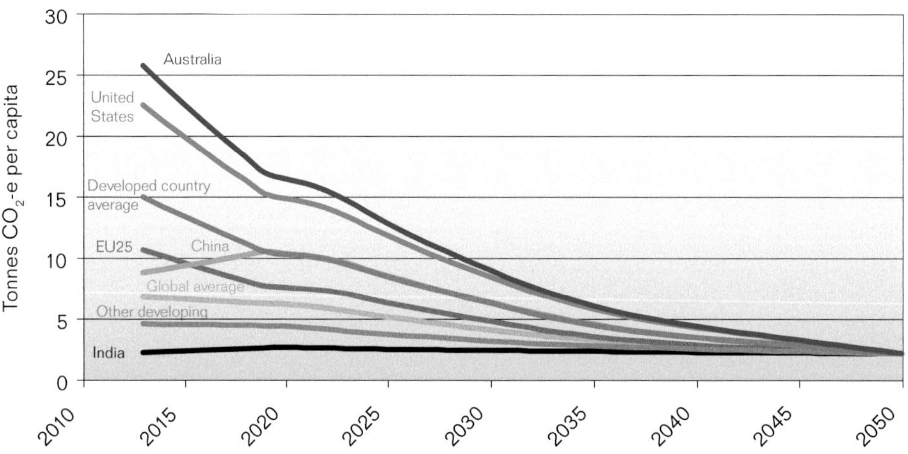

Note: The graph starts in 2012. Australia's 2012 starting value assumes Kyoto compliance, as do those for the EU25. Other countries start at their emissions level given by the reference case (the no-mitigation scenario) in 2012.

The resulting allocations of emissions entitlements to different countries and regions are shown in Table 9.2, in terms of percentage reductions over 2001 and over the Kyoto compliance commitment 2008–12 (or over 2012 for countries and regions with no Kyoto compliance commitments). Table 9.3 provides comparable data on changes in emissions entitlements in per capita terms. [6]

What is the appropriate base against which to compare commitments to reduce emissions in future? If the international community at Copenhagen were to build a set of commitments to reductions in emissions, what would be the logical base period from which to calibrate commitments? The Australian Government's current policy commitments relate to 2000. European commitments go back to 1990—the base year for the commitments under the Kyoto Protocol.

It would seem to sit more comfortably alongside the logic of the Kyoto Protocol to calibrate new commitments from the compliance period of that agreement, 2008 to 2012. This would reflect the understandings reached at Kyoto. It would reward overperformance, and not underperformance, against those understandings.

Table 9.2 shows that the Australian medium-term commitments under the Review's proposed allocation method represents a reduction of 17 per cent in absolute terms from the Kyoto commitments for the 550 reduction target, and 32 per cent for the 450 target. This is close to the developed country average in each case (16 per cent and 32 per cent respectively). It follows that Australia's per capita reductions in emission entitlements from the Kyoto compliance base to 2020 would be substantially greater than for developed countries as a whole and for each of the other developed countries (Table 9.3).

Table 9.2 Emissions entitlement allocations for 2020 and 2050 relative to 2000–01 and Kyoto/2012 (per cent)

	550				450			
	2000–01 base		Kyoto/2012 base		2000–01 base		Kyoto/2012 base	
	2020	2050	2020	2050	2020	2050	2020	2050
World	40	-13	6	-34	29	-50	-2	-62
Developed countries	-15	-76	-16	-77	-31	-86	-32	-87
Australia	-10	-80	-17	-82	-25	-90	-32	-90
Canada	-33	-80	-14	-75	-45	-89	-30	-86
EU25	-14	-69	-14	-69	-30	-82	-31	-82
Japan	-27	-75	-15	-71	-41	-86	-32	-84
United States	-12	-81	-17	-82	-28	-89	-32	-90
Developing countries	91	50	21	-5	85	-14	18	-45
China	210	-4	34	-58	195	-45	27	-76
India	98	230	35	126	97	90	35	30

Note: Australia's allocations over the 2000–01 base are relative to 2000 actuals, and are rounded. Unrounded figures (also relative to 2000) are -10%, -80%, -27%, and -89%. All other countries' allocations over the 2000–01 base are relative to the 2001 no-mitigation scenario.

The 2020 reduction figures for Japan and especially Canada relative to 2001 are large because convergence for developed countries that ratified the Kyoto Protocol begins with 2008–12 Kyoto compliance levels. For these two countries, Kyoto compliance levels for 2008 to 2012 are well below 2000 levels, whereas for EU25 and Australia the Kyoto compliance levels are at or above 2000 levels. Therefore, the same reduction below 2008–12 compliance levels is a much bigger reduction below 2000 levels for Japan and Canada. Since the United States did not ratify the Kyoto Protocol, its starting point is actual projected emissions in 2012, as for developing countries (Box 9.1).

One of the striking features of this set of allocations is that there is little variation in the 2050 reductions in emissions entitlements for developed countries. The required reductions from developed countries are in a fairly narrow range of 70 to 80 per cent for the 550 scenario and 80 to 90 per cent for the 450 scenario.

There is more variation in developed countries' 2020 targets. Canada's required reduction is exceptionally high in absolute terms from a 2000 base, but not from the Kyoto compliance period because its emissions have increased, not decreased as required by Kyoto.

The story is quite different for developing countries. The variation across developing countries reflects their different starting points, and the growth between 2001 and 2012 that needs to be taken into account. In many cases, their emissions are allowed to increase significantly, reflecting their low per capita starting points.

There is little difference between the 450 and 550 scenarios for developing countries up to 2020. They are protected in this period by the proposed transitional measures that allow their emissions entitlements to continue to grow. After 2020, developing countries' allocations under the two scenarios diverge markedly. It is likely that many developing countries would hold actual emissions below their entitlements, and that many developed countries would honour their commitments in part by purchasing permits.

The importance of population can be seen from Table 9.3, which presents the same data as Table 9.2 but in per capita terms. In per capita terms Australia is called on to do more than Europe, Japan and the United States, because the allocative

Table 9.3 Emissions entitlement allocations expressed in per capita terms in 2020 and 2050 relative to 2000–01 and Kyoto/2012 (per cent)

	550				450			
	2000–01 base		Kyoto/2012 base		2000–01 base		Kyoto/2012 base	
	2020	2050	2020	2050	2020	2050	2020	2050
World	14	-41	-2	-50	4	-66	-10	-71
Developed countries	-22	-79	-19	-78	-37	-88	-34	-87
Australia	-30	-90	-27	-88	-40	-95	-40	-93
Canada	-43	-86	-21	-80	-54	-92	-36	-89
EU25	-17	-69	-15	-69	-33	-82	-32	-82
Japan	-25	-69	-13	-64	-40	-82	-30	-80
United States	-26	-86	-23	-86	-40	-92	-37	-92
Developing countries	49	-5	10	-30	45	-46	7	-60
China	179	-13	29	-60	166	-50	23	-77
India	53	112	23	71	52	22	23	-2

Note: Australia's allocations relative to the 2000–01 base are relative to 2000 actuals, and are rounded based on absolute values. (Unrounded figures consistent with Table 9.2 values are: -31%, -88%, -42%, -94%.) All other countries' allocations relative to the 2000–01 base are relative to the 2001 no-mitigation scenario. See also notes to Table 9.2.

approach requires Australia to reduce its current high per capita emissions entitlement to the global average. As high per capita emitters, Australia, Canada and the United States have more 'distance' to move than the EU and Japan. In per capita terms, Australia's required reductions from the Kyoto compliance commitment for 2008–12 and from 2000 are similar.

At the 2007 Bali climate change negotiations, a particular range of emissions reductions received prominent attention. It was proposed that developed countries (strictly Annex I countries) consider emissions reduction targets in the range of 25 to 40 per cent by 2020 over 1990 levels. This target range stems from an IPCC (2007) analysis for a 450-type trajectory. The equivalent range for a 550 trajectory is 10 to 30 per cent. The emissions reduction targets for developed countries modelled by the Review are consistent with these Bali ranges, but at the lower end in terms of stringency, reflecting the limited progress made between 1990 and now towards mitigation.

The 1990 starting point is deeply problematic, because it does not recognise the effects of the Kyoto agreement on differentiation in emissions entitlements growth between 1990 and the Kyoto compliance period. It happens to be highly favourable for successors to centrally planned economies, whose emissions dropped sharply through the transition out of communism, and for Western European economies whose energy sectors were transformed by the easier availability of natural gas in the 1990s.

Relative to 1990, Australia's proposed targets under the Review's approach are at around the average for developed countries in absolute terms and much higher in per capita terms. They are the same percentage reduction as for 2000, as under Kyoto accounting rules Australia's emissions were almost the same in 1990 and 2000.

The allocative approach adopted takes no account of what might happen in a no-mitigation world. This would always be counterfactual, would lend itself to special pleading, and would be impossible to use as the basis for allocating emissions across countries. Nevertheless, it is a fact that Australia's rapid underlying emissions growth may require greater effort for Australia than others to comply with any comprehensive international agreement. For Australia, as a country likely to have comparative advantage in a range of emissions-intensive industries, the flexibility provided by international trading in entitlements is of considerable importance.

The Review estimates that the emissions allocations are only about 10 per cent below business as usual for developing countries, including China, by 2020. This suggests that the allocative approach adopted here is realistic. Developing countries need to be brought on board, but a transition period is required during which emissions allowances can keep growing. The relatively slow start for developing countries provides them with incentives and opportunity to reduce emissions below their allocations, and to sell surplus entitlements.

9.6 Reaching agreement on 550 or 450: is it possible?

The objective of climate change negotiations must be to define a consistent set of national allocations that would add up to a global emissions trajectory which in turn would allow the atmospheric concentration of greenhouse gases to settle at specified levels. The numbers presented in this chapter add up to 450 and 550 objectives.

Of course, other sets of allocations could also add up to 450 or 550. If some countries do less, others will need to do more. The two sets of allocations presented in this chapter are illustrative of what will be required. Do the proposed allocations suggest that global agreement around a 450 path is possible, or is a 550 path the best that can be hoped for?

It is important to distinguish between the short term, up to 2020, and the long term, up to 2050. The 2007 Bali Roadmap calls for agreement on a 'long-term global goal for emission reductions' (UNFCCC 2007). The G8 recently (in July 2008, in Japan) endorsed a long-term goal of a reduction in global emissions of 50 per cent by 2050. As noted earlier, this is consistent with the 450 path (Figure 9.4).

But is the world ready to commit to a 450 target in the short term, up to 2020? Not yet. No developed country or group of countries has indicated a willingness to cut emissions by 2020 to the extent implied by the 450 ppm target. The European Union comes closest, but even its 30 per cent conditional offer (relative to 1990) falls short of the 36 per cent that would be required of it under the 450 agreement (see Table 5.4 of the Review's supplementary draft report (Garnaut 2008)).

Canada's target is instructive: its current 2020 commitment would translate, the Review estimates, to a reduction of 10 per cent over 2000 levels—less than would be required of it in a 550, let alone a 450, world.

In another example, commitments by the US presidential candidates for 2050, if translated into 2020 targets with a starting point of 2012, convert into reduction commitments of around 10 to 15 per cent over 2000, again consistent with a 550 rather than a 450 agreement. (Similar targets are given or implied by various US climate change bills.)

Of course, smaller reductions could be asked of developed countries were the developing world prepared to commit to more, but we have already assumed that the developing world will reduce emissions by around 10 per cent below business-as-usual levels by 2020, which itself would be a significant achievement. The Review's judgment is that the contribution required of developing countries up to 2020 to achieve the 550 ppm path would exhaust what might optimistically be expected of them, and that the additional reductions to achieve 450 ppm would have to come from the developed countries.

Beyond 2020, additional and more demanding emissions reductions would have to come from developing countries. As Table 9.2 shows, the 450 target would require a 13 per cent reduction in emissions by developing countries in 2050 relative to 2001.

In the short term, therefore, a 450 agreement seems out of reach, unless developments over the next year transform the attitudes of developed and developing countries alike. Of course, major changes in the political outlook are not out of the question, with a new US President and Congress, a Chinese government beginning to make progress on its own energy efficiency and low-emissions energy goals, and the recent scientific evidence underlining the urgency of the question. Australia should encourage this possibility by announcing its preparedness to make its proportionate contribution—an absolute reduction in entitlements of 25 per cent on 2000 levels by 2020—if there is an effective global agreement around 450.

In the meantime, a post-2012 agreement consistent with stabilisation at 550 ppm seems to be possible. This would be a major achievement in itself, setting new standards of international cooperation in this area of policy, and holding promise of avoiding the worst outcomes from human-induced climate change.

Once the world has embarked on a 550 path there is a reasonable prospect that confidence would increase in all countries, and especially in developing countries, that strong mitigation was consistent with the continuation of desired rates of economic growth. The reality of income from sale of permits from developing countries, of progress with low-emissions technologies, and of developed country support for adaptation would build confidence in the international arrangements. Early progress on emissions reduction would reduce fears about the compatibility of mitigation with economic growth. Growth in confidence would make it possible to reconsider mitigation ambitions in an early successor to the Copenhagen meeting.

Notes

1 Just as with converting from concentration levels to temperature increases, so too converting from emissions to concentration levels involves uncertainty, in particular involving climate–carbon cycle feedbacks, the treatment of which can reduce permissible cumulative emissions associated with atmospheric stabilisation targets by 20 per cent or more (Jones et al. 2006).

2 The issue of aggregating over different greenhouse gases is not tackled here (see Chapter 3).

3 A price-based commitment is an example of an input-based commitment. Another variant of input-based commitments is the 'sustainable development policies and measures' approach, which would directly reward countries (normally developing countries) for implementing agreed policies (Winkler et al. 2002).

4 For the actual commitments made by these countries, see Chapter 8. UNFCCC data has been used as a baseline. All population projections are the 2006 medium variant projections from the United Nations. The exact global per capita emissions average at 2050 under the various stabilisation scenarios depends heavily on the trajectory of emissions through time, as well as on future population growth.

5 Neither report uses the term 'contraction and convergence', but both point to the need for all countries to aim for equal per capita emissions over the 'long term' (Commission on Growth and Development 2008) or by 2050 (Stern 2008). Stern (2008: 10) notes that this approach 'is a pragmatic ... one. It should not be regarded as strongly equitable since it takes little account of the developed countries' much larger per capita contribution to stocks of greenhouse gases.'

6 The year 2001 is the base year in GTEM, the computable general equilibrium model used by the Review.

References

Baer, P., Athanasiou, T. & Kartha, S. 2007, *The Right to Development in a Climate Constrained World: The greenhouse development rights framework*, Heinrich Böll Foundation, Christian Aid, EcoEquity & Stockholm Environment Institute, Berlin.

BASIC Project 2006, *The Sao Paulo Proposal for an Agreement on Future International Climate Policy*, Discussion Paper for COP12 and COP-MOP2, Nairobi, Kenya.

Baumert, K.A., Bhandari, R. & Kete, N. 1999, *What Might a Developing Country Climate Commitment Look Like?* World Resources Institute, Washington DC.

Bhagwati, J. 2006, 'A global warming fund could succeed where Kyoto failed', *Financial Times*, 16 August.

Bodansky, D. 2007, 'Targets and timetables: good policy but bad politics?' in J.E. Aldy & R.N. Stavins (eds), *Architectures for Agreement: Addressing global climate change in the post-Kyoto world*, Cambridge University Press, New York, pp. 57–66.

Commission on Growth and Development 2008, *The Growth Report: Strategies for sustained growth and inclusive development (conference edition)*, International Bank for Reconstruction and Development, World Bank, Washington DC.

Cooper, R.N. 2000, 'International approaches to global climate change', *World Bank Research Observer* 15(2): 145–72.

Council of the European Union 2007, *Presidency Conclusions*, 8/9 March 2007, 7224/1/07 Rev 1, Brussels.

Frankel, J. 2007, 'Formulas for quantitative emissions targets', in J.E. Aldy & R.N. Stavins (eds), *Architectures for Agreement: Addressing global climate change in the post-Kyoto world*, Cambridge University Press, New York, pp. 31–56.

Garnaut, R. 2008, *Garnaut Climate Change Review: Supplementary draft report*, Commonwealth of Australia, Canberra.

German Advisory Council on Global Change 2003, *Climate Protection Strategies for the 21st Century: Kyoto and beyond*, WGBU, Berlin.

Global Commons Institute 2000, 'GCI briefing: contraction and convergence', available at <www.gci.org.uk/briefings/ICE.pdf>, originally published as Meyer, A. 2000, *Engineering Sustainability* 157(4): 189–92.

IPCC (Intergovernmental Panel on Climate Change) 2007, *Climate Change 2007: Mitigation of climate change. Contribution of Working Group III to the Fourth Assessment Report of the Intergovernmental Panel on Climate Change*, B. Metz, O.R. Davidson, P.R. Bosch, R. Dave & L.A. Meyer (eds), Cambridge University Press, Cambridge and New York.

Jones, C.D., Cox, P.M. & Huntingford, C. 2006, 'Impact of climate–carbon feedbacks on emissions scenarios to achieve stabilisation', in H.J. Schellnhuber, W. Cramer, N. Nakicenovic, T. Wigley & G. Yohe (eds), *Avoiding Dangerous Climate Change*, Cambridge University Press, Cambridge, pp. 323–32.

Jotzo, F. & Pezzey, J. 2006, 'Non-binding targets for international emissions trading under uncertainty', paper presented at the Third World Congress of Environmental and Resource Economists, Kyoto.

McKibbin, W.J. & Wilcoxen, P.J. 2002, *Climate Change Policy after Kyoto: Blueprint for a realistic approach*, Brookings Institution, Washington DC.

McKibbin, W.J. & Wilcoxen, P.J. 2008, 'Building on Kyoto: towards a realistic global climate agreement', draft, prepared for the Brookings High Level Workshop on Climate Change, Tokyo, 30 May.

Murray, B.C., Newell, R.G. & Pizer, W.A. 2008, 'Balancing cost and emissions certainty: an allowance reserve for cap-and-trade', *NBER Working Paper Series, Working Paper 14258*, National Bureau of Economic Research, Cambridge, Massachusetts.

Nordhaus, W. 2008, *The Challenge of Global Warming: Economic models and environmental policy*, Yale University, New Haven, Connecticut.

Ott, H.E., Winkler, H., Brouns, B., Kartha, S., Mace, M.J., Huq, S., Kameyama, Y., Sari, A.P., Pan, J., Sokona, Y., Bhandari, P.M., Kassenberg, A., Rovere, E.L.L. & Rahman, A. 2004, *South–North Dialogue on Equity in the Greenhouse: A proposal for an adequate and equitable global climate agreement*, Wuppertal Institute (Germany) & Energy Research Centre (South Africa).

Pew Center on Climate Change 2005, *International Climate Efforts Beyond 2012: Report of the climate dialogue at Pocantico*, Arlington, Virginia.

Philibert, C. 2000, 'How could emissions trading benefit developing countries?' *Energy Policy* 28: 947–56.

Pizer, W.A. 2002, 'Combining price and quantity controls to mitigate global climate change', *Journal of Public Economics* 85: 409–34.

Roberts, M.J. & Spence, M. 1976, 'Effluent charges and licenses under uncertainty', *Journal of Public Economics* 5(3–4): 193–208.

Stern, N. 2008, *Key Elements of a Global Deal on Climate Change*, London School of Economics and Political Science.

Stiglitz, J. 2006, 'A new agenda for global warming', *Economist's Voice*, <www.bepress.com/ev>.

UK Royal Commission on Environmental Pollution 2000, 'Energy: the changing climate, presented to Parliament by Command of Her Majesty June 2000', Twenty-Second Report, London.

UNFCCC (United Nations Framework Convention on Climate Change) 1997, 'Proposed elements of a Protocol to the UNFCCC, presented by Brazil in response to the Berlin Mandate', FCCC/AGBM/1997/Misc.1/Add.3, Climate Change Secretariat, Bonn.

UNFCCC 2007, 'Report of the Conference of the Parties on its thirteenth session, held in Bali from 3 to 15 December 2007, Addendum, Part Two: Action taken by the Conference of the Parties at its thirteenth session', FCCC/CP/2007/6/Add.1.

Weitzman, M.L. 1974, 'Prices vs quantities', *Review of Economic Studies* 41(4): 477–91.

Winkler, H., Spalding-Fecher, R., Mwakasonda, S. & Davidson, O. 2002, 'Sustainable development policies and measures: starting from development to tackle climate change', in K.A. Baumert, O. Blanchard, S. Llosa & J.F. Perkaus (eds), *Building on the Kyoto Protocol: Options for protecting the climate*, World Resources Institute, Washington DC, pp. 61–87.

10 DEEPENING GLOBAL COLLABORATION

Key points

International trade in permits lowers the global cost of abatement, and provides incentives for developing countries to accept commitments.

Trade in emissions rights is greatly to be preferred to trade in offset credits, which should be restricted.

A global agreement on minimum commitments to investment in low-emissions new technologies is required to ensure an adequate level of funding of research, development and commercialisation. Australia's commitment to support of research, development and commercialisation of low-emissions technology would be about $2.8 billion.

An International Adaptation Assistance Commitment would provide new adaptation assistance to developing countries that join the mitigation effort.

Early sectoral agreements would seek to ensure that the main trade-exposed, emissions-intensive industries face comparable carbon prices across the world, including metals and international civil aviation and shipping.

A WTO agreement is required to support international mitigation agreements and to establish rules for trade measures against countries thought to be doing too little on mitigation.

It would be neither desirable nor feasible for each country separately to pursue national emissions-reduction targets. It would not be desirable because lower-cost abatement options would be forgone, and higher-cost options accepted. It would not be feasible, for there would be no financial incentive for developing countries to participate in strong mitigation, and they would not do so. These are two fatal flaws.

No ambitious system of emissions allocation among nations will work unless it allocates entitlements rather than actual emissions. And mitigation efforts will not succeed without international public funding to develop new technology and to transfer it as quickly and widely as possible. This chapter covers key aspects of international collaboration, from international public funding for research and development to emissions trading, to policy frameworks for particular sectors, trade rules and enforcement mechanisms.

10.1 International public funding for mitigation

One of the weaknesses of the world's response to climate change to date has been the limited extent of global public funding for mitigation. Levels of energy research and development, critical to enable the world to make the transition to a low-emissions future, have fallen over time.[1] Figure 10.1 illustrates the case of the United States, the world's largest investor in research and development. The low levels of energy research and development can be explained partly by the limited mitigation efforts so far. Nor has the issue been an important part of the public discussion. Research and development are barely mentioned in either the United Nations Framework Convention on Climate Change (UNFCCC) or the Kyoto Protocol.

Figure 10.1 Energy research and development expenditure by the public and private sectors in the United States

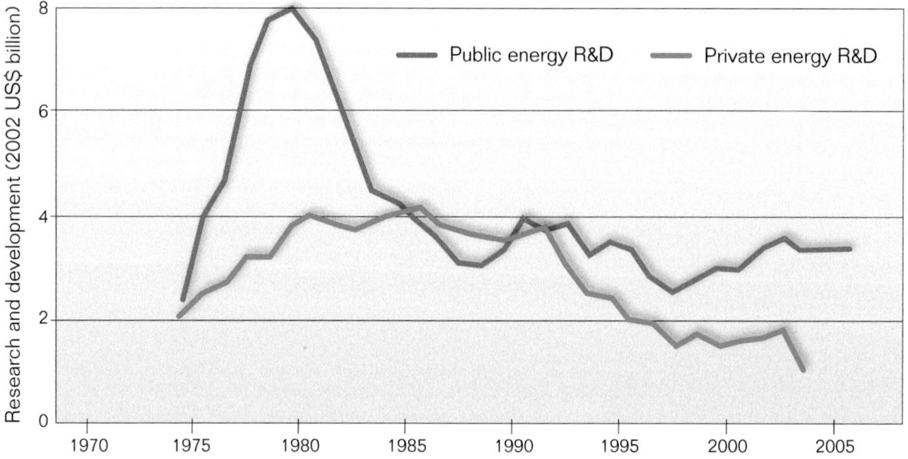

Source: Kammen & Nemet (2005a).

Unlike research and development, technology transfer features prominently under the UNFCCC and Kyoto Protocol. However, the unquantified assurances given in these treaties have not been translated into action. Some technology transfer has occurred under the Kyoto Protocol's Clean Development Mechanism, but nothing on the scale required to underpin broad-based mitigation in developing countries. The Global Environment Facility, a multilateral agency that has been designated as the 'financial mechanism' for the UNFCCC and a number of other environmental conventions, has been an active player in technology transfer, but on a limited scale; on average less than US$1 billion a year (including co-financing) was allocated to climate change projects between 1991 and 2004 (IPCC 2007).

More recently, low-emissions technology research and transfer have received increased attention, with support from prominent economists[2] and political leaders.

Venture capital funds are also starting to invest heavily in renewable energy (Pontin 2007). In February 2008, the US, UK and Japanese governments announced the establishment of 'a multibillion-dollar fund to accelerate the deployment of clean technologies and help the developing world deal with climate change' (Paulson et al. 2008). The three countries have committed about US$1–2 billion each to what is now called the Clean Technology Fund, to be administered by the World Bank. The G7[3] has issued a call to 'scale up investment in developing countries to support them in joining international efforts to address climate change' (AFP 2008). In the last few years, a number of collaborative initiatives have been launched to develop and transfer clean technologies (Box 10.1). The World Bank and regional banks have announced a new focus on energy efficiency and clean technologies, including renewable energy, with commitments to lend US$1–2 billion a year each (IPCC 2007). Australian governments have started investing in research and development on carbon capture and storage, and to a lesser extent in research on renewables. A number of Australian companies are investing large amounts of risk capital raised for the purpose in deep geothermal technologies.

Box 10.1 International research and development and technology transfer initiatives

In recent years, a number of international technology-related initiatives have been launched. Many of them are in need of additional funding. They would all be eligible for funding under the Review's proposed International Low-Emissions Technology Commitment.

- The Generation IV International Forum is a multilateral partnership fostering international cooperation in research and development for the next generation of nuclear energy systems.
- The Carbon Sequestration Leadership Forum is focused on the development of improved technologies for the separation and capture of carbon dioxide for transport and long-term safe storage.
- The Methane to Markets Partnership focuses on advancing near-term methane recovery and use of methane as a clean energy source.
- The International Partnership for the Hydrogen Economy provides a forum for its member countries for advancing policies, common codes and standards, as well as developing demonstration and commercial utilisation activities related to hydrogen and fuel cell technologies.
- The mission of the Renewable Energy and Energy Efficiency Partnership is to 'contribute to the expansion of the global market for renewable energy and energy efficiency' (REEEP 2008). The partnership's broad membership includes national governments, business, development banks and non-government organisations.
- The Asia–Pacific Partnership on Clean Development and Climate brings together governments and companies from Australia, Canada, China, India, Japan, the Republic of Korea and the United States to collaborate on the development, deployment and transfer of cleaner and more efficient technologies.

Widespread adoption of national emissions goals, as advocated in Chapter 9, would not obviate the need for international public funding for research and development related to mitigation. This follows from the external benefits generated by private investment in this area (Chapter 18). Provision of such funding will be critical to the global endeavour to live within a tightening carbon constraint to correct for international market failures relating to public goods and missing markets:

- Expenditure on the research, development and commercialisation of low-emissions technologies has international public good characteristics, as it can benefit all nations and its rewards cannot be fully captured by private investors. Delivering this international public good will be critical for lowering abatement costs and increasing confidence that the mitigation task is a feasible one. Indeed, some have argued that international climate change mitigation policy should be predominantly about the development and provision of low-emissions technologies (Schelling 2002; Barrett 2003). This view underestimates the importance of price signals for providing incentives for change, and for inducing technological change (Köhler et al. 2006). No amount of technological innovation will make some potentially important emissions-reducing technologies competitive with unmitigated release of emissions to the atmosphere. Geosequestration of emissions from fossil fuel combustion is an important example: coal-fired generation will always be more expensive with geosequestration than without it. Nevertheless, deep cuts in global emissions will require the development of new energy technologies, and international externalities in research and development provision will require international public funding.
- International trade in emissions rights will be critical for providing incentives for developing countries to participate, but it will take several years even after a new international agreement is reached for the necessary international markets to develop. In the interim, developing countries will face a problem of 'missing markets.' Developed country governments and international development finance institutions will need to provide developing countries with financing to kick-start the move to a low-emissions future (Carmody & Ritchie 2007). Such financing would provide critical technology—existing and new—to support the transition to a low-carbon economy, but could extend beyond the energy markets to other areas such as reducing deforestation.

How much public funding would be required? One estimate of global energy research and development needs for a stabilisation goal of 550 ppm carbon dioxide equivalent is in the order of US$30–100 billion per year until stabilisation (Kammen & Nemet 2005b). Another estimate for the same stabilisation level calls for annual spending of US$50 billion per year by 2050 compared to an estimated US$10 billion today (Bosetti et al. 2007).[4] Although some of this research and development will be provided by the private sector, in general these estimates understate the required public funding. They only cover energy research and development, and neglect technology transfer or other public financing for developing country mitigation. These estimates model requirements for stabilisation at 550 ppm carbon dioxide equivalent, a level that may not do enough to reduce the risk of dangerous climate change. Stabilisation at a lower concentration level will require faster and greater development and uptake of new technology, and thus larger, earlier expenditure. The International Energy Agency (2008) examined the energy technology requirements for reducing global emissions by 50 per cent by 2050. It concluded that 'a massive increase of energy technology Research, Development and Demonstration … is needed in the coming 15 years, in the order of US$10–100 billion per year' (IEA 2008: 1).

Looking beyond research and development to the financing needs of developing countries, the UNFCCC estimates that by 2030 additional global investment and financial flows of US$200 billion annually would be needed, with flows to developing countries in the order of US$100 billion annually to finance mitigation that leads to constraining emissions at 2030 to current levels (UNFCCC 2007a). While the bulk of these investment flows are expected to come from the private sector, until international carbon markets are established there will be greater reliance on public sector funding. The Group of 77 developing countries together with China have recently proposed that developed countries set aside 0.5 to 1 per cent of GNP 'to support action on mitigation and adaptation and technology development and transfer' (G-77 & China 2008).

Encouraging adequate global funding, ensuring equitable burden sharing, and deterring free-riding all contribute to making a strong case for embedding commitments to the international funding of climate change mitigation in an international agreement.

Such a commitment would apply only to high-income countries. This is consistent with the UNFCCC principle of 'common but differentiated responsibility'. It would reflect the burden the UNFCCC places on developed countries to meet mitigation costs in developing countries, and also the Bali agreement to scale up developing country mitigation action on the back of increased incentives from developed countries.

The Review therefore proposes that high-income countries support an **International Low-Emissions Technology Commitment**. This would require high-income countries to allocate a small proportion of GDP above a threshold to such purposes. They would commit to a specified funding level, but would retain flexibility in the use of funds. Funds could be spent domestically or abroad, through national or collaborative ventures.

Eligible expenditures under the Commitment would include public funding on low-emissions research and development; public funding on technology commercialisation (at a discount of, say, 2:1); and public funding to kick-start the mitigation efforts of developing countries, for example through technology transfer and support for reduction in forestry emissions. Much of the spending will be in high-income countries since that is where the technological breakthroughs are more likely to be made. A significant portion would be in developing countries, for reasons of both equity and efficiency. Research and development, and especially commercialisation of new technologies, will sometimes be less expensive in developing countries, and private investors will sometimes choose a developing country location on cost grounds. A commitment to expend some minimum portion of the funding in developing countries, perhaps up to 50 per cent, but again with flexibility over modalities and priority areas, would help accelerate technological transfer, and strengthen the incentives for mitigation in developing countries.

Funding commitments would apply as a percentage of GDP above a certain threshold level of per capita income. The threshold for funding commitments could coincide with the threshold for classification in the high-income group of countries, so that a country just entering the group would initially have only minimal funding commitments.

The size of the International Low-Emissions Technology Commitment will need to be of the order of US$100 billion to accommodate both research and development requirements, and to finance commercialisation of low-emissions technologies in developing countries. As a broad illustration, take an annual global amount of US$100 billion and the World Bank high-income threshold of US$11 000 per capita. Then for 2007 GDP levels (at then current exchange rates) under this formula, the 50 richest countries (accounting for two-thirds of global GDP) would have contributed on average 0.24 per cent of their GDP to the Commitment.

Australia's 2007 share under the above formulation would have been $2.8 billion, or 0.26 per cent of GDP. Australia would commit the public sector (federal or state) to spend at least this amount on research, development and commercialisation of new low-emissions technologies. It could acquit the commitments at home or abroad. Expenditures abroad would help developing countries finance the technologies they need to contain emissions growth.

There is a strong argument for a steep ramping up of funding for low-emissions technology, given the urgency of mitigation, the fact that research and technology transfer will bring about a permanent reduction in mitigation costs, and the need to induce developing countries to participate. Commitments might come down in future years as and when market mechanisms become effective and technological breakthroughs are made.

Developing countries would need to agree to and comply with the commitments that would be expected from them under the next climate agreement in order to qualify as recipients of funds under the Commitment.[5] For the least developed countries, the expectation would only be to put in place a carbon penalty comparable to carbon pricing in developed countries on large emissions-intensive export industries. Most other developing countries would be expected in addition to take on one-sided targets.

How funding for the Commitment is secured would be left to individual governments. Countries with an emissions trading scheme or carbon tax could choose to earmark a portion of permit sales revenue or tax revenue towards this commitment.[6]

Expenditures under the Commitment would not be restricted to the stationary energy sector. It would also cover transport, and various forms of biosequestration: research into increasing carbon content of soils, forestry management, and the use of algae for biosequestration. It would also cover research into geo-engineering.

Of particular strategic interest to Australia is the development of near-zero emissions coal technologies and their transfer to developing countries. Australia should lead a global effort to develop and implement near-zero emissions coal technologies, such as carbon capture and storage, both building on our comparative advantage and addressing key national interest concerns. First, there will be direct economic implications for Australia, through reduced demand for coal, if these technologies are not developed and commercialised in a global low-emissions future (see Chapter 20). Australia will also be affected indirectly, through reduced demand for exports in general, if our trading partners, such as China, are unable to reduce emissions without reducing their own economic growth. Second, effective participation of major developing countries such as China, Indonesia and India is critical to the success of global mitigation. The participation of these countries would be made easier by the development of such technologies. Third, such technological development has important domestic effects, as it will not only obviate the need for structural adjustment packages to coal-dependent sectors or industries, but will also greatly ease Australia's own transition to a low-emissions economy.

10.2 International public funding for adaptation

Countries will experience climate change impacts differently, and hence adaptation will mean different things for different countries. For most poor countries, an important response to the adaptation challenge is economic growth (Schelling 1997), which will put greater resources at the disposal of both citizens and governments to respond to climate change. Along with education, economic growth will take people out of the sector that is most vulnerable to climate change—agriculture. Trade liberalisation will help countries adjust to shifting production opportunities. All of these are core development assistance objectives that would benefit developing countries even if there were no climate change, but that will also help them adapt to climate change.

Given the range of potential impacts and adaptive responses, it is difficult to calculate the costs and benefits of adapting to climate change in one country, let alone across the world. There is a wide range of estimates of required expenditure on adaptation in the literature, from $4 billion to $100 billion per annum (World Bank 2006b; Stern 2007; UNDP 2007; Oxfam 2007).

The similarities between the adaptation and development agendas mean that it makes no sense to force a division between the two.

The first adaptation requirement in all of Australia's developing country neighbours is basic scientific research on potential impacts. Elementary mapping of possible impacts and vulnerability has hardly begun in any of these countries. Some aspects of the established development agenda will require an enhanced priority. For example, with projected increases in intensity of severe weather events, such as storm surges and cyclones, developing countries will need to improve their disaster mitigation and management capacities. In different contexts this may include developing building standards, early-warning systems or emergency response capacity. However, there is no need for a new adaptation architecture. The challenge is rather to make the existing aid architecture work better and to fully incorporate climate change adaptation considerations into decision making by including climate change information and impacts assessments in established development processes.

A number of funds have been established under the UNFCCC and the Kyoto Protocol that can support adaptation measures (Box 10.2), but they are small and are yet to prove themselves.

Box 10.2 Adaptation funds under the UNFCCC and the Kyoto Protocol

The **Least Developed Countries Fund** was established to assist least developed countries (as defined by the United Nations), of which there are 49, to design and implement National Adaptation Programmes of Action. It is funded through voluntary contributions, and pledges amount to US$120 million (GEF 2008a). Australia contributed $7.5 million to the fund in 2007.

The **Special Climate Change Fund** is designed to complement other funding and, while mitigation activities are within its scope, its top priority is adaptation. It is funded through voluntary contributions, and pledges currently total US$60 million (GEF 2008b). Australia is not currently a contributor to this fund.

The **Adaptation Fund** was established under the Kyoto Protocol. This fund has only just become operational. It is financed through a 2 per cent levy on certified emissions reductions (CERs) traded through the Clean Development Mechanism. The fund receives secretariat services from the Global Environment Facility, and the World Bank has been invited to become the trustee. The majority of its board members come from developing countries. Initial funding is unlikely to be available before 2010 (World Bank 2008b), when it is estimated that revenue from the levy will total US$80–300 million per year (UNFCCC 2007a). Projects in least developed countries are exempt from the levy.

The Review recommends that developed countries make a quantified commitment to providing adaptation support to developing countries. An **International Adaptation Assistance Commitment** would provide developing countries with an assurance that they will receive support in adapting, while allowing developed countries to retain flexibility over the delivery of their adaptation assistance.

Although adaptation assistance would, by its nature, be categorised as official development assistance, developed countries would need to ensure that funding to meet their shares of the International Adaptation Assistance Commitment was additional and did not displace current and planned levels of development funding.

Adaptation funding requirements, while impossible to predict and quantify, will be significant. They are likely to increase steadily over time as impacts themselves develop over time. Australia and other developed countries should be prepared to increase their development assistance significantly as the understanding of impacts improves.

The new mitigation and adaptation commitments proposed by the Review would give developing countries incentives to participate more fully in the international climate change regime. Access to such funding should be conditional on developing countries fulfilling reasonable expectations on their contribution to mitigation efforts. Note that, as set out in Chapter 9, participation conditions would be minimal for the least developed countries.

Australia's development assistance has traditionally been focused on our immediate region, and in particular our near neighbours, through the South Pacific, Papua New Guinea, Timor-Leste and Indonesia. For the same geopolitical, humanitarian and historical reasons, Australia's adaptation assistance should retain this geographic focus. Australia should remain engaged in the broader region of Southeast Asia, China and South Asia, but there and globally an indirect role will often be more appropriate, such as through funding multilateral organisations. More generally, the Review recommends that Australia retain flexibility over the deployment of its adaptation support, and consider the full range of bilateral and multilateral channels. Australia should make its decisions on funding mechanisms based on which channels can most effectively deliver adaptation services, particularly to Australia's neighbours. Australia will need to improve its own international disaster response capacity, as we are likely to be increasingly called upon to respond to major natural disasters in the region that surpass local capacities.

The non-aid policies of developed countries are also important for helping developing countries adapt. For example, developed country policies that promote free trade, especially but not only in food, and the flow of unskilled labour, can assist adaptation. Developed country policies on security, peacekeeping and disaster response will also be important to help developing countries adapt if, as some predict (Collier et al. 2008), climate change leads to increased civil conflict.

The movement of people resulting from climate change could eventually be massive. While most of it is likely to be internal, it could spill over national borders. Australia's Pacific island neighbours are already seeking assistance for moving sections of their populations that are experiencing sea-level rise and saltwater inundation. It is too early to say whether at some future point in time it might be necessary for developed countries to create special classes of entry for residents of such climate-affected nations (so-called 'climate refugees'). A more immediate response would be to increase the adaptive capacity of small developing countries. For example, countries such as Australia and New Zealand have a role to play in helping to improve labour mobility in the region. This would strengthen international private sector networks (diasporas), which in turn would help the Pacific island countries to grow and to diversify their risk. There is also likely to be an increased need for humanitarian assistance in response to projected increases in severe weather events, especially around the densely populated mega-deltas of South and Southeast Asia.

10.3 Promoting collaborative research to assist developing countries

While a price or cap on emissions will drive innovation in mitigation, and improved climate science will promote adaptation, developing countries often do not have the research capacity to turn these incentives into action. The importance of research for effective mitigation and adaptation suggests that a priority for international and Australian funding should be a collaborative research endeavour. An appropriate focus for such an effort would be the intersection between climate change, sustainable development and agriculture—for three reasons. First, agriculture is one of the sectors most vulnerable to climate change. Second, it is one of the most important sectors for developing countries. And finally, there is also significant mitigation potential in changed agricultural practices.

The Australian Centre for International Agricultural Research (ACIAR) (see Box 10.3) currently provides targeted funding to promote development-relevant agricultural research (including research on crops, livestock, fisheries, forestry, post-harvest technologies, agricultural development policy and the management of natural resources underpinning agriculture). Most of ACIAR's projects are bilateral—that is, they involve researchers from Australia and developing countries working together to solve problems of shared priority. ACIAR also funds multilateral research as part of an international research network. The coordinating body of the international research network, the Consultative Group for International Agricultural Research, recently began a process of expanding its research agenda to include climate change adaptation and mitigation. The Review recommends that the mandate of ACIAR be explicitly expanded to encompass climate change, in its biological, biophysical and social science dimensions. This would include research into, for example, the development of drought-resistant cultivars and biosequestration and could also extend to research into disaster response, and insurance. There are potential benefits in expanding the remit of ACIAR to include broader environmental issues, such as air quality and waste management, and the expansion should be reviewed by the ACIAR Commission.

Currently, a core objective of ACIAR is the development of local scientific capacity in developing countries, through collaborative research and pilot development projects with Australian scientific institutions. This should continue to be a strong focus, and the centre should consider future partnerships with other Australian research centres, such as the CSIRO Adaptation Flagship, the Climate Change Adaptation Research Facility based at Griffith University and the proposed Australian climate policy research institute (see Chapter 15).

ACIAR may also assist in translating climate projections into forms meaningful to local decision makers.

Impact assessments provide the foundation for effective adaptation planning and action. Many countries have established processes for medium- to long-term development planning, such as poverty reduction plans, and climate projections will need to be incorporated into these frameworks, and converted into policy options. Such planning should inform any research agenda. Research programs with a clear adaptation focus could be counted towards Australia's International Adaptation Assistance Commitment. Programs with a clear mitigation focus could be counted towards Australia's International Low-Emissions Technology Commitment.

The Consultative Group for International Agricultural Research is also proposing a number of governance and structural reforms. Such reforms should maintain the independence and decentralised nature of the group, while minimising bureaucracy and maximising collaboration.

Box 10.3 The Australian Centre for International Agricultural Research and the Consultative Group for International Agricultural Research

The Australian Centre for International Agricultural Research is a statutory authority that forms part of, and is funded by, Australia's aid program. ACIAR is a funding body, developing and managing bilateral and multilateral research projects relating to agriculture (including fisheries and forestry), with the goal of poverty reduction and sustainable development. The centre is also involved in the communication of research results. Its mandate includes a focus on the following five regions: Papua New Guinea and the Pacific Islands, Southeast Asia, North Asia, South Asia and Southern Africa.

ACIAR is the vehicle through which Australia funds its contribution to the Consultative Group for International Agricultural Research, an international network of 15 specialist agricultural and rural research organisations, established in 1971. The international network facilitates cross-border learning and utilises economies of scale in research.

10.4 International trade in emissions rights

10.4.1 Benefits and risks

Trading between countries in emissions rights is an integral part of the Review's proposed approach to mitigation. The agreed emissions targets would need only to hold in aggregate for the world, not at the level of each country. Some countries could emit above their allocations, buying emissions rights from other countries that in turn remain below their allocations. Indeed, it would be a natural development for countries with comparative advantage, after taking the external costs of emissions into account, in production of emissions-intensive goods, to purchase permits in international markets alongside exporting large amounts of the goods.

International trading in emissions entitlements has several advantages:

- It reduces global abatement costs by ensuring that the cheapest abatement opportunities are sought out first, wherever they occur. Cost savings are greater when there are wide differences between participants' target commitments and abatement options, as in a scheme with broad international coverage. Estimates of aggregate cost reductions from global trade are in the range of 20–80 per cent (Stern 2008).
- A broader market can reduce price volatility, dilute country-specific shocks and provide greater certainty on the domestic costs of meeting a target.
- Trade will lead to a convergence of emissions prices across countries and provide a level playing field for trade-exposed, emissions-intensive industries.
- The revenues from international trade provide financial incentives for developing country commitments. Developing countries, with some lower-cost abatement options, can expect to reduce their emissions below their allocations, and sell the freed-up emissions rights. This is the principal direct incentive for developing countries to take on national targets, with developed countries acting as purchasers.[7]

International emissions trading also carries risks. Linking internationally is a form of shared sovereignty, which will imply some loss of control over aspects of mitigation policy. Fully linking into international markets means that the speed and amount of domestic economic adjustment are determined to a significant degree by the international price. Small and medium-sized countries, such as Australia, would lose control of the domestic price of carbon. While in general free trade is welfare promoting, for a government-created market, the resulting price might be too high or low relative to domestic perceptions of the optimal rate of mitigation. Linking can also be a cause of price volatility, for example if there were external policy instability. Risks can be reduced by limiting trading, as discussed below.

From these considerations, it is clear that the spread of international emissions trading offers great opportunities, but needs to be managed in a judicious and calibrated manner. Fully linked international markets are likely to emerge only over time.

Bilateral and regional trading and other forms of cooperation are natural stepping stones towards greater international integration. Such links are already being considered between existing and proposed emissions trading systems in Europe, North America, Australia, New Zealand and Japan, but could also occur between developed and developing countries. Links between individual developed and developing countries, or among groups of countries, will be easier to achieve than comprehensive global integration, and can build on established relationships. Developed countries will need to show leadership in their regions (see Box 10.4 in section 10.8 for potential for links between Australia, Papua New Guinea and Indonesia).

10.4.2 International trading options

Imposing restrictions on trading allows countries to retain greater control over domestic prices and abatement, although with higher overall costs of complying with a given commitment. Rules for allocating trading opportunities and the profit from price differentials have to be devised in this context. Under the Kyoto Protocol, there are unquantified limits on international trade under the supplementarity principle of Article 17. The greater the trust a country has in the international system, the less it will want to resort to limits. As a result, in Chapter 14, the Review suggests limits in the Australian emissions trading scheme on the use of international offsets, but not emissions permits, from markets that meet quality standards.

Direct trading through private firms provides flexibility and is likely to lower transaction costs, especially when trade involves firms from countries with national emissions trading systems. But trading can also occur through government gateways, which would introduce the option to impose conditions on the use of international payments. For example, to make financial transfers more acceptable in permit-buying countries, buyers could require that the revenue be used for climate- and development-related purposes in permit-selling developing countries. Any such arrangements would be negotiated between the parties involved in trade.

A fundamental prerequisite for selling permits is transparent monitoring that complies with standards accepted by the international community and in particular by the main permit buyers. With international trading, incentives to under-report emissions are heightened. An international authority, possibly under the auspices of the UNFCCC, would have to assess whether minimum standards are met, similar to existing procedures under the Kyoto Protocol.

Each country would be able to determine the countries with which it would trade, to protect the integrity of its own domestic system. The scope for selectivity, however, is limited by indirect linking. For example, if Australia links to New Zealand, and New Zealand links to other countries, then Australia's market is effectively also linked to all of New Zealand's partners. Indirect linking will accelerate the tendency towards a similar permit price across countries.

Specific recommendations for how Australia should go about international linking based on consideration of these general options are provided in Chapter 14.[8]

10.4.3 International offset credits

International trading can also occur in offset credits. These involve credits for emissions reductions claimed where no overall national commitment applies. Under the Kyoto Protocol, most international trading is in offset credits derived from the Clean Development Mechanism (CDM).

The CDM has facilitated some developing country engagement in mitigation, but suffers from important limitations. Expanding the CDM beyond its project-by-project basis is currently being considered in the UN process. As discussed in Chapter 9, broad coverage of emissions sources with a safeguard for developing countries is achieved better through one-sided targets than through an expanded CDM. A one-sided target allows the quantitative commitment to be set according to agreed principles, without arbitrarily determining counterfactual baselines. One-sided targets allow for commitments below business-as-usual emissions that can nevertheless benefit developing countries through sales of emissions rights, while providing the safeguard of opting out.

Strong global mitigation will require emissions containment in developing countries in addition to (rather than in substitution for) emissions reduction in developed countries. This can only be provided by developing countries accepting national targets, and not through sales of permits within the CDM. This is increasingly recognised internationally, including by the European Union, to date the principal backer of the CDM.[9]

If this framework were adopted, offset mechanisms would only have a role where there were no national commitments. The CDM would be left as a transitional mechanism to apply in countries without one-sided targets. To remove disincentives for taking on national commitments, no new CDM projects should be accepted from countries that are expected to take on targets (see Chapter 14). In addition, implementation rules for the CDM would need to be strengthened to ensure a high standard of environmental integrity. Countries purchasing CDM credits may also decide to place quantitative or qualitative limits on purchases.

10.5 Price-based sectoral agreements for the trade-exposed, emissions-intensive sectors

Unless large producers the world over face a similar emissions price, there is a danger of artificial movement of production in emissions-intensive industries producing tradable goods from countries applying strong mitigation measures to others. This could have adverse environmental and economic effects. The fear of 'carbon leakage'—a loss of competitiveness and relocation of trade-exposed, emissions-intensive industries as a result of carbon penalties applying in some countries but not others—has been a powerful obstacle to domestic mitigation policies in many countries.

This fear can be exaggerated. Firms in different countries face very different cost structures already, in part due to differing government policies. To the extent that firms enjoy rents—and many have recently seen large increases in output prices—firms will be able to absorb carbon penalties without any adjustment (Lockwood & Whalley 2008). One country's imposition of carbon taxes or an emissions trading scheme without exclusion or compensation for trade-exposed, emissions-intensive industries would tend to lower its real effective exchange rate, offsetting the initial impact on competitiveness for some firms, and absolutely improving competitiveness relative to the prior position for others.

Australian producers of liquefied natural gas have drawn attention to the distortion that might arise from Australia but not its developing country competitors applying a price to carbon. This potential distortion would be substantially less than many differential features of fiscal regimes, for as far into the future as we can see. For example, differences in royalty-like charges affecting costs at the margin, and favourable to Australian producers, would generally be larger than any likely effects of carbon pricing.

Nevertheless, carbon leakage can be a real problem, and one that creates powerful domestic opposition to attempts to impose economy-wide carbon prices.

Countries implementing domestic policies are considering various ways to offset competitive disadvantages to their trade-exposed, emissions-intensive industries—for example, by allocating free emissions permits under emissions trading, or by applying border taxes.[10] Domestic compensation causes difficulties in implementation of domestic climate policies. Chapter 14 describes an optimal arrangement for avoiding carbon leakage without introducing new sources of distortion. This is recommended for Australia, and could usefully be applied elsewhere.

There is a good deal of current interest in developed countries in border tax adjustments. Even if compliant with World Trade Organization rules developed for the purpose (section 10.6), these can only ever be a backstop to international climate change agreements.

To avoid the need for potentially distorting domestic and trade solutions in response to the carbon leakage problem, comparable emissions pricing needs to apply to most or all of the main producers in trade-exposed, emissions-intensive industries. Effective economy-wide emissions pricing commitments for all relevant countries would be the best solution. But not all relevant countries will take on such commitments for some time. The next most straightforward mechanism to achieve a comparable carbon price is sectoral agreements that cause each government to subject the main producers in each industry producing emissions-intensive tradable goods to a carbon tax, until the country has an effective national emissions limit.[11]

An agreement about taxes does not itself allow differentiation of commitments between countries. This is not necessary in the case of the trade-exposed, emissions-intensive sectors. Producers are part of a global market. Domestic governments would keep the revenue, giving them a fiscal incentive to implement the agreement. Access to global climate funds for developing countries could be made conditional on their taking part in relevant international sectoral agreements.

Only a small number of countries would need to be involved in the key industrial sectors to achieve broad coverage. Industries that are often mentioned in the international discussion as candidates for sectoral agreements include iron and steel, aluminium, chemicals, cement, and paper and pulp. The bulk of emissions from developing countries in these sectors arise from just a few countries. To cover 80 per cent or more of developing country emissions in each sector, just three developing countries would need to be involved in iron and steel; four each in aluminium smelting and pulp and paper making; seven in cement production; and nine in chemicals and petrochemicals (Schmidt et al. 2006).[12] From Australia's perspective, additional sectors of interest are non-ferrous metals beyond aluminium, alumina, liquefied natural gas, and the products of sheep and cattle.

Price-based agreements would require agreement on the tax rate for countries not operating under UN-compliant economy-wide commitments. The tax rate would be set as an average of a basket of domestic emissions trading systems; or pegged to the price prevailing in one of the major developed country emissions trading markets (such as the European Union, Australian, or in future North American or East Asian markets).

In some industries, notably aluminium smelting and some steel production, indirect emissions in generating electricity would need to be taken into account. These emissions could be assessed according to a simple and robust approximation, based on the emissions intensity of the systems from which they draw their power, and made subject to the sectoral emissions tax. Indirect or embodied emissions that fell below a threshold would not be considered, in the interest of simplicity.

Appropriate regulatory and governance structures would need to be agreed, starting with a small number of the most important producing countries. Provisions would have to be reviewed periodically and implementation monitored by an international body.

Effective sectoral agreements could and should be struck quickly, as they are relatively straightforward and are important to help facilitate strong mitigation policies in many countries including Australia. A 2013 start date for sectoral agreements should be the goal, directly following the Kyoto Protocol's first commitment period. If coordination among candidate countries begins immediately, there is a good chance to have some agreements in place by then.

It would naturally fall to the large producing countries, in particular developed countries including Australia, to take leadership in crafting agreements among the major producers in each industry sector. To motivate the case for sectoral agreements, policy makers the world over need to understand that comprehensive emissions pricing for trade-exposed industries does not distort the optimal economic location of production of emissions-intensive tradable goods, once environmental externalities have been taken into account. If production moves elsewhere because doing so is cheaper after carbon is priced, this is economically and environmentally efficient restructuring, and should not be discouraged.

10.6 Climate change and trade policy

The links between climate and trade policy are receiving increasing attention. In December 2007, the Indonesian Government convened the first meeting on climate change of trade ministers from major economies in conjunction with the Bali Climate Change Conference.

Trade barriers to the diffusion of low-emissions technologies, and of goods and services embodying them, reduce the technologies' impact. Liberalisation of low-emissions technologies markets can be pursued unilaterally and through multilateral channels (World Bank 2008a). In December 2007, the European Union and the United States introduced a proposal to give priority in the World Trade

Organization negotiations to liberalisation of climate-friendly goods and to services linked to addressing climate change (WTO 2008). The principle is a sound one. The proposal would have been better if it had been comprehensive. The EU–US list does not include ethanol, an important exclusion, as ethanol production receives large domestic subsidies or protection in many countries, including Australia, as well as the European Union and the United States. Neither does it include motor vehicles, despite the interest that all countries have in rapid diffusion of low-emissions innovations in this sector.

The most contentious climate change issue in trade policy is whether countries should be allowed to impose border adjustments if they introduce carbon pricing ahead of others. Two rationales are suggested for such action. The first is to compensate domestic industries for a loss of competitiveness. The second is to apply pressure to other countries to impose similar policies.

The European Union proposal for the post-2012 EU emissions trading scheme and several of the climate change legislative drafts in the United States have flagged provisions for countervailing tariff measures. Economist Joseph Stiglitz (2006) has endorsed this line of action. The Review shares the concern of those who note that such moves may open the doors to protectionism and trade disputes (Bhagwati & Mavroidis 2007).

As the Director-General of the WTO, Pascal Lamy, recently commented, imposing taxes on imports to penalise countries with looser emissions controls would be a 'distant second-best to an international solution' on climate change (Point Carbon 2008). The global community has a strong interest in avoiding pressures for border taxes by moving sooner rather than later to the international agreements that avoid distortions in investment and production in trade-exposed, emissions-intensive industries. Nevertheless, if an international solution is not forthcoming, the pressure, and indeed the case, for border adjustments will grow.

Border adjustments could be imposed unilaterally. It is likely that the WTO would be open to the use of certain trade measures in support of climate change objectives (WTO 2008). Any unilateral adjustments would, however, certainly be appealed and lead to a 'long period of uncertainty and trade frictions' (Hufbauer & Kim 2008: 35).

The alternative course of action, recommended by the Review, is to work for a new WTO code on the subject (Hufbauer & Kim 2008). Such a code would provide a framework within which countries could impose border adjustments, and would greatly reduce the likelihood of the imposition of climate change–justified border adjustments degenerating into a trade war. It would give countries the right to impose adjustments on products in relation to competitors that do not impose comparable mitigation regimes (either economy-wide through national targets, or sector-specific through price-based sectoral agreements). Support for such a code would need to be unanimous. Developing countries have resisted modifications to WTO provisions on environmental grounds, but, given combined EU–US leadership, the credible threat of unilateral responses if no agreement were reached, the other incentives for cooperation on climate change, and the strong United Nations role in the emerging international mitigation regime, an agreement may be possible, though it would probably take several years to forge.

Pending such a global agreement, it would be undesirable for border adjustments to be imposed unilaterally by any country, because of the risks that they would pose to global trade. Rather, if there were a need for unilateral adjustment (due to an absence of global agreements), it would be better to provide domestic payments in WTO-consistent forms to firms.

10.7 International aviation and shipping

Emissions from international air traffic and maritime transport, or 'bunker fuel' emissions, constitute a relatively small share of global fossil fuel emissions (about 1.5 per cent and 2 per cent respectively). But emissions from international aviation grew by 2.7 per cent annually over 2000–05, and shipping emissions, though harder to measure, are estimated to have grown by 3.1 per cent per year (IEA 2007). Both, and especially civil aviation, are expected to increase rapidly their shares of global emissions as incomes and international movements of goods and people rise under business as usual. At present, emissions from the international aviation and maritime transport sectors are not regulated under the UNFCCC or the Kyoto Protocol, due to difficulties of attribution and concerns about competitiveness.

The simplest way to incorporate these two sectors into an international mitigation regime would be to treat them as emissions-intensive, trade-exposed sectors. Emissions from these two sectors should be included against national limits[13] or subject to a comparable carbon tax.[14] Emissions would be attributed to countries on the basis of fuel purchase, and the fuel-supplying country would retain the revenue raised from the tax.

Most freight ships are registered in developing countries but owned by companies in developed countries (UNFCCC 2007b). This makes a sectoral agreement particularly important for shipping. Getting broad coverage may be harder than for aviation, as ships can bunker large amounts of fuel, and have manifold options to refuel. Allowing countries to retain the revenue from any tax would give a positive incentive for enforcement.

For aviation, imposition of a fuel tax might require an amendment to the Convention on International Civil Aviation. Aviation has a range of non-carbon dioxide climate impacts, such as the emission of nitrogen oxides and the formation of condensation trails and cirrus clouds. The IPCC (1999) estimated that total radiative forcing effects from aviation are about two to four times greater than those of the carbon dioxide from burning jet fuel alone. Measurement is complex and uncertain, and this issue may best be addressed after the establishment of an initial sectoral agreement.

10.8 Land-use change and forestry

Emissions from land-use change and forestry (LUCF) include emissions from the removal of forests (deforestation) and their creation (afforestation and reforestation), as well as emissions from the management of forests (for example, through forest degradation or thickening). LUCF emissions differ in a number of ways from energy and industrial emissions. They are concentrated in the developing world because of deforestation, are difficult to estimate, and can be negative (when a forest grows and carbon is sequestered) as well as positive.

In the last few years emissions from tropical deforestation have received increased attention as a potentially important element of global mitigation, spawning interest in mechanisms for reducing emissions from deforestation and degradation (REDD).

- LUCF emissions are larger than earlier thought. The IPCC (2007) roughly estimates annual global LUCF emissions to be some 17 per cent of total emissions—more than the entire global transport sector.
- Reducing LUCF emissions in many instances would be relatively inexpensive. The Stern Review (2007: 245) found that the cost of halting deforestation in eight countries responsible for 70 per cent of LUCF emissions 'would amount to around US$5–10 billion annually (approximately US$1–2/t$CO_2$ on average)'. The World Bank (2006a) also found very low potential abatement costs, with dense tropical forests in Latin America cleared for economic gain amounting to just US$1–3 per ton of CO_2 released. Not all studies suggest such low costs. The opportunity cost of preserving forests varies greatly between sites, and is increasing with rising food and energy prices where conversion to food crops or oil palm plantations is the competing land use. However, the message that 'forestry can make a very significant contribution to a low-cost global mitigation portfolio' (IPCC 2007: 543) is sound.
- The current international regime gives limited rewards for reductions in LUCF emissions, and does little to foster sequestration. Developed countries are required to include emissions from deforestation, reforestation and afforestation (under Article 3.3 of the Kyoto Protocol) and can include other changes in land-based carbon stocks (under Article 3.4) if they choose. But most LUCF emissions are in developing countries. The CDM has no scope for credits gained by reducing deforestation. Credits can be received for establishing forests, but the rules around these are restrictive, and few forest-related CDM projects have been undertaken.

A number of proposals have been put on the table. They take either a national, a sectoral or a project-based approach.

- The simplest framework for reducing LUCF emissions in developing countries would be for those countries, like developed countries, to take on national emissions reduction commitments, and include LUCF emissions in that commitment. If developing countries bring emissions below target, as they would be expected to do, they would be able to trade their excess permits on world markets.
- The sectoral approach would establish separate baselines only for LUCF emissions. There are a large number of proposals that take such a sectoral approach to reduce LUCF emissions in developing countries (see Terrestrial Carbon Group (2008) for a recent proposal from an Australian group, and Hare and Macey (2007) for a survey). As with the national approach, countries would be rewarded if they achieved or came under their LUCF targets. The financial payments could be through either a market mechanism or public funding.
- The project approach would work along the lines of the CDM and reward developing countries for reductions in emissions from a baseline at the project level. For example, if a particular at-risk forest were conserved, an attempt would be made to calculate the saving in emissions.

The many arguments against the efficacy of the CDM approach all apply with greater force in the case of forestry (Forner et al. 2006). It appears neither desirable nor likely for such credits to gain widespread acceptance in international markets.

Sectoral approaches could be attractive to developing countries opposed to country-wide commitments. The quarantining of LUCF from other emissions is attractive given the uncertainties around reducing emissions from deforestation, but comes at the price of additional complexity. It is also far from clear which, if any, of the various competing sectoral approaches could command a consensus.

The national approach would require minimal institutional innovation, and is consistent with a simple, comprehensive approach to abatement. Chapter 9 argued that most developing countries should be given one-sided targets. Opt-out provisions could be particularly important for countries that have large LUCF emissions, and which present large but very uncertain abatement options. Bilateral or regional agreements might be required on the use of trading revenue, for example to allay concerns about displacing rural livelihoods. Averaging over time, and perhaps insurance mechanisms, would be needed to allow smoothing over base periods and commitment periods.

Note that a national approach would not commit developing country governments to introduce a domestic emissions trading scheme. Indeed, applying an emissions trading scheme to the forestry sector would probably not be appropriate for most developing countries. Instead, countries would be well advised to use a mix of regulatory and fiscal measures to help maintain or increase forest cover.

Whichever approach were taken, any serious national effort to reduce LUCF emissions would have to overcome three main challenges.

- **LUCF emissions are difficult to measure**—Measuring emissions from forest management and degradation is particularly difficult. Transparent monitoring systems would be essential if claimed emissions reductions were to provide the basis for financial flows. Ongoing emissions from cleared land (such as from the burning of dried-out peat) can also be large, but even more difficult to measure.
- **Many developing country governments lack the policy mechanisms to reduce LUCF emissions**—In many countries the government's control over the forestry sector is limited. Deforestation might be driven by subsistence agriculture or by illegal logging (for a survey of global forestry policy issues and mechanisms, see World Bank 2006a). Governments will need to develop realistic and implementable strategies for increasing forest cover, including through better forest management as well as reduced land clearing, and reforestation.
- **Logging is an export business subject to carbon leakage**—Reducing logging and LUCF emissions in one country could lead to increased logging and emissions in another. Ultimately, for success, a comprehensive approach covering all major forestry emitters is required.

Although most LUCF emissions are in developing countries, developed countries have a critical role to play. Apart from increasing sequestration within their own borders, they can help with emissions monitoring, and can provide funding to developing countries. Most importantly, they can kick-start action on a bilateral basis. Given the contentious and complex nature of the issue, it is possible that a satisfactory agreement on forest-related emissions will be several years in the making. In the interim, bilateral initiatives and regional cooperation will be particularly important. Given its neighbourhood, Australia's regional initiatives will have a focus, albeit not exclusive, on forestry (Box 10.4). Progress will require developed countries, including Australia, to commit significant resources for emissions reductions that may have no formal international status for the time being.

Box 10.4 Regional partnerships for Australia: the potential for links with Papua New Guinea and Indonesia

Chapter 8 outlined the important role regional partnerships could play in promoting international action on climate change and especially how they could build trust and confidence between developed and developing countries.

Australia has the opportunity to develop such an approach with its neighbours, in particular Indonesia and Papua New Guinea. Both have expressed interest at the highest level in cooperation with Australia on climate change policy.

Box 10.4 Regional partnerships for Australia: the potential for links with Papua New Guinea and Indonesia *(continued)*

While there is a high level of uncertainty around the data, Indonesia's emissions are thought to amount to as much as two Gt CO_2 per year, around five times Australia's total CO_2 emissions, with over three-quarters of that from deforestation (WRI 2008). According to one source, emissions from fires in peat land in Indonesia alone are estimated to be about 1800 Mt per year, about three times Australia's total emissions (Hooijer et al. 2006: 29). Papua New Guinea's annual LUCF emissions may exceed 100 Mt CO_2 (WRI 2008), a quarter of Australia's total CO_2 emissions. Both countries would have a strong interest in reducing emissions from deforestation, provided they were compensated for the loss of economic opportunity such as through the sale of rights on an international market.

Ultimately, it is desirable for both Indonesia and Papua New Guinea to be linked to Australia's emissions trading scheme and to be able to trade any reduction in emissions below their national target levels with the Australian Government or market participants. This would benefit both sides: the financial flows would benefit Indonesia and Papua New Guinea, while Australia would benefit from access to low-cost abatement options. For Indonesia, such deep integration with a large emitting country would be achieved best within larger regional arrangements involving other developed countries, with Japan and New Zealand the obvious first candidates. For such a link to become a reality, important preparatory work has to be completed. Work in several of these areas is already under way under Australia's International Forest Carbon Initiative.

- **Emissions estimation**—Current estimates of LUCF emissions are often highly contested. More accurate estimation of LUCF emissions is needed, not only from land clearing but also from forest management (and degradation) and from post-forest-clearance (for example, emissions from dried-out peat lands).
- **LUCF emissions-reduction strategies**—Reducing LUCF emissions will be a challenging task. The drivers of LUCF emissions include subsistence farming, illegal logging and poor governance. Developing a strategy that will tackle these drivers, bring benefits to local communities and promote forest regeneration and reforestation will not be straightforward.
- **Other low-emissions options**—Reducing LUCF emissions would be central to the Australia–Indonesia and Australia – Papua New Guinea partnerships, but should not be the sole goal of the partnerships. Papua New Guinea has excellent hydro potential, for example.

10.9 Enforcement mechanisms

The Kyoto Protocol has an enforcement mechanism that can be activated for breaches of greenhouse gas accounting or emissions target obligations. If a country does not meet its target, it has to make up the shortfall in the next commitment period with a 30 per cent penalty. This is a weak enforcement mechanism if subsequent commitment period targets are not defined in advance.

The problem of enforcing commitments is part of a more general problem of encouraging effective participation, regarded by some as the Achilles' heel of international efforts. In a world of sovereign states, countries cannot be forced to sign agreements, or to meet their commitments. In order to get countries to participate meaningfully, incentives must be designed so that participation is in the self-interest of each nation. All countries share an interest in reducing the risks of dangerous climate change. Several other sorts of incentives will be important.

- International trade in permits and offsets and access to international public mitigation and adaptation financing (under the proposed International Low-Emissions Technology Commitment and International Adaptation Assistance Commitment) will provide financial incentives for developing countries.
- Trade sanctions have been proposed by some as an enforcement mechanism. As discussed above (section 10.6), border adjustments to take into account differential mitigation regimes have a role to play in a world where some countries are moving faster than others on mitigation, but only once a framework has been developed and agreed under the World Trade Organization.

Enforcement requires monitoring. Under the UNFCCC and Kyoto Protocol, countries are required to produce reliable accounting (annual inventories) of their greenhouse gas emissions. Countries must also have a national registry to account for their emission credits. Teams of experts, selected from a roster of individuals nominated by the Protocol parties, check annual inventories for accuracy and completeness. If there is a dispute between the team and the party, the Protocol's Compliance Committee may intervene. The current system provides a solid foundation on which to build. As the number of countries subject to national emissions goals increases, the need for rigorous and robust monitoring will grow.

Notes

1 The Fourth Assessment Report of the Intergovernmental Panel on Climate Change (IPCC 2007: 20) found: 'Government funding in real absolute terms for most energy research programmes has been flat or declining for nearly two decades (even after the UNFCCC came into force) and is now about half of the 1980 level'. The IPCC (2007: Chapter 13) reports that OECD energy research and development has been below US$10 billion per year since the early 1990s (in 2004 prices and exchange rates) and that research and development budgets for renewable energy exceeded US$2 billion (in 2004 prices and exchange rates) in the late 1970s and early 1980s but have been well below US$1 billion since the mid-1980s.

2 Jagdish Bhagwati (2006) has called for 'subsidising the purchase of environment-friendly technologies by the developing countries'. Larry Summers (2007) has recommended 'the provision of subsidised capital for projects that have environmental benefits that go beyond national borders'.

3 The G7 comprises Canada, France, Germany, Italy, Japan, the United Kingdom and the United States.

4 Popp (2004) estimates annual energy efficiency research and development requirements at only US$13 billion in 2005, rising to US$33 billion in 2055, but the mitigation strategy he models is mild. It allows global temperatures to increase by more than 3°C.

5 The February 2008 statement by the UK, US and Japan governments indicated that the World Bank funding would support 'developing countries that undertake energy sector and climate related policy actions consistent with a low carbon growth trajectory' (Paulson et al. 2008). More explicitly, the United States has stated that it 'believes countries seeking access to the fund should be undertaking credible national plans to limit greenhouse gases and have those plans reflected in a post-2012 climate change agreement' (White House 2008).

6 It has been proposed that technology transfer commitments should be eligible for offsets (Forsyth 1999). Quite aside from its complexity, such an approach would miss the point that this commitment is in addition to an emissions reduction commitment.

7 A similar argument goes for the issue of excess Kyoto permits from Russia and some Eastern European countries, sometimes referred to as 'hot air'. Targets for these countries were negotiated knowing that emissions had fallen dramatically as a result of economic collapse and industrial restructuring in the 1990s, and were an incentive for Russia and others to join the Kyoto Protocol. Trading units from these countries for compliance elsewhere is in the logic of the agreement.

8 See Appendix 2 of Garnaut (2008) for a fuller exposition of these options.

9 The EU's Head of Emissions Trading, Yvon Slingenberg, recently signaled that the European Union wants a 'gradual shift from offsetting to cap and trade', with emissions cuts becoming 'more the contributions of developing countries' (Wynn 2008).

10 The European Union attempts to solve the problem by allocating free permits. Under the post-2012 phase III of the EU emissions trading scheme, it is proposed that affected sectors receive up to 100 per cent of their allowances for free, depending on the extent to which the industries are covered by an international agreement. The European Union has also flagged the possibility that tariffs may be used to neutralise any distorting effects from imports (European Commission 2008).

11 Sectoral agreements have received greatly increased attention recently and are explicitly mentioned in the Bali Roadmap. The various approaches raised in the international discussion (for example, Baron et al. 2007; Bradley et al. 2007; Egenhofer & Fujiwara 2008; Schmidt et al. 2006) are for more complex and less comprehensive schemes. Many revolve around best practice or technology standards, and therefore result in incentives applying only to low-efficiency operations within each technology. Choosing the benchmarks and how they should develop through time is fraught with difficulty. Other proposals are for offset credits from specific sectors, but they present the same problems as the CDM. Comprehensive coverage could be achieved by separate international emissions trading schemes for specific sectors, but this would require negotiating targets in the absence of any obvious principles for allocating them and determining what each sector's cap should be.

12 The actual analysis is in terms of countries that are not in Annex I of the UNFCCC.

13 The European Union has proposed bringing aviation into its emissions trading scheme, with coverage of emissions from flights within as well as from and to the EU. Many non-EU states, however, are opposed to this proposal. An approach agreed to by a number of countries through international negotiations would have a greater chance of success.

14 Such an approach is likely to be more achievable in the short term than negotiating the sector's own version of a trading scheme, not least because of the lack of basis for setting targets. The International Civil Aviation Organization is, however, working on an emissions trading system for international civil aviation (ICAO 2008).

References

AFP 2008, 'G7 calls for investment to fight climate change', AFP, 9 February, Tokyo, <http://afp.google.com/article/ALeqM5gc8JEW4bcoTUnJcCKemRD-iJprWg>.

Baron, R., Reinaud, J., Genasci, M. & Philibert, C. 2007, *Sectoral Approaches to Greenhouse Gas Mitigation: Exploring issues for heavy industry*, IEA Information Paper, IEA, Paris.

Barrett, S. 2003, *Environment and Statecraft: The strategy of environmental treaty-making*, Oxford University Press, New York.

Bhagwati, J. 2006, 'A global warming fund could succeed where Kyoto failed', *Financial Times*, 16 August.

Bhagwati, J. & Mavroidis, P.C. 2007, 'Is action against US exports for failure to sign Kyoto Protocol WTO-legal?' *World Trade Review* 6(2): 299–310.

Bosetti, V., Carraro, C., Massetti, E. & Tavoni, M. 2007, *Optimal Energy Investment and R&D Strategies to Stabilise Greenhouse Gas Atmospheric Concentrations*, Fondazione Eni Enrico Mattei, Milan.

Bradley, R., Baumert, K.A., Childs, B., Herzog, T. & Pershing, J. 2007, *WRI Report—Slicing the pie: sector-based approaches to international climate agreements—issues and options,* World Resources Institute, Washington DC.

Carmody, J. & Ritchie, D. 2007, 'Investing in clean energy and low carbon alternatives in Asia', Asian Development Bank, Philippines, <www.adb.org/Documents/Studies/Clean-Energy-and-Low-Carbon-Alternatives-in-Asia/clean-energy-low-carbon-alternatives-in-asia.pdf>.

Collier, P., Conway, G. & Venables, T. 2008, *Climate Change and Africa*, Oxford University, <http://users.ox.ac.uk/~econpco/research/pdfs/ClimateChangeandAfrica.pdf>.

Egenhofer, C. & Fujiwara, N. 2008, *Global Sectoral Industry Approaches to Climate Change: The way forward*, Centre for European Policy Studies, Brussels.

European Commission 2008, 'Questions and answers on the Commission's proposal to revise the EU emissions trading system', press release dated 23 January.

Forner, C., Blaser, J., Jotzo, F. & Robledo, C. 2006, 'Keeping the forest for the climate's sake: avoiding deforestation in developing countries under the UNFCCC', *Climate Policy* 6: 275–94.

Forsyth, T. 1999, 'Flexible mechanisms of climate technology transfer', *Journal of Environment and Development* 8(3): 238–57.

Garnaut, R. 2008, *Emissions Trading Scheme Discussion Paper*, <www.garnautreview.org.au>.

GEF (Global Environment Facility) 2008a, 'The Least Developed Countries Fund', <www.gefweb.org/interior.aspx?id=194&terms=ldcf>.

GEF 2008b, 'The Special Climate Change Fund (SCCF)', <www.gefweb.org/interior.aspx?id=194&terms=ldcf#id=192&ekmensel=c57dfa7b_48_60_btnlink>.

G-77 & China 2008, 'Financial mechanism for meeting financial commitments under the Convention', submission to the Third Session of the Ad Hoc Working Group on Long-term Cooperative Action under the UNFCCC, Accra, 21 August to 27 August, <http://unfccc.int/files/kyoto_protocol/application/pdf/g77_china_financing_1.pdf>.

Hare, B. & Macey, K. 2007, *Tropical Deforestation Emission Reduction Mechanism: A discussion paper*, Greenpeace.

Hooijer, A., Silvius, M., Wösten, H. & Page, S. 2006, *PEAT-CO_2: Assessment of CO_2 emissions from drained peatlands in SE Asia*, Delft Hydraulics Report Q3943.

Hufbauer, G.C. & Kim, J. 2008, 'Reconciling GHG Limits with the Global Trading System (draft)', Peterson Institute for International Economics, Washington DC.

ICAO (International Civil Aviation Organization) 2008, 'Environment', Environmental Unit: Air Transport Bureau, <www.icao.int/icao/en/env/index.html>.

IEA (International Energy Agency) 2007, *World Energy Outlook 2007: China and India insights*, IEA, Paris.

IEA 2008, *Energy Technology Perspectives 2008: Fact sheet—the blue scenario*, IEA, Paris.

IPCC (Intergovernmental Panel on Climate Change) 1999, *Aviation and the Global Atmosphere*, J.E. Penner, D.H. Lister, D.J. Griggs, D.J. Dokken & M. McFarland (eds), Cambridge University Press, Cambridge.

IPCC 2007, *Climate Change 2007: Mitigation of climate change. Contribution of Working Group III to the Fourth Assessment Report of the Intergovernmental Panel on Climate Change*, B. Metz, O.R. Davidson, P.R. Bosch, R. Dave & L.A. Meyer (eds), Cambridge University Press, Cambridge and New York.

Kammen, D.M. & Nemet, G.F. 2005a, 'Real numbers: reversing the incredible shrinking energy R&D budget', *Issues in Science and Technology*, National Academy of Sciences, National Academy of Engineering, Institute of Medicine, and University of Texas at Dallas, Dallas, pp. 84–8.

Kammen, D.M & Nemet, G.F. 2005b, 'Supplement: estimating energy R&D investments required for climate stabilization', <www.climatetechnology.gov/stratplan/comments/Kammen-1.pdf>.

Köhler, J., Grubb, M., Popp, D. & Edenhofer, O. 2006, 'The transition to endogenous technical change in climate–economy models: a technical overview to the innovation modeling comparison project', *Energy Journal: Special issue*, <www.econ.cam.ac.uk/research/imcp/technicaloverview.pdf>.

Lockwood, B. & Whalley, J. 2008, *Carbon Motivated Border Tax Adjustments: Old wine in green bottles*, Working Paper 14025, National Bureau of Economic Research, Cambridge, Massachusetts.

Oxfam 2007, *Adapting to Climate Change: What's needed in poor countries, and who should pay*, Oxfam Briefing Paper.

Paulson, H., Darling, A. & Nukaga, F. 2008, 'Financial bridge from dirty to clean energy', *Financial Times*, 7 February.

Point Carbon 2008, 'EU border tax "a distant second" to global deal: WTO chief', <www.pointcarbon.com/news/1.931067>.

Pontin, J. 2007, 'Vinod Khosla: a veteran venture capitalist's new energy', *MIT Technology Review*, March/April, <www.technologyreview.com/Energy/18299/>.

Popp, D. 2004, *R&D Subsidies and Climate Policy: Is there a 'free lunch'?* Working Paper 10880, National Bureau of Economic Research, Cambridge, Massachusetts.

REEEP (Renewable Energy and Efficiency Partnership) 2008, 'Mission', REEEP, Vienna, <www.reeep.org/511/mission.htm>.

Schelling, T.C. 1997, 'The cost of combating global warming: facing the tradeoffs', *Foreign Affairs* 76(6): 8–14.

Schelling, T.C. 2002, 'What makes greenhouse sense?' *Foreign Affairs* 81(3): 2–9.

Schmidt, J., Helme, N., Lee, J. & Houdashelt, M. 2006, *Sector-Based Approach to the Post-2012 Climate Change Policy Architecture*, Future Actions Dialogue Working Paper, Center for Clean Air Policy, Washington DC.

Stern, N. 2007, *The Economics of Climate Change: The Stern Review*, Cambridge University Press, Cambridge.

Stern, N. 2008, *Key Elements of a Global Deal on Climate Change*, LSE, London.

Stiglitz, J. 2006, 'A new agenda for global warming', *Economist's Voice*, <www.bepress.com/ev>.

Summers, L. 2007, 'Practical steps to climate control', *Financial Times*, 28 May.

Terrestrial Carbon Group 2008, *How to Include Terrestrial Carbon in Developing Nations in the Overall Climate Change Solution*, <www.terrestrialcarbon.org>.

UNDP (United Nations Development Programme) 2007, *Human Development Report 2007/2008. Fighting Climate Change: Human solidarity in a divided world*, Palgrave Macmillan, New York.

UNFCCC (United Nations Framework Convention on Climate Change) 2007a, *Investment and Financial Flows Relevant to the Development of an Effective and Appropriate International Response to Climate Change*, UNFCCC, Bonn.

UNFCCC 2007b, *Submission from Government of Norway on Emissions from Fuel Used for International Aviation and Maritime Transport*, UNFCCC, Bonn.

White House 2008, 'Increasing our energy security and confronting climate change through investment in renewable technologies', White House fact sheet, 5 March, <www.whitehouse.gov/news/releases/2008/03/20080305-2.html>.

World Bank 2006a, *At Loggerheads? Agricultural expansion, poverty reduction, and environment in the tropical forests*, Policy Research Report, World Bank, Washington DC.

World Bank 2006b, 'Clean energy and development: towards an investment framework', paper produced for the Development Committee Meeting, DC2006–0002, 23 April.

World Bank 2008a, *International Trade and Climate Change: Economic, legal and institutional perspectives*, World Bank, Washington DC.

World Bank 2008b, 'Proposal for a Strategic Climate Fund (draft)', <http://siteresources.worldbank.org/INTCC/Resources/Proposal_for_a_Strategic_Climate_Fund_April_28_2008.pdf>.

WRI (World Resources Institute) 2008, *Climate Analysis Indicators Tool (CAIT) Version 5.0*, WRI, Washington DC.

WTO (World Trade Organization) 2008, *The Multilateral Trading System and Climate Change*, <www.wto.org>.

Wynn, G. 2008, 'EU wants developing nations to do more on climate', *Reuters News Service*, 11 March.

COSTING CLIMATE CHANGE AND ITS AVOIDANCE 11

Key points

Type 1 (modelled median outcomes) plus Type 2 (estimates of other median outcomes) costs of climate change in the 21st century are much higher than earlier studies suggested. The Platinum Age emissions grow much faster than earlier studies contemplated.

The modelling of the 550 mitigation case shows mitigation cutting the growth rate over the next half century, and lifting it somewhat in the last decades of the century.

GNP is higher with 550 mitigation than without by the end of the century. The loss of present value of median climate change GNP through the century will be outweighed by Type 3 (insurance value) and Type 4 (non-market values) benefits this century, and much larger benefits of all kinds in later years.

Mitigation for 450 costs almost a percentage point more than 550 mitigation of the present value of GNP through the 21st century. The stronger mitigation is justified by Type 3 (insurance value) and Type 4 (non-market values) benefits in the 21st century and much larger benefits beyond. In this context, the costs of action are less than the costs of inaction.

Does participation in global mitigation, with Australia playing a proportionate part, and with all the costs of that part, make sense for Australia? If so, what extent of mitigation would give the greatest benefits over costs of mitigation for Australians?

The costs of mitigation come early, and the benefits of mitigation through avoided costs of climate change come later. The costs of mitigation are defined clearly enough to be assessed through standard general equilibrium modelling. The benefits of mitigation come in four types, only one of which is measurable with standard modelling techniques. This chapter applies the decision-making framework of Chapter 1 to the fundamental question before the Review. The analysis is informed by the modelling undertaken jointly with the Australian Treasury and independently by the Review.

11.1 The three global scenarios

This chapter analyses the three scenarios introduced in Chapter 4—the no-mitigation scenario, in which the world does not attempt to reduce greenhouse gas emissions; and the 550 and 450 scenarios, which represent cooperative global efforts to reduce emissions to varying degrees. To answer the question of whether Australia should support, and play its full part in, a global mitigation effort, the Review compared the costs and benefits of the no-mitigation and the 550 scenarios. To answer the question of how much mitigation Australia should support the Review compared the 450 and 550 scenarios. What is compared, through a mix of modelling and analysis, is the cost to Australia of participating in a global agreement to mitigate climate change, and the costs of climate change under the three scenarios.

In 2005, the atmospheric concentration of greenhouse gases was about 455 parts per million (ppm) of carbon dioxide equivalent (CO_2-e). In the no-mitigation world, under the view of business-as-usual emissions presented in Chapter 3, this would reach 550 ppm by 2030, 750 by 2050, 1000 by 2070, and 1600 by 2100.

The concentration of carbon dioxide (the main greenhouse gas) in this scenario would reach 1000 ppm at 2100, compared to a band of natural variability of carbon dioxide over many millennia of between 180 and 280 ppm, and 280 ppm in the early years of modern economic growth in 1840.

In the 550 scenario, concentrations of greenhouse gases stop rising by around 2060, and after slight overshooting, stabilise around 550 ppm CO_2-e, one-third of the level reached in the no-mitigation scenario, by the end of the century. In the more stringent 450 scenario, given the current concentration, significant overshooting above 450 ppm CO_2-e is inevitable. Concentrations peak at 530 ppm CO_2-e around 2050, and decline towards stabilisation at 450 ppm CO_2-e early in the 22nd century.

Atmospheric concentrations of greenhouse gases are important primarily because of their impact on global temperature. Table 11.1 shows the expected increases in global temperature associated with each of the three scenarios, as well as the temperature consistent with the highest climate sensitivity in the 'likely' range defined by the IPCC—that is, two-thirds probability of remaining within the limits (IPCC 2007). In the absence of mitigation, in the median case, the world is heading for a 2.3°C increase over 1990 levels by 2050, and 5.1°C by 2100. Temperatures would continue to rise by as much as 8.3°C by the end of the next century, or higher if the climate sensitivity were above its central estimate.

The 550 and 450 scenarios will limit median expectations of end-of-century temperature increases to 2°C and 1.6°C, respectively, above 1990 levels under the central estimate for climate sensitivity, and stabilise global temperatures at just above these levels. Even so, an end-of-century increase of 2.7°C and 2.1°C, respectively, above 1990 levels is still within the likely range of the 550 and 450 scenarios (Table 11.1).

Table 11.1 Temperature increases above 1990 levels under the no-mitigation, 550 and 450 scenarios

	2050		2100		2200	
	Best estimate	Upper end of likely range	Best estimate	Upper end of likely range	Best estimate	Upper end of likely range
No mitigation	2.3°C	2.9°C	5.1°C	6.6°C	8.3°C	11.5°C
550	1.7°C	2.2°C	2.0°C	2.7°C	2.2°C	3.1°C
450	1.6°C	2.1°C	1.5°C	2.1°C	1.1°C	1.7°C

Note: The 'best estimate' and 'upper end of likely range' temperature outcomes were calculated using climate sensitivities of 3°C and 4.5°C respectively. There is a two-thirds probability of outcomes falling within the likely range. Temperatures are derived from the MAGICC climate model (Wigley 2003). Temperature outcomes beyond 2100 are calculated under the simplifying assumption that emissions levels reached in each scenario in the year 2100 continue unchanged. They do not reflect an extension of the economic analysis underlying these scenarios out to 2100, and are illustrative only. It is unlikely that emissions in the reference case will stabilise abruptly in 2101 with no policies in place, and hence the temperatures shown underestimate the likely warming outcomes if continued growth in emissions is assumed. 1990 temperatures are about 0.5°C above pre-industrial levels.

11.2 Comparing the costs of climate change and mitigation

To understand the potential economic implications of climate change for Australia, appropriate scientific and economic frameworks must be combined to estimate impacts. This is not a trivial task. There is uncertainty in many aspects of climate change science at the climate system, biophysical and impact assessment levels. These compounding sources of uncertainty mean that quantifying the economic impacts of both climate change and its mitigation is a difficult, and at times speculative, task. The Stern Review (2007: 161) cautioned that 'modelling the overall impact of climate change is a formidable challenge, demanding caution in interpreting results'. Moreover, modelling alone will not provide an answer to the two questions posed at the beginning of this chapter. As explained in Chapter 1, many of the costs of climate change cannot be modelled.

The framework set out in Chapter 1 distinguished between four types of costs of climate change. The first type of cost (Type 1) has been measured through a computable general equilibrium model, based on measured market impacts of climate change in the median or 'average' cases suggested by the science. That is the easiest part of the problem, but still involves the most complex long-term modelling of the Australian economy ever undertaken. The requirement to model changes in the structure of the Australian economy in a general equilibrium framework to the end of the 21st century takes the models to the limits of their capacities. For details on the combination of models used, see Box 11.1.

Box 11.1 The Review's modelling

The Review's innovation was to model the cost of imposing mitigation policy alongside the benefits of the climate change avoided. This was done as follows:

- **Step 1:** A reference case of no climate change and no climate change policy was developed jointly by the Review and the Australian Treasury.
- **Step 2:** A 'no mitigation' policy scenario was developed by the Review that entailed shocking the reference case to simulate a world of unmitigated climate change.
- **Step 3:** The effect of mitigation policy was modelled by (1) imposing a carbon constraint on the model, and (2) imposing 'positive' climate shocks to simulate a lesser degree of climate change as a result of successful global mitigation policy (that is, 550 ppm or 450 ppm).

This was a highly complex and technically pathbreaking process, which required the Review to draw on a wide range of expertise and models within individual sectors. The modelling of the expected impacts of climate change by the Review included individual areas of impact, including agriculture, human health and several aspects of infrastructure.

There is currently no single model that can capture the global, national, regional and sectoral detail that was necessary for the Review's approach. As a result, the Review drew on a number of economic models to determine the costs of climate change and the costs and benefits of climate change mitigation for the Australian economy. The key models used were the Monash Multi Regional Forecasting (MMRF) model, the Global Trade and Environment Model (GTEM) and the Global Integrated Assessment Model (GIAM). The focus of the modelling of both mitigation and climate damage costs was on Australia. However, GTEM and GIAM—which extends GTEM to model the interaction between the climate system and the economy—were used to model global mitigation and climate change damages. The outcomes of this modelling (in particular, the global carbon price, and Australia's emissions entitlement and trade impacts) were fed into the MMRF model. MMRF was augmented by a series of scientific and economic models, including for the electricity, transport, and land-use change and forestry sectors. This allowed the determination of the costs of unmitigated climate change and the net costs of mitigation to Australia. GTEM was also used to derive mitigation costs for Australia, but, unlike MMRF, without calculation of avoided climate change.

In interpreting the results of the modelling, it is important to bear in mind that only one of the four types of costs of unmitigated climate change, and therefore only one of the four types of benefits of avoided climate change from mitigation, could be captured in the model.

Further details of the modelling and the climate change impact work undertaken by the Review are available in a technical appendix at <www.garnautreview.org.au>. The Review looks forward to further empirical work and refinement by others of its modelling of climate change.

The second type of cost (Type 2) involves market impacts in the median cases, for which effects cannot be measured with sufficient precision and confidence to feed into a computable general equilibrium model. By their nature, these costs and benefits are not amenable to precise quantification. The Review formed judgments about likely magnitudes, relative to the size of the impacts that were the focus of the formal modelling. These assessments were applied in a transparent way in adjustments to some of the model results, to remove the bias that would otherwise be associated with the exclusion of obviously important market impacts for which data were not available at the time of the modelling work. This is not as good as modelling these costs within the general equilibrium framework would have been if the data had been available, but it is clearly better than leaving them out altogether.

The third type of cost (Type 3) is that associated with the chance that the impacts through market processes turn out to be substantially more severe than suggested in the median cases. Type 3 costs derive their importance from the normal human aversion to risk in relation to severe outcomes, and from the possibility that the bad end of the probability distribution includes outcomes that are extremely damaging and in some cases catastrophic. Since the modelling undertaken was not of a probabilistic nature, these 'worst case' impacts could not be quantified.

The fourth type of cost (Type 4) involves services that Australians value, but which do not derive their value from market processes. Examples include deterioration of environmental amenity; loss of species and, more generally, of biodiversity; and health and international development impacts that do not necessarily have their effects through the imposition of monetary costs on the Australian community. By definition, these costs cannot be included in the modelling.

In contrast to three of the four types of climate change induced costs, all mitigation costs have a market impact and so can be measured.

The other difference between mitigation costs and climate change costs is their profile over time. The Review's modelling of the effects of climate change ends in 2100. The long time frames and large structural shifts involved in climate change analysis present considerable challenges for modelling the way the economy is likely to respond. As in most economic models, the assumed behavioural responses in models used by the Review are determined by parameters and data that have been derived from recent history. Into the second half of this century and beyond, the assumptions that must be made about economic parameters and relationships become highly speculative. And yet all of the detailed assessments of the economics of climate change indicate that the main costs of climate change, and therefore the main benefits of mitigation, accrue in the 22nd and 23rd centuries and beyond (Stern 2006; Nordhaus 2008; Cline 2004). Whereas the costs of *mitigation* can be expected to stabilise this century, the costs of *climate change* can be expected to accelerate over this century and into the next. Consideration of the long-term costs and benefits must be a feature of any evaluation of the net benefits of climate change mitigation policy.

Because of the importance of the non-quantified costs of climate change—whether they are Type 3 or Type 4 costs this century, or any of the types of costs beyond this century—a comparison of the modelled costs of mitigation and no mitigation, or even of differing degrees of mitigation, can only contribute to any comparison of two scenarios. What can be modelled are the gross costs of mitigation (purely the costs of mitigation without any of the benefits of avoided climate change) and the net costs of mitigation (the gross costs of mitigation minus the modelled Type 1 impacts and the estimated Type 2 impacts). These costs of mitigation (modelled at most out to 2100) need to be compared with judgments concerning the non-quantified Type 3 and Type 4 benefits from mitigation this century, and with the likely benefits (of all types) from climate change avoided in the next century and subsequent centuries.

11.3 Modelling mitigation

The modelling of the two mitigation scenarios is based on costs associated with Australia's adherence to an emissions allocation, derived from an international agreement commencing in 2013, to limit the concentration of greenhouse gases to 450 and 550 ppm CO_2-e respectively. As shown in Chapter 9, under the 550 scenario Australia's emissions entitlement allocation in 2050 relative to 2000 emissions in absolute terms falls by 80 per cent, and in the 450 scenario by 90 per cent. Australia's emissions can exceed that allocation if we buy permits from other countries at the global carbon price, which prevails across all sectors within Australia, and across all countries around the world. The global carbon price increases over time, along a path which ensures that emissions fall sufficiently for either of the two concentration targets to be achieved.[1] As a small emitter in global terms, Australia's emissions do not affect the global carbon price, which is taken as a given in the domestic modelling. All revenues raised from sales of carbon permits are distributed back to households. No payments are made to the trade-exposed sector, as all industries around the world face the same emissions penalty. No compensation payments are made to industry.

A critical determinant of the costs of mitigation is the assumptions made about technologies that are or will become available to reduce emissions. Technological development of any type is difficult to predict. When powerful incentives to innovation are introduced to a market environment, however, human ingenuity usually surprises on the upside. How will this ingenuity manifest itself in the face of high emissions prices and increased public support on a global scale for research, development and commercialisation of low-emissions technologies?

We do not know, but there are good reasons to believe that, if we get the policy settings right over the next few years, the technological realities later in the century will be greatly superior to those which, for good reason, are embodied in the standard technology variants of the models used by the Review.[2]

As one alternative to the *standard technology* assumptions, the Review modelled an *enhanced technology* future, embodying various assumptions of more rapid technological progress, none of which seems unlikely.[3]

As another possibility for the future, the Review examined the implications of the commercialisation of a *backstop technology* encouraged by high carbon prices, that, at a cost of $250 per tonne of CO_2-e, takes greenhouse gases from the atmosphere for recycling or permanent sequestration. In the Review's modelling of the backstop, deployment starts between 2050 and 2075.

The backstop technology has been introduced into the modelling in a stylised manner. No single technology has been modelled. Rather, the backstop technology is assumed to be available for all industries. In practice, the most likely backstop technology will not be industry specific, but will, at a substantial cost, extract carbon dioxide from the air for recycling or sequestration.

While the backstop and enhanced technologies are possibly complementary, they are assumed to be alternatives in the modelling.

Which of these three visions of the technological future, or which combination of them, or which alternative to all of them, defines the opportunities that evolve through market processes over the years ahead will be revealed in due course. Technological developments in response to a rising carbon price will have a large effect on the acceptability to the global community of 450 and 400 ppm CO_2-e mitigation strategies in future years.

Figure 11.1 shows the 450 and 550 global carbon prices in 2005 Australian dollars under different assumptions about technology.

Figure 11.1 Australia's carbon prices under different mitigation scenarios and technological assumptions

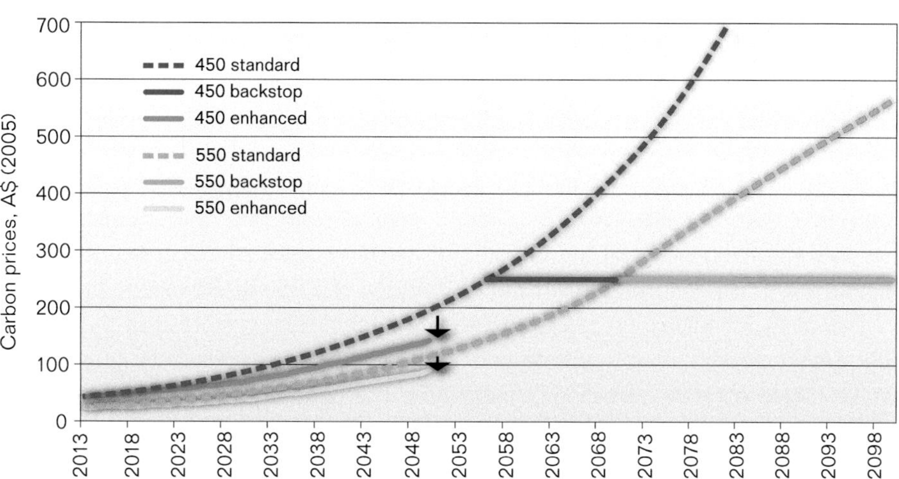

Note: The rising carbon price paths are derived in GTEM and implemented in MMRF, except for the prices derived under the enhanced technology assumptions, which are implemented only in GTEM and reported up to 2050. The 450 and 550 price paths move on to the horizontal backstop path when they reach about $250/t CO_2-e. The two arrows show the extent to which the enhanced technology assumptions reduce the carbon price relative to the standard technology assumptions.

Taking into account the deadweight costs, negative and positive, of various policies that are used to achieve reductions in emissions is just as difficult as imagining the technological future. Nordhaus's pioneering work (1994) emphasised the reductions in deadweight costs that could come from replacing distorting forms of taxation, such as income tax, by a carbon tax. Mitigation through a carbon tax—no exemptions, no shielding of trade-exposed industries—had a positive economic benefit, because the carbon tax was less economically distorting than the taxes that it replaced. A similar result could be obtained by replacing distorting Australian taxes with revenue from the competitive sale of permits from an emissions trading scheme. However, a distorted Australian emissions trading scheme that diverts management effort from commercial activities into applying pressure for political preferment could have large negative deadweight costs.

The modelling has assumed no net transactions and other deadweight costs of the mitigation regime. History will reveal whether this was an optimistic or pessimistic assumption.

The costs of mitigation depend on who bears them. Generally, an increment of money is judged to be more valuable to the poor than the rich. It follows that the costs of mitigation are higher, and the optimal amount of mitigation effort lower, the more the costs are carried by the poor. More mitigation is justified if compensation for low-income Australian households is a major feature of the policy framework. (Chapter 16 explores the distributional impacts of mitigation.) Similarly, more global mitigation can be justified if low-income countries carry a low proportion of the costs. Australia has a strong interest in the burden of mitigation being borne equitably across countries and therefore disproportionately by developed economies, as Australia's terms of trade would be damaged most by any setback to income growth in developing countries.

11.4 The decision to mitigate

Is global mitigation in Australia's interests? To test the case for action, the Review compared the no-mitigation and 550 scenarios, and compared the costs of mitigation of climate change with the benefits of avoiding climate change (the difference between the costs of climate change with and without mitigation). The costs of mitigation and the benefits of avoiding some of the costs of climate change are those associated with implementation of a 550 stabilisation strategy.

The case for mitigation rests on the large temperature increases that would be a likely outcome—not a remote possibility—of the rapid emissions growth that can be expected in the absence of mitigation. The updated, realistic projections of emissions growth developed by the Review, combined with mainstream scientific estimates of climate sensitivity, result in a best estimate of the no-mitigation scenario giving rise to a 5°C temperature increase over the course of this century. This would at best impose severe costs on the world and on Australia.

11.4.1 The cost of unmitigated climate change

Type 1 costs: modelled expected market impacts of climate change

The Review's economic modelling focused on five key areas of impact (primary production, human health, infrastructure, tropical cyclones and international trade). In each of these areas, climate change shocks were imposed reflecting the best estimates and judgments available on the likely market costs of climate change.[4] (See Table 11.2.) The modelled market impacts of unmitigated climate change relative to a world without climate change (the reference case of chapters 3 and 7) are shown in Figure 11.2.

Figure 11.2 The modelled expected market costs (median case) for Australia of unmitigated climate change, 2013 to 2100 (Type 1 costs only)

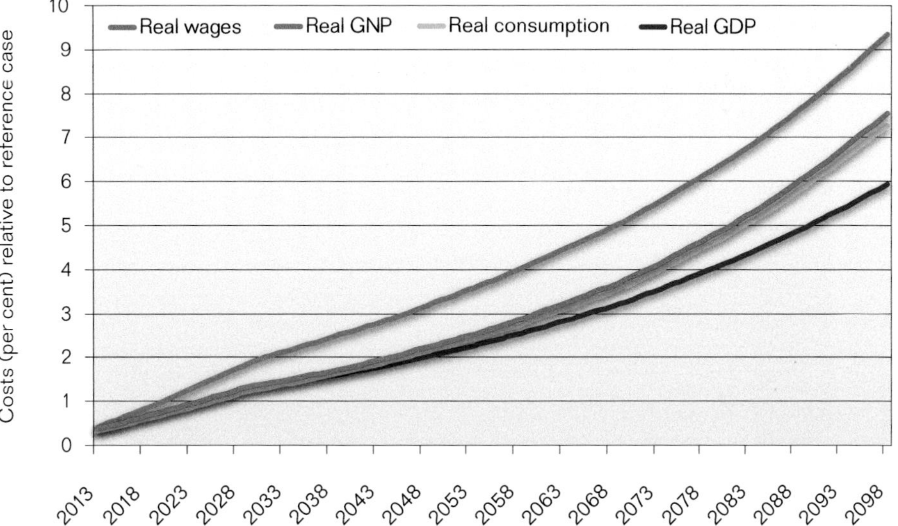

Note: All variables in this figure and throughout this chapter are in 2005 prices.

The modelled costs of climate change rise over time. Household consumption and GNP on the one hand, and GDP on the other, diverge through time due to the projected fall in Australia's terms of trade relative to the reference case.[5] If this occurs, a greater volume of exports is required to pay for the same volume of imports. Since consumption tracks GNP closely, most of the results are reported in terms of GNP.

Changes in labour demand are captured in large changes to wages, rather than unemployment, as the wage rate moves to eliminate any short-run employment effects from climate change. Unmitigated climate change causes real wages to be around 12 per cent lower than they would otherwise have been. The fall in real wages increases significantly in the second half of the century in response to reduced demand for labour as a result of climate change.

Table 11.2 Assessing the market impacts of climate change

Sector	Direct impact	Modelled	Risk	Ability to adjust or adapt	Comment	Likely economic consequence by 2100
Economy wide—international trade	Changes to import prices	Yes	High	–	Commodity-specific shocks, but methodology overlooks sectoral dimensions of climate change.	High
	Changes to world demand (commodity specific)	Yes	High	–		
Economy wide—infrastructure	Impacts on commercial buildings—changes to building codes and planning schemes	No	High	High	Capital stock of dwellings (current prices) in 2006–07 was around $1.3 trillion, 40% of total capital stocks (ABS 2007a).	Medium
	Accelerated degradation of buildings—maintenance and repair costs	Yes	High	High		Medium
Economy wide—extreme events (tropical cyclones, storms/flooding, bushfires)	Increased intensity of tropical cyclones—damage to residential infrastructure and home contents	Yes	High	Medium	The current average annual cost of tropical cyclones is estimated at $266 million, a quarter of the cost of natural disasters (BTE 2001).	Low
	Increased intensity of tropical cyclones—damage to commercial buildings and business interruption	No	High	Medium		Low
	Southward movement of tropical cyclones—infrastructure and business interruption	No	High	Medium	High uncertainty regarding southward movement of tropical cyclones.	Low
	Higher frequency of storm events (e.g. flooding from non-cyclone events)—damage to infrastructure	No	High	Medium in short run High in long run	Estimated average annual cost of floods in Australia is $314 million (BTE 2001).	Low
	Bushfires—infrastructure damage, crop loss, emergency response	No	High	Low–medium	Bushfires estimated to pose an annual average cost of about $77 million (BTE 2001).	Low
Economy wide—sea-level rise	Increase in sea levels of 0.59 m, impacts on coastal settlements	Yes	High	High	Assessment assumes there is no significant sea-level rise this century.	Low
Economy wide—human health	Heat-related stress, dengue fever and gastroenteritis—impacts on productivity	Yes	High	High	Assumes management and prevention of health impacts.	Low
	Other health impacts (productivity)	No	Low	High		Low
Agriculture	Changes in dryland crop production due to changes in temperature and CO2 concentrations	Yes	High	Medium	All crops $19.5 billion (gross value of commodities) (ABS 2007b).	High

Table 11.2 Assessing the market impacts of climate change *(continued)*

Sector	Direct impact	Modelled	Risk	Ability to adjust or adapt	Comment	Likely economic consequence by 2100
	Sheep, cattle, dairy—changes in carrying capacity of pasture from CO2 concentrations, rainfall and temperature	Yes	High	Medium		High
	Impacts on sheep and cattle from heat stress due to temperature increases	No	Medium	Medium to high		Low
	Impacts on pigs and poultry from heat stress due to temperature increases	No	Low	Medium	Limited research on potential impacts.	Low
	Irrigated agriculture—reductions in water runoff	Yes	High	Medium	Adaptation through land-use change, water conservation.	High
Fisheries	Reduced yields due to changes in water temperature	No	Medium	Low	Potential for higher adaptive capacity for aquaculture. In 2004–05, fisheries contributed 0.14% of total GDP (ABS 2008a).	Low
Forestry	Yields affected by water availability and CO2 concentrations	No	Low	Low	Possibility of using other species. In 2004–05, forestry contributed 0.12% of total GDP (ABS 2008a)	Low
Mining	Slower growth in demand due to slower increase in world income (relative to a world with no climate change)	Yes	High	Low	In 2006–07, the mining industry contributed 7% of total GDP (ABS 2007a). In the reference case, this is projected to be 10.2% in 2100.	Medium to high
	Reduction in water availability	No	High	High		Low

Table 11.2 Assessing the market impacts of climate change *(continued)*

Sector	Direct impact	Modelled	Risk	Ability to adjust or adapt	Comment	Likely economic consequence by 2100
Tourism	International tourism affected by slower growth in demand due to slower increase in world incomes (relative to a world with no climate change)	Yes	High	Low	Tourism as a share of GDP is 3.7%, which equates to $38 billion. In 2005–06, this was 10% of exports of goods and services (ABS 2008b).	Medium to high (highly uncertain)
	Reduction in international demand for Australian tourism as a result of reduced natural amenity of tourism products	No	High	Low	Requires assumptions of changes in preferences and relative amenity versus absolute amenity.	High
	Changes in domestic tourism as a result of reduced amenity of tourism products	No	Medium	Low to medium	Domestic tourism is worth $29 billion (2.6% of GDP) (relative to international tourism—$10 billion,0.9% of GDP). (ABS 2008B).	Low
Government (health)	Increased expenditure on prevention and treatment for dengue virus, heat stress and gastroenteritis	Yes	Medium	High	Public and private expenditure low to moderate.	Low
	Increased expenditure on prevention and treatment for other health impacts (air pollution, mental health etc.)	No	Medium	High		Low
Government (defence/aid)	Increase in defence and aid expenditure due to geopolitical instability in neighbouring nations	No	High	Low	Combined aid and defence budget for interventions in Timor-Leste and Solomon Islands, $900 million per year. Foreign aid 2006–07, 0.3% GNI (AusAID 2006). Defence expenditure 2006–07, $17 billion (ABS 2007c).	Medium to high (highly uncertain)
Residential dwellings	Building degradation and damage resulting from temperature, rainfall, wind etc. Changes to building codes and increased maintenance and repair	Yes	High	High	Costs of adaptation likely to be high.	High
	Impacts on buildings due to extreme events (e.g. flooding)	No	High	High	Costs of adaptation likely to be high.	Medium to high
Transport	Degradation of roads, bridges and rail due to temperature and rainfall	No	High	High	In 2004–05 maintaining and improving the total road network cost $9 billion (BITRE 2008).	Low

Table 11.2 Assessing the market impacts of climate change *(continued)*

Sector	Direct impact	Modelled	Risk	Ability to adjust or adapt	Comment	Likely economic consequence by 2100
Ports	Port productivity and infrastructure affected by gradual sea-level rise and storms	Yes	High	High		Low
Airports	Impacts on infrastructure due to sea-level rise, temperature and rainfall	No	Low	Medium		Low
Water supply	Decrease in rainfall reduces reliance on traditional water supply for urban use and increases demand for alternative water supply options	Yes	High	High	High costs associated with adaptation options.	Low to medium
	Degradation of water supply infrastructure increases maintenance costs	Yes	High	High		Low
Electricity transmission and distribution	Degradation of infrastructure increases maintenance costs Productivity losses from blackouts due to severe weather events	Yes	High	High	Net capital expenditure for electricity transmission and distribution was $6.2 billion in 2005–06 (ABS 2008e).	Low
Electricity generation	Increased demand for electricity resulting from greater use of air conditioners	No	High	High	New generation costs. Net capital expenditure for electricity generation was $2.8 billion in 2005–06 (ABS 2008e).	Low to medium

Note: High economic consequence implies 0.5–1.5% per cent loss of annual GDP by the end of the 21st century; medium economic consequence, 0.1–0.5 per cent; and low, less than 0.1 per cent.

The negative impacts of climate change on infrastructure have a significant effect on Australia's output and consumption of goods and services, and is responsible for about 40 per cent of total Type 1 GNP climate change costs.[6] The infrastructure impacts affect a wide range of assets, including commercial and residential buildings, water supply and electricity infrastructure, and ports.

By the end of the century, another 40 per cent of Type 1 GNP costs of climate change arise from the negative effect on our terms of trade. Coal demand is significantly lower than in the reference case, primarily as a result of the deceleration of global economic growth in response to climate change. The global modelling (GIAM) suggests that global GDP is likely to fall by around 8 per cent by 2100, with losses in developing countries likely to be higher than the global average. This is important for Australia as, by 2100, developing countries in Asia are projected to be overwhelmingly our major trading partners. The international modelling shows that Australian terms of trade are more adversely affected than those of any other country or region by climate change. By 2100, Australia's terms of trade are 3 per cent below the reference case, whereas for Japan, the European Union and the United States they are about 1 per cent below, and for Canada 0.5 per cent above.

About 20 per cent of Type 1 GNP costs arise from the direct climate change impacts on agriculture. The loss of agricultural productivity as a result of climate change results in agriculture drawing more resources from the rest of the economy in order to meet an assumed inelastic demand for agricultural products, and to maintain production at levels determined by domestic and world demand and prices.[7]

Agriculture is hit particularly hard by climate change. Agricultural activity is reduced by more than 20 per cent relative to the reference case. These impacts would be unevenly distributed. Without mitigation, the best estimate for the Murray-Darling Basin is that by mid-century it would lose half of its annual irrigated agricultural output (Chapter 6). By the end of the century, it would no longer be a home to agriculture.

The other sector that is hit hard is mining. Output of this sector is projected to decline by more than 13 per cent by 2100. This result is mainly driven by the deceleration of global economic growth. Most coal produced in Australia is exported. The international modelling implies that the world demand for coal falls by almost 23 per cent.[8] Iron ore activity declines for much the same reason as for coal.

The health-related impacts considered by the Review are estimated to have relatively small market effects, though this does not take into account the intrinsic value of the lives lost.

The economic effects of tropical cyclones, taken as a series of annualised losses, are estimated to be small. While a single cyclone event has the potential to create significant economic damage, particularly if it were to hit a population centre, these events are, and are likely to remain, relatively infrequent. These results may be underestimated, however, as it was not possible to consider either the impacts of flooding associated with cyclones or the impacts that might be associated with a southward shift in the genesis of tropical cyclones.

Estimating and incorporating Type 2 costs: non-modelled expected market impacts

The second type of cost, Type 2, covering the expected market costs of the median outcome of impacts for which data are too unreliable to feed into general equilibrium modelling, are estimated at about 30 per cent the size of the Type 1 GNP costs (see Box 11.2). Taken together, Type 1 and Type 2 costs amount to approximately 8 per cent of GDP, 9 per cent of GNP and higher percentages of consumption and real wages by the end of the century.

Box 11.2 Estimating Type 2 costs of climate change

Table 11.2 identifies some major expected (median case) market impacts of climate change this century, indicating which ones were included in the modelling and which were excluded, and provides a qualitative assessment of their importance in terms of market effects. The following market impacts were judged to be significant, but could not be included in the modelling, mainly because of lack of data.

There will probably be additional increases in the cost of **building construction** as a result of new building design requirements, in addition to those that have been modelled, as well as increased **road and bridge maintenance** costs. Based on the value of the building capital stock and road network in Australia, the effects of climate change on building infrastructure and roads and bridges that have not been estimated could subtract an additional 0.8 and 0.25 percentage points respectively from GDP by the end of the century.[9] The need for increased peak power usage to cool buildings could also be a significant omission.

The modelled impacts on **agriculture** are based on average changes to climate variables. It is likely that climate change will also affect the variability and predictability of the climate (especially rainfall). In the absence of forecasts describing the level of future variability, it is difficult to provide an estimate of the degree to which increased climate variability would affect the economy.

International tourism will be affected by climate change. The Review has captured the impact of climate change on tourism through incomes and relative prices, but not through the deterioration of environmental assets. Some major environmental assets that are important for Australian tourism are highly susceptible to climate change. These include the Great Barrier Reef, south-western Australia (a biodiversity hotspot) and Kakadu (see Chapter 6). International travel to Australia is projected to increase substantially in the reference case as global incomes rise strongly over the coming decades. This suggests that even small changes in demand could have significant economic implications for Australia.

Box 11.2 Estimating Type 2 costs of climate change *(continued)*

Climate change may lead to **geopolitical instability**, which will require an increase in the capability and requirements of Australia's defence force and an increase in the level of Australia's spending on emergency and humanitarian aid abroad. Previous Australian interventions in small neighbouring nations provide some indication of the potential size of future defence costs that may arise from climate change. The combined aid and defence budget for the five-year intervention in Timor-Leste has exceeded $700 million per year. Australia's intervention in Solomon Islands is estimated to cost around $200 million per year (Wainwright 2005). This level of intervention is likely to continue until at least 2013. Climate change could lead to the involvement of larger countries through geopolitical pressures, and thus may lead to much higher spending than would be indicated by recent history. A 10 per cent increase in defence spending would be a cost of 0.2 per cent of GDP. Although extra defence spending does not automatically lead to reduced GDP, the Review treats it as a cost since it represents resources that would otherwise have been available for productive use elsewhere.

To summarise, the omitted impacts on infrastructure and defence alone could subtract an additional 1.2 per cent from GDP (or GNP, the main modelling output reported) at the end of the century. The effects on tourism, variability and predictability effects on agriculture, additional impacts of geopolitical instability on Australia, and the range of other possible impacts noted in Table 11.2 need to be added to this. The total omitted market impacts could contribute an additional 1 to 2 percentage points to the loss of GNP at the end of the century, taking the estimated Type 2 loss to between 2.2 and 3.2 per cent. This would imply that the modelling has captured 77 to 70 percentage points of the 2100 no-mitigation cost to GNP. Applying only the upper bound of this range implies that Type 2 costs are about 30 per cent of the Type 1 costs of climate change.

In comparative costings of the three scenarios analysed in this chapter, it is assumed that this relationship holds not only for 2100 and the no-mitigation scenario, but for the entire century and for all three scenarios. This approach is clearly based on a significant degree of judgment and simplification. However, the inclusion of Type 2 costs is considered to be crucial to an appropriate evaluation of the expected market effects of climate change and the corresponding benefits from mitigation. The Review considers these estimates to be conservative.

These combined end-of-century Type 1 and Type 2 costs are much higher than estimates from earlier quantitative studies of the global costs of climate change during the 21st century. Stern (2007), for example, found a reduction in global GDP per capita as a result of climate change of only 2.9 per cent in 2100 after taking into account all four of the categories of costs described above, two of which (types 3 and 4) are excluded from the calculations in the preceding paragraph.

The earlier and larger costs of climate change in the Review's study derive in substantial part from the application of realistic, Platinum Age assessments of the growth in emissions in the absence of mitigation (Chapter 3).

One main theme of the Review is that the accelerated growth of the developing world, the Platinum Age, has not been factored into expectations of emissions, concentrations or temperatures. This growth, centred on but now extending well beyond China, is unprecedented, and likely to be sustained over a considerable period.

The Fourth Assessment Report of the IPCC presented a range of best-estimate temperature increases for this century from 1.8 to 4°C (or from pre-industrial levels, 2.3 to 4.5°C) (IPCC 2007: 13). The Review has generally accepted the scientific judgments of the IPCC, on a balance of probabilities, as a reasonable source of scientific knowledge on climate change. But the economic analysis of the IPCC rests on work from the 1990s, which the Review has shown to have been overtaken by events. Chapter 3 shows that the IPCC's SRES scenarios, on which its projections of climate change impacts were based, systematically underestimate the current and projected growth of emissions. Far from being alarmist, it is simply realistic to accept the conclusion from analysis that, if the mainstream science is roughly right, then 1.8 to 4°C can no longer be accepted as the central range for temperature increases in the 21st century under business as usual. Instead, that range should centre around 5°C.

Costs of catastrophic climate change (Type 3)

The rapid growth in emissions associated with the Platinum Age has an unfortunate consequence: in the absence of mitigation, it is making outcomes likely that were once seen as having low probability. The economist Martin Weitzmann justifies strong mitigation action on climate change on the basis of prevention of a possible catastrophic outcome. A recent article (Weitzmann 2007: 18) specifies a 3 per cent probability that temperatures will increase by 6°C by the end of the century, the result of which will be:

> a terra incognita biosphere within a hundred years whose mass species extinctions, radical alterations of natural environments, and other extreme outdoor consequences of a different planet will have been triggered by a geologically instantaneous temperature change that is significantly larger than what separates us now from past ice ages.

How much stronger, then, is the justification for mitigation when the probability of a temperature increase in the range of 6°C is not 3 per cent, but nearly 50 per cent (recall that the no-mitigation best estimate for temperature increase by century's end is 5.1°C)? This is the order of the change of probabilities for such a temperature increase once we move from the now-outdated SRES scenarios on which Weitzmann bases his 3 per cent calculations, to the more recent and realistic projections of emissions reported in Chapter 3.

The end-of-century temperature increase expected from the no-mitigation scenario is above the estimated range of 'tipping points' for seven of the eight catastrophic global events (enumerated in section 11.5) for which Lenton et al. (2008) present such a range, and it is at the top end of the range for the eighth.

That catastrophic events have become more likely does not make them more amenable to modelling. Table 4.1 presented results from a survey of the recent scientific literature. This indicated that, given the best estimate for climate sensitivity, the triggering of a large-scale melt of the Greenland ice sheet under temperatures expected by the end of century in a no-mitigation scenario would be a sure thing. Given the uncertainties of when the melt would start, and when it would translate into sea-level rises, the effect has been neither modelled nor included in our Type 2 estimates of the costs of climate change. As the sea level rose over a matter of centuries by 7 m, there would certainly be a large negative impact on the world, and on Australia, through the risks of severe and possibly catastrophic effects on non-market values and on the basis of median expectations of market impacts. These would also be the conditions under which irreversible melting of the west Antarctic ice sheet would be most likely to occur, so that the correlation of risks increases the chance of severe outcomes.

Non-market (Type 4) costs

The non-market risks of climate change will be significant in a no-mitigation scenario. As Table 4.1 shows, at the high levels of temperature increase expected under the no-mitigation scenario, 88 per cent of species would be at risk of extinction, and coral reefs as we know them would be destroyed. In the Australian context, under the no-mitigation scenario, by halfway through the century the Great Barrier Reef would be destroyed, and by the end of the century the Kakadu wetland system would be inundated by sea water. Non-market impacts also include the greater number of deaths due to hotter weather, the inconvenience of a greater number of extremely hot days, and much higher bushfire risk.

While there are methods through which non-market impacts can be monetised, the Review found it more useful simply to identify them, and to note that, as incomes and consumption rise, as is anticipated in the no-mitigation scenario, the relative value people assign to non-market costs and benefits will rise as well.

Costs beyond the 21st century

The lags and non-linearity of climate change impacts, even looking only at the expected market impacts, is reflected in the Review's modelling. The gradient of the modelled market impacts of the unmitigated scenario (Figure 11.2) at 2100 indicates that the costs of unmitigated climate change would grow rapidly into the 22nd century. The estimated impacts in the unmitigated scenario increase threefold from 2050 to 2075, and then threefold again from 2075 to 2100. This rate of increase in damages far outstrips the projected rate of increase in temperatures. It is obvious that if the analysis were continued into the 22nd century, estimated market impacts from climate change would be dramatically higher than for the latter decades of the 21st century.

Other studies of climate change show much higher costs in the next than in the current century. Cline (2004) used a modified version of Nordhaus's climate change model going out to the year 2300. Cline's emissions growth is lower than in the Review's modelling, but a scenario with a higher climate sensitivity yields temperature outcomes close to the Review's no-mitigation scenario at 2100, and temperature continues rising to a 15°C increase at 2300. In Cline's scenario, climate change damages as a percentage of global GDP are 9 per cent by 2100, about 25 per cent by 2200, and a remarkable 68 per cent by 2300.

The Stern Review attempted a more comprehensive assessment of global climate change damages, including market as well as non-market impacts and a probability distribution over a range of possible outcomes to 2200. Stern's analysis shows impacts on expected per capita consumption at 3 per cent at 2100, rising to 14 per cent at 2200.

As discussed in Chapter 1, under reasonable assumptions, the present value of a percentage point of Australian GNP in a century's time is about as high as a percentage point of GNP this century. The Review has not tried to model climate change impacts beyond the 21st century. It is clear that they matter.

Summary of unmitigated climate change costs

There is a risk that temperature increases, and therefore all the impacts that are related to temperature, will be much greater than anticipated in the standard cases of the modelling because of positive feedback effects. These are difficult to quantify, but they are real and potentially significant. Once temperature increases above certain threshold points, massive carbon and methane stores on earth and in the oceans may be destabilised, leading to much greater volumes of greenhouse gas release from the natural sphere, and further temperature increases.

To summarise, temperature increases of the order of magnitude associated with no mitigation—an expected increase by 2100 of 5.1°C, a 6.6°C warming at the top of the likely band, and a smaller probability of a double-digit temperature increase—would not lead to a marginal reduction in human welfare. Their impacts on human civilisation and most ecosystems are likely to be catastrophic. As the Center for Strategic and International Studies recently noted in its study of climate change scenarios, this extent of climate change 'would pose almost inconceivable challenges as human society struggled to adapt... The collapse and chaos associated with extreme climate change futures would destabilize virtually every aspect of modern life' (Campbell et al. 2007: 7, 9).[10]

To point to the devastating impact of temperature increase for this century, and of significant further increases next century, and to the possibility that such increases would leave both global and Australian welfare at the end of this century lower than at the start, is not to be alarmist. It is simply to recognise the reality of rapid emissions growth, its likely continuation in the absence of climate change mitigation, and the possibly catastrophic consequences of such large, rapid temperature increases.

11.4.2 The costs of avoiding unmitigated climate change

How much would it cost to greatly reduce the extensive climate change damage outlined in the previous section? Figure 11.3 depicts the costs of mitigation up to 2050 under stabilisation at 550 ppm, as implemented in GTEM (Box 11.1). The results are shown under both standard and enhanced assumptions concerning technological progress, as discussed in section 11.3.[11]

After an initial modelled shock to GNP growth of around 0.8 percentage points (a cost which in reality would be spread over several years), the gross costs of mitigation as modelled in GTEM typically shave a bit above 0.1 per cent per annum from GNP growth until after the halfway mark in the century under standard technology assumptions and a bit below 0.1 per cent per annum under enhanced technology assumptions. This can be seen as sacrifice of material consumption in the early decades.

The gap between the standard and enhanced cases opens up in the second half of the century. On average, annual GNP growth is 0.07 per cent faster in the enhanced case than in the standard, and the economy actually grows marginally faster in the enhanced case with mitigation than without. (The enhanced technology case is further explored in chapters 21 and 23.)

Modelling the costs of mitigation in the second half of the century is more complex, for two reasons. First, as discussed earlier, technological options become more uncertain. It is unrealistic to expect that carbon prices will continue to rise beyond many hundreds of dollars (as in Figure 11.1) without the development of new technologies to offset emissions. Accordingly, long-run cost modelling is best undertaken with the assumption that at some price a backstop technology

Figure 11.3 Change in annual Australian GNP growth (percentage points lost or gained) due to gross mitigation costs under the 550 scenario strategy compared to no mitigation, and under standard and enhanced technology assumptions, 2013–50

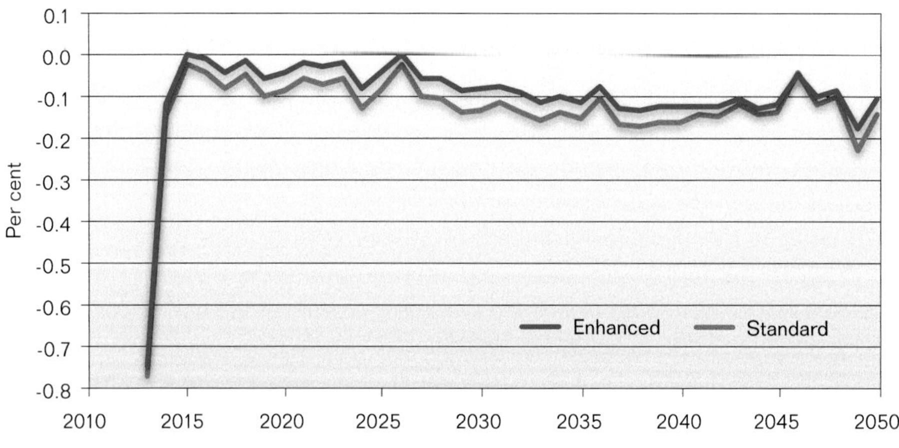

Note: Results are from GTEM (Box 11.1). For details on the standard and enhanced variants of GTEM, see section 11.3.

develops, even if there is uncertainty about the price at which that technology will develop and the precise form it will take.

Second, avoided expected climate damages become significant, so measures of costs without them (such as modelled in GTEM) become less informative. The net mitigation costs calculated in MMRF take into account both the gross costs of mitigation and the benefits of avoided expected climate change market impacts (the Type 1 costs). In addition, the net costs shown here make an adjustment for the Type 2 costs of climate change, along the lines developed for the no-mitigation scenario.

These net costs of mitigation as modelled and accounted for here are not meant to represent the full benefits of mitigation, as they do not seek to capture the Type 3 and Type 4 and post–21st century benefits of mitigation. They do, however, provide an indication of the amount Australia would need to pay to have access to the additional benefits of climate change mitigation.

Figure 11.4 shows the net cost of mitigation (including expected market costs as well as benefits) for the 550 scenario, using the MMRF model with an extension implemented by the Review to allow for a backstop technology to emerge post 2050.

Figure 11.4 shows that, in the second half of this century, mitigation towards the 550 reduction target adds to the growth rate of the economy, as, at the margin, more new climate change damages are avoided than new mitigation costs added. In fact, by the end of the century, GNP is higher than it would have been without mitigation, even when all the costs and only the expected market benefits (avoided costs types 1 and 2, but not types 3 and 4) of mitigation are taken into account.

Figure 11.4 Change in annual Australian GNP growth (percentage points lost or gained) due to net mitigation costs under the 550 scenario compared to no mitigation, 2013–2100

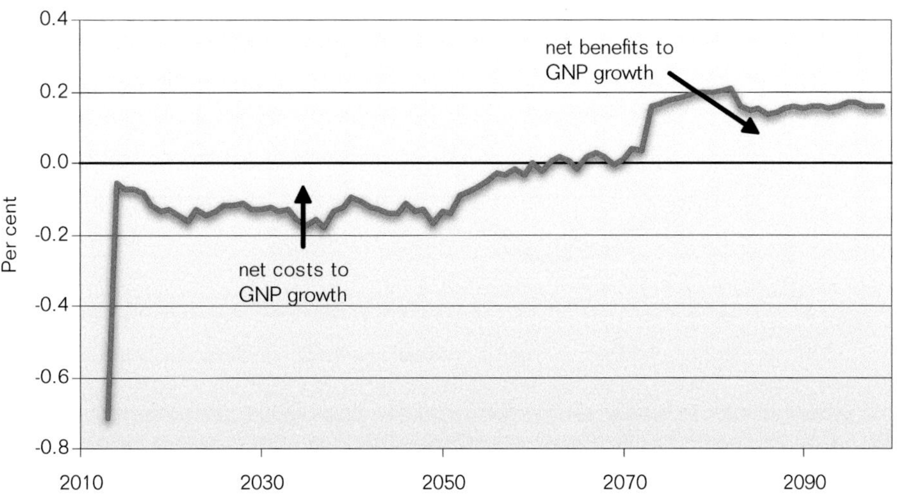

Note: The figures reflect results modelled in MMRF (Box 11.1) adjusted to incorporate Type 2 costs using the method described in Box 11.2.

Figure 11.5 tells this story by sector. At the aggregated industry level, the big winner from mitigation is the agriculture and forestry sector, which in the second half of this century is growing about 0.3 per cent faster on average every year, as forestry expands and less climate change damage is imposed on agricultural productivity. Mining does worse with mitigation in the first half of the century, but better in the second as the terms of trade improve. Growth in manufacturing and services is little affected by mitigation.

The terms of trade effects of mitigation are ambiguous. On the one hand, reduced climate change damage improves the terms of trade; on the other, a swing away from fossil fuels associated with global mitigation harms them. Under standard technology assumptions, the net impact is a further worsening of the terms of trade (from 3 per cent below the reference case in the no-mitigation scenario to 4 per cent below in the 550 scenario in 2100). But under enhanced technology assumptions, where clean coal technology is more competitive, the terms of trade in 2100 are 2 per cent above the reference case. Likewise, the backstop technology improves terms of trade to 1 per cent above the no-mitigation scenario.

Figure 11.5 Change in Australian sectoral growth rates (percentage points lost or gained) due to net mitigation costs under the 550 scenario compared to no mitigation, 2013–2100

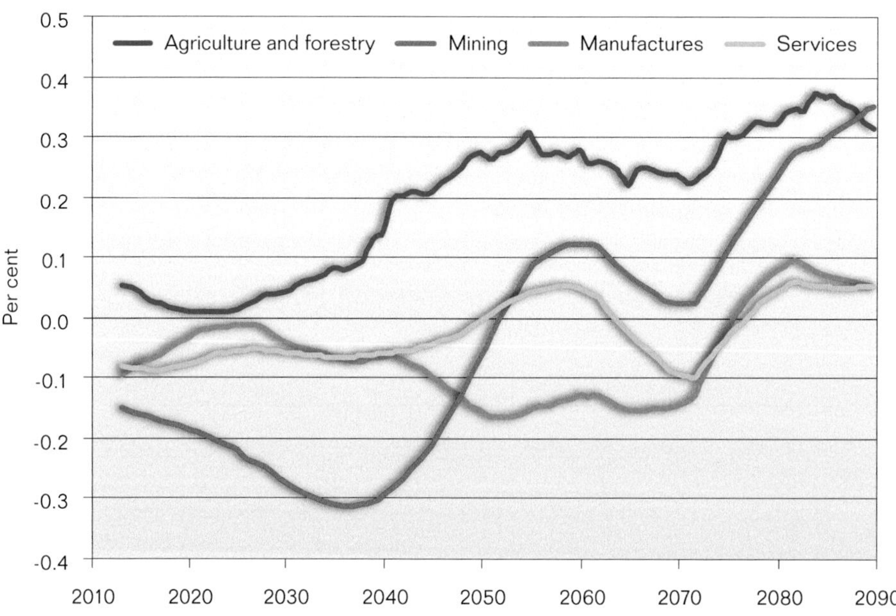

Note: Sectoral growth rates are the growth in sectoral activities (value added), as modelled in MMRF. Only Type 1 costs are shown here. Ten-year forward moving averages are used to smooth annual variability. Growth rates are presented for aggregated sectors under the backstop technology case. Uncertainty regarding the backstop technology was recognised by applying costs evenly across all sectors, and reducing emissions from all sources. Any actual backstop technology would likely have a differential impact on sectoral growth. It is therefore not possible to project sectoral changes with a high degree of certainty in the second half of the century.

Despite the boost to growth in the second half of the century, the sacrifice in the first half is substantial, even though the loss to the annual level of GNP is fully recovered with a margin by the end of the century. Of course, stabilisation at 550 ppm does not eliminate all costs associated with climate change. Temperatures would still be expected to increase by 2°C over the course of the century, with associated risks. Nevertheless, the benefits that are purchased by the cost of the 550 strategy are substantial, and take several forms.

One is insurance against the effects of severe and possibly catastrophic outcomes on material consumption during this century. Another is increased protection against loss of non-market services this century. Yet another is avoidance of all of the rapidly increasing costs throughout the 21st and into the 22nd century and beyond: the rapidly increasing negative impact on material consumption under median outcomes (types 1 and 2); the risk of outcomes much worse than the median expectations from the applied science (although throughout and beyond the 21st century the median outcomes are more severe and possibly catastrophic) (Type 3); and the impacts on non-market values (Type 4).

Figure 11.6 compares expected market damages from climate change under temperatures that would be expected for a 550 scenario with those damages associated with temperatures expected under a no-mitigation scenario (see Figure 11.2). Climate damages under the 450 scenario are also shown for later reference. This figure makes for an incomplete comparison for all the reasons

Figure 11.6 A comparison of the modelled expected market costs for Australia of unmitigated and mitigated climate change up to 2100 (Type 1 costs only)

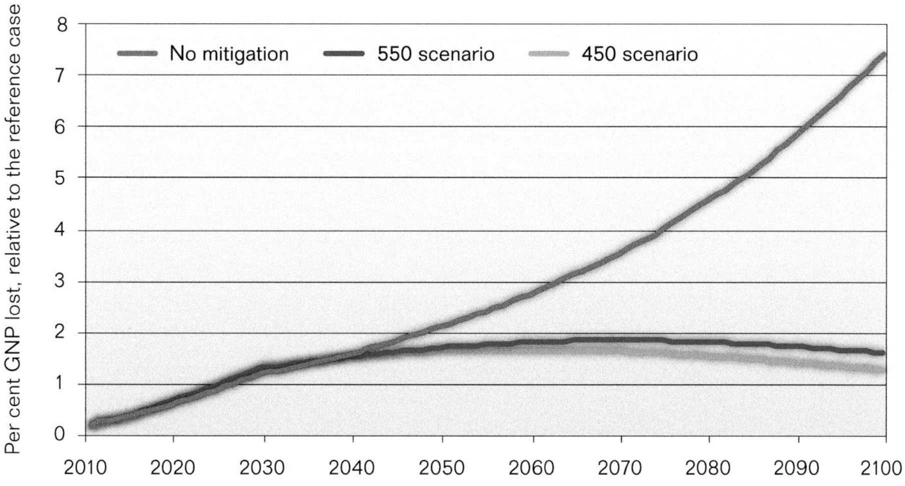

Notes: The graphs show the cost as a percentage of GNP of expected market damages from the level of climate change associated with the three scenarios. These estimates are achieved by 'shocking' the reference case with the differing levels of impact associated with the temperatures expected from the three scenarios.

already cited, but it is a telling one. Market damages under the 550 scenario flatten, stabilise and begin to decline relative to GNP by the end of the century, just as market damages under the no-mitigation scenario start to accelerate markedly. The choice is between a future with bounded expected climate change costs, but still significant risks, and one with unbounded expected costs, a high probability of severe outcomes and some chance of outcomes that most Australians would consider to be catastrophic.

The rapid growth in global emissions is increasing the costs both of mitigation (Gurria 2008) and of no mitigation. The costs of well-designed mitigation, substantial as they are, would not end economic growth in Australia, its developing country neighbours, or the global economy. Unmitigated climate change probably would.

11.5 How much mitigation?

Is it in Australia's interest to support a global goal of limiting the concentration of greenhouse gases to 450 ppm CO_2-e, or lower, rather than 550 ppm? A major portion of the Review's modelling went into weighing the relative benefits of Australia's participation in a 450 ppm and 550 ppm global climate change mitigation agreement.

Figures 11.7 and 11.8 present the same comparison in terms of growth rates as in Figures 11.3 and 11.4 above, but this time comparing 450 and 550 scenarios rather than no-mitigation and 550.

Figure 11.7 Change in annual Australian GNP growth (percentage points lost or gained) due to gross mitigation costs under the 450 compared to the 550 scenario and under standard and enhanced technology assumptions, 2013–50

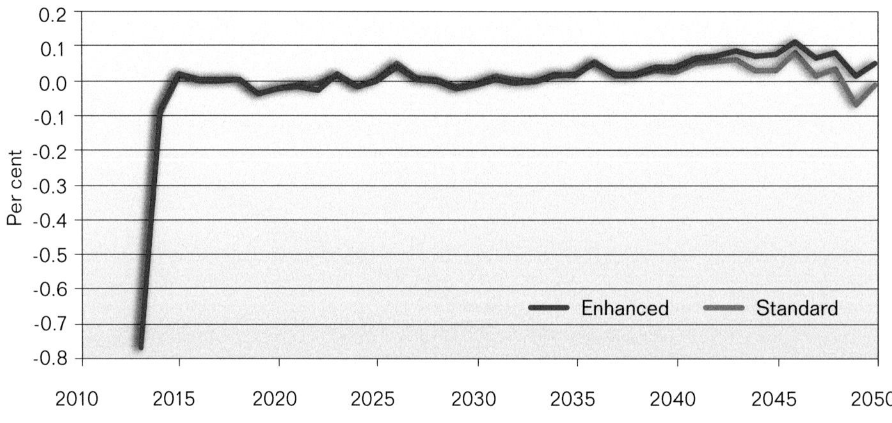

Note: Results are from GTEM (Box 11.1). For details on the standard and enhanced variants of GTEM, see section 11.3.

In GTEM, there is an additional growth penalty when the 450 regime is introduced (though again, this would be spread over several years), but after that there is no real growth differential between the 450 and 550 scenarios, whichever set of technological assumptions is made.

Net mitigation costs modelled in MMRF (Figure 11.8), with the backstop technology as before, show greater volatility in the growth differential between the 450 and 550 mitigation scenarios over the full century than GTEM does for gross mitigation costs in the first half of the century. Overall, however, the story is a similar one. After the initial shock, there is, on average, no difference in the GNP growth rates under the two scenarios over the course of the century. Sectoral differences in growth rates under the two mitigation scenarios are relatively minor under backstop technology assumptions. Gains to agriculture and manufacturing are offset by losses to mining. Sectoral differences under different technology assumptions are explored in chapters 20 to 22.

Figure 11.8 Change in annual Australian GNP growth (percentage points lost or gained) due to net mitigation costs under the 450 compared to the 550 scenario, 2013–2100

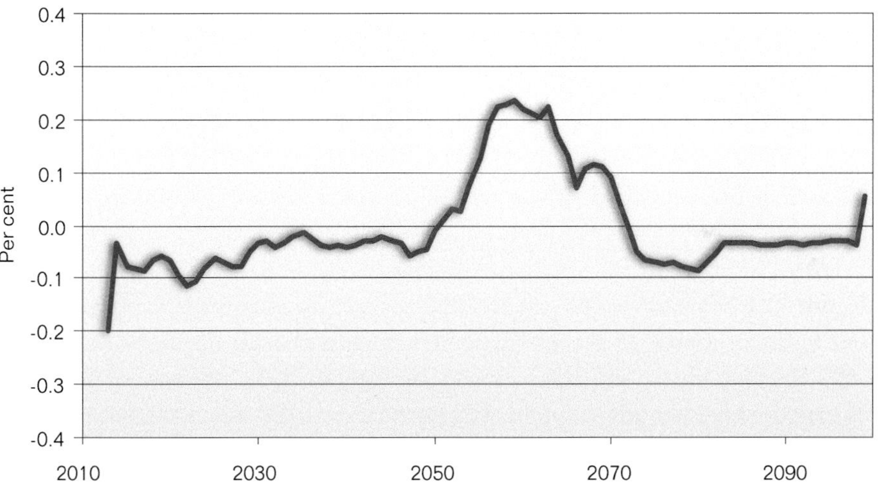

Note: The figures reflect results modelled in MMRF (Box 11.1) adjusted to incorporate Type 2 costs as per the method described in Box 11.2. Since annual differences in the 550 and 450 growth rates show considerable volatility, they are smoothed in this graph by using a three-year forward moving average.

Table 11.3 summarises the cost differences between these two scenarios. It presents results using the GTEM model for gross mitigation costs out to 2050 (where avoided climate damages are small), and the MMRF model (with the post-2050 backstop) for net mitigation costs (costs net of Type 1 and Type 2 benefits).

It calculates the net present value of these costs using two discount rates the Review considers appropriate, namely 1.4 and 2.7 per cent (as derived in Chapter 1).

At the bottom of the range of discount rates considered (1.4 per cent), and for both models, and both time periods and methods, the net present value of the excess cost of the 450 over the 550 scenario (the '450 premium') is 0.7 to 0.8 per cent of discounted GNP. At the top end of the range of discount rates (2.7 per cent), the premium is in the range of 0.7 to 0.9 per cent of discounted GNP.

Table 11.3 Net present cost of the 450 ppm and 550 ppm scenarios (in terms of no-mitigation GNP) and the '450 premium' to 2050 and 2100

Discount rate equals 1.4%	550	450	450 premium
Gross mitigation cost to 2050 (per cent)			
GTEM standard	2.6	3.3	0.7
GTEM enhanced	1.9	2.6	0.7
Net mitigation cost to 2100 (per cent)			
MMRF	3.2	4.0	0.8
Discount rate equals 2.7%	**550**	**450**	**450 premium**
Gross mitigation cost to 2050 (per cent)			
GTEM standard	2.4	3.2	0.7
GTEM enhanced	1.8	2.5	0.7
Net mitigation cost to 2100 (per cent)			
MMRF	3.3	4.2	0.9

Note: The figures give the discounted costs as a percentage of discounted GNP. The '450 premium' is the excess of the 450 ppm cost over the 550 ppm cost. Costs in GTEM are gross costs of mitigation; costs in MMRF are net costs (gross costs net of Type 1 and Type 2 benefits). MMRF modelled results are adjusted to incorporate Type 2 costs using the method described in Box 11.2.

Figure 11.6 helps to explain why the 450 scenario is always the more expensive one in terms of modelled results. Expected climate change damages are less in the 450 scenario than in the 550 scenario, but only by half a per cent of GNP. The small expected market gain from the 450 scenario to 2100 is not in itself adequate to justify the additional mitigation costs associated with it. Rather, the large difference between 450 and 550 scenarios is in terms of additional Type 3 and Type 4 avoided costs. What are the non-market (Type 4) and insurance (Type 3) benefits of a 450 relative to a 550 outcome?

Differential avoidance of non-market climate change impacts

Neither strategy will lead to the complete avoidance of non-market climate change–related impacts. Chapter 6 found that even an increase of 1°C could result in a 50 per cent decrease in the area of rainforests in North Queensland.

Nevertheless, the analysis suggests that the difference between a 450 and a 550 outcome could be of major significance for a range of environmental impacts. For example:

- A 550 outcome would be expected to lead to the destruction of the Great Barrier Reef and other coral reefs as we recognise them today. The 450 outcome would be expected to damage but not destroy these coral reefs. Under a 550 scenario, the three-dimensional structure of the corals would be largely gone and the system would instead be dominated by fleshy seaweed and soft corals. At 450 ppm, the reef would still suffer—mass bleaching would be twice as common as it is today—but its disappearance would be much less likely (Table 6.2).
- The 550 ppm outcome would lead to a greater incidence of species extinction. Under the expected temperature outcome from the 550 ppm scenario, 12 per cent of species are predicted to be at risk of extinction. This percentage is reduced to almost 7 per cent under the 450 scenario (Table 4.1).

Differential insurance value of the 450 ppm and 550 ppm scenarios

As important as these differential non-market impacts are, perhaps the decisive advantage of the 450 scenario over the 550 is its insurance value. While neither scenario would eliminate climate change risks, the 550 scenario would leave the world, and Australia, open to larger risks of exceeding threshold temperature values, even if these tipping points cannot be known in advance with certainty. Lenton et al. (2008: 1786) identified nine important 'tipping elements' ('subsystems of the Earth system that are at least subcontinental in scale and can be switched—under certain circumstances—into a qualitatively different state by small perturbations'), and conducted a survey of experts to estimate the associated temperature tipping points. For one of the nine (disruption of the Indian summer monsoon), the tipping point could not be identified. For two—melting of the Arctic summer sea ice and the Greenland ice sheet—the tipping point range was put at below 2°C (above 1990 levels). For the other six, the tipping point ranges all start at 3°C and extend to 4, 5 or 6°C. These six are melting of the west Antarctic ice sheet, disruption of the Atlantic thermohaline circulation, disruption of the El Niño – Southern Oscillation, disruption of the Sahara/Sahel and West African monsoon, dieback of the Amazon rainforest, and dieback of boreal forest.

What is the probability of reaching that 3°C degree threshold? Under the 550 scenario, 2.7°C is within the likely (two-thirds probability) range of temperatures at 2100, and 3.1°C by 2200. High-end probabilities are difficult to define for climate sensitivity, but, as reported in Chapter 2, the IPCC notes that '[v]alues substantially higher than 4.5°C [which is at the upper end of the likely range] cannot be excluded' (IPCC 2007: 12). This means there would be a smaller but still significant (say 10 per cent) probability that the 550 scenario could produce a temperature increase in excess of 3°C over 1990 levels by the end of the century. This could also happen under the 450 scenario, but even at the top end of the likely range, the increase in this scenario is 2.1°C at the end of this century.

To give just one comparison, according to the estimates in Table 4.1 the temperature increase expected from the 550 scenario would give a 26 per cent probability of initiating large-scale melt of the Greenland ice sheet. The temperature increase expected from the 450 scenario would give a 10 per cent probability.

The large temperature changes associated with the higher end of the 550 scenario probability distribution, and the tipping points that this might breach, could have far-reaching consequences. A Center for Strategic and International Studies study of climate change scenarios (Campbell et al. 2007), referred to earlier in this chapter, included a scenario of 'severe climate change', within which temperatures increased by 2.6°C over 1990 levels by 2040. In the 550 ppm scenario modelled by the Review, this is not far above the top end of the likely range by 2050 (see Table 11.1). The study found that, under this scenario:

> nations around the world will be overwhelmed by the scale of change and pernicious challenges, such as pandemic disease. The internal cohesion of nations will be under great stress ... both as a result of a dramatic rise in migration and changes in agricultural patterns and water availability. [There will be] flooding of coastal communities around the world ...

Is it worth paying less than 1 per cent of GNP more through the 21st century for the insurance value and the avoided market and non-market impacts of the 450 scenario?

This is a matter of judgment. Judgment will be affected greatly by the success of mitigation regimes and progress in research, development and commercialisation of low-emissions technologies over the years ahead. The Review thinks it likely that, with a significant and rising carbon price and support for emergence of low-emissions technologies, and confidence that the new policies are permanent features of the economic environment, there will be technological progress in areas not currently anticipated. Such developments would greatly favour a 450 outcome over a 550 outcome.

Given the benefits after 2200 of stronger mitigation and the greater risks of catastrophic consequences to the natural environment under the 550 scenario, the Review judges that it is worth paying less than an additional 1 per cent of GNP as a premium in order to achieve a 450 result.

Note, however, that Australia is not in a position to achieve 450 ppm CO_2-e on its own. Chapter 9 concluded that a credible agreement to secure the 450 scenario looked difficult for the international community as a whole in the year or two immediately ahead. Chapter 12 discusses how Australia can most effectively pursue support for a 450 global mitigation strategy in these inauspicious circumstances.

Notes

1 The global emissions path is determined within the global modelling (GTEM) using a Hotelling-style carbon price function. The start price is fixed and then increases over time at the prescribed interest rate of 4 per cent. The real interest rate is assumed to be 2 per cent. This is adjusted upwards by a 2 per cent risk premium. This approach provides a proxy for banking and borrowing, and imitates an efficient intertemporal distribution of abatement effort. The resulting emissions pathway is then used for international trading simulations. The concept of a resource price rising with the interest rate comes from resource economics. Hotelling (1931) demonstrated that profit from the optimal extraction of a finite mineral resource will increase over time at the rate of interest. Since only a finite amount of greenhouse gases can be released into the atmosphere prior to stabilisation, the optimal release of greenhouse gases into the atmosphere over time is a problem similar to the optimal extraction of a finite resource (Peck & Wan 1996).

2 The standard technology assumptions represent a best estimate of the cost, availability and performance of technologies based on historical experience, current knowledge and expected future trends. The standard scenario includes some technological cost reductions through learning by doing and improvements in existing technologies and the emergence and wide-scale deployment of some currently unproven technologies such as carbon capture and storage, hot rocks (geothermal) and hydrogen cars. It does not, however, include a backstop technology in any sector.

3 Specifically, the enhanced scenario implemented in GTEM included the following assumptions:

- Faster energy efficiency improvements of an extra 1 per cent annually from 2013 to 2030, an extra 0.5 per cent from 2031 to 2040 and no extra improvements thereafter.
- More effective carbon capture and storage in response to higher carbon prices. The share of combustion CO_2 captured increases from 90 per cent to 99 per cent as the permit price rises from zero to \$140/t CO_2-e.
- Faster learning by doing for electricity and transport technologies by increasing the parameter for the learning functions by 50 per cent relative to the standard assumptions over the whole simulation period.
- Non-combustion agricultural emissions are eliminated when the carbon price exceeds \$250/t CO_2-e.

4 For technical reasons, it was necessary for the Review to use different global average temperature changes for the assessment of domestic impacts than for the assessment of international climate change impacts. For the Australian impact analysis, median rainfall and local temperature outcomes are assumed in response to an average global temperature change of 4.5°C degrees by 2100 (above 1990 levels). This temperature change is based on the A1F1 SRES (see Chapter 3). This temperature change differs from the temperature estimated based on the Garnaut–Treasury global emissions profile used in the economic modelling. This emissions profile gives an increase in global average temperature of 5.1°C degrees by 2100 (above 1990 levels). These differences could not be avoided in the time frames available for the Review.

5 The terms of trade describe the ratio of export to import prices.

6 This decomposition is obtained by running each of the five shocks separately. Due to interactive general-equilibrium effects, the decomposition is not exact.

7 The extent to which imports replace domestic food production is limited. Two factors are influential here. First, growth in developing countries, combined with land constraints, exacerbated by climate change, is likely to result in increases in the cost of food produced overseas over the next 100 years. Second, while the Review has not undertaken detailed modelling to estimate the impacts that climate change may have on the cost of food

production in the rest of the world, any change is presumed to influence the availability of food exports to Australia.

8 Caution needs to be exercised in interpreting changes to world demand. A 23 per cent decline implies that, with prices fixed, exports will decline by 23 per cent. However, prices are not fixed in MMRF. With a typical export price elasticity of around 5, small changes to prices will change the export results.

9 An increase in the cost of constructing and maintaining buildings is equivalent to a productivity loss since more capital inputs per unit of output would be required. If buildings make up around 40 per cent of capital stocks, and capital incomes make up approximately 40 per cent of total income, then a 5 per cent reduction in productivity of the building stock would be expected to reduce GDP by approximately 0.8 per cent. In 2004–05, the cost of maintaining and improving the road network was $9 billion (Table 11.1). If the cost of maintaining the road network were to increase by 25 per cent, GDP might be reduced by around 0.25 per cent.

10 The report was prepared by, among others, former CIA Director James Woolsey, former Chief of Staff of the President John Podesta, former National Security Advisor to the Vice President Leon Fuerth, Pew Center Senior Scientist Jay Gulledge, and former Deputy Assistant Secretary of Defence for Asia and the Pacific Kurt Campbell.

11 Due to the modelling procedures followed, and the different models employed by the Review, the emissions entitlement allocations for Australia modelled in GTEM were slightly different to those presented in Chapter 9 and modelled in MMRF: over the century, a 1 percentage point greater reduction from the reference case for the 550 scenario (82 compared to 81 per cent), and a 2 percentage point greater reduction for the 450 scenario (88 compared to 86 per cent). The Chapter 9 (and MMRF) allocations are more generous early on, and less generous later than the ones in GTEM. Simulations suggested very little difference in cost over time. A partial equilibrium adjustment to account for the differential purchase of emission permits made no discernible difference in the growth rates over time. There is, however, a difference in the first year, where the GTEM allocation profile exaggerates the negative shock on GNP. See the technical appendix on the modelling at <www.garnautreview.org.au> for further discussion.

References

ABS (Australian Bureau of Statistics) 2007a, *Australian System of National Accounts, 2006–07*, cat. no. 5204.0, ABS, Canberra.

ABS 2007b, *Australian Farming in Brief, 2007*, cat. no. 7106.0, ABS, Canberra.

ABS 2007c, *Government Finance Statistics*, cat. no. 5512.0, ABS, Canberra.

ABS 2008a, *Australian National Accounts: Input – Output, 2004–05 (preliminary)*, cat. no. 5209.0.55.001, ABS, Canberra.

ABS 2008b, *Australian National Accounts: Tourism Satellite Account, 2006–07*, cat. no. 5249.0, ABS, Canberra.

ABS 2008c, *Electricity, Gas, Water and Waste Services, Australia, 2006–07*, cat. no. 8226.0, ABS, Canberra.

AusAID 2006, *Australia's Overseas Aid Program Budget 2006–07*, Commonwealth of Australia, Canberra.

BITRE (Bureau of Infrastructure, Transport and Regional Economics) 2008, 'Information Sheet 27: Public road-related expenditure and revenue in Australia (2008 update)', Commonwealth of Australia, Canberra.

BTE (Bureau of Transport Economics) 2001, *Economic Costs of Natural Disasters in Australia*, Report 103, BTE, Canberra.

Campbell, K., Gulledge, J., McNeill, J., Podesta, J., Ogden, P., Fuerth, L., Woolsey, J., Lennon, A., Smith, J., Weitz, R. & Mix, D. 2007, *The Age of Consequences: The foreign policy and national security implications of climate change*, Center for Strategic and International Studies, <www.csis.org/media/csis/pubs/071105_ageofconsequences.pdf>.

Cline, W.R. 2004, 'Climate Change', in B. Lomborg (ed.), *Global Crises, Global Solutions*, Cambridge University Press, pp. 13–43.

Dasgupta, P. 2007. 'Commentary: the Stern Review's economics of climate change', *National Institute Economic Review* 199: 4–7.

Gurría, A. 2008, 'Energy, environment, climate change: unlocking the potential for innovation', keynote speech by the OECD Secretary-General, during the World Energy Council: Energy Leaders Summit, London, 16 September, <www.oecd.org/document/18/0,3343,en_2649_33717_41329298_1_1_1_1,00.html>.

Hotelling, H. 1931, 'The economics of exhaustible resources', *Journal of Political Economy* 39(2): 137–75.

IPCC (Intergovernmental Panel on Climate Change) 2007, *Climate Change 2007: The physical science basis. Contribution of Working Group I to the Fourth Assessment Report of the Intergovernmental Panel on Climate Change*, S. Solomon, D. Qin, M. Manning, Z. Chen, M. Marquis, K.B. Averyt, M. Tignor & H.L. Miller (eds), Cambridge University Press, Cambridge and New York.

Lenton, T.M., Held, H., Kriegler, E., Hall, J.W., Lucht, W., Rahmstorf, S. & Schellnhuber, H.J. 2008, 'Tipping elements in the Earth's climate system', *Proceedings of the National Academy of Sciences of the USA* 105(6): 1786–93.

Nordhaus, W. 1994, *Managing the Global Commons: The economics of climate change*, MIT Press, Boston.

Nordhaus, W. 2008, *The Challenge of Global Warming: Economic models and environmental policy*, Yale University, New Haven, Connecticut.

Peck, S. & Wan, Y. 1996, 'Analytic Solutions of Simple greenhouse Gas Emission Models', Chapter 6 in E.C. Van Ierland & K. Gorka (eds), *Economics of Atmospheric Pollution*, Springer Verlag, New York.

Stern, N. 2007, *The Economics of Climate Change: The Stern Review*, Cambridge University Press, Cambridge.

Wainwright, E. 2005, 'How is RAMSI faring? Progress, challenges, and lessons learned', *Australian Strategic Policy Institute Insight* 14.

Weitzmann, M. 2007 'The Stern Review of the Economics of Climate Change', book review, *Journal of Economic Literature*, <www.economics.harvard.edu/faculty/weitzman/files/JELSternReport.pdf>.

Wigley, T.M.L. 2003, *MAGICC/SCENGEN 4.1: Technical Manual*, National Center for Atmospheric Research, Colorado.

12 TARGETS AND TRAJECTORIES

Key points

Australia should indicate at an early date its preparedness to play its full, proportionate part in an effective global agreement that 'adds up' to either a 450 or a 550 emissions concentrations scenario, or to a corresponding point between.

Australia's full part for 2020 in a 450 scenario would be a reduction of 25 per cent in emissions entitlements from 2000 levels, or one-third from Kyoto compliance levels over 2008–12, or 40 per cent per capita from 2000 levels. For 2050, reductions would be 90 per cent from 2000 levels (95 per cent per capita).

Australia's full part for 2020 in a 550 scenario would be a reduction in entitlements of 10 per cent from 2000 levels, or 17 per cent from Kyoto compliance levels over 2008–12, or 30 per cent per capita from 2000. For 2050, reductions would be 80 per cent per capita from 2000 levels or 90 per cent per capita.

If there is no comprehensive global agreement at Copenhagen in 2009, Australia, in the context of an agreement amongst developed countries only, should commit to reduce its emissions by 5 per cent (25 per cent per capita) from 2000 levels by 2020, or 13 per cent from the Kyoto compliance 2008–12 period.

Is it possible to secure effective international action to hold atmospheric concentrations of greenhouse gases at 550 ppm, or 450 ppm (with overshooting), or less? How should Australia define and offer its proportionate part in the global effort? What should we do in the interim if it takes time to secure effective international action?

In the remainder of the Kyoto period, ending in 2012, Australia should ensure that it meets its Kyoto targets. It should have no great difficulty in doing so. Any adverse surprise over the next few years is unlikely to be so large that it cannot comfortably be met by the purchase of international permits. During this period, Australia should work within the international community to secure a global agreement around a firm emissions concentrations goal. Australia should make it clear that it is prepared to play its full, proportionate part in achieving that goal.

Beyond the Kyoto period, Australia's central approach on targets and trajectories must be linked to comprehensive global agreement on emissions reductions, for four reasons. First international agreement is urgent and essential.

Second, agreement is possible if Australia and some other countries attach enough importance to it. Third, a comprehensive global agreement it is the only way to remove completely the dreadful political economy risks, to Australia and to the global trading system, of payments to trade-exposed, emissions-intensive industries. Fourth, international agreement lowers the cost of Australian mitigation and so allows us to be more ambitious about the reduction in emissions.

12.1 Determining our conditional and unconditional targets

The analysis presented in chapter 11 suggests that Australia's long-term interests lie in the pursuit of global action to return greenhouse gas concentrations to 450 ppm or less—even though the momentum of growth in emissions means that these concentrations can only be reached with temporary overshooting of the target concentration.

Although the goal is clear, the path to success is not.

Australia's actions now and our commitments to reducing our emissions will make a difference to whether the world has any chance of returning to 450 ppm. This global objective will not be achieved easily. There is just a chance of success. What may look improbable today may just become possible tomorrow, if we do not delude ourselves about the difficulties of the task, and are realistic about each step that we take. There is no time for complacency or for unrealistic expectations.

12.1.1 Setting conditional targets

Australia must be willing and ready to play its part in a coordinated and cooperative international effort to reduce greenhouse gas emissions. Our targets must be specified within an international framework that, when all of its parts are added up, is consistent with the desired objective. If we are not prepared to pay our fair share in the cost, then we cannot expect other countries to do so. To make an unrealistically low offer in the international negotiations is to negate the prime purpose of our own mitigation, which is to facilitate the emergence of an effective agreement.

Conversely, committing to interim targets for Australia that are unrealistically or disproportionately ambitious in the absence of an international framework (that recognises abatement and makes available opportunities for trade in emissions entitlements), is likely to be costly and difficult to achieve. It would become an example of the problems of mitigation and not of mitigation's good prospects. A vacuous commitment that denies economic reality would be as damaging to international negotiations as an unrealistically low offer that denies scientific urgency.

These issues are not unique to Australia.

The Review's modelling of the global time path to a 450 ppm objective (with overshooting), presented in Chapter 9, while closely consistent with the G8 goal agreed in July 2008 of 50 per cent reduction of global emissions by 2050, is sobering. The awful arithmetic of developing country emissions growth in the

Platinum Age, and the current state of mitigation policy in all countries, raise serious questions about whether this goal can be credibly agreed in current circumstances.

Achieving the objective of 450 ppm would require tighter constraints on emissions than now seem likely in the period to 2020. A 450 ppm objective would require an emissions reduction commitment by developed countries of 32 per cent by 2020 over Kyoto/2012 levels, or around 5 per cent reductions per year. The only alternative would be to impose even tighter constraints on developing countries from 2013, and that does not appear to be realistic at this time.

The awful arithmetic means that exclusively focusing on a 450 ppm outcome, at this moment, could end up providing another reason for not reaching an international agreement to reduce emissions. In the meantime, the cost of excessive focus on an unlikely goal could consign to history any opportunity to lock in an agreement for stabilising at 550 ppm—a more modest, but still difficult, international outcome. An effective agreement around 550 ppm would be vastly superior to continuation of business as usual, even if it were to become a final resting point for global mitigation.

An achievable agreement built around 550 ppm provides a staging platform for more aggressive reduction at a later date. In contrast, an unrealistic agreement, nominally embodying higher ambition, but with no prospects of implementation as agreed, may be an instrument of disillusionment. This conclusion is not the triumph of despair. It is the Review's appraisal of what might be achievable when the sum of the parts must come together to form a successor to the Kyoto agreement. It is based on the view that no agreement will take the world forward unless its components add up to the solution defined for each stated objective.

It is possible that the Review is wrong in its judgment about what is achievable at Copenhagen. To allow for that possibility, the Review confirms its recommendation in the supplementary draft report—that Australia should offer to play its full, proportionate part in a global agreement designed to achieve 450 ppm with overshooting. It should offer to reduce its emissions entitlements in 2020 by 25 per cent within an effective global agreement that, on realistic assessment, adds up to the 450 ppm overshooting scenario.

Pending the completion of the international discussions on post-Kyoto arrangements, it is better not to focus on a single trajectory, but to have a set of possibilities, the choice among which will be determined in an international context. This set of possibilities will be bound by Australia's 'conditional' offers:

- a 10 per cent (or 30 per cent per capita) reduction from 2000 levels by 2020 within a global agreement aimed at stabilising emissions at 550 ppm (or 17 per cent in absolute terms from Kyoto compliance over 2008–12 to 2020)
- a 25 per cent (or 40 per cent per capita) reduction from 2000 levels by 2020 within a global agreement aimed at returning emissions to 450 ppm (or by one-third in absolute terms from Kyoto compliance over 2008–12 to 2020)

- an Australian commitment between the 450 and 550 position, corresponding to a global agreeement in between.

These conditional commitments are consistent with the framework derived in Chapter 9. Over the longer term, they would respectively require 80 and 90 per cent absolute reductions (or 90 and 95 per cent per capita reductions) from 2000 levels by 2050.

The proposed targets for Australia correspond directly to the trajectories that Australia would need to adopt as its fair share of the international emissions reduction burden. They are calculated within an internally consistent framework compatible with global agreement around specified emissions concentrations objectives. The numbers expressed in absolute terms from a 2000 base turn out to look less onerous for Australia than for other developed countries in the early years, because they are based within a rigorous framework calibrated in per capita allocations of emissions rights. Australia's population, because of this country's longstanding and large immigration program, has been and will be growing much faster than populations in other developed countries. In addition, Australia's 2008–12 Kyoto targets allowed it to increase emissions. The targets are no less onerous than entitlements for other developed countries when examined within a framework of principle designed to add up to specified global mitigation outcomes, and to have a chance of success across the international community. They are no less onerous—and can be seen as being more onerous—when comparisons are made on a per capita basis, or on the absolute reduction from Kyoto compliance in over 2008–12 to 2020.

These reductions proposed for Australia would be fully consistent with the range of emissions reductions that received prominent attention at the UNFCCC Conference of the Parties in Bali in 2007 (see Box 12.1).

Box 12.1 The Bali numbers

While 2000 is a relevant comparator for Australia since it is the base year for the Commonwealth Government's announced emissions targets, 1990 has been emphasised in international discussions.

At the 2007 Bali climate change negotiations, a particular range of emission reductions received prominent attention. It was proposed that Annex I countries consider emissions reduction targets in the range of 25 to 40 per cent by 2020 over 1990 levels. This target range stems from an IPCC analysis for a 450-type trajectory. The equivalent range for a 550 trajectory is 10 to 30 per cent (see IPCC 2007: 776).

The emissions reduction targets for Annex I countries modelled by the Review are fully consistent with these Bali ranges, but at the lower end because of the limited mitigation by developed countries to date. Relative to 1990, Australia's proposed targets are at the average for developed countries.

There are advantages for Australia if the world commits itself at some time to a credible agreement that adds up to the objective of 400 ppm. This would require agreement on and progress towards a 450 objective, with a subsequent lift in ambition. The path to 450 ppm may travel through a credible agreement on and progress towards 550 ppm. The path to 400 ppm can only travel through a credible agreement on and progress towards 450 ppm.

The ultimate achievement of returning concentrations to 400 ppm is likely to depend on the commercialisation of technologies that can remove carbon dioxide from the atmosphere. This is a technical possibility at this time, notably through a range of bio-sequestration options. Such options may become commercially realistic through a combination of high carbon prices and support for research, development and commercialisation.

12.1.2 The challenges of policy setting in a world of partial mitigation

For now, there is no comprehensive international framework for reducing emissions when the Kyoto agreement ends in 2012. Each jurisdiction is left to signal its intentions in the absence of a coordinating framework.

Strong Australian mitigation outside an effective international agreement would be deeply problematic. It would impose domestic costs that are higher than they would be if similar national targets were pursued in the context of an international agreement. It has the potential to leave our traded sector at a competitive disadvantage, for no worthwhile environmental benefit. This reality opens the way to political pressure for exemptions and countervailing payments that could seriously increase the costs of mitigation.

Developed countries agreed in Kyoto that they would move first on mitigation, for reasons that have some validity or at least resonance today.

This is the context in which the world's developed countries agreed in the Kyoto discussions to take mitigation steps ahead of developing countries. Australia, and the United States, agreed to be among the developed countries that acted ahead of developing countries to reduce emissions. This is an obligation that we have already undertaken to fulfil along with the other developed countries. If there were no comprehensive global mitigation agreement out of Copenhagen, there is value in Australia playing its part in keeping the prospect of eventual agreement alive, by being prepared to act with other developed countries. It should take the first step in the expectation that this will only be necessary for a period that is short, transitional and directed at achievement of global agreement. The first step would be taken in the expectation that the ad hoc policy world can quickly be brought to an end—replaced by the cooperative arrangements that are necessary to reduce the risk of dangerous climate change to acceptable levels.

In the ad hoc world, other developed countries will be in the same position as Australia. Each country will adopt its own trajectory and implement its own policies. There is no guarantee that these policies will be well coordinated or integrated. The potentially adverse consequences for the global climate and for internationally efficient resource allocation will be significant.

What we do now will have a bearing on the likelihood of international cooperation in the near future. We, with other developed countries, help to keep the chances of eventual effective international agreement alive, by unconditional commitment to emissions reductions. We do so through the interim emissions reduction targets that we are prepared to adopt ahead of an international agreement.

12.1.3 Interim targets in the ad hoc policy world

The Commonwealth Government's policy of reducing emissions by 60 per cent from 2000 levels by 2050 provides the basis for Australia's unconditional commitment in a world of ad hoc national policies.

There is no reason to suggest that other trajectories would be superior to a linear reduction in emissions from 2013 to 2050. A well-designed market (see chapters 13 and 14) that maximises opportunities for trade among participants, at a point in time and intertemporally, will allow cost-reducing variations in annual emissions.

The Review therefore suggests that the interim target for Australia be defined as the first step along a linear path from 2012 towards meeting the Government's stated goal of reducing emissions by 60 per cent from 2000 levels by 2050. This unconditional policy commitment requires a reduction of emissions by 5 per cent from 2000 levels by 2020. This equates to a 25 per cent reduction in per capita emissions from 2000 levels. It implies an absolute reduction of 13 per cent from Kyoto compliance levels over 2008–12 and 2020. This compares with the European Union's recently announced unconditional offer, which in corresponding terms equates to reducing per capita emissions by 17 per cent from 2000 levels by 2020.

Comparable commitments should be expected from other developed countries when international negotiations reach their moment of decision in Copenhagen in late 2009.

In the modelling results presented in section 12.7, this partial mitigation scenario is referred to as the 'Copenhagen compromise'. This describes the situation in which, by December 2009 (or in meetings that follow immediately afterwards), it has not been possible to secure a comprehensive agreement on emissions reductions. Nevertheless, developed countries have endorsed a successor agreement to the Kyoto Protocol and developing countries have also adopted the kinds of approaches envisaged in the Bali Roadmap (see chapters 9 and 10).

If this was all that was achieved in a Copenhagen compromise and it was seen as an end point, it would be a disappointing conclusion. Opportunities to hold risks of dangerous climate change to acceptable levels diminish rapidly from 2013 if no major developing economies accept constraints to hold emissions significantly below business as usual by that time.

The proposed set of conditional and unconditional offers, or interim targets, are summarised in Table 12.1. The reductions required by 2050 are shown in Table 12.2 and the reductions trajectories are shown in Figures 12.1 and 12.2.

Table 12.1 Summary of interim targets in 2020 (per cent)

	Conditional offers		Unconditional offer
	450 ppm scenario	550 ppm scenario	Copenhagen compromise
Emissions entitlement reduction commitment for 2020 relative to 2000			
Reduction in total emissions	-25	-10	-5
Per capita reduction	-40	-30	-25
Emissions entitlement reduction for 2020 relative to 2008–12 Kyoto compliance			
Reduction in total emissions	-32	-17	-13
Emissions entitlement reduction commitment for 2020 relative to business as usual in 2020			
Reduction in total emissions	-39 to -44	-25 to -31	-22 to -27

Note: Two figures are used to compare Australia's allocation to a business-as-usual world: the no-mitigation scenario, as modelled in GTEM; and the lower 'with measures' projections of the Australian Government (Department of Climate Change 2008), which give a more accurate measure of additional policy effort required.

Table 12.2 Reductions in emissions entitlements by 2050 for policy scenarios (per cent)

Scenario	450 ppm scenario	550 ppm scenario	Copenhagen compromise
Emissions entitlement reduction commitment for 2050 relative to 2000			
Reduction in total emissions	-90	-80	-60
Per capita reduction	-95	-90	-75
Emissions entitlement reduction for 2050 relative to 2008–12 Kyoto compliance			
Reduction in total emissions	-90	-82	-63
Emissions entitlement reduction commitment for 2050 relative to business as usual in 2050			
Reduction in total emissions	-93	-89	-77

Note: It is unlikely that the Copenhagen compromise would be viable as a long-term outcome lasting to 2050. Even if it is the best outcome to emerge from the December 2009 meeting of the Conference of the Parties, it can be expected to be subsumed into a broader, more ambitious agreement at some future time.

The emissions reductions highlighted in tables 12.1 and 12.2 and figures 12.1 and 12.2 represent the required reduction in net emissions—that is, actual (or physical) emissions produced in Australia less any emissions entitlements purchased internationally. The comprehensive agreements that would accompany the 450 ppm or 550 ppm outcome would allow for broad trade in international permits (though opportunities would be affected by the extent to which individual countries adopt market-based mechanisms). This would, for any given level of emissions reduction, be expected to provide for lower cost abatement than could be expected under the narrower agreement represented by the Copenhagen compromise.

Figure 12.1 Australian emissions reductions trajectories to 2050 (reduction in total emissions)

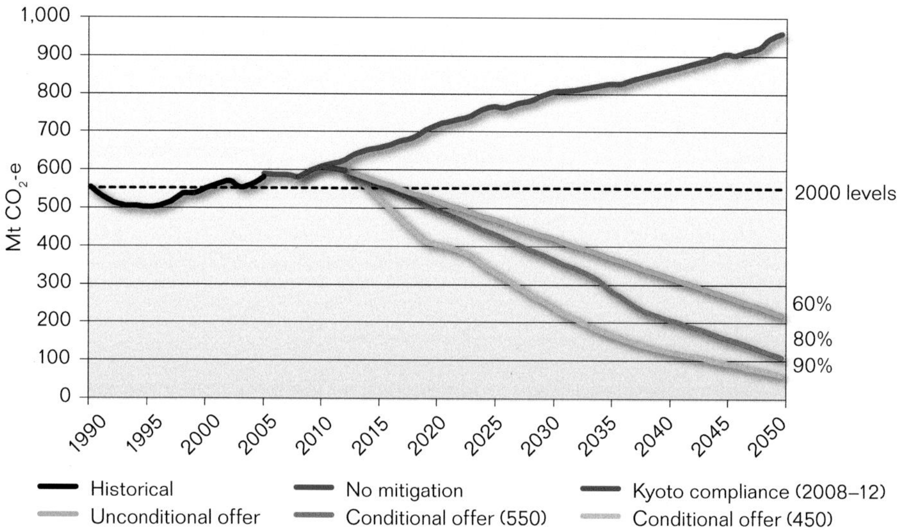

Figure 12.2 Australian emissions reductions trajectories to 2050 (per capita reduction)

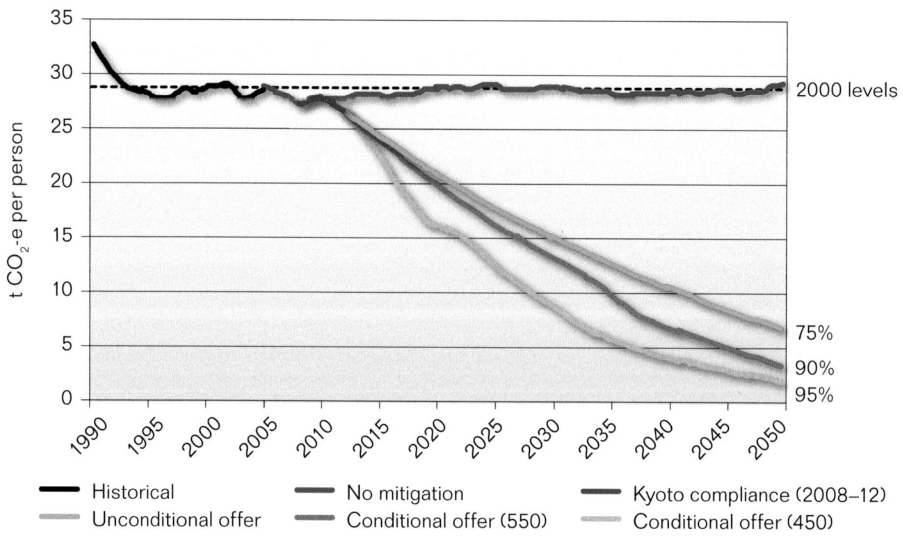

Note: The no mitigation scenario for figures 12.1 and 12.2 is as modelled in GTEM.

In the unlikely event of complete failure of agreement at Copenhagen, and in the absence of any subsequent framework agreement (even among developed countries), and therefore in the absence of clear rules and opportunities for international trade in permits, it would make little sense for Australia to impose quantitative emissions limits. Under this 'waiting game' scenario, Australia's best option would be to continue with the emissions trading scheme, and with the rising fixed carbon price of the transitional period described in Chapter 14, until international agreement or 2020. Continuing an emissions trading scheme would help to keep hopes alive of an international agreement, at reasonable cost, until all opportunities for progress had been exhausted. Current commitments by the governments of developed countries attach low probability to failure of the Copenhagen meeting even to secure an agreement among developed countries.

12.2 The benefits of global cooperation

The unilateral adoption of emissions reduction policies by individual countries has been a way of getting global mitigation started. Developments in the European countries, North American and Australian states and provinces, and in Japan, New Zealand and China have helped to establish momentum in global mitigation. We have a stronger base for moving towards effective global action than we would have had if every country and state had waited for a comprehensive global agreement. Indeed, the early actions have made it possible now to contemplate an effective global agreement. Pending international agreement, it will be helpful for individual countries to move forward unilaterally, so long as this is within policy frameworks that are designed to integrate productively with an emerging international agreement.

Nevertheless, unilateral mitigation in an ad hoc world creates problems of deep political economy for every well-intentioned government.

Unilateral mitigation in an ad hoc world is more expensive for a given degree of emissions reduction. It allows only limited international trade in emissions entitlements, and therefore does not guarantee that mitigation will be undertaken in the parts of the world at which it can be achieved at lowest cost.

Differences in carbon pricing across countries will distort the location of production and investment in trade-exposed, emissions-intensive industries. This would generate dreadful political economy problems in countries seeking to undertake mitigation, as companies seek shielding and preferment in relation to the carbon price. The domestic political economy pressures flow into the international sphere and create risks of new kinds of trade protectionism.

Once there is a comprehensive international agreement, many aspects of mitigation change for the better. Trade in entitlements between countries that have accepted emissions targets becomes possible. Countries that are able to reduce emissions below agreed trajectories are able to sell surplus entitlements to countries that are above their trajectories. This tends to equalise across countries the cost of emitting, removing distortions associated with the trade-exposed industries. Countries in which mitigation costs are high are able to buy entitlements from countries in which mitigation costs are low. This increases economic welfare in the buying and selling countries alike.

Countries with comparative advantage in emissions-intensive industries are able to acquire entitlements to allow the expansion of those industries. This will be profitable for them, so long as they still have comparative advantage after taking the carbon externalities into account. This could be especially important for Australia. With comprehensive carbon pricing, the international prices of emissions-intensive goods and services would rise. Countries with comparative advantage in an emissions-intensive industry, after taking the costs of carbon into account, and firms with competitive advantage after taking the costs of carbon into account, would remain in and expand production, and buy permits on the international market to cover any domestic shortfall. The cost of the permits would be covered, more or less depending on the range of relevant elasticities, by the increase in the international price of the final product. Countries with comparative advantage in emissions-intensive goods and services would become net importers of permits, and their domestic emissions would exceed their allocations.

Emissions-intensive export industries in which the emissions intensity of production is lower in Australia than in its main competitors in international markets may expand exports and production under comprehensive agreements. The products of the sheep and cattle industries may be examples, where Australian producers are spared the emissions costs of heated barns and grain feeding in winter.

On the other hand, in any trade-exposed industry in which production is naturally more emissions-intensive in Australia than in major competing countries, output and exports will tend to contract under arrangements that generate comparable carbon pricing across countries. Aluminium might be an example. The competitiveness of Australian production from coal-based electricity may decline for a period, relative to production from hydro-electric power and natural gas in the rest of the world. This would involve economically and environmentally efficient contraction of Australian production. If this were to occur, any attempt to slow its natural progress would increase the cost of Australian emissions reduction.

None of this happens smoothly in an ad hoc world. There is a risk of carbon leakage, from countries with strong to countries with weak mitigation regimes. In an ad hoc world the management of political economy pressures arising from trade-exposed, emission-intensive industries is a dreadful problem for every country. Chapter 14 provides a framework within the Australian emissions trading scheme for addressing this dreadful problem.

The risk is banished in a world of comprehensive agreement, even if the degree of restraint on emissions is much more severe on some countries (developed) than others (developing). Trade in permits will establish comparable carbon pricing, even if some countries face more demanding emissions reduction trajectories than others.

All countries have powerful interests in moving quickly into a world of comprehensive carbon constraints, even, within reasonable limits, one in which the commitments to emissions reductions are tighter on themselves than on some other countries.

There is one other complication in comparing costs and benefits for Australia of mitigation in an ad hoc world and under comprehensive agreements. Global mitigation will shift demand away from fossil energy sources unless and until there is commercially successful sequestration of the carbon dioxide exhaust from combustion. Australian export volumes and export prices, and therefore output and incomes, will fall with international mitigation. Countries that import fossil fuels, like the United States, face an opposite and happier prospect. For them, import prices fall and real incomes rise as a result of global mitigation. This is a significant factor in raising the cost of mitigation to Australia under a comprehensive agreement relative to the ad hoc world. Australia's status as the world's largest exporter of the world's most emissions-intensive major energy source makes it especially vulnerable to international mitigation agreements to which it is not a party.

It is in Australia's strong self-interest to engage actively in bringing about a cooperative and comprehensive international framework for reducing emissions. This will require Australia, along with other developed countries, to adopt targets and trajectories, conditionally and unconditionally, as steps along the path to resolution of the prisoner's dilemma of international action on climate change.

12.3 Solving a diabolical problem in stages

The diabolical problem of climate change has many elements. Two seem to make it intractable: uncertainties about the science of climate change and the costs of mitigation; and the prisoner's dilemma constraining international collective action.

12.3.1 The delusion of delay

Uncertainty about the benefits (in this case, the benefits of avoided climate change) and the costs of major structural changes in the economy constrain change and reform in many areas of policy. The uncertainties are particularly wide with climate change. They are compounded by the long time periods over which both the costs and benefits are expected to work their ways through the economy and society.

Uncertainties about the science of climate change and the cost of mitigation can be reduced by research, which requires time, and by new observations made available by the passing of time.

But the science, and the realities of emissions growth in the absence of mitigation, show that we do not have time. The world is rapidly approaching points at which high risks of dangerous climate change are no longer avoidable. We would delude ourselves if we thought that scientific uncertainties were cause for delay. Such an approach would eliminate attractive lower-cost options, and diminish the chance of avoiding dangerous climate change.

12.3.2 Solving the prisoner's dilemma: move by move

The prisoner's dilemma of international collective action on climate change is daunting. Time is an essential element in any resolution of the policy problem. But with every year that passes without cooperative action, the range of options diminishes.

Only an international agreement that is perceived by all parties as fair in its distribution of the burden across countries has any chance of being accepted. The analysis undertaken by this Review indicates that all major emitters, including rapidly growing developing countries, will need to be parties to such an agreement. To be widely accepted, principles to guide the allocation of a global emissions budget across countries will need to be simple, transparent and readily applicable.

As noted in Chapter 8, the incentives facing individual delegations in a single, large, multilateral negotiation are not conducive to reaching sound agreement. Each country will try to secure a 'better deal' than others, with equity concerns figuring large and incentives for free-riding working against cooperative outcomes. Countries' circumstances and interests in the negotiations differ widely, and geopolitical considerations interfere. The dominant outcome is a low common denominator.

Australians can think of many reasons why their situation is different from that of other developed countries, and why their emissions reduction targets should be less demanding. So can people from every other country. There will be no progress towards an effective international agreement if each country lays out all of the special reasons why it is different from others, and why it should be given softer targets. When climate change negotiators from any country list reasons why their country has special reasons to be treated differently, we should be quick to recognise that they, and the countries they represent, intentionally or not, are preventing effective international agreement.

The underlying free-rider problem can only be solved through a repeated game with signalling and learning (Axelrod 1984), and in agreements that are individually and collectively rational, and considered fair (Barrett 2003). This requires close communication between sovereign parties, to allow disparate perspectives to be reconciled and confidence in collective action to be developed. But this requires time, and time is running out. Without strong action by both developed and major developing countries alike between now and 2020, it will be impossible to avoid high risks of dangerous climate change.

In such circumstances, the only way through the constraints is to make a start on domestic and international action, along paths that may now be feasible, but which in themselves do not lead quickly to ideal outcomes. Early action, even if incomplete and inadequate, on a large enough scale, can buy time and begin building the foundations for effective collective action.

But any old action will not help. To buy time and to help build the foundations for effective collective action it has to be well conceived in domestic and international terms. Actions that have high costs for minimal effect are likely to inhibit rather than build domestic support for effective mitigation. In the international sphere, policy initiatives that create tensions between countries over perceptions of equity, or that set in train protectionist actions and responses, will corrode rather than build confidence in collective action.

For these reasons, the best response for now may not see immediate movement to an agreement designed to solve the global warming problem once and for all. Progress will be made by designing an interim objective large enough to keep open the better options for avoiding high risks designed well to achieve its limited goals at low cost, that builds confidence that international cooperation is possible in this difficult area, and that encourages and allows time for the accumulation of the knowledge needed to reduce uncertainty about the science and about the costs of mitigation. This is the context in which the Review has framed its recommendations on targets and trajectories.

The details of the targets and trajectories the Review recommends in section 12.1 will not be the best for all time. They are the best that are available to us now.

In the context of well-designed domestic policies on emissions reduction, encompassing correction of market failures in response to prices being placed on emissions as well as to the emissions prices themselves, and carefully conceived international policies, our interim targets will lay the foundations for effective additional steps. Those steps will become easier to take as confidence grows in the knowledge base for strong policy action and in the feasibility of effective international action.

The first step, built around immediately moving on to a path of global emissions designed to stabilise concentrations of greenhouse gases at no higher than 550 ppm, is large and far-reaching enough to keep open the possibility of avoiding high risks of dangerous climate change. It is only the first step. But it is an essential first step.

It would be wonderful if the international political constraints eased to the point that a detailed proposal that 'added up' to a credible commitment to achieve the 450 objective could be put on the table at Copenhagen, and agreed by all parties. It is more likely that a detailed proposal that 'added up' to 550 could be agreed. It is of great consequence to the prospects of strong mitigation that at least this substantial but lower outcome be achieved.

12.4 Hastening progress towards greater emissions reductions

The interim targets recommended in section 12.1, although they will not be the best for all time, represent a realistic staging post for the more ambitious reductions required to avoid dangerous levels of climate change. If an agreement based on stabilisation of atmospheric concentration at 550 ppm were realised at Copenhagen, or soon after, and came into effect at the conclusion of the Kyoto period, it is likely that clear evidence of progress would emerge within a few years.

Confidence in the regime would build as the benefits of international trade in emissions entitlements became evident; as investment appreciably increased in research and development and the commercialisation of new, low-emissions technologies; and as the measured rate of emissions growth began slowing in accordance with the global trajectory shown in Figure 9.3.

This would encourage hope and provide reason for revisiting the initial agreement and lifting the level of ambition in global emissions reductions. Progress should be reviewed regularly (annually or biennially) in order to seize opportunities for adopting more ambitious targets—possibly beginning as soon as five years from the initial agreement.

Existing and new institutional arrangements in Australia and internationally will be required to support the objective of hastening progress towards more ambitious emissions reductions. The Australian institute of climate change policy research proposed in Chapter 15 will be the obvious point of integration between Australian and international monitoring and research efforts.

Whether existing global governance structures under the aegis of the UNFCCC and the IPCC are adequate for these purposes will be tested by their ability to deliver a comprehensive agreement to replace the Kyoto agreement from 2013.

12.5 Moving from a 550 to a 450 goal

Would it be possible for the world to start on a 550 stabilisation path and then move to a path consistent with a lower ultimate stabilisation objective? Say the world agreed to a 550 stabilisation path, as in the Review's modelling up to 2020, and then switched to the 450 path. If it were to make good the slower start by 2050, to avoid additional overshooting, illustrative calculations suggest that the 450 ppm global emissions reduction target of 50 per cent by 2050 over 2000 would increase to 64 per cent. Clearly, it would be better for the world to be on the 450 ppm path from the start. It would be better to move on to the new, more ambitious part earlier (say, 2015) than later (2020). A 2015 shift would make the reduction to 2050, without additional overshooting, about 55 per cent. But being on a 550 ppm path keeps hope alive for a 450 ppm path.

Other possibilities could also be considered to improve the chances of the world eventually returning the atmospheric concentration of greenhouse gases to 450 ppm or less, even after a slow start. The 450 ppm path could allow for greater overshooting than modelled here (up to 530 ppm), and for return to the 450 ppm level early next century rather than this century. Such approaches could hold the 2050 reduction requirement to 50 per cent, but carry larger environmental risks.

12.6 Does Australia matter for global mitigation?

Only effective global action can solve the climate change problem. Australia is the source of only about 1.5 per cent of global greenhouse gas emissions. So does Australian action matter?

If our own mitigation efforts had no effect at all on what others did, we could define our own targets and trajectories, and approaches to their realisation, independently of others' perceptions or reactions. We could enjoy the benefits of reduced risk of climate change from others' actions, without accepting our share of the costs. The optimal level of Australian mitigation effort—the level that would maximise the incomes and wealth of Australians—is easily calculated. It would be zero. That is not far from the stance of Australian policy until recent times.

Australia's relevance to the international policy discussion has been apparent in the period since early 2001. The fact that Australia had joined the Bush administration in not ratifying the Kyoto agreement that we had each negotiated was a key fact in the American domestic discussion. Australia was presented as evidence that the Bush administration was not alone among developed countries.

All countries, Australia and the United States among them, agreed at United Nations meetings in Kyoto in 1997 that all developed countries would accept certain obligations. While the Review's analysis demonstrates that a substantial majority of the future growth in emissions will come from developing countries (Chapter 3), the international community has agreed that the first steps in mitigation would be taken by developed countries. This gives every developed country a veto on substantial progress on global mitigation: The failure of any one of them to do what it said it would do would make it unlikely that the necessary later steps would be taken by major developing countries. We played that veto card.

Whether we like it or not, Australia matters.

There are more general reasons why Australia may be influential to global outcomes.

There is a role for countries of substantial but moderate weight—for 'middle powers'—in taking the initiative in leading global diplomacy on issues in which they have major interests. Global warming passes the interest test for Australia, as we are likely to be the developed country that is most damaged by a failure of effective global action. Australia—at times for good and at times for ill—has demonstrated on many issues at many times in history that it is effective in a 'middle power' diplomatic role, developing ideas to shape international cooperation, and persuading others that cooperation is in their own interest. The APEC group of countries is one example.

Australia has some unusual diplomatic assets in the developing countries that are centrally important to successful global mitigation policy. Chinese policy is crucial to a successful global outcome. A history of close and productive cooperation on domestic and international policy through the reform period gives Australia a strong base for cooperation with China. Our close and well-developed relationships with Indonesia (the world's third largest emitter of greenhouse gases in absolute terms) and Papua New Guinea (a large emitter in per capita terms and one playing a global leadership role among less developed countries on greenhouse gas emissions policies) raise special opportunities.

The world, and especially developing countries, need examples of countries making successful transitions to low emissions while maintaining economic growth. Australia's established market economy and economic dynamism, with particular skills and natural resources in areas of special importance to the low-carbon economy, will be assets in making a successful transition, showing that it can be done.

Although it may miss our attention, others notice, and think it relevant, that Australia's economic strength in the early 21st century derives to a considerable degree from our higher terms of trade associated with the strong economic growth in Asian developing countries. They notice that strong growth in the Asian economies, and exceptional Australian prosperity, is the other side of the coin to the heightened urgency of the global warming problem.

Because Australia matters, we cannot contribute positively to an effective global agreement, and at the same time pick a trajectory for our own country's emissions reductions that keeps costs low for us, without assessing whether this would be consistent with a global agreement to solve the problem.

If there is to be an effective global agreement, it is not open to Australians, any more than to people from any other country, to pick and choose among principles according to what suits them best in a particular and narrow context. The corollary of the focus on per capita allocation of emissions rights for interim targets discussed above and in Chapter 9 is acceptance of long-term global allocation rules built around eventual convergence across countries in per capita entitlements. This is the source of the required 80 per cent reduction in Australian emissions (90 per cent reduction per capita) from 2000 levels by 2050 under a 550 scenario and 90 per cent reduction in Australian emissions (95 per cent reduction per capita) from 2000 levels by 2050 under a 450 scenario.

Such an approach, with these consequences, is in Australia's national interest. It is in Australia's national interest because the costs of accepting the approach are manageable and because it provides the best chance of reaching an international agreement that reduces the risks of dangerous climate change to acceptable levels.

It would help Australians to face some of these realities if we were more realistic about where we stand among developed countries in taking action to reduce greenhouse gas emissions. It is claimed by many Australians—some who want their country to be in a leadership position, and some who do not—that we are, or are about to be, ahead of other developed countries on greenhouse gas abatement.

Australia is in no danger of leading the world in greenhouse gas mitigation. In comprehensive national efforts at mitigation, it ranks behind all of the 27 countries of the European Union.

In practical policy innovation to reduce emissions, Australia ranks behind a number of states of the United States, including the largest, California, with its pervasive and costly regulatory approach. The national governments of Japan, New Zealand, the United States and the European Union have engaged in a range of partial activities to reduce emissions. In all of these countries, there are domestic debates about national abatement initiatives at similar stages to our own.

What the rest of the world notices most about Australian emissions is that ours are the highest per capita in the OECD; that over the past several decades they have been growing faster than those in other OECD countries; and that while in 1971 the emissions intensity of Australia's primary energy supply was similar to the OECD's as a whole, in recent years it has been more than one-third higher (see Figure 7.7). There are good reasons why Australia became relatively more dependent on a high-emissions source of energy—coal—while the remainder of the OECD was reducing the proportionate role of coal and increasing the contributions of low-emissions energy, including nuclear. But whatever the reasons, they are not easily reconciled with the idea that Australia is leading the world in emissions reduction.

It is often said in Australia that developing countries are strongly resistant to reductions in emissions and that it is unrealistic to expect them to participate in global constraints on emissions. This is too simple. China's selective withdrawal of export rebates within its value added tax, its export taxes on a range of energy-intensive products, its discouragement of expansion of energy-intensive industries and its specific regulatory constraints on investment in steel, aluminium and cement production add up to more substantial constraints on the most emissions-intensive industries than would occur in Australia in the early years of an emissions trading scheme. China's active encouragement of low-emissions sources of power (hydroelectric, wind, nuclear, biomass, biofuels) goes beyond current Australian efforts. These measures stand alongside a domestic policy commitment to reduce the energy intensity of economic activity by four percentage points per annum until 2020. Data released by the Chinese Government in August 2008 show the energy intensity of Chinese GDP falling by 3.7 per cent in 2007 (Xinhua 2008)—the first sign of good intentions on energy intensity being reflected in policy outcomes.

Among other developing countries, Papua New Guinea's prime minister has asked his country's newly established Climate Change Office to prepare an analysis of ambitious mitigation targets: a reduction in emissions of 50 per cent by 2020, and carbon neutrality by 2050. The Indian Government is well-known for its declamatory statements resisting commitments to reduce emissions until developed countries have gone much further. But if we listen carefully, it has also said repeatedly that it is prepared to commit itself not to increase its per capita emissions above developed country levels. India has long emphasised that convergence towards equal per capita entitlements would need to be a central part of any international agreement in which developing countries accepted constraints on emissions. Many developing countries have said they would be prepared to do more if there were commitments from developed countries to support transfer of low-emissions technologies and climate change adaptation strategies.

It is easy to be cynical about statements of good intentions by others, as it is easy for them to be cynical about ours. There is a possible path to an effective international agreement if we observe carefully what others are doing, listen to what others say they are prepared to do, and note the conditions for action. We need to listen as well to others' perspectives on our own policies and practices.

Australia matters. What we do matters. When we do it matters. It would be ill-advised to take action with costs to ourselves that is meant to assist the emergence of a good international agreement, but to do it too late to have a chance of avoiding high risks of dangerous climate change. What we do now, in time to influence the global mitigation regime from the end of the Kyoto period, is of high importance. What we do later runs the risk of being inconsequential in avoiding dangerous climate change.

12.7 Interim targets

To represent a world in which there is a more ad hoc approach to global mitigation than that assumed in Chapter 11, the Review separately and independently modelled a scenario in which Australian mitigation action is undertaken before there is a comprehensive international agreement. The modelling of this so-called Copenhagen compromise scenario was undertaken in order to assess the costs of adopting the interim target discussed in section 12.1.

An additional scenario was also modelled, the so-called waiting game. This scenario represents the unlikely event that the Australian emissions trading scheme is implemented in 2010 without any clarity about an international agreement (partial or comprehensive).

The Review's modelling assumptions are discussed in Box 12.2, with a fuller account of the modelling assumptions, methodology and results available on the Garnaut Review website at <www.garnautreview.org.au>.

Box 12.2 Modelling assumptions

In each of the scenarios all sectors were assumed to be covered by the mitigation policy from the commencement of the modelling period. All prices in the modelling are in 2005 dollars.

Under the 450 ppm and 550 ppm global mitigation scenarios of Chapter 11, globally coordinated mitigation action begins in 2013 with unlimited trading in permits between countries. In the lead-up to 2013, all countries, including Australia, are assumed to continue on business-as-usual growth and emissions. Australia meets its Kyoto commitments. From 2013, the carbon price is determined through global trade in emission entitlements, the volume of which follows the trajectory shown in Figure 9.3.

Box 12.2 Modelling assumptions *(continued)*

In the 'Copenhagen compromise' scenario, a fixed price permit system is introduced in 2010 at $20 per tonne of CO_2-e rising at 4 per cent per annum until 2013. From 2013, the carbon price is assumed to float in accordance with Australia imposing a linear reduction in emissions from 2012 to achieve a 60 per cent reduction in emissions from 2000 levels by 2050. Unlimited trading in international permits is assumed, with the international permit price set at $40 per tonne of CO_2-e in 2013 and rising at 4 per cent per annum.

In the 550 ppm, 450 ppm and Copenhagen compromise scenarios, all industries are assumed to have access to unlimited permits from 2013. This provides for domestic emissions reduction targets to be met by abatement undertaken domestically or through the purchase of international permits. The global carbon price acts as a cap on the price of domestic permits.

Under the 'waiting game' scenario, the same fixed price regime is imposed, but it remains in place as no trade is assumed available in the absence of an international framework agreement.

Shielding of Australia's major trade-exposed, emissions-intensive industries is only required under the Copenhagen compromise and the waiting game scenarios. Shielding under the Copenhagen compromise is provided through the redistribution of permit auction revenue, capped initially at 20 per cent of revenue between 2010 and 2013, and thereafter declining by 1 percentage point per annum. This decline is intended to simulate an increasing number of countries adopting emissions reduction policies between 2013 and 2020. The model assumes that all remaining revenue is transferred to households as a lump-sum payment.

Under the waiting game scenario, shielding is provided to the extent required to maintain output from shielded industries at a constant share of the economy. The value of shielding is capped in each year at 30 per cent of total permit revenue.

12.7.1 The costs of meeting our interim targets

In determining the costs of the various policy scenarios, economic outcomes were compared with the reference case of Chapters 3 and 7 that projects the global and Australian economy, assuming that there is no climate change and no climate change mitigation policy. Abstracting from the effects of climate change is a reasonable assumption in the short term. Different approaches are required when modelling the longer-term implications of policy in Chapter 11.

Table 12.3 shows the macroeconomic outcomes for all modelled scenarios to 2020.

Table 12.3 Modelling results in 2020 for policy scenarios

	Conditional offers		Unconditional offer	
	450 ppm scenario	550 ppm scenario	Copenhagen compromise	Waiting game
Emissions entitlement reduction commitment for 2020 relative to 2000 (per cent)				
Reduction in total emissions	-25	-10	-5	–
Per capita reduction	-40	-30	-25	–
Deviations from reference case in 2020 (per cent)				
GDP	-1.6	-1.1	-1.3	-0.9
GNP	-2.0	-1.5	-1.4	-0.9
Consumption	-2.4	-1.8	-1.6	-1.2
Carbon price in 2020 ($)				
Domestic	60.0	34.5	52.6	29.6

Note: Prices are denominated in 2005 Australian dollars.

These results have two clear implications.

First, the overall cost to the Australian economy from tackling climate change is manageable and in the order of one to two-thirds of 1 per cent of annual economic growth. Australia can readily afford to make unconditional and conditional policy commitments of reducing emissions by 5 per cent and 10 per cent in 2020 from 2000 levels, respectively (equivalent to per capita reductions of 25 per cent and 30 per cent, respectively).

Second, there are clear benefits from broadening the level of international cooperation in implementing mitigation policy. Australia can significantly increase its mitigation effort at negligible additional cost if the broadest possible agreement can be reached by the global community. The broader the opportunities for low-cost abatement, the lower the overall cost for Australia.

The higher carbon price under the Copenhagen compromise relative to the price observed under the 550 ppm scenario reflects more limited access to low-cost abatement opportunities in a world in which there are ad hoc arrangements. The higher permit price in the 450 ppm scenario derives from the substantially greater level of emissions reduction and the subsequent scarcity of permits.

12.7.2 The effect on the costs of interim targets of varying some major design features

The partial mitigation scenarios are predicated on a range of assumptions. From a policy perspective, the most notable of these relate to sectoral coverage, use of permit revenue, international trade in permits, and shielding of highly affected trade-exposed, emissions-intensive industries.

The modelling assumes full coverage of all sectors from the outset of the scheme. All permits are auctioned. Some of the revenue is used to assist highly emissions-intensive industries in the traded sector, with all remaining revenue returned to households.

The Review undertook preliminary analysis of varying the assumptions in relation to the ability of Australian emitters to partake in a global permit market and the consequences of using shielding policies.

Trade in international permits is found to be of particular importance and benefit to the overall economic effects of meeting an emissions reduction objective. For example, adopting emissions reduction targets such as those in the Copenhagen compromise but artificially restricting access to an international permit market would result in a much higher permit price by 2020. The higher price is explained by forcing abatement in areas of domestic activity that would otherwise have been unnecessary and could have been achieved at lower cost in other countries. This has adverse effects on macroeconomic measures such as GDP and GNP.

Preliminary analysis on the whole-of-economy implications of providing assistance to trade-exposed, emissions-intensive industries was also examined. This analysis shows that in the presence of a quantitative constraint on emissions, macroeconomic variables such as GDP and GNP in 2020 are hardly affected at all by shielding trade-exposed, emissions-intensive industries. Shielding was found to result in higher levels of activity in emissions-intensive industries, removing potentially cheap forms of abatement. This redistributes the burden of abatement across other parts of the Australian economy, at potentially higher cost, or through the purchase of more permits internationally.

The effects of shielding on households and consumption will depend on the foreign ownership structure of trade-exposed, emissions-intensive industries. In the modelling, shielded industries are assumed to have large foreign ownership shares, particularly the metals manufacturing sectors. As a result, shielding reduces income available for domestic consumption. While consumers are bearing the full cost of shielding (they receive less permit revenue as a result of the shielding), they receive only a portion of the (restored) profit as shareholders.

The Review did not model the transaction costs associated with alternative compliance arrangements for the emissions trading scheme. This could turn out to be a substantial deadweight loss on the economy, particularly in relation to the treatment of trade-exposed, emissions-intensive industries in an ad hoc policy world. If this issue is not handled well, uncertainty will affect the supply price of investment. It will lead to a diversion of management effort into rent-seeking behaviour rather than the pursuit of low-emissions production processes. It could potentially lead to a wide corrosion of good economic governance. In the worst of circumstances, it could turn out to be as expensive as the costs of mitigation itself.

12.8 Implications for an Australian emissions trading scheme

It will be important for Australia to put in place, from 2010, the architecture that will deliver emissions reductions at the lowest possible cost to the domestic economy. Great care must be taken now as consideration is given to the design of a domestic emissions trading scheme. This is the necessary centrepiece in Australia's effort to reduce emissions.

There is, however, a risk to the stability of the emissions trading scheme if the form of the post-Kyoto international agreement remains unknown at the time of the scheme's commencement. The time between the start of the domestic emissions trading scheme and a successor international agreement is best viewed as a transitional period in which the price of permits should be fixed.

In addition to avoiding unproductive interaction between the early period of a new trading system and Australia's participation in crucial global negotiations, fixing the price of permits will provide a less anxious environment for implementing the globally efficient approach to assistance to trade-exposed industries as noted in section 12.2.

The preferred principles for, and design features of, Australia's emissions trading scheme are discussed in greater detail in Chapter 14.

References

Axelrod, R. 1984, *The Evolution of Cooperation*, Basic Books, New York.

Barrett, S 2003, *Environment and Statecraft*, Oxford University Press, Oxford.

Campbell, K., Gulledge, J., McNeill, J., Podesta, J., Ogden, P., Fuerth, L., Woolsey, J., Lennon, A., Smith, J., Weitz, R. & Mix, D. 2007, *The Age of Consequences: The foreign policy and national security implications of climate change*, Centre for Strategic and International Studies, <www.csis.org/media/csis/pubs/071105_ageofconsequences.pdf>.

DCC (Department of Climate Change) 2008, *Tracking the Kyoto Target 2007: Australia's greenhouse emissions trends 1990 to 2008–2012 and 2020*, DCC, Canberra.

IPCC 2007, *Climate Change 2007: Mitigation of climate change. Contribution of Working Group III to the Fourth Assessment Report of the Intergovernmental Panel on Climate Change*, B. Metz, O.R. Davidson, P.R. Bosch, R. Dave & L.A. Meyer (eds), Cambridge University Press, Cambridge.

Lenton, T.M, Held, H., Kreigler, E., Hall, J.W., Lucht, W., Rahmstorf, S. & Schellnhuber, H.J. 2008, 'Tipping elements in the Earth's climate system', *Proceedings of the National Academy of Sciences in the USA* 105(6): 1786–93

Nordhaus, W. 1994, *Managing the Global Commons: The economics of climate change*, MIT Press, Boston.

Xinhua News Agency 15 July 2008, 'Energy consumption per unit of GDP continues to fall', <http://en.chinagate.com.cn/news/2008-07/15/content_16008506.htm>.

AN AUSTRALIAN POLICY FRAMEWORK 13

Key points

Australia's mitigation effort is our contribution to keeping alive the possibility of an effective global agreement on mitigation.

Any effort prior to an effective, comprehensive global agreement should be short, transitional and directed at achievement of a global agreement.

A well-designed emissions trading scheme has important advantages over other forms of policy intervention. However, a carbon tax would be better than a heavily compromised emissions trading scheme.

The role of complementary measures to the emissions trading scheme is to lower the cost of meeting emissions reduction trajectories, as well as adapting to the impacts of climate change by correcting market failures.

Once a fully operational emissions trading scheme is in place, the Mandatory Renewable Energy Target will not address any additional market failures. Its potentially distorting effects can be phased out.

Governments at all levels will inform the community's adaptation response. More direct forms of intervention may be warranted when events unfold suddenly or when communities lack sufficient options or capacity for dealing with the impacts of climate change.

Climate change risks are a consequence of the greatest example of market failure we have ever seen (Stern 2007). Market failure occurs when the market fails to take into account the costs (or benefits) of an action that accrue to firms or people who are not parties to the action. Market failure in relation to the pricing of a resource leads to its overexploitation (or underutilisation). The failure to place a price on greenhouse gas emissions has led to overutilisation of a scarce resource: the atmosphere's capacity to absorb emissions without risks of dangerous climate change.

The correction of this market failure is the central task of climate change policy, in Australia and in the world.

Reducing emissions is often referred to as 'mitigation'. 'Adaptation' refers to actions taken in anticipation of, or in response to, the climate change impacts that cannot be avoided by mitigation policy. These two policy areas are often treated separately. This is not necessarily helpful for the design of good policy.

To mitigate human-induced climate change effectively, a restriction must be placed on rights to emit greenhouse gases to the atmosphere. This limit must be reduced over time to the level that prevents any net accumulation in the atmosphere. This comes at a cost to the economy. But so does the alternative of unmitigated climate change. Less mitigation will, in all likelihood, require a greater adaptation effort by individuals, communities and businesses. Less mitigation may also involve costs that cannot be avoided efficiently by adaptation.

Typically, the costs of mitigation will be felt long before the costs arising from the consequences of excessive emissions. The inadequate attempts so far to address the global market failure in greenhouse gas emissions mean that some of the consequences of climate change are already unavoidable.

The optimal policy choice will involve both mitigation and adaptation.

This chapter offers a framework for considering how Australian policy makers should approach this task. It serves as a bridge between the earlier and later chapters of this report. Chapters 2 to 7 present the global and Australian impacts of climate change. Chapters 8 to 10 discuss the challenges and policy options for reaching a global agreement on limiting greenhouse gas emissions and how to approach adaptation in an international context. Based on the framework outlined in this chapter, chapters 14 to 19 outline the necessary domestic policy interventions for dealing with the causes and consequences of climate change. The report concludes with a series of chapters (20 to 23) that describe the emergence of a low-emissions economy in Australia if these policies are successfully implemented.

13.1 Confronting uncertainty: the policy challenges of climate change

13.1.1 The policy continuum

Important climate change policy decisions are required now, because delay involves cost, including the cost of lost options. These decisions must be made despite innumerable scientific, geopolitical and economic uncertainties about:

- the strength of the tendency for global emissions to continue growing
- the relationship between the accumulation of greenhouse gases in the atmosphere and global warming
- the nature, timing and extent of local biophysical impacts in Australia and elsewhere as a result of the extent of climate response
- the level of ambition and the likelihood of international cooperation to reduce greenhouse gas emissions
- the development and costs of new technologies that reduce our reliance on emissions-intensive processes
- the adaptation choices that will be available domestically and internationally, and their cost.

Decision making in the face of uncertainty is not new. Business has developed many tools for identifying and managing risk. In recent decades, some of these instruments have been adopted by governments in Australia and around the world. But the range of possible outcomes under climate change is wider than for any other challenge that we face. The probabilities that can be assigned to these outcomes remain, for now, poorly defined. The scale, scope and timing of possible outcomes at global and domestic levels are unprecedented for the consequences of human action.

We come to these problems with economic, social and political institutions (public and private, domestic and international) that may not be appropriately constituted for dealing with them. We are therefore confronted simultaneously with the uncertainties of climate change and the potential for institutional inertia in dealing with these problems.

This is not a reason for resignation. A failure to act at any point will narrow the options available at any future point. But the lives of Australians will continue beyond the point of failure and new decisions will present themselves, with the possible outcomes shaped partly by that failure.

The uncertainty of climate change must be confronted.

As with any form of uncertainty that affects decision making, there is value in methodically reducing the extent of the unknown. This is achieved in one of two ways.

First, we can seek to understand the consequences of climate change better, globally and locally. Greater information is needed in order to understand and to estimate the potential costs of different levels of climate change to our prosperity and to other things we value. Additional resources must be allocated and new institutional structures established to fill the significant gaps in Australia's climate change research program.

The quest for understanding must also cover greater knowledge about the adaptation options available for dealing with climate change as well as appropriate responses for dealing with any outstanding uncertainty.

The second strategy involves reducing the causes of the uncertainty created by climate change, namely, greenhouse gas emissions to the atmosphere. Slowing or halting the rate of climate change reduces the likelihood of extreme outcomes as well as the range of potential outcomes. This allows more confidence in decision making. But mitigation comes at a cost.

The extent to which these strategies fail to reduce uncertainty will determine the need for a third strategy, namely, the ability to operate in a world that is changing in ways that we do not now fully understand.

For now, we can only assume that neither climate change nor the uncertainty so created can be eliminated by the deployment of a single strategy.

The investment that society is prepared to make in reducing risks of climate change is dependent on the costs of taking such action, weighed against the costs of failing to do so over the period before it is expected that annual benefits will exceed annual costs. This balance has not featured prominently in the policy debate in Australia in recent years. Rather, attention has focused on the costs of reducing climate change as if doing so provided no benefit. What is true, however, is that any attempt to minimise the extent of climate change is dependent on global cooperation and not just Australian emissions reduction.

The purpose of pursuing an Australian mitigation policy ahead of a comprehensive international commitment is to maximise the chances of an effective global agreement being reached as quickly as possible. The rate at which emissions are reduced by Australia in the meantime, and the rates of reduction to which we are now prepared to commit ourselves in the context of an effective global agreement, will be among the most significant policy decisions made in this country for many years.

The cost of any given emissions constraint imposed on the Australian economy will depend on the means by which it is implemented. Poorly designed policies will result in unnecessarily high transaction costs and misallocated resources.

It is inevitable that some degree of climate change will occur and that adaptation to its impacts will be required. For our current purposes, both natural and human-induced climate change are relevant. The optimal form of this adaptation, as well as its extent and timing, will depend on the ability of communities and businesses to assess the risks they face and the options available for addressing those risks.

The relevant literature typically refers to adaptation policy as separate and distinct from mitigation policy. This has unhelpfully led to a policy approach to adaptation that is nebulous. The Review considers that mitigation and adaptation are more usefully considered within the single policy framework described in the next section.

13.1.2 A coordinating framework for climate change policy

Climate change will alter fundamentally some important relationships within the economy—for example, the relative value of different factors of production. So too will any policy interventions to mitigate greenhouse gas emissions.

The economic impacts of climate change and mitigation measures are best thought of as representing 'shocks'. A shock is defined for this purpose as an event that alters relationships within the economy.

Climate change policy is therefore most usefully considered as a set of interventions by governments to minimise the economic consequences of these shocks.

Table 13.1 summarises the nature of the shocks that direct the design of mitigation and adaptation policies.

When approached in this way, mitigation policy can be seen to consist of both the source of the shock (that is, emissions reductions) and the response to that shock through measures to minimise its adverse effects on the economy and the community. Adaptation policy also responds to shocks, but those caused by the climate change that global mitigation policy has failed to avoid.

Table 13.1 Attributes of mitigation and adaptation shocks

	Mitigation shocks	Adaptation shocks
Source	Constraint on emissions imposed by policy	Impacts from climate change
Scale	Determined by policy	Uncertain and variable[a]
Primary manifestation	Price	Productivity
Commencement	Distinct	Ambiguous
Parties directly affected	Relatively few	Localised and variable[a]
Indirect effects	Economy-wide	Variable[a]
Predictability	Relatively high	Typically uncertain
Temporal nature	Immediate and increasing	Eventual and worsening

[a] In this context, 'variable' indicates that this attribute is a direct function of the type of shock arising from climate change and so defies generalisation.

Viewing climate change policy as the management of shocks to the economy and the community suggests that properly designed mitigation and adaptation policies ought to have more in common than a simplistic policy dichotomy suggests.

Well-designed mitigation and adaptation policies only require government intervention when there is reason to believe that the effects of the shocks will not be dissipated efficiently, effectively or equitably.

As an open, flexible and market-oriented economy, Australia is well placed to deal with a wide array of events whether they are anticipated or not.

Harnessing the market in order to provide options and opportunities is central to lowering the cost to the Australian community.

In the case of mitigation, the necessary policy response is to correct for the missing market resulting in the unfettered release of greenhouse gases to the atmosphere. Section 13.2 assesses different options for doing this and concludes in favour of a well-designed emissions trading scheme, the preferred features of which are outlined in Chapter 14.

By itself, this is unlikely to be a sufficient policy response for reducing emissions. Mitigation policy must not only correct for the missing market. It must also address any market failures that inhibit the efficient operation of that new market. The main market failures are introduced in section 13.3.5 and analysed in detail in chapters 17 to 19. The case for government intervention is made where the cost of that intervention is outweighed by the reduction in the costs of the market failure being corrected.

Adaptation policy differs from mitigation policy in that there is no immediate or obvious missing market or market failure. Ongoing effort is required to enhance the capacity of existing markets, such as those for agricultural products, water and insurance, so that they may deal efficiently with the impacts of climate change. Measures that seek to promote the development of global and domestic markets for products (beyond carbon) and factors of production will assist in dissipating a wide array of shocks, whether they originate in Australia or beyond.

As with mitigation policy, correcting the market failures identified in chapters 17 to 19 will also be centrally important to an efficient and effective adaptation policy response.

Ideally mitigation and adaptation strategies would embody measures that correct the tendency for regulatory and institutional arrangements, and policy uncertainty, to create significant barriers to change.

Governments will need to review existing policies to ensure that they do not adversely interact with the objectives of successful mitigation and adaptation and, most immediately, the introduction of an emissions trading scheme. Reviews should cover federal and state taxes and subsidies, procurement policies, industry assistance programs, product and technology standards, accounting standards, taxation rules and public investment in research and development. The aim should be to identify perverse incentives that might inhibit adjustment to the effects of an emissions trading scheme or adaptation to the effects of climate change.

Commitments have already been made for reducing the regulatory burdens on business, expanding investment in infrastructure, reviewing federal tax arrangements and reforming Australia's approach to human capital formation. The successful implementation of these policy reforms would assist the introduction of an emissions trading scheme as well as the community's capacity to deal with the effects of climate change.

Beyond the establishment and enhancement of markets, governments at all levels will continue to play an important role in informing, planning and coordinating the community response to climate change. More direct forms of intervention may be warranted when events unfold suddenly or when communities lack sufficient options or capacity for dealing with the impacts of climate change.

Many of the existing instruments of government will be relevant, though their use will rarely be justified overtly in the name of climate change—which will typically be insidious rather than abrupt in its manifestation.

Health and community services, education and skills creation, quarantine and environmental protection, urban planning and transport, disaster relief and emergency services may all, at various times, be affected. It will be incumbent upon policy makers to be attentive to the changing demands on these services. The appropriate response will depend on circumstances as they emerge. It will vary between being anticipatory and reactive. Some responses will be systemic. Others will be determined by local conditions, and led and implemented by local communities and businesses. However, as climate change only forms one of innumerable considerations in the design of these policy responses, the Review does not explore these issues in detail.

The market-based discipline for tackling climate change preferred by the Review will be efficient in determining the allocation of resources across the economy, but may have undesirable distributional consequences.

A well-designed emissions trading scheme can be expected to be environmentally effective (in reducing emissions) and economically efficient. Individuals and households will be affected by its introduction to the extent to which firms pass on higher input costs in the form of higher prices. The scheme will have greater impact on regions and communities that are dependent on particular emissions-intensive industries or firms.

To the extent that climate change impacts cannot be avoided through effective global mitigation efforts, regions and communities face changing patterns of production as well as alterations to their quality of life.

Chapter 16 discusses the distributional effects of an emissions trading scheme and appropriate policy responses to them. It also outlines the limited conditions under which structural adjustment assistance by governments may be warranted, whether as a result of mitigation policy or as part of adaptation to climate change.

While human systems can ultimately respond to the impacts of climate change, at some economic and social cost, the natural environment has limited capacity to dissipate the effects of climate change through normal evolutionary and adjustment processes.

The value we attach to the natural environment is not easily quantifiable. Neither is the damage wrought by climate change, which does not readily lend itself to cost–benefit analysis. These difficulties may lead to an inefficiently low level of ambition in mitigation policy, resulting in even worse consequences for the natural environment.

Do such consequences warrant increased investment in environmental management by government as part of adaptation policy? The answer relates to the value attributed to these non-market costs by the community. (See Chapter 15 for a discussion of the challenges for the management of ecosystems and biodiversity.)

13.1.3 Modelling the effects of climate change and climate change policy

The Review has undertaken extensive modelling of the costs of climate change as well as of the costs of mitigation policy (Chapter 11). This modelling is important in informing the interim targets for Australia's emissions reduction trajectory recommended in Chapter 12.

The modelling has relied on a composite of climate models and numerous partial and general equilibrium economic models, both domestic and global. By and large, these models were not designed to be integrated into a single effort. Nor were they designed to answer the questions put to them by the Review. The economic models were certainly not built to look at the time frames that are relevant when considering climate change.

Nevertheless, the ambition and the achievements of the Review's modelling effort are unprecedented and provide invaluable insights for policy makers to consider and for the community to debate. It lays the foundation for a continuing and larger investment in the modelling of climate change and climate change policy.

As noted above, climate change and climate change policy will alter fundamentally some relationships within the economy. This does not sit comfortably with economic models that are predicated on known and measurable past behaviours.

Experience shows that once consumers and producers have accepted the inevitability of change, and face predictable incentive structures, they will alter their behaviour to account for the new conditions more efficiently and effectively than previously predicted. This experience suggests that economic models are likely to underestimate the benefits or overestimate the costs of changes in economic conditions, so long as the change is to stable institutional arrangements and predictable incentives. This bias may be further exacerbated by lack of data about the full costs of climate change impacts and a corresponding downward bias in the estimated benefits of avoided climate change.

These limitations are particularly relevant when policy makers consider the overall emissions reduction goal of mitigation policy. If they determine the goal solely on the basis of assumed technological developments and known consumer preferences at a particular moment, they will probably underestimate the true potential of the economy to reduce emissions in the future—that is, overestimate the price of permits and the economic cost of adjustment. This risks raising political resistance to new policies to tackle climate change. On the other hand, goal setting that is based on assumptions about unknown technologies and unobserved preferences runs the risk of overestimating the capacity of the economy to adjust. Economic modellers and policy makers will tend to err on the side of caution.

This is evident in the Review's own modelling. While the central assumptions of the modelling may be realistic from our perspective in 2008, history confirms the dogged recurrence of ingenuity. For this reason, the Review has also modelled alternative states of the world in which innovation is more responsive to an increasing carbon price. At first, this is applied in the energy and transport sectors (internationally and in Australia), where we have a reasonable chance of imagining how a more innovative future may unfold. However, we cannot be sure how these new technologies will manifest themselves later in the century. We can, though, foreshadow that 'backstop' technologies will remove the final vestiges of economic growth's reliance on emissions-intensive forms of production and consumption, provided there are positive economic returns for innovators from doing so.

These alternative technology scenarios were modelled by the Review, and are discussed in more detail in chapters 20 to 23. These scenarios illustrate the potential for the net costs of global mitigation to fall markedly from around the middle of the century.

As new technologies emerge, the global community will be increasingly confident in accepting more ambitious goals for reducing greenhouse gas emissions. The setting of targets and trajectories is best seen as an iterative process, with more ambitious mitigation goals being built on growing confidence that they can be reached at reasonable cost.

A study of history shows that when change is sudden, and its magnitude exceeds some hidden threshold, institutions governing the political, social and economic affairs of humanity can fracture. Things fall apart. Costs beyond previous contemplations can accumulate rapidly, as they have always done in great wars, domestic political convulsions and economic depressions. There is therefore likely to be some asymmetry between miscalculation of the costs of adjustment to moderate charges and incentives (the transition to a low-emissions economy), and to major changes in the biophysical environment (the more severe of the possible manifestations of climate change).

13.2 Avoiding the greatest market failure ever seen

The initial parameters of Australia's mitigation policy will need to be set ahead of a comprehensive international agreement to reduce global emissions. In this context, consideration needs to be given to Australia's unilateral commitment to domestic emissions reductions, the most efficient means for meeting that commitment, and the impacts that this will have on the broader economy.

13.2.1 Setting emissions limits for Australia

There would be no point in Australia introducing mitigation policy on its own. The entire purpose of Australian mitigation policy is to support the emergence of an effective global effort.

Reaching a comprehensive international agreement will not be easy, but there is a chance that Australia and the world will manage to develop a position that strikes a good balance between the costs of dangerous climate change and the costs of mitigation. The consequences of the choice are so large that it is worth a large effort to take that chance while we still can. A significant mitigation effort by Australia and other developed countries is the cost of preserving some hope of a comprehensive international agreement for avoiding dangerous climate change.

How Australia defines and implements its mitigation policy will establish its credibility and its place in negotiating an international agreement.

Nevertheless, until there is a comprehensive international agreement, there will be little difference between gross and net costs to the Australian economy from domestic mitigation policy. There will be little countervailing benefit arising from climate change avoided. Setting emissions limits will rely on a series of judgments about what value to place on Australia, with other wealthy countries, assisting movement towards a comprehensive global agreement by moving ahead of such an arrangement.

The period of Australian mitigation effort before there is an effective global effort should be short, transitional and directed at achievement of a sound global agreement.

The Review therefore proposes unconditional and conditional interim targets and trajectories that balance the requirements of developed country policy leadership with the costs of acting ahead of a comprehensive global agreement (Chapter 12). The unconditional offer needs to be broadly in line with the approaches of other developed countries. The conditional offer is determined by Australia's likely share of the burden under the most ambitious global agreement that is feasible in the current state of knowledge about the costs and benefits of mitigation and the current state of international cooperation. At present, and until any new agreement at or beyond the Copenhagen conference in 2009 is reached, the most ambitious feasible outcome may be to stabilise emissions at 550 ppm CO_2-e by 2100. In time, and with the introduction of new technologies, Australia along with other countries will feel increasingly confident in adopting more ambitious targets for emissions reductions.

Having established as a policy objective the reduction of Australia's greenhouse gas emissions according to a set of trajectories (and the conditions by which those trajectories might be changed), policy makers must choose the most efficient option for limiting emissions.

13.2.2 Domestic policy options for reducing emissions

The options for meeting the policy objective of reducing Australia's greenhouse gas emissions are either regulatory or market based. Within these two categories, numerous policy instruments can be applied.

Regulatory responses to the mitigation objective work by:

- mandating restrictions or banning particular items from the set of product choices available to consumers, and/or
- mandating, licensing or banning particular technologies or production techniques used by firms operating in the domestic economy.

Regulatory, or prescriptive, approaches to reducing emissions can be haphazard. They are inevitably informed by assessments of current and future mitigation opportunities by officials, based on expectations about the rate of technological development and the changing state of consumer preferences. Such policy mechanisms have difficulty in responding to the sometimes rapid but usually unpredictable evolution of technology and consumer preferences.

Market-based approaches seek to alter price relativities in a way that reflects the externality embedded in goods and services—that is, direct and indirect emissions arising from the production and distribution process. Consumers are left to choose whether, when and how to change from high to low carbon-intensive products. As they do so, firms begin responding to new consumption patterns by investing in alternative technologies and new products.

Under market-based approaches, governments cannot simultaneously control both the price and the quantity of emissions. The choice of approach should take into account the importance placed on having control over the level of emissions, relative to the importance attached to being able to control the emissions price.

Four market-based approaches are available.

Emissions (or carbon) taxes

Administratively, the simplest pricing mechanism is to impose a tax on emissions, typically known as a carbon tax. Carbon taxes are straightforward to apply and avoid the need for governments to take discretionary decisions about who ought to be allowed to emit. Carbon taxes also provide certainty about the marginal costs of mitigation.

However, while a carbon tax avoids the arbitrariness of regulatory interventions, the meeting of emissions reductions targets cannot be guaranteed. Compatibility with other systems internationally may also be limited. Moreover, the achievement of ongoing and increasing reductions in accordance with one of the trajectories outlined in Chapter 12 would require variation of the carbon tax rate on the basis of continuing reassessment of the relationship between the rate of the tax and the level of emissions.

Emissions trading scheme 1: cap and trade

Under a cap and trade scheme, the government issues tradable permits that allow the holder of the permit to emit a specified volume of greenhouse gases to the atmosphere. A permit is an instrument with clearly established property rights. The sum of all permits on issue equates to the total greenhouse gases that may be emitted to the atmosphere. Permits are issued according to the trajectories discussed in Chapter 12.

The issuing of permits may involve government auction, or free allocation to particular parties. The decision about how to allocate permits involves a judgment over the allocation of the rent value of the permits.

Trading between parties allows permits to move where they have the greatest economic value.

As permits are traded, the price comes to reflect the balance between scarcity of permits and options to abate. The price is the balancing variable between the supply of, and demand for, permits. The price is determined by the market, not the government. It is likely to entail some volatility, especially at the outset of the scheme when there is no or limited experience about abatement responses and costs. A well-designed scheme will not eliminate volatility in the permit price, but it can avoid the unnecessary dissipation of resources arising from second-guessing of policy makers on changing scheme parameters by market participants.

As well as providing incentives for mitigation beyond the scheme, a cap and trade scheme provides greater potential to reduce the cost of abatement opportunities though international trade in permits, which can concentrate higher levels of abatement in the countries where it can be achieved at lowest cost.

Emissions trading scheme 2: baseline and credit

Baseline and credit schemes also rely on the creation of tradable permits. These schemes differ from cap and trade schemes in that they effectively place the creation of permits in the hands of private parties (existing emitters) rather than the government.[1]

The baseline feature of these schemes involves an algorithm that provides existing emitters with some level of entitlement to emit. If their actual emissions are below this entitlement, then the surplus entitlement is converted into tradable permits (or credits). Emitters that exceed their entitlement must purchase permits to account for any emissions above their respective baseline.

Options for calculating the baseline entitlement include:

- emissions in a particular base year
- average emissions per unit of production based on installed technology in a base year
- average emissions per unit of production based on best practice technology
- any combination of these or other approaches.

The choice of algorithm introduces a high and unavoidable degree of arbitrariness into the design of a baseline and credit scheme. This would raise transaction costs and encourage rent-seeking behaviour (as the entire rent value of permit scarcity accrues to existing emitters).

Hybrid schemes

Hybrid models address the tension between wanting certainty in both price and quantity. The basic feature of these models is the establishment of an emissions trading scheme (cap and trade) with an imposed upper limit on the price of permits (McKibbin & Wilcoxen 2002; Pizer 2002). This involves initially issuing tradable permits up to a cap, but with a commitment by government to issue unlimited amounts of extra permits at a specified ceiling price.

Like the carbon tax, the hybrid approach with a ceiling price has the advantage of providing certainty about the maximum permit price while preserving some aspects of an emissions trading scheme to the extent that the market price can be expected to remain below the cap. However, it also combines the disadvantages of both approaches. In particular, the full institutional and administrative apparatus—and therefore cost—of an emissions trading scheme is required, without any guarantee of the required domestic emissions reductions. The use of ceiling prices would create a problem for Australia's role and credibility in international mitigation negotiations, since it would not allow firm commitments on levels of emissions.

A floor price for permits would require the scheme administrator to enter the market to purchase permits whenever the permit price fell below a specified value. A floor price is incompatible with international trade in permits as it would effectively create an unlimited liability for the Australian scheme administrator.

Ceiling and floor prices would dampen the incentive for development of secondary markets. They would limit intertemporality and international flexibility in use of permits. The emergence of these markets and this flexibility are important in transferring risk to the parties best able, and most willing, to manage it, to stabilising price, and to providing market guidance on future prices.

13.2.3 Australia's preferred approach

In determining the preferred approach for Australia's mitigation effort, the primary policy objective must be to meet a specified trajectory of emissions reductions at the lowest possible cost. Policy must be designed to facilitate this transition to a lower-emissions economy, with as little disruption as possible and at least cost to the overall economy.

Australian mitigation policy needs to be considered in the international context of action and commitments. The world is now some way down the track towards an international system based on emissions reduction targets, starting with developed countries. Regulatory approaches, carbon taxes, hybrid schemes and baseline and credit schemes would not be readily integrated with existing and emerging international arrangements that could provide Australia with lower-cost mitigation opportunities.

A well-designed emissions trading scheme (cap and trade) can be relied upon to constrain emissions within the specified emissions limit (or trajectory). Current as well as future prices are set by the market, without the need for bureaucratic clairvoyance in relation to prices or mitigation options and costs.

As with any policy intervention, an emissions trading scheme will involve transaction costs that represent a deadweight loss to the economy. A well-designed emissions trading scheme requires rules governing:

- the limit on emissions
- the creation and issuance of permits
- who must or can participate in the scheme
- the means by which permits are exchanged between buyers and sellers
- the timing and method of acquittal of obligations
- the consequences for non-compliance
- the treatment of sectors not covered by the scheme
- the roles of government and other bodies in operating the scheme.

With a well-designed and comprehensive emissions trading scheme in place, price signals will begin flowing through the economy reflecting the scarcity value of the emissions of greenhouse gases to the atmosphere. Consumers will begin modifying their behaviour and businesses will respond accordingly.

Conversely, poor design would put at risk the environmental effectiveness and the economic efficiency benefits that are the reason for establishing an emissions market (see section 13.3).

The superiority of an emissions trading scheme over a carbon tax depends on the former's good design. In Australia's circumstances, a well-designed emissions trading scheme is superior to a carbon tax. A carbon tax is superior to a poorly designed emissions trading scheme.

13.2.4 Understanding the impact of an emissions trading scheme

An emissions trading scheme will correct the major market failure associated with climate change by establishing the right to emit greenhouse gases to the atmosphere as a tradable commodity. It is the most direct instrument for securing Australia's emissions reductions, if properly designed and allowed to play its role without extraneous interventions (for example, by attempts to control the permit price).

The supply side of the market is represented by the government-controlled issuing of permits in accordance with an agreed emissions reduction trajectory. As such, the Australian emissions profile is capped by the force of law. No further measures are required to control national emissions in covered sectors.

On the demand side are all the goods and services whose production or consumption results in the release of emissions. There are innumerable decisions by households and firms that, when summed, determine the economy-wide demand for permits.

The demand side of the market is given force by the government requiring emitters to acquit permits if they wish to release greenhouse gases to the atmosphere. In so doing, the government must have the administrative machinery to enforce such a requirement credibly.

A fully functioning market mediates between the variety and priority of wants of consumers and the productive capacity of the economy. The price of permits will be determined by the balance between demand for, and supply of, permits.

If the sum of all decisions across the economy implies that demand for emissions permits is in excess of supply, the price of permits will increase, and continue to increase, until demand is subdued and brought into line with the quantum of permits on issue.

A credible market will establish a forward price for permits that reflects expectations about the future demand for permits. The price rises at a rate of interest corresponding to the opportunity cost of capital.[2,3] The whole price curve—the spot price and all of the forward prices, together—embodies the market's expectations of what is required to induce the necessary substitution of low-emissions alternatives for high-emissions goods and services, and for economising on the use of goods and services that incorporate high proportions of emissions.

The price curve provides fundamental stability to the market, with opportunities for hedging price risks, and adjusting quickly to new information. Any change in expectations in demand or supply or in the interest rate would see the spot and forward prices adjusting immediately.[4]

The economic effect of an emissions price

The emissions price flows through the economy in two ways.

First, it causes the substitution of higher-cost, low-emissions processes or goods and services for lower-cost established processes, goods and services. This former is a real cost to the economy as it involves the reallocation of resources to uses that would not otherwise have attracted them.

This substitution effect gradually decouples economic growth from its former reliance on processes and products with high greenhouse gas emissions. Even though the price of permits can be expected to continue increasing, as reflected by the forward price curve, the proportion of the economy exposed to this higher cost will be ever diminishing. Once a product enters the market, technological and institutional improvements and scale economies are likely to lead to relative cost reductions over time.

The second way in which the emissions price will flow through the economy is by generating rents from the scarcity of the permits. This involves a transfer of wealth (mostly from households) to whoever receives the scarcity rents of the permits. This will be established emitters if the permits are simply given to them; or the government in the first instance, and then the beneficiaries of reduced taxation or increased public expenditure, if the permits are sold competitively.

On the basis that this major environmental reform—the introduction of an emissions trading scheme—is not meant to arbitrarily increase the proportion of the economy under the control of the public sector, the proceeds of the sale of permits should be identified for return to the community, either to households or to business. Demonstration that revenues from the sale of permits had been returned to the private sector in one way or another would neutralise what could otherwise become a rallying point for opposition to effective mitigation policies.

13.2.5 Is emissions trading the next great reform agenda?

The pervasive consequences of an emissions trading scheme make it a major reform of the Australian economy.

Although it is tempting to compare the mitigation challenge to earlier Australian programs of economic reform, we should exercise caution. Previous reforms—such as trade liberalisation, financial regulation and competition policy—were designed to raise incomes by allowing the allocation of resources to their most productive uses. By contrast, the climate change reform agenda must be focused on minimising the potential for loss of income after the introduction of measures to limit the release of greenhouse gases.

In any event, Australians are well placed to deal with the challenges posed by the introduction of an emissions trading scheme. The reforms of the past have made the Australian economy more open, market oriented and adaptable than at any time in its history. We have a good record in institutional design and in establishing genuinely independent agencies to implement those arrangements. In the case of an emissions trading scheme, we have the benefit of learning from schemes that have been implemented internationally, most notably, the three phases of the European Union's scheme.

As with all reform agendas, the commitment by government and the community must be ongoing and firm. Decisions must be made even in the face of unknown prospects for an international agreement and some uncertainty about how the domestic economy will respond.

13.3 Bungling Australia's emissions trading scheme

An emissions trading scheme imposes compliance costs on businesses and administrative costs on government. These costs represent a deadweight loss on the economy that can only be justified if the scheme enables the least-cost adjustment (in terms of resource allocation across the economy) to a quantifiable and verifiable commitment to reduce emissions.

If the necessary conditions of environmental effectiveness and economic efficiency cannot be satisfied, costs will rise due to the introduction of new sources of uncertainty into business transactions. In these circumstances, policy makers should consider alternative policy interventions, possibly on a temporary basis.

A broad-based emissions tax implemented as a transitional measure would be preferable under such circumstances. Chapter 14 describes an innovative interim measure of an emissions trading scheme but with fixed price permits in the early years. This approach would minimise the transition costs of moving to a genuine market-based policy at a later date.

13.3.1 Blowing the cap: the easy but meaningless way out

The easiest path for policy makers to avoid disturbing the status quo would be to lower the level of ambition for the emissions trading scheme. Giving in to well-organised interests by adopting weaker positions on the basic design of the scheme will place at risk the benefits that justify the implementation of a market-based mitigation policy and that make the case for using an emissions trading scheme rather than a carbon tax.

Exempting some sectors or particular greenhouse gases would distort the burden of reduced emissions and shift it disproportionately onto others.

Freely allocating permits to some emitters but not others safeguards the profits of the fortunate recipients while imposing even greater adjustment costs on other emitters and on the community.

Most damaging of all would be measures that rendered ineffective the credibility of the quantitative restriction (the emissions limit) upon which the entire emissions trading scheme is predicated. There are numerous compromises in the design of the scheme that could have this effect, directly or indirectly. These include:

- caps on the permit price resulting in the issuance of additional permits for as long as the price remained above the ceiling price
- poorly defined emissions reduction trajectories and vaguely defined conditions for changing trajectories, which would lend themselves to periodic pressure on the political system—poor design features of the system would make it difficult to resist these pressures
- non-compliance measures that failed to enforce the overall constraint on emissions (known as 'make good' provisions).

Such compromises, while seeming to help secure support at the time of introduction of the scheme, would undermine the policy objective of reducing emissions. This would erode business confidence when investment decisions are being made and cause the mitigation policies to impose costs on the Australian community for little or no environmental benefit. Our international credentials on this issue would be severely damaged—putting at risk access to the benefits of global cooperation as well as our ability to influence the outcome of international negotiations.

The most costly and damaging policy for Australia would be to implement a policy that was designed to appear meaningful, but was largely meaningless in application.

13.3.2 Withstanding vested interests

The emissions trading scheme needs to be free of ongoing disputation over key parameters. It will be costly if it provides opportunities for special interests to exert political pressure for favourable treatment—most notably, in permit allocation.

Not only does this represent a risk in terms of the potential revenue forgone, but it will raise the overall cost to the Australian economy. If there is a chance that political pressure will reap rewards in the form of special treatment, then the system will promote a large diversion of management resources, away from commercially focused profit maximisation towards rent seeking from governments.

Any scheme that promotes such behaviours by rewarding pressure must be viewed as an abject failure.

Nevertheless, an emissions trading scheme will, by design, alter pre-existing relationships within the economy. This will generate winners and losers.

Consumers who are willing and able to replace higher-emissions products with lower-emissions products will adjust relatively painlessly. Firms with less dependence on emissions-intensive production processes, or that have the ability to switch production processes quickly in order to minimise their exposure to a carbon price, may find that their market share and profitability increase. Firms that have less flexible capital structures could be faced with having to choose between passing on the price (and losing market share) or absorbing the price of emissions at the expense of profitability. All things being equal, such firms may face some loss of market value.

As with all programs of economic reform, mitigation policy must be forward looking. Policy interventions and the use of scarce resources should focus on improving future economic prospects rather than reacting to past decisions by governments or the private sector.

While it is not possible to foreshadow all the demands that will be placed on the revenue raised from the sale of permits, the case for compensatory payments to shareholders in firms that lose value as a result of introduction of the scheme is a low priority for a number of reasons.

First, it will be difficult or impossible to assess the effects of the emissions trading scheme on an individual firm's profitability as the counterfactual supply and demand conditions in those markets cannot be observed. The potential information asymmetry problem would lead to disputes.

Second, there is no tradition in Australia for compensating capital for losses associated with economic reforms of general application (for example, general tariff reductions, floating of the currency or introduction of the goods and services tax) or for taking away windfall gains from changes in government policy (for example, reductions in corporate income taxes).

Third, alternative forms of assistance such as structural adjustment assistance that is targeted at the future competitiveness of firms (or in some cases, regions) is likely to provide a greater benefit to the overall economy than a backward-looking, private compensatory payment to existing emitters.

Fourth, this is a difficult reform, and a permit price that is high enough to secure levels of emissions within targets and budgets will have major effects on income distribution—including workers and communities dependent on emissions-intensive industries that may be unable to adjust readily to alternative employment. Directing scarce resources towards addressing these impacts will be a significant challenge and an unavoidable priority. There will also be large calls on the revenue from sale of permits for support of research, development and commercialisation of new low-emissions technologies, and for avoiding 'carbon leakage' through payments to trade-exposed, emissions-intensive industries.

Stationary energy, which in Australia is a particularly large source of emissions, is the dominant industry with expectations of compensation. This is the subject of further discussion in Chapter 20.

13.3.3 The dreadful problem of trade-exposed, emissions-intensive industries

Trade-exposed, emissions-intensive industries represent a special case. All other factors being equal, if such enterprises were subject to a higher emissions price in Australia than in competitor countries, there could be sufficient reason for relocation of emissions-intensive activity to other countries. The relocation may not reduce, and in the worst case may increase, global emissions. This is known as the problem of carbon leakage.

Policy makers are therefore faced with a truly dreadful problem. Shielding these industries from the effects of a carbon price either undermines attempts to limit national greenhouse gas emissions or increases the adjustment burden elsewhere in the economy. Moreover, it results in the paradoxical outcome of shielding our most emissions-intensive industries (with the exception of stationary energy) from the effects of the scheme; that is, low emitters feel the effects of the scheme, but high emitters do not.

Chapter 10 outlines the benefits of sectoral agreements in avoiding this problem, while Chapter 12 suggests that Australia will need to show global leadership in pursuing such arrangements. In the meantime, Australia is faced with implementing special domestic arrangements. These transitional arrangements should be based on efficiency in international resource allocation and not on some false premise of compensation for lost profitability.

There can be no doubt that the arbitrary nature of such assistance measures will make them the subject of intense lobbying, with potential for serious distortion of policy-making processes. Their continuation for more than a few years would be deeply problematic. The establishment of comparable carbon pricing arrangements in countries that compete with Australia in global markets for emissions-intensive products is an urgent matter.

Policy makers would be better off abandoning an emissions trading scheme in favour of a broad-based emissions tax without exemptions if they felt unable to resist pressures on the political process for ad hoc and overly generous assistance arrangements for these industries.

13.3.4 Pandering to pet solutions

Detractors of market-based mechanisms often argue that additional emissions reduction measures (be they regulatory or programmatic) are required in order to reduce greenhouse gas emissions. They are wrong.

Unless private parties contravene the law without consequence, a comprehensive and well-designed cap and trade scheme ensures that emissions will decline in line with the reduction trajectory (the 'cap').

The very purpose of a market-based approach to mitigation policy is to enable producers and consumers throughout the economy to determine the most effective response to meeting a mandated emissions limit.

Programs and other regulatory interventions—whether federal, state or territory—that seek to reduce emissions from specific activities covered by the emissions trading scheme will not result in lower overall emissions. They will simply change the mix of mitigation activities that deliver the same, required level of emissions reductions. Such interventions presuppose that government officials, academics or scientists have a better understanding of consumer preferences and technological opportunities than households and businesses. This is generally unlikely and cannot ever be guaranteed.

Within the Australian domestic policy space, a variety of policies have been discussed or put in place by various levels of governments with the aim of reducing greenhouse gas emissions from sectors to be covered by an emissions trading scheme. While some are in place for historical reasons, other schemes are being considered prospectively. The most significant of these is the expansion of the Commonwealth Government's Mandatory Renewable Energy Target (see section 14.8.1).

13.3.5 Don't pick winners. Fix market failures.

For the emissions trading scheme to have the desired effect of driving new consumption behaviour and investment decisions, it must be well integrated within the broader economy. Barriers to change must be removed or minimised in order that there may be an efficient economic response to the ever diminishing supply of permits.

Federal and state governments must avoid policies that skew investment decisions towards technologies that are currently in favour or consumption behaviours that are judged to be desirable. Existing policies—such as tax expenditures, and direct- and cross-subsidies—must be reviewed in light of the introduction of an emissions trading scheme.

Such reviews will need to extend beyond programs and policies that directly compete with the emissions trading scheme for emissions reductions. The aim should be to identify perverse incentives that might inhibit investment in low-emissions technologies or promote activities associated with high emissions.

Other policies operating alongside an emissions trading scheme can have no useful role in reducing emissions once the emissions trading scheme is in place. From that time, the only useful role for additional policies of this kind is to reduce the effect of market failures that have the potential to raise the economic cost of the structural adjustment process. Three market failures must be addressed by the relevant levels of government if the benefits of an emissions trading scheme are to be maximised.

First, there are market failures in the end use of energy, as a result of misplaced incentives, and externalities in gathering and analysing information about known technologies. Correcting these market failures would reduce energy consumption and lower the overall demand for permits. Government intervention would include mechanisms for subsidising the provision of information. Regulatory responses may be warranted if they are the most efficient means of correcting the market failure.

Second, the market failure associated with research, development and commercialisation of new technologies must be corrected. Policies are required that recognise that private investors are not able to capture for themselves the full social value of their innovations. There is therefore a need for high levels of public expenditure across a broad front, including:

- climate science
- the impacts of climate change (nationally, regionally and locally)
- technology responses to changing climatic conditions
- low-emissions technologies and processes (including energy efficiency)
- geo-, bio- and soil sequestration.

Public assistance must be introduced in different forms for different stages of the innovation process.

Third, governments must address the possibility of market failures associated with the external benefits from pioneering investment in the provision of network infrastructure related to electricity transmission, natural gas pipelines, carbon dioxide pipelines associated with sequestration, and transport infrastructure linked to urban planning. This may or may not require public expenditure.

These sources of market failure are addressed in chapters 17, 18 and 19, respectively.

Notes

1 The Greenhouse Gas Reduction Scheme (or GGAS) established by the NSW Government, which has been in operation since 1 January 2003, contains elements of a baseline and credit scheme (NSW Department of Water and Energy 2008).

2 This is because investors will be choosing between alternative investments, with an emissions permit being one possible investment. Investors will assess whether the long-term value of holding an emissions permit is higher or lower than the return from an alternative investment. This leads to selling or buying of emissions permits until a forward price curve emerges that causes the expected return from holding a permit to be equivalent to that on alternative investments. The price would therefore rise at a rate of interest corresponding to alternative investments available to holders of permits.

3 Incidentally, it is a common error to see a rising forward price curve for emissions permits as reflecting an increasing external cost of emissions as the volume of emissions rises over time. Later emissions do not impose greater costs. Rather, the rising price reflects the market's approach to optimise depletion over time of a finite resource (Hotelling 1931), in this case the resource being the atmosphere's capacity to absorb greenhouse gases without seriously adverse consequences.

4 Any new information that increased optimism about new, lower-emissions ways of producing some product, whether they were expected to become available immediately or in the future, would shift downwards the whole structure of carbon prices, spot and forward. Any new information that lowered expectations about the future availability of low-emissions alternative technologies would raise the whole structure of carbon prices, spot and forward.

References

Hotelling, H. 1931, 'The economics of exhaustible resources', *Journal of Political Economy* 39(2): 137–75.

McKibbin, W.J. & Wilcoxen, P.J. 2002, *Climate Change Policy after Kyoto: Blueprint for a realistic approach*, Brookings Institution, Washington DC.

NSW Department of Water and Energy 2008, *Transitional Arrangements for the NSW Greenhouse Gas Reduction Scheme*, consultation paper, NSW Government.

Pizer, W.A. 2002, 'Combining price and quantity controls to mitigate global climate change', *Journal of Public Economics* 85: 409–34.

Stern, N. 2007, *The Economics of Climate Change: The Stern Review*, Cambridge University Press, Cambridge.

AN AUSTRALIAN EMISSIONS TRADING SCHEME

14

Key points

A principled approach to the design of the Australian emissions trading scheme is essential if the scheme is to avoid imposing unnecessary costs on Australians.

The integrity, efficiency and effectiveness of the scheme will require:
- establishment of an independent carbon bank with all the necessary powers to oversee the long-term stability of the scheme
- implementation of a transition period from 2010 to the conclusion of the Kyoto period (end 2012) involving fixed price permits
- credits to trade-exposed, emissions-intensive industries to address the failure of our trading partners to adopt similar policies
- no permits to be freely allocated
- no ceilings or floors on the price of permits (beyond the transition period)
- intertemporal use of permits with 'hoarding' and 'lending' from 2013
- a judicious and calibrated approach to linking with international schemes
- scheme coverage that is as broad as possible, within practical constraints

Seemingly small compromises will quickly erode the benefits that a well-designed emissions trading scheme can provide.

The existing, non-indexed shortfall penalty in the Mandatory Renewable Energy Target needs to remain unchanged in the expanded scheme.

It will be important for Australia to put in place, from 2010, the architecture to deliver emissions reductions at the lowest possible cost to the domestic economy. Great care must be taken now in the design of a domestic emissions trading scheme. This is the necessary centrepiece in Australia's effort to reduce emissions.

The public debate that has accompanied the release of the Review's draft documents and the Commonwealth Government's Green Paper on a Carbon Pollution Reduction Scheme has focused attention on the need for a highly principled approach to the design of the scheme.[1] If the necessary conditions of environmental effectiveness and economic efficiency cannot be satisfied in scheme design, this will raise costs by introducing new sources of uncertainty into business transactions. The net effect will be to distort economic activity and investment in new productive capacity in ways that will be damaging in the long term and potentially disastrous for emissions reductions as well as for economic efficiency.

This is no more evident than when designing the appropriate assistance arrangements for trade-exposed, emissions-intensive industries in a world of ad hoc mitigation policy. It would be a significant failure of public policy if such assistance arrangements sought to compensate businesses for the effect of an Australian emissions trading scheme rather than for the failure of our trading competitors to implement comparable policies.

There is a risk to the stability of the emissions trading scheme if the form of the post-Kyoto international agreement remains unknown in advance of the scheme's commencement. The time between the start of the domestic emissions trading scheme in 2010 and a successor international agreement from 2013 is best viewed as a transitional period requiring special consideration. This is a period covered by Australia's established commitments under the Kyoto protocol.

Table 14.3 at the end of this chapter provides an overview of the Review's preferred design for an Australian emissions trading scheme.

14.1 The framework to guide efficient scheme design

A principled approach is required if an emissions trading scheme is to be effective and efficient in supporting transition to a low-emissions economy. Successful implementation will result in observable outcomes, such as:

- low transaction costs
- price discoverability
- emergence of forward markets and other derivatives
- investor confidence
- low-cost mitigation spread over time in a way that minimises the present value of costs.

Conversely, a poorly designed scheme will compromise some or all of these outcomes; encourage and reward rent-seeking behaviour; delay at high cost the necessary structural adjustment; and raise the overall burden incurred by households.

14.1.1 The objective of an emissions trading scheme

To mitigate climate change effectively, a limit must be placed on rights to emit greenhouse gases to the atmosphere, and this must be reduced over time to the level that prevents any net accumulation in the atmosphere. Australia's limit will represent an agreed share of a global limit.

An emissions permit represents a tradable instrument with inherent value that can be exchanged between sellers and buyers in an emissions permit market. This enables the movement of permits about the economy to their highest value (or most economically efficient) use. It does this while ensuring the integrity of the volumetric control, or emissions limit, imposed in order to satisfy the policy objectives of climate change mitigation.

After the policy objective of reducing emissions is established and it has been determined that this is most efficiently achieved by the implementation of an emissions trading scheme, the objective of the scheme should be kept as simple and focused as possible in order to avoid compromising its efficiency, namely:

> *To provide for the low-cost transmission of permits to the parties for whom they represent the greatest economic value.*

Other policy objectives—be they economic, environmental or social—should be pursued through alternative policy instruments that operate alongside the scheme.

14.1.2 Guiding principles for scheme design

The design of an emissions trading scheme is guided most appropriately and transparently by the following five principles.

Principle 1: Scarcity aligned with the emissions target

Market participants must have confidence that permits are in scarce supply and reflect the targets and trajectories for national emissions reductions discussed in Chapter 12. Where the scarcity of permits is uncertain, market participants will factor in risk premiums (if they suspect that the commodity will become more scarce) or risk discounts (if they suspect that the commodity will become more abundant). This will distort resource allocation decisions and impose unnecessarily high costs on the economy.

Principle 2: Credibility of institutions

Credibility, or faith in the enduring nature of the rules and institutions that define the emissions trading scheme, is essential for its ongoing success. Markets can quickly collapse if their credibility is shaken. This is all the more pertinent for markets that owe their existence solely to government decree.

As an emissions trading scheme exists entirely at the behest of government, market participants will be alert for any signs of shifts in policy, management protocols or operating procedures that may undermine the integrity of the market. A poorly designed scheme will also create incentives to press for change if there appears to be a chance that the rules of the scheme can be influenced by political pressure. Arbitrary changes to rules that benefit one party are likely to come at the expense of other market participants, or the community, or the environment.

Reliable, steady and transparent operating rules are a necessary condition for the credibility of the market. These rules may need to be adjusted over time. This too must be done through reliable, steady and transparent processes.

Principle 3: Simplicity of rules

Simplicity requires that rules for the scheme should be easily explained and implemented. Rules should apply consistently; and special rules, concessions and exemptions should be avoided. Rules should be unambiguous and internally consistent. Where one rule necessitates the creation of another rule to ameliorate unwanted consequences, the first rule is probably suboptimal.

Compromises to the simplicity of the scheme should not be made lightly as they will inevitably result in increased uncertainty and transaction costs for market participants.

Principle 4: Tradability of permits

If market participants have no means by which to exchange permits, the scheme's objective, of moving emissions permits to those who value them most, cannot be achieved. Tradability requires that:

- permit characteristics and the benefits they bestow are unambiguous
- the terms and conditions of trade are commonly understood
- those wanting to participate have ready access to the market
- transactions can be secured at minimal cost
- offer and bid prices are transparently available.

Principle 5: Integration with other markets

An emissions trading scheme must be able to coexist and integrate with international markets for emissions entitlements as well as with other financial, commodity and product markets in the domestic and international economy. This requires that there be no barriers to the appropriate transmission of information within and between markets.

If the scheme contains distortions that result in an emissions permit price that does not reflect its true scarcity value, this mis-priced market will adversely affect decisions about resource allocation by investors in other markets.

The converse is also true. Distortions in other markets may result in mis-priced outcomes in the scheme. However, the integrity of the scheme should not be compromised to compensate for distortions in other markets. Rather, policy makers should use the opportunity and insights gained from establishing the scheme to identify and correct distortions in other markets.

14.2 Elemental design features

This section applies the principles in order to guide the design of elemental features of the Australian emissions trading scheme. Because a comprehensive global agreement is the longer-term objective in taking mitigation action, a domestic emissions trading scheme should support Australia in moving toward this ultimate objective.

14.2.1 Establishing and changing the scheme's emissions limit

An emissions permit will enable the holder to emit, on a one-off basis, a specified quantity of greenhouse gas—one tonne of carbon dioxide equivalent (CO_2-e). The emissions reduction trajectory will determine the number of permits that can be issued in any given period. The total number of permits that can be issued (for example, in accordance with an international agreement to reduce emissions, see Chapter 9) over time specifies the 'emissions budget' for all the sectors covered by the scheme. The integrity of the trajectory and the overall emissions budget is paramount in order to satisfy the scarcity principle.

In its early years, it will not be possible for the scheme to cover all emissions from all sectors. The limit on emissions from sectors covered by the scheme should be in accordance with the following calculus in any given compliance period:

Australia's total emissions allocation under international obligations
equals
Emissions from sectors covered by the emissions trading scheme
less
Emissions from sectors not covered by the scheme[2]
plus
Emissions entitlements purchased internationally

Maintaining the credibility of the scheme (Principle 2) will require that changes to the emissions limit under the scheme are kept to an absolute minimum. Where changes are necessary, they should be predictable and carried out in accordance with clearly articulated rules and transparent processes. The targets and trajectories framework of section 12.1 provides the basis for minimising market uncertainty by defining a limited set of possible trajectories (three) for Australia's total emissions entitlement.

To ensure predictability, the conditions that would lead to movement from one trajectory to another—namely, an international agreement—would be specified in advance. If and when it was announced that the conditions had been met for movement to a tighter trajectory, five years' notice would be given to the market. This would provide the market with five years of firm 'caps' at all times.

If international obligations required Australia to move to lower emissions within five years, the government (or scheme regulator, the independent carbon bank (see sections 14.4.2. and 14.7)) would meet this commitment by purchasing international emissions entitlements. This would cushion the effect on participants in the emissions trading scheme in the period between the commencement of the agreement and the end of the five-year notice period.

Information should also be provided with sufficient lead time to market participants about any other changes that would significantly affect the scheme's scarcity constraint. These changes could include: the inclusion of new sectors or gases under the scheme; government purchases of international emissions entitlements; or changes to the rules about accepting international emissions entitlements to acquit domestic obligations. These issues are covered below in further detail.

14.2.2 Who will the scheme cover?

Coverage refers to the scope of the scheme in terms of the greenhouse gases and the sectors that come under the ambit of the scheme. Emitters of any of the six anthropogenic greenhouse gases covered by the Kyoto Protocol that contribute to climate change should have an obligation to acquit permits under the Australian emissions trading scheme.

Coverage of the scheme should be as broad as possible, within practical constraints, in order to:

- provide an incentive for emissions reductions in all sectors according to lowest-cost mitigation opportunities
- maximise market liquidity and stability
- distribute the costs of the scheme in ways that minimise distortions in resource allocation
- facilitate integration with other markets.

Sectors should be covered by the scheme unless the costs of inclusion are prohibitive due to:

- Lack of accurate estimation, measurement, monitoring and verification methodologies (see Box 14.1)—for some sectors, establishing the necessary processes may require some significant investment.
- Uncertainties in emissions measurement due to unreliable or inaccurate ways to monitor, measure or estimate, and verify emissions from operations in that sector—if a reliable proxy or 'rule-of-thumb' can be identified, then the sector should be included under the scheme. A poorly defined proxy can create distortions—failing to reward good performers and failing to penalise poor performers. Further, a major revision to the proxy introduced arbitrarily could cause significant market shock.
- Scale-related transaction costs—in some sectors large volumes of emissions come from relatively few sources (for example, electricity generators). In other sectors, there may be many small emitters. Even if emissions can be accurately measured, it may not be cost effective for all sources of emissions to take on an obligation under the scheme. Sectors comprising many small emitters may be more appropriately covered by imposing the obligation upstream or downstream rather than directly on the emitter (see 'Point of obligation' below)—provided this can be achieved cost effectively and with sufficient accuracy.

If a sector is not covered by the emissions trading scheme, policies should be developed to drive net emissions reductions from that sector, consistent with contributing to Australia's overall emissions reduction goal.

Domestic offsets from non-covered sectors

Australia's national emissions reduction commitments, as defined in Chapter 12, will relate to emissions from sectors included in the emissions trading scheme as well as those beyond the scheme's coverage.

Emissions reductions in non-covered sectors could be encouraged by recognising such reductions in the form of domestic offset credits. An offset credit could be created for each tonne of emissions removed by or reduced in non-covered sectors. It can be traded into the emissions trading scheme and would be treated as a substitute for a permit. An offset credit could be used by parties covered by the scheme to meet their obligations under the scheme. This enables lower-cost mitigation from offsets created outside the scheme to replace higher-cost mitigation options within the covered sectors.

This approach may be suitable for sectors in which emissions from some sources and activities, but not others, can be measured or estimated.

Importantly, to be eligible to create a credit, an offset project must provide an emissions reduction that is additional to that which would have occurred anyway. If it did not, allowing an offset credit to be created would undermine the overall domestic mitigation effort by introducing double counting.[3]

There may also be a role for international offsets in an Australian emissions trading scheme. This is discussed in section 14.4.3.

Point of obligation

The point of obligation defines the liable party for surrendering permits under the emissions trading scheme. The point of obligation may be anywhere in the supply chain—from those who produce goods and services that involve the release of greenhouse gases to the atmosphere, to those who consume those products. It is most reasonably imposed at the point at which monitoring and reporting of emissions is most easily, accurately and cost effectively achieved (see Box 14.1).

A natural starting point when considering the point of obligation is the emissions source. However, an alternative point of obligation may be selected when there is evidence that transaction costs can be lowered significantly by doing so, or if accuracy of emissions measurement is higher or coverage would be substantially wider.

There is no need for the point of obligation to be harmonised across schemes in different countries. Each country should adopt the most effective arrangements under local conditions.

There have been suggestions that the Australian emissions trading scheme should base the legal obligation at the point of consumption. While this has some attraction, it is not feasible for two reasons. First, the information requirements to support a consumption-based approach would be prohibitively costly. Second, there are now a number of emissions trading schemes, actual and nascent, imposing production-based points of obligation. If a consumption-based scheme were to be established in Australia, it would be extraordinarily difficult to integrate Australia's scheme with those in other countries. The integration principle highlighted in section 14.1.2 is essential in order to reduce the costs of mitigation in Australia.

Strict compliance: penalties and make-good provisions

A financial penalty must apply if a party with an obligation under the scheme fails to surrender permits equal to its emissions during a given compliance period or for failing to repay permits loaned from an independent authority (the independent carbon bank, see sections 14.4.2 and 14.7). This is a punitive measure rather than an alternative form of compliance.

To ensure the integrity of the emissions limit and credibility of the scheme, financial penalties would need to be accompanied by a make-good provision applying to the non-compliant party. That is, payment of a financial penalty does not negate an extant obligation to acquit permits. The independent carbon bank should have the necessary powers to address repeated failure by a private party to make good its obligation.

Box 14.1 Emissions monitoring, reporting and verification

The emissions trading scheme will require parties covered by the scheme to monitor and report their emissions to the scheme regulator. The system used to collect this information must be transparent, credible and efficient.

In September 2007, the *National Greenhouse and Energy Reporting Act 2007* was introduced.[4] This legislation established a national greenhouse and energy reporting system that will underpin the emissions trading scheme. Firms registered under the Act will provide information on their greenhouse gas emissions, energy production and energy consumption to the Greenhouse and Energy Data Officer. Those required to report will be facilities with over 25 kilotonnes of emissions, or production/consumption of 100 terajoules or more of energy in a given year. Thresholds have also been set at corporation level, and are to be phased in progressively during the first three years of the reporting system.

The system was in place from 1 July 2008 and the first year of reporting will be the 2008–09 financial year.

Data from the national greenhouse and energy reporting system should be the basis for making assessments about parties' obligations under the emissions trading scheme. However, additional data may be required, for example, in order to net out emissions from an upstream party's obligation.

Sector-specific issues

Australia's emissions can be classified as coming from the following sectors: stationary energy; transport; fugitive emissions from fuel production; industrial processes; waste; agriculture; and land use, land-use change and forestry.

Emissions from stationary energy, transport,[5] industrial processes and fugitive emissions from fuel production can be accurately measured or estimated at reasonable cost and should be covered by an Australian emissions trading scheme commencing in 2010.

There is a reasonably strong, although not definitive, presumption that the source of emissions is the best point of obligation for stationary energy. The possibility of allowing large energy users to opt in to accept an obligation for their (indirect) stationary energy emissions should be considered. This would require the generator to have the ability to track and net out that energy use. The existence of a power purchase agreement may support this option.

Emissions from transport are released at a much smaller scale by individual vehicles. For the transport sector, then, an upstream point of obligation may be a cost-effective way to cover a large number of smaller emitters. Many parties that produce fuel for the Australian market are located overseas, beyond the coverage of an Australian emissions trading scheme, so petroleum could logically be covered by making the point of excise the point of obligation. Large liquid fuel users, for example, fleets or freight operators, might be allowed to opt in to accept an obligation under the scheme.

A complication will arise where the relationship between fuel and emissions is not constant. For example, sometimes petroleum is used as an input in manufacturing processes (such as for plastics or petrochemicals), resulting in the release of few or no emissions. Where this is the case, fuel sales would need to be netted out of an upstream party's obligation, or a credit system established so that producers could claim back the permit price passed through to their liquid fuel purchase.

The point of obligation can be set at the facility level for oil and gas production, gas processing and fugitive emissions from coal mining. There are measurement difficulties and site-specific variability with fugitive emissions from coal mining and oil and gas fields (DCC 2008a). Overcoming these issues, with a robust methodology to estimate emissions, should be a priority, although proxy measures could be used in the interim.

The point of obligation for pipeline system fugitive emissions could be placed on pipeline systems, as defined by operational control of the physical infrastructure, such as pipes, valves and compressor stations. Generally, industrial process emissions can be measured or estimated at their source.

Emissions from waste—primarily methane emissions from organic waste—could also be covered at source—that is, the landfill facility or treatment plant. While there are difficulties associated with coverage of emissions from waste, due to the variability of these emissions and the timing of their release, the early inclusion of waste is desirable. Ahead of being covered in the scheme, other policies to encourage mitigation in the waste sector should be pursued. An offset regime may not be appropriate, because it is unclear whether additionality would apply to mitigation activities in the waste sector. Further, it may not be cost effective to implement an offset program for a sector, when full coverage of that sector will be possible in the short term.

Inclusion of forestry, agriculture and land management on the earliest possible timetable is also desirable. The treatment of these sectors is of large consequence for the Australian and global mitigation efforts. Among the many implications are prospects for large-scale participation of Indigenous land managers in the mitigation effort (NAILSMA 2008). There is considerable potential for sequestering carbon through change in land and forest management and agricultural practices (Chapter 22). However, their full inclusion in an emissions trading scheme will require issues to be resolved regarding: (1) measurement or estimation and monitoring of greenhouse gas emissions and removals, and (2) consideration of changes to current emissions accounting provisions for these sectors under the Kyoto Protocol.

Those undertaking reforestation should be allowed to opt in for coverage (that is, liability for emissions and credit for net removal from the atmosphere) from scheme commencement. Reforestation and afforestation activities should be covered, based on full carbon accounting rules, once issues regarding emissions estimation and administration are resolved. Those undertaking deforestation should be liable for resulting emissions.

Forestry is a potential source of domestic offsets, including for net sequestration, even as a covered sector. The use of these offset credits should be unlimited. The increasing carbon content of growing forests should be brought to account. Recent technological developments would seem to make that possible. As reliable estimation methods are developed, carbon stored in wood products and biochar could also be reflected in carbon accounting and under the scheme.

The same comprehensive emissions accounting approach could be applied to agriculture. However, given the magnitude and variety of difficulties associated with emissions measurement in this sector, it is worth investigating whether other policies may deliver greater emissions reductions, at lower cost, than an offset regime. Where practical difficulties interfere with measuring or estimating emissions at the source, a downstream point of obligation may be suitable. For example, under the New Zealand emissions trading scheme, a point of obligation further downstream is being considered for a subset of agriculture emissions—such as covering emissions from enteric fermentation and manure management through a point of obligation at the dairy or meat processor. For Australia, the large coverage issues in agriculture relate to accretion of carbon in soils and vegetation.

Chapters 20 to 23 provide a more general treatment of the role of these sectors in a low-emissions Australian economy.

14.3 Releasing permits into the market

The government (or its agent, the independent carbon bank, see section 14.7) will be the sole creator and issuer of permits under the proposed emissions trading scheme. How permits are released into the market will have distributional consequences with respect to the dissipation of their economic rent value.

14.3.1 Manner of permit release: auction or free allocation?

Permits can be released by allocating them freely to a range of potential recipients, selling them through a competitive process (auctioning), or through a combination of the two. Whether a permit is sold or granted freely, the recipient will acquire the full economic and financial benefit it bestows because it is a scarce and valuable resource.

The manner of permit allocation will not affect the price of permits or the costs of adjustment to the scheme. Coase (1960) demonstrated that economic efficiency will be achieved as long as property rights are fully defined and that free trade of those property rights is possible. With a well-designed emissions trading scheme in place, the price of goods and services is independent of the approach adopted for allocating permits.

Allocation of permits, however, will have large effects on the distribution of income. Costs and risks differ depending on the manner of allocation. Free permit allocation would be highly complex, generate high transaction costs, and require value-based judgments regarding who is most deserving. If permits were to be allocated freely to existing emitters, an agreed methodology would be required. This would typically involve a baseline emissions profile against which an emitter's entitlement to free permits could be determined.[6] This would involve introducing unavoidable arbitrariness.

Agreeing principles of merit, collection and application of data, and resolution of disputes would be time-consuming. The complexity of the process, and the large amounts of money at stake, encourage pressure on government decision-making processes and the dissipation of economic value in non-productive rent-seeking behaviour.

Free permits are not free. Although they may be allocated freely, their cost is borne elsewhere in the economy—typically, by those who cannot pass on the cost to others (most notably, households). This is explained in Box 14.2, which also highlights the experience of the European Union following the free allocation of permits.

Recent public wrangling in Australia over these issues is evidence enough of the undesirability and impracticality of administering a system of free permit allocation. In contrast, a competitive process (auctioning) for releasing permits will provide greater transparency and have lower implementation and transaction costs. These are important attributes for the credibility and simplicity of the Australian scheme.

Australia, with its well-established legal, regulatory and administrative structure, is in a favourable position for full auctioning of permits. A sound auction design is important to avoid introducing new inefficiencies or distortions in the market.[7]

The introduction of the emissions trading scheme will be associated with many valid claims for increased government expenditure. The full auctioning of permits maintains government policy control over the disbursement of the rent value of permits in the most transparent and accountable manner. Revenue from the auction of permits will provide government with a tool to address market failures in the development of new, low-emissions technologies and to address the scheme's income distribution effects. Permit auction revenue will provide a means of meeting these claims, without placing pressure on public finances.

The Review concludes that there are no identifiable circumstances that would justify the free allocation of permits.

Phase 3 of the European Union's emissions trading scheme and numerous states in the north-eastern Regional Greenhouse Gas Initiative in the United States are also moving to full auctioning of permits.

As discussed in section 14.5, it would be inappropriate to use freely allocated permits as part of the proposed transitional assistance arrangements for trade-exposed, emissions-intensive industries. Doing so would suggest that assistance is being provided on compensatory grounds. This would be wrong.

During the proposed transition phase (2010 to the end of 2012), permits would be sold as of right and at a fixed price rather than auctioned (see section 14.6).

Box 14.2 Pass-through of permit value

If a manufacturer is emitting as part of its production process and is required to purchase a permit via an auction, the cost will need to be recovered through the price received for the manufactured good.

Alternatively, if the manufacturer is granted a free permit, then it must decide whether the permit is of greater value if used or sold. If it is of greater value to use rather than sell the permit, the manufacturer will need to at least recover its opportunity cost. In other words, the recipient will need to attain value from the use of the permit at least as great as if the permit had been sold at the market price.

The manufacturer selling in the domestic market in the absence of international competition faces the choice of either (1) continuing to manufacture (thus emitting greenhouse gases) and using its permits to acquit its obligation, or (2) selling some or all of the freely acquired permits, and reducing its production to a level consistent with its remaining permits. If the manufacturer decides to use rather than sell the permits, then it has forgone income. Therefore, the manufacturer will recover the price of every permit not sold by the income generated from continuing to produce.

It follows that the impact on the price of goods and services of pricing carbon through an emissions trading scheme is independent of the approach adopted by governments for determining the allocation of permits. Although the price impact is independent of the allocation method, the pass-through of permit price to the price of goods and services will depend on the competitive nature of the relevant market.

Box 14.2 Pass-through of permit value *(continued)*

Studies of the power sector in certain countries under the EU emissions trading scheme indicate pass-through rates of between 60 and 100 per cent, depending on carbon intensity of the marginal production unit and other market or technology-specific factors concerned (Sijm et al. 2006).[8] There will be situations in which a firm will have to decide between passing through the cost of purchasing permits (or reducing emissions), risking a loss of market share, or absorbing those costs with a resultant loss in profit.

14.3.2 Rate of permit release

Permits, including those for post-2012, should begin to be sold into the market as soon as possible after the full details of the scheme are finalised and before the scheme commences in 2010. This will provide market participants with a guide to price before price figures directly in domestic market transactions. Liable parties could ensure that they obtained necessary permits in advance of operation of the scheme.

Auctioning should proceed on a fixed schedule—weekly, monthly, quarterly or on any other basis that best suits market participants. The frequency and timing of auctions will have implications for business cash flows and corporate balance sheets. Some parties with an obligation, such as fuel companies, will be required to purchase permits for all emissions from their fuel. Fears about this financial risk have led some fuel companies to suggest that auctions should be as frequent as weekly. The Review expects new financial services to emerge quickly around the scheme, so that the market will be able to operate effectively across a range of frequency of auctions.

Despite the transitional period of fixed-price permits, fully tradable permits for use from 2013 should be sold into the market in small quantities from 2010 (or even sooner). This will support the development of forward markets and provide guidance to market participants on future prices.

14.3.3 Flexibility in purchasing permits

There is some anxiety about potential cash flow problems associated with purchase of permits. The Review does not expect that this will be an important issue in practice if a principled approach is taken when establishing the emissions trading scheme. An elaborate system of financial services will develop for the financing of permit purchases before acquittal. Moreover, it is expected that acquittal will be after receipt of revenue from sales in most cases.

To ease anxieties about financing permits without distorting the system, the Review suggests a simple expedient for at least the early, post-transitional, years of the scheme. On request, the independent regulator could issue emitters with a number of deferred payment permits (taken from the future release trajectory). For example, some anticipated permit requirements over the next five years could be set aside for direct purchase at the time of acquittal. These would be issued up to a maximum proportion (say, one-third) of expected annual requirements—enough amply to cover permits for which corresponding sales revenue had not been received at the time of acquittal. These permits would allow payment for them to be made at the time of acquittal. The payment price would be the market price on the day of acquittal or the average price over a preceding period.

The effectiveness and need for these special measures should be evaluated at regular intervals. They should be disbanded once they are no longer necessary.

14.3.4 Accounting issues

Implementation of an emissions trading scheme will require resolution of issues relating to financial accounting standards and tax treatment, including:

- avoiding distortions between the purchase of emissions permits and other options for meeting emissions targets—that is, pursuing tax neutrality between purchasing a permit, undertaking capital expenditure to reduce or sequester emissions, investing in research and development or reducing production
- valuing permits, given that they are only valid once, but can be hoarded and loaned. Consideration needs to be given to the discount rate, or interest rate, to be applied over time (Shanahan 2007). The price of the permit will be rising over time so that the interest rate, or the expectation of it, will be built into any lending transaction. The independent carbon bank may also choose to add a margin. Valuation may be on the basis of current market values (market-to-market), but other approaches could be considered.

14.4 Lowering the costs of meeting targets

Demand for permits, and therefore the price of permits, will fluctuate over time with economic and seasonal conditions, changes in consumption preferences and technologies. Rigid adherence to annual targets would place large and unnecessary short-term adjustment strains on the economy.

This problem can be partially addressed by setting targets spanning several years, as the Kyoto Protocol has done with its 2008–12 compliance period.

Other options for helping smooth permit prices and helping parties meet obligations include: price controls, intertemporal flexibility in the use of permits and international trade in permits. The Review rejects outright the first of these three options (beyond the transition period) but is highly supportive of the latter two options in providing flexibility for better matching the rate of permit use with domestic permit release schedules. The tradability and integration principles outlined in section 14.1.2 support intertemporal flexibility in the use of permits and international trade in permits.

14.4.1 The damaging effects of price ceilings and floors

A ceiling on the price of permits would place a limit on the cost of mitigation in the period in which it is effective, but in doing so it renders unreliable the scheme's capacity to deliver emissions reductions in relation to targets.

Setting price ceilings or floors is inherently arbitrary. These controls would need to be based on predictions on all of the many variables affecting demand for permits: incomes growth; technologies; consumer preferences; seasonal climatic conditions; and others.

More importantly, price ceilings would:

- undermine Australia's role and credibility in international mitigation negotiations since it would not allow firm commitments on levels of emissions
- present a barrier to international linking (see below) because the domestic price ceiling would, through trade, become the default price ceiling for all schemes linked to the Australian scheme
- prevent the intertemporal use of permits (see below) because there would be no limit on the number of permits that could be purchased for use in later years
- dampen the incentive for development of secondary markets. The emergence of these markets is important in transferring risk to the parties best able, and most willing, to manage it. A price ceiling leaves this risk with the government (or more accurately, taxpayers).

Price floors carry the possibility of higher levels of mitigation than anticipated at a higher total adjustment cost than in the absence of the floor.

While politically expedient, the introduction of a price ceiling or floor on permits would damage greatly the normal operation of the scheme.

The specific advantages of a transitional fixed price during the last years of the Kyoto period (2010–12) are discussed in section 14.6.

14.4.2 Allowing for the intertemporal use of permits

Permits created under an emissions trading scheme are designed to allow the holder to emit a given unit of emissions within a given emissions trajectory or budget. The economic efficiency of the emissions trading scheme can be improved and the overall cost to the economy reduced by allowing permit holders to determine the most appropriate time to use a permit that is in their possession. All permits should be auctioned without restriction on the time of their use. That is, permits should not be 'date stamped'.

Hoarding of permits by market participants and lending of permits by the authorities (within prudential restrictions) introduces flexibility without breaching emissions budgets.[9] This helps to minimise volatility in permit prices and allows market participants to use permits at the time when they have greatest value. This intertemporal flexibility would cause market participants to see the issue as one of optimal depletion of a finite resource. Optimisation over time would see the market establish a forward curve rising from the present at the rate of interest, forcing increasingly deep emissions reductions, in an order that would minimise mitigation costs.

Lending by the independent regulatory authority would allow parties to use permits from the future—ahead of their scheduled release according to the trajectory—to meet current obligations. Of course, the loan must be repaid, and the borrower would need to provide security against default. The independent regulator would undertake prudential monitoring of the level of lending. It would place restrictions on the amount of lending if it became so large as to raise questions about the current or future stability of the market. Restrictions on lending could be applied in terms of:

- **Time**—Permits should not be loaned for a period exceeding five years given the plan for the permit release trajectory to be fixed for five years and the same period of notice to apply before major changes are made to scheme operations.
- **Quantity**—The independent carbon bank should lend amounts it believes will not destabilise the current or future market.
- **Eligibility**—Borrowers must be creditworthy. Criteria determining creditworthiness should be applied and communicated so participants have a clear understanding of the likelihood of their eligibiity to borrow. Financial intermediaries would provide opportunities for others to borrow, with a commercial margin.

Loaned permits should be repaid when the loan becomes due. The value of the permit at the time of repayment would generally be higher than at the time of lending, and participants would factor in that cost. The independent regulatory authority could also apply an interest rate to cover risk and costs. The interest rate would be raised at times when the authorities judged it prudent to reduce the amount of lending.

The framework of trajectories established in Chapter 12 is one in which there is an expectation that the trajectories will tighten over time. Within this framework, the market would price in the possibility of the emissions budget tightening in future. This would be reflected in a higher forward price for permits, which would be likely to encourage hoarding of permits by participants and discourage the use of the lending provision.

Flexibility in the time of use of permits, through hoarding and lending, means that actual emissions could be above or below the emissions reduction trajectory at a given point. If actual annual emissions were above the level specified in international commitment periods (and if this were not made up by reductions in the non-covered sectors), the government could purchase permits in the international market to meet these commitments.

Hoarding and lending also obviate the need for the date stamping of permits and the large administrative apparatus that would accompany a date-constrained scheme.

Box 14.3 Minimising risks associated with lending

Recent commentary has suggested that intertemporal flexibility in the use of permits, and in particular lending, might affect the overall timing of mitigation—and delay mitigation—in a way that was environmentally disadvantageous; that it might breach international commitments on emissions reduction targets; and that it would lead to breaches of emissions budgets if loans of permits were not repaid.

There are two reasons why this is unlikely to be an important issue. First, variation in the timing of permit use within firm trajectories on the scale likely to emerge in any foreseeable commercial circumstances would not be material to environmental impact. Instead, the multiple emissions trajectories proposed by the Review would create a bias towards hoarding of permits by participants and away from lending.[10] The initial budgets would be looser than the budgets that were expected to succeed them. The market would therefore tend to price in some probability of budget tightening, so that future prices were higher than those that would probably emerge from expectations that budgets would remain at their current severity. Such expectations would be likely to encourage hoarding, rather than lending.

The Review considers that, with the five-year limit on term of lending, environmental impacts due to variations in timing of acquittal of permits are not likely to be a material consideration. Such short-term lending is akin to smoothing, and would not be expected to have any global environmental impacts. This lending arrangement is similar to the five-year Kyoto commitment period and the five-year carbon budget approach in the UK Climate Change Bill. The Review's approach formalises the mechanisms by which participants can borrow, and has a five-year rolling, rather than fixed, period within which lending can occur.

In the context of international agreements on targets and trajectories, any unlikely strong tendency towards net lending in Australia would be accompanied by a requirement to buy permits abroad to meet commitments on emissions reductions. As a result, delays in reductions of emissions in Australia would be balanced by acceleration of reductions elsewhere.

On the suggestion that loans may lead to a blow-out in the emissions budget because they may not be repaid, this is a matter of governance. The authorities would need to ensure that loans of permits were made only to creditworthy borrowers, that they were backed by security, and that contracts were enforced—just as they would have to ensure that emissions were backed by permits.

14.4.3 Opportunities for international trade and links

The costs of any specified degree of mitigation can potentially be reduced substantially by international trade in permits. Ultimately, global mitigation will only be successful if countries can trade in emissions permits. However, linking with an economy that has a flawed domestic mitigation system will result in the import of those flaws. Variations in the quality of mitigation arrangements across countries will make the decision to link with particular markets a matter for judgment. Opportunities for international linking of the Australian scheme should therefore be sought in a judicious and calibrated manner.

Currently, opportunities for linking are limited, but are likely to grow. The benefits of linking centre around the potential of integrated carbon markets to:

- reduce mitigation costs and price volatility by making it easier to set and adhere to national emissions budgets
- provide financial incentives for developing countries with opportunities for low-cost mitigation to take on commitments
- provide equal treatment or a level playing field for trade-exposed industries, through convergence of carbon pricing across countries.

But linking also has risks. Since the Australian market is relatively small, if it is linked to other, bigger markets it will become a price taker. The price would be set by carbon markets in the European Union, the United States, Japan or China should they develop and Australia link to them. This exposes Australia to risk from other countries' policies and market responses.[11] Linking might lead to price volatility, for example, due to unexpected external policy change.

Given the rapid growth of emissions-intensive industries in Australia, it is expected that Australia will be a net purchaser of permits for some time. Linking opens the possibility of Australia remaining a large exporter of emissions-intensive products, to the extent that that is economically and environmentally efficient on a global basis, and balancing this with import of permits.

Note that separate approaches are required for trading in permit and offset markets, and for trading with countries that have an emissions cap but not a carbon market.

Linking with other permit markets

Determining strategic and policy parameters for linking with other permit markets should be a role for the Commonwealth Government. The independent regulatory authority would certify individual permit markets as being of a suitable standard for linking.[12] Certification would be periodic. If there were a decline in quality, then the certification could be revoked. Once a market was certified as being suitable then unlimited trading with that market—or more precisely, unlimited acquittal of permits from the overseas market—would be allowed. All private sector parties would be able to trade. There would be no limits on the amount of overseas permits that could be acquitted in fulfilment of obligations under the Australian scheme, at the individual or the aggregate level.

When making its assessment, the independent authority would assess the compatibility of the market proposed to be linked with the Australian one. Both markets need to have firm and mutually acceptable levels of mitigation ambitions. Both need to have adequate monitoring and enforcement mechanisms. And they need to have compatible market rules—for example, on the unit of emissions, and possibly on lending and hoarding (see Box 14.4).

When making its assessment, the independent authority would also need to consider indirect links. If Australia were considering linking to one market, which was itself linked to a third market, Australia would have legitimate reasons not to link to the second market if the rules governing the third market were not acceptable.

In parallel, Australia should seek to strengthen international monitoring and enforcement and to harmonise standards across markets. Deep integration with other markets (that is, joint regulation) should be sought where appropriate and where prospects for policy coordination exist.

Linking, and any resulting changes, would have a fundamental impact on the effect of the emissions limit under the scheme, and on the functioning of the scheme. Therefore, advance notice of new links should be provided in the same way, and with the same five years' notice, as a move to a different emissions reduction trajectory.

Decisions to cut links or alter quantitative limits on acceptance of international permits, however, may need to be taken more quickly if market quality elsewhere deteriorates suddenly. As with the notice for change of trajectory, it would be open to the government to move more quickly in introducing the new trade opportunity, and to balance the revealed effects of the change on the domestic market by countervailing international permit sales or purchases.

In the initial stages, it may be a useful precaution to set a quantitative limit on aggregate permit purchases from certified international schemes. Any such limit would be applied in aggregate (to all certified permits). The limit would only apply in unusual, potentially destabilising circumstances, and therefore should be set so high that it is not expected to be reached in a typical trading period.

Box 14.4 Opportunities for international linking

Given Australia's close economic links with New Zealand, and common interests on greenhouse gas mitigation, linking or even deeper integration may make sense if the New Zealand scheme is judged to be of sufficient integrity. The Review suggests that, before the indelible conclusion of scheme design in either country, the Australian and New Zealand governments meet at ministerial level to discuss linking, and to identify any impediments to linking that may warrant adjustment to one or other or both scheme designs.

Similarly, scheme design development in Japan will proceed over the next few years, and high-level consultations should take place to ensure that there are no unnecessary impediments to productive interaction.

Proposals for phase 3 (post-2012) of the EU emissions trading scheme appear well designed. Australia should explore the possibility of trading with the EU scheme, although EU views on excluding forestry and agriculture from its scheme may be a problem for two-way linking in the early stages. Australia should seek, at a minimum, agreement with the European Union to accept EU permits into the Australian emissions trading scheme, thus making the EU permit price an effective ceiling price for the Australian market.

Box 14.4 Opportunities for international linking *(continued)*

Building a regional market that encompasses (in the first instance) Papua New Guinea, other south-west Pacific developing countries, and—with greater difficulty and in the context of involvement by other developed countries—Indonesia, would also be desirable. Papua New Guinea and Indonesia have large opportunities to reduce land-use change and forestry emissions and to quickly replace coal (Indonesia) and petroleum with low-emissions fuels. To be fully engaged, these countries would need to accept national emissions trajectories, which would be set on a different basis (that is, for developing countries) than Australian trajectories. Australia should be prepared to work with these countries within the international framework and, if necessary, outside it, to accelerate progress on mitigation, and to demonstrate new modes of cooperating with developing countries.

Permit trading by and with governments

The Commonwealth Government could always trade directly with other governments and firms in other countries. This could be necessary in order to balance the actual emissions trajectory against Australia's national commitments under an international treaty. Such divergence could occur, for example, if Australia's international obligations were to change before its scheme's trajectory changed or if domestic emitters chose to hoard or lend permits.

Trading through government gateways may also be necessary in purchasing permits from countries that take on recognised national targets but do not have a domestic emissions trading scheme in place. The transition economies are currently in this category, and other developed countries may also decide not to implement emissions trading schemes domestically. Similarly, developing countries would be expected to be sellers of permits but are unlikely to have developed national emissions trading schemes.

Linking with offset markets

Offset credits arise when emissions are reduced in a country or sector not subject to an emissions limit. Under the Kyoto Protocol, international offsets can be created as certified emissions reductions under the Clean Development Mechanism. Linking with international offsets raises different issues to the acceptance of international permits. This is because of the inherent flaws in the design of offsets (see section 10.4.3). One of the objectives of the post-2012 agreement should be a much smaller role for international offsets, with countries moving instead to national targets, which are in many instances one-sided. To encourage participation by low-income developing countries that do not yet have targets, provision should be made for international offsets, but with restrictions on the source and quantity of offset credits that can be used under the Australian scheme. If the role of the Clean Development Mechanism is substantially changed or expanded after 2012, a re-evaluation would be needed of international linking in general, both to offsets and to permit markets.

The European Union has limits on the extent of the Clean Development Mechanism for use in its emissions trading scheme, expressed in terms of a share of expected reduction effort (European Commission 2008).

The Review supports a limit on international offsets fixed as a proportion of Australian permits. This would provide greater investor confidence and simplicity.

It is simplest to enforce the limit on acquittal of international offsets in a centralised way, through the regulatory authority. The authority would auction a limited number of supplementary permits, each of which would allow the holder to acquit one Clean Development Mechanism credit. Once attained, these international offset credits could be traded and used as other permits in fulfilment of obligations under the scheme. The market price of permits to acquit a certified emissions reduction would reflect the expected differential between the price paid for certified emissions reductions in the international market and the domestic permit price in the emissions trading scheme.

14.5 Addressing the distortion faced by trade-exposed, emissions-intensive industries

14.5.1 A dreadful problem

A potential distortion arises if an Australian emissions trading scheme is introduced in the absence of, and until such time that there is, an international arrangement that results in similar carbon constraints or carbon pricing among major trade competitors. If firms in the traded sector were subject to a higher emissions price in Australia than in other countries (which, as price takers, they were unable to pass through), there could be sufficient reason for emissions-intensive activity to relocate, in part or in whole, from Australia to countries with fewer constraints on emissions. This could result in carbon leakage.[13]

The concern arising out of differences in carbon constraints from those applying in our trade competitors is not that some Australian firms may reduce their level of production. Rather, the concern is that some firms may reduce their level of production too far—that is, beyond the level that would eventuate if competitor countries were subject to commensurate carbon constraints (see Box 14.5). This loss in productive capacity may not be reversible at a later stage when a carbon-inclusive world price eventuates in the relevant commodity and goods markets. In addition, new investment in trade-exposed, emissions-intensive industries may be stalled even though it may have been viable had all competitor countries adopted policies consistent with those in Australia.

Therefore, under certain circumstances, there are environmental and economic reasons for establishing special arrangements for emissions-intensive industries that are trade-exposed. However, the choice of options available to countries prepared to act to reduce their emissions ahead of a comprehensive global agreement is dreadful.

No government will be comfortable about subjecting its traded sector to an additional impost on inputs when its trade competitors are not willing to take corresponding policy measures. However, every other alternative facing policy makers means either heavily compromising a national commitment to reduce emissions or increasing the burden on other sectors (non-traded)—most notably, and ultimately, domestic households.

The inevitable consequences of such decisions about burden sharing (including the environment's share) is that the domestic discourse ahead of implementing an emissions trading scheme quickly degenerates into loud professions of support but even louder pleadings for special treatment.

These are dreadful problems for every nation's emissions trading scheme in the absence of a global arrangement. Indeed, the dilemma created for individual governments is so great that it has the capacity to destabilise public support and pervert individual domestic schemes to the point of non-viability. The sum consequence of the compromising of individual schemes could leave the world with little chance of avoiding dangerous climate change.

In the era of global trade, it takes only a handful of non-compliant countries—large or small, developed or developing, high or low emitters—to drive all other countries to implement policies that significantly compromise the overall objective of reducing emissions.

However, there are options that may avoid this destabilisation and descent into ineffective global action before a binding, comprehensive international agreement can be reached.

14.5.2 Taking a multitrack approach

Australia may well have more to lose than any other country from an internationally fractured and partial approach to dealing with trade-exposed, emissions-intensive industries. The immediacy of this problem means that Australia must simultaneously pursue three options for solving this problem. This is not an either/or choice. Two of the options rely on international agreements while the third is a domestic arrangement that could pave the way for an international approach. In order of preference, these options are:

- a comprehensive global agreement on mitigation under which all major emitters have national emissions limits (see chapters 9 and 10)

- effective sectoral climate change agreements for trade-exposed, emissions-intensive industries, placing particular industries on a more-or-less level playing field. These agreements may require backing by a World Trade Organization agreement on border adjustments (see section 10.6),[14] and, as a last resort
- domestic assistance measures for our most exposed industries that address the failure of our global competitors to act on limiting their carbon emissions.

Alongside the negotiation of a global agreement, the negotiation of sectoral agreements in priority areas for Australia (including metals, liquefied natural gas, cement, and sheep and cattle products) must be an urgent international policy priority for the Commonwealth Government. Nevertheless, despite their importance, global and sectoral agreements will not be in effect in 2010 when the Australian emissions trading scheme begins operating. The Review judges that, given effective Australian leadership and diplomatic commitment, there are reasonable prospects for international sectoral agreements for carbon pricing to be in place by the end of 2012, at least for some of the resource-based industries in relation to which the Australian economy would be at greatest risk.

In the meantime, a domestic transitional assistance arrangement for Australia's most exposed industries is required.

14.5.3 Getting it right from the outset

In recent public debate and commentary, it has been apparent that industries will seek to influence the design of any such assistance arrangements in ways that maximise their respective returns from the scheme. This is to be expected. It also signals the scale of the challenge faced by policy makers in not becoming distracted by vocal and well organised interests.

Unless government takes a principled policy approach to tackling this dreadful problem from the outset, it has the potential to undermine the efficiency and effectiveness of the emissions trading scheme and with it, Australia's commitment to reducing greenhouse gas emissions. There is a view that we do not necessarily have to get it absolutely right from day one; that we will have other chances to deal with the dreadful problem as time goes by until the rest of the world adopts similar policies. This is a myopic view.

We might not have to get it absolutely right. But if we get it wrong, we will have heavily, maybe permanently, compromised our ability ever to find our way back on to a sound path.

What does 'wrong' mean in this context? It comes down to how we view our domestic policy within the context of international efforts to deal with climate change.

All the models currently in the debate about the shielding or compensation of our trade-exposed, emissions-intensive industries, despite paying lip service to the contrary, are all predicated on beliefs that:

- nothing is happening elsewhere in the world
- nothing will happen elsewhere in the world
- nothing done by Australia makes any difference to what happens elsewhere in the world.

These three views are simply wrong. Australian policy is not being made in a world of stagnant attitudes towards climate change policy. It is a highly dynamic environment in which every country is closely monitoring the policies and actions of every other country.

Countries around the world are acting to limit their emissions and are likely to do so with increasing ambition. This is already having an effect on global markets for goods and services that are emissions intensive. This is favouring, for example, Australian exporters of natural gas and potential exporters of coal-seam gas, whose markets relative to other fossil fuels are stronger than they were a few years ago. It is favouring Australian exporters of aluminium, some of whose overseas competitors are facing higher costs as a result of other governments' measures to constrain energy use and emissions.

The Europeans, Americans, Chinese, and Japanese, among others, are all watching Australia with acute interest to see how we handle the treatment of our trade-exposed, emissions-intensive industries. If we get this wrong, it will give every country on earth another excuse also to get it wrong. In aggregate, if Australia and others get it wrong, the global outcome will be a shambles with greatly reduced scope for emissions reductions and the potential for serious damage to the global trading system.

China, the United States, the European Union, Japan—the big economies—may find ways partially and expensively to protect their own industries in a mad scramble for preferment in a world of deep and differentiated government intervention over the dreadful problem. Middle-sized countries like Australia will find it more difficult. But if we get it right, then we can help other countries to get it right. Getting it right means shifting the mindset that is currently dictating the policy debate in Australia.

14.5.4 Efficiently designed transitional assistance arrangements in an ad hoc world

It is important that we stop thinking in terms of payments to Australian firms in order to compensate them for the effects of the domestic emissions trading scheme. There is no basis for compensation arising from the loss of profits or asset values as a result of this new policy. The rationale for payments to trade-exposed, emissions-intensive industries is different and sound. It is to avoid the economic and environmental costs of having firms in these industries contracting more than, and failing to expand as much as, they would in a world in which all countries were applying carbon constraints involving similar costs to ours.

There is a strong case to be made on the basis of transitional arrangements that are based on efficiency in international resource allocation. There is a clear distinction between compensation and payments to correct for distortions in the efficiency with which resources are used. Providing assistance to address the failure of our global competitors to act on limiting their carbon emissions is not the same as compensating domestic firms for the government's decision to implement a domestic emissions trading scheme. A constructive and efficient solution must focus on policy design that assists our domestic industries to address the failure of our global competitors to act on limiting their carbon emissions.

Despite this being a complex problem, the correct response is based on the following policy prescription:

> *For every unit of production, eligible firms receive a credit against their permit obligations equivalent to the expected uplift in world product prices that would eventuate if our trading competitors had policies similar to our own.*

It is simple. It ensures that firms are encouraged to produce at levels that are sustainable in the context of a global agreement, but they are not required to bear the full cost of doing so on their own until such time as there is an agreement.

It rewards firms that might be described as early movers but does not penalise other producers. It encourages firms to invest in new low-emissions production processes rather than rewarding those who are most successful in their lobbying efforts. Unlike the input-based compensation arrangements currently dominating the debate, this approach fully accounts for the policies of our trading competitors. In this sense it is self-correcting. As long as other trade competitors do not impose carbon constraints, payments continue in full. (See Box 14.5).

As trading competitors adopt emissions reduction policies such as ours, observed world product prices will increasingly reflect their true carbon-inclusive value. As this occurs, the gap between the two will narrow and payments will decline without recourse to a political process. The assistance will simply become redundant once a global agreement is in place. Sectoral agreements for particular products will also remove the need for payments to firms in covered industries.

This formulation for calculating payments ensures non-distorted price signals for Australian businesses from the outset of the scheme. Firms will face incentives that accurately reflect those that will eventuate with global or sectoral agreements in place. Australian-based businesses will only reduce domestic production if that is consistent with the long-term loss of comparative advantage in a world of carbon-inclusive pricing.

Box 14.5 The economics of trade-exposed, emissions-intensive industries and proposed assistance arrangements

Firms will seek to produce the level of goods or services that maximises their profits, although in the short term they might deviate from this objective in order to gain or maintain market share. Some factors of production will be relatively fixed in the short term—namely, the firm's fixed capital stock such as plant and machinery. Where these firms compete in global product markets they are assumed to be 'price-takers'. Each firm's level of production has no bearing on the world price of the relevant product.

These descriptions of a trade-exposed, emissions-intensive firm can be usefully represented graphically with an upward sloping (marginal) cost curve (C_0) and a flat price curve set at the world price (P_0) (Figure 14.5a). The firm's resultant profit maximising level of production is given by q_0.

The imposition of a carbon price increases production costs for all levels of production to the extent that firms employ emissions-intensive (direct and indirect) production processes. This shifts the cost curve to the left (C_1) but has no bearing on the world product price (which remains at P_0). In response, profit maximising firms will reduce their level of production to q_1.

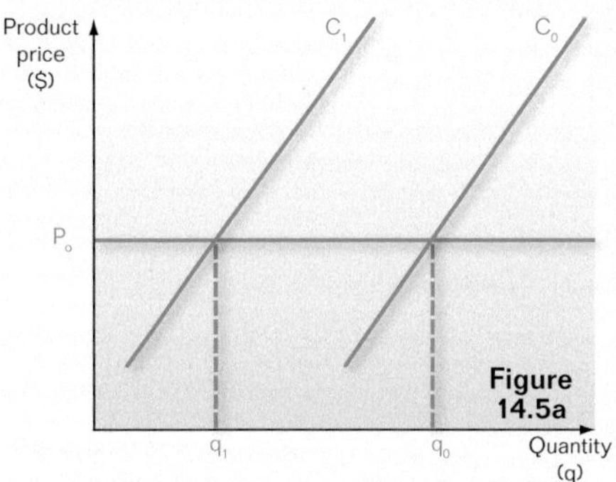

Figure 14.5a

Eventually, as more and more countries adopt a carbon pricing regime, the world price of the relevant commodity, good or service will increase to P_n and domestic firms would produce at a level q_n (Figure 14.5b). The overshooting problem is demonstrated graphically by the difference in production levels between q_n and q_1. That is, a domestic carbon price in the absence of similar schemes elsewhere will see production drop to q_1 only to increase eventually to q_n in the longer term.

Note: q_n may be greater, equal or less than q_1 depending on relative movements in domestic production costs (from C_0 to C_1) and the carbon-inclusive world product price (P_0 to P_n).

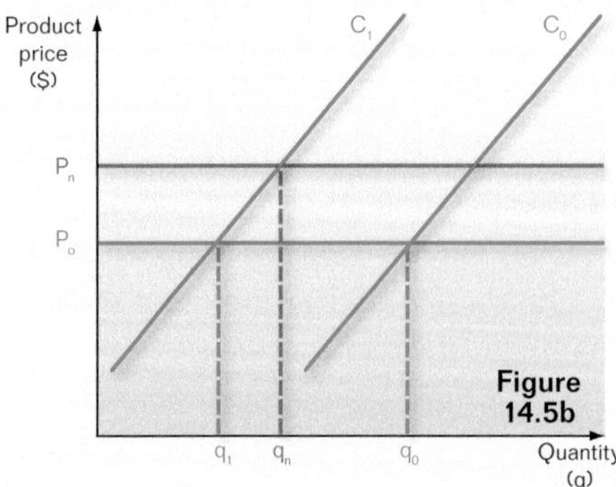

Figure 14.5b

Assistance for trade-exposed, emissions-intensive industries would most efficiently support the level of production that would be sustainable in the long run (q_n) rather than allowing production to overshoot (q_1) or trying to maintain an unsustainable status quo (q_0) (Figure 14.5c). Such assistance entails payments (or credits) to firms equal to the long-run price uplift (P_n – P_0) for each unit of production. This assistance has the effect of increasing production to Q_n. The total payment to the firm is shaded in orange in Figure 14.5c.

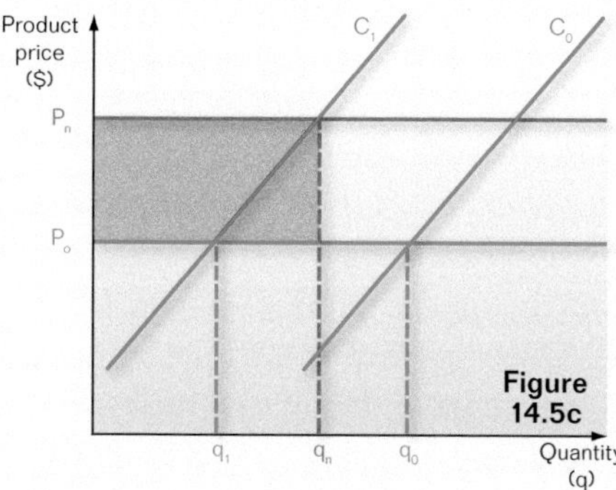

Figure 14.5c

If payments (or free permits) were to be made to compensate for higher input costs imposed by the emissions trading scheme due to a firm's direct and indirect emissions, the result would be an excess payment to the firm shown by the excess of the blue shaded area over the orange shaded area (Figure 14.5d). These excess payments would be supporting unsustainable levels of production at the expense of investment in R&D in new technologies or support for low-income households.

Moreover, input-based compensatory payments will typically reward inefficiency rather than promoting investment in new low-emissions production processes (Figure 14.5e), in which the effect of an emissions trading scheme on two firms in the same industry is shown. One firm has low emissions intensity and the other has high emissions intensity, represented by carbon inclusive cost curves C_L and C_H, respectively.

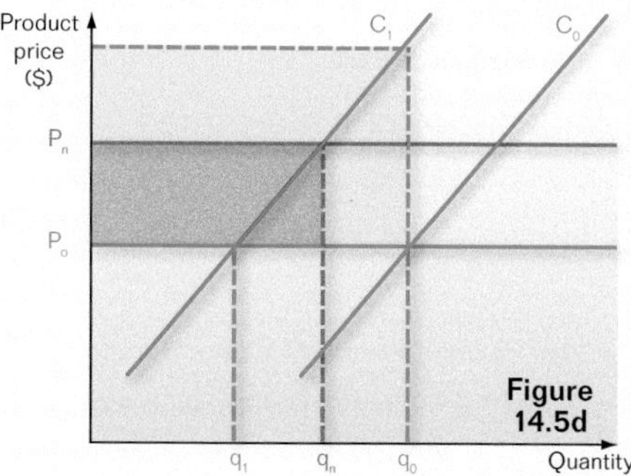

Figure 14.5d

Under an input-based compensatory payment, the high-emissions firm would be rewarded with greater payments (the area defined by AENV) than the low-emissions firm (AEMK). This is not the case under the mechanism outlined in section 14.5, which penalises high-emissions firms with lower payments (DEFG rather than BEFJ).

This analysis also demonstrates that the mechanism proposed by the Review rewards investment in new low-emissions production processes, whereas an input-based approach penalises such investment. That is, a high-emissions firm investing in new capital to reduce its emissions intensity (moving from C_H to C_L) receives an additional payment under the proposed mechanism (represented by the area BDGJ). Under an input-based compensation scheme, the firm would be penalised with a loss of payments (KMNV) following the investment.

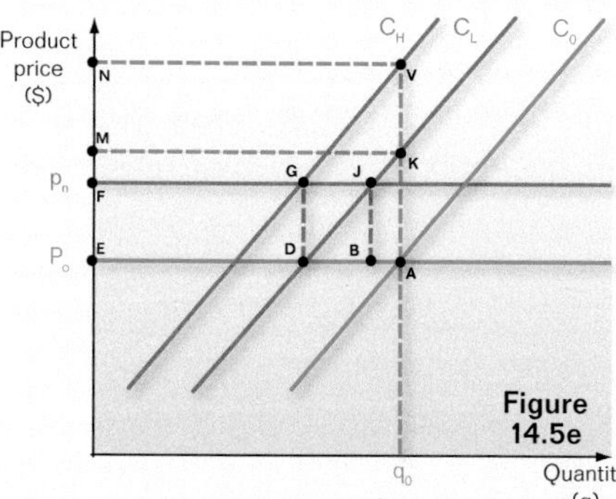

Figure 14.5e

As with the administration of customs and taxation, the size of potential revenues from the sale of permits suggests strong arguments for delegating administrative judgments to an independent entity. An independent authority will need to be established with the necessary skills to develop carbon-sensitive price models for relevant product markets.

In an open economy like Australia, there are very few non-traded products. Almost all businesses face some degree of trade exposure. It would be nonsensical to think that the above formulation could or should be applied to every product, particularly as in most instances the impact on global product prices from an international carbon price will be negligible. Once general equilibrium effects on the exchange rate are taken into account, the negative effect on trade-exposed industries will be smaller than partial analysis would suggest and some firms will actually receive net benefits.

An eligibility threshold is required. It is defined most appropriately in terms of the expected uplift in the unit price (in percentage terms) in the given compliance period such that:

- only products that are expected to increase in price by a percentage in excess of a low threshold would attract credits under the scheme
- eligible producers would receive credits for that part of the expected price uplift that is in excess of the threshold
- an eligible firm could not receive credits in excess of its permit obligations in any compliance period.

This is best demonstrated with a numerical example in which the threshold ratio has been set at, say, 3 per cent and the observed world price for a particular product is $1000. Following the introduction of the Australian emissions trading scheme, the independent authority forecasts that if a corresponding global carbon-pricing regime were in place, then the world price would rise by $90 per unit (to $1090). The firm would be eligible for credits for each unit produced in Australia because the projected price rise, at 9 per cent, is in excess of the 3 per cent eligibility threshold. The firm would receive credits against its permit obligation valued at $60 per unit of domestic production.

Special arrangements may be required, as a safeguard, where the estimated price uplift factor exceeds a relatively high proportion of gross value added per unit of production.[15] This is intended as an exceptional rather than a standard design feature.

Credits could be provided in the form of cash or as cash-equivalent reduction in a liable party's obligation to acquit permits at the end of the relevant compliance period.

The inherent arbitrariness of the eligibility threshold will make it the subject of intense scrutiny and lobbying by special interests. Yet there will be no counterbalancing representation arguing against the lowering of the thresholds—

even though doing so inevitably shifts the burden onto others in the economy and community. Policy makers will stand alone in having to resist the temptation to appease narrow interests.

The calculation of expected price uplift factors, the frequency and timing of credits being distributed to eligible firms, and the relevant accounting rules should all operate to ensure minimum disruption and maximum certainty. Expected price uplift factors would be produced by the independent authority at regular intervals (at a minimum, yearly) through a transparent and consultative process. The process used by the Productivity Commission in inquiries on protection issues is a suitable model.

Bedding down the appropriate institutional arrangements, methodologies and operating rules should be a matter of priority. As the same arrangements would be of benefit to every country tackling the dreadful problem in regard to its own trade-exposed, emissions intensive firms, Australia should, over time, look to form international institutions to provide global price indexes for use in calculating credit arrangements for eligible businesses in all countries. This would ensure globally efficient outcomes.

14.5.5 General business tax cuts versus special assistance

Materiality tests and algorithms for assistance payments will to some extent be arbitrary and will impose some additional costs on business that, in principle, would warrant assistance. Policy makers can be expected to encounter extreme pressure to err towards generosity in the setting of thresholds. This poses risks to the credibility of the entire emissions trading scheme.

If the eligibility threshold is set too low, the burden of emissions reductions will be shifted elsewhere in the economy.

This is not to say that firms that do not meet the eligibility threshold would not encounter some disadvantage from the introduction of the emissions trading scheme. However, government simply cannot efficiently administer, and the community cannot afford, a firm-by-firm assistance scheme that effectively addresses this impact at such an atomistic level of application.

If this were considered to be a material issue in development of the arrangements for trade-exposed, emissions-intensive business activities, it would be better to introduce an explicit element of efficiency-raising reduction of business taxation than to pare the thresholds to administratively difficult levels.

It is expected that direct assistance to trade-exposed, emissions-intensive industries, proposed above, would be significantly below 30 per cent of total permit revenue (though this would be tested as data becomes available to the regulatory authority). The government could notionally return to business taxpayers in an efficient way an amount of permit revenue roughly corresponding to the gap between 30 per cent and the amount actually paid as permit credits to trade-exposed, emissions-intensive firms.

While this approach sacrifices the precision of targeted assistance schemes, well-designed tax relief has the potential to reduce rather than increase the deadweight loss incurred by the economy. Obvious candidates for reduction or abolition would be input-based or transaction taxes, which are highly inefficient. Any such tax cuts should be ongoing not transitional.

14.6 Transition period: Australia's emissions trading scheme to the end of 2012

In the remainder of the Kyoto period, ending in 2012, Australia should ensure that it meets its Kyoto targets. It should have no great difficulty in doing so—and any adverse surprise over the next few years is unlikely to be so large that it cannot comfortably be met by the purchase of international permits.

However, the remainder of the Kyoto period presents challenges for the implementation of a domestic emissions trading scheme.

The timing of a global agreement to replace the Kyoto Protocol remains unclear. It is not even certain that the full details of such an agreement will be resolved ahead of the commencement of Australia's emissions trading scheme in 2010. There would be considerable benefit in avoiding the unproductive interaction between the early period of a new trading system and Australia's participation in crucial global negotiations. Otherwise, this period will be one in which every new development in the international negotiations, encouraging or adverse, could have a disproportionate and unhelpful effect on the domestic permit price in an unconstrained market.

Further, there is a chance that if Australia was to overperform on its Kyoto targets, the price of permits could be zero or close to zero. Allowing permits to be hoarded for later use may help maintain a positive price, but it is far from clear that market participants would do so if significant uncertainty remains about a future international agreement.[17]

A period of derisorily low prices would be damaging for the credibility of the scheme in its formative years. Despite this being the likely outcome, business has continued to express concern about high permit prices in an unconstrained scheme.

The remainder of the Kyoto period is best considered as a transition period in which the emissions trading scheme is established soundly for the larger task that will lie ahead of it after 2012. The Review considers the most appropriate approach to the transition period will be to fix the permit price in order to address both the likelihood of very low prices and fears of the opposite outcome.

During the transition period, permits should be sold by the independent regulatory authority at $20 per tonne in 2010 (in 2005 prices), rising each year by 4 per cent plus the percentage increase of the consumer price index (see section 12.1). This is more or less the price path that the modelling suggests would be followed if there were effective global agreement directed towards stabilisation of global greenhouse gas concentrations at 550 ppm. If a post-Kyoto agreement is struck around 550 ppm, the fixed price is likely to allow relatively seamless transition to a floating price regime.

This will have implications for the implementation and management of the scheme before and during the transition period.

- It will be necessary to separate the markets for permits before and after the end of 2012. In particular, it will be necessary to prevent the hoarding of permits acquired in the period of price controls, for use after 2012.
- It will preclude the opportunity for two-way trade of permits, as no foreign system will accept Australian permits and, on that basis, may even prevent their emissions entitlements from being sold into the Australian scheme.
- It will need to be accompanied by sale of post-2012 permits from the time that full details of the scheme had been articulated. This will be important to provide guidance on post-2012 market pricing, and to establish the credibility of post 2012 arrangements.

Fixing the price of permits will also provide a less anxious environment for implementing the assistance arrangements for trade-exposed industries proposed in section 14.5. It may even obviate the need for assistance payments during the transition period and allow time for diplomacy to work towards establishing satisfactory global or sectoral agreements.

14.7 Governance: institutional arrangements

Scheme governance has large implications for the efficiency, stability, credibility and simplicity of the scheme. New institutions will be required to operate and regulate the emissions trading scheme.

Some of the governance functions related to the scheme are, by their nature, the prerogative of government. These include decisions about establishing the scheme—setting the emissions limit and providing assistance to those whose incomes are reduced by the introduction of the scheme (for example, structural adjustment assistance and payments to low-income households). Government will undertake policy functions as distinct from an administrative role.

Legislation of key features of scheme design, such as the permit release trajectory, can assist stability—particularly in the Australian situation of qualified government control of the legislative process.

The administrative content of several of the governance functions is of a kind that lends itself to independent administration, particularly due to the large amounts of money associated with administrative decisions (for example, payments to trade-exposed, emissions-intensive industries). Government would be under pressure from particular interests to favour them in administrative decisions. There is an advantage in delegating administrative judgments to an independent entity.

The Review suggests that the administration of the emissions trading scheme be made the responsibility of an independent authority. An independent carbon bank should be established with a high degree of executive independence in the exercise of its powers. The closest analogue is the Reserve Bank of Australia. As with the Reserve Bank, the powers of the independent authority would be defined by legislation and by agreement with government. This same legislation would define the way in which government would exercise its policy responsibilities in relation to the scheme and the obligations of private parties in relation to emissions and the need for permits.

The overarching objective of the independent carbon bank should be the maintenance of the long term stability of the Australian emissions trading scheme. It must be given the necessary powers and sufficient discretion in order to fulfil this duty.

The distinctive roles of government and of the proposed independent carbon bank are summarised in Table 14.1.

Table 14.1 Governance of an Australian emissions trading scheme

Functions of scheme governance	Government responsibilities: Policy making	Independent carbon bank: Administration and regulation
Emissions trading scheme rules	All, including coverage, point of obligation, acquittal, offset rules and standards (domestic and foreign),[a] accounting rules, compliance regime. Undertake reviews of scheme rules. First review to occur two years after commencement.	Advise government on rules and their application. Manage monitoring, reporting and verification systems. Enforce compliance.
Setting emissions limit	Decide/announce: • initial target and trajectory • conditions for changing target and trajectory • nature, extent and timing (with five years' notice) of changes to the target and trajectory.	Administer movement from one emissions trajectory to another after government has certified that the conditions of change have been met.
Permit issuance Use of revenue from permit sales	Determine manner of permit issuance and setting fixed prices for permits during the transitional period (2010–12). Receive revenue for general allocation under formula.[b] Decide on the use of permit revenue.	Sell permits in line with fixed prices (2010–12) or emissions trajectories established by government. Purchase permits abroad as required to reconcile domestic emissions in particular years with international agreements or to provide for the honouring of the five-year forward commitment after a change in trajectory.
Trade-exposed, emissions-intensive industries	Negotiate global agreements and encourage effective sectoral agreements. Set rules for assistance payments and determine eligibility thresholds.	Calculate price uplift factors and assess eligibility. Make paymentsaccording to agreed thresholds and formula.

Table 14.1 Governance of an Australian emissions trading scheme *(continued)*

Functions of scheme governance	Government responsibilities: Policy making	Independent carbon bank: Administration and regulation
Hoarding and lending	Set broad policy on hoarding and lending, market participation	Make decisions on lending and interest rates.
Market supervision		Monitor integrity of the market and supervision of transactions. Monitor the creditworthiness of borrowers, and more generally the relationship between hoarding and lending and the stability of the market. Undertake stabilisation interventions when required.
Trade rules (international linking)	Establish international trade agreement and rules for international linking.	Certify that conditions have been met for linking. Monitor international trade by market participants. Purchase international permits to reconcile domestic and international obligations (for example, to meet a 2020 target). Undertake stabilisation interventions when required.

a There should be independent, expert review, amendment and approval of offset protocols and offset projects.

b Revenue would come from sale of permits, interest on loans of permits and profits from stabilisation interventions (losses would be a sign of counterproductive intervention and to be accounted transparently). The formula would leave the independent carbon bank sufficient income to cover the costs of its overhead, plus monitoring and enforcement of the system, including international permit purchases that are necessary to meet international obligations.

14.8 Addressing the relationships between an emissions trading scheme and other policies

Largely for historical reasons, a variety of policies have been put in place by the Commonwealth, State and Territory governments over the last decade with the aim of reducing greenhouse gas emissions. With the advent of a broad-based emissions trading scheme, other emissions reduction policies become largely redundant. The only useful role for additional policies of this kind is to reduce the effect of market failures, so as to reduce the cost of adjustment to the low-emissions economy.

14.8.1 The Mandatory Renewable Energy Target and the emissions trading scheme

The Mandatory Renewable Energy Target (MRET) was introduced in 2000 to drive the take-up of renewable energy and reduce emissions. MRET operates by placing an obligation on energy retailers to purchase a proportion of their energy from renewable sources in the form of renewable energy certificates. The value of a certificate is equal to the difference in the cost of producing the renewable energy and the average wholesale price of electricity. By design, MRET causes deployment of the lowest-cost eligible technologies. To date, the increase in renewable energy production has been mostly wind and solar hot water. As renewable energy production is currently more expensive than alternative sources, this higher cost is passed on by retailers to households and businesses.

An important design feature of MRET is the shortfall penalty of $40/MWh, which operates as a cap on the certificate price. The shortfall penalty is not indexed. To date, the price cap has been largely untested because, although the price of renewable energy has been increasing, the average wholesale price of electricity has been increasing at a similar rate.

MRET is set to expand from the current 9500 gigawatt hours to 45 000 gigawatt hours (around 20 per cent of energy demand) by 2020. The expanded MRET will drive increasingly expensive options for the deployment of currently favoured technologies (for example, building wind farms in more remote areas) as well as the deployment of newer and more expensive technologies (such as geothermal and solar photovoltaic). This will lead to a higher renewable energy certificate price and higher electricity prices for consumers. Conversely, the recent uplift in world energy prices (coal and gas) raises the average wholesale price of electricity and puts downward pressure on the certificate price.

The emissions trading scheme differs in objective and operation from MRET, in that it caps the level of emissions and is neutral as to how the necessary reductions will be achieved. In some cases, the emissions cap may lead to investment in renewable energy production, but in other cases it may lead to fuel switching or the deployment of more efficient operating practices among existing energy producers or other sectors of the economy. The market, rather than government, is left to find the solution. A competitive market can be expected to deliver these emissions reductions at the lowest cost to the community and business.

Implementing the expanded MRET alongside the emissions trading scheme means that these two policy instruments, with their differing objectives, will be interacting in the electricity market. This clash of objectives will potentially be detrimental to electricity users (households and businesses) and electricity producers (incumbent and new providers). Many factors will affect the extent of this adverse interaction. The most notable is the trajectory of the emissions trading scheme and the ramp-up rate of MRET (see Table 14.2). Both schemes must, by force of law, meet their mandated targets. The concerns are threefold: What is the cost? Who will bear the cost? What are the long-term consequences?

Table 14.2 Interaction between the emissions trading scheme and the Mandatory Renewable Energy Target

Emissions trading scheme trajectory	MRET ramp-up rate: Gentle	MRET ramp-up rate: Aggressive
Gentle	• Low permit price • Moderate renewable energy certificate price • Moderate impact on retail electricity prices • Mitigation activity outside MRET unlikely	• MRET cannibalises emissions trading scheme • Very low (even zero) permit price • Emissions trading scheme becomes non-functional • High renewable energy certificate price • High impact on electricity prices • Little mitigation activity outside MRET • No incentive for investment in other low-emissions technologies
Aggressive	• Permit price steadily increases over time • As wholesale electricity prices rise, renewable energy certificate prices decline, possibly to zero—implying early phase-out of the MRET • Moderate-to-high impact on retail electricity prices—depending on level of mitigation elsewhere in the economy • Investment in portfolio of renewable and other low-emissions technologies	• Permit and certificate price paths would be highly dependent on interaction of the two schemes • Prices could be range from high to very low • MRET most likely to cannibalise emissions trading scheme • High impact on retail electricity prices • Most investment likely to be in increasingly expensive renewable energy options

On balance, the Review considers that carrying forward the existing, non-indexed shortfall penalty of \$40/MWh into the expanded MRET provides the best opportunity for a smooth transition from MRET into the broader emissions trading scheme. The units of account are different for MRET and the emissions trading scheme, but it happens that \$40/kWh in MRET under current conditions roughly corresponds to \$40 per tonne of CO_2-e in the emissions trading scheme. Since the shortfall penalty is a feature of the current MRET, its retention would seem to be fully consistent with the government's commitment. As the price of permits increases above \$40–45 per tonne of CO_2-e, the emissions trading scheme would come to dominate investment decisions and the economic effects of MRET would be subsumed within the emissions trading scheme.[18]

Maintaining the shortfall penalty would place an upper limit on MRET's higher costs relative to the emissions trading scheme, and on MRET's contribution to electricity prices, while maintaining the incentive for investment in renewable energy that can be delivered below this level. Furthermore, as the emissions trading scheme took over from MRET, some of the rents previously accruing to investors in renewable energy (who will now be competitive with other forms of carbon-intensive energy) would shift to government and could be used to support research, development and commercialisation of newer technologies (see Chapter 18).

Modelling will be important to provide some indication of the likely impacts of MRET on the permit prices and mitigation occurring under the emissions trading scheme. It will be critical that the interactions between MRET and the emissions trading scheme are fully understood when the parameters of the scheme are being finalised (see Box 14.6).

Box 14.6 Expanded MRET

The Garnaut–Treasury modelling did not include the Commonwealth Government's commitment to an expanded MRET; however, a brief high-level assessment of the implications of the scheme is instructive.

By 2020, under the 550 ppm scenario in which the emissions trading scheme is the primary vehicle for delivering lower emissions, electricity generation would be 257TWh—of which 23 per cent would be gas fired and 12 per cent from renewable sources (including hydro).

The achievement of the 20 per cent expanded MRET would require an additional 21TWh of renewable energy. Technology forecasts from the Review's 550 ppm scenario suggest that this demand would be filled largely by wind-based generation (representing an additional 8000MW of installed wind capacity). If this capacity were to replace gas-fired power generation as a result of the MRET, as seems reasonable to assume, the additional cost would be around $750 million to $1.1 billion per annum by 2020.[19]

These estimates compare with estimates by CRA International (2007) of a GDP impact of $1.5 billion. The higher estimate by CRA largely arises from its assumed electricity demand of 377TWh by 2020, which is about 50 per cent higher than assumed demand in the Garnaut–Treasury model (257TWh).

There is an interesting and seemingly perverse consequence of expanding MRET at the same time as the emissions trading scheme is to be implemented. Having both schemes operating side by side could see an increase in coal-fired power generation (by more than 2000MW) as gas-fired plants are crowded out by MRET. This would not occur if the emissions trading scheme were operating without MRET.

14.8.2 Greenhouse Gas Reduction Scheme

The New South Wales Greenhouse Gas Reduction Scheme (GGAS) is one of the world's first mandatory greenhouse gas emissions trading schemes, originally designed to run until 2012.

Both the emissions trading scheme and GGAS cause a price to be applied to greenhouse gas emissions associated with energy consumption. It is not efficient or appropriate to have multiple emissions price signals. Therefore, NSW legislation provides that GGAS will cease to operate upon commencement of the emissions trading scheme (NSW Department of Water and Energy 2008).

There are several issues to consider in ensuring a smooth transition from GGAS to the emissions trading scheme, including:

- Treatment of accredited abatement providers—if emissions reduction projects under GGAS were not reaccredited at all under the emissions trading scheme, or they were reaccredited but scheme permit prices were lower than certificate prices under GGAS, this could reduce the income stream and project value.
- Forestry carbon sequestration projects—reaccreditation under the emissions trading scheme would be necessary, and depend on rules developed for the inclusion of forestry in the emissions trading scheme.
- Unused GGAS certificates, which may be held by existing providers, intermediaries or parties with an obligation—transition arrangements should not provide an incentive for oversupply of certificates, or holding of them in expectation of a higher price under the emissions trading scheme (and non-compliance with GGAS).

14.8.3 The voluntary market for emissions reductions

There is a growing market for individuals, households and businesses wishing to voluntarily purchase credits for greenhouse gas reductions, to offset emissions associated with their activities. Such measures include the purchase of GreenPower, and offset credits from the Commonwealth Government Greenhouse Friendly program.

As the emissions trading scheme develops, both in depth and breadth, it is likely to cannibalise the market for such measures, although the nature and pace of such changes are uncertain.

Voluntary demand for offsets is likely to continue even with an emissions trading scheme. For example, the South Australian Government believes offsets will play a role in meeting its commitment to be carbon neutral by 2020 (Government of South Australia 2008).

Robust standards for voluntary offsets are important. It is likely, and desirable, that the voluntary emissions market will move increasingly toward the compliance market, in terms of standards.

As well as buying domestic offset credits, under an emissions trading scheme those looking to purchase emissions reductions voluntarily may buy and surrender compliance-grade credits, including emissions permits and domestic and international offset credits.

14.9 Summary of design features of an Australian emissions trading scheme

It is very likely that an Australian emissions trading scheme will be established ahead of a comprehensive global agreement to reduce emissions. For now, it remains unclear what can be expected of international negotiations, but this should not delay the introduction of the domestic scheme in 2010.

Table 14.3 provides a summary of the Review's preferred design features for the Australian emissions trading scheme.

Table 14.3 Overview of the proposed emissions trading scheme design

Design decision	Proposal
Transition period	The two or more years between scheme commencement (2010) and the end of the Kyoto period (end 2012) should be treated as a transition period to an unconstrained, fully market-based emissions trading scheme.
Setting an emissions limit	The overall national emissions limit should be expressed as a trajectory of annual emissions targets. A number of trajectories should be specified upon establishment of the scheme. The first, up to 2012, should be based on Australia's Kyoto commitments. The others, for the post-2012 period, should reflect increasing levels of ambition. Movement between them should be based on developments in international negotiations.
Changes to the emissions limit	Movement from one trajectory to another should only occur in response to international policy developments and agreements (which should allow for new information and developments of an economic or scientific kind). Government should provide five years' notice of movement to another trajectory. Any gap between the domestic emissions trajectory and international commitments during this period would be reconciled by purchasing international permits or abatement in non-covered sectors.
Coverage	Gases: The six greenhouse gases as defined by the Kyoto Protocol. Sectors: Stationary energy, industrial processes, fugitives and transport should be covered from the outset. Waste and forestry should be covered as soon as practicable. The inclusion of agriculture should be subject to progress on measurement, administration and cost effectiveness.
Domestic offsets	Domestic offsets should be allowed from uncovered sectors if it is cost effective to do so. Unlimited offset credits for net sequestration should be accepted from forestry (and potentially soil management practices). The appropriateness of an offset regime for agriculture should be analysed further in the context of coverage of these emissions. During the transition period, the purchase of offset credits would be expected to occur only up to the value of the fixed permit price.
Point of obligation	The point of obligation should be set at emissions source when efficient. Otherwise, an upstream or downstream point of obligation should be preferred where transaction costs are lower, accuracy of emissions measurement higher, or coverage greater.
Issuing (or releasing) permits	Permits should be released according to emissions reduction trajectory, with all permits auctioned. Auctions should take place at regular intervals. There should be no limit on the use of permits (that is, no date stamping). During the transition period, permits should be sold at a fixed price as of right. Permits to be released according to demand, rather than in line with the emissions reduction trajectory. Permits for post-2012 should be auctioned as soon as possible.
Trading permits	Unlimited trading of permits should be allowed. Trading of transition period permits is unlikely. Some forward trading of post-2012 permits can be expected.

Table 14.3 Overview of the proposed emissions trading scheme design *(continued)*

Design decision	Proposal
International links	Opportunities for international linking of the Australian scheme should be sought in a judicious and calibrated manner. Government may purchase international permits or offset credits to meet its Kyoto commitment if Australia's national emissions exceed its target.
Price controls	Price controls are not supported except during the transition period to end 2012. During this transition period, permit price should be fixed, starting at $20 per tonne and increasing by 4 per cent per annum plus the rate of inflation.
Inter-temporality (flexibility in time of use of permits)	Unlimited hoarding should be allowed, except for permits issued for the transition period. Official lending of permits by the independent carbon bank to the private sector should be allowed within five-year periods. Hoarding and lending would not be expected within the transition period. Hoarding of fixed price permits not allowed for use beyond the transition period.
Treatment of trade-exposed, emissions-intensive industries	Global and sectoral agreements to achieve comparable treatment of emissions in important competitors should be pursed as a priority. If they have not been reached post-2012, assistance should be provided to account for material distortions arising from major trading competitors not adopting commensurate emissions constraints.
Governance	The emissions limit and policy framework for the scheme should be set directly by government. The scheme should be administered by an independent authority (independent carbon bank).
Compliance and penalty	A penalty should be set as a compliance mechanism. The penalty does not replace the obligation to acquit permits; a make-good provision should apply.

Notes

1 In recent years, debate in Australia has been promoted by (1) the National Emissions Trading Taskforce set up by state and territory governments in 2004 (National Emissions Trading Taskforce 2007); (2) the Task Group on Emissions Trading established by the former Prime Minister in late 2006 (Prime Ministerial Task Group on Emissions Trading 2007); (3) this Review's Emissions Trading Discussion Paper in March 2008 (Garnaut 2008a), its Draft Report in July 2008 (Garnaut 2008b) and its Supplementary Draft Report in September 2008 (Garnaut 2008c); and (4) the Carbon Pollution Reduction Scheme Green Paper released by the current Commonwealth Government in July 2008 (DCC 2008c). Much has also been learnt from experiences with other schemes such as the New South Wales Greenhouse Gas Reduction Scheme, the European Union emissions trading scheme and the movement toward emissions trading in other jurisdictions, including New Zealand, Japan, and parts of the United States and Canada.

2 It will be important to implement measures to drive emissions reductions in the non-covered sectors ahead of their inclusion in the scheme, to ensure that the task of achieving an economy-wide emissions reduction target is not borne solely by sectors covered under the scheme.

3 There are several tests of additionality. For example, regulatory additionality would require emissions mitigation to be undertaken beyond what is undertaken to comply with existing legal or regulatory requirements. Such tests are arbitrary, and potentially a source of distortion. If the tests of additionality are poorly constructed they will contravene the scarcity and credibility principles outlined in section 14.1.2.

4 Further information about obligations under the National Greenhouse and Energy Reporting Act and supporting regulations is available from the Department of Climate Change at <www.greenhouse.gov.au/reporting/> and DCC 2008b.

5 Domestic and civil aviation and sea transport should be included, with trade-exposed, emissions-intensive industry principles applied if appropriate. Bunker fuels, which are used in international aviation and shipping, are not covered by the Kyoto Protocol or included in countries' emissions targets (Article 2.2). The European Union, subject to final agreement, will include emissions from domestic and international aviation—operators of all arriving and departing flights—from 2012 (flights within the European Union are to be covered in 2011) (European Commission 2008). A sectoral agreement between international transport providers, such as a global fuel tax, should be pursued as a priority (see Chapter 10).

6 For instance, options for an emissions baseline could include emissions in a particular base year or years (say, 2008 to 2012); average emissions per unit of production, based on installed technology in a base year; average emissions per unit of production based on best practice technology; other approaches; or any combination of these.

7 See, for example, Evans and Peck (2007) for a discussion of key issues to consider in designing an emissions trading scheme permit auction.

8 During the first two phases of the EU scheme, the majority of allowances were allocated free of charge, including to established fossil fuel–fired electricity generators. Generators have generally passed on to consumers the opportunity cost of permits that they were given free (European Commission 2005; IPA Energy Consulting 2005). Taking into account the demonstrated ability of generators to pass on the notional cost of emissions allowances, the European Commission has recommended that all permits for the power sector be auctioned in the post-2012 arrangements (European Commission 2008).

9 The term 'lending' refers to transactions of permits between the independent authority and the private sector. The term is adopted in preference to 'borrowing' in order to differentiate the Review's proposal from other schemes that involve the free allocation of permits. These schemes have provided eligible parties with a guaranteed future stream of entitlements to free permits (which are date stamped). In such schemes, 'borrowing' allows the private party advanced access to their future permit entitlement—that is, to use permits ahead of their eligible (or 'stamped' use date). The Review rejects such arrangements on the grounds that all permits should be auctioned without restriction on the time of their use.

10 In order for hoarding to occur, there would have to be early and cost-effective mitigation opportunities beyond those set by the emissions reduction trajectory.

11 There is a particular issue in relation to surplus eastern European permits from the Kyoto period. Some argue that the Russian permits and some others should not be purchased because they have not arisen as a result of mitigation effort. Future treaties would not be credible, however, if countries' targets are agreed to at the time of signature, but those countries are not allowed to reap the financial rewards if they exceed them. Pre-2012 purchases of such permits in Australia could be restricted to government, and not opened to the market.

12 Ultimately, the decision to link or not is within the gift of executive government, given the international dimensions. Therefore, certification may involve an additional step in which the independent carbon bank makes a recommendation to the government regarding linking with another scheme.

13 Carbon leakage refers to a situation whereby production moves from Australia to other countries without carbon constraints and potentially with higher emissions intensity production processes. The effect is that overall emissions remain unchanged. They may even increase.

14 It is possible that even with a broad international agreement in place, trade-exposed, emissions-intensive industries in some countries may continue to operate outside of a national emissions limit. As outlined in Chapter 10, the sectoral agreements would ensure that trade-exposed, emissions-intensive industries in countries without national emissions limits would nevertheless face an emissions price comparable to those in countries which have such as limit. The WTO agreement, proposed in section 10.6, would allow countries to impose border adjustments to ensure that competitors in countries with neither national emissions limits nor sectoral agreements do not have an unfair advantage. The WTO agreement would also play the important role of preventing the use (or rather, abuse) of border adjustments as instruments of protectionism.

15 For example, if the ratio of price uplift (in dollar terms) to gross value added per unit of production exceeds a proportion in the order of, say, 10 per cent, then eligible producers could receive credits for that part of the expected price uplift that is in excess of the threshold when the latter is calculated in dollar terms.

16 Detailed advice on tax reductions could be provided within the Henry Tax Review, due to report in December 2009.

17 There have been suggestions that hoarding could be allowed to overcome the risk of a near-zero permit price while a price ceiling could also be set to prevent unacceptably hight prices. As outline in section 14.4.1, a price ceiling is incompatible with hoarding provisions.

18 The assumed equivalence of $40/MWh shortfall penalty and permit prices of $40–45 per tonne is calculated on an assumed average emissions intensity ratio of 1.0 to 0.9, respectively, for the electricity supplied beyond MRET.

19 If the expanded MRET was to be implemented as a minimum of 60 000GWh of renewable energy by 2020, then this would add approximately $300 million per annum to this estimate. This target is identified as a policy outcome in a consultation paper released in July 2008 by the COAG Working Group on Climate Change and Water (COAG 2008).

References

COAG Working Group on Climate Change and Water 2008, 'Design options for the expanded national renewable energy target scheme', consultation paper, July.

Coase, R.H. 1960, 'The problem of social cost', *Journal of Law and Economics* 3: 1–44.

CRA International 2007, 'Implications of a 20 per cent renewable energy target for electricity generation', prepared for Australian Petroleum Production & Exploration Association, November.

DCC (Department of Climate Change) 2008a, *Fugitive Sector Greenhouse Gas Emissions Projections 2007*, Commonwealth of Australia, Canberra.

DCC 2008b, *National Greenhouse and Energy Reporting System*, Regulations Policy Paper, Commonwealth of Australia, Canberra.

DCC 2008c, *Carbon Pollution Reduction Scheme Green Paper*, Commonwealth of Australia, Canberra.

European Commission 2005, 'Review of the EU Emissions Trading Scheme—survey highlights', <http://ec.europa.eu/environment/climat/emission/review_en.htm>.

European Commission 2008, 'Questions and answers on the Commission's proposal to revise the EU Emissions Trading System', MEMO/08/35, Brussels, 23 January.

Evans & Peck 2007, *Further Definition of the Auction Proposals in the NETT Discussion Paper*, report to the National Emissions Trading Taskforce, <www.emissionstrading.net.au>.

Garnaut, R. 2008a, *Emissions Trading Discussion Paper*, <http://garnautreview.org.au>, March.

Garnaut, R. 2008b, *Draft Report*, <http://garnautreview.org.au>, July.

Garnaut, R. 2008c, *Supplementary Draft Report*, <http://garnautreview.org.au>, September.

Government of South Australia 2008, submission to the Garnaut Climate Change Review, <http://garnautreview.org.au>.

IPA Energy Consulting 2005, *Implications of the EU Emissions Trading Scheme for the UK Power Generation Sector*, report to the UK Department of Trade and Industry, 11 November, <www.berr.gov.uk/files/file33199.pdf>.

National Emissions Trading Taskforce 2007, 'Possible design for a national greenhouse gas emissions trading scheme: final framework report on scheme design', submission to the Garnaut Review, <www.garnautrevi ew.org.au>.

NAILSMA (North Australian Indigenous Land and Sea Management Alliance) 2008, submission to the Garnaut Review, <www.garnautreview.org.au>.

NSW DWE (New South Wales Department of Water and Energy) 2008, *Transitional Arrangements for the NSW Greenhouse Gas Reduction Scheme*, consultation paper, NSW Government.

Prime Ministerial Task Group on Emissions Trading 2007, *Report of the Task Group on Emissions Trading*, Department of the Prime Minister and Cabinet, Canberra.

Shanahan, J. 2007, 'A carbon conundrum', *CFO Professional*, July.

Sijm, J., Neuhoff, K. & Chen, Y.2006, 'CO_2 cost pass-through and windfall profits in the power sector', editorial, *Climate Policy* 6, 49–72.

Government of South Australia 2008, submission to the Garnaut Climate Change Review.

ADAPTATION AND MITIGATION MEASURES FOR AUSTRALIA

15

Key points

Every Australian will have to adapt to climate change within a few decades. Households and businesses will take the primary responsibility for the maintenance of their livelihoods and the things that they value.

Information about climate change and its likely impacts is the first requirement of good adaptation and mitigation policies. This requires strengthening of the climate-related research effort in Australia. The Australian Climate Change Science Program should be provided with the financial resources to succeed as a world-class contributor to the global climate science effort from the southern hemisphere.

A new Australian climate change policy research institute should be established to raise the quality of policy-related research.

Flexible markets using the best available information are the second essential component for successful adaptation and mitigation policies. It will be important to strengthen markets for insurance, water and food.

Government regulatory intervention and provision of services will be required in relation to emergency management services and preservation of ecosystems and biodiversity.

Mitigation will come too late to avoid substantial damage from climate change. Uncertainties in the science mean that we do not know with certainty the extent of damage from successful 550 ppm CO_2-e, or even 450 ppm CO_2-e strategies. Weak mitigation will result much worse outcomes and much greater uncertainty about how bad it could get.

It is likely that Australians and Australian institutions will be adapting to climate change within a few decades. Before that time, Australians will be dealing with the price effects of an emissions trading scheme.

The experience of climate change will vary between households, and across communities, businesses, sectors and regions. Geographic location, degree of exposure and the capacity of those affected to reduce their vulnerability will all influence the Australian experience. The appropriate adaptation response will always depend on a range of local circumstances. Therefore, unlike the mitigation effort, adaptation is best seen as a local, bottom–up response. Households, communities and businesses are best placed to make the decisions that will preserve their livelihoods and help to maintain the things they value.

Some may expect that government can, and should, protect the community from climate change by implementing the right strategy, program or initiative to allow Australians to maintain established lifestyles. This is not a realistic expectation for four reasons. First, climate change will require adjustment of innumerable, locally-specific customs and practices over time. Second, the range and scale of impacts that is likely across Australia is such that it is not feasible for governments to underwrite maintenance of established patterns of life for all people in all places. Third, the uncertainty surrounding climate change impacts makes it impossible to predict their timing, magnitude or location with precision.

Finally appropriate responses to climate change impacts will be specific to circumstances. In many instances, centralised government will lack the agility to orchestrate a differentiated response with the necessary precision to address local needs. The informational requirements of government would be extreme and costly. It unlikely that an intrusive or directive approach to adaptation would be as effective as one motivated by local interests.

The Review favours strong reliance on local initiative in determining how Australia as a whole adapts to climate change. Government, in its roles as manager of public land, national water and infrastructure assets, regulator of markets and other activities, and manager of equity issues, can provide support for this approach by creating the necessary conditions for effective and efficient decision making by communities, households and businesses as they begin (and continue) to adapt to climate change.

Chapter 13 laid out an overarching framework that suggested that it was unhelpful to think about climate change policy in terms of the simple dichotomy of adaptation and mitigation. Typically, the policies required to support the community's adaptation effort have much in common with those required to support adjustment to the effects of a carbon price arising from mitigation policy—though the nature of the shocks being addressed are different (see Table 13.1).

Direct intervention by government in developing the national policy response should involve:

- deepening our understanding of climate change, its impacts and the options available to respond
- developing the capacity of the community to use this information and take advantage of available options
- dealing with events that unfold suddenly or require resources that are of a scale beyond a community's capacity to address.

This entails developing a climate change response, coordinated across all levels of government, that is focused on:

- producing and disseminating information and advice that is useful and useable by a wide range of interests
- utilising markets and market-based policies to create options for individuals and businesses to manage the uncertainties associated with climate change
- building capacity for dealing with events that can overwhelm individual communities or the natural environment.

15.1 Information and understanding

Information on the possible impacts of climate change is essential for determining the most appropriate means of adapting to climate change and the timing and scale of the response. But, as noted in section 12.4, hastening progress towards more ambitious global mitigation will also require a better understanding of climate change, its impacts and our progress in dealing with it.

There is, of course, inherent uncertainty in the basic climate science. The climate is an immensely complex system and long-term projections are inherently fallible. The current range of projections is too wide to be usefully applied by households, businesses and governments in preparing for climate change. For example, our current understanding of potential rainfall outcomes for Australia indicates a range that includes both an extreme dry outcome (10th percentile) and a wet outcome (90th percentile). There are similarly significant uncertainties in projections of sea-level rise.

Added to the uncertainty is the lack of certainty about when and how quickly the impacts will arrive.

An enormous domestic and international effort is required if we are to acquire the information needed for dealing with decision-making in the face of this uncertainty—as well as taking the necessary steps to reduce the cause of the uncertainty and the likelihood of extreme events (by hastening global agreement on more ambitious targets). Of course, a greater global mitigation effort alleviates, at least to some extent, the adaptation challenge that lies ahead.

Integrated assessments of economic and social effects will provide bioclimatic, social and economic analysis and information on the possible economy-wide, regional and sectoral impacts of climate change. New institutional arrangements are required to inform policy development as well as decision making by individuals, communities and business. This will involve public and private efforts in undertaking research and analysis.

The mix of benefits, and the demonstrated willingness of groups to seek information without government involvement, will require government to be disciplined in ensuring that it only funds activities with public benefits without crowding out private funding or providers.

15.1.1 Institutional arrangements to support the creation and dissemination of information

As well as assisting the development of policy, the creation of scientific knowledge and its dissemination have the characteristics of a public good. A private provider of the information will not be able to capture all the benefits of that information once it is in the public domain. Such information is therefore likely to be underprovided without government involvement.

Confident decisions made with intellectual authority will require a strong base of knowledge across all areas of climate change science—from global science and modelling, to the localised impacts of climate change. Figure 15.1 summarises the various areas of research and demonstrates how they form a system of information creation. Information from earlier stages is built on in later stages, and information from later stages feeds back to inform earlier stages.

Figure 15.1 Areas for further support and investment in the climate change research system

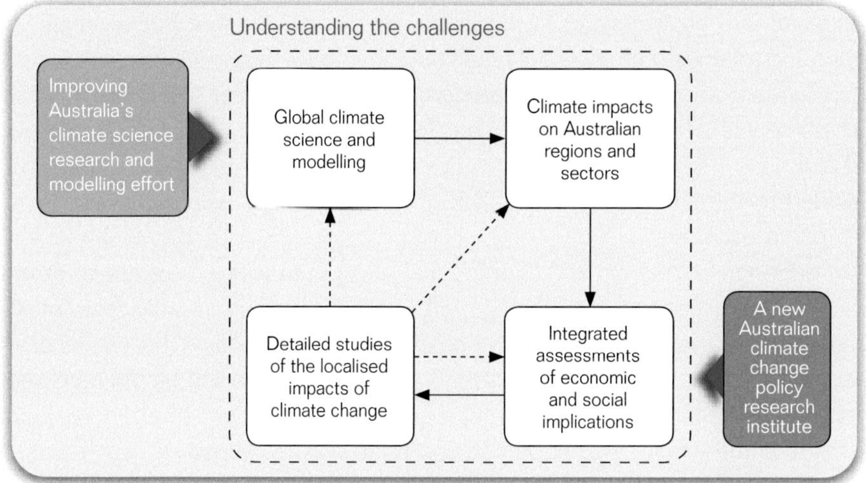

Australia's climate science research and modelling effort

Understanding the earth as a complete system requires an understanding of the various dimensions of that system, including the terrestrial surfaces, oceans, biological systems, atmospheric chemistry, and carbon cycles. The ultimate aim is to model as much as possible of the whole of the earth's climate system, using robust models supported by sufficient computing power, to allow all the components to interact. The predictions of climate change can then take into account the effect of complex feedback between components (IPCC 2001: 48).

Australia has fallen behind in key areas of research such as global and regional climate modelling (DCC 2008). Recent studies have found the evidence and modelling base for climate change in Australia to be sparse (Rosenzweig et al. 2008), and many global climate models have a bias towards the northern hemisphere due to the influential capabilities of the European and North American research institutes.

While pluralism is always valuable in basic research, it is likely that, because of the sheer size and complexity of the task, Australia can only afford one strong coordinated effort in its contribution towards the international climate modelling work. Australia can maintain no more than one dedicated supercomputing capability cost-effectively. The relative size of Australia's population and economy makes the case for a single consolidated approach.

The most recent attempt to improve and consolidate Australia's climate modelling capability has been the establishment of the joint CSIRO – Bureau of Meteorology Centre for Australian Weather and Climate Research (see Box 15.1). This appears to be a step in the right direction.

Addressing the uncertainty and gaps in knowledge at the most fundamental levels of climate science and at the intermediate stages of analysis of climate impacts has large public good dimensions. There is strong justification for Australian governments expanding funding to these research areas on a long-term basis. This suggest that greater effort and funding be allocated as part of the renewal of the Australian Climate Change Science Program (which aims to improve our understanding of the causes, nature, timing and consequences of climate change).

To strengthen this effort, governments should ensure that there is adequate funding for world-class performance in high-priority areas, including through the ACCESS model (see Box 15.1) and investment in substantial supercomputing infrastructure. Such support should be focused on areas where Australia has both a national interest in answering particular research questions and the international comparative advantage in being able to do so. This should be considered as part of the renewal of the Australian Climate Change Science Program.

Some support should be targeted at increasing Australia's participation in international research and modelling efforts (such as the World Climate Research Program and the International Geosphere Biosphere Program) and review and assessment processes (like those undertaken by the Intergovernmental Panel on Climate Change). First, Australia needs to ensure that it is able to play its role as the leading country of science in the southern hemisphere in many of these areas. Second, the ongoing participation of the Australian scientific community in international processes is necessary to enable global progress in climate modelling to be interpreted and absorbed by Australia.

Box 15.1 The Centre for Australian Weather and Climate Research

The Centre for Australian Weather and Climate Research was established in recognition of problems in the historically fragmented approach to climate modelling in Australia, and the need for a unified national effort.

The centre is a new joint venture between CSIRO and the Bureau of Meteorology on climate change research. It will develop a next-generation unified modelling and assimilation capability for Australia known as the Australian Community Climate and Earth System Simulator (ACCESS).

The ACCESS model will incorporate some components of the model used by the UK Met Office Hadley Centre, with a focus on further developing the components that are of national interest to Australia and where Australian researchers have a comparative advantage. The ACCESS model will be developed in collaboration with Australian universities and the Cooperative Research Centre for Earth Sciences, and will supersede all existing Bureau of Meteorology and CSIRO models.

The Review sees it as a matter of importance that the Centre for Australian Weather and Climate Research be given the secure, long-term financial resources and the scientific independence it requires to succeed in its important task and to play its role as the southern hemisphere's anchor for the world's scientific effort on climate change. It is too early to tell whether recent developments will secure the objectives of the centre. An early review should examine progress.

A new Australian climate change policy research institute

Climate change policy is a wide-reaching issue requiring many disciplines and capabilities to come together to understand the full extent of the problem and to analyse and develop potential solutions.

There is no continuing independent centre of research on the range of policy issues the Review has examined. There is a gap in the national research capacity, covering the implications of climate change for various areas of government policy—beginning with targets and trajectories for Australian mitigation policies and extending through to a wide range of adaptation issues.

An Australian climate change policy research institute should be established to inform public discussion and to strengthen the intellectual context for policy development. The roles and responsibilities of the proposed institute should include, but not be limited to, six key areas:

- interacting with climate science and modelling institutions, including the Centre for Australian Weather and Climate Research
- working with the applied science and broader research community
- developing models and analytic frameworks to provide integrated analysis
- linking Australia to international thinking about climate change mitigation and adaptation policy
- enriching the policy debate
- pursuing linkages and joint appointments and training with other academic institutions.

The proposed research institute would need to have disciplinary strengths in the physical (climate) and biological sciences, in applied biological and engineering sciences, and in economics and the relevant social sciences.

It is envisaged that the institute would interact with various elements of the scientific and research communities and in the process expand their sensitivity to issues that are relevant to policy analysis.

As climate change policy will have significant economic, social and environmental consequences with long time frames, it is important that the proposed institute is independent. To maintain independence, the institute should have its own governance structure. The type of governance and institutional arrangement would need to be explored further. An independent stand-alone institute would ensure that institutional capture is minimised, but a research consortium approach as adopted by the Tyndall Centre in the United Kingdom, would facilitate the inclusion and participation of Australia's geographically dispersed base of policy-oriented researchers.

15.1.2 Limits to the quantity and quality of information

While high-quality and high-resolution information on the projected impacts of climate change are critical inputs to adaptation decisions by households and businesses, there are inherent limits to the availability of such information.

First, there are likely to be elements of irreducible uncertainty in the basic climate science. Nevertheless, efforts must be made to reduce uncertainty, where possible, for adaptation and mitigation decisions. Some level of uncertainty in the climate science, and gaps in the relevant knowledge base, will persist. One of the most significant adaptation challenges for the Australian community is to respond to the immense potential impacts of climate change on the basis of imperfect information.

Second, while it is desirable that the information available on the impacts of climate change be at a high level of detail, cost considerations will mean that the resolution of studies will vary between regions.

Third, scarcity of relevant scientific resources will mean that there is competition for them. Groups faced with the greatest potential exposure and loss will be prepared to invest most in research into impacts. It is important that those who lack the resources—or the capacity to cooperate with others with similar interests to pool funds—are engaged.

Fourth, the creation of information does not guarantee its optimal uptake. Information must be prepared and released in a form that is usable by its intended beneficiary (see Box 15.2).

Box 15.2 Improving the communication of information

Correcting gaps in the public knowledge base rests not just on the research effort, but also on the interpretation and presentation of scientific projections in a meaningful and relevant form that can be factored into local risk management and decision making.

Even when soundly researched information is widely communicated, it may be of limited utility if users have problems comprehending it or using it in making their decisions. The behavioural economics literature acknowledges that people making decisions, whether in households or businesses, deal poorly with probabilistic information.

15.2 The role of markets and market-based policies

The objective of adaptation policy is to facilitate the ability of households, communities and businesses to respond effectively to the impacts of climate change. Markets provide the most immediate and well-established avenue for addressing many of the uncertainties posed by climate change. Fortunately, flexible markets also provide the most efficient mechanism for dissipating the price impacts of an emissions trading scheme.

Australia's prime asset in responding to the adaptation and mitigation challenges that lie ahead is the prosperous, open and flexible market-oriented economy that has emerged from reform over the last quarter century. Government can facilitate adaptation by continuing to promote broad and flexible markets, and seeking to correct remaining barriers to efficient exchange.

Markets are well placed to transfer risk to those best placed to deal with it and disperse concentrated risks across a wide base of industries, communities, regions and countries. This is achieved through insurance and financial markets, and through dispersed but interconnected domestic and international product markets. The smooth flow of goods and services, and factors of production, increases the ability of the Australian economy to respond at least cost to abrupt shocks and anticipated changes over the longer term.

Broad and flexible markets allow scarce resources to move to where their economic value is highest, at a time when new information is continually changing our understanding of value. When resources are 'stuck' in an area of declining productivity, the growth of the economy is hampered by the exacerbation of scarcity. For example, as agricultural yields decline in some regions due to increases in temperature or declining precipitation, capital, labour and remaining water resources will produce better outcomes for their owners, for regional communities and for the national economy if they are able to move to other more productive crops, industries or regions.

The benefits of flexible markets are evident even in the absence of climate change. However, the requirements of adaptation to climate change, and adjustment to mitigation policies, increase the importance of efficient markets.

Some particular domestic and international markets will be especially important to Australia's adaptation response, and potentially to our capacity to smoothly adjust to the effects of a carbon price. These markets may require increased policy attention. Included in this category are markets for insurance, water and food.

15.2.1 Insurance markets

Households and businesses are able to manage many risks effectively through the insurance and financial markets. As the frequency and intensity of severe weather events increase with climate change, demand will rise for related insurance and financial services.

In the context of climate change adaptation, there are two types of benefits of insurance and related financial instruments.

First, the insurance industry, and financial markets more broadly, provide financial instruments that enable the market to moderate exceptionally bad outcomes for particular groups of people. They do this through sharing risk across a broad base of parties facing different degrees of risk from varying sources. They also transfer risk between those who are more risk averse and those more willing and able to accept risks. The financial markets' capacity to reallocate costs and risks globally to those most willing and able to bear them will help reduce the costs of adaptation for the wider society (IMF 2008).

Second, adjustments to insurance premiums for risk-reducing behaviour will provide incentives for households and businesses to adopt risk-reduction measures (IMF 2008). Of course, insurance and other financial products do not remove risk altogether. They are least suitable for reducing large and highly correlated risks, such as large climate change outcomes with global effects.

Climate change poses new challenges for insurance markets by widening the probability distribution of possible losses and increasing the severity of damages and payouts.

Primary insurers are able to provide coverage for risks that are not strongly correlated and where the exposure is limited relative to the total size of any one insurer. Traditional indemnity-based insurance has been able to deal with climate variability and weather risks for centuries (OECD 2008). However, the insurance industry will have to deal with an increase in the more 'exceptional' risks. These involve the possibility of extremely large losses, with the risks being highly correlated across many households, businesses or even regions.

If insurance claims greatly increase as a result of severe weather events, there is a possibility that such risks may overwhelm the financial sector's capacity to provide insurance coverage (Association of British Insurers 2005). The challenge for the industry will be to develop new ways to spread risks so that it can offer coverage to more people exposed to a wider range of more costly outcomes. As the probability distribution of impacts widens, more risk capital is needed to bridge the gap between expected and extreme losses (Association of British Insurers 2005). Figure 15.2 depicts the higher capital needs that result from climate change impacts.

The global market for reinsurance has supported primary insurers by providing a range of financial instruments to transfer catastrophic, low-probability, highly correlated risks between parties with different preferences or appetites for risks at a global scale. Over the past decade, the market for global catastrophe reinsurance has grown strongly in volume and in the variety of its financial structures, while its geographic coverage has expanded to a more limited degree (IMF 2008).

Figure 15.2 Impact of climate change on probability loss distribution and implications for risk capital requirements

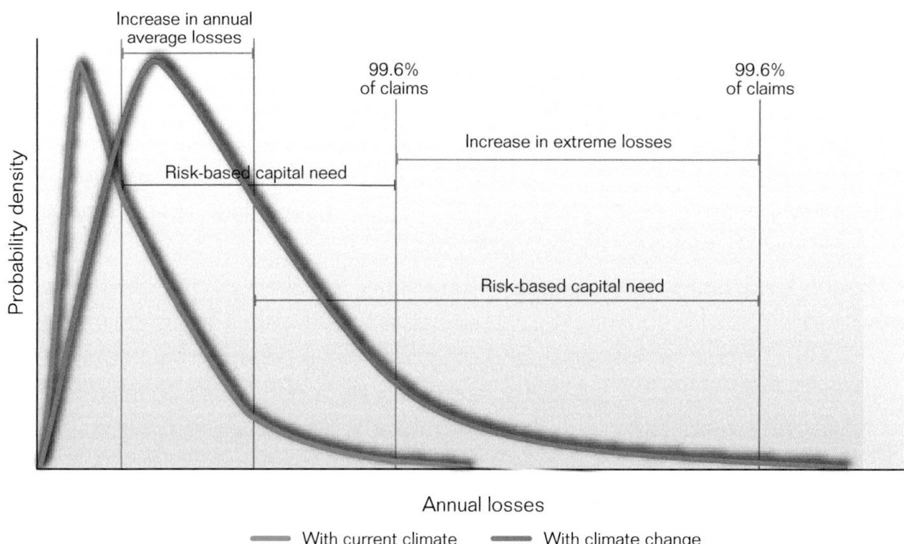

Source: Association of British Insurers (2005).

The recent innovation and deepening in these markets shows their considerable potential to promote adaptation to climate change (IMF 2008). By its nature, however, conventional insurance is of limited value when an adverse event (for example, damage from increases in sea level) is likely to have similar impacts over wide areas of the world. Nor is conventional property insurance of much help when the uncertainty mainly involves the timing rather than the extent of an impact. An example is sea-level rise if it were to become clear that the melting of the Greenland ice sheet had become irreversible. In this case, there might be scope for developing new property insurance products that share characteristics with traditional life insurance: life insurance covers the risk of timing of death, although the fact of eventual death is itself certain. The development of innovative products along these lines would occur within the commercial insurance sector.

Government can promote a deep and flexible insurance sector by:

- **Improving the provision of basic climate science information**—With improved information, insurance markets can be expected to deepen in response to need.[1] The insurance industry is equipped to analyse and respond to new risks that flow from climate change, including in risk assessment capacity, diversification of risks, and pricing.
- **Promoting insurability through appropriate approaches to policy**—Moderate levels of climate change should not raise any significant challenges for the industry, but there will always be some residual risks that are highly correlated and sometimes uninsurable. In the case of more severe climate change, such risks will increase, as will the expectation that government will play the role of insurer of last resort. Appropriate approaches to regulation, particularly in land-

use planning and zoning decisions, will improve insurability (Insurance Council of Australia 2008), and minimise pressures for the expansion of this role for government.

- **Minimising regulation**—In the light of the required rapid expansion of the industry, government will need to avoid discouraging rapid growth and innovation in the sector while maintaining standards.[2, 3] As the Henry Review has noted, insurance products are subject to a range of insurance transaction taxes and direct contributions to the funding of fire services, which leads to inefficient outcomes. The interaction of these taxes and levies increases the cost of premiums, which may reduce insurance uptake (Australian Treasury 2008; IPART 2008). Such regulations may ultimately inhibit the necessary expansion of the industry and the emergence of new products. The revenue benefits of such taxes need to be evaluated against their inefficiencies and economic costs, particularly given the role of insurance in encouraging firms and households to adapt to climate impacts. The Henry Review is well placed to undertake such analysis.

15.2.2 The role of water markets

The challenges for rural and urban water supply are partly a consequence of emerging climate change, partly of increased demand with population and economic activity, and partly of uneconomic pricing leading to unsustainable consumption patterns. The pressures associated with climate change can be expected to intensify, increasing the importance of removing unnecessary market distortions.

Australia's rural water market is the result of many years of reform, but some barriers to efficient operation remain. While extraction of in-stream flows has been priced, access to groundwater and surface flow is less regulated, which has led to perverse incentives.

While water trading and the use of price signals have been in place in some parts of the rural sector for over 20 years, in others water remains free. Where water trading has been introduced, the initial allocation of property rights in water has locked in, and even exacerbated, the overallocation that prevailed under previous systems of licensing.

Artificial jurisdictional boundaries have resulted in restrictions on trading between rural consumers, and have largely prohibited trading between the rural and urban water sectors. While this has been done to manage the pace of the transition that a region exporting water undergoes, it has resulted in a distorted price signal in some areas, and no price signal in others. This effect has also occurred in the urban sector and been compounded by the absence of a competitive water market.

Climate change provides an added impetus to accelerate and raise ambitions for water market reform. Water markets that are transparent, broad and flexible, and based on clearly defined property rights, are better able to manage shocks.

By 2100 under a best-estimate no-mitigation scenario, the impact of climate change on urban water supply infrastructure is expected to be moderate to extreme, depending on location. Under the hot, dry extreme scenario, the impact is expected to be extreme in all capital cities. An effective market could minimise ad hoc infrastructure investment decisions, promote optimal timing of large infrastructure investments, and assist in bringing a broader range of supply options to the market.

One major benefit of water reform is securing environmental flows. This involves recognising the need for waterways to have sufficient flows to maintain the ecosystem services they provide, by supporting freshwater flora and fauna, or improving water quality. Traditionally the environment was not considered a 'productive' use of water, but contemporary thinking recognises its significant benefits for the wider community.

The establishment of a well-functioning water market that delivers the best possible outcomes in the context of climate change will require the active involvement of government. Government is required to establish the most effective administrative and regulatory arrangements for the functioning of the market. Once the water market has matured, the role of government moves to one of adequate monitoring and enforcement.

The Review endorses a set of common principles, developed from a set proposed by PricewaterhouseCoopers (2006), that would go a long way to deliver effective management of Australia's water resources and assist in the adjustment to a new climate future if supported in practice, namely:

- Water rights need to be clearly defined, with exclusive ownership, and be separable from other resources, such as land.
- The total quantity of water rights allocated in each catchment and over time needs to be flexible enough to accommodate new scientific information on climate change and sustainable water use.
- Security levels for water rights need to be defined in a way that is consistent with the variability and uncertainty of aggregate supplies.
- Clearly defined rules are needed to set the boundaries of the market and acceptable behaviour by market participants, and reduce transaction costs by providing greater certainty.
- Reliable and timely information is needed to allow buyers and sellers to make informed decisions.
- Clear administrative processes are necessary for effective trade and to support the enforcement of the trade.
- Arrangements must secure low transaction costs compared to the value of the trade, with costs known in advance.
- The market should have few limitations on who can participate.

In addition to these principles, the Review considers that providers of new sources of water should not be prevented or discouraged through regulatory and other barriers (provided they meet health and safety standards). Market participants should be able to introduce new, including manufactured, sources of water to each market and to transfer water between markets, when they consider it commercially viable to do so.

With a well-designed and comprehensive water market in place, price signals will reflect the scarcity value of water across rural and urban Australia. It can reasonably be expected that households, businesses and other consumers will modify their water use accordingly.

15.2.3 Food markets

The global agricultural sector will be adversely affected by climate change in the absence of effective mitigation. It is likely that the levels and volatility of prices on world markets will increase. The recent global food price crisis provides an indication of how things could develop (ACIAR 2008). In future, it is likely that international food markets will face multiple supply shocks with resulting impacts not unlike those recently witnessed.

If Australia becomes increasingly dependent on food imports, as projected under a best-estimate no-mitigation case, rising global food prices and price volatility could become increasingly important issues for consumers.

Global food markets that are functioning efficiently can absorb shocks. However, existing food markets are characterised by arbitrary and variable interventions by governments of both developing and northern hemisphere developed countries. The responses of governments around the world to recent price shocks have exacerbated price increases and instability.

Traditionally, trade tariffs, subsidies and quarantine restrictions have been the primary obstacles to efficiently functioning food markets. With food prices soaring and in the face of mounting political pressure, recent policy changes by major food-exporting countries have increased barriers to trade. The most notable of these have been increased export restrictions (or similar restrictions) and domestic price controls (von Braun 2008b).

- **Export restrictions**—Worldwide production of cereals has not kept pace with demand. Droughts, bad weather and general rises in the domestic price of food in the key food-exporting regions of Asia have led many countries to restrict exports to protect domestic supplies. This has in turn increased international food prices sharply (von Braun 2008a, 2008b).
- **Domestic price controls**—In many other countries, governments have sought to soften the impact of price rises on domestic markets by resorting to price controls and government subsides to curb inflation and manage the political backlash. However, holding down prices through regulations or subsidies reduces the incentive for producers to increase supply, thereby exacerbating shortages and pushing global prices even higher. Price controls may yield short-run political benefits, in countries imposing them, but not in the rest of the world. In the longer term, these benefits are likely to be outweighed by the damaging effects on economic efficiency (World Bank 2008).

It is in Australia's long-term interest to pursue the liberalisation of international food markets by removing the distortionary policies of the world's future food importers and suppliers. Australia would benefit from broader and more open and reliable global food markets as an exporter. It would benefit as a neighbour of developing countries that are from time to time reliant on supplies from international markets. And it would benefit in food security if, as it may, climate change makes Australia a large importer of food.

It is also in our long-term interest to ensure that barriers to import such as unnecessarily restrictive quarantine measures do not unnecessarily prevent goods from overseas entering our domestic markets. The banana market in the aftermath of Cyclone Larry in North Queensland in March 2006 illustrates how restrictions on trade can lead to extreme price fluctuations in response to supply checks. Banana prices rose by up to 500 per cent in the wake of the cyclone (Watkins et al. 2006). Australia needs to be sure that the restrictions are justified—especially in circumstances in which supply shocks are likely to be more important.

15.2.4 Beyond markets

Some services that are of high value to Australians do not lend themselves to provision through a market (see the discussion of Type 4 costs of climate change in Chapter 1). The foremost example is the conservation of ecosystems and biodiversity. The irreversibility of ecosystem collapse and species extinction warrants regulatory intervention.

In addition to the priority areas for domestic policy outlined above, government will have a role to play in two other policy areas relevant to both adaptation and mitigation.

- There are second-order market failures common to adaptation and mitigation that will not be corrected through the provision of climate information and the development of broad, open and flexible markets. These market failures—in information; research, development and innovation; and network infrastructure—can be dealt with through specific measures that aim to lower the costs of adaptation. They are discussed in chapters 17, 18 and 19.
- Government also has a broader role to play in the areas of income distribution and related questions of structural adjustment to avoid regressive and inequitable outcomes that may arise from climate change adaptation. These issues are discussed in Chapter 16.

15.3 Scaling the challenges: five examples

Chapter 6 identifies four areas—irrigated agriculture in the Murray-Darling Basin; urban water supply infrastructure; buildings in coastal settlements; and ecosystems and biodiversity—for which the impacts of climate change are anticipated to be large. Reference to these four sectors illustrates the diversity of considerations and issues inherent in the adaptation policy challenge. A fifth area of emergency management is also examined.

15.3.1 Irrigated agriculture in the Murray-Darling Basin

Declining runoff in southern Australia is a significant threat to the continuation of irrigated agriculture in the Murray-Darling Basin. Extensive development of water storage and distribution systems during the 20th century has encouraged excessive extraction of water from the environment. This has resulted in significant adverse impacts on streamflows and groundwater supplies. Existing strategies for managing water supplies were developed in the second half of the 20th century, during a period of higher rainfall, and are not suited to a progressively drying climate (CSIRO 2008).

Historically, and for a variety of reasons, the price of irrigation water does not often reflect its true economic value. Not surprisingly, when viewed independently of these historical considerations, we observe the inefficient allocation of this resource.

The effect of declines in water supplies for irrigated agriculture could be reduced by improving the efficiency of the existing infrastructure for water delivery. Currently, between 10 and 30 per cent of the water diverted from rivers into irrigation systems is lost before it reaches the farm gate, and up to 20 per cent of the delivered water may be lost in on-farm distribution channels (CSIRO 2008). It is unlikely that this situation would be perpetuated if the scarcity of the water was accurately reflected in its price. The combination of current water market reforms and the ongoing drought conditions experienced in much of the Murray-Darling Basin has prompted large-scale infrastructure commitments by state and federal governments.

Agriculture in Australia has developed around its capacity to adapt to natural climate variability. However, the scale and accelerated impacts of human-induced climate change with limited mitigation are likely to breach the capacity of the sector to continue operations at even a tiny proportion of its current scale. By 2100 under a best-estimate no-mitigation case, the economic production of irrigated agriculture in the Murray-Darling Basin is projected to fall by 92 per cent; under the hot, dry extreme case, production falls by 97 per cent. Under such outcomes, the variety and volume of irrigated agricultural crops will diminish. Significant structural change of the sector would be necessary from 2050 when impacts from reduced precipitation are projected to start to become critical (Quiggin et al. 2008). The only viable adaptation response might be to abandon established patterns of agriculture.

Even with effective mitigation aimed at stabilisation of CO_2-e concentrations at 550 ppm, and possibly 450 ppm, major adaptive changes are likely to be required. Commercial-scale and technologically sophisticated farming operations, with the ability to increase water efficiency and compete in water markets, may replace smaller operators that fail to invest in efficiency improvements.

15.3.2 Urban water supply infrastructure

Australia's urban water supply infrastructure is old, inadequate for current population levels, and not designed to cope with changing climate conditions (Marsden Jacob Associates 2006; Productivity Commission 2008).

By 2100 the Garnaut–Treasury reference case points towards an Australian population of 47 million people. Population growth alone is likely to place significant additional stress on urban water supply infrastructure. With projected increased temperature and evaporation and reduced rainfall, under the best-estimate no-mitigation case, climate change adds greatly to the stress on urban water supply infrastructure later in the century.

For Australia's coastal cities, which include the major population centres, an array of water supply and demand management options are available to allow diversification of supply in response to climate change (Marsden Jacob Associates 2006). The expansion and opening of water markets would allow the emergence of the lowest-cost supply options and the optimal balance between reduction of use and expansion of supply.

In addition to improved use of existing water sources, it is likely that new forms of supply that are not climate dependent will be required. New water sources can initially be expected to have higher capital and operating costs. These should fall over time because of 'learning by doing' and economies of scale, although it is likely that the cost of water will be permanently higher than it is today (J. Quiggin 2008, pers. comm.).

For Australia's major inland cities and towns, recycled water and purchase of irrigation entitlements may be among the few alternative water supplies available (Marsden Jacob Associates 2006).

A drying climate would be likely to also result in increased ground movement, which would damage water distribution infrastructure such as mains pipes. In the next two to three decades, operational and maintenance costs for existing water supply infrastructure will probably increase (Maunsell 2008).

By mid-century a degree of adaptation to the new climate will be under way. Technological development and engineering standards will have been modified to reflect the changing climate, and ageing assets replaced. However, as the century continues, the significant increase in severe weather events anticipated under a best-estimate no-mitigation case would lead to further increases in operational expenditure.

15.3.3 Buildings in coastal settlements

As the climate changes in the absence of strong effective global mitigation, Australia's coastal communities and associated infrastructure would be subject to increasingly frequent and severe weather events, as well as impacts from sea-level rise, storm surges and associated coastal flooding. The broad preference of

Australians for living on the coast makes the ability of the coastal built environment to withstand climate change impacts a determining factor in the distribution of future human settlements.

Domestic and public infrastructure tends to be long-lived—for example, residential buildings typically have a design life of about 40 years (Maunsell 2008), although average actual lives are longer. Infrastructure planning for new and existing settlements should consider the potential climate change impacts on the entire life cycle of the proposed infrastructure. This can reduce future maintenance requirements, the need for premature replacement or abandonment, and the need for relocation of entire settlements in the case of increasingly severe weather events.

Steps can be taken to decrease the vulnerability of new buildings to climate change. These generally fall into three categories: changes in design, changes in materials, and changes in location (BRANZ Ltd 2007; CSIRO et al. 2006; Engineers Australia 2008).

For existing settlements, changes in both building design and materials can provide effective options for adaptation at reduced cost if retrofitting aligns with asset renewal. For new settlements, the foremost consideration is to avoid placing infrastructure in highly exposed positions.

By 2100 under a best-estimate no-mitigation case, measures for coastal protection may not be adequate to withstand the damaging impacts of climate change on buildings. The relocation of industries, activities and households away from certain coastal areas may be the only available adaptation response.

15.3.4 Ecosystems and biodiversity

Climate change is a significant and additional stressor on ecosystems and biodiversity in Australia. It will affect ecosystems and biodiversity by shifting, reducing and eliminating natural habitats. In Australia, many species of flora and fauna are at risk from rapid climate change because of their restricted geographic and climatic range. Where ecosystems and species have low tolerance for change, altered climatic conditions can trigger irreversible outcomes such as species extinction.

Climate change impacts will reduce the availability of various ecosystem services.[4] Opportunities for medical advances from natural sciences research may be lost. Natural resource–based industries such as snow season tourism and tourism in the Great Barrier Reef, wet tropics and Kakadu regions will be adversely affected.

Given that net losses are currently occurring with only the initial effects of climate change, significant resources will be required to minimise future loss (Australian State of the Environment Committee 2001; Beeton et al. 2006).

Natural resource management networks and programs have been established in Australia to conserve our natural environments. With climate change, additional efforts will be required to build the resilience of the Australian environment. This can be achieved by reducing existing non-climatic stressors such as land-use change, overallocation of water, and pollution (Howden et al. 2003). Similarly, expanding the existing system of land reservation and exploring new methods for engaging private landholders will facilitate species migration, encourage conservation and promote resilience.[5]

It is important to avoid perverse outcomes for ecosystems in the implementation of policy in related areas, such as agriculture, forestry, fisheries and fire management. For example, water markets must ensure that environmental needs are adequately met in water allocations. The incentives for plantation forestry introduced by an emissions trading scheme must sit alongside adequate valuation of native vegetation.

Maintaining viable, connected and genetically diverse populations increases their likelihood of survival (IPCC 2007; WWF–Australia 2008). Conserving Australia's ecosystems will also assist in greenhouse gas mitigation due to their large cumulative sequestration capacity (see Chapter 22).

Now and in the future, natural resource managers will need to consider geographical shifts in habitats, the resulting new species assemblages, and the effect of these developments on, for example, the location and management of conservation reserves. Future natural resource management practice will need flexibility to allow managers to respond quickly to a dynamic environment and new information.

An enhanced research effort is required to improve knowledge of ecosystem function under differing levels of climate change. Functionally critical species and habitats need to be identified and appropriate actions taken to manage them. Much less is known about marine than terrestrial ecosystems. This information divide will need to be addressed (Beeton et al. 2006). Systematic effort is required to improve statistical information on environmental qualities including water quality, populations, endangered species and land degradation.

The development of environmental markets, through which incentives are provided for private landholders to assist in conserving and restoring ecosystems, has potential to assist in the adaptation and mitigation effort. These mechanisms can be designed to reveal the necessary price incentive for private landholders to change land-use practice. Pilots such as the BushTender project in Victoria (see Box 15.3) have been successful across a variety of ecosystems (DSE 2008); however, the deployment of market-based mechanisms is relatively new and is yet to be applied on a broad scale.

Box 15.3 Victoria's BushTender project

The Victorian Department of Sustainability and Environment has conducted a trial of an auction-based environmental market—the BushTender project—since 2001. Through the project, landholders bid for contracts to conduct conservation activities on their land. The land in each tender is assessed for its conservation value and landholders nominate their costs, which presumably include opportunity costs from reduced productivity. Bids are then assessed based on (1) estimated change in the on- and off-site environmental outcomes; (2) the value of the assets affected by these changes; and (3) the cost (determined by the landholder's bid). Successful bidders enter into contracts with government under which they receive periodic payments for conservation. The project is part of a range of experimental markets which include BushBroker (for native vegetation land clearance credits), CarbonTender (for carbon offsets) and EcoTender (combining multiple environmental objectives).

15.3.5 Emergency management services

Australia's emergency management systems and services have been operating for many years and are relatively robust and well developed (Department of Transport and Regional Services 2004). The provision of these services is largely within the jurisdictional responsibility of state and territory governments but is heavily supplemented by the work of volunteers, most notably those in the rural fire services and state emergency services.

With the mainstream science projecting increases in the intensity and frequency of severe weather events across Australia in the absence of mitigation, and to a lesser extent with mitigation, there will be increased demands on emergency services. Understanding the implications of climate change at a local level will be centrally important in the future planning of these services.

In some instances, at least for the time being, investing in emergency services will be the most cost-effective response for dealing with the uncertainties of climate change. One example is rising sea levels, the effect of which will become manifest gradually over many decades. Storm surges accompanying deep low pressure systems may become more frequent long before it is possible to justify investment in sea walls or the abandonment of low-lying coastal property. They may overwhelm the capacity of affected communities to respond.

Notes

1 Before Hurricane Hugo in 1989, the insurance industry in the United States had not experienced any losses from a single disaster of over US$1 billion, whereas today such disasters are relatively common, predictable and manageable. Most insurance experts now agree that it is the potential US$100 billion-plus catastrophic event which is now a challenge for the industry (King 2005).

2 Increases in average and extreme losses will tend to increase the amount of risk capital needed to satisfy the requirements of the regulator.

3 Insurance remains the only industry within the Australian financial services sector operating under both state and federal supervision.

4 Ecosystem services transform natural assets (soil, plants and animals, air and water) into benefits that people value for financial, ecological or cultural reasons (Binning et al. 2001). Ecosystem services with direct use values are quantifiable. Other values are difficult to calculate and highly subjective. Attempts to quantify some ecosystem service values include (1) an estimated $2 billion per annum contribution of the Great Barrier Reef to the Australian economy through tourism (Hoegh-Guldberg & Hoegh-Guldberg 2003), and (2) an estimated $1.2 billion per annum derived in benefit to the Australian agricultural sector from natural crop pollination (PMSEIC 2002).

5 Government regulation or acquisition may be justified where land is of significant conservation value, or where certainty of outcome is required. Currently, around 10 per cent of Australia's land mass is under some form of government conservation through the National Reserve System (DEWHA 2004). This compares with 64 per cent under private ownership (ABS 2008). National statistics are not available, but estimates suggest that the Indigenous estate accounts for approximately 20 per cent of the Australian land mass (Altman et al. 2007). Approximately 10 per cent of the Indigenous estate, or 2 per cent of Australia's land mass, is within the boundaries of the National Reserve System (DEWHA 2004).

References

ABS (Australian Bureau of Statistics) 2008, *Year Book Australia, 2008*, cat. no 1301.0, ABS, Canberra.

ACIAR (Australian Centre for International Agricultural Research) 2008, 'Food prices on the rise', <www.aciar.gov.au/node/8549>.

Altman, J.C., Buchanan, G.J. & Larsen, L. 2007, *The Environmental Significance of the Indigenous Estate: Natural resource management as economic development in remote Australia*, Australian National University, Canberra.

Association of British Insurers 2005, *Financial Risks of Climate Change*, <www.abi.org.uk/Display/Display_Popup/default.asp?Menu_ID=1090&Menu_All=1,1088,1090&Child_ID=552>.

Australian State of the Environment Committee 2001, *Australia State of the Environment 2001*, Independent report to the Commonwealth Minister for the Environment and Heritage, CSIRO Publishing on behalf of the Department of the Environment and Heritage, Canberra.

Australian Treasury 2008, *Architecture of Australia's Tax and Transfer System*, <http://taxreview.treasury.gov.au/content/downloads/report/Architecture_of_Australias_tax_and_transfer_system.pdf>.

Beeton, R.J.S., Buckley, K.I., Jones, G.J., Morgan, D., Reichelt, R.E. & Trewin, D. (2006 Australian State of the Environment Committee) 2006, *Australia State of the Environment 2006*, independent report to the Australian Government Minister for the Environment and Heritage, Department of the Environment and Heritage, Canberra.

Binning, C., Cork, S., Parry, R. & Shelton, D. (eds) 2001, *Natural Assets: An inventory of ecosystem goods and services in the Goulburn Broken catchment*, CSIRO, Canberra, <www.ecosystemservicesproject.org/html/publications/docs/Natural_Assets_LR.pdf>.

BRANZ Ltd 2007, *An Assessment of the Need to Adapt Buildings for the Unavoidable Consequences of Climate Change*, report to the Australian Greenhouse Office, Commonwealth of Australia, Canberra.

CSIRO (Commonwealth Scientific and Industrial Research Organisation) 2008, *Climate Change Adaptation in Australian Primary Industries*, report prepared for the National Climate Change Research Strategy for Primary Industries, CSIRO, Canberra.

CSIRO, Maunsell Australia Pty Ltd & Phillips Fox 2006, *Infrastructure and Climate Change Risk Assessment for Victoria*, CSIRO, Victoria.

DCC (Department of Climate Change) 2008, *Submission to the Review of the National Innovation System*, DCC, Canberra.

DEWHA (Department of the Environment, Water, Heritage and the Arts) 2004, 'National Summary (Terrestrial)', Collaborative Australia Protected Area Database, <www.environment.gov.au/parks/nrs/science/capad/2004/index.html>.

Department of Transport and Regional Services 2004, *Natural Disasters in Australia: Reforming mitigation, relief and recovery arrangements*, report for the Council of Australian Governments (COAG) High Level Group on the Review of Natural Disaster Relief and Mitigation Arrangements, Department of Transport and Regional Services, Canberra.

DSE (Department of Sustainability and Environment) 2008, *BushTender: Rethinking investment for native vegetation outcomes. The application of auctions for securing private land management agreements*, Victorian Department of Sustainability and Environment, East Melbourne.

Engineers Australia 2008, *Inquiry into Climate Change and Environmental Impacts on Coastal Communities*, submission to the House of Representatives Standing Committee on Climate Change, Water, Environment and the Arts, <www.engineersaustralia.org.au/shadomx/apps/fms/fmsdownload.cfm?file_uuid=FC2504E8-E9C7-594D-7A67-823FB95D6886&siteName=ieaust>.

Hoegh-Guldberg, H. & Hoegh-Guldberg, O. 2003, *The Implications of Climate Change for Australia's Great Barrier Reef*, WWF–Australia and Queensland Tourism Industry Council, <www.wwf.org.au/publications/ClimateChangeGBR>.

Howden, M., Hughes, L., Dunlop, M., Zethoven, I., Hilbert, D. & Chilcott, C. 2003, *Climate Change Impacts on Biodiversity in Australia*, outcomes of a workshop sponsored by the Biological Diversity Advisory Committee, 1–2 October 2002, Commonwealth of Australia, Canberra.

IMF (International Monetary Fund) 2008, *World Economic Outlook: Housing and the business cycle*, <www.imf.org/external/pubs/ft/weo/2008/01/index.htm#ch4fig>.

Insurance Council of Australia 2008, *Improving Community Resilience to Extreme Weather Events*, <www.insurancecouncil.com.au/Portals/24/Issues/Community%20Resilience%20Policy%20150408.pdf>.

IPART (Independent Pricing and Regulatory Tribunal) 2008, *Review of State Taxation: Report to the Treasurer: Other Industries—Draft Report*, IPART of New South Wales, Sydney.

IPCC (Intergovernmental Panel on Climate Change) 2001, *Climate Change 2001: The scientific basis. Contribution of Working Group I to the Third Assessment Report of the Intergovernmental Panel on Climate Change*, J.T. Houghton, Y. Ding, D.J. Griggs, M. Noguer, P.J. van der Linden, X. Dai, K. Maskell & C.A. Johnson (eds), Cambridge University Press, Cambridge and New York.

IPCC 2007, *Climate Change 2007: Impacts, adaptation and vulnerability. Contribution of Working Group II to the Fourth Assessment Report of the Intergovernmental Panel on Climate Change*, M.L. Parry, O.F. Canziani, J.P. Palutikof, P.J. van der Linden & C.E. Hanson (eds.), Cambridge University Press, Cambridge.

King, R.O. 2005, *Hurricane Katrina: Insurance losses and national capacities for financing disaster risk*, Congressional Research Service Report for Congress, <www.fpc.state.gov/documents/organization/53683.pdf>.

Marsden Jacob Associates, 2006, *Securing Australia's Urban Water Supplies: Opportunities and impediments*, discussion paper prepared for the Department of Prime Minister and Cabinet, Marsden Jacob Associates Pty Ltd, Victoria.

Maunsell Australia Pty Ltd, in association with CSIRO Sustainable Ecosystems 2008, *Impact of Climate Change on Infrastructure in Australia and CGE Model Inputs*, report commissioned by the Garnaut Climate Change Review.

OECD (Organisation for Economic Co-operation and Development) 2008, *Economic Aspects of Adaptation to Climate Change: Costs, benefits and policy instruments*, OECD, Paris.

PMSEIC (Prime Minister's Science, Engineering and Innovation Council) 2002, *Sustaining Our Natural Systems and Biodiversity*, Department of Education, Science and Training, Canberra.

PricewaterhouseCoopers 2006, *A Discussion Paper on the Role of the Private Sector in the Supply of Water and Wastewater Services*, Department of the Prime Minister and Cabinet, Canberra.

Productivity Commission 2008, *Towards Urban Water Reform: A Discussion Paper*, Productivity Commission Research Paper, Melbourne.

Quiggin, J., Adamson, D., Schrobback, P. & Chambers, S. 2008, *The Implications for Irrigation in the Murray–Darling Basin*, report commissioned by the Garnaut Climate Change Review.

Rosenzweig, C., Karoly, D., Vicarelli, M., Neofotis, P., Wu, Q., Casassa, G., Menzel, A., Root, T.L., Estrella, N., Seguin, B., Tryjanowski, P., Liu, C., Rawlins, S. & Imeson, A. 2008, 'Attributing physical and biological impacts to anthropogenic climate change', *Nature* 453: 353–7.

von Braun, J. 2008a, 'Rising food prices: What should be done?' *IFPRI Policy Brief*, April, International Food Policy Research Institute, Washington DC.

von Braun, J. 2008b, 'Supply and demand of agricultural products and inflation: how to address the acute and long-run problem', paper prepared for the China Development Forum, Beijing, March 22–24.

Watkins, A.B., Diamond, H.J. & Trewin, B.C. 2006, 'An Australian season of extremes: yes, we have no bananas', in A. Arguez (ed.), 'State of the Climate in 2006', *Bulletin of the American Meteorological Society* 88: S61.

World Bank 2008, *China Quarterly Update, June 2008*, <http://go.worldbank.org/NZLENK2LK0>.

WWF–Australia 2008, *Australian Species and Climate Change*, WWF–Australia, Sydney.

SHARING THE BURDEN IN AUSTRALIA

16

Key points

Low-income households spend much higher proportions of their incomes than other households on emissions-intensive products. The effects of the emissions trading scheme will fall heavily on low-income households, so the credibility, stability and efficiency of the scheme require the correction of these regressive effects by other measures.

At least half the proceeds from the sale of all permits could be allocated to households, focusing on the bottom half of the income distribution. The bulk could be passed through the tax and social security systems, with energy efficiency commitments to low-income households in the early years.

To assist in early adjustment of low-income households, a system of 'green credits' should be introduced to help with funding of investments in energy efficiency in housing, household appliances and transport.

It is possible but not certain that regional employment issues could arise in coal regions. They would not emerge in the early years of an emissions trading scheme. Up to $1 billion in total should be made available for matched funding for investment in reducing emissions in coal power generation, as a form of preemptive structural adjustment assistance.

Climate change and its mitigation can both have significant effects on the distribution of incomes in Australia. Both can lower the income of poor households relative to others if government policy is not well designed to counteract some underlying tendencies. The largest negative distributional effects of mitigation come early, and the main effects of climate change itself after many decades.

The main guarantor of equity during rapid structural change is maintenance of economic growth and full employment within a flexible economy. This helps to ensure that there are ongoing economic opportunities for those displaced by the differential impacts of an emissions trading scheme and climate change. Contemporary Australia is well placed to absorb major structural change, given the current high demand throughout the country for skilled labour and the shortages of unskilled labour in many regions.

While sustaining these favourable circumstances is the hope of all citizens and the focus of policy, these hopes and intentions may not be continuously realised.

Even if labour displaced by the structural change associated with the mitigation regime were quickly employed elsewhere, there would still be important income distribution effects to be considered.

Alongside high employment, the most important guarantor of equity through a period of changing relative prices and structural change is the general social safety net, comprising social security arrangements, and public provision or funding of health and educational facilities. Australia is relatively well endowed in these respects, and will have opportunities to improve income transfer arrangements following the completion of the Henry Tax Review. Adjustments to the social security and taxation systems provide an opportunity for effective responses to the negative income distribution effects of an emissions trading scheme. In due course they would provide an efficient avenue of response to income distribution consequences of reforms in water management arrangements that become part of Australia's adaptation to climate change.

In the short term, the impacts of climate change on household income are not as significant as the likely impacts of the emissions trading scheme. This chapter focuses on the short-run income distribution issues associated with the introduction of an emissions trading scheme.

The emissions trading scheme will be only one of several contributors to rising electricity, gas and transport fuel prices over the next decade. In each case, rising prices from the emissions trading system are likely to be smaller than those from other market developments in the first decade of the scheme.

16.1 Effects of mitigation policy in the short term

16.1.1 How will an emissions price flow through the economy?

Although the cost of reducing emissions or purchasing permits will rest with certain parties, such as electricity generators, the cost will be passed down the demand chain. Pass-through will be quick and complete in some industries and incomplete in others, depending on the nature of the competitive environment, and especially competition from imports. It is likely to be complete for petroleum products, and substantial but possibly less than complete for electricity.

With a price on emissions, production costs will increase, with the cost of electricity, natural gas, petrol, diesel, chemicals, fertiliser and other inputs all increasing. These costs will be reflected in higher-priced goods, from cement and steel to paper and plastic. This will have an impact on the input costs for a range of industries, including construction and retail. Through the supply chain, those disposing of waste will also pay more, and transport costs will be higher. The pass-through of permit prices to users of goods and services whose production requires permits will be constrained by international competition, where there is opportunity for export and import, until there is comparable international pricing of emissions. The way this price flows through the economy is illustrated in Figure 16.1.

Figure 16.1 How will an emissions price flow through the economy?

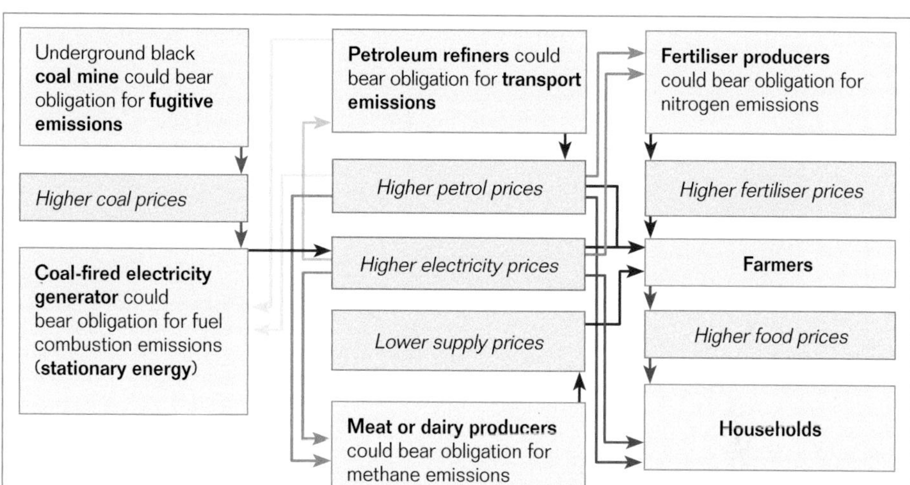

16.1.2 Effects on Australian households

Consumers will pay more for a range of goods and services as businesses pass on the emissions price. A major part, if not all, of the costs faced by electricity generators will be passed down the chain from electricity generators, distributors and retailers and finally to households through higher prices for electricity. Lower-emissions sources of electricity will receive an unrequited benefit from the high electricity prices that emerge from the high permit prices paid by emissions-intensive competitors. Petrol and food prices will rise as a result of the emissions trading scheme's coverage of emissions from transport, energy and eventually fertiliser and livestock.

These higher prices will require households to spend a greater proportion of their incomes to obtain the same goods and services purchased before the introduction of an emissions price. This will reduce households' real incomes and purchasing power. Under the arrangements proposed for the emissions trading scheme in this Review, the Commonwealth Government would receive large amounts of revenue from the competitive sale of permits, and this could be passed back to households to offset a major part of (and, for a proportion of households, the whole of) the reduction of real purchasing power. Moreover, over the time prices are projected to increase, average incomes are also expected to increase. Thus, the extent to which expenditure as a share of income changes over time will depend on the increases of prices relative to incomes and on the extent and nature of the return of permit sales revenue to households.

Low-income households spend a greater proportion of their income on basic necessities than households with higher incomes, and will therefore be disproportionately affected (Figure 16.2).

Figure 16.2 Expenditure on basic goods as a share of disposable income

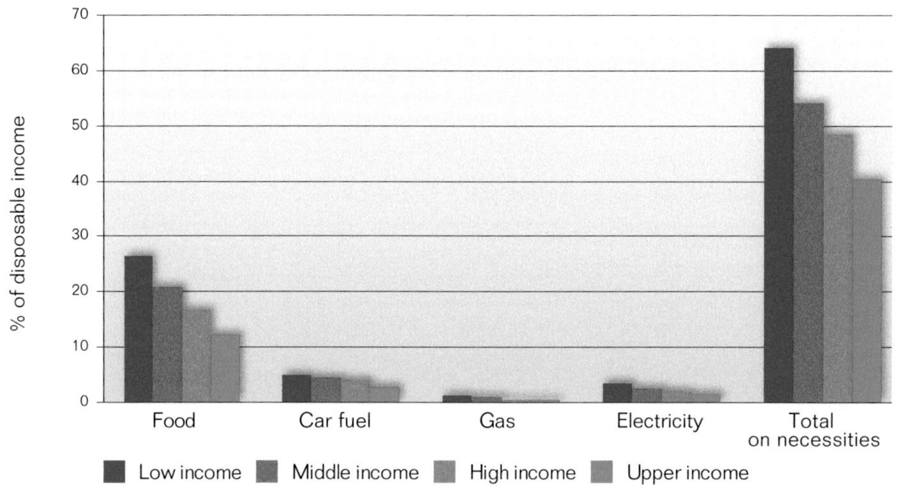

Note: Income deciles used to group households into low, middle, higher and upper income households are based on total current weekly household income from all sources divided by the (modified OECD) equivalising factor, and weighted using sample weights. The lower and upper percentiles for low, middle, high and upper income households are 10 and 29, 30 and 49, 50 and 69, and 70 and 100, respectively. Reported are the mean shares of household expenditure on necessities expressed as a percentage of disposable income.

Source: Unique expenditure codes from the Household Expenditure Survey 2003–04 (ABS 2006) grouped as 'necessities' by the Melbourne Institute of Applied Economic and Social Research.

These findings suggest that the effects of a carbon pricing regime, such as an emissions trading scheme, will fall disproportionately on households on low and modest incomes. This effect will be exacerbated, in absolute terms and relative to higher income households, by the financial constraints faced by lower income households in switching to less emissions-dependent lifestyles—such as energy efficient appliances, household retrofits and vehicle type and fuel use.

The modelling presented in earlier chapters of this report finds that average real household income rises significantly during the 21st century, even with the higher carbon price that would accompany more ambitious levels of mitigation. Thorough distributional analysis, supplemented by empirical research informed by fieldwork, will be imperative in designing policies that most appropriately redress these otherwise regressive impacts.

There is likely to be spatial variability of effects, although modelling does not disaggregate income effects finely enough to illustrate this. Electricity prices already differ among regions and, due to variability in emissions intensity, an emissions trading scheme will affect residential electricity prices more in some regions than in others (DCC 2008). Some analysis suggests the difference in effects will be significant, with New South Wales and Victorian households experiencing electricity price rises almost double those experienced by households in the Northern Territory (Australian Conservation Foundation, ACOSS & Choice, 2008: 10).

More broadly, the impacts will differ between people living in and outside capital cities and, within cities, between inner and outer suburbs. Higher product prices will be influenced by higher transport costs, disadvantaging rural or outer suburban dwellers.

More significantly, however, because of their dependence on private transport and their need to drive longer distances to access services—such as shopping, medical care, and schooling—rural and outer suburban dwellers will be particularly vulnerable to the rising fuel prices brought on by an emissions trading scheme. Low-income households are often over-represented in the urban fringe, where there is limited access to public transport (Randolph & Holloway 2005; Baum et al. 2005).

Remote Indigenous communities in northern and central Australia are likely to be particularly affected, given their reliance on diesel fuel for power supply as well as transport.

The impact of rising fuel prices, combined with low incomes and limited access to public transport, means that for some households, reducing their use of private transport will be a primary means of reducing exposure to costs. Reduced mobility has negative flow-on effects, such as feelings of isolation and social exclusion (Currie et al. (eds) 2007; Dodson et al. 2006). This makes expanded access to public transport, and the introduction of more energy- and emissions-efficient private vehicle choices, important equity issues.

The longer-term burden for households will depend on the extent to which they can reduce their exposure to emissions prices. Key questions are:

- What low-emissions substitutes are available?
- Are there any constraints on the uptake of those substitutes?

Ability of low-income households to respond to higher energy prices

Reducing energy use by using energy more efficiently would reduce households' exposure to higher prices. Chapter 17 discusses information and agency barriers that inhibit the best use of known technologies to use energy more efficiently. These barriers and underlying economic considerations cause demand by the household sector for energy—particularly electricity—to be inelastic in the short term (IPART 2003; Kamal & Stern 2001; Owen 2007). The National Institute of Economic and Industry Research (2007) estimates residential demand elasticity in Australia's National Electricity Market to be 0.25.

However, over the longer term there is likely to be greater price elasticity of demand for energy, as consumer preferences change, price increases are considered to be permanent, and assets are turned over for more energy-efficient appliances and houses.

Two factors may constrain this response in low-income households. First, low-income households are less able to fund the use of energy-efficient technologies. While such technologies might be cost effective over the longer term as lower energy bills recoup the initial outlays, people on low incomes may not be able to meet the capital cost of low-emissions technologies.

Second, due to principal–agent problems, households paying rent for housing, including those living in public or community housing, have limited incentive to pay for the capital costs of energy-saving insulation, space heating, hot-water systems and cooking appliances. Around 29 per cent of Australian households rent their homes, with a disproportionate number of these being low-income households (ABS 2007).

Ability of low-income households to respond to higher transport fuel prices

Demand for private transport has a low price elasticity where there are limited substitutes.

As fuel prices rise, responses will be determined by the proximity of public transport services. Where substitutes are available, there will be a switch in demand away from private transport and fuel. However, for many households public transport is not available. As noted above, this is particularly the case in outer suburban and regional areas, which tend to have a higher concentration of low-income households (Wulff & Evans 1699; Baum et al. 2005; Dodson & Sipe 2007). In these locations, private transport must continue to some extent, and people will be forced to pay a higher price for it.

The emissions trading scheme will raise community interest in and pressure for extension and upgrading of public transport infrastructure and services. In the best of circumstances, change in services can be expected to be slow. Noticeable improvements in public transport for many communities are likely to occur over decades rather than years. This transition is discussed in Chapter 21.

For many people on relatively low incomes in the established outer suburbs of large cities and in rural areas, there is no likely public transport substitute for the private motor vehicle. For them, an early transition to low-emissions motor vehicles with reasonable capital and operating costs is the only reliable offset to rising petroleum and fossil fuel prices. Exploration of opportunities for the rapid deployment of cost-effective low-emissions vehicles, with the electric or hydrogen-fuelled car the ultimate solution, has a strong rationale in equity as well as economic efficiency.

16.1.3 Effects of an emissions price on industry

Industries most likely to be affected by an emissions price are those:

- with a high emissions intensity
- with access to few or no substitutes to reduce their emissions intensity or exposure to increased costs under the emissions trading scheme (access by firms to substitutes may be constrained by lack of information, capital, and physical location)
- with limited capacity to pass through the emissions price (firms trading in the domestic sector will generally be able to pass through costs, while firms producing traded goods and services—with overseas competitors not subject to a commensurate emissions price—may not).

Households' reliance on emissions-intensive industry

Some emissions-intensive industries will be placed under considerable pressure by the introduction of the emissions trading scheme. At some point, their viability may depend on their capacity to adjust to use of low-emissions technologies. Their capacity to make these adjustments can determine the fate of regions that are heavily reliant on emissions-intensive industries. If adjustment is not possible, affected workers may need to pursue alternative employment. A community's capacity to manage such change may depend on the extent of its reliance on the affected industry (the number of people it employs in the local community), the availability of alternative employment, and whether workers are able to take up other vocations (Box 16.1).

Box 16.1 Labour mobility

The capacity of the labour market to adjust to changes in the demand for labour and accommodate changes in the job preferences of workers is influenced, in part, by the extent and ease of labour mobility or turnover.

Mobility has several elements. It can involve a change of job between firms in the same industry, a movement between firms in different industries, or a change in occupational type. It may also involve a change in location. Where industries and/or regions are subject to general decline, the potential for interindustry and locational mobility is important for accommodating adjustment (Productivity Commission 1998).

Workers in declining industries who have skills and experience that are valued in other activities (and ideally, can find employment in those activities in their local area) are likely to be much less disadvantaged by structural change than workers with a narrower range of re-employment opportunities, or those who must move to another location to find work.

Impacts may be concentrated in particular regions or towns that rely heavily on the affected industries. They are likely to be felt more in rural and provincial communities, where there are fewer alternatives than in cities. The two industries likely to be most affected—coal-fired electricity production and agriculture—are discussed below.

Australia's coal-based regions

The success of Australia's emissions trading scheme will be inextricably intertwined with the future of the coal industry. The coal industry underpins Australia's domestic electricity supply sector, and is by far our biggest export commodity. The domestic coal-fired generation sector, the domestic and export mining sectors and, most importantly, the communities who live and work in centres of coal mining and coal-fired power generation, all have a vital stake in the long-term viability of the industry.

The prospects for the coal regions depend on the resolution of several powerful cross-currents. In the immediate future, the high price of black coal—metallurgical and thermal—on the Asian and world markets will underwrite prosperity and expansion for regions producing this commodity. High export prices for black coal (and for gas, now potentially exportable from the east coast) have been improving the competitive position and profitability of power generation based on brown coal in the Latrobe Valley of Victoria. These tendencies are likely to be more powerful than an emissions price in the prosperity of black and brown coal regions, at least through to the conclusion of the Kyoto period at the end of 2012.

As discussed at length in Chapter 20, developments after that depend on, among much else, the success of technological change in reducing emissions in coal-based electricity generation, and on whether there is a global mitigation agreement. Within 550 mitigation (and for a while probably 450 as well), existing coal-based generators would remain competitive, with higher electricity prices balancing some modest contraction of volume. Effective investment to reduce emissions using known technologies could greatly improve the competitive position of Victorian generators (brown coal) in particular. The early retirement of one of the older and economically and environmentally less efficient generators would greatly improve the commercial prospects of others—for a while maintaining old volumes at higher prices.

Effective mitigation would open up some new opportunities for investment and employment in the Victorian coal fields.

In the end, the future of coal depends on successful carbon capture and storage, through geosequestration or biosequestration. Global mitigation commitments would make this important for coal exports as well as domestic markets. Successful partial sequestration (say 90 per cent capture), perhaps based at first on retrofitting old facilities, would provide expansion opportunities for several decades. After that, coal's future, export and domestic, will depend on commercial success of near-complete capture.

Any large negative impacts in the coal regions are many years away. With effective application of known technologies to reduce emissions in the immediate future, and commercially successful carbon capture and storage after that, the future prospects are for continued expansion.

This view of the future of the coal regions has clear implications for structural adjustment assistance to coal communities. Support for transformation to lower emissions and then near-zero emissions use of coal is the main priority in the years immediately ahead. Only if that fails will it be necessary to plan support for coal regions that are disadvantaged by major reductions in local employment.

Targeted transitional assistance for the coal regions is discussed in section 16.2.4.

Australia's agricultural regions

From the commencement of an emissions trading scheme, costs of agricultural inputs—electricity, liquid fuel and fertiliser—will rise. This will particularly affect parts of the sector where energy costs and energy-dependent costs are a large proportion of total costs. In the event of coverage of agriculture in the scheme, those parts of the sector with high direct emissions—for example, methane emissions from sheep and cattle—will incur costs.

Over time, as the emissions price increases, a range of new low-emissions commodity options—such as biosequestration and bioenergy—will become increasingly cost effective. Most landowners will have the option to continue with traditional forms of farming that remain profitable, or switch to lower-emissions forms of production. Low-emissions substitutes are already technically feasible for the agriculture sector, and will be economically feasible with a sufficient carbon price.

16.2 A framework for government intervention

16.2.1 Why income distribution effects warrant government intervention

The price imposed by an emissions trading scheme is not intended to result in large, arbitrary transfers of wealth, especially regressive changes in income distribution. There is a clear role for government in ensuring distributive efficiency and addressing the social welfare implications of climate change mitigation policy on those people who are most affected by an emissions price and least able to respond (see Box 16.2).

Box 16.2 Distributive efficiency and social welfare implications of an emissions trading scheme

Economic efficiency is of the utmost importance in designing an emissions trading scheme but distributive efficiency is also an important consideration. Distributive efficiency occurs when goods are distributed to those who gain the most utility from them (Lerner 1944).

It is accepted that income has diminishing marginal utility—that is, an extra dollar has more utility to the poor than to the rich. Income distribution is a key dimension of welfare. The introduction of an emissions price without consideration and assistance to low-income households will reduce social welfare.

The initial transfer of wealth as a result of the emissions trading scheme will have impacts on the distribution of income—some of them inequitable. The way in which the wealth transfer is handled in the longer term—that is, the use of permit auction revenue—will determine whether or not that income distribution is corrected. Therefore, in responding to the impacts of the emissions trading scheme, equity must be considered.

The first lines of defence during rapid structural change are maintenance of economic growth and full employment within a flexible economy, and the maintenance of the general social safety net. Contemporary Australia is well placed to absorb major structural change, and it is important to the success of adaptation to climate change that this continues to be the case. The main focus of this section is the effects not covered by these two defences: regressive income effects and the concentrated employment effects in some regions.

To address these effects, the Review proposes a package of measures limited to three key elements where the rationale for assistance is strongest:

- general assistance to most households through efficiency-raising improvements to the social security and tax systems
- targeted assistance to address the regressive income effects of an emissions trading scheme on low-income households and facilitate a more efficient transition
- targeted assistance for regions grossly affected by the loss of livelihood as a result of the implementation of an emissions trading scheme.

Each of these elements has a strong but different underlying policy rationale, which should guide the timing of implementation and policy design details.

16.2.2 General assistance: social security and tax systems

Many market imperfections and policy-related distortions affect the adjustment process across the economy rather than having effects that are specific to particular industries or regions.

An emissions trading scheme will affect a broad range of goods and services throughout the economy—including, food, fuel, and housing—with a flow-on effect across almost all aspects of daily living. These higher-priced necessities will particularly affect low-income households (see section 16.1.2). The degree of impact across low-income groups will depend on the structure of each individual's consumption bundle, which will itself depend on a range of factors.

This diversity of effects and differences in consumer preferences means that it is difficult for government to determine household trade-offs between these goods. This argues against assistance measures that are prescriptive or directive in their nature, and suggests that households can be most efficiently assisted through the taxation and social security system, particularly by the reduction of distortionary taxes.

Reductions in taxation rates, such as the lowering of marginal effective tax rates at the lower end of the income range, have the dual benefit of increasing household disposable income as well as stimulating labour supply and increasing the efficiency of the taxation system.

Increases in social security payments and/or amendments to social security tapers can address equity issues for households and individuals out of the labour force or in retirement, while also increasing labour supply and the efficiency of the welfare system.

Direct financial support through the taxation and social security systems will have a more important role as impacts become greater. Adjustments through the taxation and social security systems would need to be permanent.

If the emissions permits are auctioned, as recommended in Chapter 14, the sales revenue will provide an important source of funding for assistance. It is proposed that half of the permit sales revenue be allocated to payments to households, focusing on the lower half of the income distribution.

Changes to the tax and social security systems to assist low-income households following the introduction of an emissions trading scheme should be integrated with the Henry Tax Review.

16.2.3 Targeted assistance for low-income households

While there are benefits in providing assistance through general measures, there are some equity issues that cannot be addressed with broad improvements to the social welfare or tax systems. In particular, general financial support may fail to assist households without sufficient capital or information to make the change to energy-efficient technologies; rental households or those living in public or community housing with limited incentive to access substitutes; and households for whom public transport is not available. In these cases, government may need to provide more targeted assistance for these households to make the transition to a lower-emissions future.

In designing such assistance, it is important that the emissions price signal, designed to encourage a shift toward a lower emissions economy, not be blunted or eroded. Instead, in addition to addressing equity concerns, such measures should be tailored to improve the efficiency of the adjustment process by addressing market-based impediments to adjustment (Productivity Commission 2001). This provides a strong rationale to use a proportion of permit revenue to fund this assistance.

To assist low-income households with the upfront capital and information for the purchase and installation of energy-efficient services, products and appliances, the Review proposes that the federal government, with assistance from the state and territory governments, establish a system of 'green credits'. This system would combine an upfront once-off grant for households in the bottom quarter of the income distribution with third-party audits where requested.

Households could choose to use the grants of $1000 per person in the household to assist in the purchase of a list of approved items. To address information barriers, the list would identify products that would improve energy efficiency in housing or transport.

Alternatively, households could elect to have a third-party audit. Auditors would identify energy efficiency opportunities in these households, and provide a wider range of options for these households to use their credits, including household appliances. The Review considers the audits could be provided most efficiently by the private sector, at a cost of around $150 per household. Governments could provide additional funding for low-income households that are particularly vulnerable to rising energy prices for a more detailed audit, as they have significantly higher exposure to high energy prices and greater opportunities for energy efficiency savings—savings that would easily exceed the cost of the audit.

Should households prefer not to select an eligible technology from the list provided, the grant would be made available in cash after a period of five years.

The total cost of the system would amount to approximately one-quarter of the permit revenue allocated to assisting households in the lower half of the income distribution for the first five years of the emissions trading scheme. The grants should be provided by the federal government, with state and territory governments providing administration of the scheme.

The early years of the emissions trading scheme—indeed, in the period leading up to it—would be the appropriate time for implementation of the green credits system to reduce the impacts on low-income households when the emissions trading scheme is introduced. Given an emissions trading scheme start date of 2010, the scheme should begin in 2009 and run for five years to 2013, with the value of the system rolling into the general assistance following this.

To assist low-income renters in public housing, the government should invest in increasing the energy efficiency of its public housing stock. For example, in Victoria, the Energy and Water Taskforce program retrofits public housing estates and low-income housing with energy efficiency improvements, resulting in notable decreases in gas and electricity consumption (Sustainability Victoria 2008).

To assist outer suburban households, improving access to public transport will also be important in adjustment to a low-emissions future. Rolling out additional public transport services will take time, and may not be viable or cost effective in some locations. Therefore, in the shorter term the assistance in the green credits system for low-emissions, energy-efficient private transport will be the main form of assistance for these households.

16.2.4 Targeted assistance for coal-generation regions

Many impacts of an emissions trading scheme are specific to certain sectors, industries or regions, but few of these warrant government assistance. A flexible labour market, with sufficient employment opportunities and a strong social safety net, will preclude the need to provide targeted assistance for all sector-specific impacts.

An emissions price may seriously affect some business and asset owners, and shareholders, through its effect on industry. Claims by these groups for special consideration on equity grounds should be assessed by government alongside the equity claims of others. It is unlikely that effects on these groups would result in inequitable outcomes, relative to effects the emissions trading scheme will have on other groups in the community.

Additional assistance measures that target particular sectors, industries or regions therefore are only likely to be appropriate where there are wider income distribution considerations, notably regional income effects. Inadequate arguments for special compensation are discussed in Box 16.3.

Box 16.3 Inadequate arguments for government assistance

A number of false arguments are sometimes advanced to justify compensation for the impacts on an emissions trading scheme on the most affected industries:

- There will be a large and rapid decline in the value of emissions-intensive assets, and their future profitability will decline. These negative impacts will be disproportionate compared to the rest of the economy; compensation could offset these disproportionate income effects.
- An 'unanticipated' regulatory change, such as an emissions trading scheme, could undermine investor confidence; compensation would increase investor confidence about market operation in the face of future change.
- Adverse affects on the investment climate could jeopardise or delay required investment; compensation would reassure current and future investors and help facilitate necessary investment.

The Review considers that these arguments are inadequate to justify government intervention, for a number of reasons. First, most domestically traded industries will be able to pass on the costs of the emissions trading scheme, and thus will not require assistance to recover costs or avoid potential losses.

Second, the emissions trading scheme does not represent a sovereign risk issue, but a policy risk. Individuals and firms win or lose from market-based changes every day; this is part of the normal operation of markets. Governments do not provide guarantees that regulation will remain unchanged or that asset values will be immune from policy changes (Productivity Commission 2001), and industry is generally cognisant of the risk. Industry has been aware of a carbon price for some time; the fact that the industry has been citing uncertainty on climate change policy as a deterrent to new investment would suggest that it has been recognised for many years.

Third, though impacts of emissions trading are likely to be significant, policy changes that adversely affect asset values, without compensation, are not unusual. Government does not, as a matter of course, compensate asset owners when environmental or social externalities are internalised (for example, compensation was not provided to the tobacco or asbestos industries).

Finally, even if some existing investors are deterred from future investment, it is expected that a clearly communicated and credible policy response to climate change will provide significant investment opportunities and that these will be attractive to an adequate range of both new and existing investors.

For a case to be made for public support on those grounds, it must be shown also that such measures can cost-effectively improve the efficiency of the adjustment process. Special transitional assistance is only valid under exceptional circumstances where there are clear market barriers to an efficient transition.

The Review recognises that market imperfections and policy-related distortions can make transitional unemployment and production losses associated with policy-induced changes larger and more sustained than they might otherwise be (Productivity Commission 1998).

Assistance for adversely affected coal regions

In considering the various impacts of an emissions trading scheme on different sectors, industries and regions, there is one geographic area the Review identified where such targeted transitional assistance may turn out to be warranted—the brown coal region of the Latrobe Valley. It will be a number of years before it is clear whether there is likely to be a regional problem.

The Latrobe Valley satisfies the dual criteria in that:

- Brown coal electricity generation is one of the most emissions-intensive industries in Australia, and the expected consequences may be severe, depending on the range of factors affecting future competitiveness (see section 16.1.3), and concentrated in the region.
- There would be limited opportunities for the employment of people who may be made redundant in the event of industry decline.

Other areas that may be considered comparable, such as the black coal mining, exporting and power-generating regions of the Hunter Valley would seem to face less severe adverse impacts due to the ongoing strength of exports and other employment options; this may change in the event of a successful international mitigation agreement. Regions with emissions-intensive agriculture may be severely affected by the emissions price but have the options to diversify towards less emissions-intensive production or to seek alternative employment for their labour force.

Once appropriate recipients have been identified, government must then consider the most appropriate form for targeted assistance. In doing so it should adhere to three principles:

- **Assistance should be non-distorting**—It is important that the form of assistance provided does not distort the incentives to move away from emissions-intensive practices.
- **Assistance should benefit households and communities**—The form in which any transitional assistance is provided needs to ensure that the benefits of that assistance accrue to households and communities in the form of ongoing employment in, and the continued economic viability of, the region concerned.
- **Assistance should be transitional**—It should move industry toward a low-emissions future, not provide compensation for loss or in order to continue the status quo. The long-term viability of the domestic coal industry as an employer and source of income in regional Australia hinges on the successful transition

to low-emissions technologies that will allow the industry to keep operating. This is, in turn, ultimately dependent on whether a near-zero emissions future is feasible. If it is not, then Australia needs to know as soon as possible, so that all who depend on the coal industry can begin the process of adjustment, and so that adequate and timely investments can be made in other industries.

If the coal industry in Australia is to have a long-term future in a low-emissions economy, it will have to be transformed to near-zero emissions. A rising permit price will provide increasingly strong incentives for investment in low-emissions technology. Innovation in the sector will also benefit from the early research funding and investment in demonstration and commercialisation from the matched funding scheme suggested in Chapter 18.

In addition, the Review proposes a specific one-off allocation of additional funding, in the order of $1 billion, to facilitate structural adjustment in the industry. It is suggested that this be made available preemptively, so as to reduce the probability that other regional support measures may be required in future. This 'targeted transitional assistance' would match industry funding on an equal basis, and would be available to support investment in reducing emissions in coal-based generation in existing and new plant through the adoption of emissions-reducing technologies. For example, funding could be provided to retrofit a facility with low-emissions technologies to support drying of brown coal and restructuring of boilers to use the product, or to improve use of materials and logistics. For these structural adjustment purposes, it would not be necessary for the technology to be new.

Such investments will maintain the short-term viability of communities such as those in the Latrobe Valley during the early period when the necessary new technologies such as carbon capture and storage are being developed. If the expectation in the industry is that carbon capture and storage is likely to allow the sector to continue competing in the low-carbon economy, this targeted transitional assistance will help generators to operate more efficiently in the interim, thereby allowing a less volatile transition.

If a time comes when the industry considers that neither carbon capture and storage nor any other technology is likely to allow continued profitable operation in the long term, firms will not invest their own capital in low-emissions technology, and they will not make demands on matching funds. Governments will then need to provide measures such as the retraining of workers for new employment (as with textile and steel workers in the 1980s after reductions in protection); and provide grants to communities to support improvements in infrastructure that would be helpful in attracting alternative industries to affected regions. Any remaining funds in the targeted transitional commitment should be directed towards such programs.

16.3 Long-term impacts and structural change

Over the next decade, the distributional implications for Australian households that flow from the impacts of climate change are mild compared to the likely effects of an emissions trading scheme as described in section 16.1. Over the longer term, the magnitude and scope of climate change impacts, and consequent impacts on household incomes, will be determined by the ambition and effectiveness of the global mitigation agreement.

If an effective global mitigation agreement is not reached, climate change will have significant income distribution effects. The possible impacts of climate change on Australia beyond 2030 will be varied and extensive, and will be unevenly spread across the country. Climate change impacts will mostly be gradual but could also involve shocks, either in the form of severe weather events, a rapid decline of productivity in certain areas, or other sudden socio-economic or environmental shifts. Some communities in severely affected regions will experience disadvantage from structural change.

The redistribution of income will initially be the result of direct climate impacts such as increases in temperature or declining precipitation. These impacts will result in an increase in the scarcity of some resources, such as water and land, a decline in the productivity of some industries, and an overall decline in domestic or international competitiveness. The clearest example of this is the impact on households and communities that rely on agriculture for income and employment.

Storm surges, bushfires and other severe weather events may have severe effects on economic and social life in some regions. They will, at a minimum, raise the price of insurance in affected areas. In the most extreme cases, some regions will be rendered less habitable, to the point that there will be a need for communities and industries to relocate.

A significant proportion of the income distribution effects of climate change and climate change policy will come from changes in the industrial make-up of the economy over the longer term. Regional communities and industries are likely to be more vulnerable to these impacts than urban centres, due to their reliance on agriculture and other natural resource-based industries, and low levels of infrastructure stock. Regional communities, in particular farming regions, have already been subject to structural change to a much greater extent than metropolitan centres in recent history (Productivity Commission 1998).

These are issues for policy in the longer-term future.

References

ABS (Australian Bureau of Statistics) 2006, *Household Expenditure Survey, Australia: Summary of results, 2003–04 (reissue)*, cat. no. 6530.0, ABS, Canberra.

ABS 2007, *Housing Occupancy and Costs, Australia, 2005–06*, cat. no. 4130.0.55.001, ABS, Canberra.

Australian Conservation Foundation, ACOSS & Choice 2008, *Energy and Equity—Preparing households for climate change: efficiency, equity, immediacy*, <www.acoss.org.au/upload/publications/papers/4204__EnergyEquity%20low%20res.pdf>.

Baum, S., O'Connor, K. & Stimson, R. 2005, *Faultlines Exposed: Advantage and disadvantage across Australia's settlement system*, Monash University Press, Melbourne.

Currie, G., Stanley, J. & Stanley, J. (eds) 2007, *No Way to Go: Transport and social disadvantage in Australian communities*, Monash University ePress, Melbourne.

DCC (Department of Climate Change) 2008, *Emissions Factors for Consumption of Purchased Electricity by End Users*, DCC, Canberra.

Dodson, J., Buchanan, N., Gleeson, B. & Sipe, N. 2006, 'Investigating the social dimensions of transport disadvantage—I. Towards new concepts and methods', *Urban Policy and Research* 24(4): 433–53.

Dodson, J. & Sipe, N. 2007, 'Oil vulnerability in the Australian city: assessing socioeconomic risks from higher urban fuel prices', *Urban Studies* 44(1): 37–62.

IPART (Independent Pricing and Regulatory Tribunal) 2003, *Inclining Block Tariffs for Electricity Network Services*, Secretariat Discussion Paper DP64.

Kamal, M. & Stern, D. 2001, *The Structure of Australian Residential Energy Demand*, Working Paper in Ecological Economics, no. 0101, Australian National University, Canberra.

Lerner, A.P. 1944, *The Economics of Control*, Macmillan, New York.

National Institute of Economic and Industry Research 2007, *The Own Price Elasticity of Demand for Electricity in NEM Regions*, prepared for the National Electricity Market Management Company.

Owen, A.D. 2007, *Report of the Owen Inquiry into Electricity Supply in NSW*, New South Wales Government, Sydney.

Productivity Commission 1998, *Aspects of Structural Change in Australia*, Research Report, AusInfo, Canberra.

Productivity Commission 2001, *Structural Adjustment: Key policy issues*, Commission Research Paper, AusInfo, Canberra.

Randolph, B. & Holloway, D. 2005, 'Social disadvantage, tenure and location: an analysis of Sydney and Melbourne', *Urban Policy and Research* 23(2): 173–201.

Sustainability Victoria, 2008, Energy and Water Taskforce, <www.sustainability.vic.gov.au/www/html/1464-energy-task-force.asp>.

Wulff, M. & Evans, S. 1999, 'The spatial impacts of Commonwealth rent assistance on Australia's low income households', in J. Yates & M. Wulff (eds), *Australia's Housing Choices*, University of Queensland Press, Brisbane.

17 INFORMATION BARRIERS TO KNOWN TECHNOLOGIES

Key points

There are potentially large and early gains from better utilisation of known technologies, goods and services, including energy efficiency and low-emissions transport options.

Externalities in the provision of information and principal–agent issues inhibit the use of distributed generation and energy-saving opportunities in appliances, buildings and vehicles.

A combination of information, regulation and restructuring of contractual relationships can reduce the costs flowing from many of the market failures blocking optimal utilisation of proven technologies and practices.

The introduction of an emissions trading scheme will increase returns for business and households from adopting opportunities to lower their direct and indirect emissions. The opportunities will often involve adopting existing technologies and practices. However, market failures will impede adoption of opportunities that may be profitable once applied. Policies that tackle these market failures would lower the cost of mitigation across the economy.

As climate change impacts begin to be felt, there will also be cost-effective opportunities to adapt to these impacts by changing economic behaviour, with some changes requiring investment. As with migitation, reducing information and agency barriers will lower the costs of adaptation across the economy.

17.1 The impact of information and agency barriers

Two kinds of market failures are especially important in inhibiting the adoption of established technologies and practices. One relates to externalities in the supply of information and skills. The other involves a principal–agent problem—where the party who makes a decision is not driven by the same considerations as another party who is affected by it.

These market failures are most important for small to medium enterprises and households, where the benefits of reducing emissions are small relative to the transaction costs of securing them. Large firms are more likely to overcome information and principal–agent barriers, but may still miss opportunities where the benefits are relatively small or diffuse.

For mitigation, the barriers largely affect the adoption of energy efficiency (see Box 17.1), fuel switching and small-scale generation in buildings, industry and transport.

For adaptation, they will affect the adoption of water efficiency measures, and improvements to buildings to withstand the impacts of climate change (see Box 17.2). Private adoption of some water efficiency measures can occur over relatively short time frames. This is not the case for improvements to buildings, which typically have a life of 40 to 60 years or more, meaning that those built and modified today will need to be able to withstand the impacts of future climate change. The success of this will be inhibited somewhat by the availability of reliable information regarding climate change impacts and appropriate responses. This chapter focuses on changes required in the short to medium term and therefore discusses buildings alone in considering barriers to adaptation.

Box 17.1 What is energy efficiency?

Energy efficiency generally refers to reducing the amount of energy required to deliver an amount of a service, such as kilowatts per unit of heat. The International Energy Agency (2006) has estimated that increased energy efficiency could account for 45 to 53 per cent of global emissions reductions in projections to 2050.

Energy efficiency does not always correspond to economic efficiency, which involves maximising the efficiency of use of all resources (Sutherland 1994). Where efforts to improve energy efficiency require more input of capital, labour and other resources than is saved in energy, economic efficiency would be reduced.

Nevertheless, the evidence indicates that there are significant opportunities for increased energy efficiency in Australia that are economically beneficial (Allen Consulting Group 2004), despite methodological issues in accurately determining the quantum of the opportunity (Productivity Commission 2005).

Box 17.2 Impacts of climate change on buildings

The Commonwealth Government established a project in 2005 to investigate the capacity of Australia's building stock and building practices to maintain current levels of amenity in the face of a changing climate and the scope to consider changes in building practices to adapt to climate change.

The main impacts of climate change with implications for Australian buildings were found to be:

- increased energy consumption due to higher temperatures
- increased risk of damage from more intense tropical cyclones, storms and winds
- damage to foundations and pipe work from increased ground movement due to reductions in soil moisture
- increased flood damage from intensified weather events
- increased bushfire risk due to higher temperatures (BRANZ 2007; Holper et al. 2006).

It was also established that the Building Code of Australia, which already addresses the issue of minimum structural performance standards in buildings, will have a significant role to play to ensure that public health, safety and amenity are not put at risk by the impacts of climate change.

17.1.1 Mitigation potential

Various studies attempt to estimate the extent of mitigation opportunities in different sectors. Work by the IPCC (2007: 9, 409) suggests that the majority of global mitigation potential to 2030 at under US$20 per tonne of CO_2-e would occur in areas affected by information and principal–agent market failures, with around 5 billion tonnes of mitigation potential in the building sector alone out of a total abatement potential of 9–18 billion tonnes in all sectors.

Similarly, work by McKinsey & Company (2008) estimates that in 2020 Australia's emissions could be reduced by around 11 per cent below business as usual levels through zero and negative net cost mitigation opportunities.

Some studies of energy efficiency are overly optimistic as they do not include potentially unavoidable transaction costs from the uptake of more efficient products, such as time spent in information gathering and decision making, policy costs and appropriate discount rates (Stavins et al. 2007). Conversely, many studies are also conservative in limiting the potential for future technology development.

17.1.2 Rationale for additional policies

Some of the reasons given for government intervention to improve energy efficiency lack a sound economic basis. An effective emissions trading scheme would address the issues of reducing greenhouse gas emissions and urgency of action. The rationale for policies to support the uptake of low-emissions technologies and practices should be the correction of market failures that increase the cost of mitigation or market failures related to other issues. If these market failures cannot be tackled cost-effectively then there is no case for action.

Reasons for energy efficiency policy that do not have a sound economic basis include:

- Energy efficiency policy is needed to ensure sectors meet their targets. Once an emissions trading scheme is in place the cap will prevent emissions from increasing in covered sectors.
- We need to invest in energy efficiency to lessen the impact of the carbon constraint. Investing in energy efficiency when there are no requisite market failures requiring correction is likely to lead to greater economic cost, not less.

These reasons should be rejected.

17.2 Information barriers

Individuals will rarely have perfect information relevant to a decision they are making. However, efficient adoption of established technologies and practices requires individuals to know:

- the options available
- the approximate costs and benefits of the different options
- how to deploy the options (including hiring experts)
- the cost of investigating the options.

Governments should not be expected to fill the gap in every situation where individuals lack sufficient information to make good decisions. However, where information barriers are caused by market failures, governments may sometimes be able to improve the efficiency of the market.

These market failures have their origins in the public good nature of some information, information asymmetry and bounded rationality. They are discussed below, together with policies to address them.

17.2.1 Public good information and spillovers

Some information is a pure public good as it is not possible to exclude individuals from using it, and one person's use of that information does not prevent others from using it.

Where information has public good characteristics, it is likely to be underprovided by the private sector (Jaffee & Stavins 1994a). The private sector may disseminate information with public good characteristics, for example through consumer magazines. However, as firms are not able to capture all the benefits from public good information, there is insufficient incentive to make information as extensive and widely available as consumers may demand.

Training and education have positive benefits to society and support the use of available information. Even if individuals have access to information regarding established technologies and practices, they, or commercial agents supplying services to them, may require new skills or a wider body of knowledge to use that information (Consumer Affairs Victoria 2006). Given the wide range of technical issues associated with energy efficiency, gaps in the skill sets of specialists such as engineers or tradespeople could prevent the uptake of these options across a range of sectors.

17.2.2 Information asymmetry

Information asymmetry occurs when two parties to a transaction do not have equal access to relevant information.

There are potentially significant information asymmetries for appliances, vehicles and houses as it is extremely difficult for non-experts to determine the ongoing energy used by, for example, an appliance without outside assistance. This allows opportunism, as a product manufacturer could mislead a buyer on the efficiency and efficacy of a product, which the buyer is unable to verify.

As noted by the Productivity Commission (2005), market participants may attempt to gather or verify information to reduce information asymmetries through such expedients as obtaining an assessment of a product before they buy it. This can be costly and may only be done for large purchases such as houses or cars.

Some features of a good can increase the likelihood of information asymmetries. Where the quality of the good can be determined before purchase, there will be limited information asymmetry. Where the quality of the good can be determined only after purchase, repeat purchasing will overcome information asymmetries where the good is purchased regularly. Where the quality of a good cannot be determined even after purchase, it is difficult to overcome information asymmetries (Sorrel et al. 2004).

Adverse selection

Information asymmetry can lead to adverse selection, which can occur where sellers are better informed than buyers, resulting in lower-quality goods dominating a market (Akerlof 1970).

In a market where it is difficult for buyers to verify whether a product is of good or bad quality, they may be unwilling to pay a premium for goods that are actually of good quality. Even if manufacturers voluntarily give information on a product's quality, buyers may be wary of this information (Aronson & Stern 1984). Where this occurs, there would be limited incentives for manufacturers or developers to produce higher quality products (Jaffee & Stavins 1994b).

In the markets for appliances and houses, tenants and users of appliances have a strong incentive to reduce ongoing energy costs. Developers and manufacturers do not have this incentive unless they can command higher prices for more efficient buildings and appliances (Golove & Eto 1996) and in fact have a strong incentive to lower the upfront costs, usually by avoiding energy-saving features. Unless buyers can confidently assess the energy efficiency of buildings and appliances, most goods for sale on the market will be less energy efficient than if buyers could be sure of their quality.

17.2.3 Bounded rationality

Even where people have access to sufficient information, they may make decisions that are suboptimal. Situations of suboptimal decision making or 'bounded rationality' have been observed and documented in the behavioural economics literature (Camerer et al. 2004).

First, people faced with complex decisions, often use rules of thumb to aid decision making. Some rules of thumb deliver broadly accurate results. However, when Kempton and Montgomery (1982) examined how people estimated savings from investments in insulation, they found households significantly underestimated its cost-effectiveness.

Second, people often assign a budget in their own mind for a particular class of expenditure, hoping to constrain their expenditure (Thaler 1999). The implication is that if people have assigned a low budget to capital improvements in their home and a high budget to variable utility costs, they may be unwilling to reallocate their budget to undertake a capital upgrade that would lower their overall expenses.

Third, there are some predictable biases in human decisions that could result in decisions that are both personally and socially suboptimal (Kahneman & Tversky 2000). Particularly important biases include:

- biases towards the status quo
- high rates of discounting of future costs and benefits (IEA 2005).

Finally, information can be difficult to use, which may prevent people from weighing up the costs and savings of various options. Even where savings are known, households may pay them limited attention compared to their perceptions of upfront costs, effort, comfort and social norms (Komor & Wiggins 1988). In particular, people may have difficulties in making use of information that is probabilistic in nature (Camerer & Loewestein 2004). This factor will have particular implications for those faced with investment decisions necessary to avoid risks from the projected impacts of climate change.

17.2.4 Tailored information, education and training

Information, education and training programs can tackle the undersupply of public goods directly. Information and education programs have strong synergies with an emissions trading scheme, as they can help individuals to identify the energy and other costs affected by a carbon price and respond to it. This is particularly important during the scheme's initial phase, when the costs of many goods and services will change.

Public information programs

Basic media campaigns and pamphlets are often neither targeted nor tailored and there is considerable evidence that their effectiveness is limited (Cone & Hayes 1980). Information programs for households are more effective if they consider social and attitudinal issues and involve alternative communication techniques such as audits, community-based programs and diffusion through social networks (Shipworth 2000). Developing these types of programs generally requires:

- identifying target groups and assessing their knowledge, attitudes and behaviours
- developing communications, possibly using social networks
- testing, evaluating and improving the program before rolling it out.

When governments lead by example, such as undertaking energy efficiency audits, this can support the credibility of such programs (Bjornstad & Brown 2004).

Programs need to be targeted and tailored to ensure that the right individuals receive suitable information. This seems to be done particularly well in the Western Australian Government's TravelSmart program (Box 17.3). Programs relevant for the general public include those that raise awareness of the benefits of energy efficiency, provide basic information on low-emissions practices and educate consumers on how to identify the costs and benefits of different low-emissions options.

A well-designed information program should:

- attempt to overcome biases by providing a simple comparison between current and future costs and current and future benefits
- use familiar language (such as payback periods)
- be located as close to the point of sale as possible.

In designing information programs for households, governments should tailor the program to their target audience, draw on the extensive literature on bounded rationality, and not rely on basic media campaigns.

Box 17.3 Tailored information: TravelSmart

Some individuals do not have ready access to basic information about the transport options that are available to them and the costs and benefits of those options. Interviews in Perth suggested that information failures may have prevented 24 per cent of all trips being switched from car to other modes of transport. The TravelSmart Household Program in Perth aims to overcome these information failures through tailored information provision, including:

- localising and simplifying information to make it relevant to people's needs
- providing motivation through dialogue and personalised communication
- assisting new users of public transport to navigate the system.

Education and training for specialists and industry

Where extensive knowledge and skills are needed, education and training programs will be more effective than public information programs. In industry, formal education and reskilling courses are generally suitable for addressing the lack of skilled professionals, such as engineers. There are also gaps in organisation-wide skills that support energy management, such as energy reporting (Paton 2001). Here companies may need to be engaged directly, as general information provision may be limited in its effectiveness (Energy Consult 2002).

Like public information programs, education and training programs need to be targeted and tailored. Target groups for programs should include:

- market intermediaries such as retailers and estate agents—for basic education programs
- managers and other non-specialists in business—for programs that raise awareness of practices for energy and carbon management
- specialists—for programs that cover practical skills in the installation and maintenance of low-emissions options for trades such as building and plumbing, and a mixture of theory, knowledge and skills for professions such as engineering (Desha et al. 2007).

Pprograms should be tailored around the information needs and structures of sectors and should use existing approaches where suitable, such as extension programs in the agricultural sector. New structures, such as the independent Carbon Trust that was established in the United Kingdom to specialise in delivering knowledge and skills to firms, may also be required.

As a general rule, participation in programs should be voluntary, allowing firms and individuals to make decisions based on the benefits of the program to them. Where certain costs and uncertain benefits confuse this decision, certification programs could provide an incentive for specialists to learn new information or mandatory requirements may be necessary. These requirements should only exist early in the transition to the carbon-constrained economy, as, in future, these new energy management processes will become integrated into standard business practices.

The Commonwealth Government's Energy Efficiency Opportunities program provides an example of combined mandatory and voluntary activities. The program requires businesses using more than 0.5 petajoules of energy per annum to undertake an audit to identify areas for efficiency gains. To assist them, the government provides instructions and free training. Implementation of the audit recommendations is voluntary. The final review of the precursor to this program, the Energy Efficiency Best Practice program, found that it had been cost effective, and that projects planned under the program could save $74 million by 2010. The regulatory impact statement conducted before the introduction of the Energy Efficiency Opportunities program estimated a net present value of $760 million (Parliament of Australia 2005).

The Review has formed a favourable view of the Energy Efficiency Opportunities program. Governments should remove overlapping and mandatory programs.

It would be productive for the Ministerial Council on Education, Employment, Training and Youth Affairs to investigate the support required to enable universities and other education institutions to deliver training, education and certification in low-emissions and climate change resilience options to specialists, particularly engineers, tradespeople and business managers.

In the building sector this could specifically entail:

- developing retraining courses and incorporating energy efficiency components into vocational and university courses
- providing tools such as design guides and advisory services
- fostering on-site training through demonstration programs
- introducing accreditation to provide an incentive for specialists to learn.

17.2.5 Third-party programs

Information programs may be less effective when they attempt to convey complex information or where habits or practices are entrenched. Specialists, such as energy service companies, can use economies of scale in gathering and processing information. These companies are paid by firms to make decisions about which technology to buy, thus spreading the cost of gathering information across several parties.

Unfortunately, transaction costs make current forms of energy service contracting less suitable for smaller parties with significant information and bounded rationality problems, such as households and small businesses (Sorrell 2005).

Various countries have attempted to foster the market for energy service contracting and auditing although with limited success to date (Eoin Lees Energy 2006). For example, energy retailers could offer contracts to households for 'services' such as heating, hot water and appliances, creating an incentive for the retailers to improve households' energy efficiency.

Governments have tried two approaches to overcoming this problem. One subsidises third parties to provide advice or directly install low-emissions options in houses and businesses. The other creates obligations or incentives for parties, such as energy retailers, to deliver energy efficiency improvements in households and firms. There are a number of problems with the approaches.

First, if the number of audits and subsidised installations is limited and schemes rely heavily on households to make the decision to take up these options, the schemes will tend to favour informed individuals who are already motivated to save energy and so create adverse distributional impacts. Therefore, if these approaches are used they should focus on low-income households.

Second, there are challenges in the obligations or incentives approach in estimating the energy savings from these programs.

Finally, there are problems in requiring that retailers undertake activities to improve energy efficiency when their primary incentive is to sell more energy.

There is evidence that these programs can be effective in changing household energy use (Nadel & Geller 1996; Eoin Lees Energy 2006). If they can be developed to be cost effective they could lead to changes in both household behaviour and building efficiency that benefit households and the economy more widely.

Overall, there seems to be a case for removing requirements for retailers to improve energy efficiency in households and instead subsidising a limited amount of energy efficiency audits targeted at low-income households.

17.2.6 Expanding mandatory disclosure

Ensuring that both parties in a transaction have access to sufficient information will generally be the most effective way to address information asymmetry. Disclosure schemes, such as energy efficiency ratings, complement an emissions trading scheme as they assist individuals to act on the price signal.

Disclosure schemes will be far more effective if they are mandatory, as sellers are only likely to apply voluntary labels to high-performing products, leaving consumers unable to select among average and poorly performing products (Productivity Commission 2005).

The disclosure mechanism should be designed to should show the ongoing running costs of the good, use familiar language (such as payback periods), and should be located at the point of sale, as for public information programs.

Mandatory disclosure should be applied to goods where it is cost effective to do so. This will be determined largely by the administrative cost of the scheme, its accuracy and the potential savings to consumers. The potential for accurately and cheaply rating energy use will vary between goods. For refrigerators, it is relatively cheap to assess their energy use—most households' patterns of using a refrigerator will have limited effect on the comparative efficiency of different models. For vehicles, the situation is more complex, as a driver's behaviour may influence the efficiency of some cars relative to others, but even partially accurate ratings are likely to be valuable.

Australia already has a labelling program in place for appliances. It is argued that labelling programs for appliances are successful in assisting the uptake of more energy-efficient products in Australia and other countries (George Wilkenfeld and Associates & Energy Efficient Strategies 1999: 49).

Governments should continue to implement the energy label program for appliances where energy consumption is substantial and there is significant variation in performance. These include refrigerators, freezers, washing machines, water heaters, televisions and air conditioners.

Australia also has a labelling program for new vehicles. Arguably, labelling for second-hand vehicles, which are likely to be older and in some cases less efficient, is even more important. This, however, would be more costly than for new vehicles. New vehicles could have one test per production run, whereas second-hand vehicles would need individual tests. The lower value of the car would raise the cost of the test as a proportion of the total cost of the vehicle. Overall, labelling of used cards is unlikely to be cost effective.

The Australian Capital Territory (ACT) has introduced a mandatory energy efficiency rating scheme for houses at the point of sale. A recent study suggests that there is a statistically significant correlation between house prices and energy efficiency ratings (Department of the Environment, Water, Heritage and the Arts 2007). Modelling results suggest that, for a house worth $365 000, increasing the rating by half a star would, on average, increase its market value by $4489.

Concerns have been raised about the cost of this scheme, particularly the cost of rating assessments. Administrative simplicity and cost are obviously key features in scheme design and building owners should have the option of electing to undertake a more detailed assessment if they feel this would give a more accurate rating and that this would have value for them.

Although the ACT scheme does not apply ratings at the point of lease, doing so would assist in overcoming some of the principal–agent problems discussed below.

The success of the ACT scheme suggests that a national building rating program could be useful throughout Australia at the point of sale and the point of lease.

Concerns with the accuracy of building rating schemes (Williamson 2004) rightly raise the issue that efforts need to be made to ensure that rating tools are as accurate, flexible and useful as possible.

17.3 Principal–agent problems

17.3.1 Principal–agent market failures

Principal–agent problems can occur when one person (the principal) pays an agent for a service, but the parties face different incentives and the principal cannot ensure that the agent acts in her best interest. For example, landlords (agents) selecting fixed appliances for their rental property do not face the same incentive as renters (principals) to lower the ongoing energy cost of the appliances (IEA 2007a).

The kind of principal–agent relationship can influence both the nature of the problem and the appropriate policy response. The International Energy Agency has categorised four kinds of principal–agent relationships (see Table 17.1), using energy use as an example, depending on:

- who chooses the energy-using equipment
- who pays the energy bills.

In all four kinds the principal uses the equipment.

Table 17.1 Four kinds of principal–agent problems

	Principal chooses technology	Agent chooses technology
Principal pays the energy bill	**1:** The principals select the energy-using equipment and pay the energy bill. They have an incentive to select efficient equipment and lower their energy use. There is no principal–agent problem.	**2:** The agents select equipment on behalf of the principals, and the principals pay the energy bill. As a result, the agents may not have an incentive to select efficient equipment. This type of relationship occurs between landlords and tenants.
Agent pays the energy bill	**3:** The principals select the equipment, but do not pay for the energy bill. As a result the principals have no incentive to select efficient equipment or lower their energy use. For example, staff select company cars but do not pay ongoing fuel costs.	**4:** The agents select the equipment on behalf of the principals, and pay the energy bill. As a result, the agents have an incentive to select efficient equipment, but the principals do not have an incentive to lower their energy use. This occurs in hotels.

Source: Derived from IEA (2007a).

Principal–agent problems may entirely insulate some decisions from a carbon price, potentially reducing the adoption of low-emissions options. For example, since residential tenants pay energy bills, landlords may not install energy-efficient appliances (IEA 2007a).

Principal–agent relationships have repercussions throughout the wider market for goods. For example, the new car market dictates which cars are available in the second-hand car market. Therefore, the principal-agent problem that arises from company car purchases could have significant repercussions on Australia's car fleet.

Principals and agents may be able to negotiate to align their incentives more effectively. In the landlord–tenant example above, the rental contract could stipulate that the landlord install a fixed appliance meeting particular efficiency requirements. Principal–agent problems persist when:

- it is difficult to enforce contracts, or
- the costs of negotiating and establishing a better contract exceed the benefits. For example, while residential tenants can attempt to renegotiate leases, offering to pay more rent if landlords improve energy efficiency, the effort of doing so is likely to be substantial (Sanstad & Howarth 1994).

In the Australian rental market, a mixture of principal–agent problems arise. Problems arise with regard to the thermal and energy efficiency of a building: landlords are generally responsible for the purchase and maintenance of fixed appliances, such as water heaters, insulation and air conditioners, and tenants pay the energy bills and are subject to the thermal discomfort. During the period of the lease there is no incentive for landlords to invest in improving the energy efficiency of their properties, even if energy prices rise (IEA 2007a) and temperatures change.

This appears to affect the energy efficiency of the 29 per cent of homes that are rented in Australia (ABS 2007). A survey in South Australia supported this conclusion, finding that, for example, low-flow shower heads were installed in over 42 per cent of owner-occupied households but in only 25 per cent of private rental homes (ABS 2004).

In the commercial sector, industry sources suggest that at least 70 per cent of offices are leased rather than owner occupied. Commercial tenants are generally more aware of energy costs and are often in a better position to negotiate with landlords.

17.3.2 Linking principals and agents

Where possible, principal–agent problems can be tackled directly by fostering new standard contracts that are readily available and better align the interests of principals and agents.

New contracts have been mandated in Japan to tackle problems in the vending machine market (IEA 2007a). Previously, Japanese beverage companies rented space from building owners for vending machines, but building owners paid the electricity bill, resulting in a principal–agent problem of the third kind. To address this barrier, the Japanese Government stipulated that contracts for vending machines should make beverage companies responsible for both selecting the appliance and paying the energy bill. In combination with standards for vending machines, this policy appears to have driven a 34 per cent increase in energy efficiency in vending machines between 2000 and 2005, in contrast to similar but unregulated display cabinets.

The Commonwealth Government has developed 'Green Leases' that set out obligations for landlords and tenants to cooperate in reducing energy and water use (Christensen & Duncan 2007). It demonstrates and promotes the viability of these leases by using them when it leases commercial property or leases out its property to commercial tenants. Other governments should consider similar measures.

17.4 Minimum performance standards

There may be limits to the extent to which providing information can overcome information barriers and improving the links between principals and agents can eliminate principal–agent problems.

In these situations, minimum standards are usually considered to provide some level of protection for individuals and firms. However, they also:

- reduce flexibility
- reduce the opportunity for individuals to make choices
- operate on the presumption that governments are better informed than market participants, both now and in the future.

If standards are designed appropriately, with good knowledge of the costs and benefits and sufficient lead time for industry to respond, experience from both

Australia and abroad has indicated that they can be cost effective in supporting the uptake of low-emissions options (IEA 2007b).

California has been held up as a success story in improving energy efficiency, with electricity sales per capita remaining steady at the same time as output per person grew strongly. Although this is likely to have been partly driven by California's industry structure and higher electricity prices, recent work indicates that energy efficiency policies account for a substantial proportion of the state's successes (Kandel et al. unpublished). Building and appliance standards account for around half of these savings (Geller et al. 2006).

Figure 17.1 Residential per capita electricity consumption in the United States, California and as predicted for California

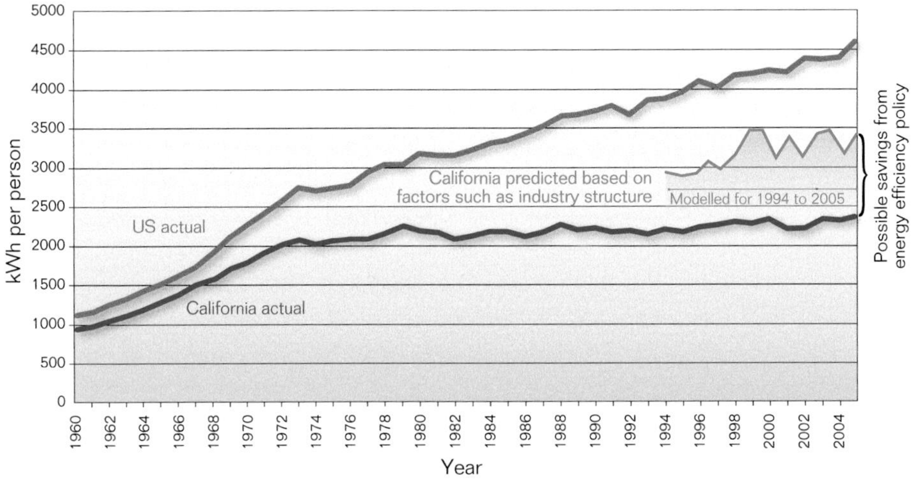

Note: The area between California predicted and California actual (modelled for 1994 to 2005) indicates possible savings from energy efficiency policies.

Source: Kandel et al. (unpublished).

17.4.1 Approach to minimum performance standards

Given the likely limits on information available to governments, standards should focus on:

- performance rather than specifying a particular technology
- features that are unlikely to affect consumers' amenity, such as energy efficiency, rather than features that consumers may value, such as the size of appliances
- removing poorly performing products, as it will be generally easier to identify the products that are the least cost-effective for the majority of users, than the products that are the most cost-effective options for all parties in all circumstances.

17.4.2 Minimum performance standards for buildings

Existing building standards in Australia

Building standards have been in place in Australia for many years. Historically they have been required to ensure the safety and structural integrity of buildings, though in recent years their scope has broadened to incorporate other issues including disability access.

The primary vehicle for delivery of building standards, the Building Code of Australia, is administered by the Australian Building Codes Board, a joint government and industry body. The code stipulates minimum building standards at the point of construction or major refurbishment, which are required to be met prior to building approval. The code often refers to stipulated requirements for particular building elements and materials established by Standards Australia. The code is administered at a national level, but its standards are brought into law by the states and territories.

In recent years the Building Code of Australia has become a vehicle for supporting greenhouse gas mitigation through the implementation of minimum energy performance standards. Energy Performance Standards were introduced into the code in 2003 for houses (with increased stringency incorporated in 2006), and 2006 for commercial and office buildings.

For houses, these requirements pertain to the heating and cooling energy required to maintain the building's thermal comfort. They refer to the building's fabric only. For commercial and office buildings, the standards have a slightly broader coverage and, in addition to the fabric, include the building's heating, ventilation and air-conditioning systems, as well as lighting.

While the Building Code of Australia adopts 'deemed to satisfy' prescriptive building design standards, flexibility is provided to allow equivalent performance standards to be met in a non-prescriptive manner through the achievement of a particular performance level using allowable design simulation software tools.

The Australian Building Codes Board is undertaking an ongoing work program to ensure that building materials are appropriately resilient to climate change impacts.

Are current building standards adequate?

The standards currently focus on the performance of the building, not particular technologies. They provide flexibility for meeting the minimum standards in a range of ways that impose minimal restrictions on the building's design in regard to materials or aesthetics. They do not force people to reduce the utility of their dwellings by, say, living in a smaller house. Software tools do not discriminate against larger buildings, provided they meet the per area energy performance requirements.

Having said this, some improvements could be made.

First, the standards, particularly for houses, do not include the appliances within the building envelope and therefore fail to guarantee energy performance for the household as a whole. While there appears little reason for incorporating non-fixed appliances (such as washing machines and refrigerators), which can be switched by occupants, there are strong arguments for incorporating fixed appliances (such as water heaters).

Second, changes in building standards can create uncertainty for industry and increase costs. This can be overcome by providing an indicative pathway for the standards, which may be introduced in the future, to assist the sector in adapting its practices. Such a pathway, updated as new information becomes available, could be a powerful tool for providing information on the implications for buildings of possible future trends in energy prices and implications of projected climate change impacts to developers and building owners, who may not otherwise consider such issues.

In relation to climate change adaptation, it remains to be seen whether the existing approach of enhancing structural performance in the building code is sufficient to ensure buildings will be resilient in the face of future climate change. The challenge will be to achieve reasonable accuracy in the predicted impacts for particular climate zones and locations. Priority should be given to locations most at risk from climate change.

17.4.3 Minimum performance standards for appliances

The National Mandatory Efficiency Performance Standards for refrigerators and freezers were introduced in Australia in 1999 and revised in 2005. This set of standards removes appliances from sale that do not meet minimum benchmarks of energy efficiency. Retrospective analysis in 2006 estimated that these policies saved more than 3000 gigawatt-hours of energy by 2005, savings that were 34 per cent higher than was forecast in the original Regulatory Impact Statements (Energy Consult 2006).

Estimates of the costs and benefits of appliance standards have been contested, particularly in the United States (see for example, Meyers et al. 2002; Sutherland 2003; Nadel 2004). This debate does not suggest that standards are unsuitable, but underlines the importance of using robust methodologies in assessing the benefits of appliance standards and regularly updating standards to ensure that they remain relevant.

There is a risk that applying excessively stringent standards will have consequences for those on low incomes. That is, it may remove from the market products that are attractive for those whom it suits to pay higher ongoing costs rather than higher upfront costs. For this reason, the Minimum Energy Performance Standards scheme for appliances should focus on removing poorly performing products with considerable energy consumption and significant variation in performance, without eliminating features that consumers value.

References

ABS (Australian Bureau of Statistics) 2004, *Domestic Use of Water and Energy, South Australia*, cat. no. 4618.4, ABS, Canberra.

ABS 2007, *Housing Occupancy and Costs, Australia, 2005–06*, cat no. 4130.0.55.001, ABS, Canberra.

Akerlof, G. 1970, 'The market for lemons: quality uncertainty and the market mechanism', *Quarterly Journal of Economics* 89: 488–500.

Allen Consulting Group 2004, *Economic Impact Analysis of Improved Energy Efficiency Phase 2 Report*, report to the Sustainable Energy Authority of Victoria, Allen Consulting Group, Canberra.

Aronson, E. & Stern, P.C. 1984, *Energy Use: The human dimension*, Freeman, New York.

Bjornstad, D.J. & Brown, M.A. 2004, *A Market Failures Framework for Defining the Government's Role in Energy Efficiency*, Joint Institute for Energy and Environment, Knoxville, Tennessee.

BRANZ Ltd 2007, *An Assessment of the Need to Adapt Buildings for the Unavoidable Consequences of Climate Change*, Department of the Environment and Water Resources, Australian Greenhouse Office.

Camerer, C.F. & Loewestein, G. 2004, 'Behavioral economics: past, present, future', C.F. Camerer, G. Loewestien & M. Rabin (eds), in *Advances in Behavioral Economics*, Princeton University Press, pp. 3–51.

Camerer, C.F., Loewestien G. & Rabin, M. (eds) 2004, *Advances in Behavioral Economics*, Princeton University Press.

Christensen, S. & Duncan, W.D. 2007, 'Green leases: a new era in landlord and tenant cooperation', *Australian Property Law Journal* 15(54): 1–10.

Cone, J.D. & Hayes, S.C. 1980, *Environmental Problems/Behavioral Solutions*, Brooks/Cole Publishing Company, Monterey, California.

Consumer Affairs Victoria 2006, *Information Provision and Education Strategies*, Research Paper No. 3, Consumer Affairs Victoria, Melbourne.

Department of the Environment, Water, Heritage and the Arts 2007, *Modelling the Relationship of Energy Efficiency Attributes to House Price: The case of detached houses sold in the Australian Capital Territory in 2005 and 2006*, a statistical consultancy report, Australian Bureau of Statistics, Canberra.

Desha, C., Hargroves, K., Smith, M., Stasinopoulos, P., Stephens, R. & Hargroves, S. 2007, *State of Education for Energy Efficiency in Australian Engineering Education: Summary of questionnaire results*, The Natural Edge Project, Adelaide, <www.naturaledgeproject.net/Documents/Energy_Efficiency_Survey_-_Summary.doc>.

Energy Consult Pty Ltd 2002, *Full Term Review of the Energy Efficiency Best Practice Program*, Department of Industry, Tourism and Resources, Canberra.

Energy Consult Pty Ltd 2006, *Retrospective Analysis of the Impacts of Energy Labelling and MEPS: Refrigerators and freezers*. prepared for the Australian Greenhouse Office, Energy Consult Pty Ltd, Jindivik.

Eoin Lees Energy 2006, *Evaluation of the Energy Efficiency Commitment 2002–05*, Department for Environment, Food and Rural Affairs, London.

Geller, H., Harrington, P., Rosenfeld, A.H., Tanishima, S. & Unander, F. 2006, 'Policies for increasing energy efficiency: thirty years of experience in OECD countries', *Energy Policy* 34: 556–73.

George Wilkenfeld and Associates & Energy Efficient Strategies 1999, *Regulatory Impact Statement: Energy labelling and minimum energy performance standards for household electrical appliances in Australia*, report prepared for the NSW Department of Energy and the Australian Greenhouse Office, George Wilkenfeld and Associates, Killara, <www.energyrating.gov.au/library/detailsris-modelregs.html>.

Golove, W.H. & Eto, J.H. 1996, *Market Barriers to Energy Efficiency: A critical reappraisal of the rationale for public policies to promote energy efficiency*, Lawrence Berkeley National Laboratory, Berkeley, California, <http://eetd.lbl.gov/EA/EMS/ee-pubs.html>.

Holper, P., Lucy, S., Nolan, M., Senese, C. & Hennessy, K. 2006, *Infrastructure and climate change risk assessment for Victoria*, CSIRO, Aspendale.

IEA (International Energy Agency) 2005, *The experience with energy efficiency policies and programmes in IEA countries: Learning from the critics*, IEA, Paris.

IEA 2006, *Energy Technology Perspectives 2006: Scenarios and strategies to 2050*, IEA, Paris.

IEA 2007a, *Mind the Gap: Quantifying principal–agent problems in energy efficiency*, IEA, Paris.

IEA 2007b, *Experience with Energy Efficiency Regulations for Electrical Equipment: IEA Information Paper in support of the G8 Plan of Action*, IEA, Paris.

IPCC (Intergovernmental Panel on Climate Change) 2007, *Climate Change 2007: Mitigation of climate change. Contribution of Working Group III to the Fourth Assessment Report of the Intergovernmental Panel on Climate Change*, B. Metz, O.R. Davidson, P.R. Bosch, R. Dave and L.A. Meyer (eds), Cambridge University Press, Cambridge and New York.

Jaffee, A.B. & Stavins R.N. 1994a, 'The energy paradox and the diffusion of conservation technology', *Resource and Energy Economics* 16: 91–122.

Jaffee, A.B. & Stavins R.N. 1994b, 'The energy-efficiency gap: what does it mean?', *Energy Policy* 22(10): 804–10.

Kahneman, D. & Tversky, A. 2000, *Choices, Values, and Frames*, Cambridge University Press, Cambridge.

Kandel, A., Sheridan, M. & McAuliffe, P. unpublished, 'A comparison of per capita electricity consumption in the United States and California', paper submitted to the American Council for an Energy Efficient Economy Annual Conference 2008.

Kempton, W. & Montgomery, L. 1982, 'Folk quantification of energy', *Energy* 7(10): 817–27.

Komor, P.S. & Wiggins, L.L. 1988, 'Predicting conservation choice: beyond the cost-minimisation assumption', *Energy* 13(8): 633–45.

McKinsey & Company 2008, *An Australian Cost Curve for Greenhouse Gas Reduction*, McKinsey & Company, Sydney.

Meyers, S., McMahon, J.E., McNeil, M. & Liu, X. 2002, 'Impacts of US federal energy efficiency standards for residential appliances', *Energy* 28: 755–67.

Nadel, S. 2004, 'Critique of the CATO Institute study "The high costs of federal energy-efficiency standards for residential appliances" by Ronald Sutherland', American Council for an Energy Efficiency Economy, Washington DC, <http://aceee.org/buildings/policy_legis/stnds_info/cato.pdf>.

Nadel, S. & Geller, H. 1996 'Utility DSM: What have we learned? Where are we going?' *Energy Policy* 24(4): 289–302.

Parliament of Australia 2005, *Energy Efficiency Opportunities Bill 2005, Explanatory Memorandum*, House of Representatives, Canberra.

Paton, B. 2001, 'Efficiency gains within firms under voluntary environmental initiatives', *Journal of Cleaner Production* 9: 167–78.

Productivity Commission 2005, *The Private Cost Effectiveness of Improving Energy Efficiency*, Productivity Commission, Canberra.

Sanstad, A. & Howarth, R. 1994, 'Normal markets, market imperfections and energy efficiency', *Energy Policy* 22: 811–18.

Shipworth, M. 2000, *Motivating Home Energy Action: A handbook of what works,* prepared for the Australian Greenhouse Office, Australian Greenhouse Office, Canberra.

Sorrell, S., O'Malley, E., Schleich, J. & Scott, S. 2004, *The Economics of Energy Efficiency*, Edward Elgar Publishing Ltd, Cheltenham, United Kingdom.

Sorrell, S. 2005, *The Economics of Energy Service Contracts: Tyndall Centre working paper No. 81*, Tyndall Centre for Climate Change Research, Norwich, United Kingdom.

Stavins, R.N., Jaffee, J. & Schatzki, T. 2007, *Too Good to Be True: An examination of three economic assessments of California climate change policy*, Resources for the Future, Washington DC.

Sutherland, R.J. 1994, 'Energy efficiency or the efficient use of energy resources?' *Energy Sources* 16: 257–68.

Sutherland, R.J. 2003, 'The high costs of federal energy-efficiency standards for residential appliances', *Cato Institute Policy Analysis* 504: 1–15.

Thaler, R.H. 1999, 'Mental accounting matters', *Journal of Behavioral Decision Making* 12: 183–206.

Williamson, T. 2004, *Energy-Efficiency Standards in Residential Buildings: A Plea for evidence-based policy making*, submission to Productivity Commission Public Enquiry into Energy Efficiency, <www.pc.gov.au/inquiry/energy/docs/submissions>.

THE INNOVATION CHALLENGE

18

Key points

Basic research and development of low-emissions technologies is an international public good, requiring high levels of expenditure by developed countries.

Australia should make a proportionate contribution alongside other developed countries in its areas of national interest and comparative research advantage. This would require a large increase in Australian commitments to research, development and commercialisation of low-emissions technologies, to more than $3 billion per annum by 2013.

A new research council should be charged with elevating, coordinating and targeting Australia's effort in low-emissions research.

There are externalities associated with private investment in commercialising new, low-emissions technologies.

To achieve an effective commercialisation effort on a sufficiently early time scale, an Australian system of matching funding should be available automatically where there are externalities associated with private enterprise investment in low-emissions innovation.

Research in adaptation technologies is required. Existing arrangements are well placed to meet immediate priorities.

The successful development and deployment of new technologies across sectors will be critical to minimising the costs of adjustment to the emissions trading scheme.[1] This will be a global effort. Australia will contribute proportionately, focusing on areas of research, development and commercialisation of new technologies, according to its comparative advantage and national interest.As other countries adopt similar constraints on emissions, there will be new opportunities for expansion in sectors where Australia can develop an international comparative advantage. Specifically, there may be significant sequestration opportunities in forestry and agriculture, and in parts of the low-emissions energy sector.

A variety of new technologies and practices are potential contributors to Australia's mitigation task. They include:

- **energy efficiency**—electrical equipment, fixed appliances, and building materials and design
- **electricity generation**—geothermal (hot rocks), improved generation efficiency (e.g. coal drying), and solar (photovoltaic and thermal)
- **transport**—lower-emissions vehicles, second- and third-generation biofuels (including from mallee and algae) and biomass, and electric cars
- **agriculture and forestry**—anti-methanogen technologies for livestock producers, altered savanna management, and nitrification inhibitors
- **sequestration**—soil sequestration (biochar and mallee), geosequestration and algal sequestration.

The role of new technologies will also be important in lowering the cost of adapting to climate impacts. Commercial agriculture in Australia often requires imported agricultural technologies to be adapted to Australian conditions, with high levels of government participation in research and the dissemination of information (Raby 1996).

Reliable information about the impacts of climate change will be needed for the continued development of new adaptation technologies. Those areas that will play a direct and significant role in Australia's adaptation challenge include:

- **agriculture**—use of improved seasonal forecasts, heat tolerant crop cultivars, and different methods of crop and livestock management
- **the built environment**—more resilient building materials, climate-appropriate building design and more efficient heating, ventilation and air-conditioning systems
- **biodiversity**—connectivity corridors and conservation methods.

Some technologies, such as those that improve the thermal properties and energy efficiency of buildings, contribute to both the adaptation and mitigation efforts.

18.1 What is innovation?

Although it entails simplification, the 'innovation chain' concept can help identify policies appropriate for different stages of development (Foxon et al. 2008).

For economic analysis and policy development purposes, the Review adopts a simplified model of the innovation chain, as shown in Figure 18.1, in which there are three distinct phases.

Figure 18.1 The innovation chain

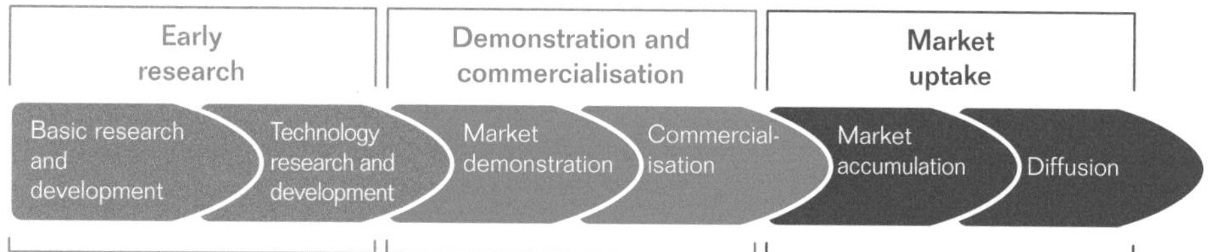

Source: Adapted from Grubb (2004).

Early research—Contributions are made to basic science and knowledge, usually at research institutions at a laboratory scale, with few immediate commercial returns. The knowledge and information generated tend to be of benefit globally, are difficult to keep secret, and can be easily disseminated at low cost.

Demonstration and commercialisation—The new knowledge is applied to the real world through pilot, demonstration and first commercial-scale projects. These activities require research bodies or firms to take on substantial risk as the technology requires proof in the intended operating environment and may not be cost competitive at first—even in cases that later turn out to be commercially successful. Some studies call this phase 'the valley of death', where most technologies fail either technically or financially (Grubb 2004; Murphy & Edwards 2003).

Market uptake—Once new knowledge becomes embodied in a tested product or service, it is sold to the open market. Technologies at the market uptake stage are able to compete with other mature products in the marketplace, with successful instances being associated with falling costs as market share expands.

18.1.1 How will an emissions trading scheme affect technological development?

As the emissions trading scheme raises the costs of greenhouse gas–emitting activities, new and existing low-emissions technologies will become more profitable. Mature technologies will be most immediately affected by the demand–pull effects of an emissions trading scheme.

An emissions trading scheme will spur private sector research and development activities by creating the long-term demand–pull for more low-emissions products and processes. However, there may be only limited impacts on early research activities since most early research is publicly funded. Changes to funding are dependent on how quickly funding resources are allocated to new research areas.

Both public and private research and development will have a large impact on the economy-wide cost of emissions reductions in the medium to long term. Over time, technological change and development will bring down the cost of various low-emissions technologies.

18.1.2 Barriers to an efficient market response

While an emissions trading scheme and projected climate impacts will both drive the development and uptake of new technologies, market failures that impinge on the efficient and competitive functioning of markets for new ideas and technologies are likely to result in suboptimal levels of investment in innovation. This could lead to unnecessarily expensive substitutes being deployed to reduce emissions and to a carbon price that is higher than it would otherwise be. Similarly, inadequate investment in developing superior adaptive technologies could result in a more costly adaptation response.

These market failures are most important in the early research and demonstration and commercialisation phases of the innovation chain (see Figure 18.2). There are some externalities associated with the early adoption of proven technologies, but the emissions trading scheme and better research and dissemination of knowledge about impacts will create enough demand–pull for new technologies so that generally there will be no need for any additional support for innovation at the market uptake stage.

Figure 18.2 Market failures along the innovation chain

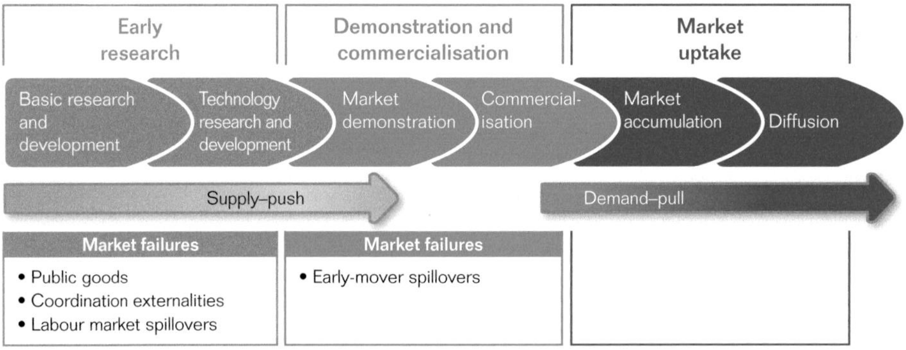

Early research	Demonstration and commercialisation	Market uptake
Basic research and development; Technology research and development	Market demonstration; Commercialisation	Market accumulation; Diffusion
Supply–push		Demand–pull
Market failures • Public goods • Coordination externalities • Labour market spillovers	**Market failures** • Early-mover spillovers	

Correction of market failures justifies government policy intervention. Economic studies have emphasised the role of innovation policy in delivering least-cost emissions reduction (Stern 2007; Productivity Commission 2007a; Jaffe et al. 2005). This rationale for government intervention holds true even in the absence of climate change, but as the emissions trading scheme delivers quick and profound shifts in the economic context, there will be a special requirement for high rates of technological improvement in low-emissions technologies. The emissions trading scheme will raise the opportunity cost of an inadequate market response to incentives for new technologies.

The required adaptation response is less likely to face the same time pressures. The impacts of climate change are likely to be felt more gradually, suggesting that the adaptation effort can be managed within established research and development funding practices. The cost of market failures in the development of new adaptation technologies should not be expected to add significantly to Australia's total adaptation costs.

There is a risk that undisciplined innovation policies will become the focus of strong pressures on the political process for unjustified payments to industries and firms (Banks 2008). Government must therefore ensure that innovation schemes address material market failures that yield net benefits to society and that processes of resource allocation are insulated from political pressures. Any justifications for policy outside the market failure rationale should be rejected (see Box 18.1).

Such a rationale, however, is not sufficient to dictate government intervention—two additional requirements should also be met. The proposed measures need to target the problem and the cost of a market failure needs to be more than the cost of government intervention, with all of its political economy and other risks and costs.

Box 18.1 Wrong arguments for innovation policy in the context of an emissions trading scheme

Some rationales for government intervention in the area of innovation do not have a sound economic basis. A credible emissions trading scheme would address the issues of environmental integrity and urgency of action. The case for public support for new technologies related to climate change should require the correction of material market failures that would otherwise increase the cost of mitigation where benefits of intervention clearly exceed costs.

Some mistaken arguments for innovation policy in this context include:

- *There will not be enough innovation or time to develop new technologies for Australia to meet its national targets successfully.*
  The cap on emissions in an appropriately designed emissions trading scheme is binding, such that emissions reductions will have to be delivered regardless of the technologies available. Ultimately, this cap may be met through reductions in consumption of emissions-intensive goods and services, if need be, in the short term.
- *The permit price will be low initially and therefore will not drive much innovation.*
  Where there is a domestic emissions trading scheme supported by a global agreement, there is no reason for the permit price to be uneconomically low with the anticipated future scarcity of permits. If this seems to be the case, then it may reflect market optimism that suitable new technologies will be available in the future. If this is not the expectation, then the incentive would be to hoard permits for future use when scarcity, and therefore prices, were higher.
- *We need to invest in innovation to lessen the impact of the carbon constraint.*
  Investing in innovation when there are no requisite market failures requiring correction is likely to lead to greater economic cost, not less.

18.2 Ensuring optimal levels of early research

18.2.1 Market failures in early research

Public good nature of early research

Early research is characterised by substantial spillovers arising from the non-excludable or non-appropriable nature of knowledge. In most cases, once new basic knowledge is created, it is impossible to exclude others from sharing the benefits (Arrow 1962). This is the strongest rationale for government intervention at the early research phase of the innovation chain (Productivity Commission 2007b). Excludability is further diminished by the typically low incremental costs of diffusion (Stephan 1996), especially when knowledge is transparently embodied in a product or process, readily codified and easily diffused (Productivity Commission 2007b). In addition, knowledge is non-rivalrous—once created, many individuals or firms can use and apply it, thus making it a public good.

Externalities from early research

There is evidence both nationally and internationally that collaboration yields measurable benefits for participating individuals and organisations. In Australia, an analysis of Australian Bureau of Statistics data found that businesses that engage in collaboration are more likely to achieve higher degrees of innovation (Department of Industry, Tourism and Resources 2006). However, in some cases, the costs of negotiating an agreement are sufficient to prevent a coordinated approach from happening.

Early research activities often have the added benefit of being vehicles for education and training because those who conduct research also usually teach. Therefore a shortfall in early research and development funding could also result in medium- to long-term skills shortages. The Productivity Commission (2007b) found that the benefits of research and development in both universities and public sector research agencies are high, due to their orientation to public good research and their role in the development of high-quality human capital for the Australian economy.

18.2.2 Are current policies sufficient?

Like most developed countries, Australia has an established institutional framework for allocating research funding across the economy.

Early research in low-emissions technologies

The Productivity Commission (2007b) recently highlighted the expanding need for public good research in the light of future environmental, energy and climate challenges as a potential stress on the Australian innovation system.

Despite being among Australia's stated national priority research areas (DEEWR undated), funding for energy supply research has increased only marginally off a low base (ABS 2008a). Australia's expenditure on energy supply technologies ranks low among OECD countries (based on OECD 2008). Internationally, energy-related research and development is dominated by just a few countries (based on OECD 2008) whose research activities and priorities are likely to determine the global range of future low-emissions technologies.

Falling levels of investment in early research in mitigation have been a global phenomenon (see Figure 10.1). For example, the Fourth Assessment Report of the Intergovernmental Panel on Climate Change (2007: 20) found:

> Government funding in real absolute terms for most energy research programmes has been flat or declining for nearly two decades (even after the UNFCCC came into force) and is now about half of the 1980 level.

The low levels of government expenditure on research and development in key areas like energy supply, juxtaposed with the rising importance of low-emissions energy technologies for Australia's mitigation effort, suggest that current funding levels do not reflect the priority required to meet the rapidly changing pattern of demand established by an emissions trading scheme.

It is important that this issue be looked at from an international perspective since research has global spillovers. The Review proposes that high-income countries support an International Low-Emissions Technology Commitment, suggesting an indicative global figure for this fund of $100 billion per year, and an indicative Australian expenditure of $2.8 billion (section 10.1).

Early research in adaptation technologies

Research in adaptation technologies is critical for the agriculture sector as impacts will be increasingly severe in a future without mitigation. New technologies and practices will also be required to support the private adaptation response of households and businesses in Australia.

Early research in agriculture is an area of strength in Australia. In 2006–07, 22 per cent of all government expenditure on research and development could be attributed to research in plant and animal production and primary products, while environmental management accounted for a further 18 per cent (ABS 2008a). Australian research publications in plant and animal sciences, ecology and environment and agriculture were relatively numerous, accounting for more than 4 per cent of the total world publication in these fields in the period 2000 to 2004 (Productivity Commission 2007b).

The Review has not examined the adequacy of the early research effort in other sectors relevant to adaptation, such as the built environment sector, and areas likely to be affected by the increasing scarcity of water. Various research organisations are already undertaking work on improving our technological responses to the effects of climate change,[2] but better outcomes in the resilience of buildings, energy efficiency and water efficiency will require greater uptake of existing technologies rather than further research and development of new technologies. Market failures associated with the uptake of available solutions by households and businesses are discussed in Chapter 17.

18.2.3 New institutions to drive early research

The significant challenge of deep cuts to emissions suggests that Australia's research agenda needs to focus more strongly on early research into low-emissions technologies, to shorten the lag between the introduction of the emissions trading scheme and the response of the research community.

In the past, government has managed the issue of priority by establishing institutions, including the Australian Research Council, that allocate resources according to strategic importance and national capability. There is a case for a specialist research body related to low-emission technology, to elevate, coordinate and target Australia's effort in this field. Such a body could operate in a similar way to the National Health and Medical Research Council (NHMRC),[3] overseeing a large expansion of effort in early research for low-emissions technologies, and operating independently to correct market failures.

Like the NHMRC, the proposed research council could have three mutually reinforcing core functions:

- allocating public funding for early research based on clearly established criteria (see section 18.2.5)
- promoting linkages across relevant early research activities within Australia and with research activities in the Asia–Pacific region, and being alert to opportunities for international cooperation
- guiding training in low-emissions technologies throughout Australia, including the development of research training through higher degrees.

18.2.4 Sources of additional funding

One possible source of additional funding for allocation by the proposed research council is the consolidation and reallocation of existing funds now allocated to related research areas. However, there are good reasons for not reallocating all existing funding. First, it is important to maintain continuity in the allocation of existing funds. Second, it is beneficial to maintain some plurality of funding sources. General institutions can co-exist with specific funding bodies, so that no one body holds the purse strings for all the funding in any one area. Some existing funding arrangements may continue on this basis, but a review of all programs may be warranted to identify those that are yielding below-average returns on investment.

Additional funds for early research in low-emissions technologies could come from the revenue from the auctioning of emissions permits. The allocation of a consistent level of annual permit revenue towards this public good research could form the major portion of funds to be allocated by the proposed research council. It is sensible to use permit revenue to fund early research because there are potentially strong links between the early research effort, the long-term cost of mitigation and the carbon price.

All commitments of funds for early research would qualify under the International Low-Emissions Technology Commitment proposed in section 10.1, alongside expenditure on matching funding for investment in commercialisation of new, low-emissions technologies.

18.2.5 Criteria for funding early research

The allocation of resources to innovation in general is complicated by two trade-offs. First, there is a trade-off between the desire to provide technology-neutral support in order to avoid distorting the selection of technologies by the market; and the competing desire to concentrate resources on more promising areas. Policies to assist innovation must find the right balance between providing technology-neutral and technology-specific support, and between encouraging options and maximising returns.

Second, funding decisions must balance the role of knowledge generation within Australia and the adoption of ideas and technologies from the global research effort. Technologies with broad application and commercial potential are likely to be developed outside Australia, and it will sometimes be preferable for Australia to be a technology taker rather than duplicating the international research effort.

Despite the desire to avoid 'picking winners', there is inevitably a good deal of discretionary judgment in decisions on allocation of public funding for early research. The proposed research council should be guided by criteria that ensure funds are allocated to areas likely to result in the highest economic value. Two important criteria should underlie any funding decisions in early research: (1) Is this area of research of national interest? (2) Is this an area of early research where Australia has a comparative advantage?

The criteria for both national interest and comparative advantage can be expected to shift over time. The funding allocation should be subject to a transparent and independent process of periodic evaluation and review (Productivity Commission 2007b), with swift termination of funding for projects that no longer meet the criteria.

Criterion 1: National interest

Australia should only fund early research aligned with its national interest. In the case of climate change mitigation, considerations should be based on both current circumstances and future projections, and could include:

- **Australia's emissions profile**—The high emissions intensity of electricity generation and the high levels of emissions from agriculture are two examples of unusual characteristics of Australia's emissions profile.
- **Technological solutions particular to local conditions**—Many technologies can be adopted from overseas and applied to the Australian context. The deployment of wind turbines from Europe and any future use of nuclear power are examples. However, some technologies will be subject to local factors, including geography, geology and climatic conditions.
- **Sources of Australia's economic prosperity**—Sectors that are important sources of economic prosperity today or could become sources of economic advantage in the future have a broad strategic value for Australia.

- **Technologies that build on Australia's natural resources**—Australia is in a unique position among developed countries of having an abundance of a wide range of natural resources that are relevant to low-cost transition to a low-carbon economy (for example, solar, wind and geothermal sources of energy).

Box 18.2 Examples of areas of national interest

Technological solutions in carbon capture and storage, soil sequestration, solar technologies, algal biosequestration and geothermal energy are among the areas in which Australia has disproportionately strong opportunities and interests. The successful development of these technologies could be expected to have exceptional value within Australia.

Table 18.1 Brief assessment of two technology categories against criteria for national strategic interest

	Carbon capture and storage for coal-fired electricity generation	Algal biofuels
Australia's particular emissions profile	Coal-fired electricity generation is a major contributor to Australia's high emissions intensity of energy.	Some algal biosequestration processes could absorb emissions from coal-fired electricity generation and metals smelting.
Technological solutions particular to local conditions	Australia has a variety of geological formations that are suitable for long-term geosequestration.	Few other developed countries have the required natural conditions.
Sources of Australia's economic prosperity	Any proven technology that cost-effectively reduces the emissions from coal-burning will be in high demand in the future when climate change mitigation becomes a global priority. Carbon capture and storage will also maintain the value of Australia's coal resources as a commodity both for domestic consumption and export.	Algal biofuels could provide energy security and economic growth as they have a higher yield per hectare than traditional crops, with much higher energy returns. Algal biofuels could also prove competitive with fossil fuels in light of increasing global scarcity.
Technologies that build on Australia's natural resource advantage	The abundant availability of coal, and subsequently low energy prices are sources of comparative advantage for Australia. The export of coal itself is a significant contributor to Australian GDP.	There are several regions around Australia that could potentially provide the intense insolation and saline and other non-productive land needed to cultivate algae for biofuel production at a large scale.

Criterion 2: Comparative advantage

Australia should only undertake early research in areas where it has a comparative advantage. There are no perfectly objective measures for comparing different fields and disciplines. The proposed research council would therefore need to consider a range of proxy indicators of comparative advantage when making funding allocation decisions.

In some instances, the absence of any comparative advantage should be clear. For example, although the export of uranium is one source of economic prosperity in Australia (and therefore an area of national interest), Japan and France outspend Australia by a factor of 300 and 150 respectively on nuclear energy research (Commonwealth of Australia 2006), suggesting that early research in nuclear generation is not an area in which Australia is likely to have a comparative advantage.

Australia's demonstrated strength in agricultural research is an example of an area of clear comparative advantage.

18.3 Rewarding early movers

The early movers of a new industry are those that undertake the first demonstration and commercialisation projects. The spillovers from these early-mover activities mean that in the absence of government intervention, there will be suboptimal levels of private investment in demonstration and commercialisation projects.

18.3.1 Spillovers from early movers

In most new industries, the early movers bear all the costs of demonstrating and bringing a new technology to market, while later movers share in all the associated benefits that spill over directly from the early movers' investments. These spillovers can result in a strong disincentive for any firm to be a pioneer and result in an undersupply of demonstration and commercialisation activities. For some new industries, multiple spillovers may result in no activity at all.

There may be secondary mechanisms through which these spillovers are internalised. Early movers may reap the benefit from early gains in the form of brand reputation, product recognition and early leads in market share. These benefits may provide sufficient incentives to bear the upfront costs if the remaining spillovers are relatively small.

Research and development spillovers are both prevalent and important (Griliches 1992). There are five main types of spillovers that result from early mover activities.

- **Knowledge externalities**—Early movers who make the initial high-cost investment to demonstrate or apply new technologies can generate substantial contributions to the knowledge base of an industry, which later benefits the industry more widely. These knowledge and information benefits over the long run have been observed in the steep decline in the costs of new technologies during the demonstration and commercialisation stages.

  While knowledge spillovers can be internalised through the creation and enforcement of intellectual property rights under the patent system, not all knowledge lends itself to patent protection (Jaffe et al. 2005; Fri 2003).

Furthermore, patent rights are not self enforcing. Remedying breaches often requires costly legal action (Martin & Scott 1998).

- **Skills spillovers**—Early movers contribute to the future of all firms in an industry by bearing the upfront costs to develop appropriate technical skills and capacity and associated training courses. This has a positive lingering effect in the labour market and later movers are able to draw on this pool of skilled labour.
- **Regulatory and legal spillovers**—Early movers may bear the large upfront costs of working with government and other industries to develop new regulations and standards. This could include significant costs associated with resolving legal disputes regarding new regulatory frameworks with government and other industries. Later movers benefit from regulatory clarity and have established avenues for secure agreements and contractual arrangements.
- **Support sector externalities**—The development of supporting industries inevitably requires some additional investment by early movers—for example, to identify suppliers with appropriate manufacturing capabilities, develop suitable products and product standards with those suppliers, and test new parts and components. Firms that enter the market at a later stage are able to benefit from an established supply chain without having had to bear the upfront costs.
- **Social acceptance spillovers**—Communities can be apprehensive of new technologies that are intrusive, potentially dangerous, or simply not yet fully understood. An early-mover firm looking to commercialise such a technology will often bear the costs of demonstration projects and communication and information exercises to increase people's confidence in the safety and effectiveness of its particular technology. The higher level of social acceptance is enjoyed at no cost by later movers promoting similar technologies.

18.3.2 Are current policies sufficient?

Current programs for the demonstration and commercialisation of low-emissions technologies

The Productivity Commission (2007b) noted that the emphasis on innovation in Australia has moved towards demonstration and commercialisation projects. Many recent low-emissions research and development policies have targeted the market uptake stages of the innovation chain (see Table 18.2). This could be because the scarcity of funds intensifies the pressure to focus resources 'downstream' on shorter-term applied research aimed at the deployment of mature and commercial technologies and because of political economy distortions.

Table 18.2 Research and development programs in Australia targeting low-emissions technologies

Policy/fund name	Description	Funding
Low Emissions Technology Demonstration Fund	Supports the commercial demonstration of technologies that have the potential to deliver large-scale greenhouse gas emissions reductions in the energy sector.	$410 million over 11 years
Renewable Energy Development Initiative	A competitive merit-based dollar-for-dollar grants program supporting renewable energy innovation and commercialisation.	$100 million over seven years
Solar Cities	Demonstrates how solar power, smart meters, energy efficiency and new approaches to electricity pricing can be combined.	$93.8 million over nine years
Energy Technology Innovation Strategy (Victorian Government)	Assists the commercialisation of coal drying, coal gasification and geosequestration technologies, distributed generation energy efficiency, and renewable and enabling technologies. This funding supports some Low Emissions Technology Demonstration Fund projects.	Up to $369 million
Queensland Future Growth Fund	Supports the deployment of low-emissions coal and renewable energy technologies. Will operate separately from the Queensland state budget.	$350 million
Green Car Innovation Fund	Aims to support the manufacturing of low-emissions vehicles in Australia. Will operate on a matched funding basis at a ratio of 1:3 public:private.	$500 million over five years
National Low Emissions Coal Fund	Aims to reduce greenhouse gas emissions and secure jobs in the coal industry by stimulating investment in clean coal technologies with matched funds at a ratio of 1:2 public:private.	$500 million over seven years
Renewable Energy Fund	Targets renewable energy demonstration projects with private sector funds matched at a ratio of 2:1 public:private. Funding distributed through competitive grants, based on the goal of encouraging a range of technologies across a range of geographic areas. Fifty million dollars has been earmarked for dollar-for-dollar matched funding for private investors in the geothermal industry.	$500 million over seven years
Energy Innovation Fund	Investments targeted equally towards the Australian Solar Institute (solar thermal), photovoltaic research and development, and general clean energy research and development, including energy efficiency, energy storage technologies and hydrogen transport fuels.	$150 million over four years

Sources: Prime Ministerial Task Group on Emissions Trading (2007); Australian Treasury (2008).

Many of these industry support programs have the effect of providing incentives for early movers, but there is a conspicuous absence of a targeted technology-neutral program for dealing with the spillovers discussed in section 18.3.1.

The Productivity Commission (2007b: 371) found that this issue of poorly targeted policy was characteristic of technology programs in Australia more generally:

> Australia's current suite of business programs do not target rationales for public support (additionality and spillovers) effectively and, as a consequence, involve substantial transfers from taxpayers to firms without attendant net benefits. The need to raise taxation revenue to fund these transfers creates large efficiency losses.

Current programs for the demonstration and commercialisation of new adaptation technologies

Rural research and development corporations and companies are a major vehicle for driving the development of new adaptation technologies in the agriculture sector. Under the Rural R&D Corporations program, such corporations commission agricultural research and development on a competitive basis among public and private providers using funds from levies on production and matching Commonwealth grants.

This basic model was established in 1989 with the aim of correcting the spillovers by collecting compulsory industry levies for industry research and development (CRRDCC 2008). Without these levies, it is not likely that individual and voluntary agricultural associations would be able to capture enough of the spillover benefits and they would therefore fail to justify the level of research and development investments currently undertaken by the rural research and development corporations. This model allows for a targeted approach to fund allocation by industry and promotes accountability, allowing levy payers to contribute to decisions on the rural corporations' strategies, including on the amount of the levy collected.

The Rural R&D Corporations program is in essence a matched funding scheme with two main sources of funding: (1) levies from producers and (2) matching funds from government at an average ratio of 1.5:1 private to public investment.[4] The government contribution, in part, assists in overcoming some of the market failures associated with demonstration and commercialisation activities, which would not otherwise occur.

The investment of $511 million in this program in 2004–05 is expected to deliver gross private and social returns of around $1.65 billion to Australia over five years (CRRDCC 2006). Benefits include improvements in on-farm production; the development of new products for emerging markets; better management and use of water and natural resources; building and developing rural skills; building research and development capacity; and improving biosecurity (CRRDCC 2006).

Agricultural research in Australia is undertaken by public sector research organisations, notably the CSIRO, cooperative research centres, universities and agencies within the primary industries portfolios at federal and state levels. The demand for new technological solutions in the light of future climate change impacts will test the research capabilities of these institutions.

Overall, the shortfall in demonstration and commercialisation activities for new adaptation technologies is likely to be much smaller than that for low-emissions technology, given that there has been longstanding demand for adaptation technologies.

18.3.3 Supporting early movers with a matched funding scheme

The externality benefits from early-mover activities will vary widely and on a case-by-case basis. Different projects in different contexts will generate different types of spillovers at various levels. The prohibitively high administration and compliance costs of quantifying spillovers on a case-by-case basis means that such an approach would not be viable.

There are a variety of vehicles through which compensation for these spillovers could be provided. These can be classed into three broad categories, as set out in Table 18.3.

Table 18.3 Mechanisms for directly subsidising positive externalities in demonstration and commercialisation

Category	Instrument	Description
Tax instruments	Tax rebates or concessions	Tax concessions allow companies to claim a deduction of R&D-related expenditure, usually for a proportion beyond the actual expense incurred (i.e. more than 100 per cent).
	Accelerated depreciation	For research projects with high capital costs, approved accelerated depreciation of assets is an alternative tax concession.
Niche market creation	Technology target schemes	Policies such as mandated targets may establish guaranteed markets for particular categories of goods or services.
	Guaranteed revenue	Policies such as regulated prices or tariffs can provide innovators with revenue certainty.
	Government patronage	Government may itself provide a niche market for new products through its internal procurement policies or through advance purchasing contracts.
Direct funding	Competitive grants	Competitive grants are a common means by which government subsidises specific projects selected by merit based on defined criteria.
	Income-contingent loans	Income-contingent loans compensate innovators for spillovers through government assuming some of the short-term exposure to risk.
	Matched funding	Matched funding stimulates demonstration and commercialisation activities by lowering the costs associated with being the first mover by some fixed proportion.

Among these many instruments, matched funding is the preferred option based on a range of criteria.[5]

- **Simple and targeted**—Matched funding can be simple and directly targeted in its design to address specific spillovers.
- **Technology neutral**—Matched funding can be technology neutral across the whole range of economic sectors and has the potential to allow for all technological possibilities.
- **Maintains risk exposure**—By leveraging private funds, a matched funding scheme would ensure that applicants continue to bear and manage the potential risks associated with bringing a new technology to market. Moreover, it is unnecessary for applicants to demonstrate technical feasibility, commercial competitiveness or the pathway to uptake and diffusion as these criteria are implicit in the matched funding approach. The private investment would not be made unless a project is expected to earn a return in the long run.
- **Capped expenditure**—Matched funding can be designed so that the total expenditure does not exceed an imposed budget constraint.
- **Transparency, impartiality and independence**—The body or institution that administers a matched funding scheme should operate at arm's length from government to ensure that it is insulated from the political process.

18.3.4 Funding sources for a matched funding scheme

The preferred matched funding scheme could be paid for from three sources.

- **Reallocation from existing funds**—This would occur if, after review, existing programs were shown to be inefficient in compensating firms for the external benefits they generated. If consolidated, the schemes listed in Table 18.2 could benefit from (1) more efficient administration, (2) easier access for business, and (3) more consistent application of criteria.
- **Industry levies**—Matched funding schemes could be augmented by funds collected by individual industries. Given that most of the spillover benefits from early movers are likely to accrue to later movers within the same industry, these levies should be set aside strictly for new technologies within the source industry. Compulsory levies have been the established way of funding research and development in many rural industries for several decades.

  Levies on the coal industry for investment in the development and commercialisation of carbon capture and storage technologies are another example. The futures of Australia's domestic-oriented and export coal industries are both dependent in the long term on the success of carbon capture and storage. The Coal 21 Fund will raise an estimated $1 billion over the next decade from voluntary levies of 20 cents per tonne on black coal production.[6] While this is not an insignificant sum, it is relatively low when compared to the voluntary investment by some agricultural industries. If this fund were to be extended to be comparable to the contributions from sales revenue to research and development in the agricultural sector through the rural research and development corporations, and if the levy were confined only to exports, it would generate around $250 million per year. This would not seem to be an excessive allocation to research into technologies that are going to be necessary to secure the future of the domestic and export elements of the coal industry.

- **Auction revenue from the sale of emissions permits**—There is a strong policy rationale for a substantial amount of permit revenue to be allocated to matched funding of demonstration and commercialisation projects for low-emissions technologies.

18.3.5 Criteria for early-mover support

In any policy that targets early-mover spillovers, an accurate and simple set of criteria for a project to qualify is required. If the criteria are satisfied, funding should follow automatically. The criteria must be based on the answers to three key questions:

- Will the technology contribute to lowering the cost of mitigation?
- Does the project qualify as an early-mover innovation?
- Are there expected spillovers associated with the project?

In assessing a project against the criteria, government needs to balance the accuracy of the assessment process against the associated complexity and transaction costs.

The balance of considerations strongly favours simplicity and low transaction costs. The more complex the criteria, the more dependent the assessment process will be on the subjective judgments of the assessing panel. Simple criteria would be more objective and transparent.

Criterion 1: Lowering the cost of mitigation

Applicants must demonstrate the relevance of their technology to the mitigation challenge. Technologies that contribute to the delivery of existing goods and services at lower emissions intensity would qualify, even if emissions reductions are not the primary aim of the new technology, as long as the potential contribution to emissions reductions are material.

Choosing the appropriate cut-off level so as to select only those technologies that can be expected to make material differences at a reasonable economic cost will require specialist technical advice.

Criterion 2: Early-mover innovation

Pilot, demonstration or first commercial-scale projects should qualify for support. Determining whether or not a project falls into one of these three categories is not a straightforward exercise. Project proponents have the incentive to expand the scope of non-innovative projects at the margins to increase the chances of qualifying for funding, while other projects using non-novel technologies may in fact be making a significant contribution to the state-of-the-art knowledge at a highly technical level.

Given these difficulties, this determination should be made by an independent panel of experts that would assess whether a particular project is materially different from current available technology. Assessment would involve two stages:

- **Selection of an appropriate comparator**—'Current available technology' can be defined as a technology that is currently contributing to the production of commercial goods or services in Australia or overseas.
- **Technical judgment of material difference**—The panel should consider the characteristics, scale and context of the technology or technologies being proposed and assess these against the comparator.

Criterion 3: Expected spillovers of the project

Given the difficulty in attempting to quantify the size of different spillovers on a case-by-case basis, government will need to base its assessment on a proxy measure. The straightforward proxy is the assessment of whether or not a particular project, if successful, would be a genuine early mover.

A second method for identifying early movers would be to adopt a scalar measure of quantity, and an associated cut-off point for the 'first fleet' of early movers. For example, to determine whether a centralised electricity generation plant is part of a first fleet, the panel could assess whether the proposed plant is part of the first five of its kind or within the first 1000 megawatts of its kind, whichever is less.

18.3.6 What is an appropriate ratio for matched funding?

For a matched funding scheme, the difference between the private and social rates of return[7] may be a good proxy indicator for the estimated spillovers from demonstration and commercialisation activities in general. Table 18.4 shows that estimates of the private rate of return on research and development spending by firms tends to be much lower than the social rate of return, which is often more than twice that of the private rate.

This suggests that it could be appropriate for the proposed matched funding scheme to be based on a ratio of between $0.50 to $1.50 of public funding per dollar of private funding. Many matched funding schemes currently use a ratio of between 1:1 and 1:3 (see Table 18.4). Dollar-for-dollar matched funding is consistent with the evidence base.

The ratio of matched funds should not be varied based on criteria such as the level of expected emissions reduction. Doing so would reward investors on the basis of fine judgments about matters that in their nature are difficult to quantify.

It is likely that in the early years of the emissions trading scheme the funds allocated from the permit sales revenue towards research, development and commercialisation will not be exhausted, as the market will need time to assess and put forward appropriate candidate technologies. There will also be lags in the approval process. In this scenario, funds should be allowed to accumulate for use in future years.

Table 18.4 Estimates of private and social rates of return to private research and development spending

Studies	Private rate of return (%)	Social rate of return (%)
Minnasian (1962)	25	–
Nadiri (1993)	20–30	50
Mansfield (1977)	25	56
Terleckyj (1974)	27	48–78
Sveikauskas (1981)	10–23	50
Gotto & Suzuki (1989)	26	80
Mohnen & Lepine (1988)	56	28
Bernstein & Nadiri (1988)	9–27	10–160
Scherer (1982, 1984)	29–43	64–147
Bernstein & Nadiri (1991)	14–28	20–110

Source: Griliches (1995: 72).

On the other hand, it is also likely that in at least a few years, demonstration and commercialisation activities will be at a peak and the claims for funds will be above the annual allocation, even after allowing for the surplus of funds accumulated in the early years. The funding scheme should include measures that automatically reduce the rate of matching once the budgeted level of expenditure has been exceeded.

18.4 Overcoming barriers from technological lock-in

If the deep cuts necessary for the stabilisation of atmospheric greenhouse gas concentrations are to be achieved, far-reaching innovation will be needed. Technological lock-in however is an obstacle to such innovation (Foxon et al. 2008).[8]

Analysis of innovation systems suggests that it is important to create a long-term, stable and consistent strategic framework to promote investment in low-emissions technologies (Foxon et al. 2008; Stern 2007). High policy uncertainty on the other hand can create the incentive to delay investment and raise investment thresholds in an already high-risk environment (Blyth & Yang 2006). A clear, credible and consistent policy framework will provide investors with long-term signals, and incentives to deal with the challenge of technological lock-in, accelerating Australia's technological transition to a low-carbon economy (Foxon et al. 2008). The most important overarching policies for creating investor confidence and overcoming technological lock-in are the long-term emissions trajectory and the emissions trading scheme. Policy certainty and long-term investment signals can be backed up by strengthened international policy action that enhances domestic policy credibility (Blyth & Yang 2006).

Notes

1 In this chapter, the term 'low-emissions technologies' refers to those technologies that reduce the emissions intensity of existing technologies, reduce the need for emissions, or capture and sequester greenhouse gases.

2 For example, the CRC for Construction and Innovation (soon to become the Sustainable Built Environment National Research Centre) has developed resources such as 'Your Building'. In addition, work is being undertaken by the Australian Building Codes Board and Standards Australia to ensure building materials are manufactured to be resilient to climate change impacts.

3 The NHMRC is a national organisation with diverse responsibilities in health and medical research, including the allocation of research funding, fostering medical and public health research and training, and the development of health policy advice.

4 In 2004–05, the rural research and development corporations invested $511 million, of which about 60 per cent was funded by industry. Gross value of production for agriculture in 2006–07 was $36.1 billion (ABS 2008b). Therefore the proportion of industry expenditure on research and development was 0.85 per cent of gross value of production. Note that matched government funding is typically limited to 0.5 per cent of gross value of production (CRRDCC 2006).

5 For a discussion of a range of other key design principles for business research and development programs, see Productivity Commission (2007b: Chapter 10.2).

6 The Coal 21 Fund is the Australian black coal mining industry's funding commitment to research, development and demonstration of clean coal technologies.

7 The private rate of return is the benefit a firm receives on its investment, while the social rate of return is the broader benefit that accrues to both the firm and society more generally. The difference is therefore the spillover benefit that the firm is unable to appropriate.

8 Technological lock-in occurs when incumbent technologies benefit from positive feedbacks that come from being the status quo to the extent that superior technologies struggle to displace inferior incumbents.

References

ABS (Australian Bureau of Statistics) 2008a, *Research and Experimental Development, Government and Private Non-Profit Organisations, Australia, 2006–07*, cat. no. 8109.0, ABS, Canberra.

ABS 2008b, *Value of Agricultural Commodities Produced, Australia, 2006–07,* cat. no. 7503.0, ABS, Canberra.

Arrow, K. 1962, 'Economic welfare and the allocation of resources for invention', in R. Nelson (ed.), *The Rate and Direction of Inventive Activity*, Princeton University Press, pp. 609–26.

Australian Treasury 2008, Budget Measures, *Budget Paper No. 2 2008–09*, Commonwealth of Australia, Canberra.

Banks, G. 2008, Colin Clark Memorial Lecture, University of Queensland, Brisbane, 6 August.

Blyth, W. & Yang, M. 2006, *Impact of Climate Change Policy Uncertainty in Power Investment*, document no. IEA/SLT (2006) 11.

Commonwealth of Australia 2006, *Uranium Mining, Processing and Nuclear Energy: Opportunities for Australia?* Department of the Prime Minister and Cabinet, Uranium Mining, Processing and Nuclear Energy Review (UMPNER), Canberra.

CRRDCC (Council of Rural Research and Development Corporations) 2006, *The Benefits of Rural R&D*, prepared by the Council of Rural Research and Development Corporations' Chairs, Rural R&D Corporations, Canberra, 5 September.

CRRDCC 2008, *Council of Rural Research and Development Corporations' Chairs, Submission to the National Innovation System Review*, Rural R&D Corporations, Canberra, April.

Department of Industry, Tourism and Resources 2006, *Collaboration and Other Factors Influencing Innovation Novelty in Australian Businesses: An econometric analysis*, Commonwealth of Australia, Canberra.

DEEWR (Department of Education, Employment and Workplace Relations) undated, *The National Research Priorities and Their Associated Priority Goals*, DEEWR, Canberra.

Foxon, T., Gross, R., Pearson, P., Heptonstall, P. & Anderson, D. 2008, *Energy Technology Innovation: A systems perspective*, report for the Garnaut Climate Change Review, Imperial College Centre for Energy Policy and Technology, London.

Fri, R.W. 2003, 'The role of knowledge: technological innovation in the energy system', *The Energy Journal* 24(4): 51–74.

Griliches, Z. 1992, *The Search for R&D Spillovers*, National Bureau of Economic Research Working Paper Series, No. 3768, NBER, Stanford, California.

Griliches, Z. 1995, 'R&D and productivity', in P. Stoneman (ed.), *Handbook of Industrial Innovation*, Blackwell, London, pp. 52–89.

Grubb, M. 2004, 'Technology innovation and climate change policy: an overview of issues and options', *Keio Economic Studies* 41(2): 103–32.

IPCC (Intergovernmental Panel on Climate Change) 2007, *Climate Change 2007: Mitigation of climate change. Contribution of Working Group III to the Fourth Assessment Report of the Intergovernmental Panel on Climate Change*, B. Metz, O.R. Davidson, P.R. Bosch, R. Dave & L.A. Meyer (eds), Cambridge University Press, Cambridge and New York.

Jaffe, A.B., Newell, R.G. & Stavins, R.N. 2005, 'A tale of two market failures: technology and environmental policy', *Ecological Economics* 54(2–3): 164–74.

Martin, S. & Scott, J. 1998, *Market Failures and the Design of Innovation Policy*, <www.mgmt.purdue.edu/faculty/smartin/vita/csda4.pdf>.

Murphy, L.M. & Edwards, P.L. 2003, *Bridging the Valley of Death: Transitioning from public to private sector financing*, National Renewable Energy Laboratory, Colorado.

OECD 2008, OECD Stat Beta Version, 'Gross domestic expenditure on R&D by sector of performance and socio-economic objective', <http://stats.oecd.org/wbos>.

Productivity Commission 2007a, *Productivity Commission Submission to the Prime Ministerial Task Group on Emissions Trading*, Productivity Commission, Canberra.

Productivity Commission 2007b, *Public Support for Science and Innovation*, research report, Productivity Commission, Canberra.

Prime Ministerial Task Group on Emissions Trading 2007, *Report of the Task Group on Emissions Trading*, Department of the Prime Minister and Cabinet, Canberra.

Raby, G. 1996, *Making Rural Australia: An economic history of technical and institutional creativity, 1788–1860*, Oxford University Press Australia & New Zealand, Melbourne.

Stephan, P.E. 1996, 'The economics of science', *Journal of Economic Literature* 34(3): 1199–35.

Stern, N. 2007, *The Economics of Climate Change: The Stern Review*, Cambridge University Press, Cambridge.

NETWORK INFRASTRUCTURE

19

Key points

There is a risk that network infrastructure market failures relating to electricity grids, carbon dioxide transport systems, passenger and freight transport systems, water delivery systems and urban planning could increase the costs of adjustment to climate change and mitigation.

The proposed national electricity transmission planner's role should be expanded to include a long-term economic approach to transmission planning and funding. The Building Australia Fund should be extended to cover energy infrastructure. A similarly planned approach is necessary to facilitate timely deployment of large-scale carbon capture and storage.

There is a limited case for carefully calculated rates for feed-in tariffs for household electricity generation and co-generation.

The need to reduce the costs of mitigation reinforces other and stronger reasons for giving higher priority to increasing capacity and improving services in public transport, and for planning for greater urban density.

Opportunities to reduce costs as the emissions price rises will require good network infrastructure. So will effective adaptation to climate change.

Good infrastructure will not always be provided in a timely manner and adequate scale by the market. Network infrastructure is vulnerable to market failure. Effective government action may be necessary for its provision in relation to electricity transmission, transport of combustible gas and carbon dioxide, freight and passenger systems, water storage and transport, and planning of urban settlements. One of these—gas—provides an example, from eastern Australia, of the market finding ways to provide adequate network infrastructure in a timely manner. The others may require planning and regulatory and sometimes wider roles for government to correct market imperfections.

There are several sources of potential market failure that can block private sector provision of adequate infrastructure.

- **Public goods**—Infrastructure that is a pure public good (that is, non-rival and non-excludable) may be underprovided because the infrastructure provider is unable to capture the full benefits of its investment.
- **Natural monopoly**—Where infrastructure is best provided by a single firm, the firm may, without competition or regulation, underprovide and overcharge for use of the infrastructure.
- **Externalities**—Where infrastructure has positive or negative spillovers to third parties, the level of infrastructure provided may not be socially optimal. These spillovers include *early-mover spillovers* where the first party to invest in infrastructure may face all of the costs, while some of the benefits accrue to later movers; and *coordination externalities* where private companies may not coordinate provision of infrastructure where trust is low or the cost of reaching agreement is high.

There may be circumstances in which, with well-directed and minimal government intervention, private activity can overcome market failures. However, these market failures can mean that there is less than economically optimal investment in network infrastructure. If left unattended, this will increase the cost of adjustment to an emissions trading scheme and inhibit an effective response to the impacts of climate change. If the cost of a market failure exceeds the cost of government intervention, with all of its political economy and other risks and costs, then regulatory or fiscal intervention by government may be required.

19.1 The transmission of electricity

19.1.1 Public good aspects of interconnectors

Interconnectors are the high-voltage transmission lines that transport electricity between adjacent regions. An interconnector's ability to transfer electricity is constrained by the extent of its physical transfer capacity.

The adequacy of interstate interconnection will be a key infrastructure issue for the National Electricity Market[1] in the near future. There are public good arguments for reducing constraints in light of the expected changes required for Australia's transition to a carbon-constrained future.

Both the emissions trading scheme and international prices for fuel source commodities such as tradable coal and natural gas will result in changes to regional comparative advantage associated with different fuel sources. Adequate interconnection will allow the National Electricity Market to accommodate structural change in the electricity sector as costs and demand change rapidly and differentially across the power sector.

Without a network of interconnectors with enough capacity to cope with the potentially large shifts in interstate flows of electricity over time, much of the generation capacity must remain within a region, even if there are more economic sources elsewhere. Confidence in the capacity of a national system will be particularly important for the period of transition. Interconnector constraints will be reflected in unnecessarily high, and more regionally differentiated and volatile, energy and emissions permit prices.

While it may seem inefficient to have permanent abundant excess capacity in the interconnectors between regions, in the world of structural change that Australia is entering, generation cost differences will exceed the distribution losses and infrastructure costs for higher levels of capacity.

Adaptation to climate change and more frequent disruptions of electricity supply will require deeper interconnection capacity. Climate impacts and pressures on electricity infrastructure are forecast to increase and include changes to demand for electricity (particularly daytime peaks from increased air conditioner use), more rapid deterioration of assets, and increased network failures resulting from severe weather events (see Box 19.1) (Maunsell 2008). The operator of the National Electricity Market (NEMMCO 2008) has also identified water scarcity as a factor that could affect generation capacity.

These pressures cumulatively threaten the overall security and reliability of electricity supply. Adaptation to climate change and more frequent disruptions of electricity supply will require deeper interconnection capacity that can provide additional security for the system as a whole by allowing electricity to be supplied from alternative areas if one section of the network is damaged.

Having excess capacity in interconnectors provides additional security for the system as a whole in the light of the pressures likely to arise from both climate change and an emissions trading scheme.

Box 19.1 Benalla bushfire blackouts

On 16 January 2007, a bushfire tripped the two Dederang to South Morang 330 kilovolt lines in northern Victoria, leading to the electrical separation of South Australia, Victoria/Tasmania and Snowy/New South Wales/Queensland into three 'islands'.

The power failure hit as Victoria was experiencing high temperatures leading to record high demand of over 9000 megawatts of electricity to run air conditioners and fans, and was drawing extra power from New South Wales through the interconnector transmission lines.

The sudden loss of 2000 megawatts of power—a quarter of the state's supply—caused an automatic load-shedding system to kick in, shutting down power to large areas of Victoria. Customer demand on the Victorian system was reduced by an estimated 2490 megawatts (NEMMCO 2007).

The blackout across Melbourne, Geelong and northern and eastern Victoria left an estimated 200 000 homes without electricity. Melbourne's public transport system and road network were also adversely affected.

Adequacy of current arrangements

At present, interconnector constraints do not appear to significantly affect short-term operations, in the absence of shocks to supply, demand or transmission infrastructure. The most constrained interconnector is DirectLink from New South Wales to Queensland, which was constrained for 285 hours in 2005–06 (Energy

Supply Association of Australia 2007). But the extent of current constraints is not the test of whether there is optimal interconnection capacity. The test is whether there would be a different pattern of investment in new generation capacity, and greater net value in greater insurance from shocks, if there were more interconnection capacity.

The current regulatory arrangements provide for the sharing of interconnection costs between the regions involved, subject to a dual test of reliability and market benefits. While the benefits of reliability often accrue to both regions, there can be differences in perceptions of the relative size of benefits. This can lead to difficult negotiations. Additionally, state governments may place limits on interconnectors to ensure that local generators are able to maintain market share within their region.

Reforms to the regulatory and institutional arrangements for the planning and funding of improvements to interconnector capacity are under way. The key focus of reform should be the facilitation of new private interconnection capacity to allow flexibility in the amount of interstate electricity trade.

19.1.2 Market failures in transmission network extensions

Rising average temperatures are likely to increase demand for energy, while an emissions trading scheme will make higher-emissions forms of energy generation more expensive, shifting demand towards lower-emissions sources. There are clear differences between the location and character of supply and demand today and into the future. Current transmission networks are geared to handle increments of supply that are near the established grid; have consistent supply; are on a large scale; and are highly centralised. The new technologies tend to be far from the grid (geothermal, thermal solar and wind), have intermittent supply (wind and solar), operate on a smaller scale (tidal), and be decentralised or embedded (photovoltaic solar and biomass). Without major changes in the transmission infrastructure, new technologies will find it difficult to compete, even in circumstances in which they are expected to be highly competitive once compatible infrastructure has been established. There are two barriers to successful network augmentation that could significantly slow or even halt the progressive deployment of lower-emissions generation technologies.

Free-rider problems and first-mover disadvantage

The current regulatory regime requires those seeking connection to cover the cost up to the point of connection. For a single remotely located generator the additional cost of connection is likely to be insurmountable. If the costs can be shared between multiple generators, the likelihood of a successful network extension increases. But the extension may not eventuate due to the strong incentive to free ride on the efforts of early movers.

The first party that connects to the network is faced with all the cost of extending the network. Later parties are then able to connect to the expanded network at a substantially reduced cost. The incentive is for potential larger-scale generators to delay investment in the hope that others will take the first step, or to select plant sizes and locations that simply 'use up' existing capacity in sections of the grid.

Barriers to achieving optimal scale in network extensions

Current processes for extending the electricity network may result in extensions without adequate capacity to carry future generation load. At present, regulatory arrangements stipulate that additional network capacity can only be funded by the broader customer base if it is judged to be the best alternative to meet reliability requirements or provides net market benefits. From this perspective, it will usually be better to install network capacity that is only adequate for current needs. When the next project to develop a resource in close proximity is proposed, the transmission network will have to be augmented, with the total cost greater than if the network had been built to that capacity from the outset.

These tendencies are exacerbated by the long lead times for transmission investment.

19.1.3 Expanded role for proposed national transmission planner

Current electricity market reform proposals involve the introduction of a national transmission planner to promote the development of a strategic and nationally coordinated transmission network.[2] The core function of the national transmission planner will be to prepare and publish an annual national transmission network development plan. The planner would have regard to 'the most efficient combination of transmission, generation, distribution and non-network options that will deliver reliable energy supply at minimum efficient cost to consumers under a range of credible future scenarios' (Australian Energy Market Commission 2008: 10). It would also take into account demand side, embedded generation and fuel substitution alternatives.

These new arrangements are expected to deliver a coordinated and efficient national transmission grid that meets local and regional reliability and planning requirements, and is flexible enough to respond to generation and load changes.

The Review endorses the recommendations for national transmission planning arrangements in the draft report by the Australian Energy Market Commission (2008). It suggests that the role of the national transmission planner be extended to incorporate an economic approach to transmission planning, and financial incentives for priority projects.

An economic approach to transmission planning

The Review endorses the Australian Energy Market Commission's recommendation that the national transmission network development plan should 'present a broad and deep analysis of different future supply and demand scenarios ... taking account of various policy, technology and economic assumptions and looking out at least 20 years into the future' (Australian Energy Market Commission 2008: 23). The Review favours the national transmission planner adopting an economic approach to transmission planning that covers more forward-looking demand and supply scenarios, rather than simply focusing on technical feasibility.

The Renewable Energy Transmission Initiative in California provides some important lessons for such an approach (see Box 19.2). The national transmission planner could undertake a similar process to that followed in California's Renewable Energy Transmission Initiative. Unlike the California initiative, the planning process should be technologically neutral and consider potential projects for both renewable and non-renewable fuels. The process would start with a resource assessment that analyses the resources considered in previous studies and identifies the most cost-effective potential power resources in areas throughout relevant parts of Australia. In particular, when analysing the need for new infrastructure, the national transmission planner must consider the effects of climate change on demand (higher temperatures) and supply (severe weather events, water scarcity and bushfires). Among other things, this analysis should take into account engineering feasibility and environmental factors.

Box 19.2 California's Renewable Energy Transmission Initiative

The Renewable Energy Transmission Initiative is a statewide initiative of the California Energy Commission that aims to identify the transmission projects needed to accommodate the state's renewable energy goals. The purpose of the initiative is to bring together all of the renewable transmission and generation stakeholders in the state to participate in a consensus-based process to identify, plan and establish a rigorous analytical basis for regulatory approvals of the next major transmission projects needed to access renewable resources.

There are five core steps to the process:

- identifying competitive renewable energy zones having densities of developable resources that best justify building transmission to them
- ranking zones on the basis of environmental considerations, development certainty and schedule, and cost and value to Californian consumers
- developing conceptual transmission plans to the highest-ranking zones
- supporting the California Independent System Operator Corporation, investor-owned utilities and publicly owned utilities in developing detailed plans of service for commercially viable transmission projects
- providing detailed analysis regarding comparative costs and benefits to help establish the basis for regulatory approvals of specific transmission projects.

The Office of Gas and Electricity Markets in the United Kingdom undertakes a similar exercise with its long-term electricity network scenarios.

Source: RETI Coordinating Committee (2008).

This analysis would be informed by a comprehensive stakeholder consultation process with private sector generation companies. Firms would submit proposals and estimates of the costs of developing the generation resources within an area and delivering that energy to consumers. These project and technology costs would by necessity be estimates, intended primarily to provide information to compare areas. The open and transparent process would support the emergence of a consistent set of assumptions.

Ultimately, based on analysis of comparative economics and other factors, potential power supply areas would be grouped into high-demand zones. These areas would then be allocated priorities on the basis of economic contributions.

Financial incentives for priority projects

The Australian Energy Market Commission (2008: ix) states that 'the [national transmission planner] will be required and resourced to produce its own development strategies, including its own transmission investment options'.

The role will need to be developed as the location and structure of Australian electricity generation and demand change rapidly. The national transmission planner will need to be alert to market failure leading to slow and suboptimal response to changing supply and demand.

The role of the planner should be expanded to include advising the Commonwealth Government on whether there is a need for initial public funding for transmission investments. The objective would be to ensure that optimal extensions of transmission capacity were not inhibited by first-mover problems, and that extension and expansion of the network were designed at optimal scale. Advice could include processes for recovery of investments as utilisation expands over time. Care would need to be taken to ensure that there was no crowding out of private provision of transmission capacity.

The Review proposes that funds be made available for this purpose from Infrastructure Australia, and its Building Australia Fund. The Building Australia Fund is currently earmarked for national transport (roads, rail and ports) and communications infrastructure (broadband) that cannot be delivered by the private sector or the states. It would be appropriate for the Building Australia Fund to be extended to finance high-value national electricity transmission infrastructure.

19.2 The distribution of electricity

19.2.1 Externalities of embedded generation

There are two main positive externalities created by embedded generation that may not be adequately priced. These could lead to inefficient investment decisions.

Network externalities arise from:

- **Deferred augmentation of the transmission and distribution systems**—The community expects the electricity distribution (and transmission) networks to be engineered to meet requirements in periods of peak demand. At sufficient scale, embedded energy generation (particularly during peak periods) can reduce the engineering requirements of the system to the extent that this allows deferral of network augmentation. This would lower the overall cost burden for end users.

- **Reduced transmission losses**—Energy losses from electrical resistance in transmission cables are significant when electricity is transported over long distances.[3, 4] Losses are exponentially related to load. By siting a generator near a load, the amount of energy required to be imported from the network is reduced. The non-linear relationship between load and loss means that all customers benefit from reductions in system losses. Current rules do not recognise the reduction in losses that embedded generation brings to the system as a whole.[5]

The market failure arises because the investor in the embedded energy infrastructure cannot appropriate the benefits created for others from deferred network augmentation or transmission losses.

The current regulatory framework prevents these externalities from being internalised. Distribution businesses receive revenue based on the value of the asset base, creating the incentive to build more distribution infrastructure. Rewarding embedded generators for the benefits of deferred network augmentation is in direct conflict with this arrangement. The first best solution would be reform of the regulatory framework for distribution businesses. The existing regulatory frameworks are the result of many years of reform and therefore the first best solution may not be achievable in the short term.

Feed-in tariffs can be used to internalise the positive externality for investors in embedded generation, though they can only do so at the margin.

19.2.2 What should the value of a feed-in tariff be?

Metering

There are two methodologies for calculating feed-in tariffs: gross metering and net metering.[6] Gross metering pays the embedded generator for all electricity it produces and does not discriminate between embedded and centralised generation. Net metering pays only for the energy exported to the grid (gross generation minus local energy consumed).[7]

The two externalities from embedded generation are present for every unit of electricity produced, not just the amount sold—implying that gross metering is the more appropriate approach for addressing this market failure.

Rate

The rate embedded generators receive per unit of electricity should be based on a rigorous quantification of the externalities described above and must include full accounting of implementation costs. This may result in a lower feed-in tariff than is currently being applied in most schemes. If governments opt for a higher tariff, then the rest of the customer base will be cross-subsidising embedded generators. The reintroduction of a cross-subsidy would run counter to the reforms of the last decade.

Where the network externalities of embedded generation (less implementation costs) are found to be positive and material, a consistent approach should be adopted across jurisdictions. This is not currently the case. The policy objective should be a consistent *methodology*, not necessarily the same tariff rate (which will depend on local conditions in each region).

19.3 Gas transmission infrastructure

Australia's gas transmission system is privately owned, and today serves the dual purpose of connecting gas fields to gas markets and interconnecting regional systems. Interconnections provide a degree of supply diversity and security.

While the potential impediments to private provision of optimal amounts of network infrastructure, such as first-mover and free-rider barriers, are not absent from the gas market, there is evidence that the market has been able to overcome them.

Australia's east coast gas transmission system has expanded rapidly over the last 30 years through private sector investment, with little government intervention. The construction of the SEA gas pipeline connecting the Victorian and South Australian gas systems through its link between Port Campbell and Adelaide provides a recent example. The pipeline connected the joint interests of gas producers in Victoria and a gas generator and gas retailers in South Australia, and was ultimately constructed as a three-way joint venture.

The majority of Australia's gas transmission pipelines are not regulated. Pipeline developers and owners, who can contract directly with shippers, use pricing structures that have avoided such a requirement. This contrasts starkly with the electricity market.

There is no reason to suggest that existing impediments would be any more significant following the introduction of an emissions trading scheme. This is an example of a network infrastructure market working efficiently without government intervention.

19.4 The transportation of carbon dioxide

19.4.1 Infrastructure challenges

Because of the relative immaturity of the technology for geosequestration of carbon dioxide, current projects in Australia are located close to storage sites of varying capacities, eliminating the costs of transportation over long distances.

As the number of sources of carbon dioxide and identified sequestration sites increases, there will be a corresponding increase in the need for pipeline networks to transport carbon dioxide between locations, some of which may be relatively isolated (see Figure 19.1). There may be good arguments for locating a point source far from a sequestration site. The current location of many coal-fired power plants close to coal seams is an example.

Figure 19.1 Major sequestration sites and carbon dioxide sources in Australia

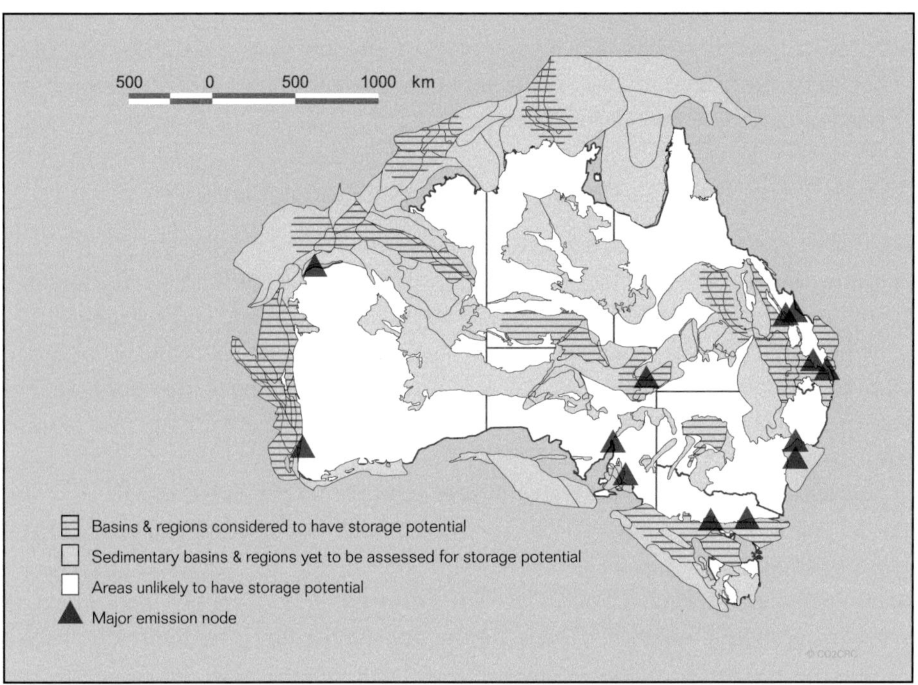

Source: Image courtesy of CO2CRC.

Carbon dioxide gas is most efficiently transported when compressed to a supercritical state (a temperature and pressure at which it shows properties of both liquids and gases). Pipelines are the most economic mode for transporting large amounts of carbon dioxide over distances of up to 1000 km. This method of transporting pressurised carbon dioxide is already a mature technology in the United States, where about 40 million tonnes per year travels through a 2500 km network of high-pressure pipelines (mainly in Texas) for the purpose of enhanced oil recovery (IEA 2001).

19.4.2 Potential roles for government

The provision of a system of pipelines transporting carbon dioxide from points of capture to points of storage has the potential for market failure in three phases.

Pre-commercial planning

While carbon dioxide sequestration technology matures and approaches commercial feasibility, an appropriate independent body established by government could start assessing appropriate carbon dioxide sources, sequestration sites, existing projects and potential future projects with the aim of highlighting some of the possible long-term priorities for key pipeline infrastructure. This process could go beyond a study of technical feasibility and explore economic competitiveness based on consultation and proposals from the carbon capture and storage industry.

Government should not act on these assessments until substantial demand has been confirmed.

Establishment

Once the industry has matured to the point of being potentially commercially competitive, government will need to be prepared with efficient mechanisms for initial development and funding of a pipeline grid. As has been the experience with the gas industry in Australia (see section 19.3), it is possible that the physical infrastructure for carbon dioxide transport could be successfully provided by the private market, thereby requiring minimal intervention by government.

However, the potential for delays in overcoming inhibitions to private cooperation in a new market may warrant government intervention. This could involve supporting the construction of the main pipelines at a socially optimal scale, regulating pipeline construction, providing a contingent subsidy, or providing adequate information regarding sites and sources. If government funding were required in the establishment phase then early outlays could be recovered from future users of the spare capacity. Government could divest itself of the asset by sale to a private operator as the pipeline approaches full utilisation.

As with the electricity transmission infrastructure discussed in section 19.1.3, a program (also based on the Californian Renewable Energy Transmission Initiative) could provide an efficient mechanism to determine the initial coverage and scale of a carbon dioxide pipeline grid, and to fund identified carbon dioxide pipeline priorities if this proves to be necessary. Arrangements for cost recovery and eventual sale to the private sector should be structured so as to maintain incentives for purely private pipeline investment.

Long-term management and access

Since the pipeline system could be a natural monopoly, access arrangements for multiple users may be required. The gas industry has privately established these arrangements. The carbon dioxide sequestration industry may be able to do the same. If not, the Australian Competition and Consumer Commission would need to establish an appropriate regime.

19.5 The transport of passengers and freight

An emissions trading scheme will make higher-emissions forms of vehicles and, modes of transport more expensive, shifting demand to lower-emissions forms. The extent to which consumers can express these preferences will be strongly dependent on the availability of the appropriate network infrastructure.

In some cases, the private sector can deliver low-emissions options, such as inter-regional passenger coach services and private rail freight systems. In other cases, particularly in urban passenger transport, market failures will justify the involvement of government and affect the efficient provision of infrastructure and services.

The Review identified the following areas in which market failures would seem to warrant corrective action:

- The quasi-public good nature of road, bicycle and walking infrastructure.
- The natural monopoly characteristics of hard rail infrastructure.
- The coordination externalities of integrated service provision—Where two or more services combine to provide a passenger trip (such as a bus then a train), benefits accrue to the passenger if the infrastructure, ticketing, provision of information, and timing of these services are well integrated. This coordination does not always occur, resulting in a suboptimal outcome for passengers.
- The positive externalities associated with new transport infrastructure and services—New infrastructure, such as rail lines, can increase the value of local properties but the party providing the train line does not capture this benefit.
- Externalities in land use and transport—If the price of new housing does not reflect the cost of providing new infrastructure and services to that location, it can encourage development further away from current infrastructure than would otherwise occur.

To correct these sources of market failure, governments have traditionally funded transport infrastructure, funded and provided transport services, regulated pricing of natural monopolies, and regulated where people can develop land and build houses. Australian governments have attempted to introduce a larger role for markets in decisions by entering into public–private partnerships for some infrastructure development, corporatising or privatising the service provision, and increasing the reliance of service providers on ticketing and/or revenue. Despite this, most decisions regarding the location, timing and extent of infrastructure investments for public transport services are ultimately made by governments.

Are current arrangements suitable for managing the changing needs for transport infrastructure into the future? The increased demand for low-emissions transport options reinforces other and more powerful reasons for increased public policy focus in this area. Other reasons include rapid escalation of congestion costs and related equity issues at a time of rapid growth of population and incomes.

19.5.1 Funding

The current arrangements for transport funding may create biases in infrastructure spending in favour of roads relative to other modes.

As highlighted by the Victorian Competition and Efficiency Commission (2006), road bias could occur if funding for roads is 'more flexible, more accessible or gives greater autonomy to road project managers compared with project managers for other modes'. The Review proposes that state and territory governments investigate their current transport funding arrangements, including dedicated road funding.

Second, of the $12.3 billion the Commonwealth Government allocated to transport through Auslink in 2004–05 to 2008–09, the majority was directed to roads, including urban roads and grants to local governments (Department of

Infrastructure, Transport, Regional Development and Local Government 2008). Less funding has been directed to rail, and urban public transport was excluded.

While this could be understood if the intent were to distribute the fuel excise that is levied to pay for road development and to concentrate on roads or rail lines of significance for the national economy, it runs the risk of creating incentives for state and territory governments to give priority to road (where they can achieve matched funding) over rail projects (which they must fully fund) and non-urban over urban projects.

For this reason, the Review considers that federal funding for transport infrastructure should be broadened to include contributions to all modes of transport, in urban and non-urban areas. The establishment of the Building Australia Fund and recent commitments to contribute to road and rail projects to alleviate urban congestion in Melbourne are steps in this direction.

19.5.2 Prices for transport use

Charges that reflect mode use are important in ensuring optimal decisions on allocation of investment in various modes of transport. There are many sources of divergence between private and public benefits affecting investment in various transport modes:

- the equity benefits of some modes of transport and the many externalities generated, related to noise, air pollution, accidents and congestion
- the equity benefits of some modes of transport in some locations
- incomplete price signals in land-use decisions, as some costs associated with new developments do not reflect the full extent of subsequent government investment and private transport costs
- the difficulty of quantifying all these externalities and equity benefits
- the many charges levied by all levels of government and the private sector, including fuel excise, road tolls, registration fees and heavy vehicle duties
- the direct and indirect provision of subsidies from government.

The Review has not investigated to what extent users pay for the costs of use (including externalities) across modes. However, some specific externalities can be analysed and taken into account fairly precisely. Congestion charging has been estimated to provide multiple benefits in reducing road congestion independently of any mitigation considerations (Bureau of Transport and Communications Economics 1996). State governments should investigate congestion charging in major cities.

It is desirable to have closer links between pricing structures and the full cost of providing infrastructure and services. Then pricing structures would take account of such factors as distance travelled, mass of cargo (especially for trucks) and place of travel (especially to take account of congestion). This would enable users to maximise the efficiency of their travel and providers (road agencies and public transport service providers) to respond in areas and times of high demand. This would also enable people to respond to an emissions price more flexibly.

There are moves in this direction, such as trials on mass–distance–location pricing for freight, new roads with private investment that have tolls, and point-to-point fares for some public transport systems. Reform in this area should be accelerated. The Productivity Commission's recommendation that incremental pricing form a precursor to mass–distance–location pricing for freight is worth another look.

19.5.3 A coordinated transport policy

Delivering an effective national transport system requires balancing a wide range of objectives on a local and regional scale. As a result, although many jurisdictions have set out principles to guide transport planning, there is much ad hoc decision making. The lack of explicit principles in transport planning results in the implicit use of less desirable principles, which can create a bias towards some modes. The absence of principles can lead to systematic discrimination in favour of continuation of established trends, and in favour of expansion of modes at the time experiencing congestion. In contemporary Australia, this has favoured road infrastructure over rail, cycling and walking infrastructure.

Governments should plan transport infrastructure and land-use change with a horizon of 40 years or more. Transparent long-term planning will undoubtedly create controversy, as both higher urban densities and some new areas of development will be required. However, failing to make long-term plans will create a burden of poorly functioning cities that is difficult to unwind and will last for many decades.

Given the clear need for strategic policy to be coordinated across modes to make the whole system more efficient, the Review suggests that a single body in each state and territory should be responsible for transport policy and coordination across the transport portfolio. The institution could be supported by a number of service delivery agencies, each responsible for a single mode.

19.6 Water supply infrastructure

The eastern, south-east and south-west regions of Australia house the majority of Australians and produce most of Australia's irrigated output. These areas are likely to experience a decline in rainfall and higher temperatures causing increased evaporation. This is likely to diminish inflows to local, rainfed water supplies, compounding longstanding problems associated with a dry and variable climate and strong population growth in Australia.

Natural monopolistic characteristics in water supply, and equity dimensions in its use, have led governments traditionally to manage the supply and regulate the price of water in urban and rural environments. However, in the past 20 years, the water sector has undergone significant change. Governments have attempted to account for some of the environmental externalities by progressively capping extraction levels on most surface water systems, and there has been progressive introduction of market forces to water supplies. There is now a water market that allows irrigation water to flow to the highest value use. The majority of water provision has been corporatised or privatised. Water business providers increasingly rely on revenue from water sales.

19.6.1 Adequacy of current arrangements

Australia's water markets are, nevertheless, far from fully competitive and there is scope for greater integration of contiguous water systems. Extant physical and regulatory barriers lead to differential pricing and underinvestment in infrastructure. There are restrictions on the volume of trade between irrigation regions and there is only limited trade between urban and irrigation water systems. As a result, the price of water in different regions often does not reflect its scarcity value, with the result that it may be allocated to lower-value uses.

Urban water supply remains centralised, with government agencies generally undertaking the roles of planner, regulator, wholesaler, distributor and retailer (Productivity Commission 2008). This restricts the diversity of supply options, including decentralised local supply options, to those considered by governments. It has also left the market reliant on government infrastructure provision in most cities which, until the recent long drought, experienced low levels of investment, resulting in the need for stringent restrictions on water use in most major cities (Quiggin 2007).

Finally, current pricing arrangements do not reflect the long-run marginal cost of supply. This inhibits timely and efficient investment decisions.

19.6.2 Continuing water reforms

Future reforms in the water market should focus on developing an integrated and competitive market. By removing or reconfiguring barriers to trade and supply, and encouraging prices based on long-run marginal cost, price signals would accurately reflect and mediate between demand and supply. Coupled with reform of institutional arrangements to allow for private investment in supply, this would lead to efficient, timely and effective competition among a greater diversity of supply options.

In such a market, government's role would be to ensure that equity, security and environmental considerations were met transparently. Equity considerations have led to many government interventions in water supply and pricing, but they need not do so. Community service obligations are a transparent method of ensuring that low-income households have access to basic services. The effects of higher pricing can be addressed efficiently through the income tax and social security systems (see Chapter 16). Water security can be provided through diversity of supply options, particularly where uncertainty about climate change makes it more difficult to forecast supply and demand. Regulatory approaches are necessary to secure environmental objectives.

19.7 The planning of urban settlements

Climate change is likely to affect all Australian communities, with coastal areas under particular pressure. This is largely associated with sea-level rise, storm surge and coastal flooding, and increased frequency and intensity of severe weather events. Acknowledging the strong preference of Australians for living on the coast, the relationship of climate change to coastal infrastructure will be a determining factor in the distribution of future settlements for a rapidly growing population.

The location of developments, the strength and stability of built structures, the thermal comfort of buildings, and the capacity and physical resilience of electricity networks and stormwater systems will be affected. Coastal communities experiencing rapid population growth will experience pressure for rapid development approval, with the risk that this will occur in advance of climate change considerations being factored into planning and assessment frameworks (Gurran et al. 2008).

Given the long functional life of most built infrastructure, ensuring resilience to anticipated future climate change impacts will be crucial. Poor decisions will result in outcomes that are costly to fix.

Across Australia, relevant planning policy and legislation has been slow to incorporate climate change considerations (Planning Institute of Australia 2007), both in the context of reducing the carbon intensity of settlements and developments (mitigation) and of building resilience to climate change impacts (adaptation). This section focuses on adaptation.

19.7.1 Current arrangements for strategic and statutory urban planning

The statutory framework for planning and development in Australia is set at the state and territory government level and administered by local government. The manner in which climate change adaptation is included in the high-level parameters set by state and territory governments is generally limited or in the early stages of development (Walsh et al. 2004). Similarly, few municipal planning schemes include specific provisions for climate change adaptation (Gurran et al. 2008).

Municipal councils have noted the presence of various barriers to the effective integration of climate change within local policy and planning schemes. Councils have expressed concern about:

- access to reliable data on potential climate risks
- the need for advice on the best way to reflect these matters in a planning scheme
- their capacity to develop an assessment of impacts that is defensible if reflected in planning permit conditions and subject to appeal.

Compounding these challenges are competing priorities, financial constraints and a chronic shortage of skilled planners across the country. The combination of factors has resulted in a wide failure to incorporate climate change into planning (Burton & Dredge 2007).

19.7.2 Minimising climate change impacts through improved planning

To reduce the impacts of climate change on future coastal settlements three factors should be considered:

- the location of new or infill developments and associated public and private infrastructure
- the resilience of the infrastructure in the face of new climatic conditions such as higher temperatures, wind and floods
- the capacity of the infrastructure to service demand given the expected climate.

Each of these can be addressed through the planning system.

Location

The fundamental role for planners in relation to minimising climate change impacts will be decisions on the allocation of land for development.

Impacts such as anticipated storm surge and inundation of coastal areas already present a challenge to planning in Australia. This is likely to be exacerbated by the speed with which governments will be expected to release land for new developments to keep pace with population growth.

Land-use zoning and other controls have historically prevented or discouraged development in areas where the risk of damage to infrastructure and buildings is thought to be too great because of the prospective incidence of severe weather events. Maintaining current approaches to zoning and development control may no longer achieve this end in a future altered by climate change.

There has been a growing call for local government to consider the risks posed by climate change in coastal development approval processes (Burton & Dredge 2007).

Where applications for development have been affected by assessments related to climate change, many have ended up before a civil appeals process (McDonald & England 2007; Thom 2007).

Resilience

Climate change will affect the resilience of infrastructure in new and existing settlements. Although improved resilience can be achieved by measures independent of the planning system, such as through building codes, sound planning can still play a useful role (see Box 19.3). For example, fixed-line networks and towers for electricity and telecommunications will need to be able to withstand more intense wind and storms (Maunsell 2008). Strategic decisions about the location of new installations can minimise exposure to the elements.

Box 19.3 The limits of planning: infrastructure to protect coastal settlements

There is already considerable coastal infrastructure in areas that may become prone to storm surge, coastal flooding and other climate change impacts in the future. For existing infrastructure the planning system offers little opportunity for adaptation to climate change.

A range of adaptation actions independent of planning may be successful in managing such situations. Examples of private actions include adjusting insurance risks or retrofitting buildings to improve resilience against particular threats such as wind or flood. However, if these options are exhausted and protection is still not adequate, there is a possibility that more intensive actions such as building sea walls or other protective structures may be necessary.

Sea walls are only likely to be successful if they are built over a length of coastline. They are not suited to protecting individual properties (New South Wales Government 1990) and they are costly to construct. Consequently, this kind of infrastructure is unlikely to be provided by individuals or markets. In cases where a sea wall is likely to produce a net benefit, a public good market failure arises and government intervention may be appropriate.

Assessments of what constitutes a net benefit in relation to the development of sea walls will be difficult, and will depend upon judgments of how the benefits should be distributed among the relevant community. In addition, there is ultimately the question of whether protective infrastructure in a particular location is actually likely to produce greater benefit than abandonment, relocation or simply accepting risk. It will also be difficult to establish the most appropriate location for a sea wall, and the scale of climate change impacts it should be designed to withstand. These are complex issues that will have to be resolved under considerable uncertainty. There is a significant risk of large government expenditure with little actual benefit (Dobes 2008). Also, public provision of protection through sea walls would raise complex distributional issues.

Planning schemes already consider the risk of bushfire, which is expected to increase in both frequency and intensity with climate change. Development controls can stipulate building envelopes, access, height, design, materials and landscaping requirements (SMEC Australia 2007: 35). Such controls require revision to take climate change into account.

Capacity

The impacts of climate change may affect the scale of infrastructure required to service human settlements. For example, higher temperatures may result in greater use of peak-load electricity for the cooling of buildings. Drier conditions may create a need for new water supply infrastructure. Flooding from storm surge may increase the need for greater stormwater drainage capacity.

Decisions about supplementary infrastructure requirements are complex and will need to be made in the context of uncertainty about climate change impacts.

19.7.3 Changes to Australia's planning regime

Planning is facing some new and challenging land use and development issues because of climate change; nowhere more so than in our growing coastal settlements. To bring climate change adaptation into mainstream planning practices and long-term strategic directions a number of enhancements to the current planning system are required.

Councils must be able to demonstrate a sound evidence base for identifying and justifying planning responses to climate change. Many coastal councils, particularly those with a small rate-base, will need assistance in accessing, interpreting and applying consistent and reliable sources of scientific information about climate change scenarios (Gurran et al. 2008). The federal, state and territory governments have roles to play in:

- supporting local government to access information and build expertise
- undertaking vulnerability assessments[8] at a regional or local scale to support strategic land-use planning decisions and significant development assessment.

At a more detailed level, state and territory planning legislation and policy must clarify local governments' obligations in relation to developments that may present an unacceptable exposure to climate change. This can be done through a strengthening of the state planning policy frameworks. There is a case for developing nationally consistent planning guidelines, recognised within planning policy, that codify standards of acceptable risk for development approvals (A. Kearns 2008, pers. comm.). The principal aim of such guidelines would be to provide:

- clear guidance for councils as to how climate change should be factored into a development approval
- policy support for the decisions they make
- transparency for the community and developers on how planning applications as affected by climate change are likely to be determined.

Such guidelines may reduce the number of cases that appear before civil appeal.

Guidelines should encourage and enable consistency in decision making. This is particularly relevant for cross-boundary settlement planning and coordinated growth management along our coastline. However, they must be flexible enough to account for the variability of expected climate impacts in different regions. They must also be able to respond to unexpected changes, new technologies or new scientific information as it comes to hand (Gurran et al. 2008).

The federal government can play a role in ensuring that states, territories and local government have ready access to authoritative national and international data and information on climate science and the impacts of climate change.

Notes

1 The National Electricity Market is a wholesale market for electricity supply covering the Australian Capital Territory and the states of Queensland, New South Wales, Victoria, Tasmania and South Australia. In 2005–06, approximately 88.6 per cent of electricity generated was sent out in the National Electricity Market.

Because the vast majority of electrical energy in Australia is traded on the National Electricity Market, the Review's analysis of electricity infrastructure provision will focus on barriers to that market. The report commissioned by the Review from McLennan Magasanik Associates contains detailed discussion of market failures in the National Electricity Market (see <www.garnautreview.org.au>). That said, the analysis of potential problems and solutions is relevant to the other Australian electricity markets.

2 The Council of Australian Governments and the Ministerial Council on Energy have provided some guidance and prescription on the characteristics of the new arrangements.

3 The average weighted distribution loss in Australia in 2005–06 was 5.9 per cent, with the highest loss factor of 7.2 per cent in Tasmania (Energy Supply Association of Australia 2007).

4 There are technological solutions to transmission losses such as lower-resistance power lines, but the capital costs are currently prohibitive.

5 While National Electricity Market rules currently require network businesses to pass on these savings to larger embedded generators (known as the avoided transmission use of system charge), there is no requirement to similarly compensate the smaller embedded generators.

6 The selection of the type of tariff will depend on the technological capabilities of the meters installed.

7 Some argue that a gross-metered feed-in tariff is undesirable because, from a sustainability perspective, it does not encourage embedded generators to consume less electricity, whereas under a net-metered scheme profits can only be made by exporting more to the grid. This reasoning is erroneous because the incentives to consume should come through the retail tariff paid for electricity, not through the feed-in tariff system.

8 In this context 'vulnerability assessment' is used to describe the consideration of the environmental, social, cultural and economic issues that relate to land use and settlement patterns and as anticipated to be affected by climate change. For example, analysis of affected assets, population and natural landscapes.

References

Australian Energy Market Commission 2008, *National Transmission Planning Arrangements*, draft report, Australian Energy Market Commission, Sydney.

Bureau of Transport and Communications Economics 1996, *Transport and Greenhouse: Costs and options for reducing emissions*, Report 94, Bureau of Transport and Communications Economics, Canberra.

Burton, D. & Dredge, D. 2007, 'Framing climate: implications for local government policy response capacity', *Proceedings of the State of Australian Cities National Conference 2007* Adelaide, S. Aust.: SOAC : Causal Productions, 2007. ISBN 978-0-646-48194-4

Department for Infrastructure, Transport, Regional Development and Local Government 2008, *Auslink Funding Allocations Consolidated*, <www.auslink.gov.au/funding/allocations/index.aspx>.

Dobes, L. 2008, 'Getting real about adapting to climate change. Using "real options" to address the uncertainties', *Agenda: A Journal of Policy Analysis and Reform* 15(3), <http://epress.anu.edu.au/titles/agenda.html>.

Energy Supply Association of Australia 2007, *Electricity Gas Australia 2007*, Energy Supply Association of Australia Ltd, Melbourne.

Gurran, N., Hamin, E. & Norman, B. 2008, *Planning for Climate Change: Leading practice principles and models for sea change communities in coastal Australia*, report prepared for the National Sea Change Taskforce, University of Sydney.

IEA (International Energy Agency) 2001, *Putting Carbon Back into the Ground*, IEA Greenhouse Gas R&D Programme.

Maunsell Australia Pty Ltd 2008, *Impact of Climate Change on Infrastructure in Australia and CGE Model Inputs for the Garnaut Climate Change Review*, Maunsell Australia Pty Ltd, Melbourne.

McDonald, J & England, P. 2007, 'A risky climate for decision making: the legal liability of development authorities for climate change impacts', paper presented to QELA 2007 Conference Your System or Mine?, Kingscliff, 16–18 May.

NEMMCO (National Electricity Market Management Company) 2007, *Market Event Report: System separation and load shedding*, <www.nemmco.com.au/opreports/180-0074.pdf>.

NEMMCO 2008, *Drought Scenario Investigation, June 2008 Update*, NEMMCO, Melbourne.

New South Wales Government 1990, *NSW Coastline Management Manual*, 'Appendix D6: Protective Works Options', New South Wales Government, September, <www.environment.gov.au/coasts/publications/nswmanual/appendixd6.html>.

Planning Institute of Australia 2007, The Delivery of Training Seminars to Planning Practitioners on the Impacts of Climate Change: Final report, report prepared for the Australian Greenhouse Office, Planning Institute of Australia, Kingston, ACT.

Productivity Commission 2008, Towards Urban Water Reform: A discussion paper, Productivity Commission Research Paper, Productivity Commission, Melbourne.

Quiggin, J. 2007, 'Urban water: markets and planning', presentation at the Australian Academy of Science's 2007 Fenner Conference on the Environment on Water, Population and Australia's Urban Future, Canberra, 15 March.

RETI Coordinating Committee 2008, Renewable Energy Transmission Initiative: Phase 1A, Final Report, Black & Veatch Corporation, California

SMEC Australia 2007, *Climate Change Adaptation Actions for Local Goverments*, Australian Greenhouse Office, Department of the Environment and Water Resources, Canberra.

Thom, B. 2007, 'Climate change and the coast: the institutional challenge', paper presented to Climate Change Forum, Department of Primary Industries & Southern Rivers Catchment Management Authority, Bega & Nowra, 19–20 June. http://www.wentworthgroup.org/docs/Climate_Change_&_the_Coast.pdf>

Victorian Competition and Efficiency Commission 2006, *Making the Right Choices: Options for managing transport congestion, Final Report*, Victorian Competition and Efficiency Commission, Melbourne.

Walsh, K.J.E., Betts, H., Church, J., Pittock, A.B., McInnes, K.L., Jackett, D.R. & McDougall, T.J. 2004, 'Using sea level rise projections for urban planning in Australia', *Journal of Coastal Research* 20(2): 586–98.

20 TRANSFORMING ENERGY

Key points

Australians have become accustomed to low and stable energy prices. This is being challenged by rapidly rising capital costs and large price increases for natural gas and black coal. These cost effects will be joined by pressures from rising carbon prices, and will be larger than the impact of the emissions trading scheme for some years.

Australia is exceptionally well endowed with energy options, across the range of fossil fuel and low-emissions technologies.

The interaction of the emissions trading scheme with support for research, development and commercialisation and for network infrastructure will lead to successful transition to a near-zero emissions energy sector by mid-century.

The future for coal-based electricity generation, for coal exports and for mitigation in developing Asia depends on carbon capture and storage becoming commercially effective. Australia should lead a major international effort towards the testing and deployment of this technology.

As emissions fall in electricity, it will become an increasingly important source of energy to other stationary energy and transport.

If the world is to meet the challenge of climate change, there will need to be a transformation in Australia's stationary energy sector as it adjusts to mitigation policies. Over the next 40 years, we will see the emergence of something close to a zero-carbon energy sector in Australia and around the world—an energy transformation.

This will be part of a wider set of big changes for Australians. We have become accustomed to low and stable energy prices. These have underpinned aspects of our economic structure and lifestyles. The cheap energy is being challenged by rapidly rising capital costs, large increases for natural gas and black coal prices on world markets, and a big lift of gas prices in eastern Australia to international levels as export facilities are established. On top of all this—smaller in magnitude in the early years, but increasingly important towards and beyond 2020—will come carbon pricing within the emissions trading scheme.

There will be three broad phases to the transformation to a low-emissions energy sector:

- an initial adjustment phase involving a transition from high-emissions growth to greater use of known lower-emissions technologies
- a technology transition phase as new technologies, some of which may be important through this phase only, emerge and then facilitate and accelerate the restructuring of the sector
- a long-term emergence phase to sustainable, low- and zero-emissions technologies.

The electricity sector is projected to play a central role in the way in which the Australian economy achieves an abatement commitment within a global agreement. The role emerges from the 35 per cent contribution that electricity makes to greenhouse gas emissions today. It is magnified by the capacity for other sectors, notably other stationary energy and transport, to achieve lower emissions by changing from high-emissions fossil fuels to lower-emissions electricity. The transformation described in this chapter draws on and expands upon the results of Garnaut–Treasury economic modelling to describe the role and structure of the energy sector in a low-emissions economy.

For this reason, electricity is the major focus of this chapter, although broader issues for energy, and directly affected sectors such as coal and aluminium, are also described.

20.1 The energy sector today

The current low and stable price of energy has been largely taken for granted by the Australian community. The realities are changing rapidly.

The energy sector, driven by the reforms of national competition policy and progressive privatisation, is now a physically and financially sophisticated and increasingly national sector, delivering security of supply, competitive prices and new investment. This evolution remains unfinished, with regulatory responsibility for monopoly subsectors still to make the full transition to national bodies. Government ownership remains dominant in several states, and price and service regulation remains in areas where competition should be capable of delivering greater consumer benefits.

The energy sector makes a larger contribution to greenhouse gas emissions in Australia than in other developed countries. An energy sector that addresses mitigation will therefore need to establish a balance between driving change towards a low-emissions future built on the underlying national reform agenda, and preserving as much as possible of the energy sector's positive contribution to the Australian economy.

20.1.1 Australia's energy sector in the economy

Growth in energy consumption has historically moved closely with GDP in Australia, with a tendency to be slightly lower since the early 1990s. The stationary energy sector is dominated by electricity generation and manufacturing processes. The size of the sector and its fuel mix vary across the country and reflect different regional economic structures and local fuel sources (see Figure 20.1).

Figure 20.1 Installed electricity generation capacity, 2005–06

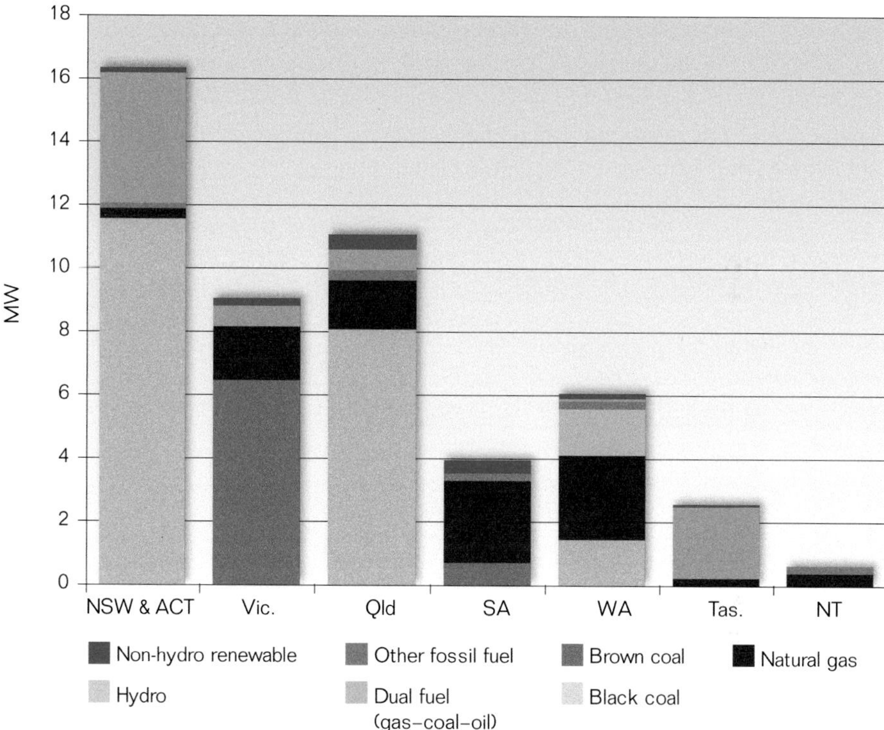

Source: ESAA (2007).

The availability of large and accessible coal and gas resources has delivered electricity and gas prices that have been low in Australia relative to those in comparable countries. Australia and other developed countries have experienced substantial increases in domestic electricity prices in recent years (see Figure 20.2).

20.1.2 Recent developments

Since the early 1990s, there have been remarkable changes in Australia's energy sector. Privatisation and competition policy have led to a fundamental restructuring of the retail and generation sectors, with the integrated generator–retailer across both electricity and gas taking an increasingly dominant role.

The distribution and transmission pipeline sectors in gas are now entirely, and in electricity partially, privately owned. The electricity transmission grid remains largely government-owned. Distribution and transmission assets generally exhibit natural monopoly characteristics and are subject to economic regulation, for which responsibility is gradually passing to the national Australian Energy Regulator.

The result is a highly competitive, increasingly national market with considerable regional interconnection, sophisticated financial structures and flexible fuel substitution. This era of change has delivered choice and broadly stable prices for customers and an attractive climate for investors, while maintaining and increasing supply security.

Power generation based on black and brown coal for base-load supply, transmission interconnection for flexibility and additional security, and gas-fired plant to meet the growing demand for peak and intermediate capacity have all been important in this period of rapid change. Almost 5000 MW or approximately 12 per cent of net additional generation capacity was added between 1999 and 2006, with further capacity either under construction or planned in response to the consistent growth in demand.

Figure 20.2 Comparison of industrial electricity prices

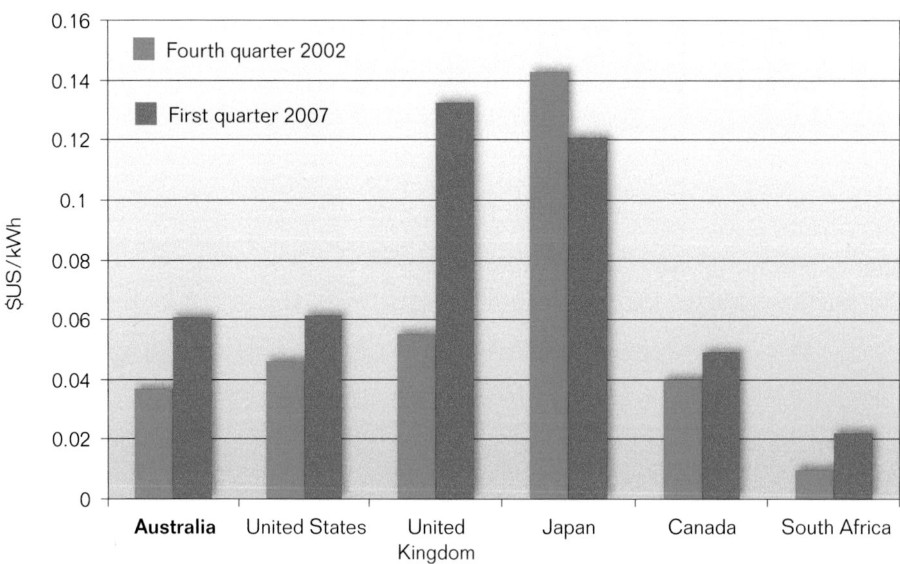

Note: Exchange rate movements were significant sources of changes in relativities between 2002 and 2007.
Source: IEA (2003, 2008).

From the mid-1990s until around 2006, prices for both electricity and gas were generally stable. Domestic prices for thermal coal were relatively low and steady. In the case of gas, the market has witnessed a growth in the depth and breadth of supply sources in response to:

- renewal cycles of long-term contracts
- the requirements of the Queensland Government that a minimum proportion of electricity be generated from gas

- the increasing role for gas in meeting peak electricity demand
- the development of new gas fields
- the emergence of coal-seam gas as a major new supply source.

The electricity market has been characterised by strong growth, which has been met by new capacity, increased operability of existing plants and new inter-regional transmission lines.

Since 2006 there has been some upward tendency in prices (see Figure 20.3). Announced price increases have been large in 2008, and are likely to continue independently of any emissions trading scheme impacts.

The most significant remaining step towards establishing competitive markets is the removal of retail price regulation. State and territory governments have been cautious about relaxing the regulatory process. At the same time, the market has been evolving through competitive activity, as evidenced by the significant number of customers switching their retailer as full retail competition has been introduced.

Through the Ministerial Council on Energy, there is now agreement for the Australian Energy Market Commission to review the status of competition in each jurisdiction with a view to opening up the market to full competition, while maintaining structures that protect consumers who are in financial hardship. There has been a general move away from cross-subsidies and towards directly funded community service obligations. Subject to this review, the commission will make recommendations on the removal of remaining retail price controls.

Figure 20.3 Average electricity market prices, 1999–2008

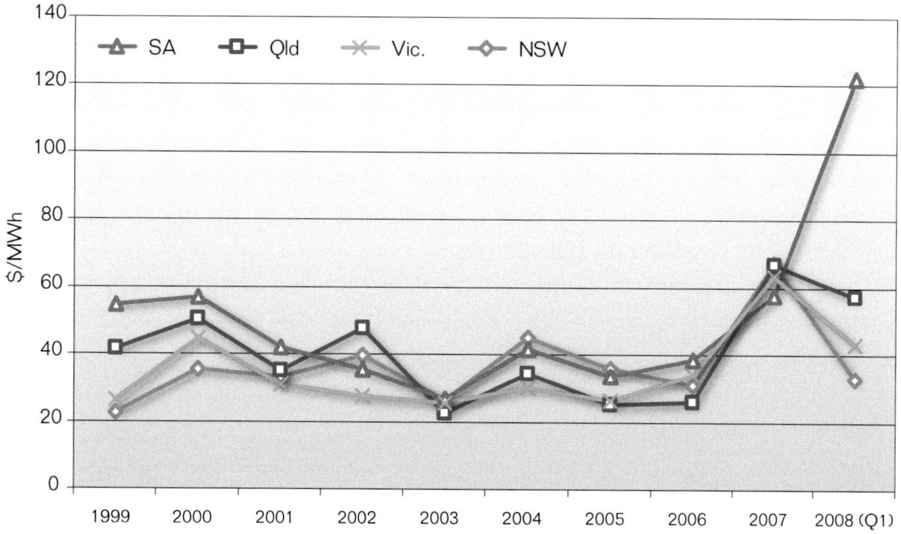

Source: NEMMCO (2008).

These structures and processes generally allow the private sector's assessment of supply and demand to determine the need for additional capacity and to deliver this capacity in a timely fashion. They have generally been successful for gas exploration, production and transportation and for electricity generation. There are mixed views across the industry as to whether established mechanisms are able to deliver the most efficient and timely investment in electricity transmission.

20.1.3 Energy and existing climate change policy

In the last several years, there have been three major developments in the Australian energy sector's response to the climate change challenge.

First, federal and state policies have been catalysts for investment in a specific set of lower-emissions technologies. The federal government's Mandatory Renewable Energy Target (MRET), the NSW Greenhouse Gas Abatement Scheme and the Queensland Gas Scheme have been the principal instruments. Second, the increasing public awareness of climate change has seen the emergence of a voluntary market exemplified by the continuing rapid growth in GreenPower, an Australia-wide program whereby consumers use renewable energy sourced from the sun, wind, water and waste that is purchased by their energy company on their behalf. Finally, there has been a general hesitancy to invest in new power generation assets outside existing schemes in the absence of a clear, broad and stable policy framework.

20.2 Drivers of the transformation

The energy industry is not new to change in the face of external pressures. Different societies at different times have moved away from gathering trees for firewood and charcoal; from burning coal in homes and commercial buildings for heating; and from 'town gas' made from coal to natural gas. The key to changes in sources of energy has always been the interaction between economic and environmental factors. With the challenge of climate change, the introduction of a price on greenhouse gas emissions will accelerate the change by increasing the cost of fossil fuels relative to alternatives.

The pace and direction of the energy transformation will be dictated by the dynamics of the supply and demand surrounding key fuel sources, and the global and domestic policy response to mitigation of climate change.

20.2.1 Global fuel dynamics

In the last few years, the domestic energy sector has been challenged by three developments:

- **Capital costs** have risen markedly with particular impact on capital-intensive industries. Industry advice to the Review indicates that there have been increases of up to 60 per cent in construction costs per installed kilowatt of power plants since 2004, across all technologies.

- A major uplift in **global coal prices**—180 per cent over the past three years and over 100 per cent over the past year—has been driven by recent strong demand in China and India and a supply system that takes some time to respond with new infrastructure and transport capacity.
- Increases in **global prices for traded natural gas** have not kept pace with even larger increases in oil prices. These movements in energy commodity prices are illustrated in Figure 20.4.

Figure 20.4 International energy commodity prices, indexed to 1994

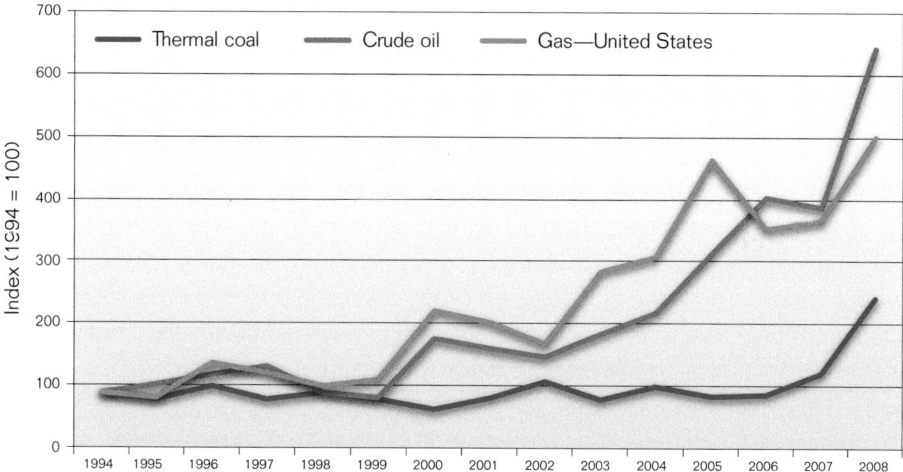

Sources: ABARE (2008); State of Nebraska (2008); Government of Western Australia (2008).

In Australia, rising capital costs are starting to affect electricity prices. This has been compounded by the effect of drought on the availability of water for hydroelectric generation and power station cooling. The existence of long-term domestic contracts for black coal, the unsuitability of brown coal for export and the absence of liquefied natural gas export infrastructure on the east coast have largely cushioned Australian prices from the other two factors.

This position is not sustainable: contracts will be renegotiated, new coal export infrastructure is being developed and several east coast liquefied natural gas export projects have been announced.

To put these price increases into perspective, a $20 per tonne price on carbon dioxide emissions could add $16–20 per MWh to the average wholesale electricity price. An increase of $3 per gigajoule in the gas price, to somewhere closer to but still short of export parity would add more than $20 per MWh to the price of gas-fired electricity, while an increase of $100 per tonne to black coal prices would add approximately $53 per MWh to the price of coal-fired electricity.

There is considerable scope for substitution across fuel types. In the discussion of the four broad fuel types in the Australian energy sector that follows, the outlook for oil—as traditional sources become increasingly constrained by resource availability and increasing extraction costs—will tend to act as an external determinant of prices for the other fuels, at least in the long periods required for large expansion of production and transport capacity for the more abundant fossil fuels.

Coal

Coal generally gets to the point of consumption in one of three ways:

- as part of an integrated coal supply–power generation entity
- through commercial contracts with generators
- through export contracts.

The first way recognises the economic advantages of integrating coal production and power generation within a single business, especially for low-quality coal and when international prices are low. This model generally applies in Victoria and Queensland, including at the recently established plant at Kogan Creek.

The second is characterised by relatively long-term contracts. The most important example is in New South Wales, where the government separated mining rights from electricity generation when it corporatised the energy sector some years ago. In the past, this structure has provided some flexibility for competitive market forces to optimise outcomes and keep domestic prices low. However, firms operating under such supply contracts are highly vulnerable to the continuation of high export prices combined with tendencies towards internationalistion of low-quality black coal markets. Coal generators relying on the purchase of exportable coal will be at an increasing disadvantage in competing with generators with tied access to coal in Victoria and Queensland.

Third, in export markets annual contracts are common and prices reflect the dynamics of global supply and demand.

Coal demand is experiencing a period of unprecedented growth, driven by China and India. As contracts for the domestic supply of thermal coal expire, and if suppliers have the potential to access this export market, there is upward pressure on domestic prices. Such prices, if they hold up for an extended period, have the potential to encourage coal-drying technologies and enhance the economic outlook for Australia's brown coal industry.

In the absence of any expectation of a significant slowing of growth in Asian developing economies, prices in real terms are likely to remain well above the levels of the late 20th century for the foreseeable future, but to come down from current peaks as supply expands. There is not likely to be a supply constraint on coal resource availability for many decades, although the exploitation of deeper and more distant deposits is likely to keep prices at a high level.

Gas

Natural gas in Australia has two geographically separate main markets. The east coast has gas coming from the Cooper Basin in central Australia and the Gippsland Basin in the south-east. The North-West Shelf supplies the local West Australian market but also provides large volumes of liquefied natural gas for export. Recently, a third market of smaller domestic significance has developed in northern Australia, centred on Darwin. For more than 20 years, gas prices in Australia have been relatively stable, with no significant new sources of supply or demand being developed. Gas for power generation has remained confined to a relatively small peaking demand, where the higher fuel cost could be offset by lower capital costs.

As long-term contracts came to an end during the 1990s, new sources of supply emerged, as did an expansion in the gas transmission network to provide greater security of supply and competition between basins. The take-up of gas for power generation to meet increasing peak and intermediate demand and the introduction of the Queensland Gas Scheme have influenced these developments. Most importantly, the coal seam gas sector, an unintended consequence of the Queensland Gas Scheme in the first instance, now has a vigorous life of its own.

The increasing confidence in large coal seam gas reserves in Queensland has led to recent announcements about the development of liquefied natural gas export infrastructure in the east. Major global energy companies have been taking a stake in the sector, having identified the potential to supply a growing Asian demand for gas.

The recent strength of global demand and higher world prices have resulted in a lift in gas prices in Western Australia. Other domestic gas prices can be expected to rise towards export parity and remain at that level over the longer term. International prices for gas are likely to rise relative to other fossil fuels, as its advantages in lower emissions are increasingly recognised and valued.

The Garnaut–Treasury modelling assumes this rise will take place over a period of around 10 years. If recently announced liquefied natural gas developments were to accelerate that movement, then it would affect the relative cost of gas in Australia for all applications.

Nuclear

The global uranium industry has recently experienced a surge in demand. This is set to go further. China has committed to a program that is immense by any previous global standards, and India has high nuclear power ambitions. In China, nuclear will still account for no more than 6 per cent of total power in 2020. Some countries, including the United Kingdom, have decided that nuclear power stations should continue to play a role alongside other low-carbon electricity sources (HM Government 2008), and that should lead to investment in new capacity. This expanded demand for nuclear energy arises from a combination of influences, from climate change, energy security and relative costs. With more than one-third of currently estimated global resources of uranium, Australia is well placed to benefit from this growth.

Nuclear power stations will have been disproportionately affected by the recent increases in capital costs on account of their exceptional capital intensity, although the latest nuclear technologies indicate potentially lower costs.

Australia has better non-nuclear low-emissions options than other developed countries, especially (but not only) if carbon capture and storage is commercialised within the range of current cost expectations. Australia is a major net exporter of a wide range of energy sources, notably coal, liquefied natural gas and uranium. Transport economics should favour local use of those fuels in which the gap between export parity and import parity price is greatest (first liquefied gas, then coal). As a consequence, Australia is not the logical first home of new nuclear capacity on economic grounds.

In Australia, as well as in most other developed and developing countries, the level of public acceptance of nuclear power is an important barrier. Any government committed to implementation of a nuclear power program would need to recognise public opposition as a constraint—a potential source of delays and increased costs. The Australian Government is firmly against Australian nuclear power generation, and the Coalition parties retreated quickly from nuclear advocacy in the face of community antipathy during the 2007 federal general election. It would be imprudent, indeed romantic, to rely on a change in community attitudes as a premise of future electricity supply for the foreseeable future.

Given the economic issues and community disquiet about establishing a domestic nuclear power capacity, Australia would be best served by continuing to export its uranium and focusing on low-emissions coal, gas and renewable options for domestic energy supply. Modelling by the Review of the effects of allowing nuclear power generation in Australia (Figure 20.12) supports this assessment. However, it would be wise to reconsider the constraints if:

- future nuclear costs come in at the low end of current estimates
- developments in technologies reduce the need for long-term storage of high-level radioactive waste
- there is disappointment with technical and commercial progress with low-emissions fossil fuel technologies.

In these circumstances, there would be reason for the Australian Government to engage with community disquiet on the issue, and to seek a change of policy.

Renewable energy

In recent years, power generated from non-hydro renewable sources has increased as a result of MRET and, to a lesser extent, GreenPower demand. However, by 2007 it represented only around 3.3 per cent of capacity and 2.5 per cent of delivered electricity (ESAA 2008). This capacity has been dominated by wind, with contributions from solar hot water and biomass.

There is little likelihood of large expansion in storage-based hydroelectric generation in Australia, although there is scope for much better use of existing storage capacity in the current environment, in which renewable power has greatly increased value. The anticipated growth in intermittent supply technologies (wind and solar) and ongoing, above-average growth in peak demand mean that existing hydroelectric infrastructure will play an enhanced role as a provider of flexible and readily available stored energy to meet short-term demand peaks.

This role could be substantially expanded through judicious investment aimed at making the hydroelectricity assets important balancing components in the eastern Australian system. Australia's main hydroelectric assets—in the Snowy Mountains and Tasmania—will have increased value, far beyond that suggested by their installed capacities (3676 MW and 2278 MW respectively) alone. The value comes initially from their zero-emissions status and low underlying operating costs. This is enhanced by their potential for counteracting the intermittent supply from wind and solar power. If market conditions can be effectively exploited, power from intermittent sources at times of low demand and price could be used to pump water into hydroelectric storage for use at times of greater demand and value. Public ownership, and in the case of Snowy Hydro ownership by three governments, has applied constraints on the supply of capital to the optimisation of the value of these major national assets. These constraints have high opportunity costs in the emerging environment. It is important that they be removed.

20.2.2 Things might have been different

In the absence of climate change, global and domestic forces would lead to major adjustments in the Australian energy sector in particular.

One key force for change is strongly increasing Asian and global demand for Australia's commodity exports, driving strongly rising terms of trade in the longer term, and a GDP that roughly triples by 2050 in the Garnaut–Treasury reference case. Population and Australian household incomes both rise strongly. In the longer term, the depletion of relatively low-cost resources leads to increases in energy prices. Australia's energy supply continues to be dominated by coal, while globally, nuclear energy and gas play lesser but significant roles, dictated by fuel availability and the public acceptance of nuclear. In this scenario greenhouse gas emissions grow without constraint, with Australia's emissions projected to double by 2050 through rising energy consumption and ongoing dependence on fossil fuels.

Climate change adds a major dimension to the future of the energy sector, with direct implications from the impacts of climate change and even greater implications as mitigation responses are adopted. Unmitigated climate change is predicted to cause greater storm, wind and bushfire damage and increased levels of materials degradation. This will mean additional transmission and distribution losses across the gas and electricity networks.

The specific risk to electricity transmission and distribution networks that arises from the increased frequency of extreme weather events is illustrated by the power supply outages of January 2007, when a bushfire caused disruption to the transmission system between New South Wales and Victoria.

The most significant impact that will require adaptation planning in the energy sector is that on urban water supply. In 2007, the drought exposed the obvious dependence of part of the market, the hydro generators, on water supply. However, it also exposed the extent to which most fossil-fuel generators depend on water for cooling. This is likely to lead to a move towards air-cooled plant in future, with an associated reduction in efficiency. One consequence of water shortages is additional electricity demand for recycling, desalinisation or longer-distance pumping.

There will also be an impact on energy infrastructure demand through compounding growth in the peak summer period.

These challenges amplify the need for governments to maintain momentum towards a truly national energy market at the same time as they respond to the structural adjustment imposed by the mitigation task.

20.2.3 The domestic emissions trading scheme

The implementation of an emissions trading scheme, as the central element in the policy recommendations of the Review, will unleash far-reaching change, as the market responds to the emissions constraint and delivers an assessment of consequent pricing expectations. In the electricity market, the short-term price implications will cause a direct adjustment in marginal cost structures and asset values. This market response is also expected to be associated with a more certain framework to underpin contracting behaviour across the sector. The long-term price expectations will provide long-needed clarity to frame major investment decisions for new energy infrastructure, including base-load power generation. In addition to investment in technologies with known operating and cost characteristics, this longer-term perspective is expected to facilitate research, development and commercialisation of technologies assessed to have greater mitigation potential in the future. The Review's recommended support for research, development and commercialisation of low-emissions technologies would have a powerful effect in accelerating innovation. The overall suite of domestic mitigation policy measures recommended by the Review provides the necessary and sufficient policy conditions for the transformation described below.

For much of the 21st century energy, including transport, is projected to provide the major source of reductions in emissions. The marginal cost of energy supply, interacting with the permit price, will determine the balance of domestic mitigation and international permit trading. The role of technology developments and associated assumptions on availability and costs forms a central theme in the Review's analysis of the energy sector in a low-emissions economy.

20.3 The transformation

In considering the way forward, three broad phases can be identified. These are neither prescriptive nor precise, but separate the ebb and flow of particular developments as they might unfold, especially future changes in technology.

From the perspective of 2008, the first phase could be expected to apply for the initial 5–10 years after the scheme is introduced, the second over the next 10–15 years and the third beyond that.

Australia is ideally placed for this transformation, with its abundant coal, gas, uranium, geothermal, solar and other renewable sources, and exceptional opportunities for geosequestration and biosequestration of carbon dioxide.

Therefore, while major structural change always presents challenges, energy supply security is not likely to be one of them. Furthermore, Australia has a strong recent history of supporting its resources and engineering industries with appropriately skilled people. However, that skills base may be challenged as the transformation accelerates.

Within Australia and through the application of the emissions trading scheme, the size, structure, greenhouse gas emissions and, ultimately, the cost of the electricity sector will be determined by:

- the adoption of supply-side energy efficiency, retrofitting of CO_2 capture and other in-plant abatement opportunities
- rebalancing the use of current generation plant in favour of plant with lower emissions
- demand reduction through demand-side energy efficiency and price elasticity
- adoption of new and replacement plant with lower emissions, driven by post–permit price economics and leading to progressive retirement of existing higher-emissions plant.

Section 20.3 describes the phases of the energy transformation over the century, while section 20.4 provides an analysis of this transformation as informed by the Garnaut–Treasury modelling.

20.3.1 Phase 1—commitment and adjustment

The Review has recommended that the emissions trading scheme commence in 2010 at about $20 per tonne (2005 prices), and rise at 4 per cent per annum in real terms. This is roughly what the Garnaut–Treasury modelling indicates would be generated by a global agreement around stabilisation of greenhouse gas concentrations at 550 ppm CO_2-e. Tighter global trajectories than that implied by the 550 scenario would generate a higher price after 2013. Even with the initial fixed price of the Kyoto period, the primary and secondary markets would be expected to quickly establish a spot and forward price curve for emissions permits beyond 2012.

As the constraints tighten from early 2013, low-cost mitigation opportunities and expectations of tightening of trajectories in response to an international agreement are likely to lead to some hoarding of permits.

As this phase evolves, and the trajectory diverges from the business-as-usual path, the next set of responses is likely to involve some fuel switching. Constraints will include transmission interconnection for new (and possibly remote) capacity, and gas availability and cost, involving existing gas-fired open-cycle plants being operated more intensively. Competitive tensions will arise from the relative emissions intensities of existing coal-fired plants as the permit price is incorporated into short-run marginal costs. Increased price volatility is likely to be a feature of this period—around a tendency for prices to be driven by factors outside the emissions trading scheme, but augmented by the emissions price.

The fuel mix and cost implications will be strongly influenced by the extent to which new black coal contracts in the domestic electricity sector are negotiated at higher prices and the speed with which domestic gas prices move towards global price parity. The implications for brown coal generators will, in the short term, be dominated by the effect of these factors on their competitors and east coast electricity prices, and therefore their capacity to recover lost volume in prices.

It is likely that some coal-fired generators with captive coal supply will stand to reap significant increases in profits from the higher price environment driven by increases in capital costs and gas and black coal prices. There will be a vigorous search for in-plant mitigation including partial fuel substitution. Beyond the commercial limits of in-plant emissions reduction, it is likely that it will be economical for some time for such generators to purchase domestic offset credits or international permits to maintain substantial production despite their high emissions intensity, in an environment in which high gas and black coal prices are underpinning higher electricity prices.

In this phase, new baseload generation capacity is likely to be based on established, combined-cycle gas turbine technology, ideally designed for post-combustion capture of carbon dioxide.

Offsets, trade in permits, and hoarding and lending of permits would all provide the flexibility necessary to modify the high permit prices that could flow through to delivered energy prices as demand and supply factors, including any short-term demand surprises, adjust to the emissions constraint.

The Review recognises that this period will generate acute pressures for owners and operators of existing coal-fired plants, some of which have been optimised to run efficiently in a mode that will be challenged in this new world. It has concluded that compensation for the cost of permits has no priority in circumstances in which there are stronger calls on permit revenue. Chapter 16 addresses in some detail the specific arguments raised in relation to emissions-intensive electricity generation. However, other factors will tend to ameliorate the otherwise negative consequences for well-managed coal-based generators:

- There will be opportunities for some relatively low-cost reductions in emissions, including through coal drying.
- There will be capacity to recover volume loss through price. The strong upward pressure on competitors' costs for reasons beyond the mitigation regime will strongly favour established producers with sources of non-tradable coal including some of these generators most affected by the emissions trading scheme. Some of these generators will not see a loss in cash flows for several years, and may well see opportunities for increasing profit in the current circumstances.

Opportunities for increased energy efficiency are envisaged to begin slowly, with the support of the programs described in chapters 18 and 19, and accelerate as the rising permit price provides an increasing incentive for their adoption. Research, development and innovation funding across the development cycle will drive substantial investment in a range of technologies with the potential to be competitive over time, as the emissions cap tightens and the price rises. This dynamic will be strongly influenced in the early years, while emissions permit prices may be relatively low, by the proposed expansion of MRET, even with the price limitations proposed in Chapter 14.

20.3.2 Phase 2—transition

The second phase of the transformation will see resolution of the tension between the pull of global gas prices and successful deployment of the first coal-fired power stations with carbon capture and storage. Either way, this scenario plays out to Australia's advantage due to its diversity of fuels, its favourable sites for geosequestration and biosequestration, and its wide range of relatively low-cost renewable generation opportunities.

This phase is likely to be dominated by technology shifts as the investment in research, development and commercialisation delivers the commercial-scale models of new generation capacity across several technologies. New baseload fossil fuel generation plant is likely to incorporate coal drying and coal gasification technologies. It is expected that retrofitting of oxy-firing and carbon dioxide capture will be added to existing coal and gas-fired plants, accompanied by carbon dioxide pipelines and commercial-scale geosequestration operations.

For other coal-fired plant, where such changes are not economically feasible, this phase will see increasing cost pressure as the permit price rises. This phase will be characterised by investment in technologies for which the electricity costs have been demonstrated at commercial scale through the investments in research, development and commercialisation of the first phase. Victoria's brown coal resource, unsuitable in its natural state for export, could be expected to have a strong future in this scenario.

At the same time, it is expected that various factors—the rising permit price, the results of programs such as the large-scale solar energy program called Solar Cities, and funding for research, development and innovation in renewable technologies such as geothermal, solar thermal and solar photovoltaic—will be delivering favourable trends in the deployment of such technologies at a commercial scale.

A challenge for wind power could be that costs struggle to remain competitive due to site availability, wind quality and community restrictions. Energy storage technologies, including through effective use of the stored hydroelectric potential in the Snowy Mountains and Tasmania, can be expected to be available on a commercial basis to support the intermittent nature of solar and wind, meaning that these sources could act as baseload sources. The marrying of such technologies to demand that matches their availability will enable a more comprehensive approach to infrastructure planning. This phase may see the validation of the potential for technologies such as biochar and algal conversion of carbon dioxide as a form of recycling. Market developments in vehicle fuels and motor technologies will strongly influence whether such biomass material realises greater value as a liquid transport fuel or for stationary electricity generation.

The combined impacts of rising energy prices, capital replacement cycles and complementary measures to deploy cost-effective energy efficiency changes will contribute to major changes in the energy technology portfolio in this phase, driven primarily by the increasingly stringent emissions trading scheme trajectory, with considerable pressure from the supply side.

20.3.3 Phase 3—emergence

In the third phase of the transformation, the energy sector will move close to a position of zero carbon emissions. The balance of supply and demand that will achieve this outcome will ultimately be determined by the economics of technology developments, which cannot be forecast with certainty. The transport sector, both public and private, is also likely to be based largely on this zero-emissions electricity generation supply.

The success of near-zero emissions coal technologies would lead to the retention of coal as the main fossil fuel energy source, while Australia continues to gain as an exporter from the ongoing high global gas prices. Gas is likely to be most valuable to countries without local coal resources and for which near-zero emissions coal technologies are neither physically nor economically feasible.

The development of storage technologies and the reduction in solar costs—driven by larger-scale deployment and ongoing technological innovation—are expected to combine with geothermal energy to begin to replace fossil fuels as the long-term solution to our energy needs. Near-zero emissions coal technology will have carried out its primary role and remain a significant energy source for some time. An alternative possibility could be the successful development of biosequestration technologies. Such a development could deliver a more favourable long-term future for coal in the energy sector, allowing it to compete with renewable energy technologies as resources and geography dictate.

As in the earlier phases, Australia will be in the fortunate position of being able to monitor the global competitive dynamics of coal, gas, nuclear and renewable technologies and to apply economically superior options flexibly as they emerge.

20.4 Modelling results for the energy sector

This section provides a quantification assessment of the transformed electricity sector based on the Garnaut–Treasury economic modelling, followed by a description of the main modelling sensitivities.

Results from three scenarios of Chapter 11 are discussed—a no-mitigation scenario (the reference case), and 550 and 450 ppm CO_2-e global stabilisation scenarios—drawing on three of the models used by the Review: the Australian general equilibrium model, MMRF; the global general equilibrium model, GTEM; and the bottom–up modelling of the Australian electricity sector, by McLennan Magasanik Associates using the Strategist model. As in Chapter 11, MMRF is implemented with the post-2050 emergence of a fixed-cost backstop technology, and GTEM is implemented with both standard and enhanced technology assumptions. Unless otherwise stated, the modelling results presented in this section and in section 20.5 are based on standard technology assumptions.

20.4.1 The central policy scenarios

The net result of the factors discussed above on electricity demand over the modelled period is shown in Figure 20.5. Initially this demand is determined within the electricity sector, but progressively it rises above the reference case as the sector becomes decarbonised and fuel substitution towards electricity occurs in other sectors. Transport is the major contributor to this development; its role is described more fully in Chapter 21.

Figure 20.5 Australia's electricity demand

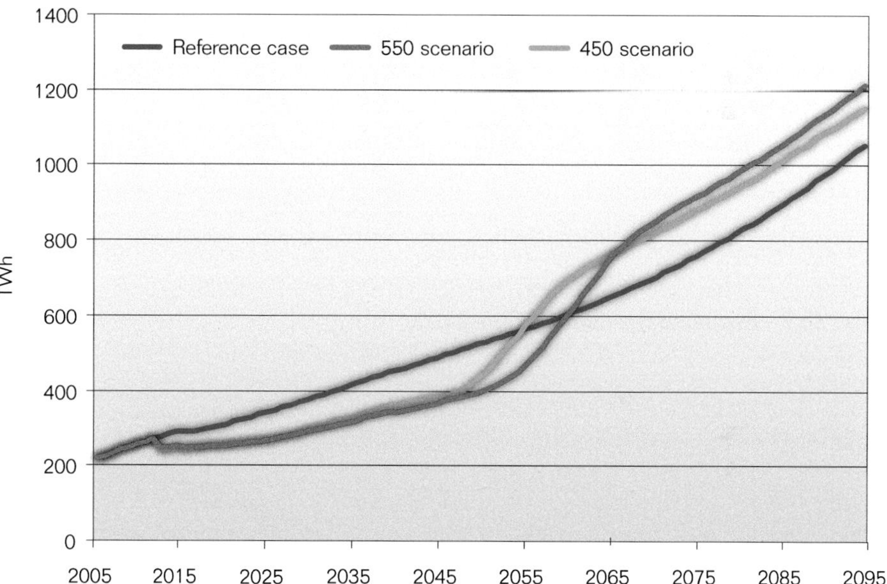

Note: These results were generated using MMRF and standard technology assumptions.

Within the constraints and assumptions of the modelling, the initial response to the carbon constraint is a reduction in electricity demand. The major contributors (aluminium, residential and commercial) are shown in Figure 20.6. While the real world could smooth out this initial price impact, the longer-term change is for a demand reduction, with residential demand falling 11 per cent below the reference case by 2020.

The implications of the above for the aluminium sector are clearly significant; these are explored more fully in section 20.5.4. After the initial price response, residential demand on a per capita basis grows steadily, as shown in Figure 20.7, until this growth is strongly augmented by the switching to electricity for private transport.

Figure 20.6 Electricity demand reduction in selected sectors, 550 scenario

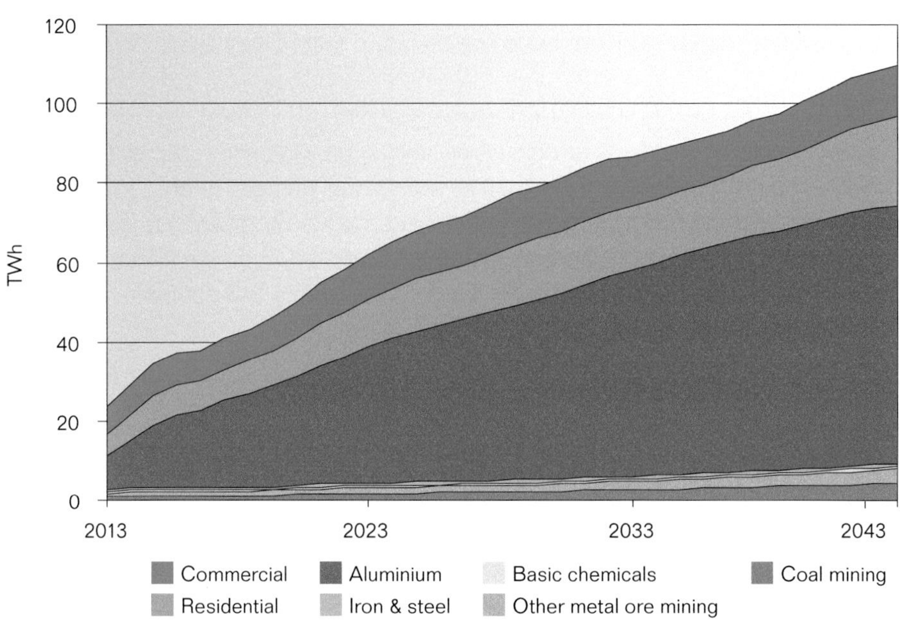

Note: These results were generated using Strategist and standard technology assumptions.

Figure 20.7 Residential demand, 2005–2100

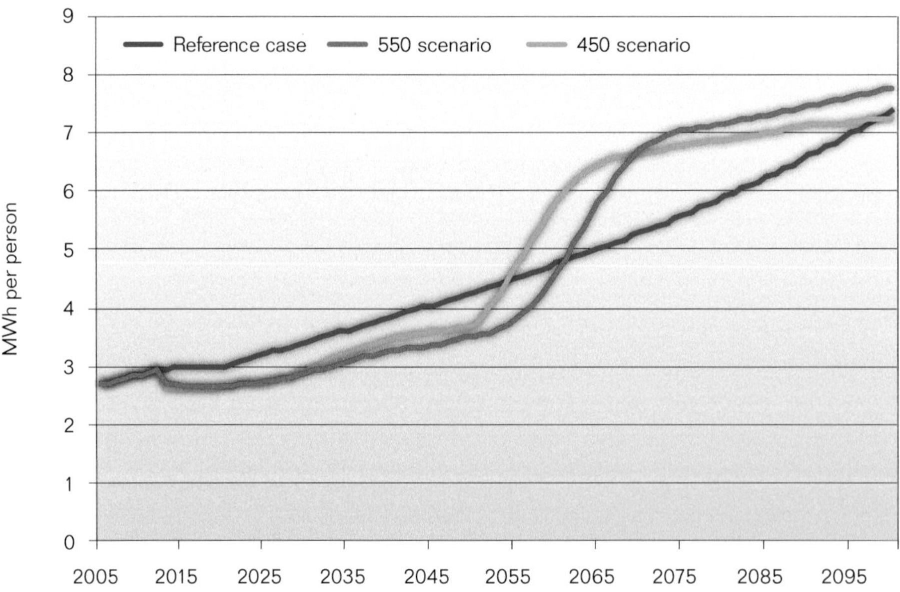

Note: These results were generated using Strategist and standard technology assumptions.

The structure and technology mix is illustrated in Figure 20.8. The transformation begins with an increase in the role of gas as a relatively low-cost and immediately available source of lower-emissions electricity. Progressively, the availability and relative cost of carbon capture and storage technology (either as geo- or biosequestration) sees it emerging to play a strong role by mid-century with additional growth in renewable supply sources. As the abatement target becomes more restrictive, the residual CO_2 emissions associated with carbon capture and storage would cause it to become increasingly less competitive and renewable sources dominate.

The relative contributions of fossil fuels and renewable sources are driven by the cost assumptions (as described in the technical report available at <www.garnautreview.org.au>) and technology-specific constraints (for example, for intermittent sources). The modelling excludes any additional policy instrument such as the expanded MRET, that acts to 'force in' specified technologies.[1] Accordingly, non-hydro renewables' share of the electricity mix remains below 10 per cent until the mid-2020s, but then rises strongly to reach 30 per cent by 2050.

Figure 20.8 Australia's electricity technology shares, 550 scenario

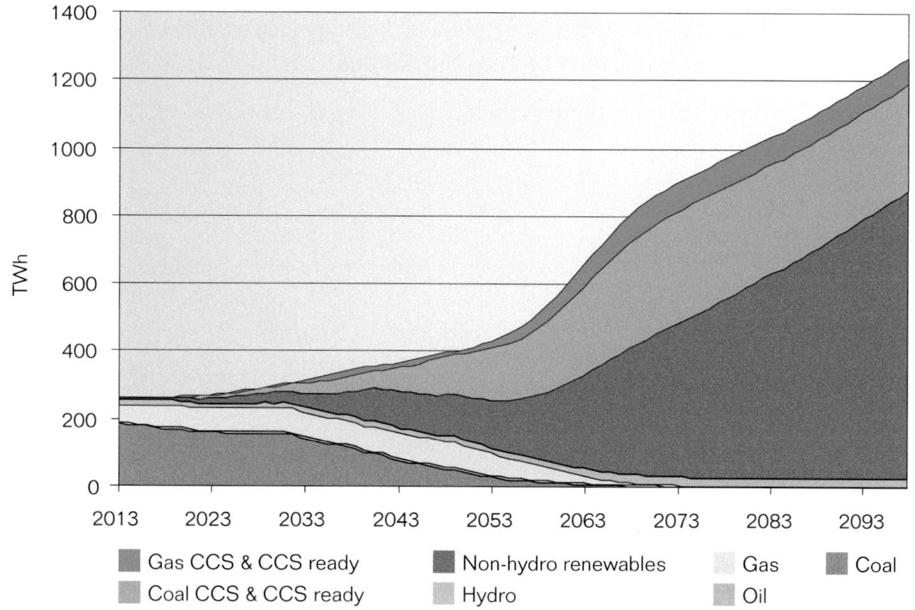

Note: These results were generated using MMRF and standard technology assumptions.

Over the longer term, the critical factor that is likely to determine the structure of the electricity supply sector and the future of fossil fuels, both in Australia and internationally, is achieving near-zero emissions carbon capture and storage. The implications are described in section 20.5.3 and are illustrated in Figure 20.9, which shows the much greater share taken by coal with carbon capture and storage when such a technological change is included.

The major difference between the Australian outlook in these figures and the global picture is that the modelling of the former has excluded nuclear generation. The effects of removing this constraint are described in section 20.4.2.

In the 450 standard technology scenario, the constraints described above would act much more tightly, such that the role of carbon capture and storage diminishes and non-hydro renewable energy dominates Australia's electricity supply, rising to 13 per cent by 2020 and towards 50 per cent by mid-century. This is illustrated in Figure 20.10.

It is clear that whatever specific technology mix emerges, it is likely to deliver a progressive decarbonisation of electricity generation by mid-century to an extent ultimately determined by the cost of near-zero and zero emissions coal technologies. This is shown in Figure 20.11.

Figure 20.9 Australia's electricity generation technology shares, 550 scenario with zero-leakage carbon capture and storage

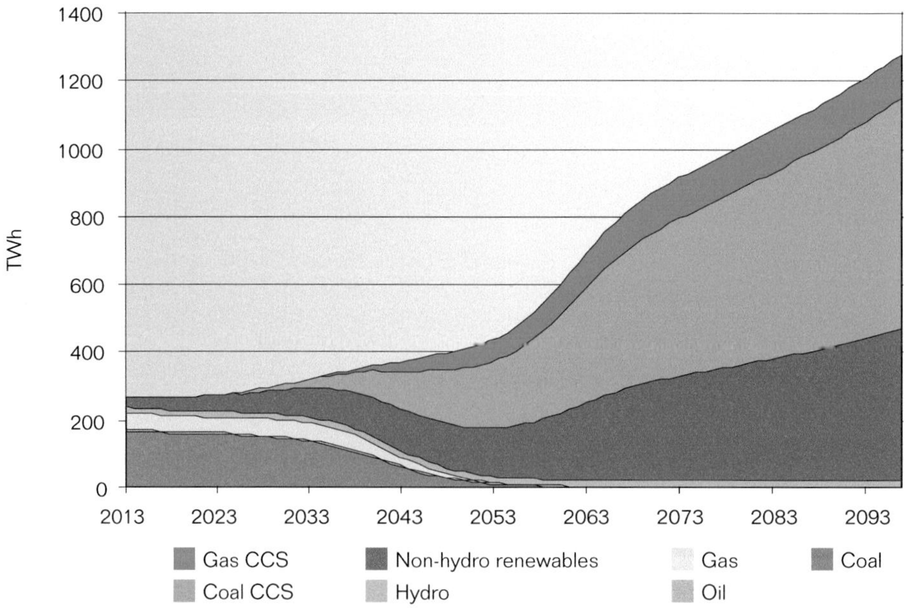

Note: These results were generated by applying the technology shares from GTEM under the zero-leakage CCS scenario to the electricity demand from MMRF under the 550 standard technology scenario. This illustrates the impact of zero-leakage CCS in MMRF.

Figure 20.10 Australia's electricity generation technology shares, 450 scenario

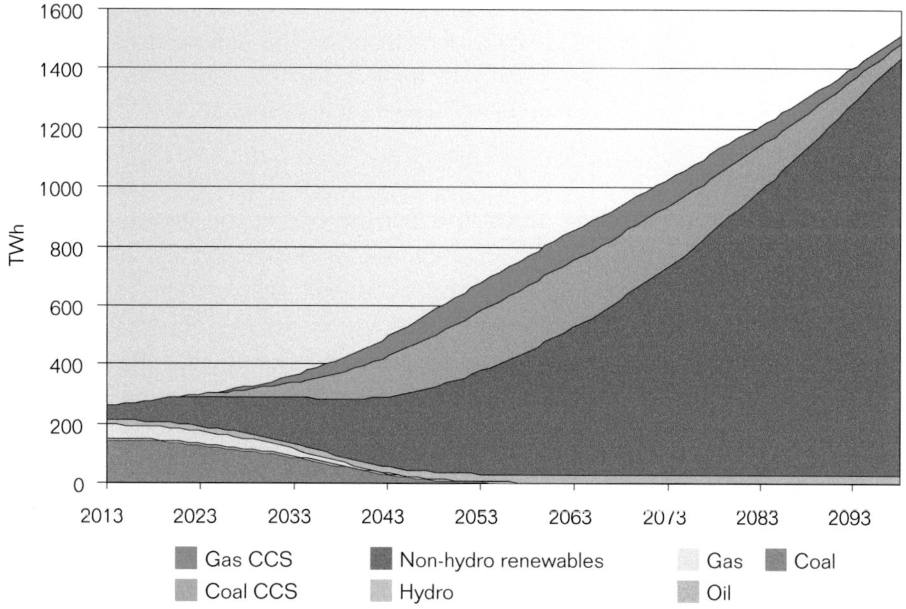

1600
1400
1200
1000
800
600
400
200
0
TWh
2013
2023
2033
2043
2053
2063
2073
2083
2093
Gas CCS
Coal CCS
Non-hydro renewables
Hydro
Gas
Oil
Coal

Note: These results were generated using GTEM and standard technology assumptions.

Figure 20.11 Electricity emissions intensity

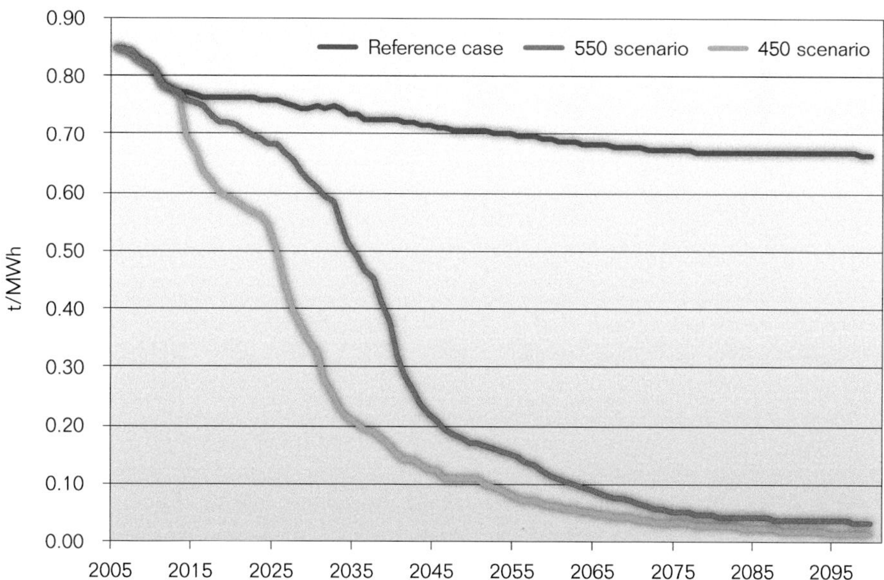

0.90
0.80
0.70
0.60
0.50
0.40
0.30
0.20
0.10
0.00
t/MWh
Reference case
550 scenario
450 scenario
2005
2015
2025
2035
2045
2055
2065
2075
2085
2095

Note: These results were generated using MMRF and standard technology assumptions.

20.4.2 Modelling sensitivities

The structure of the energy sector and the future for specific technologies as projected in the modelling are critically dependent on the assumptions of future technology developments. While many of these could be subject to sensitivity analysis, the Review focused on three of these for assessment.

Nuclear energy

When the economic modelling includes a nuclear option for Australia, nuclear is adopted, supplying 27 per cent of total electricity demand by 2050 in the 550 scenario, and primarily replacing coal combined with carbon capture and storage. This outcome is particularly sensitive to relative technology cost assumptions. Within this framework, and as shown in Figure 20.12, the impact on electricity costs is modest. As discussed in section 20.2.1, other issues are likely to be more important for determining the future of nuclear energy in Australia's energy mix.

Figure 20.12 Total wholesale electricity costs, with and without nuclear, 550 scenario

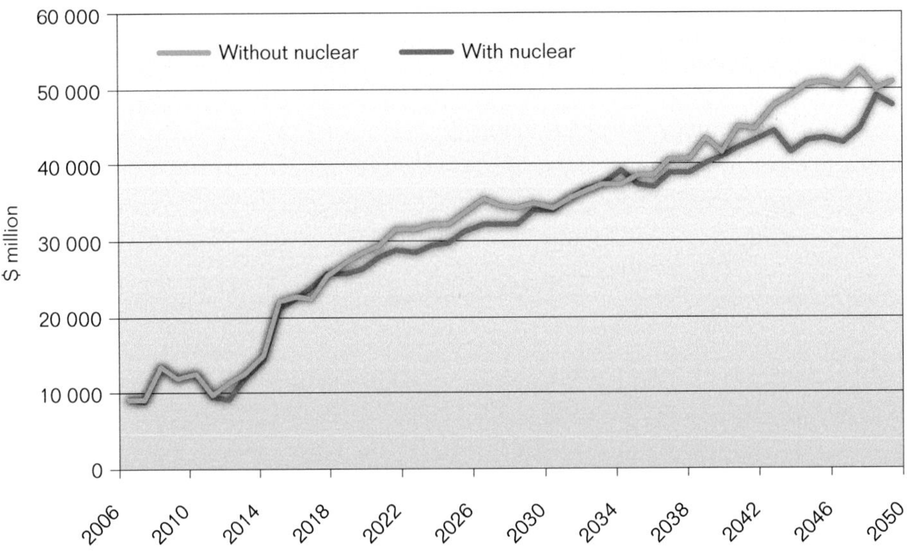

Note: These results were generated using Strategist and standard technology assumptions.

High rate of technology development: enhanced technology assumptions

The Garnaut–Treasury modelling ran an enhanced technology scenario using GTEM, for which the macroeconomic impacts are discussed in chapters 11 and 23. The lower permit price, relative to the standard technology scenario combined with high rates of carbon dioxide sequestration by carbon capture and storage, to shift the technology mix somewhat towards carbon capture and storage at the expense of non-hydro renewables. This is illustrated in Figure 20.13.

Figure 20.13 Technology mix under an enhanced technology scenario

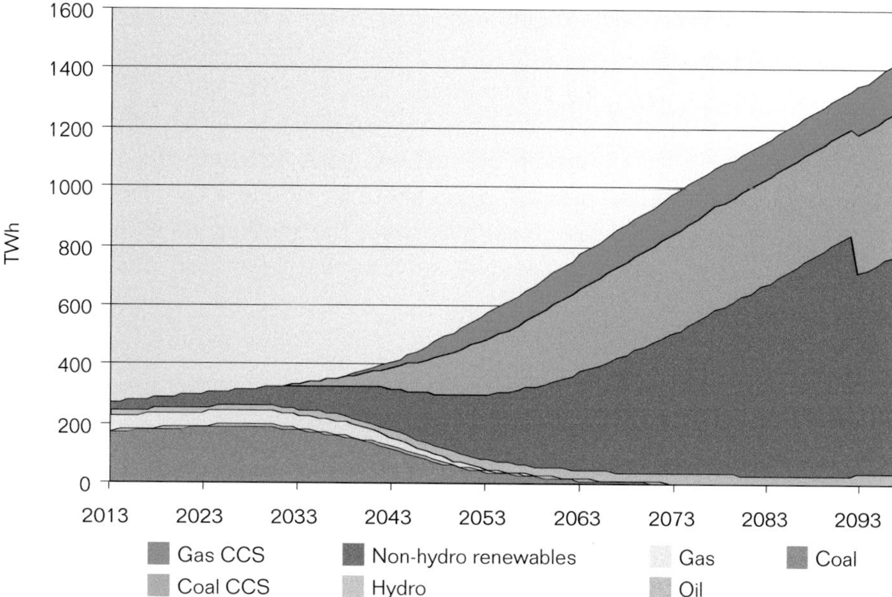

Note: These results were generated using GTEM.

Near-zero emissions coal technologies

Global and Australian dependence on coal, with its high combustion emissions, for power generation means that the future of the energy sector is particularly sensitive to assumptions regarding the way in which this subsector is affected by an emissions constraint. This sensitivity is described in section 20.5.3.

20.5 Major economic impacts

The transformation as described through economic modelling will have widespread impacts, both within and beyond the electricity sector.

20.5.1 Electricity costs

The transformation is projected to be accompanied by increases in the cost of electricity influenced by assumptions for future technology costs. The result is an increase in electricity wholesale costs, as shown in Figure 20.14, increasing steadily as existing and lower-cost/higher-emissions technologies are displaced and then reflecting future assumptions regarding the relative costs of the range of near-zero emissions technologies described in the technical appendix on the Review's website.

These wholesale electricity prices would then flow through to retail prices, with the proviso that there are no regulatory impediments. In the 550 scenario, Australian retail electricity prices in the first few years following the introduction of the emissions trading scheme are projected to be around 40 per cent higher in real terms than they would have been otherwise. The implications for households in regard to the affordability of essential services are considered in chapters 16 and 23.

Figure 20.14 Wholesale electricity prices, 2005–50

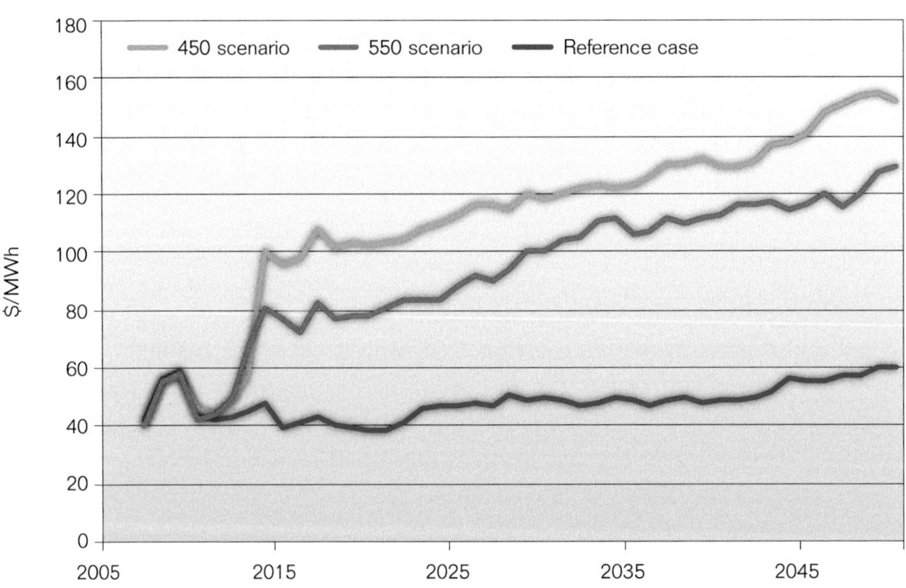

Note: These results were generated using Strategist and standard technology assumptions.

20.5.2 Existing electricity generation

The changes in demand and technology mix, described earlier in this chapter, carry with them significant implications for established generation with its concentration on black and brown coal. Most of the existing generators are projected to continue in profitable operation beyond 2020, although growth is either curtailed or supplied by gas and renewables as described earlier in this chapter. Figure 20.15 indicates the impact on electricity generation from coal.

The dynamics of price projections for black coal and gas have an influence on the permit price at which switching occurs. A result is that coal generators are still frequently marginal suppliers, thereby setting the wholesale price, more often than under the reference case. This leads to the impact on electricity prices described in the preceding section, with permit prices being passed through by generators.

Figure 20.15 Electricity generated from coal

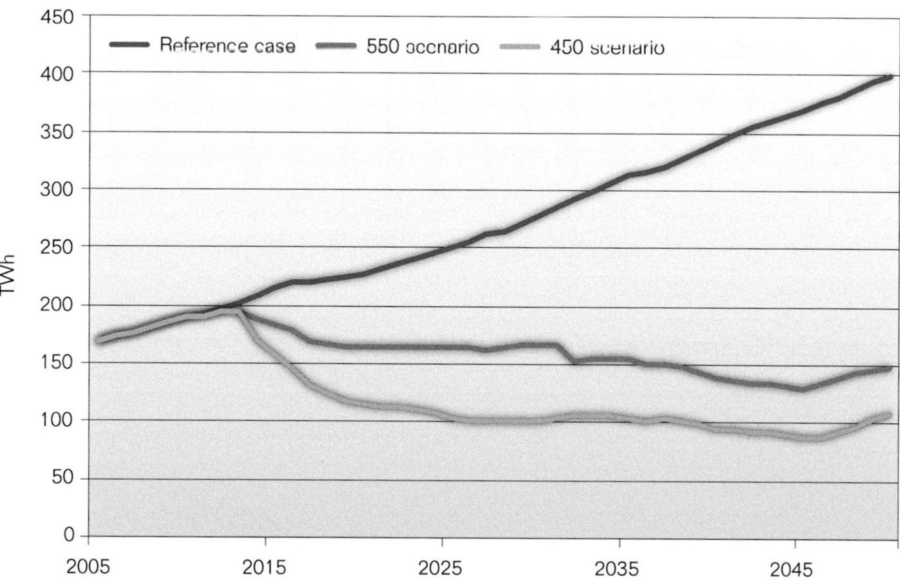

Note: These results were generated using Strategist and standard technology assumptions.

In the 550 scenario, there is an impact on existing coal generators as shown in Figure 20.16, although most of the current capacity remains profitable and in place beyond 2020. Even for brown coal, 93 per cent of today's capacity is projected to still be in place by 2020.

Figure 20.16 Generation capacity, 550 scenario

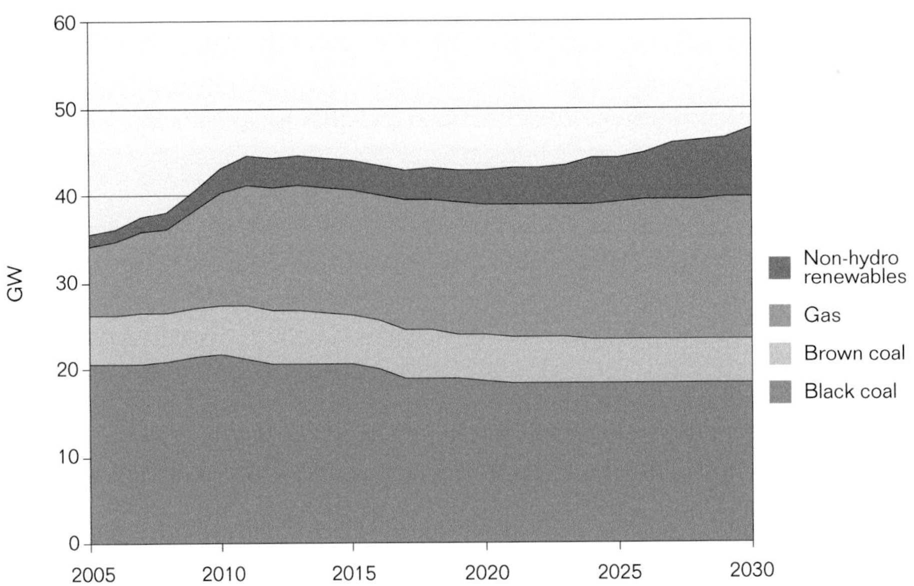

Note: These results were generated using Strategist and standard technology assumptions.

In the 450 scenario (Figure 20.17), the impacts are magnified—the installed brown and black coal capacities are 37 per cent and 26 per cent, respectively, below today's level by 2020.

Figure 20.17 Generation capacity, 450 scenario

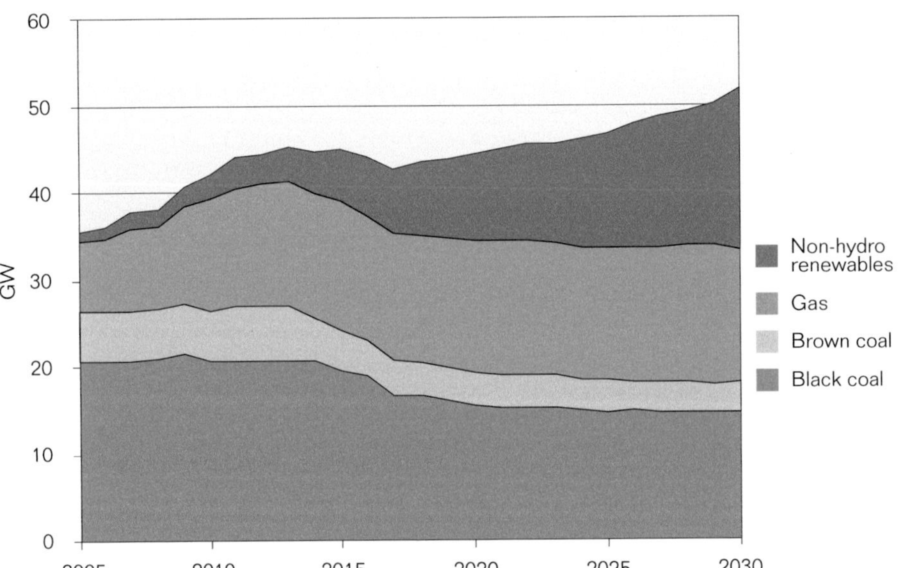

Note: These results were generated using Strategist and standard technology assumptions.

In addition to the gains made by some coal generators in both scenarios, existing gas generators generally benefit substantially from the higher permit prices, as do renewable generators such as hydro and wind. The pump storage units associated with the existing hydro generators are projected to see some reduction in profitability through the higher price of their electricity purchases.

20.5.3 The future for coal

For Australia, reducing emissions from coal combustion is of national importance. Under any realistic scenario, Australia's response to climate change, both internationally and domestically, will be inextricably intertwined with the long-term future of the coal industry.

First, coal underpins Australia's domestic electricity supply sector. An early resolution of the future role of coal-fired generation will be important in shaping a smooth transition for Australia's energy sector into a low-emissions economy.

Second, the significant role that coal plays as Australia's biggest export means that we have a major economic interest in working with other coal-exporting countries and the importing countries themselves to determine as quickly as possible how and when low-emissions coal technologies can assist these countries to follow lower-emissions development paths. The $22 billion contribution to exports in 2007 is set to leap dramatically in 2008–09 to almost $50 billion, mainly through much higher prices. These contributions will be at risk sooner or later, unless low-emissions ways of using coal are applied successfully in our major Asian markets.

Third, the effective participation of China, Indonesia, India and other major developing countries in ambitious emissions reduction is essential for the success of the global mitigation effort. These counties could find such participation in an effective global agreement easier if zero-emissions technologies for using coal are available.

Last but not least, there are possible scenarios in which significant adverse impacts arise for communities dependent on coal and coal-fired power generation, notably those in the Latrobe Valley. The continued health of the industry could obviate the need for assistance measures.

If the coal industry is to have a long-term future in a low-emissions economy, then it will have to be transformed to near-zero emissions, from source to end use, by mid-century. A range of technical, environmental and economic challenges must be addressed effectively to achieve this objective, in a time frame consistent with a global agreement on climate change and Australia's own domestic commitment (Box 20.1).

Priority should be given to the resolution of whether a near-zero coal future is even feasible, either partially or in total. If it is not, then Australia needs to know as soon as possible, so that all who depend on the coal industry can begin the process of adjustment, and so that adequate and timely investments are made in other industries.

The Review has identified three key initiatives that would contribute to an efficient transition.

- Technology innovation in the sector will benefit from the early research funding described in Chapter 18, as will investment in demonstration and commercialisation from the suggested matched funding scheme.
- The Review explores the need for structural adjustment assistance under very specific circumstances in Chapter 16, and proposes an additional allocation of funds, of the order of $1–2 billion, to match industry investments in the adoption of proven technologies during the period of transition.
- In Chapter 11, the Review identifies the leadership role that Australia could play in coordinating a major global effort to develop and deploy carbon capture and storage technologies, and to transfer those technologies to developing countries.

Box 20.1 The technologies of zero-emissions coal

At its simplest, the challenge is to develop technologies that allow coal combustion with zero, or near-zero, carbon dioxide emissions while maintaining its relative competitive position as a fuel. Carbon dioxide is an unavoidable product of combustion but must not be released into the atmosphere if coal is to achieve near-zero emissions status. This can be achieved by capturing the carbon dioxide and then either converting it to some environmentally benign end product or consigning it to permanent storage (sequestration).

Carbon capture technologies

There are broadly two groups of technologies related to carbon dioxide capture.

Members of the first seek to capture the carbon dioxide from an existing gas stream, such as the exhaust stack of a generation plant. These are of most relevance to existing plants and have the advantage of extending the life of such assets. The most significant challenge in this area is that such plants were not designed with carbon dioxide capture in mind and the exhaust gas stream is generally low in carbon dioxide, making the capture more expensive and energy intensive.

Members of the second deploy fundamentally different new approaches to create, at some point in the process, a concentrated stream of carbon dioxide that is more readily suitable for large-scale capture. The challenges here are those of technology commercialisation and cost in the early stages.

Clearly, those technologies that apply after coal has been gasified, or a carbon dioxide stream created, are equally applicable to carbon dioxide capture and sequestration from gas-fired power generation plants. With large gas reserves, including coal-seam gas, Australia also has a strong strategic interest in such applications. Process streams involving coal gasification can potentially be applied to the production of transport fuels as an alternative to electricity generation.

Box 20.1 The technologies of zero-emissions coal *(continued)*

Transport technologies

If the source point of the carbon dioxide is physically distant from the final destination, then some form of transport will be required. Carbon dioxide transport is relatively well developed as a technology. The issues associated with the provision of appropriate transport infrastructure are discussed in Chapter 19.

Sequestration technologies

There are two categories of sequestration. The first, geosequestration, involves storing carbon dioxide permanently underground or below the seabed in depleted oil or gas reservoirs or in deep saline aquifers. Another possibility is sequestration in deep coal seams, where the injection of carbon dioxide could enhance coal-seam gas recovery. The challenges in this area mostly involve geology and geophysics, including seismic mapping and developing a robust regulatory regime that may have to coexist with the extraction of gas or petroleum products.

There are several projects under way or proposed in Australia that will test various aspects of these technologies, including the CO2CRC project in the Otway Basin in Victoria.

A more intriguing, and potentially highly valuable, approach is biosequestration. There are, for example, proposals to produce biofuels from algae, the growth of which is enhanced by access to a constant stream of carbon dioxide from power stations or industrial process exhaust.

New technologies: what are the issues?

A focused approach will be needed to address the range of technical, regulatory, environmental and economic issues associated with each area of technology described in Box 20.1.

Many of the individual technologies are technically proven. Issues of scale, integration and economics are likely to be the greatest challenges.

The challenge posed by the scale of the task is the most significant of these. It will ultimately involve the annual capture and sequestration of several hundred million tonnes of carbon dioxide in Australia alone. An operation on this scale will be a substantial new industry in itself, and will include export of services and possibly export of storage reserves to countries which lack low-cost geo-sequestration sites. The power requirements of the geo-sequestration process itself will add substantially to the demand for power generation. These developments will place a considerable strain on regulatory processes and human resources and be the source of considerable growth in associated activity, employment and incomes for the regions in which they are located.

If the challenges of low-emissions coal technologies can be successfully addressed, a profoundly different medium- and longer-term future emerges. Those same forces of high capital costs, high world gas prices and relatively strong export coal prices will favour retrofitted (post-combustion capture) coal plants with captive coal supplies and low-emissions profiles and, ultimately, near-zero emissions plants that use integrated coal drying and gasification technology.

Under the 550 and 450 standard technology scenarios modelled by the Review, the demand for Australia's coal largely depends on the ability of coal generation to capture a share of an expanding electricity market in a rapidly growing world. The modelling assumes that carbon capture and storage technologies from 2020 onwards are able to capture 90 per cent of coal-fired electricity generators' emissions. While this assumption causes global demand for coal to remain relatively high, global mitigation causes the rate of growth to moderate, such that Australian coal exports fall by around 25 per cent by 2050 and 20 per cent by 2100, relative to the base case.

As the carbon price rises to high levels, zero-emissions electricity generation becomes increasingly competitive against coal generation, even where 90 per cent of carbon capture and storage is assumed. It is likely that the development of zero-emissions technologies would increase demand for coal-fired energy generation and hence maintain global demand for coal. Illustrative global modelling undertaken by the Review shows that the introduction of a zero-leakage carbon capture and storage technology could significantly increase the demand for coal-fired electricity generation and hence increase demand for Australian coal, relative to a scenario with only 90 per cent carbon capture and storage. Figure 20.18 shows the impact of zero-leakage carbon capture and storage relative to 90 per cent carbon capture and storage.

Figure 20.18 Carbon capture and storage scenarios

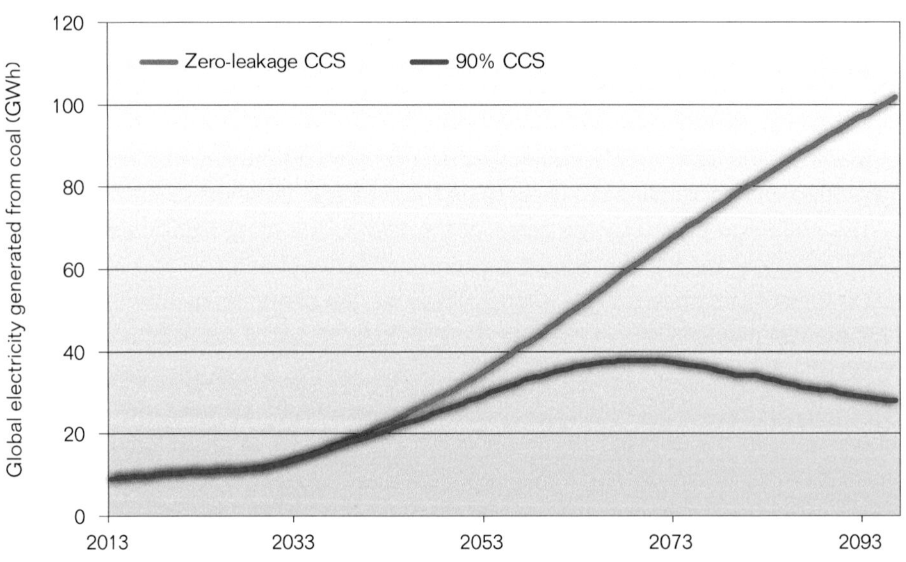

Note: These results were generated using MMRF and standard technology assumptions.

Australian coal production moves in line with the global demand for Australia's coal exports. A future scenario in which Australia stands aside from a strong global mitigation effort is more likely to damage than to assist the circumstances of the Australian coal industry.

20.5.4 The impact on aluminium

The aluminium industry comprises two distinct subsectors: alumina production and aluminium smelting. Each is modelled separately in MMRF.

Figure 20.19 shows the projected changes in activity for the two sectors under the no-mitigation and the 550 standard technology scenario. They show that both industries are projected to grow to mid-century in an unmitigated world. Beyond that, global productivity convergence within the modelling framework influences the results. One of the consequences is that the aluminium industries in developing countries gain competitiveness and gain market share relative to the Australian industry.

While both industries are energy intensive, they use energy from different sources. The process of refining bauxite to form alumina (alumina production) uses energy to generate heat, for example, mainly from coal and gas. Aluminium smelting, on the other hand, uses mainly electricity. Under an emissions trading scheme, the aluminium smelting industry loses its advantage in cheap electricity and hence is affected more than the alumina industry.

Figure 20.19 shows that the Australian coal-based aluminium industry is projected to decline in absolute terms for a while, as sources of low-emissions electricity, such as stranded hydro in Africa, the island of New Guinea and South America, and standard natural gas in a number of countries become increasingly competitive. The ultimate scale of this tendency will be determined by the ongoing availability of hydro sites. The alumina industries' growth is projected to be significantly moderated under an emissions reduction scheme.

Figure 20.19 Projections for aluminium and alumina industries

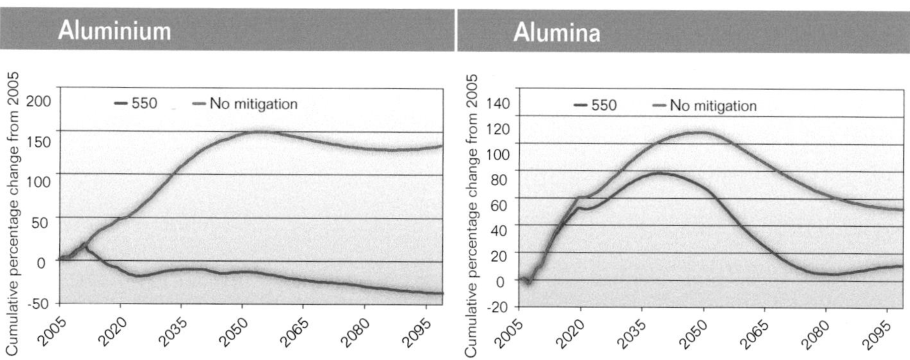

Note: These results were generated using MMRF and standard technology assumptions.

20.5.5 The role of gas

As described above, the introduction of an emissions trading scheme is projected to lead to an increasing role for gas in power generation, as shown in Figure 20.20, reaching 23 per cent and 27 per cent by 2020 in the 550 and 450 scenarios, respectively. In the process, gas is projected to meet 21 per cent of Queensland's demand, more than would be required under the existing Queensland Gas Scheme. Mitigation policies overseas also expand the opportunities for exports of natural gas at higher prices. Gas's role becomes constrained in later years as coal with carbon capture and storage and renewable sources become more competitive under the rising permit price, even when combining carbon capture and storage with gas is included. As with coal, this outcome can be strongly influenced by relative movements in future technology costs and global commodity prices.

Figure 20.20 Electricity from gas sources

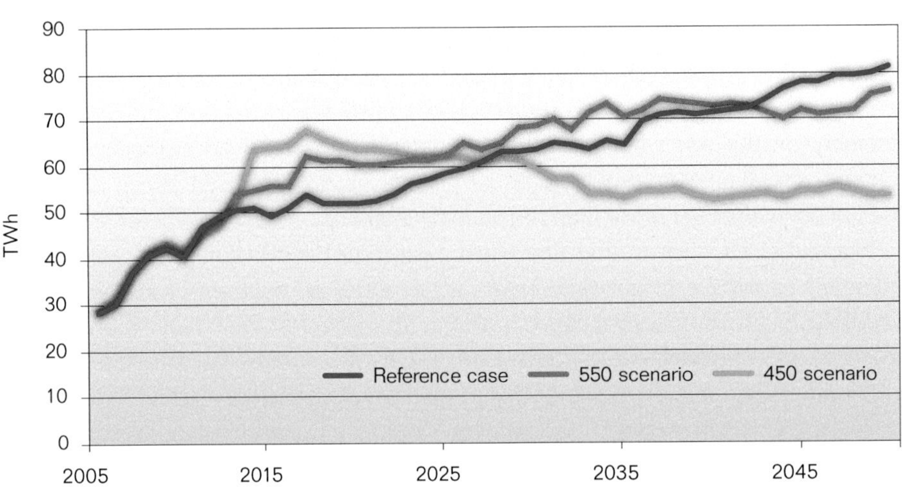

Note: These results were generated using Strategist and standard technology assumptions.

Australia's competitive position in the global gas market will be driven by fuel-specific dynamics as described earlier in this chapter. As with coal, global gas demand in the 450 scenario is constrained later in the century by the increasing permit price and would also be favoured by the development of near-zero emissions carbon capture and storage technology.

20.5.6 The role of other stationary energy

Emissions from fuel combustion for energy, other than electricity or transport, include onsite energy production for the agriculture, mining, manufacturing, communications, services and residential sectors. Key sources of other stationary energy emissions in 2005 were the steel and alumina sectors, which use large amounts of coal; the gas and private heating sectors, through their use of gas; and the coal, agriculture and services sectors through their use of diesel.

Current levels of emissions from other stationary energy sources are of a similar magnitude to transport and, like the transport sector, there is a high potential for abatement through switching towards zero- or low-emissions fuels. In the manufacturing sector, for larger plants in industries such as cement, lime and plaster, iron and steel, and alumina refineries, there is also the potential for the application of capture and storage technologies.

Emissions reductions from other stationary energy contribute 12 per cent of the overall abatement task in both the 450 and 550 standard technology scenarios. The majority of abatement occurs after 2050, reflecting a switch to low-emissions energy sources and the adoption of carbon capture and storage technologies in parts of the manufacturing sector. The carbon capture and storage technologies are assumed to become commercially viable with a lag to their rate of adoption in the electricity generation sector. The switching towards low- or zero-emissions fuels occurs at the same rate as the electrification of the transport sector.

By the end of the century, emissions from other stationary energy account for 12 per cent and 18 per cent of overall emissions for the 550 and 450 scenarios respectively, and are increasing very gradually in the 450 scenario. This reflects an assumed natural limit to the degree to which the process of fuel switching can occur, which has been set at 90 per cent, reflecting, for example, the lack of availability of grid-generated electricity in remote areas. It also reflects the assumed limit to the capture of emissions from carbon capture and storage technologies.

As outlined above, emissions in the 'other stationary energy' category are sourced from a range of different activities. Stationary energy emissions represent some or all of the fuel combustion emissions from almost every sector considered, with the exception of electricity generation and a subset of the transport sectors. The potential for fuel switching will vary considerably within the combustion activities in any one sector—some of these combustion activities, such as residential heating, would be especially conducive to a switch to electricity, as there is already widespread use of electricity in these activities.

Such switching would amplify the increased demand for electricity and its central role as described earlier in this chapter. While reductions in emissions were included as described above, this was not accompanied, in the modelling detail, by price-based fuel switching.

20.6 Risks to the transformation

20.6.1 Inertia

In recent years, there has been much discussion regarding new, integrated coal gasification technologies, featured in such projects as ZeroGen in Australia and FutureGen in the United States. These projects are complex and costly and have struggled to make real progress. Their complexity is partly responsible for the generally held view that clean coal technologies will not be commercially viable until after 2020.

While such projects remain critical for the longer term, the work of the Review suggests that there are other possibilities with shorter time frames. Having considered the economics of the technologies, the urgency of making major inroads into our emissions, and the other fuel cost pressures, the Review concludes that there is a strong case for accelerated work on the retrofitting of technologies applied in existing plants. This could facilitate the capture of much of the carbon dioxide from such plants, even if it does not involve complete capture. In some areas, such developments could also be associated with carbon capture and storage from gas-fired plant, at least in the medium term.

There is a compelling case for Australia to play a major role in accelerating the international research effort on carbon capture and storage across the range of technological change.

20.6.2 Second-guessing the market

There is a risk that Australia is not bold enough to rely on a market-based emissions trading scheme, supported by mechanisms to remove defined market failures, as proposed by the Review. This hesitancy could arise from the general Australian business and community distrust of market mechanisms. It may also arise from anxiety that the cap and trade system will not drive new technologies, even with the support for research, development and commercialisation of new technologies proposed by the Review. There will be pressure from interests that stand to lose from high permit prices for caps on price that would compromise the emissions reduction objectives. Political resistance to the implications of carbon pricing on costs for some products may drive demands for truncation of sectoral coverage.

20.6.3 Reform fatigue

The energy sector has been on a path of continual reform since the mid-1990s, and that journey is not yet complete. These reforms are consistent with the aims of the emissions trading scheme and, in cases such as removal of retail price regulation, may be important in facilitating it. However, there is a risk that the added complexity associated with the introduction of the emissions trading scheme may introduce unexpected delays into energy market reform. It will therefore be important that an effective linkage is created between the energy market and emissions reduction reforms, through the agenda of the Ministerial Council on Energy of the Council of Australian Governments.

20.6.4 Short-term instability

The energy market is likely to experience wholesale price volatility in the short term as the impact of pre-existing cost pressures, the emissions constraint and the full emissions reduction policy suite works its way through the economy. Price volatility can be an important and essential feature of an effective market. It is therefore important that governments and their regulators work closely with industry to monitor the causes and effects of any such price volatility, and allow the normal mechanisms of the market to operate. Adverse effects of price fluctuations on the living standards of low-income Australians should be managed through fiscal arrangements outside the markets for electricity or emissions permits.

20.6.5 Insufficient people and inadequate skills base

The depth and breadth of the transformation described in this chapter carry significant implications for human resource requirements. The transformation will be evolving as the economy in general, and the resources sector in particular, is suffering from an acute skills shortage in engineering, management, finance, and a range of trades. Maintaining strong investment in appropriate education and training will be an important element in the success of the transition to a low-emissions Australian energy sector.

Note

1 The reference case includes existing federal and state and territory policies on climate change mitigation. The modelling assumes that the measures will be phased out when the emissions trading scheme begins.

References

ABARE (Australian Bureau of Agricultural and Resource Economics) 2008, *Australian Mineral Statistics 2008*, ABARE, Canberra.

ESAA (Energy Supply Association of Australia) 2007, *Electricity Gas Australia*, ESAA, Melbourne.

ESAA 2008, *Electricity Gas Australia*, ESAA, Melbourne.

Government of Western Australia 2008, *Western Australian Mineral and Petroleum Statistics Digest 2007*, Department of Industry and Resources.

HM Government 2008, *Meeting the Energy Challenge: A White Paper on Nuclear Power*, Department for Business Enterprise and Regulatory Reform, Norwich, United Kingdom.

IEA (International Energy Agency) 2003, *Key World Statistics*, IEA, Paris.

IEA 2005, *Projected Costs of Generating Electricity 2005 Update*, IEA, Paris.

IEA 2008, *Statistics by Country and Region*, <www.iea.org/Textbase/stats/index.asp>.

NEMMCO (National Electricity Market Management Company) 2008, *Operational Market Data*, <www.nemmco.com.au/data/market_data.htm>.

State of Nebraska 2008, *Henry Hub Natural Gas Spot Prices 2008*, <www.neo.ne.gov/statshtml/124.htm>.

TRANSFORMING TRANSPORT

21

Key points

Transport systems in Australia will change dramatically this century, independently of climate change mitigation. High oil prices and population growth will change technologies, urban forms and roles of different modes of transport.

An emissions trading scheme will guide this transformation to lower-emissions transport options.

Higher oil prices and a rising emissions price will change vehicle technologies and fuels. The prospects for low-emissions vehicles are promising. It is likely that zero-emissions road vehicles will become economically attractive and be the most important source of decarbonisation from the transport sector.

Governments have a major role to play in lowering the economic costs of adjustment to higher oil prices, an emissions price and population growth, through planning for more compact urban forms and rail and public transport. Mode shift may account for a quarter of emissions reductions in urban passenger transport, lowering the cost of transition and delivering multiple benefits to the community.

Under strong mitigation scenarios, emissions from Australian land transport will fall rapidly around the middle of the century. It will be more difficult—and more expensive—to reduce emissions radically in civil aviation, where the early emphasis will be on improvements in conventional energy use. Late in the century, civil aviation will account for a high proportion of residual emissions, despite people making proportionately more use of other modes of transport.

The path to low-carbon transport will be driven by variations in rates of technological progress across and within transport modes. Early emergence of low-cost biofuels that do not compete with food for agricultural land would reduce the need for structural change. The early emergence of a low-cost electric car, alongside the decarbonisation of the electricity sector, would secure a large place for the private car. Greater use of rail within and between our large cities is being driven by other factors, which will now be reinforced by carbon pricing.

More generally, the transformation of the transport sector, like that of stationary energy, will be driven by interactions of the emissions trading scheme with a range of other factors. These factors include:

- higher global oil prices
- research and development in vehicle and fuel technology
- population growth
- government decisions on transport infrastructure, public transport services and land-use planning, induced in part by the other factors.

This transformation will take place through three main processes, which may operate in parallel:

- vehicles becoming more fuel efficient and shifting to low-emissions fuels, such as electricity
- a shift to lower-emissions modes, such as rail and public transport, accompanied by changes in the structure of towns and cities (urban form)
- reduction in travel frequency and distances, facilitated by changes in consumption, production and distribution patterns and changes in urban form, and driven by changing relative prices.

21.1 The role of transport and its current structure

Transport can be subdivided into six sectors, categorised by purpose (passenger or freight) and by the distance and destination (local, inter-regional or international) (see Table 21.1).

Table 21.1 Transport sectors

	Passenger	Freight
Local	Short-distance passenger travel, such as trips within cities and in regional towns	Short-distance freight trips, including urban deliveries and moving grain from farms to silos
Inter- regional	Longer-distance passenger travel between regions, cities and states and territories	Longer-distance freight movement between regions, cities and states and territories
International	Tourism and other international travel	Imports and exports

Despite the large differences between these sectors in the purpose of the trip, the length of journey and the mode of travel, all are currently dependent on petroleum-based fuels. Petroleum-based fuels currently account for around 97 per cent of Australian transport energy use in Australia (CSIRO 2008). These fuels have been relatively cheap in Australia in the recent past, largely due to low global oil prices in the 1990s and relatively low fuel taxes by international standards (Joint Transport Research Centre 2008).

Low fuel prices, in combination with patterns of urban development and the low priority given to public transport, are a key factor behind the extensive use of fuel-intensive modes of transport in Australia, including trucks and cars. These modes accounted for over 85 per cent of Australia's transport emissions in 2006 (see Figure 21.1). Long-distance freight is the exception, where rail and shipping play major roles. The efficiency of rail and shipping means that they produce a fraction of Australia's domestic emissions, despite carrying a high proportion of loads.

Figure 21.1 Australian domestic transport emissions, 2006

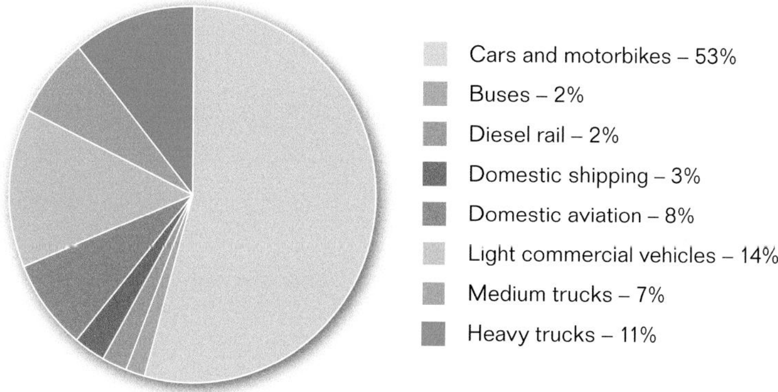

Note: Excludes electric rail and trams.
Source: DCC (2008a).

Emissions from the transport sector have grown rapidly with the increase in demand for transport, particularly higher-emissions forms of transport (see Figure 21.2). Passenger travel per person has increased, with incomes growing faster than the costs of car use and aviation (ABS 2008c, 2008d). The amount of freight carried in Australia has doubled over the last 20 years in tonne-kilometres,[1] largely caused by declining real freight rates and economic growth (BTRE 2006).

There have recently been changes in these trends. In some transport sectors there appears to have been a shift of travel to lower-emissions modes, and the rate of improvement in the fuel efficiency of the passenger car fleet has accelerated.

21.2 Causes of the transformation

The transport system is likely to undergo a profound transformation in this century, irrespective of mitigation. The main causes of these changes will be higher oil prices, new transport technologies, rising incomes and population growth. These factors will interact with an emissions trading scheme and will be mediated by market forces and government decisions on public transport and urban planning.

Figure 21.2 Passenger travel per capita by various modes, 1970–71 to 2006–07

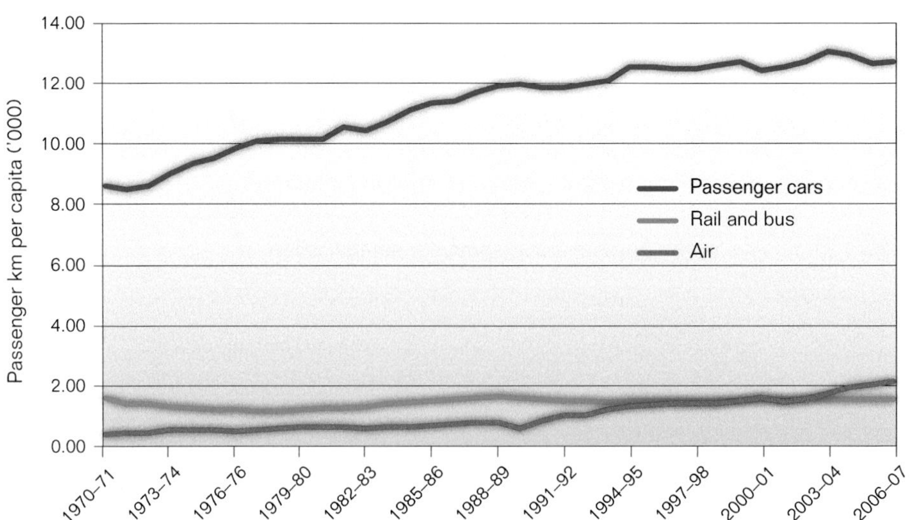

Sources: Derived from ABS (2008a); BTRE (2007); BITRE (2008, pers. comm).

21.2.1 Global oil prices

Increases in the scarcity and price of oil will profoundly affect the costs of our current transport patterns and the relative price and competitiveness of different fuels, vehicle technologies and modes of transport.

Oil prices have risen steeply over the last few years and are likely to remain well above those of the late 20th century. World oil prices have more than tripled from an average of under US$30 per barrel in the 1990s to an average of over US$90 so far in 2008, with a peak of nearly US$150 in July 2008 (US Energy Information Administration 2008a, 2008b). The modelling undertaken for the Review assumes that oil prices will remain above $75 (US$60). However, prices could rise much further if conventional oil reserves come under pressure, partly because the cost of extracting and processing oil from other sources such as tar sands is higher than from conventional sources (CSIRO 2008: 14).

Higher oil prices will improve the competitiveness of alternative fuels, such as synthetic diesels, biofuels and electricity. They will provide incentives to travel less or by more fuel-efficient vehicles and modes.

21.2.2 Technology and fuel development

Higher oil prices have already stimulated investment in research and development for a variety of alternative fuels and vehicle technologies, including liquid fuels (produced from shale, tar sands, natural gas and coal), biofuels, fuel-cell hydrogen vehicles and electric vehicles. The take-up of these fuels and technologies will depend on:

- when commercial feasibility is demonstrated
- relative costs

- the size of the resource base for the fuel
- compatibility with existing commercial and public infrastructure.

The prospects for several of these technologies appear excellent, although it is uncertain when they will become fully competitive and which technologies will do relatively well. However, it is likely that a relatively low-cost, low-emissions vehicle technology will become competitive in the coming decades, and possibly within the next 10 years.

21.2.3 Population growth and economic growth

Australia's population and income growth (see Chapter 11) will dramatically increase transport demand, requiring major expansions to transport systems. Population growth will increase the competitiveness of modes such as public transport and intercity rail, as their costs per passenger decrease with scale (Mills & Hamilton 1994). Similarly, economies of scale in infrastructure provision and costs of traffic congestion mean that it will be increasingly cost effective to meet the needs of a growing population through denser urban settlements and greater use of rail and public transport.

Income growth is likely to increase the demand for long-distance transport, and in particular for aviation. Based on trends in the relationship between income and per capita demand for transport in Australia, the Bureau of Transport and Regional Economics suggests that the demand for local land transport in Australia's capital cities may be approaching a saturation point, with increases in income beyond 2020 not leading to further increases in per capita urban car use (2007: 27).

Economic growth will increase the demand for freight transport, heightening congestion pressures.

21.2.4 The emissions trading scheme

In the early years of the emissions trading scheme, it is likely that high global oil prices will have a larger impact on the cost of petroleum-based transport than an emissions price. The tripling in global oil prices from 1997 to 2008 more than doubled the average cost of petrol in Sydney, from 74 cents to $1.52 per litre (ABS 1997, 2008b). This increased the cost of travel in an average medium-sized car by around 10 per cent if the fixed costs of car ownership are considered.[2] An emissions price of $20 per tonne CO_2-e would increase the cost of petrol by around 5 cents a litre, and the cost of travel in a medium-sized car by less than 1 per cent. The impact of an emissions price will become more substantial as it rises over time. For example, an emissions price of $200 per tonne of CO_2-e would increase the cost of petrol by around 50 cents a litre.

Applying an emissions price that reflects the full contribution of aviation to climate change will be complex but is required for effective mitigation. Radiative forcing from aviation may be two to four times the impact of its carbon dioxide emissions alone, due to the complex effects of emissions such as nitrogen dioxide and cloud formation at high altitude, although there is uncertainty around these figures (IPCC 1999). Based on the UK Government's methodology for calculating emissions from flights for the purpose of offsets, a carbon price of \$20 per tonne CO_2-e would increase the cost of a one-way flight from Sydney to Brisbane by \$2.50 per passenger, and a price of \$200 per tonne of CO_2-e by around \$25 (Department for Environment, Food and Rural Affairs 2008).

21.2.5 How the factors interact through market forces

Higher oil prices and an emissions price will increase the price of petroleum-based fuels, potentially lowering demand for them.

The response to higher fuel prices strengthens over time. The Bureau of Infrastructure, Transport and Regional Economics (2008b) estimates that a 10 per cent increase in fuel prices leads to a 1.5 per cent reduction in car fuel use within one year, but around 4 per cent in the longer run. Goodwin et al. (2004) estimate reductions in fuel consumption associated with a 10 per cent increase in fuel prices of 2.5 per cent within one year and 6 per cent in the longer run. This is because the options to reduce fuel consumption increase with time, through use of more fuel-efficient vehicles, development of public transport infrastructure and changes to urban structure.

Individuals' responses to higher fuel prices will be affected by their incomes—which are expected to rise strongly over the coming century.

The higher the oil price, the lower the emissions price will need to be to make the transition to lower-emissions options competitive.

However, an emissions price and higher oil prices do not have identical effects. First, higher oil prices will improve the competitiveness of all alternative fuels—including liquid fuels produced from shale, tar sands and coal, which all involve much higher emissions per unit of energy than fuels from conventional oil sources. By contrast, an emissions price will selectively encourage lower-emissions fuels. The Garnaut–Treasury modelling projects that fuels such as coal-to-liquids would have a significant place in the market by 2050 if there were no mitigation, but not if an emissions price were introduced. Second, an emissions price will increase the incentive for reducing the use of all fuels that produce emissions, not just petroleum-based fuels.

There are three categories of response to higher oil prices and an emissions price:

- using vehicles that are more fuel efficient or run on alternative fuels
- switching to other modes such as rail and public transport
- reducing the demand for transport.

Vehicle emissions efficiency

The combined effect of rising oil prices and an emissions price will drive greater fuel efficiency, including through the take-up of hybrid petrol–electric vehicles, smaller cars and fuel substitution. The latter includes the substitution for petrol by fuels that can be used more efficiently, such as diesel, and by fuels that produce fewer emissions per litre, such as liquefied petroleum gas and ethanol.

The demand for travel

The amount of travel is likely to fall in response to higher fuel prices, including through:

- improved freight logistics, such as consolidating loads
- substitution of travel with other options, such as telecommunications
- travelling less often and staying longer
- travelling shorter distances by adjusting the origin or destination.

Travelling shorter distances will be supported by changes in urban form, such as increased urban density, which brings destinations such as homes and work, shops and recreation closer together. Population growth is likely to increase the density of some urban areas, facilitating reductions in per capita travel. However, government decisions will have a significant impact on urban density, as there are a range of externalities that affect land use decisions.

Changes in mode

Higher oil prices and an emissions price will encourage switching to more fuel-efficient and lower-emissions modes of transport, such as rail, shipping, public transport, walking and cycling. For example, shifting bulk freight from road to rail could reduce emissions and fuel use by 60 per cent (BITRE 2008, pers. comm.).

The Bureau of Infrastructure, Transport and Regional Economics produced estimates for the Review of typical emissions intensities of various modes of transport (Figure 21.3).

The emissions intensities of modes will vary with changes in vehicle technology and fuel sources. As low-cost, low-emissions road vehicles are developed, the

Figure 21.3 Emissions intensity of passenger modes, 2007

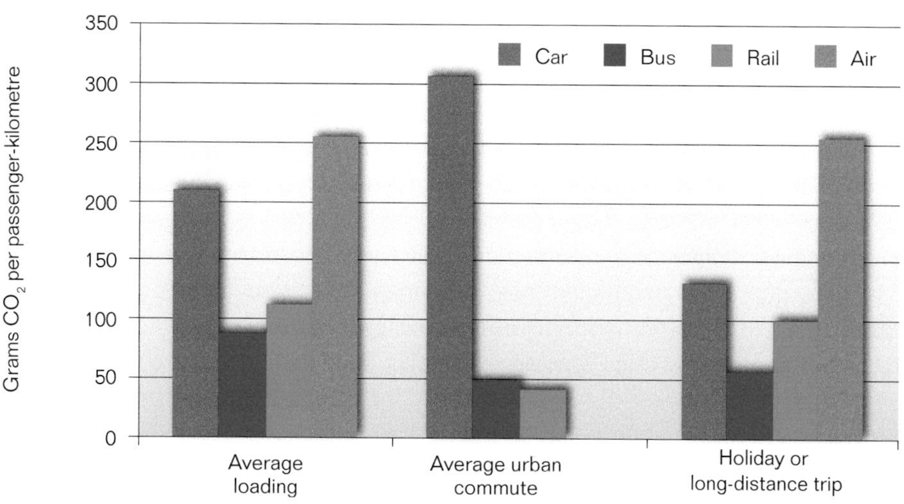

Note: Includes a modest rate of global warming potential for aviation of 168 per cent of carbon dioxide emissions.

Source: BITRE (unpublished data).

competitive advantage of more fuel-efficient modes of transport will decline. This will moderate the pressure for mode shift from higher oil prices and an emissions price.

There are substantial opportunities for mode shift in local passenger transport, particularly in urban areas. Over the last two decades, only around 20 per cent of commuters in Australian capital cities travelled to work by public transport, walking and cycling. In contrast, in many European and Asian cities these modes account for more than 50 per cent of trips. The potential for mode shift is discussed in more detail in section 21.4.2.

If population growth increases urban density, this could mean that more people use public transport, walk and cycle in their daily lives. While increased urban density does not automatically lead to the take-up of these modes, there is evidence that well-planned higher-density environments facilitate them (Newman & Kenworthy 1999). In addition, the cost of public transport per passenger declines with higher patronage, and higher density can support increased demand.

Some major changes in the proportion of the transport task undertaken by different modes will be determined by the market on its own. In freight, it is likely that a portion of the task will be transferred to rail and shipping. In passenger transport, public transport use could increase where services either already exist or can be expanded by the private sector, for example through intercity coach services.

In many sectors, governments' responses in delivering infrastructure and services, and influencing urban form will have a critical effect on the extent of mode shift.

21.2.6 The role of governments

Firms and individuals will only be able to express their demand for mode shift if there are suitable services and infrastructure. Surveys suggest that the main reasons that people do not currently use public transport relate to the lack of suiTable quality infrastructure and services (ABS 2003). Governments have a role in delivering these infrastructure and services.

Governments are also responsible for land use planning. The types of transport infrastructure in which governments invest can further influence urban form. Public transport services, for example, may attract residents to higher-density environments. Conversely, investing in highways can lead to lower urban density. A growing population means that major changes to Australia's cities and transport systems are inevitable, and governments have an opportunity to influence fundamentally the way that urban areas develop and function into the future.

Even in the absence of an emissions price, there are many good reasons for governments to improve infrastructure and services for public transport, walking and cycling and to increase urban densities. Following business-as-usual trends, avoidable traffic congestion would cost Australians $20.4 billion by 2020

(BTRE 2007). The same amount of space dedicated to one bus or rail line can carry 10 times or more the number of people per hour as one lane of freeway (Vuchic 1981, 2005).

In addition to efficiency reasons for changing transport patterns, there are also major equity considerations. Individuals and communities are economically and socially disadvantaged if they lack access to employment, services and social opportunities. Currently, 14 per cent of Australians who are over 18 years of age do not have access to a car (ABS 2006c) and may be disadvantaged if they do not have alternative transport options. People who do have access to cars face major risks of social exclusion and financial hardship if they live in car-dependent areas and oil prices rise (Dodson & Sipe 2007).

In locations where the co-benefits of investing in more compact urban form and mode shift are significant, mode shift will be a low-cost mitigation option. For example, an assessment by the Bureau of Transport and Communications Economics (1996) estimated that upgrading rail lines between cities to support the transfer of some freight from truck to rail would be a no-regrets measure, providing economic benefits and reducing emissions. In locations where the co-benefits are small, mode shift will be a high-cost mitigation option.

The relative merits of different urban structures and transport patterns will vary between regions and even suburbs. However, the rising price of fossil fuels and emissions, and increasing population, add to the case for investment in more compact cities and better public transport, walking and cycling infrastructure.

21.3 Economic modelling results: a possible future?

The economic modelling undertaken by the Review jointly with the Australian Treasury paints one picture of a possible future for the transport system, considering the interaction of some of the factors discussed in section 21.2. The modelling assessed the impact of a carbon price on a range of different transport forms, including private cars, road freight, rail passenger, rail freight, shipping and aviation. CSIRO and the Bureau of Infrastructure, Transport and Regional Economics used a partial equilibrium model to assess the take-up of different fuels and technology in road transport to 2050, and the outputs were fed into the economy-wide mitigation modelling conducted jointly by the Review and Treasury and the independent modelling of climate change impacts conducted by the Review. As discussed in Chapter 11, three technology scenarios were assumed: 'standard', 'enhanced' and 'backstop'. This section focuses on the outcomes of the standard technology assumptions. A scenario was also modelled that examined the impact of higher global fossil fuel prices.

Changes in urban form, which could have major effects on future transport patterns, were excluded by assumption in the modelling.

21.3.1 Economic growth and emissions from transport

In the no-mitigation scenario, demand for all types of transport rises throughout the century due to increasing incomes, population and economic activity. Aviation is the fastest-growing transport type, reflecting growing incomes and an increased proportion of international and domestic spending on tourism. Bus, taxi and passenger rail use also increase with increased income. Private car use grows more slowly, in line with population growth, due to saturation of per capita demand early in the modelling period.

Road freight and shipping increase roughly in line with overall economic growth, and the growth in rail freight is influenced by the expansion of mining output.

Total transport activity under the 550 standard technology scenario is little different from the no-mitigation scenario, reflecting the relatively small impacts on transport costs from the carbon price compared to the impacts on sectors such as electricity. Demand for aviation and shipping grows slightly slower than in the no-mitigation scenario. Road transport shows a larger drop in activity, and rail transport activity increases relative to the no-mitigation scenario. This indicates that the carbon price induces a modest shift from road to rail for both freight and passenger transport, although the model structure does not allow the extent of mode shift to be quantified in passenger-kilometres or tonne-kilometres.

With no carbon price in place, transport emissions in the no-mitigation scenario nearly quadruple by 2100. In the 550 standard technology scenario, transport emissions are around 70 per cent below their 2006 levels by 2100 (Figure 21.4).

Figure 21.4 Projected emissions from the domestic transport sector with standard technology assumptions, 2006–2100

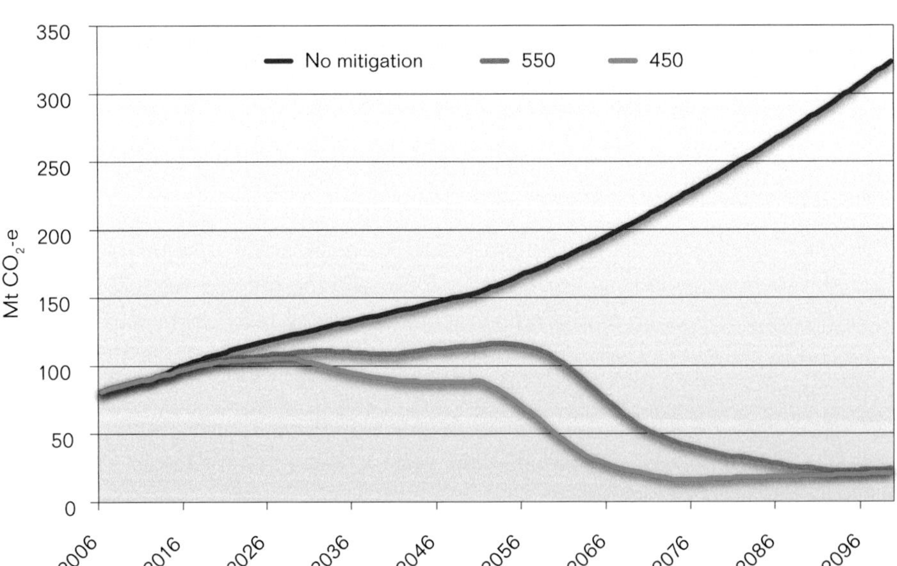

Source: These results were generated using MMRF.

In the 550 standard technology scenario, following the introduction of the carbon price in 2013, much of the increased road transport activity is offset by improvements in vehicle technology. Emissions increase by around 40 per cent from 2006 levels to 2050, but are 25 per cent lower than in the no-mitigation scenario. After 2050, road transport emissions decline rapidly with the adoption of electric road vehicles. Emissions intensity from aviation, shipping and rail transport declines gradually over the modelling period (Figure 21.5).

Figure 21.5 Breakdown of transport sector emissions in the 550 standard technology scenario, 2006–2100

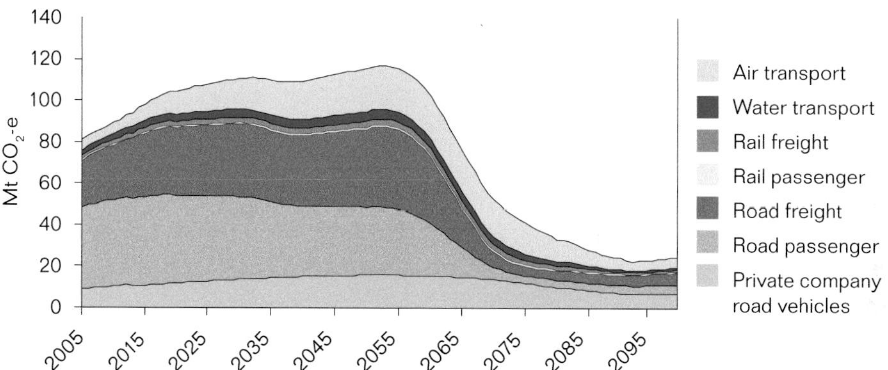

Note: Emissions data do not take into account the impacts of climate change.

Source: These results were generated using MMRF.

21.3.2 Road transport technology

The Garnaut–Treasury modelling undertook more detailed 'bottom–up' modelling of road transport for the period out to 2050, as this sector currently makes the largest contribution to transport emissions. The majority of emissions reductions from transport in the 550 standard technology scenario come from changes in road vehicle technology.

The bottom–up modelling covered a wide variety of technologies and fuels, including petrol, diesel, gas, coal-to-liquids, ethanol, biodiesel and electricity. In the short term, the number of fuels and engine types used in road transport increase in both the no-mitigation and 550 standard technology scenarios. Under the 550 standard technology scenario, non-conventional oil sources such as coal-to-liquids do not become competitive due to their high emissions. Lower-emissions fuels such as ethanol and natural gas increase their share of the market, and more efficient diesel engines displace petrol engines. The shares of many fuels then declines, with a shift back to petrol as hybrid petrol–electric and plug-in vehicles become cost-competitive around 2020 and 2030 respectively (see Figure 21.6). By 2050, hybrid petrol–electric vehicles and plug-in hybrids account for 50 per cent and 13 per cent, respectively, of the road transport undertaken in Australia.

In the long run, the emissions price and technology development result in a more limited range of fuels and technologies remaining competitive. In the 550 standard technology scenario, in the next 40 years fully electric vehicles are taken up primarily for passenger travel, and account for 14 per cent of the transport task in 2050. After 2050, electric vehicles become the predominant technology, and there is a rapid increase in electricity demand from the road transport sector under standard technology assumptions (see Figure 21.6).

Figure 21.6 Modelling of road transport fuel use in a 550 standard technology scenario

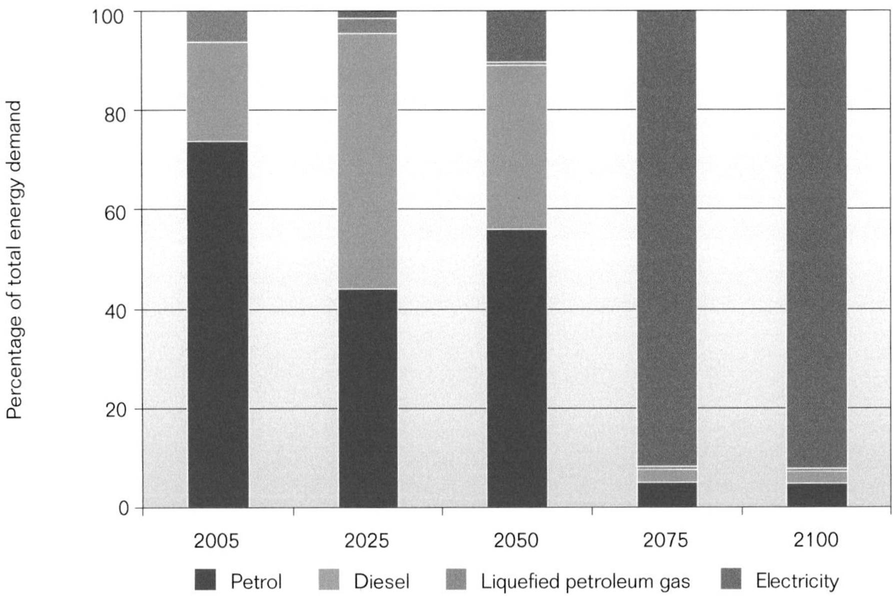

Source: These results were generated using MMRF.

21.3.3 Other transport modes

Other transport modes include rail, shipping and aviation. The modelling did not include specific fuel-switching and technology options for these modes of transport. Instead, under standard technology assumptions, the model assumes a gradual shift towards low- and zero-emissions fuels at approximately the same rate as road transport moves towards electric vehicles. It was assumed that by the end of the century zero-emissions fuels account for 90 per cent of fuel use in these transport modes, and that petroleum-related fuels account for the remaining 10 per cent.

Different results would emerge with more specific technology and fuel assumptions in these sectors. For example, it is possible that the aviation sector may still be producing some emissions in 2100, as there are currently more limited prospects for zero-emissions fuels in that sector, although the inclusion of biofuel elements in aviation fuel is proceeding more rapidly than had been considered likely a few years ago. There appear to be fewer constraints in adopting zero-emissions fuels in the ancillary road, rail and shipping sectors.

21.3.4 Technology assumptions

The technology assumptions in the 550 standard technology scenario represent an estimate of the cost, availability and performance of technologies based on historical experience, current knowledge and a cautious view of future trends. The assumptions incorporate improvements in existing technologies and the gradual, slow emergence and wide-scale deployment of new technologies, such as electric cars. The assumptions also incorporate learning effects, which result in substantial reductions in the cost of new technologies over time.

Strong globally coordinated mitigation action creates a large and sustained shift in relative prices (and therefore relative competitiveness) of high-emissions and low- or zero-emissions technologies. It would be surprising if these developments did not result in faster technology improvements and stronger changes in consumer preferences than might be expected on the basis of past experience and current knowledge. Faster improvements in fuel and engine efficiency and low- or zero-emissions fuel sources in the transport sector would bring forward the decarbonisation of the transport sector.

In addition to faster decarbonisation, technology improvements could also result in deeper and lower-cost decarbonisation. The standard technology scenarios assumed that there are limits to the decarbonisation possible for various modes of transport. As noted, the modelling assumed that the maximum take-up of zero-emission fuels for the rail, shipping and aviation sectors was 90 per cent in 2100. As a result, a rising carbon price imposes ever-increasing costs on the economy from the remaining 10 per cent of carbon-intensive transport. Technologies that enabled further decarbonisation could therefore significantly lower the cost of mitigation.

21.3.5 Modelling faster technology development with 'enhanced' and 'backstop' technology assumptions

The Review modelled the possible impacts of more rapid low-emissions technology development in the transport and electricity generation sectors globally. In this 'enhanced technology' scenario (see Chapter 11), technology improvements lead to a decrease in the demand for emission permits from transport and electricity generation, which lowers permit prices. This shows that faster development of road transport technology could bring forward the decarbonisation of the transport sector and reduce aggregate economic costs. However, under a fixed emissions cap, accelerating emissions reduction in transport would not necessarily accelerate decarbonisation of the economy as a whole. Instead it could allow higher emissions in other sectors and/or reduce the purchase of international emission permits.

In the enhanced technology scenario, electric vehicles become cost-competitive 20 years earlier than in the standard technology scenario but, as a result of lower carbon prices, they are taken up more slowly than in the standard technology scenario. The combined effect of faster development and slower take-up sees electric vehicles become a major component of the fleet around a decade earlier in the enhanced technology scenario than in the standard technology scenario.

Hybrid vehicles also become a major component of the fleet earlier in the enhanced technology scenario, but then are replaced earlier by electric vehicles.

Rates of technology development also affected mode shift in the modelling. Non-road transport activity (including rail and pipelines) is lower in 2100 in the enhanced technology scenario than in the standard technology scenario. This suggests that the faster that cost-competitive low-emissions road transport technologies are developed, the lower the pressure for mode shift.

The Review independently modelled a scenario in which a technology is developed that could provide unlimited emissions reductions at an emissions price of $250 per tonne of CO_2-e (the 'backstop technology'). The effect of introducing a backstop technology is that emissions continue to fall but the carbon price no longer rises. As a result, the introduction of a backstop technology halts further transition to low-emission vehicles in the transport sector.

While this indicates that a backstop technology would reduce the pressure from the carbon price on the transport sector, in all likelihood once a transition to a particular type of road vehicle has built up momentum it would continue.

21.3.6 Modelling of higher fossil-fuel prices

A sensitivity run was undertaken that examined the potential impacts of sustained high fossil fuel prices. In this scenario, the extraction costs for oil, gas and coal were increased by 50 per cent in both the no-mitigation and 550 standard technology scenario. This raises the price of coal in 2050 in the no-mitigation scenario by 16 per cent, gas by 28 per cent and oil by 42 per cent relative to the standard no-mitigation scenario. This oil price would still be lower in 2050 than the peak of almost US$150 in July 2008. These price rises result in general substitution away from fossil fuels, but within the energy sector there is substitution towards coal, which is the most emissions-intensive fuel.

The net effect is a small reduction in global emissions in the no-mitigation scenario—5 per cent lower at 2050—which reduces the overall scale of the mitigation task in the 550 standard technology scenario. It also narrows the cost gap between conventional and low-emissions technologies, so that advanced vehicle technologies become cost-competitive at lower carbon prices. As a result, in the high fossil fuel price 550 standard technology scenario, the global carbon price is 15 per cent lower in 2100 than in the 550 standard technology scenario. Higher fossil fuel prices result in an immediate shift to non-road modes of transport and distribution (including rail and pipelines) compared to the standard scenario, and electric and hydrogen vehicles are developed and adopted around five years earlier.

21.4 The path to transformation: a picture of future transport

The Garnaut–Treasury modelling explored a number of scenarios for the transport system in the future. Other outcomes are possible, including faster or slower emissions reduction at higher and lower cost. Nevertheless, emission reductions are likely to follow the pattern suggested by the modelling, with emissions stabilising over the coming decades, dropping rapidly as low-emissions road vehicles become competitive and tailing off as emissions intensity falls in aviation and shipping. However, it is possible that transport emissions could:

- grow more slowly over the coming decades if lower-emission vehicles and modes are taken up more extensively than the modelling suggested
- fall a decade or more earlier than projected in the 550 scenario if zero-emissions vehicles become cost-competitive more quickly
- fall to lower levels in 2100 if limits to reductions in emissions intensity at high carbon prices were not so tight.

Across the whole transport sector, changes in vehicle technology are likely to account for the vast majority of transport emissions reductions. Shifts to lower-emissions modes of transport are likely to account for a smaller proportion of early emissions reductions, but deliver multiple benefits as they occur gradually over the next five decades and beyond. Reductions in transport activity will probably result in a relatively small reduction in emissions in the long term.

A combination of some or all of changes in vehicle technology, mode shift and demand reduction will occur in each of the six transport subsectors discussed in section 21.1. The Garnaut–Treasury modelling is economy-wide and does not allow disaggregation of local, inter-regional and international travel, so discussion of the potential to reduce emissions faster than modelled in each sector must be qualitative.

In local passenger transport there are significant opportunities for rapid reductions in emissions over the next 30 years, with potential both for mode shift and for early adoption of low-emissions vehicles. There are also prospects for low-emissions vehicles in inter-regional passenger transport, but these may take longer to be realised, and there is significant opportunity for mode shift.

In both local and inter-regional freight there are good prospects for significant emissions reductions in the next two decades. In local freight there are better short-term prospects for vehicle changes than in inter-regional freight, but in inter-regional freight there is greater potential for mode shift.

Emissions from international passenger and freight transport are likely to fall later than in the other sectors, due to more limited opportunities for mode shift and longer useful life of existing vehicles. However, significant emissions reductions in these subsectors are still likely within this century.

Together, the changes in each sector will add up to significant reductions in transport emissions. Given the pressures from higher oil prices, major changes in transport emissions could be achieved at relatively low cost.

21.4.1 Local passenger transport

Changes in car technology are likely to be relatively rapid in local passenger transport and could be faster than suggested by the Garnaut–Treasury modelling. Smaller cars, hybrids and short-range electric vehicles are well suited to local transport, which involves short-range and stop–start driving. There are also substantial opportunities for mode shift and reductions in transport demand.

First, in the 550 standard technology scenario the share of car travel by 'light and small' cars was assumed to increase from 35 per cent to 45 per cent over the 44 years from 2006 to 2050. A faster shift is possible. Over just the last five years, with rising oil prices the proportion of new cars purchased that were small and light vehicles increased from around 35 per cent to 45 per cent of the market (FCAI 2008b), resulting in the emissions intensity of new vehicles decreasing by around 10 per cent (Figure 21.7). If these purchasing patterns continue over the next decade, the efficiency of the whole fleet would improve substantially.

Figure 21.7 Average new car emissions and oil price, January 2002 – April 2008

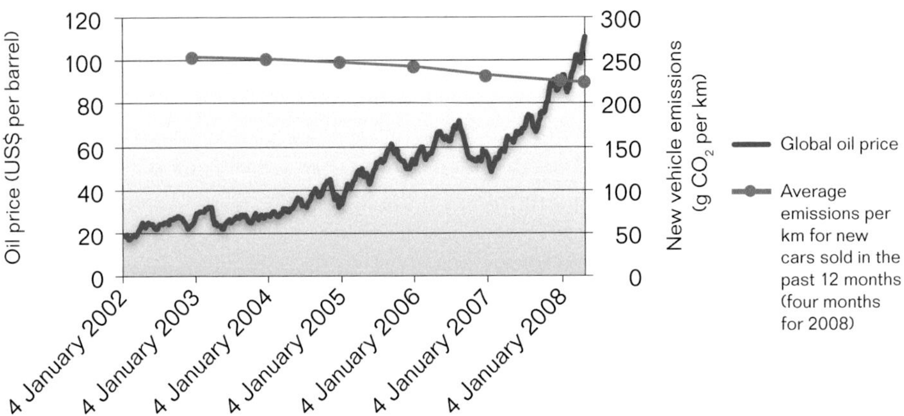

Sources: FCAI (2008a); US Energy Information Administration (2008a).

Second, partially and fully electric vehicles appear set to become cost-competitive in the near future, potentially faster than projected by the modelling (see Box 21.1).

Third, there are substantial opportunities for mode shift in local passenger transport if governments invest in infrastructure and urban planning. The modelling allowed for reductions in car activity and growth in demand for bus and train activity, but did not explicitly model the transfer of passenger journeys from car to bus or rail, or consider walking and cycling.

Box 21.1 Hybrid, electric and hydrogen vehicles

Production of hybrid petrol–electric vehicles has grown rapidly since the release of the Toyota Prius in 1997. By 2008, more than one million units of this model had been sold worldwide (Toyota Motor Corporation 2008). Sales of hybrids in Australia grew from just 0.2 per cent of cars sales in 2005 to 0.78 per cent of sales in January to July 2008 (FCAI 2008a; BITRE 2008b).

Electric vehicles are likely to become cost-competitive for local travel earlier than for long-distance travel. Battery capacity is a major part of an electric vehicle's cost, and short-range driving requires less battery capacity than longer-range driving (IEA 2008). The cost and reliability of batteries has improved dramatically over the last decade, making plug-in hybrid and fully electric vehicles that can refuel on electricity appear increasingly viable. Electric vehicles that are suitable for short-range use are already available, and both Toyota and General Motors have announced that they will release plug-in hybrids, with ranges of several hundred kilometres, by late 2010.

The emissions of electric vehicles will depend on the source of the electricity. To illustrate the point, an electric car today would generate about 30 per cent more emissions than a petrol-fuelled car of similar dimensions if the electricity had the average emissions intensity of the Australian grids. It would generate about 85 per cent less emissions than the equivalent petrol car if it drew its power from the average supplies to Tasmania. It would generate about 60 per cent more emissions than the equivalent petrol car if it drew its power from the average supplies to Victoria. If there were a widespread shift to electric vehicles, transport emissions would be tied to the emissions from the stationary energy sector. As the stationary energy sector decarbonised, emissions generated by transport activity would decline.

There is active research and development in a number of countries on the use of hydrogen in vehicles, including for energy storage in electric vehicles. Hydrogen can be produced by several means, including using electricity to split the water molecule and reforming natural gas, and so could be produced using low-emissions electricity. Its widespread use for motor vehicles would require larger investment in commercial infrastructure than a car directly recharged with electricity. Nevertheless, it could win a place in the future if its costs fall significantly.

Public transport, walking and cycling accounted for 12 per cent to 26 per cent of trips in Australian cities in 1995, but in many high-income European and Asian cities these modes accounted for more than 50 per cent of trips (Figure 21.8). Even within Australian cities there are significant variations in mode share between suburbs. For example, public transport, walking and cycling make up almost 50 per cent of trips to work in some local government areas in inner Melbourne but less than 10 per cent in many of the outer suburbs, where recent growth has been concentrated (ABS 2006b).

Figure 21.8 Trip mode, population and emissions in 57 high-income cities, 1995–96

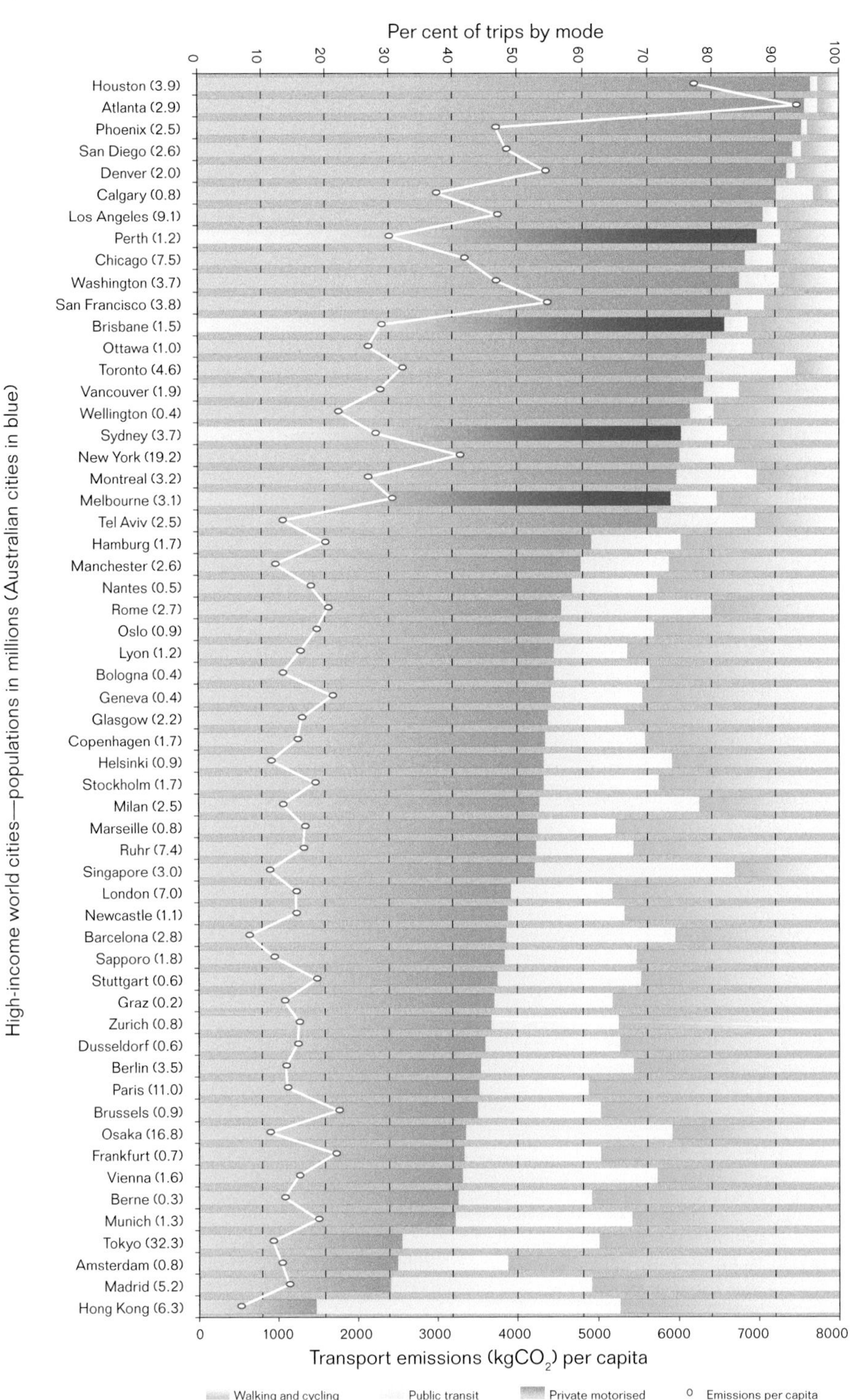

Note: Population figures are for study areas in 1995–96, which may differ from other definitions of urban areas.

Sources: Derived from Kenworthy & Laube (2001) and Kenworthy (2008).

A small shift from cars to other modes is already occurring in Australia. The census, which provides national data on mode share in Australia, indicates the share of commuting by car in Australian capital cities stabilised and started to decline slightly between 1996 and 2006 (Figure 21.9).

Figure 21.9 Mode share for journeys to work in Australian capital cities 1976–2006

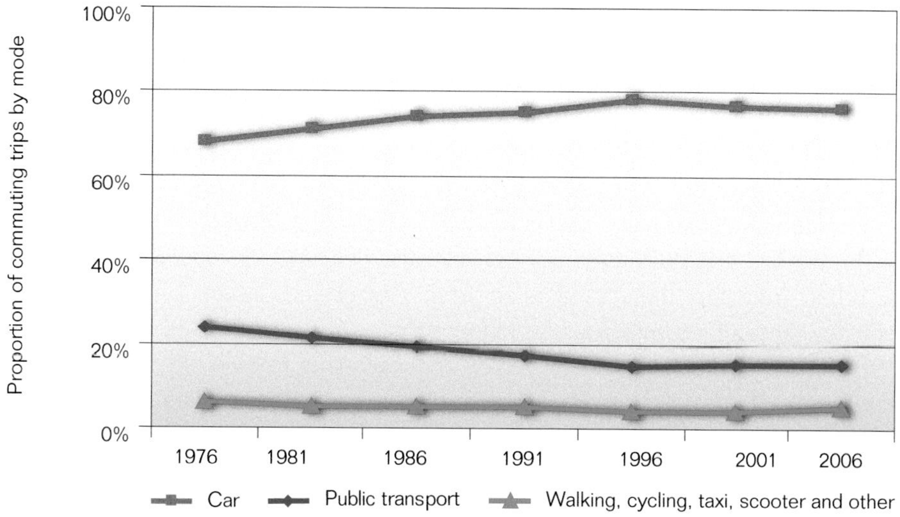

Proportion of commuting trips by mode
100%
80%
60%
40%
20%
0%
1976 1981 1986 1991 1996 2001 2006
Car
Public transport
Walking, cycling, taxi, scooter and other

Source: Derived from Mees et al. (2007).

Since the last census in 2006 there has been a resurgence in public transport use in many parts of Australia after several decades of stagnation, although it is not yet clear whether this represents a mode shift. In a recent survey, 19 per cent of respondents stated that they had used public transport more in the last year in response to higher petrol prices (Nielsen 2008). In 2007–08 public transport patronage in south-east Queensland was 5 per cent higher than in the year before (Queensland TransLink Authority 2008, per comm.). In both Melbourne and Perth patronage was 7.7 per cent higher (Victorian Department of Transport 2008, pers. comm.; Public Transport Authority of WA 2008, pers. comm.).

The patronage of rail services grew even faster. In 2007–08, rail patronage in Sydney was 5.1 per cent higher than the year before (Rail Corporation New South Wales, pers. comm.), and in Melbourne 12.7 per cent higher, stretching some services to capacity. Perth was 19 per cent higher, partly reflecting extension of rail services. Increases in patronage are also occurring in some regional areas. Following improvements in the frequency and affordability of Victorian regional rail services, patronage grew by 29 per cent between 2005–06 and 2006–07, and a further 21.8 per cent in 2007–08 (V/Line 2007, 2008 pers. comm.).

The number of commuters travelling to work solely by foot in major Australian cities increased from 3.5 per cent of trips to 4 per cent of trips between 2001 and 2006 (2006 census data reported in Mees et al. 2007). In some cities and suburbs it makes up a much larger proportion of trips, with 7.6 per cent of commuters walking to work in Hobart (Mees et al. 2007).

In most Australian capital cities, around 1 per cent of people cycled to work in 2006. Cycling is increasing rapidly in some Australian cities. The number of cyclists on measured routes in Sydney increased by 11.4 per cent per annum between 2003 and 2007, and on key routes in Melbourne by 20 per cent per annum from 2006 (Road Transport Authority 2008; Bicycle Victoria 2008).

In Canberra, 2.5 per cent of commuters already cycle to work (Mees et al. 2007). In the United States, where travel is dominated by car, cycling accounts for 3.5 per cent of trips to work in Portland, a city of over half a million people (US Census Bureau 2007). In many European cities cycling rates are much higher—for example, bicycles account for 12 per cent of traffic in Berlin, a city of around 3.5 million people (Senate Department for Urban Development 2007). Thirty-six per cent of trips to work are made by bicycle in Copenhagen, a city with a population of around half a million (City of Copenhagen 2007).

The Review's modelling did not consider the potential for reduction in demand for urban transport that could arise from changes in urban form, as people substitute shorter distance trips for longer trips.

Governments have a strategic choice about whether to invest in mode shift and more compact urban forms. The need to plan our towns and cities for population growth provides us with an opportunity to plan for different densities and public transport structures.

Changes in urban structure and mode could be implemented immediately in new suburbs and settlements. In established areas, it will take longer for the necessary system-wide changes to transport and land use to evolve. Significant mode shift will not occur overnight, or within one electoral cycle. However, a steady improvement in transport facilities and urban form over several decades accumulate to major effects on total travel demand and shifts in mode.

The combination of changes in vehicle efficiency, mode and demand could have significant impacts on emissions even without further technological development. On average, the European cities in Figure 21.8 have 60 per cent of the per capita emissions of the Australian cities, and wealthy Asian cities less than 40 per cent. The largest sources of difference relate to public transport infrastructure and use.

21.4.2 Local freight transport

The vast majority of local freight is carried by trucks and light commercial vehicles (BTRE 2006). Technological changes in these vehicles will account for almost all of the emissions reductions in this sector. Local freight vehicles could change rapidly, as short distances and stop–start conditions make hybrid engines and electric vehicles highly suitable. In addition, the use of central refuelling depots reduces their dependence on the development of widespread alternative fuel infrastructure. The modelling projected slow but in the end substantial take-up of hybrid and electric freight vehicles. Emissions from this sector could fall faster than modelled if, as seems likely, these technologies became commercially attractive sooner.

There is likely to be limited mode shift. Short travel distances and the need to travel to a wide range of destinations mean that road vehicles are likely to continue to dominate local freight. However, some demand reduction would occur almost immediately in response to higher fuel prices through more efficient practices, such as greater consolidation of deliveries.

21.4.3 Inter-regional passenger transport

Inter-regional passenger travel is largely made by car and air. Cars account for 87 per cent of all intrastate trips and 38 per cent of interstate trips. Aviation accounts for 55 per cent of interstate trips (Tourism Research Australia 2007). Other modes, including coaches and ferries, account for a smaller proportion of inter-regional transport. Reductions in emissions are likely to arise from a combination of changes in vehicle technology, mode shift and demand reduction.

There will be changes in the emissions intensity of the full range of passenger vehicles, including cars, coaches, rail and aeroplanes. Changes could occur faster than projected in the modelling if a suitable low-emissions long-distance road vehicle technology were developed in the near future, such as biodiesel from sources that did not compete with food production for land.

Aviation increased in efficiency significantly in the last century. The fuel efficiency of the US air fleet increased by 60 per cent between 1971 and 1998 (IEA 2008). High oil prices are already driving investment in research and development on a number of alternative fuels, including biofuels and hydrogen. There appear to be prospects for use of biofuels in the coming decades, but in the short to medium term the prospects for a near zero-emissions fuel in aviation appear more limited than in other modes. Unless there is a breakthrough with alternative fuels, higher oil prices and an emissions price are likely to increase the cost of flying relative to other forms of transport.

If long-distance road and aviation technologies were to improve slowly, higher oil prices and an emissions price could slow the growth in per capita inter-regional travel. Video-conferencing could supplant some business trips, and some discretionary transport, such as holidays, could be reduced or replaced by shorter trips.

However, growing incomes over this period may offset rising long-distance travel costs, and there are options to switch to other modes rather than reduce travel. For medium-distance journeys, which are currently largely made by car, passengers could shift to bus services and regional rail. For longer-distance journeys now made by car or air, passengers could shift to high-speed rail.

High-speed rail is a major component of long-distance travel in Europe, Japan, Korea and China, linking cities that are several hundred to a thousand kilometres apart. While the prospects for competitive high-speed rail for intercity journeys in Australia have seemed limited in the past, high oil prices, an emissions price, rising incomes and a growing population on the east coast improve the prospects of cost-competitive high-speed rail links between major cities.

Now is a good time for the Commonwealth Government and the governments of New South Wales, Victoria, Queensland, South Australia and the Australian Capital Territory to examine why intercity passenger train services in Australia are inferior to those in European and high-income Asian countries, with a view to removing barriers to the emergence of high-quality inter-regional rail services in Australia.

21.4.4 Inter-regional freight

Long-distance freight transport is made by a range of modes. In 2000, around 28 per cent of non-urban freight tonne-kilometres was carried by road, 37.6 per cent by rail, 33 per cent by ship and 0.05 per cent by air (BTRE 2006). Inter-regional freight may experience less rapid improvement in the emissions intensity of vehicle technology than local freight, but more mode shift.

The development of near zero-emissions trucks for inter-regional freight may take longer than for local freight, due to the additional energy storage required for long-distance travel. Accordingly, the Review's modelling placed some limitations on the adoption of hybrid and electric vehicles for long-distance travel prior to 2050, and assumed that biodiesel would not become commercial. Emissions could fall faster from this sector if biodiesel or another low-emissions fuel were developed.

Rail and shipping also have opportunities for fuel switching. Gains may take several decades to realise fully after the development of new technologies, due to the long working lives of these vehicles.

There are immediate and growing opportunities for mode shift, particularly from road to rail, given that the cost of road freight has been increasing faster than the cost of rail and shipping (ABS 2008d). The majority of bulk freight, such as mining products and grain, is already carried by rail and shipping; the Bureau of Transport and Regional Economics (2006) projected that this will continue even in the absence of an emissions price.

Around 75 per cent of non-bulk freight, such as manufactured goods, is currently carried by road, due to its ability to deliver door to door (BTRE 2006). Road is likely to retain a significant component of the non-bulk task, but part of the 15 per cent of road freight that is currently used for intercity trips could be transferred to rail (BTRE 2006: 158). In the longer term, the development of a more substantial rail freight network, along with intermodal terminals that allow the rapid transfer of goods between trucks and trains, could permit an even greater share of freight to be transferred from road to rail. The Review's modelling allowed for some limited shift of freight from road to rail.

The demand for inter-regional freight will be affected by changes in production and distribution patterns. Producing goods at several sites for local consumption, rather than at one site with wider distribution, can reduce transport costs. As both production and transport costs will involve some greenhouse gas emissions, a broad-based emissions trading scheme that includes stationary energy, industrial processes and transport is necessary to achieve the lowest-cost emissions reduction over the whole production and distribution process.

21.4.5 International passenger and freight

International passenger travel is now almost exclusively by air. Cruise ships account for only a small section of the tourist market. Conversely, shipping accounted for over 99 per cent of international freight by weight in 2006–07, with aviation carrying a small but valuable load (less than 0.1 per cent of freight by weight but 13 per cent by value) (BITRE 2008a). Vehicle technology improvements and demand changes are likely to account for the majority of emissions reductions in this sector.

There are prospects for improving the emissions efficiency of shipping and aviation. Some of these changes will take several decades to develop and implement, with the result that the cost of international transport could rise with emissions pricing in the short term. The fuel consumption and emissions of existing ships can be lowered by reducing their speed and retrofitting them with technologies such as kite-sails and biofuels (IEA 2008). New ships will be able to take advantage of additional options, including changes to hull design. As noted in section 21.4.3, there are good prospects for improved aviation efficiency, but more limited prospects for zero-emissions fuels than for other modes. The Review's modelling assumed that emissions in these sectors would decline slowly over the century. A breakthrough in technology in the coming decades would result in faster and lower-cost emissions reduction in this sector.

The demand for international freight transport will grow more slowly if shipping and aviation prices increase. Higher international transport costs would increase the advantage of producing goods closer to their site of consumption, which may result in some goods being produced domestically rather than imported. In addition, relative costs of processing some bulk mineral and energy commodities before export may decline, promoting processing in Australia.

Depending on how high oil prices rise and how long alternative fuels take to come to market, international passenger transport may grow more slowly than it otherwise would have done. In the short term, the emissions price is likely to have a smaller impact than higher fuel prices on aviation costs. Later in the century the emissions price could have a more significant impact on the demand for aviation if low-emissions fuels have not emerged. Very high oil and emissions prices would result in people choosing to travel less often and by slower modes such as ships. Any such tendency would be moderated if alternative, low-emissions aviation fuels became cost-competitive.

21.4.6 Faster, deeper, cheaper

The Garnaut–Treasury modelling concludes that the cost of the transition to a lower-emissions transport system will be relatively modest. Transport emissions could be reduced faster and at lower cost if:

- governments plan for more compact cities and invest in a shift from high-emissions modes to rail, public transport, walking and cycling
- a range of road vehicle technologies become commercial sooner than assumed

- technologies are developed in the coming decades that substantially reduce emissions from aviation, shipping, rail, ancillary road transport and the residual emissions from road.

21.5 Fostering the transformation

There are many opportunities for decarbonising the transport system. Low-cost transformation of the transport sector will require policy steps beyond the introduction of an emissions trading scheme.

Two key issues stand out. First, governments will need to respond to changing economic conditions by focusing on denser urban form and investing more in public transport infrastructure. Second, there are a range of policies that distort the costs of vehicle ownership and use. Government also has a useful role in dissemination of information, as discussed in Chapter 17.

21.5.1 Cities for the new cost environment

Australian governments have an opportunity to choose how settlements will develop. With population set to increase by about two and a half times in a century, the continued development of settlements on their current patterns would generate infrastructure, traffic congestion and equity problems. The alleviation of these problems through planning for more dense development around better public transport infrastructure would have incidental advantages for reducing the cost of adjusting to an emissions constraint.

21.5.2 Distortions in the prices of vehicle ownership and use

A number of established policies impede the take-up of more fuel-efficient, lower-emissions vehicles. The introduction of an emissions trading scheme provides an occasion for governments to consider whether the distortions arising from these policies outweigh their benefits.

There is a strong case for reducing imposts on purchasing and owning vehicles and replacing them with charges on using a vehicle. One way to do this would be to allow the full cost of the emissions permits to flow through to fuel prices, and to use the permit revenue to fund reductions in sales taxes, import duties and vehicle registration charges.

Several taxes apply to the upfront price of vehicles but not other goods, including import tariffs and stamp duty. Tariffs on vehicles manufactured abroad distort the market for vehicles, at the margin slowing the diffusion of new, low-emissions vehicles. Given that 81 per cent of vehicles sold in Australia in 2007 were manufactured abroad (Commonwealth of Australia 2008) and at the present time lower-emissions vehicles tend to be imported, this biases purchase patterns against low-emissions vehicles. In addition, four-wheel drive vehicles have a 5 per cent import tariff advantage over passenger cars, reducing the cost of these more emissions-intensive vehicles relative to other imported cars. There is a strong case for the Commonwealth Government reducing and equalising import tariffs as soon as possible.

There are also charges applied to vehicle ownership, such as registration fees and insurance, where the costs that the charges are sometimes supposed to cover are affected by the distances travelled. These charges increase the cost of owning a vehicle and, as the costs do not increase with use, provide no incentive for people to use their cars less. Options for relating vehicle charges more closely to use include shifting the third-party personal injury insurance component of registration fees to fuel excise. This is a use-related cost that should rise as use rises. This shift could be done in a revenue-neutral way and would require an agreement on revenue distribution between state and territory governments (which currently collect these charges) and the Commonwealth Government (which collects fuel excise).

Some policies reduce the cost of vehicle use or create incentives for use. The fringe benefits tax provisions attempt to value benefits provided by employers to employees as part of salary packages in order to appropriately tax them. However, the current treatment of vehicles and parking spaces distorts decisions towards private vehicle use and greater demand of transport overall (Commonwealth of Australia 2008). These provisions could be improved by:

- ensuring the salary sacrifice arrangements are mode neutral
- amending the statutory fraction method to ensure it is distance neutral.

Notes

1 Freight activity is typically measured in 'tonne-kilometres', the tonnes of freight carried multiplied by the distance it is carried. Passenger transport activity is measured in 'passenger-kilometres', the number of people carried multiplied by how far they travelled.

2 Based a medium-sized car driven 15 000 kilometres a year over five years (RACV 2008).

References

ABS (Australian Bureau of Statistics) 1997, *Average Retail Prices of Selected Items, Eight Capital Cities, June Quarter 1997*, cat. no. 6403.0, ABS, Canberra.

ABS 2003, *Environmental Issues: People's Views and Practices*, cat. no. 4602.0, ABS, Canberra.

ABS 2006a, *Census of Population and Housing Australia*, cat. no. 2001.6, ABS, Canberra.

ABS 2006b, *Census of Population and Housing Australia*, Census community profile series, ABS, Canberra.

ABS 2006c, *General Social Survey: Summary Results, Australia*, cat. no. 4159.0, ABS, Canberra.

ABS 2008a, *Australian Historical Population Statistics, 2008*, cat. no. 3105.0.65.001, ABS, Canberra.

ABS 2008b, *Average Retail Prices of Selected Items, Eight Capital Cities*, cat. no. 6403.0.55.001, ABS, Canberra.

ABS 2008c, *Average Weekly Earnings, Australia, May 2008*, cat. no. 6302.0, ABS, Canberra.

ABS 2008d, *Producer Price Indexes, Australia, June 2008*, cat. no. 6427.0, ABS, Canberra.

Bicycle Victoria 2008, *Super Tuesday 2008 Count*, Bicycle Victoria, Melbourne.

Bureau of Transport and Communications Economics 1996, *Transport and Greenhouse: Cost and options for reducing emissions*, Report 94, Bureau of Transport and Communications Economics, Canberra.

BTRE (Bureau of Transport and Regional Economics) 2006, *Freight Measurement and Modelling in Australia*, Report 112, BTRE, Canberra.

BTRE 2007, *Estimating Urban Traffic and Congestion Cost Trends for Australian Cities*, Working Paper 71, BTRE, Canberra.

BITRE (Bureau of Infrastructure, Transport and Regional Economics) 2008a, *Australian Transport Statistics 2008: Pocket booklet*, BITRE, Canberra.

BITRE 2008b, *How Do Fuel Use and Emissions Respond to Price Changes?* BITRE Briefing 1, BITRE, Canberra.

City of Copenhagen 2007, *Bicycle Account 2006*, City of Copenhagen, Copenhagen.

Commonwealth of Australia 2008, *Review of Australia's Automotive Industry: Final report* (Bracks Review), Commonwealth of Australia, Canberra.

CSIRO (Commonwealth Scientific and Industrial Research Organisation) 2008, *Fuel for Thought: The future of transport fuels: challenges and opportunities*, CSIRO, Newcastle.

DCC (Department of Climate Change) 2008, *Australia's National Greenhouse Accounts*, Australian Greenhouse Emissions Information System, <www.ageis.greenhouse.gov.au>.

Department for Environment, Food and Rural Affairs 2008, *Code of Best Practice for Carbon Offset Providers: Methodology paper for new transport emission factors*, Department for Environment, Food and Rural Affairs, London.

Dodson, J. & Sipe, N. 2007, *Shocking the Suburbs: Urban location, housing debt and oil vulnerability in the Australian city*, Urban Research Program, Research Paper 8, Griffith University, Brisbane

FCAI (Federal Chamber of Automotive Industries) 2008a, *National Average Carbon Emissions Fact Sheet*, FCAI, Canberra.

FCAI 2008b, VFACTS data, FCAI, Canberra.

Goodwin, P., Dargay, J. & Hanly, M. 2004, 'Elasticities of road traffic and fuel consumption with respect to price and income: a review', *Transport Reviews* 24(3): 275–92.

IEA (International Energy Agency) 2008, *Energy Technology Perspectives: Scenarios and strategies to 2050*, IEA, Paris

IPCC (Intergovernmental Panel on Climate Change) 1999, *Aviation and the Global Atmosphere*, J.E. Penner, D.H. Lister, D.J. Griggs, D.J. Dokken & M. McFarland (eds), Cambridge University Press, Cambridge.

IPCC 2007, *Climate Change 2007: Mitigation of climate change. Contribution of Working Group III to the Fourth Assessment Report of the Intergovernmental Panel on Climate Change*, B. Metz, O.R. Davidson, P.R. Bosch, R. Dave & L.A. Meyer (eds), Cambridge University Press, Cambridge and New York.

Joint Transport Research Centre 2008, *Greenhouse Gas Reduction Strategies in the Transport Sector: Preliminary report*, Joint Transport Research Centre of the Organisation for Economic Co-operation and Development and the International Transport Forum, Paris.

Kenworthy, J. & Laube, F. 2001, 'The millenium cities database for sustainable transport (CDROM database)', International Union (Association) of Public Transport, Brussels & Institute for Sustainability and Technology Policy, Perth.

Kenworthy, J.R. 2008, 'Energy use and CO_2 production in the urban passenger transport systems of 84 international cities: findings and policy implications', in Droege, P. (ed), *Urban Energy Transitions*, Elsevier, Oxford.

Mees, P., Sorupia, E. & Stone, J. 2007, *Travel to Work in Australian Capital Cities, 1976-2006: An analysis of census data*, Australasian Centre for the Governance and Management of Urban Transport, University of Melbourne, Melbourne.

Mills, E. & Hamilton, B.W. 1994, *Urban Economics*, 5th edn, Harper-Collins, New York.

Newman, P. & Kenworthy, J.R. 1999, *Sustainability and Cities: Overcoming automobile dependence*, Island Press, Washington DC.

Nielsen 2008, *Nielsen Omnibus*, Nielsen, Sydney.

RACV 2008, *Vehicle Operating Cost Results 2008*, RACV, Melbourne.

Road Transport Authority 2008, *Cycling in Sydney: Bicycling ownership and use: April 2008*, Road Transport Authority, Sydney.

Senate Department for Urban Development (Senatsverwaltung für Stadtentwicklung) 2007, *Mobility in the City: Berlin transport in figures, 2007*, Senate Department for Urban Development, Berlin.

Toyota Motor Corporation 2008, 'Worldwide Prius sales top 1 million mark', media release, Toyota Motor Corporation, Tokyo.

Tourism Research Australia 2007, *Interstate vs Intrastate: Domestic Tourism facts, year ending December 2007*, Tourism Research Australia, Canberra.

US Census Bureau 2007, 'Most of us still drive to work—alone: public transportation commuters concentrated in a handful of large cities', media release, US Census Bureau, Washington DC.

US Energy Information Administration 2008a, *Weekly All Countries Spot Price FOB Weighted by Estimated Export Volume (Dollars per Barrel)*, <http://tonto.eia.doe.gov/dnav/pet/pet_pri_wco_k_w.htm>.

US Energy Information Administration 2008b, *International Energy Outlook 2008*, Report # DOE/EIA-0484, Energy Information Administration, Washington DC.

V/Line 2007, *Annual Report 2006/07*, V/Line, Melbourne.

Vuchic, V.R. 1981, *Urban Public Transportation: Systems and technology*, Prentice-Hall, Englewood Cliffs, New Jersey.

Vuchic, V.R. 2005, *Urban Transit: Operations, planning and economics*, John Wiley and Sons, Hoboken, New Jersey.

TRANSFORMING RURAL LAND USE

22

Key points

Rural Australia faces pressures for structural change from both climate change and its mitigation.

Effective mitigation would greatly improve the prospects for Australian agriculture, at a time when international demand growth in the Platinum Age is expanding opportunities.

Choices for landowners will include production of conventional commodities, soil carbon, bioenergy, second-generation biofuels, wood or carbon plantations, and conservation forests.

There is considerable potential for biosequestration in rural Australia. The realisation of this potential requires comprehensive emissions accounting.

The realisation of a substantial part of the biosequestration potential of rural Australia would greatly reduce the costs of mitigation in Australia. It would favourably transform the economic prospects of large parts of remote rural Australia.

Full utilisation of biosequestration could play a significant role in the global mitigation effort. This is an area where Australia has much to contribute to the international system.

Land-use change—the alteration of management practices on a certain type of land cover—has the capacity to transform Australia's, and to a lesser extent the global, mitigation effort. Outside Australia, it is of powerful significance for Australia's immediate neighbours, Indonesia, Papua New Guinea and the other countries of Southeast Asia and the South Pacific. Getting the incentive structures right at home and abroad to realise the enormous potential for biosequestration is a major challenge, and potentially Australia's most important contribution to the global mitigation effort.

This chapter looks more speculatively at some future possibilities that have been given an unreasonably small place in Australian and international discussions of mitigation.

Climate change and climate change mitigation will bring about major structural change in the agriculture, forestry and other land use sectors. With effective global action, climate change mitigation would become the more important force for change. A rising carbon price will alter the cost of land management practices and commodities, depending on their emissions profiles.

On the other hand, without mitigation, and in the next few decades in any case, projected temperature increases and decreased rainfall in some important centres of Australian agriculture are likely to reduce water availability. This will particularly affect industries that rely on irrigation and those that are currently operating near the margins of profitable cultivation. In the longer term, land managers will respond to these dual challenges by pursuing new opportunities in carbon removal (or sequestration), energy production from biomass and low-emissions livestock production. Such opportunities could significantly lower the economy-wide cost of the emissions trading scheme—far below those suggested in the Review's modelling of the costs of mitigation.

Agriculture and forestry will experience the effects of climate change differently, and their prospects for adaptation and emissions mitigation also differ. While these sectors warrant separate consideration, they are inextricably linked. Both provide products and services based on natural systems. The issues they face can be relevant to a single landowner or business. They sometimes compete with each other for land and water. Indeed, the *IPCC Guidelines for National Greenhouse Gas Inventories* (IPCC 2006) incorporate emissions from what is known as the 'agriculture, forestry and other land use sector' into a single reporting framework. As many of the overarching issues relate to these interactions, agriculture, forestry and other land use are considered together in this chapter.

22.1 Drivers of a transformation towards lower emissions

22.1.1 Existing pressures on the agriculture and forestry sectors

At the end of the 20th century, several factors coincided to place pressure on most Australian agricultural industries. A long period of relatively low real prices for agricultural products was continuing, while costs of established patterns of cultivation were rising. Environmental limits to production came to the fore in the form of dryland and irrigated salinity, soil acidification, soil fertility and structural decline, soil erosion, and increasingly stressed water systems. Governments responded by introducing regulations and establishing environmental markets. The most notable was the 1995 cap and trade system for water, which enabled high-value uses to compete for water in the Murray-Darling Basin. This competition has led to moves away from water-intensive agriculture in some areas. Further reforms to the existing water allocation systems will be implemented in the near future, increasing the impetus for change.

Over the last two decades, agricultural subsectors have been increasingly deregulated, including the dairy industry in the 1980s, the pork industry in the 1990s and the sugar industry in a series of steps. The increase in competition has affected local communities as the geographic location of production and employment has shifted.

Since 2000, several factors have acted in concert to increase commodity prices received by Australian producers. Incomes in developing countries are rising rapidly, leading to higher consumer demand for meat, dairy products and oil seeds. This growth in demand for animal products has increased demand for grain and oilseeds for stockfeed (ABARE 2006). With growth in major developing economies expected to continue for the foreseeable future, a continuation of this strong demand is likely. A series of droughts in Australia and drought and flood in other grain-producing regions of the world have placed further upward pressure on prices. Distortionary biofuels policies in North America and Europe have also contributed to increases in food prices. Some governments have responded to increased food prices by restricting food exports, setting limits on food prices, or both (von Braun 2008). Such controls have exacerbated global price increases and volatility in the rest of the world, and placed stress on developing countries that are dependent on imports.

Higher costs for agricultural inputs, particularly for fuel, fertiliser and chemicals, have been observed in recent years, driven primarily by high global prices for petroleum and higher demand for these inputs (Figure 22.1). After a long period in which average farm sales prices fell relative to costs, through the early 21st century, the increase in the cost of inputs has almost been matched by an increase in prices received (Figure 22.2). So far in 2008, a large price increase in farm goods has improved farmers' terms of trade.

Figure 22.1 Prices paid by Australian farmers, 1998–2007

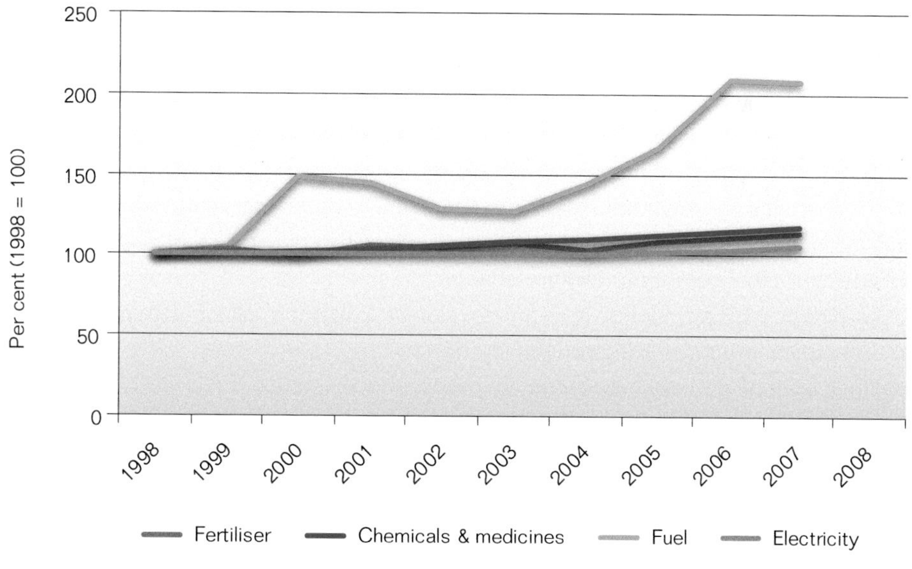

Source: ABARE (2008a).

Figure 22.2 Australian farmers' terms of trade, 1998–2007

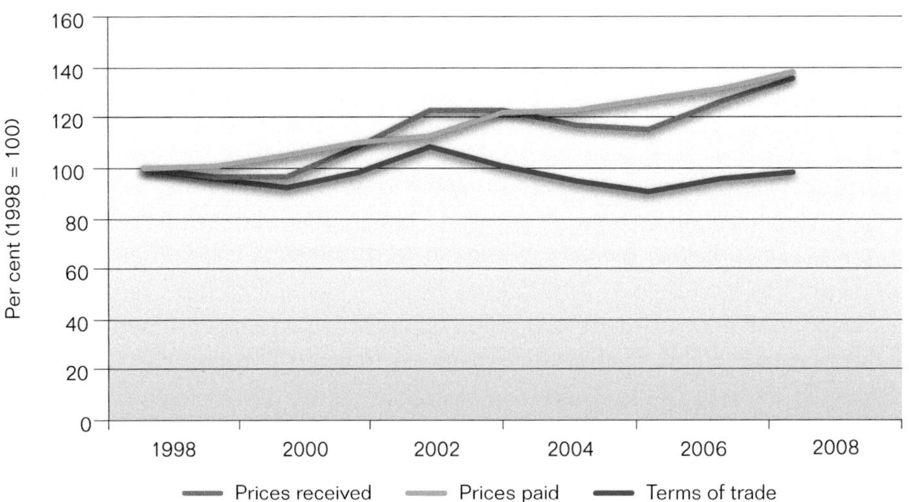

Source: ABARE (2008a).

22.1.2 Effects of climate change on food demand and supply

Effects on global food supply and demand

Even with 550 ppm global mitigation, a global average temperature increase of 2.5°C above 1990 levels is the median equilibrium outcome. Temperature increases of between 1°C and 3°C above 1990 levels, and increases in carbon dioxide concentration and rainfall, are associated with an increased potential for global food production, but above this range potential production is expected to decrease relative to current levels (IPCC 2007: 274). The melting of glaciers, leading to sea-level rise, and changes in river flow and monsoon rainfall, are likely to severely affect agricultural production, particularly in Asia. South Asia, sub-Saharan Africa and Australia have been identified as having agricultural sectors that are especially vulnerable to the impacts of climate change.

Domestic food production in many developing countries will be at immediate risk of reductions in agricultural productivity due to crop failure, livestock loss, severe weather events and new patterns of pests and diseases (FAO 2007). Climate change could disrupt ocean currents, which would have serious ramifications for the availability of fish, a major protein source.

Farmers in developing countries are less able to adapt to and effectively manage these risks due to the higher proportion of small-scale and subsistence farms, poorly developed infrastructure and lesser access to capital and technology.

These impacts, together with the considerable increases in population and food demand expected in developing countries, will lead to an increase in global food prices.

Effects on domestic food supply

In Australia, some agricultural industry subsectors will be more vulnerable than others to climate change impacts (see Table 22.1). Enterprises already close to the edge of the ideal climatic range for their dominant agricultural activity will be particularly at risk.

Changes to local climate and water availability will be key determinants of where agricultural production occurs and what is produced. Climate change is expected to reduce yields for many crops and place upward pressure on Australian food prices (Quiggin 2007). Climate change impacts will also drive a range of adaptation measures.

Table 22.1 Vulnerability of Australia's agricultural industry to the biophysical impacts of climate change, by subsector

Industry subsector	Vulnerability to biophysical impacts of climate change
Sheep (dryland)	High
Sheep (irrigated)	Very high
Grain (dryland)	High
Grain (irrigated)	Very high
Beef cattle (dryland)	High
Dairy cattle (irrigated)	Very high
Pigs (intensive)	Low
Poultry (intensive)	Low
Other (horticulture & viticulture)	Moderate (high for wine quality)
Forestry	Moderate
Fisheries	High for some species, but largely unknown

Note: Vulnerability is a measure of impacts (exposure + sensitivity) and adaptive capacity (see also Figure 6.1).

22.1.3 Drivers introduced by climate change mitigation

Existing mitigation policies

In Australia, emissions mitigation has been pursued for several years, particularly in the forestry sector, as many existing mitigation policies and agreements recognise and provide credit for carbon removal by forests. Land clearing has slowed significantly since 1990, primarily due to regulatory controls.

Forests and plantations established after 1990 accounted for net removal of about 23 Mt CO_2-e in 2006, and will make an important contribution towards meeting Australia's commitment under the Kyoto Protocol.

Mitigation through forest sinks has been encouraged by demand for emissions reduction certificates or offset credits under a number of domestic programs, including the New South Wales Greenhouse Gas Reduction Scheme, the West Australian Government's requirement for some project approvals to involve carbon offsetting, and the Commonwealth-administered Greenhouse Friendly program.

At the same time, there has been increasing interest in a range of low- to negative-cost emissions reduction activities in the agriculture sector, which generally also provide productivity benefits, such as fertiliser management.

It is important that an emissions trading scheme with comprehensive coverage replaces and expands incentives for mitigation in the agriculture and forestry sectors. For activities not included in the scheme, other policies will be required to drive mitigation.

An emissions trading scheme

When it is introduced by the Commonwealth Government, an emissions trading scheme will be the primary instrument driving emissions mitigation in Australia. The effect of the scheme on the agriculture and forestry sectors will depend on several factors:

- rules for their coverage or inclusion under the scheme
- direct and indirect emissions intensities
- availability and cost of mitigation options
- availability of alternatives for commodity production.

In relation to the treatment of forestry and agriculture under an emissions trading scheme, the Review proposes the following approach:

- Those undertaking reforestation should be allowed to opt in for coverage (that is, liability for emissions and credit for net removal from the atmosphere) from scheme commencement.
- Those undertaking deforestation should be liable for resulting emissions.
- There should be full coverage of the agriculture, forestry and other land use sector, based on full carbon accounting once issues regarding emissions measurement, estimation and administration are resolved.
- Policies should apply to the agriculture sector to drive mitigation until it is covered under the scheme.

The over-riding idea should be one of providing incentives for net sequestration within a comprehensive carbon accounting framework.

Full coverage of the agriculture, forestry and other land use sector would involve accounting for all greenhouse gas emissions and removal on managed land, including soil carbon, forests and wooded lands (regardless of the date of establishment) and life-cycle emissions from, and carbon storage by, harvested wood products. The 2006 IPCC Inventory Guidelines provide a useful framework for the development of a comprehensive approach to accounting. However, emissions reported do not necessarily have to align exactly with emissions liabilities or credits under an emissions trading scheme.

The mitigation policy modelled by the Review (Chapter 11) does not reflect this emissions trading scheme design. Consequently, the analysis in the following section and in Chapter 11 cannot be taken as a reflection of what would occur under the Review's recommended emissions trading scheme design. They take account of only a small part of Australia's biosequestration potential.

22.2 Economic modelling results: a possible future?

The modelling presented in Chapter 11 considered possible outcomes for Australia's economy without mitigation, and also considered the impacts of an emissions trading scheme under three technology assumptions: 'standard', 'backstop' and 'enhanced'. The focus of this section is on the transition for the agriculture and forestry sectors in Australia in a world with effective global action on mitigating emissions (stabilisation at 550 ppm CO_2-e or 450 ppm CO_2-e under standard technology assumptions). It does not take into account some of the large opportunities for biosequestration discussed later in this chapter. It reflects continuing application of current Kyoto Protocol rules as adopted by Australia—including Australia's decision so far not to opt in to the more expansive coverage of Article 3.4 of the Kyoto Protocol, and relevant clauses of the Marrakesh Accords.

22.2.1 Overview of emissions outcomes

Projected non-combustion emissions from agriculture, forestry and land-use change for the no-mitigation and 550 standard technology scenarios are presented in Figure 22.3.

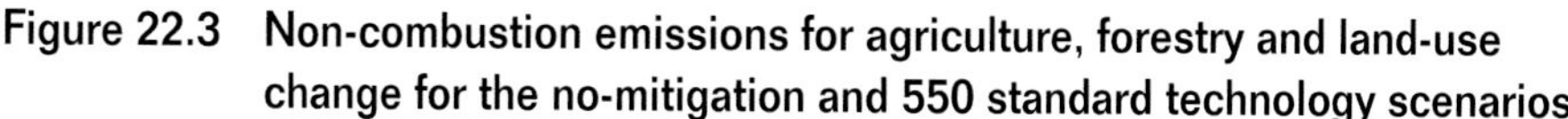

Figure 22.3 Non-combustion emissions for agriculture, forestry and land-use change for the no-mitigation and 550 standard technology scenarios

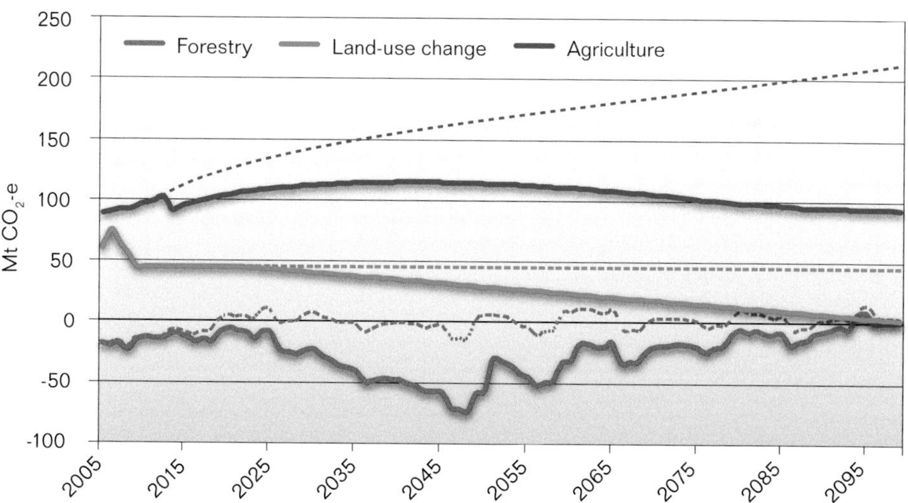

Note: These results were generated using MMRF. Emissions under no mitigation are shown with the dashed lines. Emissions from fuel combustion in the agriculture and forestry sectors, such as the on-farm use of petrol and diesel in farm machinery, are not included.

With the 550 standard technology scenario, non-combustion emissions from all three sources are lower than under the no-mitigation scenario; however, agricultural emissions still increase slowly to around 30 per cent above 2005 levels by mid-century, before declining to 5 per cent above 2005 levels in 2100.

By 2100, in the 550 standard technology scenario, the agriculture sector is responsible for more than 41 per cent of total Australian emissions and is by far the largest source of emissions under the standard technology scenarios. The agriculture sector as a whole has a lower known technological and economic potential to reduce emissions intensity than other sectors of the economy. There is currently a lack of well-quantified and well-costed mitigation methods available to agriculture. While the modelling exercise allows reductions in emissions intensity, it does not identify individual mitigation methods and technologies. Rather, an assumed marginal abatement cost curve was used (US EPA 2003, 2006).

Known and expected opportunities account for the mitigation observed in the agriculture sector, and emissions intensity progressively falls throughout the century (Figure 22.4). The cost of this was attributed to the subsector through a marginal abatement cost curve. Following the literature, emissions intensity is assumed to improve more rapidly for grains and other agriculture than for the animal subsector, reflecting greater potential for mitigation at a given carbon price. The level of aggregation of subsectors in the MMRF model means that low-emissions agricultural products are not individually identified. However, the marginal abatement cost curves have been developed to broadly reflect the potential for substitution of high-emissions with low-emissions agricultural products.

Figure 22.4 Change in emissions intensity over time in response to carbon price, 550 standard technology scenario, 2006–2100

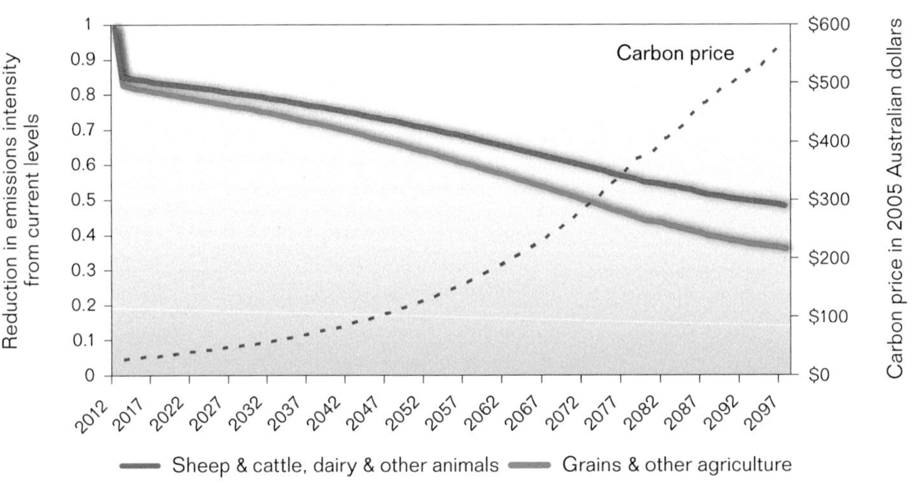

Note: These results were generated using MMRF.

The rate of emissions intensity improvement in the agricultural sector in the first half of the 21st century under the 550 standard technology scenario reflects the limited mitigation options available at the prevailing carbon price. As the carbon price increases, it becomes efficient to reduce agricultural emissions further. After 2050, the higher carbon price leads to emissions reductions that would require a widespread change in agricultural practices and/or consumer tastes, or the implementation of new technologies. By 2100, output from the sector is almost four times larger than in 2005, but agricultural emissions are just above 2005 levels and emissions intensity levels have decreased by more than half relative to current levels.

A scenario was run through the model assuming emissions reductions from forestry activities were not eligible under the global emissions trading scheme. To achieve the same level of mitigation without forestry activities, the carbon price is consistently 30 per cent higher than when forestry activities are included and when standard technology assumptions are used. The higher carbon price leads to higher gross world product costs. In Australia, GNP in 2100 is half a percentage point lower in 2100 when forestry is not included, compared to the same mitigation scenario where it is included. This sensitivity illustrates the potential impact of excluding forestry activities from emissions accounting. It also demonstrates how the availability of large, low cost sources of mitigation can reduce the global costs of mitigation policy. Such sources could include soil carbon or biochar, or a new technology to reduce emissions from livestock.

If new, low-cost mitigation options emerge for the agriculture sector, there will be greater reductions in the agriculture and forestry sectors' emissions, at a lower cost than the model predicts. An alternative, lower-emissions future for the agriculture and forestry sectors is described in section 22.3.

22.2.2 Productivity and macroeconomic effects on agriculture and forestry

After 2050 the benefits of mitigation (avoided climate change) start to become evident when the 550 standard technology scenario is compared to the no-mitigation scenario. With unmitigated climate change, all agricultural subsectors experience a reduction in output by the end of the century as temperatures increase and water availability decreases as a result of climate change. Water-intensive or water-dependent industries such as grain, farm dairy and horticulture are particularly affected.

Under the 550 standard technology scenario, a large proportion of climate change impacts are avoided. Those sectors that benefit more from the avoided impacts of climate change, and are affected less by rising carbon prices, show higher levels of output in 2100 than in the no-mitigation scenario.

The forestry sector, which includes environmental plantings, is stimulated by the introduction of an emissions trading scheme. Demand for offset credits (from carbon removals) increases demand for forestry output, which include logging and services associated with plantations, in the 550 standard technology scenario. The forestry sector increases by more than 20 times by 2100 and its share of overall activity more than doubles relative to 2005 levels.

While activity in the agriculture sector increases by 2100, agriculture has a falling share of total output. The composition of the agriculture sector changes: output from sheep and cattle, grains and dairy decreases relative to the no-mitigation scenario, while the share of other animal products and other agriculture, including horticulture, increases.

22.2.3 Changes to production of livestock and other animals

The sheep and cattle industries are highly emissions intensive, and there are currently limited opportunities for the reduction of methane emissions. Other meat products, such as pork and chicken, are less emissions intensive. While the model allows for substitution between existing meat products in response to the carbon price, there is no explicit consideration of alternative sources of animal protein that are not currently widely consumed, such as kangaroo meat.

In response to a carbon price on the agricultural sector, households move away from meat and meat products because of the higher price of these commodities under an emissions trading scheme. Households also move away from beef and lamb towards less emissions-intensive meat, such as chicken and pork. A similar pattern of change is observed in Australia's export of meat and meat products under the mitigation scenarios.

While output in the sheep and cattle industries is reduced in comparison to the no-mitigation scenario, real production in the 550 standard technology scenario still increases by around 150 per cent from current levels by 2100.

22.2.4 Land use, land-use change and forestry

Modelling land use, land-use change and forestry emissions is complex and difficult, and the results should be seen as a guide only to the possible implications of the forestry sector's response to a carbon price.

The forestry sector responds to the carbon price by establishing new plantations. In the modelling, three types of forestry activity were assumed to be available—softwood and hardwood timber plantations and environmental (carbon sequestration) plantations. All types have establishment costs, but carbon plantings do not have transport or harvesting costs.

The forestry modelling for Australia was incorporated into MMRF (see Box 11.1). The analysis took in land currently used for all forestry and agricultural activities, including minimally adjusted pastures used for livestock production in remote areas of Australia. The extent of new land dedicated to forestry is determined by the relative value of forestry activities compared to the value of agricultural activities competing for the land.

The model did not explicitly consider possible restrictions on forestry expansion for conservation reasons, the potentially negative environmental impacts of forestry expansion (such as reduced water runoff), the potential implications arising from climate change, or regional capacity constraints in timber processing or other factors leading to landholder resistance to land conversion. However, there are assumed restrictions on potential take-up rates, which may limit the potential increase in forestry activity.

Forestry experiences a significant change between the no-mitigation and effective global mitigation scenarios. Under the no-mitigation scenario, emissions from forestry rise over the study period, to the point where it is a net source of emissions in some years (Figure 22.3). By contrast, under the 550 standard technology scenario forestry is consistently an emissions sink, with removal from the atmosphere increasing particularly after the late 2020s, and reaching almost 60 Mt CO_2-e in 2050.

The fluctuations in forestry emissions are due to assumptions regarding harvesting periods for timber plantations and the maturing of environmental plantations. Carbon plantations are assumed to reach maturity after 45 years, after which no further carbon removal occurs. After 2050 in the 550 standard technology scenario, net sequestration from forestry activities declines and approaches zero by 2100.

After 2050, few new plantations are established due to rising land prices and competition with higher-value agricultural uses. By the end of the century, just over half of the new land under forestry is dedicated to carbon plantings.

Far more land goes to forest sinks in the 450 standard technology scenario; this reflects the higher carbon price. In the 450 scenario, higher carbon prices are reached earlier in the century when land values are lower, so that forestry activities, especially carbon plantations, are more competitive.

In the Review's modelling, land use and land-use change emissions—for instance, a liability for landowners for emissions from clearance, or the opportunity costs of reduced clearance—were not included. Rather, land use and land-use change emissions are imposed in the models. Land use emissions for Australia largely represent emissions from clearing of regrowth as part of agricultural management—rather than clearing for new land. In the no-mitigation scenario, emissions from land clearing were assumed to remain at 44 Mt CO_2-e per year throughout the modelling period, based on a simple extrapolation from projections in the most recent national emission projections (DCC 2008c). Under the modelled policy scenarios, clearing emissions are assumed to decline in a linear fashion in response to the carbon price, to 28 Mt CO_2-e by 2050 and reaching zero by 2100.

22.2.5 Biofuels and bioenergy

The modelling exercise assumed that the emissions intensity of fuels such as petrol and diesel would decrease over time through an increase in the share of biofuels. However, the potential impacts of increased domestic demand for first-generation biofuels is not reflected in competition between different uses of land. Due to the difficulty in making predictions about second-generation biofuel technologies and costs, the modelling did not include any progress in these technologies under a carbon price. If domestic production of bioenergy were to increase, there could be greater competition for land that is currently assigned in the model to food or forestry production.

22.3 An alternative future

There are major opportunities to reduce emissions and increase greenhouse gas removal in the agriculture, forestry and other land use sectors. Not all of these are incorporated in the modelling results. Some combination of them could reduce radically the cost of mitigation in Australia and transform the economic prospects of rural Australia, especially of remote areas. Options include reducing emissions from major sources (sheep and cattle), and carbon dioxide removal in forests, other types of vegetation and soil. Producing biomass as a feedstock for biofuels and other forms of energy could also reduce emissions. These biosequestration activities appear to offer the largest emissions reduction potential.

These sectors could reduce emissions and exposure to an emissions price through other means too—improved management of manure, changed methods of rice cultivation and reduced fuel and electricity consumption are all promising options. However, because these options are likely to offer relatively small emissions reduction benefits, they are not considered in this chapter.

Land managers will choose among mitigation options depending on the nature of their land, the price and availability of water, carbon prices, and the development of new markets (for example, for biofuels). For some commodities, proximity to markets and commodity and input prices will also determine patterns of production. Estimates of some technical potential for emissions reduction and removal in the agriculture, forestry and other land use sector are summarised in Table 22.2 and Box 22.2. It is recognised that these potentials are calculated in a context of uncertainty and will in many cases not be easy to realise without substantial investments in proving and developing the systems. Further, since some of the identified processes overlap, their mitigation potential is not intended to be aggregated. Rather, they are listed to provide a broad sense of the mitigation possibility if policy, program and research efforts were more heavily focused on endeavours that recognised the integration of climate change mitigation with the management of agriculture, forests and other land use issues.

Table 22.2 Potential for emissions per annum reduction and/or removal from Australia's agriculture, forestry and other land use sectors

Process	Potential	Key assumptions
Land clearing (deforestation)	Emissions reduction potential of 63 Mt CO_2-e per year on an ongoing basis	Land clearing ceases (resulting in zero emissions from deforestation).
Enteric emissions from livestock	Emissions reduction estimated at 16 Mt CO_2-e per year on an ongoing basis	Based on either deployment of anti-methanogen technology for ruminant livestock, or shifting of meat production from a minority proportion (7 million cattle and 36 million sheep) of ruminant livestock by kangaroos.

Table 22.2 Potential for emissions per annum reduction and/or removal from Australia's agriculture, forestry and other land use sectors *(continued)*

Process	Potential	Key assumptions
Removal by soil—cropped land	Removal potential of 68 Mt CO_2-e per year for 20–50 years	Conservative management changes assumed (e.g. conservation tillage), not pasture cropping. Changed practices implemented on all cropped land (38 million ha).
Removal by soil—high-volume grazing land	Removal potential of 286 Mt CO_2-e per year for 20–50 years	Based on the Chicago Climate Exchange for changed practices to rehabilitate previously degraded rangelands. Changed practices implemented on all grazing land (358 million ha).
Restoration of mulga country	Up to 250 Mt CO_2-e per year for several decades	Comprehensive restoration of degraded, low-value grazing country in arid Australia.
Nitrous oxide emissions from soil	Reduction potential of 0.3 Mt CO_2-e per year for 20–50 years	Improved fertiliser management practised on all agricultural soils.
Reduction in emissions from savanna burning	Reduction of 5 Mt CO_2-e per year on an ongoing basis	Assumes annual emissions from savanna fires are 10 Mt (average from period 1990–2006). Complete reduction of savanna fire is not desirable or feasible; a 50% reduction is assumed, through management.
Removal by post-1990 forests	Emissions removal potential of 50 Mt CO_2-e by 2020	Assumes Australia will have around 2 million ha of Kyoto-compliant (post-1990) plantations (including wood production plantations and specific carbon plantations) by 2020.
Removal by pre-1990 eucalypt forests	Emissions removal potential equivalent to 136 Mt CO_2-e per year (on average) for 100 years	Current carbon stocks in logged forests are about 40% below carrying capacity. Timber harvesting and other human disturbances cease in study area (14.5 million ha). Landscape growth potential has not been degraded by land use activities.
Carbon farming (plantations)	Emissions removal potential of 143 Mt CO_2-e per year for 20 years	Using 9.1 million ha of land where returns would be more than $100 per ha per year better than current land use, with water interception less than 150 mm per year and permit price of $20 per tonne CO_2-e.
Biofuel production	Up to 44 Mt CO_2-e per year on an ongoing basis	Replacement of all fossil fuel diesel with biodiesel. More than 550 000 ha required for production (cultivating algae as a feedstock) or more than 10 million ha (using other plants).

Sources: Beeton (pers. comm.); Chicago Climate Exchange (2008); DCC (2008a); de Klein & Eckhard (2008); Grace et al. (2004); IPCC (2006); Mackey et al. (2008); NAFI & TPA (2007); Chan (unpublished); Polglase et al. (2008); Russell-Smith et al. (2004); Wilson & Edwards (2008).

22.3.1 Livestock production

In Australia, enteric fermentation emissions from livestock (mainly sheep and cattle) account for about 67 per cent of agricultural emissions (DCC 2008b). Cattle and sheep production also accounts for a significant proportion of emissions from agricultural soils, and beef production is responsible for some emissions from savanna fire and land clearing. Agricultural emissions, allocated by subsector and not including emissions due to land clearing, are presented in Figure 22.5.

Figure 22.5 Contribution to Australia's agricultural emissions, by subsector, 2005

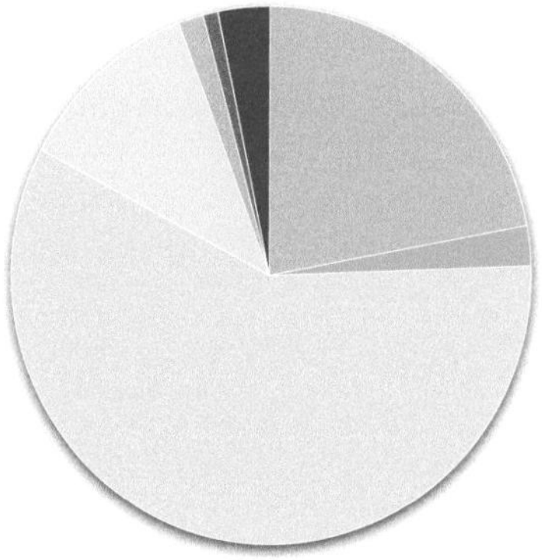

Sheep – 22.0%
Grains – 2.5%
Beef cattle – 58.1%
Dairy cattle – 11.6%
Pigs – 1.8%
Poultry – 1.0%
Other – 3.2%

Source: DCC (2008d).

Figure 22.6 shows the ratio of total emissions permit costs to the value of production by subsector in the agricultural industry for a range of permit prices. Clearly, among the agricultural subsectors, a carbon price will affect cattle and sheep producers most heavily.

Over time, increasing permit prices will encourage reduced use of energy and emissions-intensive inputs and drive mitigation of livestock emissions. Current options for the mitigation of methane emissions include:[1]

- **Practices to increase productivity**—Breeding pattern manipulation, better location of watering points and greater use of products that promote growth can all increase productivity without increasing food consumption and resultant emissions (Eckhard 2008; Howden & Reyenga 1999). These activities are already widely practised.
- **Nutritional management**—The addition of monesin, dietary fats and lipids can reduce ruminant emissions by 20 to 40 per cent (Beauchemin et al. 2008; Howden & Reyenga 1999). Nitrous oxide emissions from livestock can also be reduced through dietary changes (Miller et al. 2001; van Groenigen et al. 2005). These options are technically feasible, but are generally not yet cost effective.
- **Vaccination, biocontrols and chemical inhibitors**—Trials of immunisation on sheep found methane emissions reductions of almost 8 per cent (Wright et al. 2004). This is a longer-term category—it may be decades before this is feasible on a large scale.

Figure 22.6 Ratio of emissions permit costs to value of production, by subsector, 2005

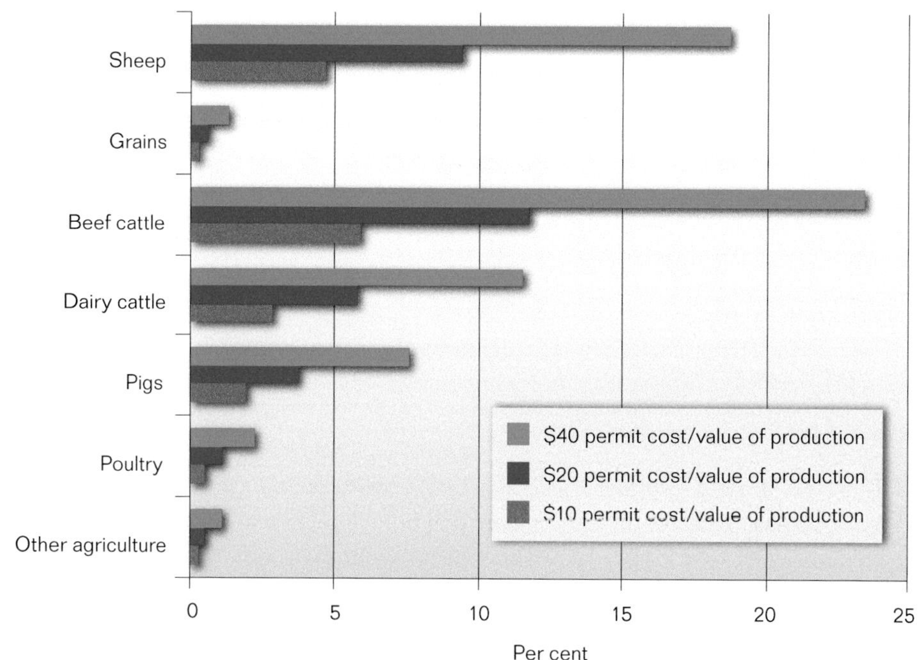

Note: Emissions due to land clearing have not been attributed to the agricultural subsectors responsible. Includes indirect emissions from purchase of electricity, does not include transport.

Source: DCC (2008d); ABS (2008).

The effectiveness of these mitigation options may be limited and research and development is likely to continue for some time.[2] To the extent that there were no cost-effective mitigation options, under an emissions trading scheme methane-emitting livestock producers would be required to purchase permits for their emissions, which would lead to an increase in costs of sheep and cattle production. In the short to medium term, the impact on meat prices and consumption may not be large, given that global demand is expected to remain strong and permit prices will be a relatively small component of the cost of animal products (see Table 22.3).

According to the Australian Bureau of Agricultural and Resource Economics (ABARE) (2008b), since 1960:

- real retail prices for beef have remained roughly steady, as has per capita consumption
- lamb prices have increased by about 30 per cent and per capita consumption has fallen by 70 per cent
- pork prices have fallen by about 10 per cent and consumption has nearly tripled
- poultry meat prices have fallen by more than 75 per cent and there has been an almost ninefold increase in per capita consumption.

Table 22.3 Impact of emissions permit prices on cost of meat production

Commodity	Kg CO_2-e emitted per kg of produce[a]	Cost increase at $40/t permit price ($/kg)	2006 retail prices ($/kg)	Price increase at $40/t permit price (%)
Lamb & mutton[b]	16.8	0.67	12.20	5.5
Beef & veal	24.0	0.96	15.38	6.2
Pork	4.1	0.16	11.87	1.3
Poultry meat[c]	0.8	0.03	3.16	0.9

a This does not take into account any emissions resulting from deforestation, which are largely attributable to the beef cattle industry.

b Emissions from sheep production were allocated between sheep meat and wool in proportion to the gross value added by each commodity.

c Emissions from poultry production were allocated between poultry meat and eggs in proportion to the gross value added by each commodity.

The ABARE data suggest that, over time, consumption patterns in Australian households are highly responsive to changes both in price and conditions of supply (see figures 22.7 and 22.8). Australian consumer preferences have changed over time, and will continue to change into the future.

Figure 22.7 Australian real retail prices for meat, 1960–2006

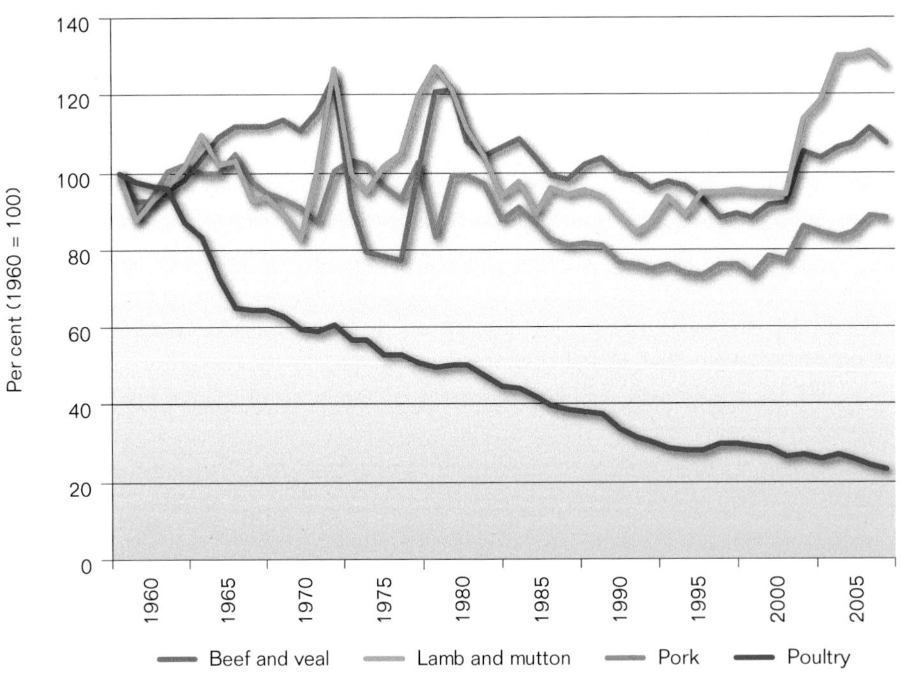

Source: ABARE (2008b).

Figure 22.8 Australian per capita consumption of meat, 1960–2006

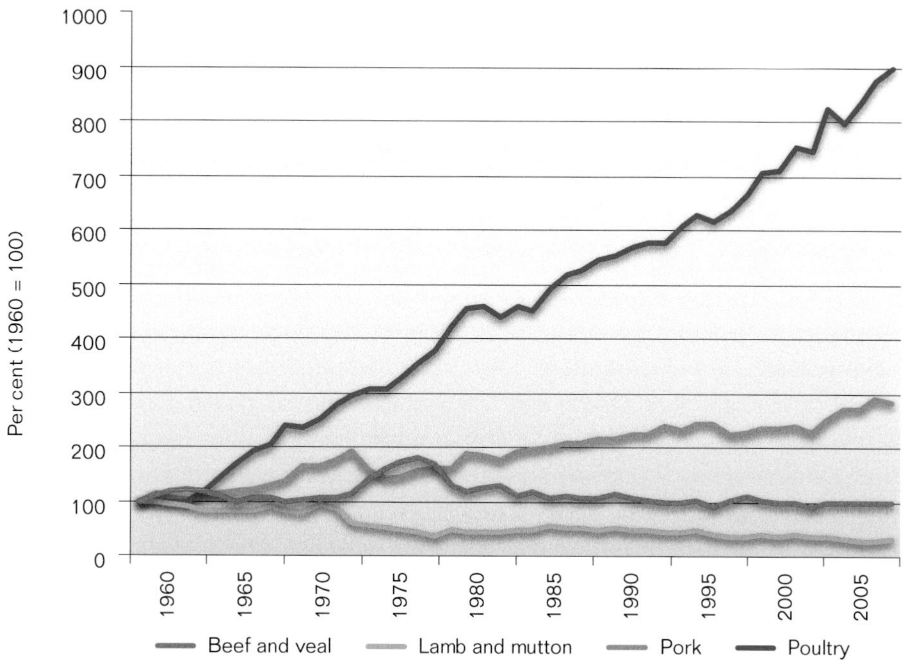

Source: ABARE (2008b).

As permit prices increase, higher prices for some meats are likely to lead to further changes in consumption patterns.

Sheep and cattle production is highly vulnerable to the biophysical impacts of climate change, such as water scarcity (see Table 22.1). This factor, combined with increased costs for methane emissions, could hasten a transition toward greater production and consumption of lower-emissions forms of meat, such as chicken, fish and pork. Demand for these products is projected to remain strong.

Australian marsupials emit negligible amounts of methane from enteric fermentation (Klieve & Ouwerkerk 2007). This could be a source of international comparative advantage for Australia in livestock production. For most of Australia's human history—around 60 000 years—kangaroo was the main source of meat.[3] It could again become important. However, there are some significant barriers to this change, including livestock and farm management issues, consumer resistance and the gradual nature of change in food tastes.

Edwards and Wilson (2008) have modelled the potential for kangaroos to replace sheep and cattle for meat production in Australia's rangelands, where kangaroos are already harvested. They conclude that by 2020 beef cattle and sheep numbers in the rangelands could be reduced by 7 million and 36 million respectively, and that this would create the opportunity for an increase in kangaroo numbers from 34 million today to 240 million by 2020. They estimate that meat production from 175 million kangaroos would be sufficient to replace the forgone lamb and beef meat production, and that meat production from kangaroos would become more profitable than cattle and sheep when emissions permit prices exceed $40 per tonne CO_2-e. The net reduction in greenhouse gas emissions would be about 16 Mt CO_2-e per year.

22.3.2 Soil management

Carbon dioxide removal by soil

Atmospheric carbon dioxide is removed from the atmosphere by plants and transferred to soil through active plant roots or the decomposition of plant and animal matter.

Soil carbon is both a source and a sink of greenhouse gases. Soil carbon can be restored and increased through active management of the biological system. It can be affected by employing conservation tillage; increasing the use of mulch, compost and manure; and changing the vegetation cover on soil. Soil carbon can be built with the use of further soil additives, including calcium-bearing silicates (Engineering and Physical Sciences Research Council 2008) and biochar (see section 22.3.4). Tests are now being conducted using lignite as a catalyst for accumulation of soil carbon (LawrieCo 2008).

Soil carbon can be lost—for example, as a result of land clearing, erosion or drought (Lal 2004). Soil carbon built up by conventional cropping with reduced tillage (such as 'zero-till' methods) may only affect soil close to the surface, and is often returned to the atmosphere within months (J. Baldock 2008, pers. comm.; Lal 2004; Chan unpublished). By contrast, carbon dioxide removed by actively growing roots of living plants and stored in soil humus can provide long-term storage. Increased soil microbe activity associated with increased vegetation is essential for soil carbon sequestration. This promotes plant availability of soil minerals and other nutrients, improves soil structure and humus content, increases water retention and increases oxygen respiration to the atmosphere (Jones 2008; Parr & Sullivan 2005; Post & Kwon 2000).

All things being equal, the potential for removal of carbon by soil is in the following order (least to greatest): degraded soils and desertified ecosystems, cropland, grazing lands, forest lands[4] and permanent crops. Some estimates are available for the removal potential of soil (Table 22.4).

There are other benefits from building soil carbon. It increases oxygen and retention of moisture when combined with other nutrients and minerals, leading to improved soil health (Grace et al. 2004; Jones 2007; Lal 2007; Wentworth Group of Concerned Scientists 2008). As a result, a number of Australian land managers are already making on-farm changes to build soil carbon (Jones et al. 2008). Australia is well positioned to further increase carbon dioxide removal by soil, due to the sheer size of its land mass and the ability of its farming sector to adopt new management practices.

Table 22.4 Technical potential for CO_2 removal by soil—selected estimates

Activity	Location	Carbon dioxide removal estimate[a]	Source
Conservation and sustainable land management practices	Worldwide	3.3 Gt (± 1.1 Gt) CO_2 per year (for 50 years)	Lal (2004)
		1.6–3.2 Gt CO_2 per year (for 50 years)	Paustian et al. (2004)
Adoption of sustainable stocking rates, rotational grazing and seasonal use grazing practices	United States rangelands (previously degraded)	0.3–1.3 t CO_2 per ha per year	Chicago Climate Exchange (2008)
Zero till or minimum tillage	United States	0.5–1.5 t CO_2 per ha per year	Chicago Climate Exchange (2008)
Conservation tillage	Australia-wide	25% increase in carbon retained, compared to conventional tillage	Valzano et al. (2005)
Conservation tillage	South-eastern Australia	43.3-46.6 t CO_2 per ha over 20 years in high rainfall region (on average between 2.1 and 2.3 t per year)	Grace et al. (2004)
Changes in cropping practice	New South Wales	2 t CO_2 per ha per year	Chan (unpublished)
Sowing of crops into perennial pastures (growth of perennial grasses alongside crops)	Trial sites in New South Wales with good soils	5-10 t CO_2 per ha per year (up to 20–30 t under good conditions)	C. Jones (2008, pers. comm.)
Pasture cropping (pasture type: Rhodes, Lucerne, Siratro, Bambatsi)	Loam soil, 'northern agriculture region', Western Australia	5.2 t CO_2 per ha per year	T. Wiley (2008, pers. comm.)

a Soil carbon will eventually reach a new equilibrium, and carbon removals in soil will cease. Removals may continue for 20–50 years before a new equilibrium is reached.

Notes: Technical potential refers to what is physically possible and does not take into account the influence of emissions reporting requirements or cost effectiveness. Some units have been converted from original source data.

A range of biophysical, economic and social constraints must be overcome in order for this potential to be realised on a large scale, although it is already technically feasible. To be pursued on an optimal scale, carbon removal by soil would require recognition under an emissions trading scheme, a sufficient carbon price and potentially the assurance that those undertaking it will not be held liable for soil carbon emissions that result from non-anthropogenic release.[5]

The barriers to recognising carbon dioxide removal by soil could be overcome within decades, presenting soil carbon as a new commodity for landowners. Though the potential is not as great as in high-quality soil, removal by soil may offer an alternative to other forms of biosequestration in areas of low rainfall or scarce water supply.

Nitrous oxide emissions

Nitrous oxide emissions that result from soil management can be reduced through currently feasible activities—fertiliser management, soil and water management, and fertiliser additives (de Klein & Eckhard 2008). These mitigation activities can significantly reduce costs. Organic additives are low-emissions alternatives to conventional fertiliser that are already available. Further research and development may help to identify new biological products that are appropriate for fertiliser production, and could also improve the efficiency of chemical fertilisers (Hargrove 2008).

Nitrification inhibitors on fertiliser have been shown to reduce nitrous oxide emissions by up to around 80 per cent (de Klein & Eckhard 2008). Nitrification inhibitors for livestock also have potential, although data are limited (de Klein & Eckhard 2008; Whitehead 2008).

Building soil carbon may have implications for nitrous oxide and other emissions, for example increases that may arise from chemical fertilisers (Changsheng et al. 2005; Grace et al. 2004). There needs to be a comprehensive and robust carbon market, and a market for other environmental externalities (such as forms of pollution), to ensure sustainable decision making and to avoid suboptimal outcomes.

22.3.3 Plantations and production forests

Forests and plantations established after 1990 already contribute to reportable mitigation of Australia's emissions, consistent with the provisions of the Kyoto Protocol. In 2006, afforestation and reforestation accounted for net removal of about 23 Mt CO_2 from the atmosphere. This could increase to about 50 Mt CO_2 per year by 2020 (NAFI & TPA 2007).

The profitability of harvested forestry systems can be improved if carbon is included as an additional, saleable product. Analysis by Polglase et al. (2008) concludes that carbon payments could increase the profitability of hardwood and softwood sawlog systems, but not of pulpwood. Carbon revenue has a lower impact upon pulpwood production because rotation periods are relatively short and these systems have less opportunity to store carbon compared with longer rotation sawlog systems.

There will be significant financial opportunities for landholders who intend to maintain permanent forest cover. However, participation in the carbon market will also carry risks, especially for landholders who intend to change from forestry to another land use and, to a lesser degree, for those who intend to harvest their forests. Permits or credits generated as a growing forest removes carbon dioxide from the atmosphere will need to be surrendered when the forest is harvested.

There is scope to reduce the carbon liability incurred when trees are harvested if inventories, and the emissions trading scheme, recognise carbon stored in harvested wood products. The provisions of the Kyoto Protocol do not account for carbon in harvested wood products. However, the 2006 IPCC Inventory Guidelines provide detailed guidance on how to estimate the contribution of harvested wood products to emissions and removals. The approach requires estimation of emissions from the decay of all wood products in the 'products in use' pool and would be likely to result in an increase in Australia's reported greenhouse gas emissions (G. Richards 2008, pers. comm.).

There are flaws in the approach. This is an important issue that warrants further analysis and then international discussion. The objective should be to credit genuine, multiyear sequestration of carbon in harvested wood products.

A large switch in land use toward production forestry would have additional consequences that might be negative (such as impacts on water supply) or positive (for example, mitigating dryland salinity and assisting with habitat restoration), depending on the type of forestry and the land use it replaces. These externalities should be addressed through the creation of market-based instruments for other ecosystem services, such as water quantity and quality, biodiversity, air filtration, and abatement of salinity and erosion.

22.3.4 Biofuels

If a biofuel is to have environmental and economical value it must be produced sustainably and contain more energy than was used to produce it. The net reduction in emissions must be secured at a cost that is competitive with alternative mitigation opportunities. Perverse incentives allow production of some biofuels that do not meet these criteria (Oxburgh 2008). A poorly conceived biofuel production process could:

- produce less liquid fuel energy than is used in production
- over the life cycle of the process, emit the same quantity of greenhouse gases per unit of liquid fuel energy as fossil fuels
- place upward pressure on food prices through competition with food production for arable land.

Subsidies and mandated targets for biofuels distort the market. The correct way to support mitigation through biofuels involves placing a price on all greenhouse gases arising from the production process and the combustion of the biofuel. This is achieved through including inputs into and the use of biofuels comprehensively in the emissions trading scheme.

Global production of biofuels in 2005 amounted to roughly 1 per cent of total road transport fuel consumption (Doornbosch & Steenblik 2007). Satisfying the global demand for liquid fuels with current (first-generation) biofuel technologies would require about three-quarters of the world's agricultural land (Oxburgh 2008). First-generation biofuels can therefore never amount to more than a minor supplement. In the future, second-generation biofuels, using resources that are not applied to food production, will be valuable.

Box 22.1 Biofuel production methods

Biofuels are produced in three main ways: through fermentation (ethanol), extraction and chemical processing of oils (biodiesel), and gasification (syngas and biochar). Biodiesel has several advantages over ethanol, among them that it requires less energy for production (Durrett et al. 2008).

Ethanol accounts for more than 90 per cent of current global biofuel usage. It is produced by fermentation of material rich in sugar and starch, such as sugar cane, corn, sugar beets, potatoes, sorghum and cassava. Compared with gasoline, use of ethanol from Brazilian sugar cane is estimated to reduce emissions by 90 per cent, while use of corn ethanol yields an estimated emissions reduction of only 15 to 25 per cent (IEA 2007).

Biodiesel can be produced from vegetable oils, used cooking oils and animal fats. Oily seeds that can be used for biodiesel include palm oil, rape (canola), soy and sunflower. Oxburgh (2008) has noted that *jatropha curcas* can be used as a feedstock for biodiesel and is cultivated on marginal land in Southern Africa, India and Southeast Asia. An assessment of the costs and benefits of cultivating *jatropha curcas* in arid regions of Australia unsuitable for food production is warranted. Considerations would need to include the degree to which it presents a pest risk to Australian native fauna or habitat. In some parts of Australia *jatropha curcas* is a declared noxious weed and growing or importing it is illegal (Western Australia Department of Agriculture and Food 2007).

Gasification involves conversion of biomass into 'syngas'—a mixture of hydrogen, carbon monoxide and other gases. Syngas can be converted directly to electricity, hydrogen or other chemicals, including liquid fuels. The biochar left over after gasification is high in carbon. Biochar degrades very slowly and has been proposed for use as a fertiliser and to build soil carbon (J. Baldock 2008, pers. comm.; Johnson et al. 2007).

Biofuels can be produced using second-generation technologies from waste biomass, lignocellulosic materials or algae. Australian native trees offer a wide range of possibilities. Mallee eucalypts, for example, can be grown on marginal arid and semi-arid lands, including land that seems to be in the process of conversion out of wheat growing by the warming and drying of southeast Australia. The mallee green top can be harvested perennially as a biofuel feedstock. The growing of mallee can lessen dryland salinity and assist in habitat restoration without competing directly with fibre production from the forestry sector. Mallee also contributes directly to mitigation through storage of carbon in its massive root system.

Biofuel production using algae can be concentrated in terms of land use (see Table 22.5). Its essential requirement is energy from sunshine. Algae can absorb carbon dioxide from the atmosphere and thrive on concentrations of the gas from combustion wastes. They do particularly well in saline environments, which are abundant in Australia and have no alternative commercial uses.

Trials of the production of second-generation biofuels are already proceeding, although there are as yet no full-scale production facilities in operation. It could qualify for commercialisation support under the innovation proposals of Chapter 18, and will be encouraged by a rising carbon price. Commercial production of second-generation biofuels could reasonably be anticipated before 2020.

Table 22.5 Estimated oil yield per ha for biodiesel production

Biofuel feedstock	Estimated oil yield (litres per ha per annum)[a]
Cottonseed	200–400
Soybean	400–600
Sunflower	900–1100
Groundnut/peanut	1000–1200
Canola/rape	1100–1300
Jatropha curcas	1200–1400
Coconut	2200–2400
Palm	2400–2600
Algae	> 30 000[b]

a Yield numbers are subject to considerable geographic and temporal variation. In the cases of palm and coconut, yield numbers represent production from mature plants and do not reflect periods of lower production during plantation establishment.

b Theoretical calculation based on photosynthetic efficiency and growth potential.

Sources: Durrett et al. (2008); Hu et al. (2008).

22.3.5 Other forms of bioenergy

Biomass can be converted to other forms of energy, such as heat and electricity. Biomass could be the basis for 'negative emissions' energy if it is coupled with carbon capture and storage or secure storage of biochar. While biomass offers the only promising way of making clean liquid fuels for vehicles, there are other ways of generating electricity cleanly, so that biofuel is likely to be the early target of commercialisation (Oxburgh 2008).

Polglase et al. (2008) assessed the potential economic outcomes and environmental impacts across Australia of agroforestry for dedicated bioenergy and integrated tree processing (that is, integrated production of bioenergy, activated carbon and eucalyptus oil), based on various species of mallee and other eucalypts. They conclude that dedicated bioenergy and integrated tree processing systems are unlikely to be profitable unless they are close to processing facilities. This is due to the high cost of production (harvesting and transport) relative to the low product price for wood energy. Lehmann (2007) suggests that, in the United States, biochar production in conjunction with bioenergy from pyrolysis could become economically attractive at an emissions permit price above US$37 per tonne.

22.3.6 Environmental carbon plantings

Australia has large areas of land, much of which would be suitable for carbon plantings and revegetation (see Table 22.6).

Table 22.6 Area of selected land uses in Australia[a]

Land use	Area (million ha)
Agricultural land	425
Grazing land	358
Cropping land	38
Other agricultural purposes	2
Agriculturally unproductive	28
Forests and wooded lands	571
Plantation forests	2
Native forests	147
Wooded lands	422
Savanna[b]	190
Settlements	3

a Data are taken from a number of sources and some categories of land use classifications are overlapping (e.g. between wooded lands and grazing land, and also between those categories and savanna).

b Savanna is defined as biogeographic regions in northern Australia that are dominated by a wet–dry climate and have landscapes dominated by grasslands or woodlands, not forests (Tropical Savanna Cooperative Research Centre 2008).

Sources: Gavran & Parsons (2008); DAFF (2008); ABS (2008); FAO (2008); Savanna Explorer (2008).

There are about 28.8 ha of forest and wooded land for every person in Australia (FAO 2008). This is the largest area of forest and wooded land per person among OECD countries and the second largest globally (see Figure 22.9).

In addition to land area, the amount of additional carbon dioxide that can be removed from the atmosphere by existing forests and woodlands and through revegetation of cleared lands is determined by the local climate, the fertility of the substrate, the characteristics of the plant species and the impact of land use history in reducing carbon stocks below the land's carrying capacity.

Polglase et al. (2008) have undertaken extensive analysis of the opportunities for carbon farming across Australia, taking account of climatic and soil suitability, species characteristics, the likely profitability of carbon farming compared with current land use and the potential impact on rainfall interception and biodiversity benefits. They modelled environmental plantings of mixed species with an open woodland structure, as well as monocultures of eucalypts and pines. Taking account of climatic and soil suitability, they find that there is about 200 million ha of land suitable for carbon plantings (see Figure 22.10) with potential revenue of up to $40 billion per year. This does not suggest that all land will be planted, rather it will depend on land availability, social attitudes, investment and farming models, and intersection with other policy (for example, on water and planning).

Figure 22.9 Per capita area of forested and wooded land, 2005

Australia
OECD average
World average
0 5 10 15 20 25 30 35
Ha per person

Source: FAO (2008).

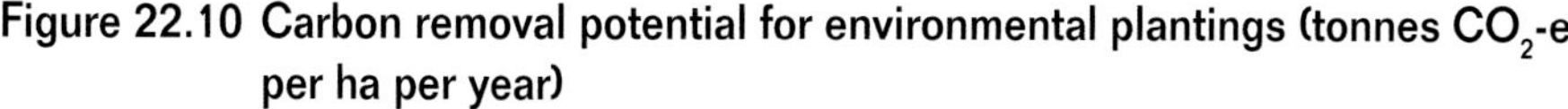

Figure 22.10 Carbon removal potential for environmental plantings (tonnes CO_2-e per ha per year)

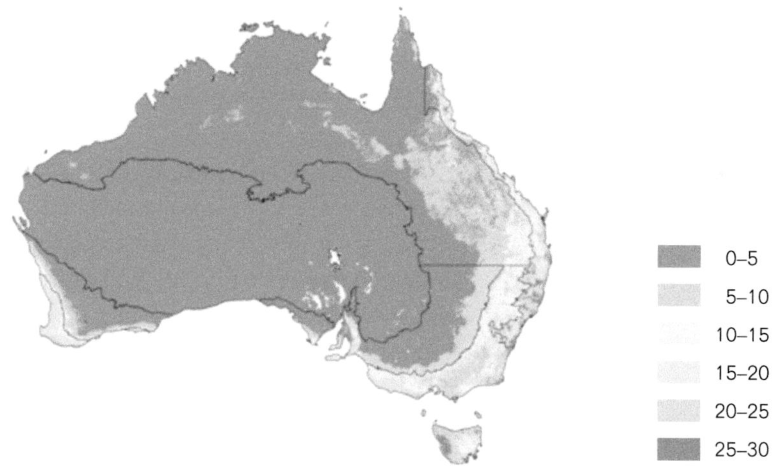

Source: Polglase et al. (2008).

Opportunities for profitable carbon farming have no harvesting or transportation costs and location is not constrained by proximity to processing facilities. Moreover, the carbon payment can be an annuity and financial returns are not delayed by years or decades until trees are harvested.

Polglase et al. (2008) found that carbon farming in Australia could remove about 143 Mt of carbon dioxide from the atmosphere annually, based on the following assumptions:

- There is a carbon price of $20 per tonne CO_2-e.
- Carbon farming would take place in areas where it would generate sales revenue of at least $100 per ha per year more than current land use.
- Carbon plantings would be restricted to areas where rainfall interception would be less than 150 mm per year.

A total of 9.1 million ha were identified as being suitable for carbon farming under these criteria and, compared with current land uses, the additional revenue that would be realised is estimated at $1.9 billion per year. The most significant opportunities are in south-eastern Australia (west of the Great Dividing Range and extending through Victoria and New South Wales up to the Queensland border), southern and south-eastern South Australia and parts of Tasmania and south-west Western Australia. (Polglase et al. note that their results are based on a particular set of assumptions in relation to variables that will differ across regions.)

22.3.7 Conservation forests

As provisions for carbon accounting become more comprehensive, carbon dioxide removal from the atmosphere could be a substantial new source of revenue for managers of national parks and forests set aside for conservation. However participation in carbon markets would also entail risks. Liability for emissions arising from fires would be the most significant risk, and would require management responses. Some but not most forests are already close to their carbon carrying capacity.

The IPCC default values for temperate forests are a carbon stock of 217 tonnes carbon per ha and net primary productivity of 7 tonnes of carbon per ha per year (IPCC 2000). However these IPCC estimates may be conservative, particularly for intact forests. Mackey et al. (2008) have shown that the stock of carbon for intact natural forests in south-eastern Australia is about 640 tonnes per ha, with an average net primary productivity of 12 tonnes per ha per year. They estimate that the eucalypt forests of south-eastern Australia could remove about 136 Mt CO_2-e per year (on average) for the next 100 years.

22.3.8 Savanna

About a quarter of Australia is covered by savanna woodlands and grasslands, and much of this land is owned and managed by Indigenous Australians (Tropical Savanna Cooperative Research Centre 2008). The upgrading of savanna management has substantial mitigation potential, and would also have positive effects for biodiversity conservation and for Indigenous land managers.

Under its implementation of the Kyoto Protocol, Australia can only account for carbon dioxide removed on land that was cleared at 1 January 1990, and most savanna areas do not satisfy this provision. Future carbon accounting provisions should include all greenhouse gases removed by and emitted from managed lands.

This would provide significant revenue opportunities for land managers. It would also require the management of risks, especially if liability for emissions resulting from non-anthropogenic activities—such as fire and the effects of drought—were brought to account.

Savanna fires are the principal source of greenhouse gas emissions in the Northern Territory, and a significant source of Australia's agricultural emissions. Ignitions of savanna fires are frequently anthropogenic (Russell-Smith et al. 2004). Reducing savanna fires can significantly increase biosequestration and protect carbon stored in vegetation sinks. Actions to reduce the area burnt include seasonally targeted management strategies such as fire breaks, early and seasonal burning, and fuel reduction burns. For example, the West Arnhem Land Fire Abatement project, which applies traditional Indigenous burning practices to 28 000 km^2, has reduced annual emissions by an average of 145 000 tonnes CO_2-e over the three years of the project, at a cost of around $15 per tonne (excluding the cost of establishing the project) (Tropical Savanna Cooperative Research Centre 2008; Whitehead et al. in press).

Box 22.2 The potential for biosequestration in arid Australia

Australia has the largest area of woodlands and forest per capita among OECD countries. These vast areas have a large and varied potential for biosequestration. To date, the focus of Australia's biosequestration efforts has been on plantation forests, encouraged, and distorted, by transportation of the limited Kyoto rules into various Australian arrangements. However, Australia enjoys a potential international comparative advantage in carbon trading through its large areas of marginal pastoral country.

Arid and semi-arid rangelands currently make up about 70 per cent of Australia's land mass, or around 5.5 million km^2. Eighteen per cent of this area consists of chenopod shrublands, native tussock grasslands, and woodlands and shrublands that are dominated by mulga (*Acacia aneura*) in eastern Australia, within the 200 to 500 mm annual rainfall zone.

Considerable degradation of these rangelands has been caused by marginal sheep and cattle grazing. It is estimated that these rangelands could absorb at least half of Australia's current annual emissions or some 250 Mt for several decades. A carbon price of $20 per tonne would provide up to a tenfold increase in income for property holders in this region if current practices were replaced by land restoration through a strategic property management program. The mitigation gains are potentially so large that it is important for Australia to commence work on program design and implementation even before the issues of coverage, national and international, are fully resolved.[6]

22.4 Barriers and limits to a low-emissions future

22.4.1 Ability to estimate or measure emissions and removals

More reliable and cost-effective ways to measure or estimate net emissions are needed in the land use sector. Without reliable estimation, it is difficult to include the sector in an emissions trading scheme.

Estimation of emissions and removal by soils is particularly difficult. There are models—such as the Rothamsted soil carbon (RothC) and GRC-3 (DCC 2008a)—but actual samples often provide different results. Soil carbon is characterised by spatial, seasonal and annual variation. Sampling is intensive and costly, and data are limited. Emissions estimation is also difficult for nitrous oxide and native forests.

Resources should be directed, as a priority over the next few years, to overcoming gaps in emissions data and measurement issues for the agriculture, forestry and other land use sectors, in order to include all of the sector's emissions in accounting and potentially in an emissions trading scheme. In addition, training will be needed to ensure that Australia has the skills needed for monitoring and verification.

The same issues arise in relation to Australia's developing country neighbours. Australia has been helpful in sharing knowledge of carbon measurement techniques in Papua New Guinea, Indonesia and elsewhere. Extending this work can be a large Australian contribution to the global mitigation effort.

22.4.2 Emissions accounting rules

If emissions removal processes are not recognised in accounting protocols, they cannot assist in meeting emissions obligations—which reduces the incentive to pursue them.

Accounting should be as broad as possible in the land use sector, particularly if it is included in an emissions trading scheme. This would minimise the likelihood of perverse incentives. With incomplete coverage (for example, exclusion of emissions from deforestation), a carbon price could provide a financial incentive to clear land for biosequestration or bioenergy even though this could result in a net increase in emissions.

The accounting framework for Australia's emissions under the Kyoto Protocol is not comprehensive. The dampening effect this has on the take-up of biosequestration was evident in the Review's modelling results, which assumed continuation of existing emissions accounting rules. As new global emissions accounting methods are developed, alternative technologies and forms of biosequestration should be considered. Australia should advocate movement towards comprehensive monitoring, reporting and recognition of emissions from land use.

It is also important that Australia take full advantage of whatever international accounting rules are in place. The Marrakesh Accords (UNFCCC 2002) of the

Kyoto Protocol determined that any Annex I party (such as Australia), in addition to claiming emission reductions from afforestation, reforestation and deforestation (under Article 3.3 of the Protocol), could (under Article 3.4 of the Protocol) 'choose to account for anthropogenic greenhouse gas emissions by sources and removals by sinks … resulting from … revegetation, forest management, cropland management, and grazing land management'. Australia has opted not to account for emissions in these areas. While there are valid concerns about the impact of bushfires on emissions, and these need to be addressed, it is in Australia's interests to implement as wide-ranging a definition of human-induced greenhouse gas emissions as possible.

It will be important to account for all emissions—including those caused by natural processes—although it may not be appropriate to include all emissions in an emissions trading scheme. Clear rules will be needed about how non-anthropogenic emissions, such as those caused by drought and fire, might be managed.

The potential contribution of biosequestration, much of it at relatively low cost, to the mitigation task is immense. This is true for the world, and particularly true for Australia. Comprehensive emissions accounting as a basis for the emissions trading scheme's application to agriculture, forestry and related sectors could meet a major part of Australia's mitigation effort. The exclusion of comprehensive accounting from the modelling of the Review's costs of mitigation is a large source of conservatism in the estimates of the costs of mitigation in Australia.

22.4.3 The high cost of mitigation in the agriculture sector

The agriculture sector lacks cost-effective mitigation options for some major sources of emissions. Most methane abatement options are costly—for example, reducing emissions through nutritional management can result in higher costs for animal feed, farm labour and animal health. The biosequestration options, if properly accounted and rewarded, are quantitatively much more important in the mitigation story.

Nevertheless, there are promising research avenues for reductions in agricultural emissions. Large-scale, and widely shared, public good research in this area is warranted.

The transaction costs of full inclusion of agriculture in an emissions trading scheme would be high. There are around 130 000 agricultural establishments in Australia (ABS 2007), each with a diverse emissions profile. Inclusion of agriculture in an emissions trading scheme will involve a trade-off between accuracy and cost. Both will be significantly influenced by the threshold set for coverage and the point of obligation. There will be a large role for collective action among farmers, or private broking functions, to reduce the costs of individual farmers' interaction with an emissions trading scheme.

Notes

1 As discussed in section 22.2, not all of these mitigation options were assumed to be available in the modelling.

2 First, there are doubts about whether reductions from nutritional management and biological treatments persist over time (Guan et al. 2006; McAllister & Newbold 2008). Second, assuming these interventions reduce methane emissions, these savings could be offset or even exceeded by the indirect emissions involved in implementing the mitigation measure—for example, the cultivation and transport of high-quality feed (Beauchemin et al. 2008; Howden & Reyenga 1999).

3 Kangaroo meat is currently used for human consumption and pet food. By volume, total kangaroo meat production declined by about 22 per cent between 2002 and 2007. This was due to a reduction in harvest quotas necessitated by a decline in the kangaroo population under severe drought conditions. Over the same period the proportion of kangaroo meat used for human consumption increased from about 40 per cent to about 60 per cent, resulting in an increase in the value of kangaroo meat production of about 15 per cent in real terms (ABARE 2008b).

4 Australian research has found that converting land to forest (afforestation or reforestation) can have variable outcomes in terms of soil carbon (Guo & Gifford 2002; Paul et al. 2002).

5 Soil emissions resulting from drought or fire would be potentially large. It may be appropriate to consider rewarding and penalising landowners only for changes in emissions that result from anthropogenic activities (for example, changed practices). Government may also seek to avoid such liability and be reluctant to agree to reporting arrangements such as those set out under Article 3.4 of the Kyoto Protocol.

6 Based on information supplied by Professor Bob Beeton, Chair of the Commonwealth State of the Environment Committee.

References

ABARE (Australian Bureau of Agricultural and Resource Economics) 2006, *Agricultural Economies of Australia and New Zealand*, <www.abareconomics.com/interactive/ausNZ_ag/htm/au_change.htm>.

ABARE 2008a, *Australian Commodity Statistics 2008*, ABARE, Canberra.

ABARE 2008b, unpublished data provided to the Garnaut Climate Change Review.

ABS (Australian Bureau of Statistics) 2007, *Australian Farming in Brief*, cat. no. 7106.0, ABS, Canberra.

ABS 2008, *Natural Resource Management in Australian Farms, 2006–07*, cat. no. 4620.0, ABS, Canberra.

Beauchemin, K.A., Kreuzer, M., O'Mara, F. & McAllister, T.A. 2008, 'Nutritional management for enteric methane abatement: a review', *Australian Journal of Experimental Agriculture* 48: 21–27.

Chan, K.Y. unpublished, 'Scoping paper: soil organic carbon sequestration potential for agriculture in NSW', Department of Primary Industries, Richmond, New South Wales.

Changsheng, L., Frolking, S. & Butterback-Bahl, K. 2005, 'Carbon sequestration in arable soils is likely to increase nitrous oxide emissions, offsetting reductions in climate radiative forcing', *Climatic Change* 72(3): 321–38.

Chicago Climate Exchange 2008, 'CCX Offsets Program', <www.chicagoclimatex.com/content.jsf?id=23>.

CSIRO (Commonwealth Scientific and Industrial Research Organisation) 2004

DAFF (Department of Agriculture, Fisheries and Forestry) 2008, *Australia's State of the Forests Report*, Commonwealth of Australia, Canberra.

DCC (Department of Climate Change) 2008a, 'Soil carbon measurement and modelling', <www.climatechange.gov.au/ncas/activities/soils.html>.

DCC 2008b, *Agriculture Sector Greenhouse Gas Emissions Projections 2007*, DCC, Canberra.

DCC 2008c, *Tracking to the Kyoto Target 2007: Australia's greenhouse emissions trends 1990 to 2008–2012 and 2020*, DCC, Canberra.

DCC 2006d, unpublished data provided to the Garnaut Climate Change Review.

de Klein, C.A.M. & Eckard, R.J. 2008, 'Targeted technologies for nitrous oxide abatement from animal agriculture', *Australian Journal of Experimental Agriculture* 48: 14–20.

Doornbosch, R. & Steenblik, R. 2007, 'Biofuels: is the cure worse than the disease?' document prepared for the OECD Round Table on Sustainable Development, OECD, Paris, 11–12 September.

Durrett, T., Benning, C. & Ohlrogge, J. 2008, 'Plant triacylglycerols as feedstocks for the production of biofuels', *The Plant Journal* 54: 593–607.

Eckhard, R. 2008, 'Greenhouse emissions from livestock and fertiliser and implications for a National Emissions Trading Scheme', *Proceedings of the Agriculture, Greenhouse and Emissions Trading Summit*, Australian Farm Institute.

Engineering and Physical Sciences Research Council 2008, 'Specially designed soils could help combat climate change', media release, 31 March, United Kingdom.

FAO (Food and Agriculture Organization of the United Nations) 2007, *Adaptation to Climate Change in Agriculture, Forestry and Fisheries: Perspective, framework and priorities*, <ftp://ftp.fao.org/docrep/fao/009/j9271e/j9271e.pdf>.

FAO 2008, FORIS database, <www.fao.org/forestry/32185/en>.

Gavran, M. & Parsons, M. 2008, *Australia's Plantations 2008 Inventory Update*, Bureau of Rural Sciences, Canberra.

Grace, P.R., Antle, J., Aggarwal, P.K., Andren, O., Ogle, S., Conant, R., Longmire, J., Ashkalov, K., Baethgen, W.E., Valdivia, R. & Paustian, K. 2004, *Assessment of the Costs and Enhanced Global Potential of Carbon Sequestration in Soil*, IEA/CON/03/95.

Guan, H., Wittenburg, K., Ominski, K. & Krause, D. 2006, 'Efficacy of ionophores in cattle diets for mitigation of enteric methane', *Journal of Animal Science* 84: 1896–1906.

Guo, L.B. & Gifford, R.M. 2002, 'Soil carbon stocks and land use change: a meta analysis', *Global Change Biology* 8: 345–60.

Hargrove, T. 2008, 'TVA fertilizer technology used worldwide—but few new products since 1970s', International Center for Soil Fertility and Agricultural Development, <www.eurekalert.org/pub_releases/2008-08/i-tft082508.php>.

Howden, S.M. & Reyenga, P.J. 1999, 'Methane emissions from Australian livestock: implications of the Kyoto Protocol', *Australian Journal of Agricultural Research* 50: 1285–91.

Hu, Q., Sommerfield, M., Jarvis, E., Ghirardi, M., Posewitz, M., Seibert, M. & Darzins, A. 2008, 'Microalgal triacylglycerols as feedstocks for biofuel production: perspectives and advances', *The Plant Journal* 54: 621–39.

IEA (International Energy Agency) 2007, *Biofuel Production*, IEA Energy Technology Essential, ETE02, January, IEA, Paris.

IPCC (Intergovernmental Panel on Climate Change) 2000, *Land Use, Land-Use Change and Forestry: A special report*, R. Watson, I.R. Noble, B. Bolin, N.H. Ravindranath, D.J. Verardo & D.J. Dokken (eds), Cambridge University Press, Cambridge.

IPCC 2006, *2006 IPCC Guidelines for National Greenhouse Gas Inventories*, <www.ipcc-nggip.iges.or.jp/public/2006gl/index.html>.

IPCC 2007, *Climate Change 2007: Impacts, adaptation and vulnerability. Contribution of Working Group II to the Fourth Assessment Report of the Intergovernmental Panel on Climate Change*, M.L. Parry, O.F. Canziani, J.P. Palutikof, P.J. van der Linden & C.E. Hanson (eds), Cambridge University Press, Cambridge.

Johnson, J.M-F., Franzluebbers, A.J., Lachnicht Weyers, S. & Reicosky, D.C. 2007, 'Agricultural opportunities to mitigate greenhouse gas emissions', *Environmental Pollution* 150: 107–24.

Jones, C. 2007, 'Australian Soil Carbon Accreditation Scheme (ASCAS)', presentation at the Managing the Carbon Cycle Workshop, Katanning, Western Australia, 21–22 March.

Jones, C. 2008, 'Liquid carbon pathway unrecognised', *Australian Farm Journal* 8(5): 15–17.

Jones, C., Wilson, R. & Wiley, T. 2008, 'Climate change and the Australian agricultural sector', evidence to the Senate Standing Committee on Rural and Regional Affairs and Transport, Proof Committee Hansard, 30 June, Canberra.

Klieve, A. & Ouwerkerk, D. 2007, 'Comparative greenhouse gas emissions from herbivores', in Q.X. Meng, L.P. Ren & Z.J. Cao (eds), *Proceedings of the VII International Symposium on the Nutrition of Herbivores*, China Agriculture University Press, Beijing, pp. 487–500.

Lal, R. 2004, 'Soil carbon sequestration to mitigate climate change', *Geoderma* 123(1–2): 1–22.

Lal, R. 2007, 'Carbon management in agricultural soils', *Mitigation and Adaptation Strategies for Global Change* 12: 303–22.

LawrieCo 2008, editorial, 'Carbon farming: stored water security', *Sustainable Update*, winter 2008, p. 1.

Lehmann, J. 2007, 'A hand full of carbon', *Nature* 447: 143–4.

McAllister, T.A. & Newbold, C.J. 2008, 'Redirecting rumen fermentation to reduce methanogenesis', *Australian Journal of Experimental Agriculture* 48: 7–13.

Mackey, B., Keith, H., Berry, S. & Lindenmayer, D. 2008, *Green Carbon: The role of natural forests in carbon storage. Part 1. A green carbon account of Australia's south-eastern eucalypt forests, and policy implications*, Australian National University, Canberra.

Miller, L.A., Moorby, J.M., Davies, D.R., Humphreys, M.O., Scollan, N.D., Macrea, J.C. & Theodorou, K.M. 2001, 'Increased concentration of water-soluble carbohydrate in perennial ryegrass (*Lolium perenne* L.). Milk production from late-lactation dairy cows', *Grass and Forage Science* 56: 383–94.

NAFI (National Association of Forest Industries) & TPA (Tree Plantations Australia) 2007, submission to the Garnaut Climate Change Review, December.

Oxburgh, R. 2008, 'Biofuels of the future', presentation made to the Biofuels, Energy and Agriculture: Powering towards or away from food security? seminar, Canberra, 15 August.

Parr, J.F. & Sullivan, L.A. 2005, 'Soil carbon sequestration in phytoliths', *Soil Biology & Biochemistry* 37: 117–24.

Paul, K.I., Polglase, P.J., Nyakuengama, J.G. & Khanna, P.K. 2002, 'Change in soil carbon following afforestation', *Forest Ecology and Management* 168: 241–57.

Paustian, K., Babcock, B., Hatfield, J., Rattan, L., McCarl, B., McLaughlin, S., Mosier, A., Rice, C., Robertson, G.P., Rosenberg, N.J., Rosenzweig, C., Schlesinger W.H. & Zilberman D. 2004, 'Agricultural mitigation of greenhouse gases: science and policy options', Council on Agricultural Science and Technology, Report R141.

Polglase, P., Paul, K., Hawkins, C., Siggins, A., Turner, J., Booth, T., Crawford, D., Jovanovic, T., Hobbs, T., Opie, K., Almeida, A. & Carter, J. 2008, *Regional Opportunities for Agroforestry Systems in Australia: A report for the RIRDC/L&WA/FWPA/MDBC Joint Venture Agroforestry Program*, RIRDC, Barton, ACT.

Post, W.M. & Kwon, K.C. 2000, 'Soil carbon sequestration and land-use change: process and potential', *Global Change Biology* 6: 317–27.

Quiggin, J. 2007, *Drought, Climate Change and Food Prices in Australia*, report for the Australian Conservation Foundation, <www.afconline.org.au/articles/news>.

Russell-Smith, J., Edwards, A., Cook, G.D., Brocklehurst, P. & Schatz, J. 2004, *Improving Greenhouse Emissions Estimates Associated with Savanna Burning in Northern Australia: Phase 1. Final Report to the Australian Greenhouse Office*, Tropical Savannas Cooperative Research Centre, Darwin.

Savannna Explorer 2008, <www.savanna.org.au/all/index.html>.

Tropical Savanna Cooperative Research Centre 2008, <http://savanna.ntu.edu.au/information/arnhem_fire_project.html>.

UNFCCC (United Nations Framework Convention on Climate Change) 2002, *Decision 15/CP.7: The Marrakesh Accords*, 21 January, <http://unfccc.int/2860.php>.

US EPA (United States Environmental Protection Agency) 2003, *International Analysis of Methane and Nitrous Oxide Abatement Opportunities: Report to Energy Modeling Forum, Working Group 21*, Office of Atmospheric Programs, Washington DC.

US EPA 2006a, *Global Anthropogenic Non-CO2 Greenhouse Gas Emissions: 1990–2020*, Office of Atmospheric Programs, Washington DC.

US EPA 2006b, *Global Mitigation of Non-CO2 Greenhouse Gases*, Office of Atmospheric Programs, Washington DC.

Valzano, F., Murphy, B. & Koen, T. 2005, 'The impact of tillage on changes in soil carbon density with special emphasis on Australian conditions', technical report 43, Australian Government, Canberra.

van Groenigen, J.W., Van der Bolt, F.J.E., Vos, A. & Kuikman, P.J. 2005, 'Seasonal variation in N2O emissions from urine patches: effects of urine concentration, soil compaction and dung', *Plant and Soil* 273: 15–27.

von Braun 2008, *Rising Food Prices: What should be done?* policy brief for the International Food Policy Research Institute.

Wentworth Group of Concerned Scientists 2008, 'Proposal for a National Carbon Bank', submission to the Garnaut Climate Change Review.

Western Australia Department of Agriculture and Food 2007, 'Physic nut (*Jatropha curcas*) and bellyache bush (*J. gossypiifolia*)', *Farmnote*, note 244, July.

Whitehead, D. 2008, 'Mitigation technologies for agriculture—now and in the future', presentation to the Australian Farm Institute, Maroochydore, April.

Whitehead, P.J., Purdon, P., Russell-Smith, J., Cooke, P.M. & Sutton, S. in press, 'The management of climate change through prescribed savanna burning: emerging contributions of Indigenous people in Northern Australia', *Public Administration and Development*.

Wilson, G.R. & Edwards, M.J. 2008, 'Native wildlife on rangelands to minimize methane and produce lower-emission meat: kangaroos versus livestock', *Conservation Letters* 1(3): 119–28.

Wright, A.D.G., Kennedy, P., O'Neill, C.J., Toovey, A.F., Popovski, S., Rea, S.M., Pimm, C.L. & Klein, L. 2004, 'Reducing methane emission in sheep by immunization against rumen methanogens', *Vaccine* 22: 3976–85.

TOWARDS A LOW-EMISSIONS ECONOMY

23

Key points

Australian material living standards are likely to grow strongly through the 21st century, with or without mitigation, and whether 450 or 550 ppm is the mitigation goal. Botched domestic and international mitigation policies are a risk.

Substantial decarbonisation by 2050 to meet either the 450 or 550 obligation is feasible. It will go fastest in the electricity sector, then transport, with agriculture being difficult unless, as is possible, there are transformative developments in biosequestration.

There is considerable technological upside. This could leave Australian energy costs relatively low, so that it remains a competitive location for metals processing.

Australia's human resource strengths in engineering, finance and management related to the resources sector are important assets in the transition to a low-emissions economy. They will need to be nurtured by high levels of well-focused investment in education and training.

The introductory impact of the Australian emissions trading scheme will not be inflationary if permit revenue is used judiciously to compensate households.

Within the policies recommended by the Review, the Australian economy has good prospects of reducing emissions by 80 per cent (under a 550 stabilisation scenario) or 90 per cent (under a 450 scenario) by mid-century, alongside continued strong growth in living standards. Decarbonisation will be associated with changes in industry structure, changing contributions by various sectors to steady falls in emissions, and a changing Australian economic relationship to the rest of the world.

While the prospects are generally favourable, there are three main downside risks in the outlook that are associated with climate change or its mitigation. The first relates to failure to achieve international agreement on mitigation. The second is associated with climate change impacts turning out to be at the bad end of the probability distribution. The third would come from mismanagement of mitigation policy. There are significant possibilities that technological outcomes will be superior to those assumed in the Review's quantitative analyses.

23.1 The dynamics of economic adjustment with global mitigation

The choice of 400 ppm, or 450 ppm, or 550 ppm, or for that matter 850 ppm, is a choice of different time paths to low emissions rather than different end points. If the mainstream science is roughly right, human net emissions will need to be close to zero by late in this century, whatever the level of concentrations. Steeper paths through mitigation to stabilisation, or to 'peaking', will have larger transition costs, but will be associated with less damage and fewer risks from climate change.

Mitigation coordinated on a global scale will be less costly and more effective than independent national action. Independent national action could not add up to a good global outcome. The difference in costs and effectiveness between comprehensive global agreement and separate national actions is so large that it is reasonable to presume that, sooner or later, the world will find its way onto that path.

Sooner or later, a small number of global rights to emit greenhouse gases will be divided among countries. It is hard to see any basis being accepted except that of equal per capita emissions. Along the way, a global system of carbon pricing can be established so long as participating countries have clearly defined trajectories for emissions entitlements.

With carbon pricing, the prices of emissions-intensive goods and services will have increased relative to those of other products. Comparative advantage in trade in goods and services will be determined by a comprehensive set of costs that includes the price of emissions.

So long as there is comprehensive pricing, countries with a comparative advantage in emissions-intensive products, and firms with a competitive advantage, will find that the prices have risen enough to cover the cost of purchasing emissions permits. These countries and firms will tend to buy permits on international markets. When all of the costs are taken into account, some countries will have a comparative advantage in emissions-intensive goods and services. These countries will tend to be net purchasers of permits, and their emissions will exceed their shares of the global population and permit allocations.

If emissions permit markets are working well, the carbon price will rise steadily at the interest rate. If it is $20 in 2010 (2005 dollars), forward prices stretching into the future will be, in real terms, around $40 in 2030, $80 in 2050 and $320 in 2090 (all measured in 2005 dollars). The whole forward price structure will rise with any disappointment of expectations of technological improvement, and fall on any good surprises.

23.1.1 Innovation and incentives

Comprehensive emissions pricing on a global scale, and large-scale fiscal support everywhere for research, development and commercialisation of low-emissions technologies, will provide incentives for exploring ways to economise on the use of emissions-intensive goods and services, and emissions-intensive ways of producing goods and services.

These incentive structures will direct the attention of generations of people over the world who are more numerous, better educated, better informed, better connected and better equipped for productive innovation than any of their predecessors. In its first 150 years, the modern global economy was built on the genius of a small corner of humanity. There was relatively little contribution to the new inventions that drove industrial modernity to ever-rising material standards of living from China, or South Asia, or Southeast Asia, or the Middle East, or Africa, or Latin America—despite the seminal and disproportionate contribution that people from several of these regions had made to the intellectual content of human civilisation in earlier and formative eras. Over the 21st century, out of the expansion of opportunity from the rising incomes and global integration of the Platinum Age, the whole of the genius of humanity will be contributing to innovation in the global economy. Invention will occur in more places and, more than ever before, invention in one part of the world will move quickly to enrich the whole.

Soundly based global mitigation policies will provide incentives for an expanded world population's talent for innovation to be focused especially on low-emissions technologies.

The Australian mitigation effort will be the beneficiary of innovation everywhere. National economic performance will be related closely to how quickly and how well the best new knowledge is absorbed from abroad. Australian prosperity will be related as well to the efficiency with which it develops new and adapts established technology that is especially suited to Australian natural and human endowments. National economic performance will be related closely to how well an economy adjusts to changes in costs and opportunities associated with carbon pricing and emissions-related technological change.

23.1.2 Backstop technologies

At some time, there will be breakthroughs that fundamentally lower the costs of producing goods and services in the low-emissions economy. One or more of these will be 'backstop technologies' that become commercially viable at one or two or three or four hundred dollars per tonne of carbon dioxide equivalent. These backstop technologies will take carbon dioxide from the air at some cost, without relevant limit, and so end the inexorable rise in the carbon price. Research is currently proceeding around the world on a number of possibilities, both industrial and biological.

In Australia, the most interesting work on what could become backstop technologies are in the applied biological sciences—areas of traditional Australian scientific and economic strength. One of these, the use of algae to convert carbon-rich wastes or carbon dioxide from the air into stable carbohydrates, would utilise the biological processes that converted an earlier carbon-rich atmosphere of earth to the oxygen-rich air that made life possible for mammals and therefore humans. They would enhance natural processes by selecting organisms and their growing conditions specifically for the sequestration task.

The emergence of backstop technologies will mark the end of history for the decarbonisation of particular and various sectors of economic activity.

Until these technologies emerge, the rising carbon price will apply ever-increasing pressure for reduction in demand for emissions-intensive goods, and for substitution away from carbon-intensive ways of producing goods and services. Decarbonisation will occur earlier in some goods and services than in others. But occur it will. After half a century, emissions will be confined to a small number of highly valuable goods and services, which have no close substitutes in demand or supply. Beyond this time, the increase in the cost of mitigation is determined by the resistance to substitution in supply and demand of a few goods and services that people continue to value highly even when their prices have risen way beyond old relativities. This will all be happening in a world much richer in purchasing power for goods and services than it is today.

23.1.3 Carbon pricing and inflation

The introduction of carbon pricing will generate a once-and-for-all increase in the general price level. The inexorable rise in the carbon price, beyond the initial shock, will not contribute comparable continuing pressure on the price and cost structure. At a $20 per tonne initial price (2005 dollars), the subsequent annual increases in carbon prices would be about one dollar in the early years, and would gradually increase in annual impact after that. If the direct price effect of the increase in the emissions price were initially to raise the general price level by about one percentage point, the next year's impact would be about one twenty-fifth of a percentage point. As the process of decarbonisation proceeds, the products of newly competitive low-emissions processes can be expected to experience more rapid technological improvement than established high-emissions products and processes, so that their relative prices fall over time. At some point, part of the way through the decarbonisation process, falling economy-wide costs of newly competitive products and processes would outweigh the effects on costs through the economy as a whole of the rising carbon price.

A starting price in 2010 of around $20 per tonne in 2005 prices would raise the consumer price index by about one percentage point. As the Governor of the Reserve Bank of Australia observed in September 2008, this would not require a tightening of monetary policy in response if it were a once-for-all increase that did not then trigger subsequent increases in the general cost and price level.[1] On the other hand, if introduced at a time of strong demand and a tight labour market, it could raise inflationary expectations and generate pressures for compensating wage increases.

One important element in a set of policies to prevent a flow-on into the general cost structure is a household compensation package that at least maintains real purchasing power of people with incomes at levels covered by the regulated wage system. The proposed household adjustment package, directing half the permit revenue to the bottom half of the income distribution, would provide this element. Normal prudence in management of monetary conditions would avoid the transmission of the price increases associated with the introduction of the emissions trading scheme to other income levels.

Thus the household compensation package is not only important for equity, but a critical element in maintaining price stability and avoiding monetary tightening that would not otherwise have occurred.

Household incomes are projected to increase at rates comparable to historical trends in the reference case and also in the 550 and 450 mitigation scenarios as shown in Table 23.1. After-tax income per capita in 2020 would be 17 per cent higher than in 2006 with no mitigation, 15 per cent higher with 550 mitigation and also 15 per cent higher with 450 mitigation. By 2050, average income per capita after tax would be 66 per cent higher than in 2006 with no mitigation, 57 per cent higher with 550 mitigation and 55 per cent higher with 450 mitigation. In the second half of the century, the gap closes then reverses.

Table 23.1 Total after-tax per capita income (2005 dollars)

	2006	2020	2050
Reference	$36 548	$42 676	$60 622
Reference—relative to 2006	–	+17%	+66%
550 standard technology scenario	–	$42 207	$57 304
550 standard technology scenario—relative to 2006	–	+15%	+57%
450 standard technology scenario	–	$42 044	$56 675
450 standard technology scenario—relative to 2006	–	+15%	+55%

Source: MMRF reference case and policy simulations without climate change impacts.

The greatest impact on consumption is projected to be on electricity, as described in Chapter 20. Residential electricity prices for the 550 and 450 backstop technology scenarios would be 21 per cent and 37 per cent, respectively, higher by 2020 than they would have been in the reference case. Beyond electricity, there would also be minor reductions in consumption of air transport, heating and some chemicals. However, these changes are small when compared with the changes in overall consumption that occur over time as a result of normal changes in consumer preferences and incomes.

23.2 The economy to and at 550 ppm

The Review's joint modelling with the Australian Treasury focused on an emissions constraint to meet Australia's proportionate share of either a global 550 or 450 objective. Average output and consumption of Australians at the end of the century are more than three times as high as now under any of the modelled scenarios.

The no-mitigation scenario from the Garnaut–Treasury modelling has GNP growth much slower than GDP growth over the next decade, as Australia's terms of trade fall from extraordinary and unsustainable heights to high but more moderate levels. From 2013 to 2020, average annual growth is expected to be 2.7 per cent for GDP and 2.1 per cent for GNP. Incidentally, the assumed fall in the terms of trade deducts almost twice the amount of incomes growth as does the more ambitious of the mitigation scenarios.

Average annual GNP growth rates from 2013 to 2020 are expected to fall from 2.1 per cent with no mitigation, to around 1.9 per cent under a 550 scenario with backstop technology assumptions (see Chapter 11) and 1.8 per cent under a 450 backstop technology scenario.[2] These initial costs are not small. The growth costs of mitigation fall after the carbon pricing has been established, and lessen over time. The difference between no-mitigation and 550 growth rates falls to a tenth of a percentage point between 2020 and 2050 (Table 23.2).

Table 23.2 Annual average growth rates for GNP and GDP under the no-mitigation, 550 and 450 scenarios with backstop technology (Type 1 and Type 2 benefits of mitigation) (per cent)

Average annual growth rates	2013–20	2021–50	2051–2100
Unmitigated climate change (GNP)	2.1	1.7	2.0
Unmitigated climate change (GDP)	2.7	1.7	2.0
Stabilisation to 550 ppm backstop technology (GNP)	1.9	1.6	2.1
Stabilisation to 550 ppm backstop technology (GDP)	2.5	1.7	2.1
Stabilisation to 450 ppm backstop technology (GNP)	1.8	1.6	2.2
Stabilisation to 450 ppm backstop technology (GDP)	2.5	1.6	2.1

Source: MMRF adjusted to include Type 2 costs of climate change (see Chapter 11).

The modelling has the Type 1 and estimated Type 2 benefits from avoided climate change significantly reducing the net economic costs of mitigation from 2030, and increasingly so with passing decades. Over the second half of the century, average growth rates are a tenth of a percentage point higher under 550 than with no mitigation, and higher still under 450. By 2100, GNP is slightly higher with mitigation than it would have been without it.

The achievement of this mitigation task involves a major change in the structure of the economy. Australia's total emissions entitlement in the 550 scenario will be 35 per cent below business-as-usual emissions in 2020 and 90 per cent by 2050. Note that the comparison is with business-as-usual for 2020 and 2050, and not with a base year from the MMRF modelling. As shown in Figure 23.1, Australia purchases a significant volume of emissions permits from other countries. Beyond a certain point, it is cheaper for Australia to buy permits from abroad at the prevailing international price than to reduce emissions from the domestic economy. International purchases cease once a backstop technology is adopted.

The proportion of domestic mitigation relative to purchases of permits is influenced by the specific parameters of the models used. For example, GTEM shows higher levels of domestic mitigation than MMRF for the same permit price.

Figure 23.1 Australia's emissions in the 550 backstop scenario (global entitlement, net of trading)

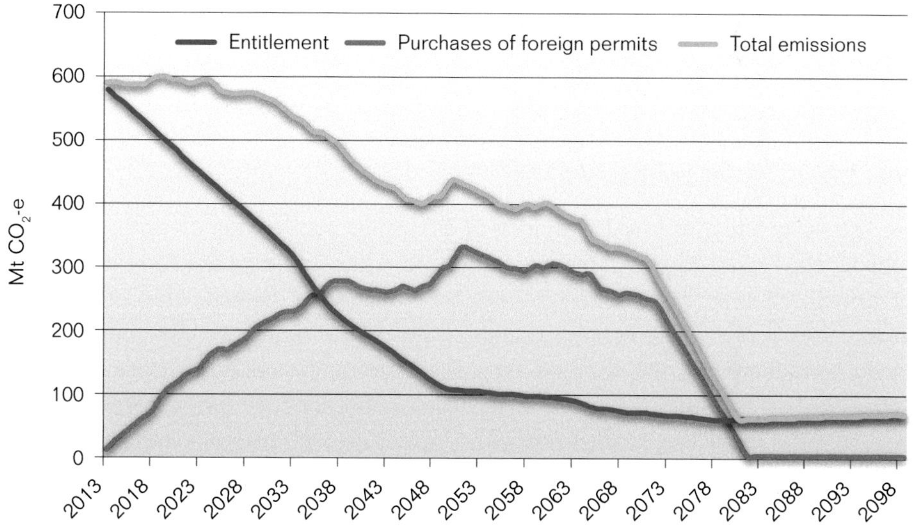

Note: These results were generated using MMRF.

Figure 23.2 shows how mitigation under the 550 backstop scenario is achieved in Australia. Electricity generation makes the largest contribution, with the sector largely decarbonised by mid-century. The modelling indicates that there will be little growth and, over time, some pressure for contraction of highly energy-intensive, trade-exposed sectors such as non-ferrous metals in Australia. This follows from the exceptionally high emissions intensity of Australian energy. However, as discussed further below, this is one area of uncertainty, as there is some prospect that Australian energy costs will remain low by world standards. Other determinants of Australian competitiveness in processing minerals and energy commodities will strengthen.

Figure 23.2 Sources of mitigation under the 550 backstop scenario

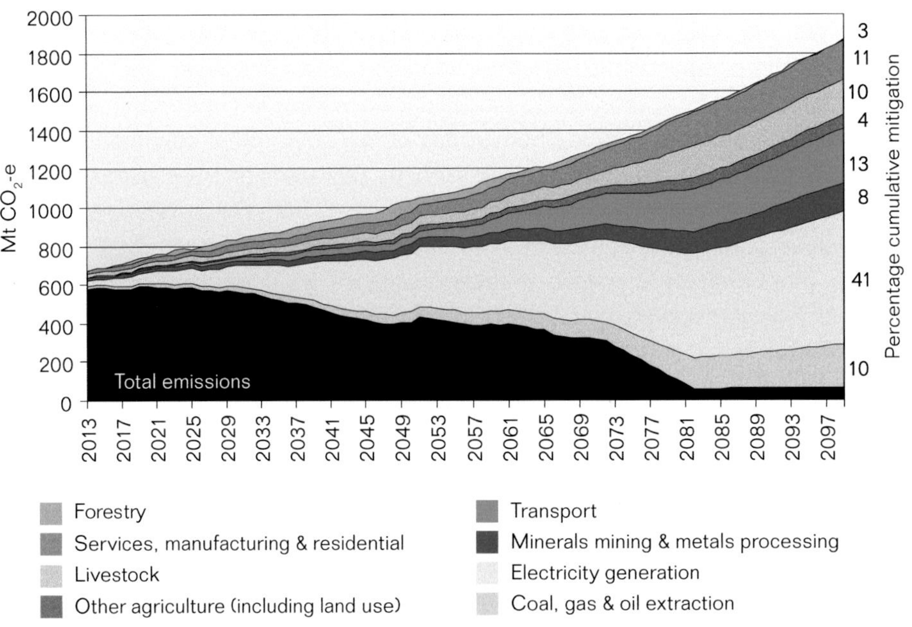

Note: These results were generated using MMRF.

The rising and eventually high price of emissions accelerates a move away from fossil fuels, first of all coal. This would have to occur in the longer term regardless of climate change. It will come sooner with mitigation. Beyond this period, and in line with achieving emissions reductions of 80 or more per cent below 2000 levels by mid-century, the sources and cost of mitigation will be determined by the emergence of technologies either to provide net greenhouse gas absorption, or to deal with the remaining fugitive emissions from ruminant livestock and some other activities, probably including civil aviation.

The sectoral implications of climate change mitigation

The responses of various sectors to the rising emissions price are triggered by the emissions intensities of wider supply chains. The immediate impact is on sectors with large direct emissions, as illustrated in Figure 23.3 for 2005. Electricity, gas and agriculture stand out for their high direct exposure to emissions pricing.

Electricity is a primary input to many industries. The impact on electricity generation flows through to consumers and also to electricity-intensive downstream sectors such as aluminium. This is shown in Figure 23.4, which includes both direct and indirect emissions.

The ratios in the figures are sensitive to the relevant commodity prices. In the resource-based sectors such as coal, liquefied natural gas, non-ferrous, iron and steel, the rising prices of recent years will have significantly reduced these intensities.

Figure 23.3 Direct emissions per million dollars value added, 2005

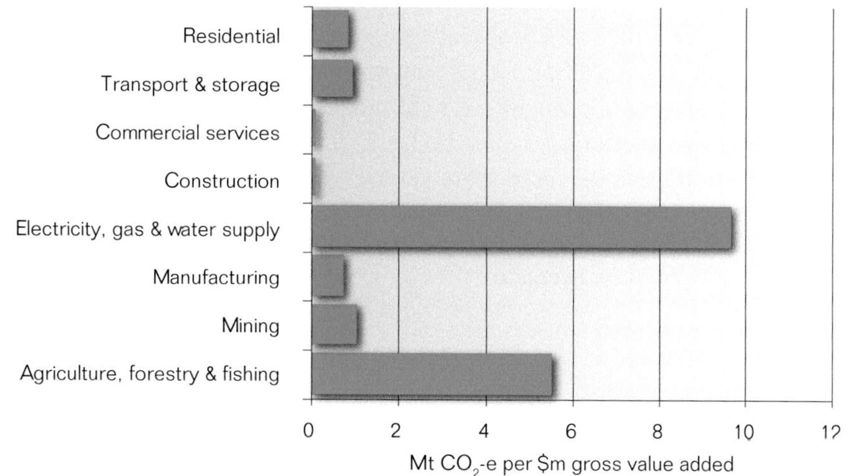

Sources: ABS (2008); DCC (2008b).

Figure 23.4 Direct and indirect emissions per million dollars value added, mining and manufacturing, 2005

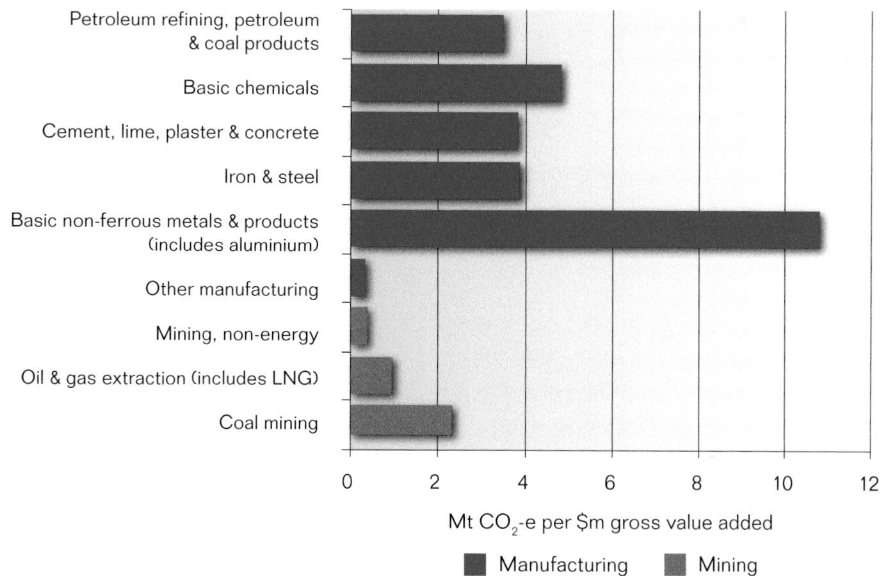

Sources: ABS (2008); DCC (2008a).

The direct impact on petroleum products will be felt through the transport sector, and subsequently in sectors where transport is a significant input cost.

Figure 23.5 illustrates how emissions are projected to fall across major sectors as the rising carbon price forces change. It highlights differences in the incidence of pressures for structural change across various Australian industries. Whereas Australian emissions are heavily concentrated in electricity in the early years, by the later decades of the century a high proportion is associated with livestock. This reflects the substantial known technology opportunities to reduce electricity sector emissions, and the more limited known opportunities for livestock.

Figure 23.5 Emissions sources (not including forestry) in the 550 backstop technology scenario

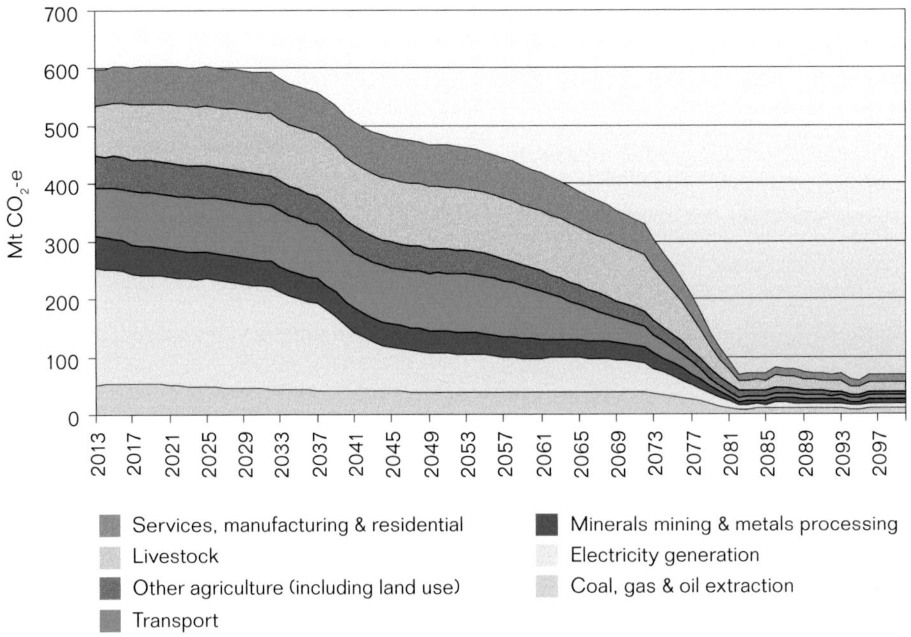

Note: These results were generated using MMRF.

Mitigation results in strong expansion of some industries, in particular agriculture, forestry and renewable electricity generation. Agriculture is particularly exposed to the effects of climate change. Global mitigation of climate change will therefore result in significant benefits to agriculture as the more severe outcomes are avoided.

23.3 The difference between 550 and 450

Stronger mitigation deepens and accelerates the changes described in the 550 scenario. In the 450 scenario, Australia's emissions entitlement would be 47 per cent below the levels that it would have been in 2020 in the absence of mitigation, and 94 per cent lower by 2050.

As the permit price rises in each of the scenarios, but more quickly and powerfully in the 450, deeper utilisation of domestic mitigation opportunities becomes cost effective. The international price of permits is substantially higher in the 450 backstop technology scenario, and Australians buy fewer permits abroad despite the greater abatement task (Figure 23.6).

Figure 23.6 Australia's emissions in the 450 backstop technology scenario (global entitlement, net of trading)

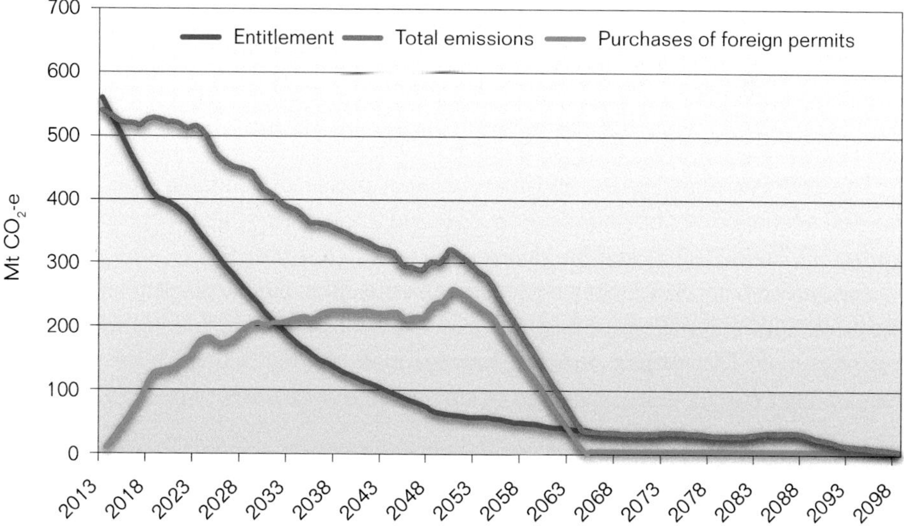

Note: These results were generated using MMRF.

Australia's emissions, even after substantial permit acquisition through trading, will be significantly below what they would have been in a scenario without mitigation, as illustrated in Figure 23.7.

Many of the changes under the 450 backstop technology scenario are qualitatively similar to those in the 550 scenario, but they are larger and occur sooner. Retail electricity prices increase more and continue increasing (approximately 70 per cent higher in real terms by 2020, compared with 40 per cent for 550). Decarbonisation of the electricity sector and electrification of the transport sector occur earlier, with consequences for the roles of fossil fuels and renewable technologies, as described in Chapter 20. Some existing high-emissions electricity generation plant retires earlier.

Figure 23.7 Total emissions for the no-mitigation, 450 and 550 backstop scenarios

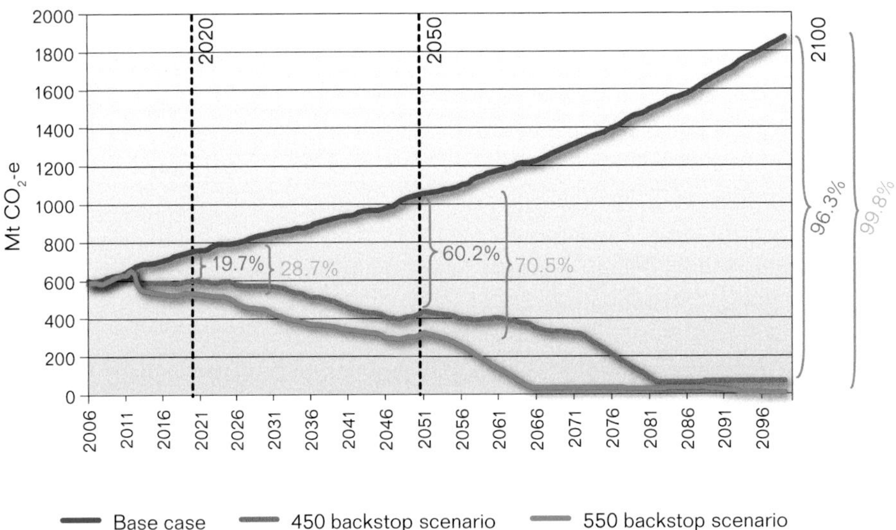

Note: These results were generated using MMRF.

The sources of mitigation and the changing profile of Australia's emissions in the 450 scenario are shown in figures 23.8 and 23.9. These can be compared with figures 23.2 and 23.5 to see the differences across the scenarios.

An overarching conclusion from the above is that the availability of superior low-emissions technology becomes more urgent as more ambitious targets are contemplated. This places an even heavier load on early support for research, development and commercialisation of such technologies (see Chapter 18).

23.4 Australia in the low-emissions world energy economy

Global mitigation is likely to lower Australia's terms of trade through its effects on demand for fossil fuels. Australia would experience these costs whether or not it was itself engaged in strong mitigation. In this, Australia's position is different from that of most other developed countries. The United States, the European Union, Japan, New Zealand and Korea are all net importers of fossil fuels. Their terms of trade would all rise with strong global mitigation.

Will Australia continue to have low (carbon-price-inclusive) energy costs on an international scale? The answer to this question will be the main determinant of whether Australia continues to have a comparative advantage in metals processing, and exports metals and other processed mineral products. Countries with superior natural opportunities for low-cost biosequestration or geosequestration of emissions will have an advantage in producing goods and services with large

Figure 23.8 Sources of mitigation under the 450 backstop technology scenario

Mt CO_2-e

Percentage cumulative mitigation

Total emissions

Forestry
Services, manufacturing & residential
Livestock
Other agriculture (including land use)
Transport
Minerals mining & metals processing
Electricity generation
Coal, gas & oil extraction

Note: These results were generated using MMRF.

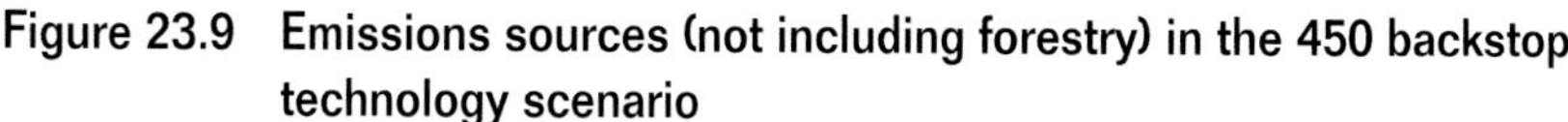

Figure 23.9 Emissions sources (not including forestry) in the 450 backstop technology scenario

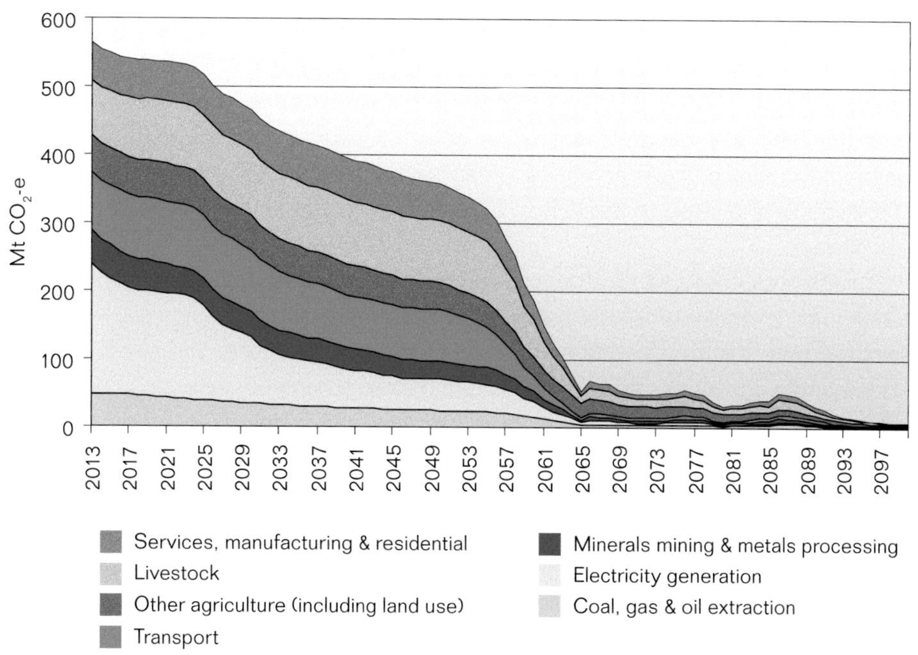

Note: These results were generated using MMRF.

emissions from fossil fuels. If geosequestration with high capture efficiency (near 100 per cent) is commercially successful, the answer will be in the affirmative. It would be in the affirmative for energy based on fossil fuels even if its geosequestration sites were of average quality in global terms. Australia seems to be relatively well endowed with sites that are suitable for geosequestration. If, as may be the case, biosequestration based on the use of algae to convert carbon dioxide wastes to stable carbohydrates becomes commercially viable, Australia will also be a favourable location. Our sequestration sites seem to be of superior economic quality across the range of possible technologies. Australian energy costs will be lower than international costs to the extent that carbon capture and storage or biosequestration is commercially successful, as the export parity price of gas, especially, and coal, will be much lower than the price in importing countries.

For a number of years, new investment in energy-intensive processing, for example aluminium smelting, will be concentrated at sites with low-cost, low-emissions energy supply that do not have established outlets in major urban and industrial centres. Hydroelectric and conventional geothermal capacity in developing countries will be the first prize, and gas deposits that are not easily developed for export through pipelines or liquefaction second. But it will not take long to exhaust this capacity, and whether or not Australia re-emerges as an attractive location for energy-intensive processing beyond that point will depend on the many factors that will determine whether it has relatively low energy costs in a low-emissions world.

Natural gas, from traditional geological formations but increasingly from coal seams, will receive a powerful competitive boost in the early years of a rising emissions price. The growth and prices of coal exports will soon be damaged by other countries' mitigation in the absence of carbon capture and storage in our major markets and beyond (Chapter 20). Partial capture and storage of carbon emissions from coal combustion will provide a long reprieve. But a time will come, over the next several decades, when only full capture and storage of carbon wastes from coal and gas combustion can hold a role for coal in power generation. Metallurgical coal exports would hold up better for longer in the absence of capture and storage.

If carbon capture and storage and commercial-scale biosequestration of carbon wastes fail, so that fossil fuels become unimportant in the global energy equation, Australia may still be a country of relatively low energy costs. We seem to have exceptionally low-cost resources, in abundance compared with population, for most renewable energy: deep hot rocks (geothermal), solar, wind, wave, biomass and second-generation biofuels.

Although Australia does not have abundant hydroelectric capacity, the world's low-cost capacity is likely soon to be fully utilised. Global prices for energy-intensive products are likely still to be determined by the cost of other sources of power, in which Australia is relatively well placed on a global scale.

23.5 The downside risks

There are three types of risks that could seriously and adversely change economic outcomes in the process of transition to a low-carbon economy. The first is that there may not be an early comprehensive global agreement. The second is the possibility that, even with an effective international agreement to hold concentration at around 550 or even at 450 with overshooting, the inevitable climate change with which it is associated will turn out to be at the bad end of the probability distribution of possible outcomes. The third relates to unexpectedly high costs of mitigation emerging from technological disappointments, or unnecessarily high costs from bungling of mitigation policy design or implementation.

The biggest risk is that there will be no comprehensive global agreement at Copenhagen. This, at this stage, looks more than a small risk. The framework of international negotiations established at Kyoto in 1997, and continued in Bali in 2007, contains weaknesses that, if not removed over the year ahead, make successful agreement at Copenhagen unlikely.

The weaknesses have been described in chapters 8 to 10. The UN discussions have enshrined the idea that developing countries for the foreseeable future will not be required to enter commitments on emissions reductions. Developing country commitments are needed for the numbers to add up to effective international agreement to reach a 550 outcome.

The most likely alternative to comprehensive agreement in Copenhagen is partial, ad hoc agreement. This is a problematic second best. Strong global mitigation is impossible in an ad hoc world.

The second risk relates to uncertainty about the climate change effects of increases in greenhouse gas concentrations, and in their impacts on Australian economic activity and life (chapters 2 to 6). At concentrations of 550 ppm and 450 ppm, there is a possibility of effects on economic activity and life that are so much greater than the median prospects upon which the modelling is based that they could negatively transform the outlook, even after successful international cooperation around strong mitigation objectives.

No feasible degree of global mitigation from this time, at the end of the first decade of the 21st century, would remove the possibility of potentially destabilising outcomes. Some risks would remain even if the international community effectively committed itself now to 450 ppm atmospheric concentrations of carbon dioxide equivalent—which would require a peaking profile, and therefore periods in which concentrations went higher than 450 ppm, and then contracted.

There is some small risk that the suite of technologies available for the economic transition to low emissions will turn out to be less favourable than assumed even in the standard technology case in the modelling of costs of mitigation discussed in chapters 11, 20, 21 and 22. The risk of mitigation turning out to be much more expensive than expected for reasons outside technology is more important. It has its origin in human behaviour rather than in laws of nature. This is mostly about frictions in the introduction and operations of the mitigation regime. Three sources of friction are potentially important.

23.5.1 Frictions in the mitigation regime

There could be a long period of uncertainty about the rules of the game, and in particular about targets and trajectories going forward. Uncertainty would raise the supply price of investment to the emissions-intensive industries, and miscalculations arising out of it could greatly increase the costs of adjustment to the low-emissions economy.

Some uncertainty is inevitable, given the good reasons why trajectories will need to be adjusted from time to time. The challenge will be to economise on uncertainty. This can be done best by specifying as closely as possible and as far into the future as possible, the possible trajectories, international and national, and the conditions under which there will be movement from one to another.

Any perception that parameters of the arrangements remained open for negotiation with business after its commencement will be counterproductive. Here the greatest vulnerability would arise out of payments to trade-exposed industries, and, should the advice from this Review not be followed, compensation to business for the introduction of the emissions trading scheme. There could also be high costs from diversion of managerial effort into competition for government preferment through the interaction of national mitigation regimes with the international trading system. The differential impact across countries of arrangements affecting trade-exposed, emissions-intensive industries would lead to pressure for countervailing payments, in the form of 'free permits' or in other ways. Inevitably, this would come to interact with protectionist pressure of a conventional kind. Protectionist responses in some countries would increase pressures for the transfer of value to special interests in other countries.

The costs to good governance of continuing negotiation and encouragement of rent-seeking behaviour could be high—in the extreme, possibly as high as the unavoidable costs of mitigation itself.

There are three elements of the mitigation regime that are important in avoiding a blow-out in costs. First, any transfer of value to special interests must be undertaken transparently on the basis of clear and sound principles. Second, no elements of the regime involving transfer of value to particular interests can be left open for negotiation after the commencement of the scheme. Third, any transfers of value to particular firms must be made through an independent agency, operating at arms length from political government. Internationally, it is important that payments and trade interventions in relation to trade-exposed industries be made within a principled framework agreed within the World Trade Organization.

23.6 The upside in technology assumptions

The quantitative analysis undertaken by the Review has extended models into time frames and levels of mitigation not previously explored. In doing so, the limits of assumptions regarding technology developments and substitution possibilities become a further constraint on the timing and costs of projected mitigation in the second half of the century. Together, these assumptions become the critical determinants of the preferred sources of mitigation and the residual emissions that are shown in figures 23.2, 23.5, 23.8 and 23.9.

The consequence of these assumptions around development of mitigation technologies is that the overall cost of mitigation is kept implausibly high, even before the introduction of a backstop technology at an emissions price of $250 per tonne. A more plausible future would be one in which the increasing carbon price delivers a range of new technologies across many sectors that have not been modelled in detail at this stage. This unfolding technological reality may diverge from the modelling assumptions in ways that transform for the better the costs of change. Such developments would also transform ambitions for strong mitigation, raising perceptions of what is possible, and in all likelihood raising mitigation achievements on a global scale. A number of broad areas of opportunity have been identified that would lead to this outcome.

23.6.1 Broad opportunities

The first area of opportunity involves better use of known and established technologies. The realisation of potential for greater energy efficiency may turn out to be much less costly than anticipated in the model. The standard technology assumptions used in the joint modelling with the Australian Treasury allowed for modest gains from this source, but much less than some detailed studies have suggested were available (most recently in Australia, from the McKinsey Consultancy (2008) and by McLennan Magasanik Associates for the Climate Institute (2008)).

Chapter 17 discussed the reasons for incomplete use of established technologies for improving economic outcomes by using energy more economically. They are of two kinds. One relates to the economics of information. The other relates to flaws in incentives structures, most importantly through the presence of contractual arrangements that separate the parties who must pay the costs of change from those who benefit from change.

Can these sources of market failure be overcome at reasonable cost? We do not know what will turn out to be possible with a rising emissions price and a culture of adaptation to an emissions constraint. The standard technology scenarios assume that they can be overcome to some extent. However, it is possible that the combination of rising carbon and energy prices, some well-designed public programs on energy efficiency, and changes in the habits of communities that are focusing strongly on the changes necessary to make greenhouse mitigation work, will generate stronger results.

In energy, the range of existing primary energy sources such as coal, gas and uranium, the development of low-emission fossil fuel technologies, and the development of renewable technologies such as geothermal and solar thermal could occur faster than projected and, possibly, at lower costs than are assumed in the standard cases. A critical uncertainty in this area relates to the costs and effectiveness of geosequestration of carbon dioxide wastes. This is a source of downside as well as upside risks. It is possible that highly effective carbon capture and storage could come earlier and at lower cost than assumed in the modelling. There is also a chance, albeit assessed as much less likely, that it will come later, be more expensive or prove to be less effective. Much hangs on the outcome. A large disappointment to outcomes embodied in the models would increase the medium term costs of power, and undermine the prospects of what is now and prospectively Australia's largest export industry.

The broad area of biosequestration is a potentially much larger source of opportunity for upside surprises on technology beyond that assumed in the modelling. The upside here is so large that it has the potential to transform both the Australian and the international mitigation tasks (Chapter 22). There are many barriers to overcome if this potential is to be realised. There must be changes in the accounting regimes for greenhouse gases. Investments are required in research, development and commercialisation of superior approaches to biosequestration. Adjustments are required in the regulation of land use. New institutions will need to be developed to coordinate the interests in utilisation of biosequestration opportunities across small business in rural communities. Special efforts will be required to unlock potential in rural communities in developing countries, nowhere more importantly than in Australia's near neighbours.

In transport, Australia is likely to be a fast follower. However, the Australian market and its location relative to fast-growing Asian economies could facilitate a role as a testing and development market for new, low-emissions fuel and vehicle technologies. For example, the introduction of hybrid and fully electric cars may proceed more rapidly than assumed in the models under the influence of high prices for fossil fuels and a rising carbon price.

Chapters 20 to 22 drew attention to other ways in which the standard technology assumptions of the modelling were conservative on the application and potential for cost reduction in technologies that are currently close to large-scale commercial applications.

An 'enhanced technology' scenario was developed to capture some of these opportunities and to assess their impact on the mitigation scenarios. This scenario includes a greater level of energy efficiency across the economy and considers how the energy, transport and agriculture sectors might respond to stronger assumptions around technology development. These details are described in chapters 20 to 22, while the scenario and its macroeconomic implications are described in Box 23.1.

Box 23.1 Enhanced technology assumptions

The Garnaut–Treasury modelling work conducted a simulation of an 'enhanced technology' scenario in the global general equilibrium model (GTEM) to assess the potential impact of a more optimistic outlook for technology development. This scenario incorporates the combined effects of greater energy efficiency and faster technology learning in the early years, and more effective carbon capture and storage technology, as well as the elimination of non-combustion emissions in agriculture at higher carbon prices (see Chapter 11).

Under this enhanced technology scenario the effects on Australia's GDP in both the 550 and 450 ppm scenario are reduced by around one-third on average throughout the century. The cost savings are particularly strong in the second half of the century. In the 450 ppm scenario, Australia's GDP costs are reduced by almost 45 per cent. In the latter half of the century, Australia's relative cost reductions exceed the global average, as agricultural emissions and carbon capture and storage play a relatively greater role.

The effects on GNP for Australia are even more pronounced than on GDP, because of reduced expenditure on permit purchases in international markets. Under the enhanced technology scenario, Australia achieves greater levels of mitigation domestically and remaining permits that need to be bought come at a lower price because mitigation options are better in other countries as well. As a result, Australia's GNP costs in the second half of the century are reduced by over 50 per cent.

The near-zero-emissions carbon capture and storage assumption within the enhanced technology scenario plays an important role in the future composition of power generation in both scenarios, as described in Chapter 20. The share of electricity generated in Australia using fossil fuels with carbon capture and storage is substantially higher in the enhanced technology scenario, especially under 450 mitigation and under high carbon prices.

This in turn has major implications for Australian coal production and exports. Near-zero carbon capture and storage would allow a continued expansion of coal use both domestically and internationally, even under a tight global carbon constraint. By contrast, if carbon capture and storage does not move beyond 90 per cent, then it would be poised to become a transitional technology on the way to very low emissions levels, which ultimately would be achieved almost exclusively through renewable technologies and/or with the introduction of nuclear energy in circumstances described in Chapter 20.

23.7 The importance of flexible global and national markets

The maintenance and enhancement of Australian prosperity through and after the transition to a low-carbon economy will depend mainly on the same factors, plus one, that determined success in earlier times, before the realisation that global warming was an issue to be managed.

That one is the effective introduction of efficient means of constraining Australian greenhouse gas emissions, within a global system of constraints that holds concentrations to levels that avoid dangerous climate change.

Continuing and expanding prosperity will depend on sound economic governance, now including governance of an inevitably intrusive system of mitigation. These issues have been discussed at length in relation to the emissions trading scheme, the correction of market failures associated with adjustment to a carbon constraint, and adaptation to inevitable climate change (chapters 14, 15, 17, 18 and 19).

Success will depend on the presence of flexible and efficient markets for goods, services, factors of production and now emissions entitlements, all deeply integrated with international markets. It will depend on high and well-focused investment in education and the development of skills, now with even greater concentration than before on the areas in which Australia has both longstanding strengths and current deficits, related to engineering, management and financing in the resources, energy, construction and transport sectors.

In circumstances of changing relative prices of goods and services in response to the mitigation regime, there will be pressures for government to constrain the increases in prices that are a necessary part of adjustment to the emissions constraint. There will be pressures to protect particular interests and asset values from the changes that are inevitably associated with successful adjustment to the carbon constraint. There is a danger that governments will respond to these pressures in ways that greatly increase the costs of adjustment and which seriously diminish prospects for economic growth. Other governments will be under similar pressures, and there is therefore substantial risk that international markets for goods and services will be distorted by protectionist interventions.

The risks will be greatest in markets for the most emissions-intensive goods and services: energy, transport and food.

23.7.1 Energy markets

In energy markets, two innovations are critical to provide the necessary flexibility. One is the removal of price caps on retail power prices accompanied by reliance on competition to secure reasonable pricing. For this to be possible, government would need to introduce explicit, alternative approaches to fund equity measures through the budget, in ways that do not distort market pricing of power.

The second innovation required is the establishment of a genuinely national energy market across all of the regions that can be connected at reasonable cost to the main centres of generation and demand in eastern and south-eastern Australia—the electricity markets of New South Wales, Victoria, Queensland, South Australia, Tasmania and the Australian Capital Territory. This will require deeper and more dense interconnectivity between subregional markets, allowing electricity production to expand from sources at which it can be supplied at lowest cost in the new, carbon-price-inclusive environment. Deeper interconnectivity will support efficient expansion and use of potential for pump storage in Australia's main hydroelectric assets in the Snowy Mountains and Tasmania, thus easing the constraints associated with intermittency of wind and solar power. The test of the emergence of a national market is economically low differentials in electricity prices, and close correlation in changes in electricity prices, across regions.

Effective markets will require removal of constraints on investment in response to opportunity for profitable expansion of power generation and distribution capacity. This adjustment could be made through extension of the privatisation of power business assets owned by state, territory and federal governments. Where major facilities continue to be owned by governments, it is important that these assets be managed within corporate structures that allow commercially efficient responses to opportunity. To artificially constrain investment in expansion of the main hydroelectric assets, for example, would remove a major opportunity for overcoming constraints on the use of intermittent renewable energy technologies.

On managing the risks to open international markets, the main challenge is to ensure that there are principled constraints on assistance to trade-exposed industries in the period of partial, national mitigation prior to comprehensive global agreement. The formula for support of trade-exposed, emissions-intensive industries set out in Chapter 14 would be extendable internationally in ways that minimised risks of distortion.

23.7.2 Transport

In transport markets, too, there will be resistance to price changes in response to the emissions constraint. Recent Australian experience illustrates the sensitivity of this issue, with the Australian Government's proposal in the Green Paper on a Carbon Pollution Reduction Scheme effectively to exclude transport fuels from the early operations of the Australian emissions trading scheme. Such exclusions create arguments for other exclusions, and can quickly lead to a plethora of market-distorting interventions. They are to be avoided.

23.7.3 International food markets

International food markets are deeply distorted by subsidies, restrictions and tariffs on imports and price controls in many countries, and most of all in the developed countries of the northern hemisphere. Such distortions will create major problems for efficient response to changing climatic conditions, competition between food and mitigation (biofuels) for land resources, and costs of inputs in the years ahead. The extent of the problem is illustrated by international responses to high food prices in 2007 and 2008, when new restrictions on exports exacerbated fluctuations in global prices, and reduced confidence in international markets in food importing countries.

The world, and Australia, will need deep and flexible global food markets to manage the fluctuations and changes in supply conditions in response to climate change and its mitigation. Population growth, greater variability in climatic conditions, and the effects of climate change on agriculture will make Australia economically a food importer from time to time, perhaps even under normal conditions. The new global challenge of climate change adds to the good reasons for giving priority to resolving the old problem of distorted global agricultural markets.

23.8 The importance of education and training

The structural changes that will emerge in a low-emissions, growing economy will change requirements for human capital. In Australia, a history of skills development has been inherent in a globally successful resources sector. Australia should be structurally well placed to apply such skills to new activities.

At the same time, the additional requirements will place added strain on an education and training system already under pressure from the resources boom and the associated shifts in employment between key sectors.

The work of the Dusseldorp Skills Forum (Hatfield-Dodds et al. 2008) has indicated that more than 2.5 million jobs will need to be filled over the next two decades. Many of these jobs will be in areas either directly or indirectly influenced by the climate change response. In addition to the construction and energy sectors, areas of potential employment change include transport, agriculture and a range of services. Many of these jobs will be in industry subsectors that barely exist today and some that lie within the imagination of farsighted entrepreneurs. The need to supply appropriately skilled people for these jobs is in addition to the need to develop new knowledge and skills in existing roles and sectors around the issues that emerge from the implementation of climate change policies. There will be few sectors left untouched.

The implications of these changes for Australia's education and training sector are yet to be fully appreciated. These implications, and the necessary response from government, business, labour and our educational and training institutions, need to be comprehensively understood and integrated into the long-term planning of these bodies.

23.9 Global mitigation and ongoing prosperity

The exceptional prosperity of Australia through the early 21st century has emerged through the productive interaction of a flexible, internationally oriented domestic economy with the opportunities generated by rapid growth in the major Asian countries. Will the international environment continue to be conducive to Australian prosperity in an era of climate change mitigation and climate change?

Sustained, strong economic growth now has deep roots in China. It is well established, at a less frenetic pace, in India and Southeast Asia. It is gaining institutional sinews but remains vulnerable in other developing and transitional economies.

Throughout the developing and transitional economies, the beneficent processes of rapid economic growth that have emerged or strengthened in the current century are vulnerable to large shocks that disturb established social and political relationships and institutions and break the momentum of growth.

The risk in China, in the two decades that are required to complete the major part of its remaining transition to a developed economy, derive mainly from the dynamics of the complex political and social change set in train by sustained rapid economic growth. Large shocks from climate change could interact with these dynamics in destabilising ways. More immediately, differences in approach to mitigation between China and the developed world, especially the United States, would add fuel to the reactions to Chinese export expansion that are already smouldering. A protectionist conflagration fuelled by differences over climate change is as likely to hurt the developed world as China. But all would be damaged—China and its trading partners, and none more than Australia.

In the developing world beyond China, shocks from climate change are amongst the potential threats to growth, as risks of large impacts increase through the 21st century.

Climate change mitigation in the developing world will have large effects on Australian opportunity. Australia has a strong interest in mitigation not slowing growth in developing countries. The proposals for global mitigation put forward in this Review are consistent with sustained strong growth in the developing world.

Partial mitigation would define a difficult world for Australia. This is a world in which there are large risks of trade interventions in developed countries, prompted initially by valid concerns about 'carbon leakage', but spilling over into crude protectionism. This would be directly damaging for Australian trade, and probably even more damaging through its effects on developing countries whose economic health is important for Australia.

More than other developed countries, Australia has a comparative advantage, properly assessed to take carbon externalities into account, in industries vulnerable to carbon leakage. Australia therefore has a particularly strong interest in early movement to comprehensive global mitigation. It is important for Australia that major developing countries participate in international trade in emissions entitlements, or else accept participation in sectoral agreements, applying domestic carbon taxes, in the main trade-exposed, emissions-intensive industries.

There is a large risk that governments will apply domestic mitigation policies that favour domestic over international coal sources. China and Indonesia in particular, but also India, have large domestic coal resources, while being substantial and rapidly growing importers. In a shift away from reliance on coal for power generation, there would be strong tendencies first to exclude imported coal. The only safeguards would be provided by strong international trading rules and efficient domestic mitigation regimes. Metallurgical coal is less vulnerable.

Effective mitigation regimes would affect Australian metal exports in complex ways, but would mostly be advantageous. It is likely that effective mitigation would be associated with reduction of subsidies for domestic processing of metals—such as is already important in Chinese policy development. Again, Australia has a strong interest both in an effective global trading system and in efficient mitigation.

Australia is a global leader in the wide range of skills and technologies that are relevant to high performance in the resource industries. Businesses with headquarters functions in Australia are playing major roles in development of the resource sectors in all inhabited continents. In addition, Australia is a major exporter of engineering, management, financial and legal services related to the resources sector.

The skills and capacities that are important in the resources sector, more than others, will be crucial to the global transition to a low-emissions economy. The development in Australia of low-emissions technologies in the energy and agriculture sectors in particular will provide a basis for Australian businesses to play leading roles in innovation associated with the low-emissions transition in many countries, and especially in the Asia and the Pacific. The importance of export of related services from Australia would be enhanced by early establishment in Australia of efficient and extensive mechanisms in support of research, development and commercialisation of the new technologies.

23.10 Australia in a successful world of change

The Review has recommended a necessary and sufficient mitigation policy package that will facilitate the reliable, efficient and equitable transformation for Australia to a low-emissions economy. Australia has a strong national interest in a global agreement and the development of competitive, low-emissions energy technologies such as large-scale renewables and carbon capture and storage. The former will ensure that currently emissions-intensive commodities, such as livestock and fossil fuels, do not face the risk of carbon leakage, while the latter offers the best chance for sectors such as aluminium to be competitive—except against specific locations with availability of large-scale, otherwise stranded hydro power.

The results of our economic modelling also indicate that a comprehensive global agreement will mean that Australia can play to its strengths, importing permits when relatively high domestic abatement costs cause that to be the most effective and efficient result.

Given these outcomes, Australia's advantages in resource availability and sophisticated skills in critical areas of resource development, engineering and financial management are likely to ensure that we remain a strongly competitive suppler of commodities. As described in Chapter 20, Australia is well-placed to maintain a competitive domestic energy supply sector as well as to continue as a major exporter of primary energy. The shifting global economic balance towards our near neighbours over the 21st century will also tend to act in Australia's favour as the benevolence of proximity erodes the tyranny of distance, particularly as international freight costs rise as a result of the emissions constraint. These factors, and rising international transport costs, may also provide an incentive for greater domestic, value-adding processing of raw materials.

A world of effective global mitigation, with Australia playing its proportionate part, would be a congenial world for continued Australian prosperity.

A world of partial and inadequate mitigation, overtaken from time to time with rushes to catch up with lost mitigation opportunity, would be deeply problematic.

It is worth Australia's effort to invest now, when there is time still to obtain a good result, in the best of the possibilities.

Notes

1 In evidence to the Parliamentary Committee on Monetary Policy on 8 September 2008.

2 The results presented in this chapter are for the mitigation scenarios discussed in Chapter 11, which include an assumption of a backstop technology at a carbon price of around $250 per tonne of CO_2-e.

References

ABS 2008, *Australian National Accounts: Input–Output, 2004–05 (preliminary)*, cat. no. 5209.0.55.001, ABS, Canberra.

DCC (Department of Climate Change) 2008a, Australia's National Greenhouse Accounts, *Australian Greenhouse Emissions Information System*, <www.ageis.greenhouse.gov.au>, data as at 26 May 2008.

DCC 2008b, Australia's National Greenhouse Accounts, *Australian Greenhouse Emissions Information System*, <www.ageis.greenhouse.gov.au>, data as at 23 September 2008.

Hatfield-Dodds, S., Turner, G., Schandl, H. & Doss, T. 2008, 'Growing the green collar economy: skills and labour challenges in reducing our greenhouse gas emissions and national environmental footprint', report to the Dusseldorp Skills Forum, CSIRO Sustainable Ecosystems, Canberra, June.

McKinsey & Company 2008, *An Australian Cost Curve for Greenhouse Gas Reduction*, McKinsey & Company, Sydney.

McLennan Magasanik Associates 2008, *Defining a National Energy Efficiency Strategy, Stage 1 Report*, Report to the Climate Institute, Sydney.

FATEFUL DECISIONS 24

There are times in the history of humanity when fateful decisions are made. The decision this year and next on whether to enter a comprehensive global agreement for strong action on climate change is one of them.

Or rather, in this case, a fateful series of decisions. The world will not arrive at a satisfactory single settlement in one meeting in Copenhagen, or in one meeting after that.

If things go well, the decisions of many governments will lead into a comprehensive global agreement in Copenhagen. That agreement will lead to the world taking major new steps on mitigation in all major countries. Substantial financial flows to developing countries for mitigation and adaptation will expand beyond recognition. Structures and incentives will have been established to support a large increase in investment in the new technologies necessary for mitigation to occur at reasonable cost.

If things go well, very well, Copenhagen will be the end of one process, and the beginning of others that will lead, over time, to effective global mitigation at a level that reduces risks of dangerous kind to an extent that seems acceptable to most informed people.

If things go badly, they could go very badly.

When human society receives a large shock to its established patterns of life, the outcome is unpredictable in detail but generally problematic.

Things fall apart.

The initial financial shocks that hit Australia in the 1890s, central Europe and the industrial world in the 1930s, or Indonesia in the 1990s, were in themselves substantial, but turned out to be small in comparison to the chain of events that followed. In themselves, these shocks could have been expected to cause a pause in growth, but not one that would throw history from its course. But each shock was large enough to exceed some threshold of society's capacity to cope with change. In each case, what might have been a recession of substantial but ordinary magnitude became a great depression. Total output fell by a fifth or more. The associated social convulsions changed political institutions fundamentally and as permanently as human institutions can be changed. They shifted the whole trajectory of economic growth.

Unmitigated climate change, or mitigation too weak to avoid dangerous climate change, could give human society such a shock.

The case for strong mitigation is a conservative one. Even at the levels of mitigation that now seem to be the best possible, the challenges could be considerable. In the absence of mitigation, we can be reasonably sure that they would be bad beyond normal experience.

We know that immense shocks unsettle basic institutions, with unfathomable consequences.

We know that the possibilities from climate change include shocks far more severe than others in the past that have exceeded society's capacity to cope, and moved societies to the point of fracture.

Here we are talking about global fracture.

If sea level rises by a metre or more this century and as much again in the first half of the next, and displaces from their homes the people of the low-lying coasts and river banks of the island of New Guinea, it will not be a problem for Papua New Guinea and Indonesia alone.

If sea level rises and displaces from their homes a substantial proportion of the people of Bangladesh and West Bengal, and many in the great cities of Dhaka, Kolkata, Shanghai, Guangzhou, Ningbo, Bangkok, Jakarta, Manila, Ho Chi Minh City, Karachi and Mumbai, it will not be a problem for Bangladesh, India, Pakistan, China, Thailand, Indonesia, the Philippines and Vietnam alone.

If changes in monsoon patterns and the flows of the great rivers from the Tibetan plateau disrupt agriculture among the immense concentrations of people that have grown around the reliability of water flows since the beginning of civilisation, it will not just be a problem for the people of India, Bangladesh, Pakistan, Vietnam, Myanmar and China.

There will be no islands of normality in Melbourne or Mildura, even if the same forces on climate have not displaced the people around the edges of Port Phillip Bay and scorched the economic life from the Murray-Darling Basin.

The problems of unmitigated climate change will be for all humanity.

During the discussions following the release of the Review's draft report in early July 2008, some critics said that my descriptions of impacts had been 'alarmist'. I responded that I was simply telling the story as it fell out of the analysis, when the emissions growth suggested by the Review's own work was applied to 'centre of the road' scientific judgments on the relationship between CO_2 concentrations and temperatures.

I was talking then about impacts in the middle of the probability distributions that come, as best we can judge, from contemporary science.

I did not then talk about some of the possible shocks that I am discussing now: shocks that until recently were a fair way along the 'possible but not very likely' end of the probability distribution, but have been moved closer to the centre by the Review's work on business as usual scenarios. Some shocks that would be severe and damaging that were once near the edges of the distributions are now near the middle. In the absence of mitigation, as we move beyond this century, some of these shocks move to the higher probability ends of the distributions. As noted in Chapter 11, without strong mitigation, the melting of the Greenland ice sheet, sooner or later, becomes something close to a sure thing.

In Chapter 2, the Review accepts the views of mainstream science 'on a balance of probabilities'. That formulation allows the possibility that the views on climate change of the IPCC and the learned academies in all of the main countries of scientific achievement are wrong.

There is a chance that they are wrong. Just a chance. But to heed instead the views of the minority of genuine sceptics in the relevant scientific communities would be to hide from reality. It would be imprudent beyond the normal limits of human irrationality.

It is prudent to give the major weight to the mainstream science. This is fully compatible with investing more in improvement of knowledge to narrow the dispersion of the probability distributions. The improvement of knowledge, the narrowing of uncertainty, the sharpening of predictions: all these can and should proceed alongside the commencement of international collective action in pursuit of strong mitigation.

The annual costs of strong mitigation continue to increase over the first half of the century. The mitigation process can be cut short, with due notice to those who have committed their capital to a new economy of low emissions, if at any time the international community comes to the view that new scientific knowledge establishes that the concerns of 2008 were erroneous to the extent that mitigation judgments based on them have become obsolete. Mitigation could come to a stop in 2020, for example, on the basis of new knowledge that it was unnecessary, after mitigation had been put in place to return to concentrations of 450 ppm.

In this case, Australia would have paid 2 per cent of GNP as insurance against what would otherwise have been a high risk of immense damage. It would be a high price, but one that was reasonable on the basis of the evidence available at the time when decisions had to be made.

The consequences of inaction now are not similarly reversible. The arithmetic of Chapter 3 about the new patterns of global growth takes away the time we may once have thought we had for experiment, talk, and leisurely decision making. It tells us that business as usual is taking us quickly towards what the science tells us are high risks of highly disruptive climate change.

So fateful decisions are to be taken at Copenhagen.

The analysis of the current international situation in chapters 8, 9 and 10 tells us that a good outcome is not assured.

The international community is on a course plotted before the implications of the current era of growth we call the Platinum Age had been absorbed into its decision-making framework. It is on a course plotted before humanity had absorbed the implications of the acceleration of economic growth in the early 21st century; the concentration of that growth in economies at the stage of development when growth absorbs huge amounts of energy; and in countries where coal is the cheapest and most convenient energy source. New knowledge changes the calculus.

The old calculus said that there was time—time for all developed countries to take the early steps in mitigation, and then for all developing countries to join at a later unspecified date. The old calculus said that it was good enough for the developing countries to begin to contribute through the Clean Development Mechanism and in other ways that made no additional contribution to the global mitigation effort, beyond commitments that the developed countries had already made.

The Review's updated projections show that approaches based on the old calculus will not hold the risks of dangerous climate change to acceptable levels.

Success at Copenhagen is not an agreement along the lines of the Bali Roadmap. Success will need to build on the foundations of Bali and earlier UNFCCC agreements, because there is no time to start again. But the content of any agreement will need to go beyond what had been contemplated at Kyoto and Bali.

Success at Copenhagen requires agreement to large emissions reductions from developed countries, plus agreement on a framework for early contributions to mitigation from China and as soon as possible from other successful developing countries.

This formulation underplays the importance of another part of the contemporary reality. It is much more likely that effective mitigation from developed countries will be achieved within a comprehensive global mitigation regime. Developing country participation would remove competitive distortion in trade-exposed industries. It would demonstrate to the polities of the developed countries that their contributions are not pointless self sacrifice, but part of a solution to the global problem of climate change.

So success at Copenhagen, or at subsequent meetings convened for the purpose, must encompass inclusion of developing countries in a global mitigation regime. The arithmetic of Chapter 3 shows that the participation of China is urgent. Comprehensive participation, beyond China, is necessary for the political and economic viability of the regime.

So the fateful decision at Copenhagen is not just about whether there will be a comprehensive regime.

It has to be a *credible* agreement. This means that the sum of national commitments must 'add up' to the environmental objective.

Chapter 9 set out a principled basis for global agreement that meets these objectives and places manageable obligations on developing countries with a reasonable chance of acceptance. The Chinese constraints, which the arithmetic says must be binding, are consistent with domestic goals that the government of China has set for itself. For other developing countries, acceptance of constraints would not be binding, but there would be large advantages for them in participating. The trajectories for emissions constraint, based on modified contraction and convergence, would provide opportunities for them to do better, and to sell surplus permits, providing new economic opportunities. Acceptance of constraints would allow developing countries access to the low-emissions technology and adaptation funding commitments of the developed countries. They would avoid the disruption to trade that might come to be associated with standing aside from international cooperation on mitigation.

Let us be clear about the contemporary reality of global mitigation, and of the gap between where we are and where we need to be. There are few countries in which mitigation policies have yet had a substantial effect on emissions reduction. The large reductions that have occurred in some countries have come from structural change that was not associated with mitigation policies. Global expenditure on low-emissions technologies has been at a low ebb—much lower than had been induced by the high oil prices of the 1970s. No developed country has yet put in place policies that can be reasonably expected to achieve its share of the reductions in emissions necessary for 550 ppm concentrations objectives, let alone something more ambitious. While China and some other developing countries have implemented policies that are moderating the growth in emissions, no developing country has been willing to concede that binding emissions constraints should also apply to its own economy.

The first essential step at Copenhagen is a comprehensive global agreement that adds up to the environmental objective to which it is directed.

Achievement of a comprehensive agreement around a 550 ppm objective would be a step forward of historic dimension. Such an achievement and its effective implementation would avoid the worst outcomes from unmitigated climate change. It would give confidence to the international community that cooperation is possible in this difficult sphere. Once in effect, alongside a low-emissions technology commitment, it would unleash forces for innovation and structural change that would demonstrate that strong mitigation was consistent with continued economic growth, and bring more ambitious goals into the realm of the possible. It would bring the next step to 450 closer to reach.

Effective comprehensive global agreement around a 450 ppm objective, if it were realistic in conception and implementation, would be better still, for Australia and for the international community. It would be 450 ppm with overshooting, because we are already at around 450 ppm and this level will go much higher before the momentum of emissions growth is slowed, halted and then turned around.

The numbers that add up to a 550 ppm global objective seem not to be impossible for the separate sovereign nations that will have to form the view, one by one, that acceptance and compliance is in their own interests.

The numbers that add up to 450 ppm are not yet within the decision frames of the people who will need to commit to them. More ambitious numbers may become feasible, and sooner rather than later, through the building of confidence from the early years of successful implementation of a 550 ppm regime.

There is much that we do not know about the future costs or possibilities of low emissions technologies, as there is much that we do not know in climate science itself. Chapters 20, 21, 22 and 23 describe the possibility that the incentives provided by a substantial and rising carbon price, and public fiscal support for investment in innovation, could lead to large reductions in the cost of structural transformation. In the nature of things, we will only learn by doing. It is important to start taking the measures we need to take within carefully designed institutional frameworks.

The difference in environmental outcome between successful achievement of a 550 ppm objective and of a 450 ppm objective is substantial for Australia, as demonstrated in chapters 6 and 11 in particular. But it is small compared with the difference between 550 ppm and the complete failure of mitigation. The difference between 550 ppm and 450 ppm is small compared with the difference between 550 ppm and emissions growth remaining anywhere near its current course.

The fateful decisions at Copenhagen will be for all sovereign nations. But the fates will be set long before December 2009. They will be set in the earlier national policy decisions taken by many countries, including Australia.

The Review's conclusion that it would be in Australia's interests for the world to agree on commitments that add up to a 450 ppm objective, and which is capable of implementation, is the basis of our recommendation that Australia express its willingness to do its full, proportionate part in such a global agreement. But Australia and the world have an even bigger interest in ensuring that realistic and comprehensive global mitigation is begun around some attainable mitigation objective as a result of Copenhagen.

Australia will matter to the international community's fateful decision. We can make a difference by announcing at an early stage that we are prepared to play our full proportionate part in an ambitious global mitigation effort. We can take the lead in a global effort to commercialise carbon capture and storage technologies, that would, if successful, greatly ease the adjustment to low-emissions economies of the developing world, and incidentally preserve a future for Australia's coal industry. We can take the lead in promotion of the International Low-Emissions Technology Commitment, and by commencing with a national commitment.

Australia can take the lead by building on what we have already begun, in establishing productive cooperation on climate change issues with our neighbours, first of all Papua New Guinea and Indonesia. An example of successful cooperation that was advantageous for development, including in low-emissions technologies, adaptation and permit trading, the latter covering forestry and all other emissions, would be influential.

Australia can take a lead by placing in the international marketplace for ideas proposals that add up to realisation of a global environmental objective. Australia would do this not with any arrogant insistence that this is the only proposal to be considered, but making it clear that we would also want to consider alternative proposals, developed in other countries, that add up to specified objectives.

Australia could suggest that a number of heads of government with commitments to a strong outcome at Copenhagen each appoint a representative to a group of experts. This group would be given the task of coming up with a practical approach that adds up to defined environmental objectives for consideration by leaders in the lead-up to Copenhagen.

Australia can take a lead by preparing to implement at reasonable cost the full range of mitigation programs necessary to meet the commitments that we make to reduce emissions. Australia will be more effective if we are introducing a well-designed emissions trading system. It will help if Australia makes it clear that we are proceeding in any case with national mitigation, within the parameters suggested in this report.

It is sometimes said that Australia's influence would be greater and more positive if, in the absence of comprehensive international agreement, we would unilaterally implement much more radical reductions in emissions than those put forward in Chapter 12. This neglects the economic reality. A world of partial mitigation, in which individual countries do their own thing, is a world in which mitigation is more difficult and more expensive. To go it alone beyond certain levels of ambition would be to demonstrate the problems rather than the feasibility of mitigation. It is doubtful that this would encourage the global mitigation effort. It may deter it.

So the fateful decision at Copenhagen will follow many decisions in Australia and elsewhere between now and then.

And after Copenhagen, there will be more big decisions to be made. If there is a comprehensive and effective global agreement, the scene will be set for reconsideration of ambition once it has been demonstrated that mitigation is consistent with continued economic growth.

If there is no such agreement, the outlook is an unhappy one.

On a balance of probabilities, the failure of our generation would lead to consequences that would haunt humanity until the end of time.

List of figures and tables

Figures

Figure 1.1	The risk–uncertainty spectrum	8
Figure 1.2	A probability distribution	8
Figure 1.3	The four types of climate change impacts	10
Figure 1.4	Utility with and without mitigation	16
Figure 1.5	Utility under a more ambitious level of mitigation	16
Figure 1.6	Utility with more climate change impacts taken into account	17
Figure 2.1	Trends in atmospheric concentrations of carbon dioxide, methane and nitrous oxide since 1750	26
Figure 2.2	A stylised model of the natural greenhouse effect and other influences on the energy balance of the climate system	28
Figure 2.3	Contribution of human and natural factors to warming since 1750	30
Figure 2.4	Steps in the causal chain of greenhouse gas emissions leading to climate change	31
Figure 2.5	Effect on extremes of temperature from an increase in mean temperature, an increase in variance, and an increase in both mean temperature and variance	41
Figure 2.6	Inertia in the climate system	42
Figure 2.7	Response of different carbon sinks to the rate of emissions over time	44
Figure 2.8	Different pathways of emissions reductions over time to achieve the same concentration target	45
Figure 2.9	Temperature outcomes of varying levels of overshooting	46
Figure 2.10	Emissions pathways required to achieve a low concentration target following an overshoot	47
Figure 3.1	The 20 largest greenhouse gas emitters: total emissions and cumulative share (%) of global emissions, c. 2004	54
Figure 3.2	The 20 largest greenhouse gas emitters: per capita emissions including and excluding emissions from land-use change and forestry, c. 2004	55
Figure 3.3	CO_2 emissions/GDP, energy/GDP and CO_2 emissions/energy for the world, OECD and non-OECD, 1971–2005 (1971 = 1)	57
Figure 3.4	Energy intensities of GDP for China and other developing countries, 1970–2005	57
Figure 3.5	The reference case: global population, GDP and GDP per capita, 2001 to 2100	59
Figure 3.6	The reference case: global population, GDP, GDP per capita, and CO_2-e emissions, 2000 to 2100—average growth rates by decade	60
Figure 3.7	Shares in global output of various countries and regions, 2001 to 2100, under the reference case	61

Figure 3.8	Global CO_2 emissions growth rates from fossil fuels and industrial processes to 2030: a comparison of Garnaut Review no-mitigation projections with SRES and post-SRES scenarios and historical data	62
Figure 3.9	Global greenhouse gas emissions growth rates to 2030: a comparison of Garnaut Review no-mitigation projections, SRES and post-SRES scenarios, and historical data	63
Figure 3.10	Global greenhouse gas emissions to 2100: a comparison of Garnaut Review no-mitigation projections and various SRES scenarios	64
Figure 3.11	China total energy consumption, levels and growth, 1978 to 2006	66
Figure 3.12	Oil, gas and coal prices, 1970 to 2008	67
Figure 3.13	Global energy use and CO_2 emissions, 1970 to 2007	68
Figure 4.1	Selected regional climate change observations	76
Figure 4.2	Average global air temperature anomalies, 1850–2005	78
Figure 4.3	Global average sea-level rise, 1870–2005	80
Figure 4.4	Concentrations of greenhouse gases in the atmosphere for the three emissions cases, 1990–2100	86
Figure 4.5	Global average temperature outcomes for three emissions cases, 1990–2100	88
Figure 4.6	Spatial variation in temperature change in 2100 for the three emissions cases	90
Figure 4.7	Temperature increases above 1990 levels for the three emissions cases	92
Figure 4.8	Abrupt or rapid climate change showing the lack of response until a threshold is reached	97
Figure 5.1	Australian annual average temperature anomalies, 1910–2007	107
Figure 5.2	Annual streamflows into Perth's dams (excluding Stirling and Samson dams)	110
Figure 5.3	Best estimate (50th percentile) of Australian annual temperature change at 2030, 2070 and 2100 under three emissions cases	114
Figure 6.1	Vulnerability and its components	125
Figure 6.2	State and territory impacts of climate change by 2100 under the no-mitigation case	126
Figure 7.1	Per capita greenhouse gas emissions	154
Figure 7.2	Greenhouse gas emissions by sector, 1990 and 2006	155
Figure 7.3	Greenhouse gas emissions by sector: 1990, 2006 and reference case scenarios	156
Figure 7.4	Per capita emissions due to energy use, 2005	157
Figure 7.5	Factors underlying per capita energy emissions, 2005	157
Figure 7.6	Fuel mix contributing to total primary energy supply, 2005	158
Figure 7.7	Trends in average emissions intensity of primary energy supply, Australia and OECD, 1971–2005	159
Figure 7.8	Primary energy consumption in Australia, by sector, 2005–06	159
Figure 7.9	Per capita emissions due to electricity, 2005	160

Figure 7.10	Factors underlying per capita electricity emissions, 2005	160
Figure 7.11	Per capita emissions due to transport, 2005	161
Figure 7.12	Factors underlying per capita transport emissions	162
Figure 7.13	Per capita emissions due to agricultural production	163
Figure 7.14	Per capita area of forested and wooded land, 2005	164
Figure 7.15	Emissions attributable to Australian industry by sector, 2006	166
Figure 7.16	Emissions attributable to the Australian mining and manufacturing industries, disaggregated by sector, 2005	167
Figure 7.17	Ratio of permit costs to value of production, 2005	169
Figure 7.18	Direct emissions intensity of Australia's agriculture industry compared with selected OECD countries, 2006	170
Figure 8.1	Kyoto targets and 2005 emissions, relative to 1990	181
Figure 9.1	Different concentration goals: stabilisation, overshooting and peaking	193
Figure 9.2	Different cumulative emissions from the same end-year target (y)	195
Figure 9.3	Emissions trajectories for the no-mitigation, 550 and 450 scenarios, 2000–2100	206
Figure 9.4	Per capita emissions entitlements for the 550 scenario, 2012–2050	208
Figure 9.5	Per capita emissions entitlements for the 450 scenario, 2012–2050	208
Figure 10.1	Energy research and development expenditure by the public and private sectors in the United States	218
Figure 11.1	Australia's carbon prices under different mitigation scenarios and technological assumptions	251
Figure 11.2	The modelled expected market costs (median case) for Australia of unmitigated climate change, 2013 to 2100 (Type 1 costs only)	253
Figure 11.3	Change in annual Australian GNP growth (percentage points lost or gained) due to gross mitigation costs under the 550 scenario strategy compared to no mitigation, and under standard and enhanced technology assumptions, 2013–50	264
Figure 11.4	Change in annual Australian GNP growth (percentage points lost or gained) due to net mitigation costs under the 550 scenario compared to no mitigation, 2013–2100	265
Figure 11.5	Change in Australian sectoral growth rates (percentage points lost or gained) due to net mitigation costs under the 550 scenario compared to no mitigation, 2013–2100	266
Figure 11.6	A comparison of the modelled expected market costs for Australia of unmitigated and mitigated climate change up to 2100 (Type 1 costs only)	267
Figure 11.7	Change in annual Australian GNP growth (percentage points lost or gained) due to gross mitigation costs under the 450 compared to the 550 scenario and under standard and enhanced technology assumptions, 2013–50	268

Figure 11.8	Change in annual Australian GNP growth (percentage points lost or gained) due to net mitigation costs under the 450 compared to the 550 scenario, 2013–2100	269
Figure 12.1	Australian emissions reductions trajectories to 2050 (reduction in total emissions)	284
Figure 12.2	Australian emissions reductions trajectories to 2050 (per capita reduction)	284
Figure 15.1	Areas for further support and investment in the climate change research system	366
Figure 15.2	Impact of climate change on probability loss distribution and implications for risk capital requirements	372
Figure 16.1	How will an emissions price flow through the economy?	387
Figure 16.2	Expenditure on basic goods as a share of disposable income	388
Figure 17.1	Residential per capita electricity consumption in the United States, California and as predicted for California	416
Figure 18.1	The innovation chain	425
Figure 18.2	Market failures along the innovation chain	426
Figure 19.1	Major sequestration sites and carbon dioxide sources in Australia	454
Figure 20.1	Installed electricity generation capacity, 2005–06	469
Figure 20.2	Comparison of industrial electricity prices	470
Figure 20.3	Average electricity market prices, 1999–2008	471
Figure 20.4	International energy commodity prices, indexed to 1994	473
Figure 20.5	Australia's electricity demand	483
Figure 20.6	Electricity demand reduction in selected sectors, 550 scenario	484
Figure 20.7	Residential demand, 2005–2100	484
Figure 20.8	Australia's electricity technology shares, 550 scenario	485
Figure 20.9	Australia's electricity generation technology shares, 550 scenario with zero-leakage carbon capture and storage	486
Figure 20.10	Australia's electricity generation technology shares, 450 scenario	487
Figure 20.11	Electricity emissions intensity	487
Figure 20.12	Total wholesale electricity costs, with and without nuclear, 550 scenario	488
Figure 20.13	Technology mix under an enhanced technology scenario	489
Figure 20.14	Wholesale electricity prices, 2005–50	490
Figure 20.15	Electricity generated from coal	491
Figure 20.16	Generation capacity, 550 scenario	492
Figure 20.17	Generation capacity, 450 scenario	492
Figure 20.18	Carbon capture and storage scenarios	496
Figure 20.19	Projections for aluminium and alumina industries	497
Figure 20.20	Electricity from gas sources	498
Figure 21.1	Australian domestic transport emissions, 2006	505
Figure 21.2	Passenger travel per capita by various modes, 1970–71 to 2006–07	506

Figure 21.3	Emissions intensity of passenger modes, 2007	509
Figure 21.4	Projected emissions from the domestic transport sector with standard technology assumptions, 2006–2100	512
Figure 21.5	Breakdown of transport sector emissions in the 550 standard technology scenario, 2006–2100	513
Figure 21.6	Modelling of road transport fuel use in a 550 standard technology scenario	514
Figure 21.7	Average new car emissions and oil price, January 2002 – April 2008	518
Figure 21.8	Trip mode, population and emissions in 57 high-income cities, 1995–96	520
Figure 21.9	Mode share for journeys to work in Australian capital cities 1976–2006	521
Figure 22.1	Prices paid by Australian farmers, 1998–2007	533
Figure 22.2	Australian farmers' terms of trade, 1998–2007	534
Figure 22.3	Non-combustion emissions for agriculture, forestry and land-use change for the no-mitigation and 550 standard technology scenarios	537
Figure 22.4	Change in emissions intensity over time in response to carbon price, 550 standard technology scenario, 2006–2100	538
Figure 22.5	Contribution to Australia's agricultural emissions, by subsector, 2005	544
Figure 22.6	Ratio of emissions permit costs to value of production, by subsector, 2005	545
Figure 22.7	Australian real retail prices for meat, 1960–2006	546
Figure 22.8	Australian per capita consumption of meat, 1960–2006	547
Figure 22.9	Per capita area of forested and wooded land, 2005	555
Figure 22.10	Carbon removal potential for environmental plantings (tonnes CO_2-e per ha per year)	555
Figure 23.1	Australia's emissions in the 550 backstop scenario (global entitlement, net of trading)	571
Figure 23.2	Sources of mitigation under the 550 backstop scenario	572
Figure 23.3	Direct emissions per million dollars value added, 2005	573
Figure 23.4	Direct and indirect emissions per million dollars value added, mining and manufacturing, 2005	573
Figure 23.5	Emissions sources (not including forestry) in the 550 backstop technology scenario	574
Figure 23.6	Australia's emissions in the 450 backstop technology scenario (global entitlement, net of trading)	575
Figure 23.7	Total emissions for the no-mitigation, 450 and 550 backstop scenarios	576
Figure 23.8	Sources of mitigation under the 450 backstop technology scenario	577
Figure 23.9	Emissions sources (not including forestry) in the 450 backstop technology scenario	577

Tables

Table 2.1	Sources of greenhouse gases	31
Table 2.2	Estimates of the amount of carbon stored in different sinks in 1750 and how they have changed	36
Table 3.1	Growth in CO_2 emissions from fuel combustion, GDP and energy	56
Table 3.2	Shares of total greenhouse gas emissions by country/region in the Garnaut–Treasury reference case	65
Table 3.3	Time to exhaustion of current estimates of reserves and reserve base for various metals and minerals, and fossil fuels	71
Table 4.1	Summary of extreme climate responses, high-consequence outcomes and ranges for tipping points for the three emissions cases by 2100	102
Table 5.1	Projected changes to statewide annual average rainfall, best-estimate outcome in a no-mitigation case (per cent change relative to 1990)	115
Table 5.2	Projected changes to statewide average rainfall, dry and wet outcomes in a no-mitigation case (per cent change relative to 1990)	116
Table 5.3	Projected increases in days over 35°C for all capital cities under a no-mitigation case	117
Table 5.4	Projected per cent increases in the number of days with very high and extreme fire weather for selected years	118
Table 6.1	Sectors and areas considered in this chapter	122
Table 6.2	Climate cases considered by the Review	124
Table 6.3	Differences between probable unmitigated and mitigated futures at 2100—median of probability distributions	127
Table 6.4	Decline in value of irrigated agricultural production in the Murray-Darling Basin out to 2100 from a world with no human-induced climate change	130
Table 6.5	Percentage cumulative yield change from 1990 for Australian wheat under four climate cases	132
Table 6.6	Magnitude of impacts to water supply infrastructure in major cities under four climate cases	136
Table 6.7	Infrastructure impacts criteria	137
Table 6.8	Magnitude of impacts on buildings in coastal settlements under four climate cases	138
Table 6.9	Change in likely temperature-related deaths due to climate change	140
Table 7.1	Comparison of the highest per capita emissions among OECD countries (tonnes per person per year)	154
Table 7.2	Agricultural emissions and land use, land-use change and forestry emissions, by commodity and economic sector, 2005	168
Table 9.1	2020, 2050 and 2100 global emissions changes for the two global mitigation scenarios, relative to 2001 (per cent)	205

Table 9.2	Emissions entitlement allocations for 2020 and 2050 relative to 2000–01 and Kyoto/2012 (per cent)	209
Table 9.3	Emissions entitlement allocations expressed in per capita terms in 2020 and 2050 relative to 2000–01 and Kyoto/2012 (per cent)	210
Table 11.2	Assessing the market impacts of climate change	254
Table 11.3	Net present cost of the 450 ppm and 550 ppm scenarios (in terms of no-mitigation GNP) and the '450 premium' to 2050 and 2100	270
Table 12.1	Summary of interim targets in 2020 (per cent)	283
Table 12.2	Reductions in emissions entitlements by 2050 for policy scenarios (per cent)	283
Table 12.3	Modelling results in 2020 for policy scenarios	296
Table 13.1	Attributes of mitigation and adaptation shocks	303
Table 14.1	Governance of an Australian emissions trading scheme	352
Table 14.2	Interaction between the emissions trading scheme and the Mandatory Renewable Energy Target	355
Table 14.3	Overview of the proposed emissions trading scheme design	358
Table 17.1	Four kinds of principal–agent problems	414
Table 18.1	Brief assessment of two technology categories against criteria for national strategic interest	432
Table 18.2	Research and development programs in Australia targeting low-emissions technologies	435
Table 18.3	Mechanisms for directly subsidising positive externalities in demonstration and commercialisation	437
Table 18.4	Estimates of private and social rates of return to private research and development spending	441
Table 21.1	Transport sectors	504
Table 22.1	Vulnerability of Australia's agricultural industry to the biophysical impacts of climate change, by subsector	535
Table 22.2	Potential for emissions per annum reduction and/or removal from Australia's agriculture, forestry and other land use sectors	542
Table 22.3	Impact of emissions permit prices on cost of meat production	546
Table 22.4	Technical potential for CO_2 removal by soil—selected estimates	549
Table 22.5	Estimated oil yield per ha for biodiesel production	553
Table 22.6	Area of selected land uses in Australia	554
Table 23.1	Total after-tax per capita income (2005 dollars)	569
Table 23.2	Annual average growth rates for GNP and GDP under the no-mitigation, 550 and 450 scenarios with backstop technology (Type 1 and Type 2 benefits of mitigation) (per cent)	570

List of shortened forms

ABARE	Australian Bureau of Agricultural and Resource Economics
ABS	Australian Bureau of Statistics
ACCESS	Australian Community Climate and Earth System Simulator
ACIAR	Australian Centre for International Agricultural Research
APEC	Asia–Pacific Economic Cooperation
BoM	Bureau of Meteorology
CCS	carbon capture and storage
CDM	Clean Development Mechanism
CFCs	chlorofluorocarbons
CO_2-e	carbon dioxide equivalent
CO2CRC	Cooperative Research Centre for Greenhouse Gas Technologies
CSIRO	Commonwealth Scientific and Industrial Research Organisation
DCC	Department of Climate Change
GDP	gross domestic product
GEF	Global Environment Facility
GGAS	New South Wales Greenhouse Gas Reduction Scheme
GIAM	global integrated assessment model
GNP	gross national product
GTEM	global trade and environment model
HCFCs	hydrochlorofluorocarbons
IEA	International Energy Agency
IMF	International Monetary Fund
IPCC	Intergovernmental Panel on Climate Change
LNG	liquefied natural gas
LUCF	land use, land-use change and forestry
MMRF	Monash Multi Regional Forecasting
MRET	Mandatory Renewable Energy Target
NEMMCO	National Electricity Market Management Company
OECD	Organisation for Economic Co-operation and Development
OPEC	Organization of the Petroleum Exporting Countries
R&D	research and development
SRES	Special Report on Emissions Scenarios of the IPCC
UNFCCC	United Nations Framework Convention on Climate Change
WTO	World Trade Organization

Units of measurement

Gt	gigatonne (one billion metric tonnes)
kWh	kilowatt hour
Mt	megatonne (one million metric tonnes)
MWh	megawatt hour
PJ	petajoule (1015 joules)
ppb	parts per billion
ppm	parts per million
TJ	terajoule (1012 joules)
TWh	terawatt hour

Glossary

Definitions are taken from the Intergovernmental Panel on Climate Change wherever possible. A list of sources is provided at the end of the glossary.

Terms in a definition that appear elsewhere in the glossary are italicised.

abatement. Activity that leads to a reduction in *greenhouse gas* emissions.

abrupt climate change. The nonlinearity of the *climate system* may lead to abrupt *climate change*. The term 'abrupt' often refers to time scales faster than the typical time scale of the responsible *forcing*.

adaptation. Adjustment in natural or human systems in response to actual or expected climatic stimuli or their effects, which moderates harm or exploits beneficial opportunities.

adaptive capacity. The ability of a system to adjust to *climate change* (including climate variability and extremes) to moderate potential damages, to take advantage of opportunities, or to cope with the consequences.

additionality. Reduction in net emissions by sources or enhancement of removals by *carbon sinks* that is additional to the reduction that would occur in the absence of an incentive provided through a program.

aerosols. A collection of airborne solid or liquid particles, with a typical size between 0.01 and 10 micrometres (a millionth of a metre) that remain in the atmosphere for a relatively short time.

afforestation. Planting of new forests on lands that historically have not contained forests.

albedo. The amount of *solar radiation* reflected by a surface or object, often expressed as a percentage.

Annex B countries/parties. Industrialised countries and economies in transition countries listed in Annex B to the *Kyoto Protocol* that have emissions reductions targets for the period 2008–12.

Annex I countries/parties. Industrialised countries and economies in transition listed in Annex I to the *United Nations Framework Convention on Climate Change*. They include the 24 original OECD members, the countries of the European Union, and 14 countries with economies in transition.

anthropogenic. Resulting from or produced by human beings.

Bali Roadmap. The key decisions agreed at the 2007 Bali Climate Change Conference, charting the way for the UN negotiations on a post-2012 UN climate agreement.

base case. In the Review's modelling, the evolution of the global and Australian economies and associated *greenhouse gas* emissions to the end of the 21st century taking into account the impacts of *climate change*.

biochar. A charcoal product made through anaerobic combustion of biomass (for example, farm or wood waste) at high temperatures.

biosequestration. The removal from the atmosphere and storage of *greenhouse gases* through biological processes, such as growing trees and practices that enhance soil carbon in agriculture.

business as usual. A scenario of future *greenhouse gas* emissions that assumes that there would be no major changes in policies on mitigation.

carbon budget. The amount of carbon (or emissions, expressed as *carbon dioxide equivalent*) allowed to be released over a number of years, by a given party or parties.

carbon–climate feedback. See *feedback*.

carbon cycle. The term used to describe the movement of carbon in various forms (for example, as carbon dioxide or methane) through the atmosphere, ocean, plants, animals and soils.

carbon dioxide equivalent (CO_2-e). A measure that allows for the comparison of different *greenhouse gases* in terms of their *global warming potential*.

carbon dioxide equivalent concentration. The concentration of carbon dioxide (measured in parts per million) that would lead to the same amount of *radiative forcing* as a given mixture of carbon dioxide and other *greenhouse gases*.

carbon dioxide equivalent emissions. The amount of carbon dioxide emissions that would cause the same integrated *radiative forcing*, over a given time horizon, as an emitted amount of a well-mixed *greenhouse gas*. The equivalent carbon dioxide emission is obtained by multiplying the emission of a well-mixed greenhouse gas by its *global warming potential* for the given time period.

carbon dioxide fertilisation. Increasing plant growth or yield by elevated concentrations of atmospheric carbon dioxide.

carbon price. The price at which *emissions permits* can be traded, nationally or internationally.

carbon sink or **reservoir.** Parts of the *carbon cycle* that store carbon in various forms.

Clean Development Mechanism (CDM). A flexibility mechanism under the Kyoto Protocol that allows *Annex I countries* to meet part of their obligation to reduce emissions by undertaking approved emissions reduction projects in developing countries. Emissions reductions under the CDM can create tradable permits offset credits, called certified emission reductions or CERs.

climate change. A change in the state of the climate that can be identified (for example, by using statistical tests) by changes in the mean and/or the variability of its properties, and that persists for an extended period, typically decades or longer.

climate sensitivity. A measure of the *climate system*'s response to sustained *radiative forcing*. Climate sensitivity is defined as the global average surface warming that will occur when the climate reaches equilibrium following a doubling of carbon dioxide concentrations.

climate system. A highly complex system consisting of the atmosphere, the water cycle, ice, snow and frozen ground, the land surface and plants and animals, and the interactions between them.

CO_2-e. See *carbon dioxide equivalent.*

commitment period. The period in which *Annex B countries* are required to meet their emissions reduction commitments. The first commitment period is 2008 to 2012. The dates of the second commitment period have not yet been determined.

committed warming. Warming of the climate which, due to the thermal inertia of the ocean and slow processes in ice sheets, biological sinks and land surfaces, would continue even if the atmospheric composition were held fixed at today's values.

contraction and convergence. A model for allocating a global emissions budget among nations. Allocations of emissions entitlements start at current emissions levels, and converge over time to equal per capita allocations in all countries. At the same time, the global emissions budget, and thus the global per capita average, contracts toward lower levels.

deforestation. Conversion of forest to non-forested land.

direct emissions. Emissions from sources within the boundary or control of an organisation's or facility's (or individual's) processes or actions. They can include emissions from fuel combustion (for example, in transport) and non-combustion emissions arising from physical or chemical processes (for example, in agricultural production or industrial manufacturing).

ecosystem. A distinct system of interacting living organisms, together with their physical environment. The extent of an ecosystem may range from very small spatial scales to, ultimately, the entire earth.

El Niño – Southern Oscillation. A coupled fluctuation in the atmosphere and the equatorial Pacific Ocean that has a large influence on Australia's climate.

emissions (or **carbon**) **intensity.** A measure of the amount of carbon dioxide, or other *greenhouse gases*, emitted per unit of, for example, electricity, energy output or kilometre of travel.

emissions limit or **emissions cap.** A limit on the number of tonnes of *greenhouse gases* that can be emitted under an *emissions trading scheme*. The limit could apply to the whole economy, or to the sectors covered under the scheme.

emissions permit. See *permit*.

emissions trading scheme. A market-based approach to reducing emissions. An emissions trading scheme places a limit on emissions allowed from all sectors covered by the scheme. It allows those reducing *greenhouse gas* emissions to use or trade excess emissions *permits* to offset emissions at another source. Also referred to as a 'cap and trade scheme'.

energy efficiency. The ratio of energy required to produce a certain level of a service, such as kilowatt hours per unit of heat or light.

energy intensity. A measure of the amount of energy supplied or consumed per unit of, for example, gross domestic product or sales.

enteric fermentation. Part of the digestive process of ruminant animals, such as cows and sheep, that results in the release of methane.

evapotranspiration. The sum of evaporation and plant transpiration from the earth's land surface to the atmosphere.

exposure. The nature and degree to which a system is exposed to significant climatic variations.

feedback. An interaction mechanism between processes, where the result of an initial process triggers changes in a second process and that in turn influences the initial one. A positive feedback intensifies the original process, and a negative feedback reduces it.

fluorinated gases. Hydrofluorocarbons (HFCs), perfluorocarbons (PFCs) and sulphur hexafluoride (SF_6). See *greenhouse gas*.

forcing. An induced change to a system.

geo-engineering. Technological efforts to reduce global warming by stabilising the *climate system* through intervention in the energy balance of the earth.

geosequestration. Injection of carbon dioxide directly into underground geological formations.

global warming potential. The index used to translate the level of emissions of *greenhouse gases* into a common measure in order to compare the relative *radiative forcing* of different gases without directly calculating the changes in atmospheric concentrations.

greenhouse effect. The effect created by *greenhouse gases* in the earth's atmosphere that allow short-wavelength (visible) *solar radiation* to reach the surface, but absorb the *long-wavelength radiation* that is reflected back, leading to a warming of the surface and lower atmosphere.

greenhouse gas. Any gas that absorbs infrared radiation in the atmosphere. This property causes the *greenhouse effect*. With the exception of Chapter 2, where a wider range of greenhouse gases are discussed, the term 'greenhouse gases' in this report relates to those gases covered by the Kyoto Protocol, which are carbon dioxide, nitrous oxide, methane, sulphur hexafluoride, perfluorocarbons (PFCs) and hydrofluorocarbons (CHFCs).

ice sheet. A mass of land ice that is sufficiently deep to cover most of the underlying bedrock, so that its shape is mainly determined by the flow of the ice as it deforms internally and/or slides at its base.

indirect emissions. Emissions that are a consequence of the activities of an organisation (or individual) but originate from sources owned or controlled by another. Indirect emissions can refer to the emissions attributable to the purchase of electricity, heat or steam from another party, and also from activities such as outsourcing and waste disposal.

intertemporal flexibility. The ability to use *emissions permits* at different points in time, made possible through the flexibility mechanisms of hoarding and lending.

Kyoto Protocol. An agreement adopted under the *United Nations Framework Convention on Climate Change* in 1997. It entered into force in 2005.

long-lived greenhouse gases. A term used to identify the selection of *greenhouse gases* covered by the Kyoto Protocol to distinguish them from ozone and water vapour, both of which are removed from the atmosphere relatively quickly.

long-wavelength radiation. Thermal radiation, or heat, emitted by the earth's surface, the atmosphere and the clouds. It is also known as 'infrared radiation'.

Marrakesh Accords. A series of agreements signed in Morocco in 2001 on the rules of meeting the targets set by the *Kyoto Protocol*.

mitigation. A reduction in the source of, or enhancement of the sinks for, greenhouse gases.

Montreal Protocol. The Montreal Protocol on Substances that Deplete the Ozone Layer, adopted in 1987. It controls the consumption and production of chemicals that destroy stratospheric ozone, such as chlorofluorocarbons.

offsets. Reductions or removals of greenhouse gas emissions that are used to counterbalance emissions elsewhere in the economy.

overshoot scenario or **profile.** A mitigation scenario where concentrations of a *greenhouse gas* (or a mix of greenhouse gases) peak at a higher atmospheric concentration than the eventual target, and then reduce over time to achieve *stabilisation.*

passenger-kilometre. A measure of passenger transport activity, equal to one passenger carried one kilometre. For example, two individuals in a car travelling 50 kilometres is equal to 100 passenger-kilometres.

peaking scenario or profile. A mitigation scenario where concentrations of a *greenhouse gas* (or a mix of greenhouse gases) stabilise or peak, and then continue to reduce.

permit or **emissions permit.** A certificate created under an *emissions trading scheme* that enables the holder to emit a specified amount of *greenhouse gases*, generally one tonne of *carbon dioxide equivalent.*

price ceiling. An upper limit on *carbon prices*. Once the price ceiling is reached, an unlimited amount of permits are issued at that price.

floor price. A lower limit on *carbon prices*. When the floor price is reached, authorities would intervene to reduce the supply of permits, in order to keep prices at or above the floor.

primary energy. Energy in the forms obtained directly from nature, for example coal, natural gas or solar energy.

radiative forcing. A measure of the influence that a factor has on the energy balance of the climate system. Positive forcing tends to warm the surface, while negative forcing tends to cool it.

reference case. In the Review's modelling, the evolution of the global and Australian economies and associated greenhouse gas emissions to the end of the current century in the absence of climate change.

reforestation. Replanting of forests on lands that once contained forests but were converted to some other use.

secondary market. In the context of an emissions trading scheme, a financial market for trading of permits, whether by auction of some other method of allocation. It may also include markets in physical or financial contracts for the future purchase or sale of permits (forward contracts).

sensitivity. With respect to the *climate system*, the degree to which the system is affected, either adversely or beneficially, by climate-related stimuli. With respect to modelling, a sensitivity analysis may be used to assess how the variation of model assumptions affect the output of that model.

sequestration. Removal of carbon from the atmosphere by, and storage in, terrestrial or marine reservoirs.

severe weather event. An event that is rare within its statistical reference distribution at a particular place. The characteristics of what is called 'severe weather' may vary from place to place. An 'extreme climate event' is an average of a number of weather events over a certain period of time—an average that is itself extreme (for example, rainfall over a season).

sink. See *carbon sink*.

solar radiation. Electromagnetic radiation emitted by the sun. It is also referred to as 'short-wavelength radiation'.

stabilisation. In the *climate change* context, keeping constant the atmospheric concentrations of one or more *greenhouse gases* (such as carbon dioxide) or of a *carbon dioxide equivalent concentration* of a mix of greenhouse gases.

storm surge. A temporary increase, at a particular location, in the height of the sea due to extreme meteorological conditions (low atmospheric pressure and/or strong winds). A storm surge is the excess above the level expected from the tidal variation alone at that time and place.

stratosphere. The highly stratified layer of the atmosphere above the *troposphere* extending from about 10 km (ranging from 9 km at high latitudes to 16 km in the tropics on average) to about 50 km in altitude.

sunspot cycle. Periods of high activity observed in numbers of sunspots (small dark areas on the sun), as well as radiative output, magnetic activity and emission of high-energy particles.

temperature reference point or **baseline.** Unless otherwise specified, temperature changes discussed in this report are expressed as the difference from the period 1980–99, expressed as '1990 levels' in the Fourth Assessment Report of the Intergovernmental Panel on Climate Change. To compare temperature increases from 1990 levels to changes relative to pre-industrial levels, 0.5°C should be added.

thermal expansion. In connection with sea level, the increase in volume (and decrease in density) that results from warming water. A warming of the ocean leads to an expansion of the ocean volume and hence to sea-level rise.

thermohaline circulation. Large-scale circulation in the ocean driven by high densities at or near the surface, caused by cold temperatures and/or high salinities, in addition to mechanical forces such as wind and tides.

threshold or **tipping point.** The point in a system at which sudden or rapid change occurs, which may be irreversible.

tonne-kilometre. A measure of freight activity, equal to one tonne of freight carried one kilometre. For example, 20 tonnes carried 5 kilometres is equal to 100 tonne-kilometres.

trade-exposed, emissions-intensive industries. Industries with product prices that are set by world markets and that produce significant emissions during their production processes.

transaction costs. Costs associated with a market exchange (which may include indirect costs of market participation, such as information gathering).

transition countries. Countries in central and eastern Europe and the former Soviet Union defined in the *United Nations Framework Convention on Climate Change* and the *Kyoto Protocol* as 'undergoing the process of transition to a market economy'.

troposphere. The lowest part of the atmosphere, from the surface to about 10 km in altitude at mid latitudes (ranging from 9 km at high latitudes to 16 km in the tropics on average), where clouds and weather phenomena occur.

ultraviolet radiation. The high-energy, invisible part of the spectrum of light emitted by the sun. The majority of ultraviolet radiation is absorbed by the layer of ozone in the *stratosphere*.

United Nations Framework Convention on Climate Change (UNFCCC). The international treaty that sets general goals and rules for confronting *climate change*. It has the goal of preventing 'dangerous' human interference with the *climate system*. Signed in 1992, it entered into force in 1994, and has been ratified by all major countries of the world.

upstream point of obligation. Designating the point of obligation at a point higher or earlier in the supply chain. For example, the obligation for emissions from petrol can be placed upstream at the point of excise tax collection.

utility. Personal satisfaction or benefit derived by individuals from the consumption of goods and services.

vector-borne disease. A disease that is transmitted between hosts by a vector organism (such as a mosquito or tick—for example, dengue virus).

volumetric control. The imposition of a restriction on the amount of something allowed. For example, a cap and trade **emissions trading scheme** sets a limit on the amount of emissions that may be released over a given period of time without incurring a penalty. By contrast, a price control policy would set the cost of emissions or permits, but not the amount.

Sources

IPCC 2007, *Climate Change 2007: The physical science basis. Contribution of Working Group I to the Fourth Assessment Report of the Intergovernmental Panel on Climate Change*, S. Solomon, D. Qin, M. Manning, Z. Chen, M. Marquis, K.B. Averyt, M. Tignor & H.L. Miller (eds), Cambridge University Press, Cambridge and New York.

IPCC 2007, *Climate Change 2007: Impacts, adaptation and vulnerability. Contribution of Working Group II to the Fourth Assessment Report of the Intergovernmental Panel on Climate Change*, M.L. Parry, O.F. Canziani, J.P. Palutikof, P.J. van der Linden & C.E. Hanson (eds), Cambridge University Press, Cambridge.

IPCC 2007, *Climate Change 2007: Mitigation of climate change. Contribution of Working Group III to the Fourth Assessment Report of the Intergovernmental Panel on Climate Change*, B. Metz, O.R. Davidson, P.R. Bosch, R. Dave, L.A. Meyer (eds), Cambridge University Press, Cambridge and New York.

IPCC 2007, *Climate Change 2007: Synthesis report. Contribution of Working Groups I, II and III to the Fourth Assessment Report of the Intergovernmental Panel on Climate Change*, Core Writing Team, R.K. Pachauri & A. Reisinger (eds), IPCC, Geneva.

IPCC 2007, *Climate Change 2007: Synthesis report. An assessment of the Intergovernmental Panel on Climate Change*, A. Allali, R. Bojariu, S. Diaz, I. Elgizouli, D. Griggs, D. Hawkins, O. Hohmeyer, B. Pateh Jallow, L. Kajfez-Bogataj, N. Leary, H. Lee & D. Wratt (eds), Cambridge University Press, Cambridge.

Melbourne Water 2006, *Port Phillip and Westernport Region, Flood Management and Drainage Strategy*, Melbourne Water, Melbourne.

National Emissions Trading Taskforce 2007, *Final Framework Report*, submission to the Garnaut Climate Change Review, 2008.

Prime Minister's Science, Engineering and Innovation Council 2007, *Climate Change in Australia: Regional impacts and adaptation—managing the risk for Australia*, Independent Working Group, Canberra.

Index

Note: Page numbers in **bold** text denote boxed text; those in *italics* denote glossary definitions. References to figures, tables and endnotes are indicated by *fig*, *tab* and *n* after the page number.

450/550 mitigation scenarios
  description 86
  feasibility of reaching international agreement on 212–13
  greenhouse gas concentration projections 86
  likelihood of extreme climate outcomes 101
  modelling results
    450 backstop technology 575–76
    global emissions changes over 2001: 205–06
    per capita emissions, 2012–2050: 208
    purchases of foreign permits (450 and 550) 571, 575
    sources of emissions (450 and 550) 574, 577
    sources of mitigation (450 and 550) 572, 577
    total Australian emissions 571, 575
  summary of projected climate change 87–96
450 case *see* 450/550 mitigation scenarios
550 case *see* 450/550 mitigation scenarios
2020 targets *see* interim targets

abrupt climate change 96–97, *608*
adaptation
  adaptation challenges
    buildings in coastal settlements 378–79
    ecosystems and biodiversity 379–80
    emergency management services 381
    irrigated agriculture 377–78
    urban water supply infrastructure 378
  attitudinal aspects 14
  community protection as non-government responsibility 363–64
  cost–benefit impact 13–14
  defined 302, *608*
  role of food markets 375–76
  role of insurance markets 370–73
  see also agency barriers; climate information; information barriers
Adaptation Fund 224
adaptation policy
  creation and dissemination of climate information 366–69
  defined 302, 303
  funding climate research 367–68
  intervening to prevent ecosystem collapse 379–80
  liberalising food markets 375–76
  preventing inequitable responses 376
  promoting insurability 372–73
  Review's institutional proposals 366–69
adaptation technologies
  Australian early research 429
  current demonstration and commercialisation programs 434–36
  market failures in developing 426
adverse selection 407
aerosols
  and fossil fuels 89
  defined **29**, *608*
  in geo-engineering 49
  influence on climate system **29**, 29–30, 38, 45, 85, 108
agency barriers 404
agriculture
  adaptation to climate change 12
  Australian early research 429
  Australia's emissions, relative to OECD countries 167–70
  barriers to a low emissions future 558–60
  economic impact of climate change on **259**
  effect of emissions price on 393
  existing mitigation policies 535
  impact of climate change on 129, 534–35, 535*tab*
  in an emissions trading scheme 330
  potential for reducing emissions 538, 542–43*tab*
  projected future emissions 537–39, 541, 547–48
  projected productivity and composition of sector 539–40
  recent changes 532–33
  transaction costs of including in emissions trading scheme 540, 544–45
air traffic emissions 234
albedo effect 29, 49, *608*
algae, in carbon-cycle 35, 48
algal biofuels **432**, **495**, 552
algal biosequestration 424, **432**, 567–68, 578
alpine tourism **134–35**
alumina production 497
aluminium industry 497
Annex B countries/parties 187*n2*, 198, *608*
Annex I countries/parties 174, 175, 201, 206–07, *608*
Antarctic ice melt 94
anti-methanogen technology 424, 542

appliances
energy efficiency rating 413
minimum performance standards 417–18
Arctic warming
observed 76–77*fig*, 78, 82
projected 89, **95**
sea-ice decline 76, 82, 83, 271
Asia–Pacific Economic Cooperation (APEC) 176
Asia–Pacific geopolitical stability 145–50
Asia–Pacific Partnership on Clean Development and Climate 176, 219
atmosphere
accumulation of greenhouse gas concentrations in 25–26
as component of climate system 27
changing temperature over time 25–26, 78
components and temperatures 24, 27, 28
natural greenhouse effect of 24
Australia
as highest per capita OECD emitter 293
as laggard in mitigation efforts 292–93
as projected low-emissions economy 565–78
emissions entitlement 571
emissions reduction targets 209–11, 278–81, 307–08
policies to mitigate climate change 177–79
policy leadership role 223, 302
potential for backstop technology 567–68, 571
preferred climate change policy 310–11
projected effects of mitigation
on industry sectors 571–74
on inflation 568–70
on terms of trade and comparative advantage 576–78
ratification of Kyoto Protocol 180
Australian Capital Territory (ACT) 413
Australian Centre for International Agricultural Research (ACIAR) 226–27, **227**
Australian Community Climate and Earth System Simulator (ACCESS) **366**
Australian climate change policy research institute (proposed) xxxiii, 366*fig*, 368–69
Australian Solar Institute 435
Australia's International Forest Carbon Initiative **238**
aviation emissions 234, 360*n5*, 507, 523, 525

backstop technologies 251, 516, 567–68
Bali Roadmap 175, 280, *608*
Bayesian decision theory 8
Benalla bushfire blackouts **447**
benefits of mitigation xxiii–xxiv, 14–18, 264–68, 301–2
biochar 48, 424, 481, **552**, 553, *609*
biodiesel **552**
biodiversity, climate change impact on 101, 141–44, 377
bioenergy 551–54
biofuels 541, 542–43*tab*, 551–53, **552**
biomass 35, 48, 156, 165, 481, 542, 552, 554
biosequestration
algal 424, **432**, **495**, 552, 567–68, 578
Australia's potential xxxii, 164–65, 542, **557**, 578, 582
barriers 582
defined *609*
effect of Kyoto Protocol on uptake 558
importance to emissions mitigation 559, 582
see also soil sequestration
bounded rationality 408
Brazil, policies to mitigate climate change 180
brown coal electricity generation 398
building and construction industry **259**
Building Australia Fund 451, 457
Building Code of Australia 417
buildings
impact of climate change on 128*tab*, 137–39, 138*tab*, 376–77, **405**
minimum performance standards 417–18
bushfires
emergency management 377
impacts 134, **135**, 139, 262, 559
liabilities for emissions 557
observed increase 112
projections for Australia 118
BushTender project **381**

California
energy efficiency policies 416
Renewable Energy Transmission Initiative 450
Canada, emissions reduction targets 177
cap and trade schemes **178**, 195–97, 309
carbon accounting, for agriculture and forestry xxxii, 165, 558, 582
carbon budget *609*
carbon capture and storage
and future of coal industry 392, 399, 438
capture from the air 48, 567–68
in Australia's national interest 431–32, 567
investment in 219, 223, 399, 438
modelling of 485–89, 496, 499, **583**
research on 432, 500
scale of the task 495
carbon cycle
carbon–climate feedbacks 37
carbon sinks and 35–36
changes since 1750: 36
defined 35, *609*
future reduced absorptive capacity 44*fig*

carbon dioxide
as greenhouse gas 24, 31
as part of carbon cycle 35–36
carbon intensity of energy use 56
concentration in atmosphere 25–26, 33, 45
contribution to atmospheric warming 30*fig*, 37
emissions 31
from fossil-fuel use 55–58
long lifetime 32–33, 43
OECD vs non-OECD countries 56–57
projected growth to 2030: 62–63
recent accelerated growth 55–58
removal by soil 548–49
removal from air 48, 567–68
carbon dioxide equivalent (CO_2-e) 35, 37, *609*
carbon dioxide transportation 453–55
carbon intensity of energy 57, 61, *611*
'carbon leakage' problem 231–32
carbon plantings 542–43*tab*, **552**, 554–56
carbon sequestration *see* biosequestration; geosequestration
Carbon Sequestration Leadership Forum 219
carbon sinks
defined 35–36, *609*
rate of carbon storage over time 36
carbon taxes
advantages 196
border adjustments to compensate domestic industries 233–34
limitations 309
catastrophic events
examples 99–101, 118
likelihood 101–2, 271–72
catastrophic fire weather 118
cattle industry *see* livestock industry
Centre for Australian Weather and Climate Research **367**
China
energy intensity 57–58, **65**
mitigation efforts 179, 293
projected energy consumption growth **65–66**
projected share of global emissions **65**
projections of emissions growth 64, **65–66**
chlorofluorocarbons (CFCs)
accumulation in atmosphere 34
and the Montreal Protocol 34
as a greenhouse gas 31, 34
sources of emissions 32*tab*
Clean Air Act (US) 178
Clean Development Mechanism (CDM) 174–75, *610*
international trade in offset credits under 229–30, 341
limitations and flaws **182**, 183, 229–30
Clean Technology Fund 219
climate–carbon feedbacks
defined 37
uncertainty in projecting 98–99
climate change
abrupt 96–97, *608*
as diabolical public policy challenge xviii–xxi, 287–89
Australia's vulnerability 124–25
challenge for insurance markets 370–73
definitions 27, *610*
extreme outcomes 97–99
high-consequence outcomes 99–101
high risk from xvii
human impact on 26–30, 82–83
impact of aviation on 507
impact on buildings 137–39, **405**, 462
importance of rate of temperature change 39
income distribution effects 400
public attitudes to xviii–xix
climate change (Australia)
observed impact
air and surface temperature 106–07
bushfires 112
cyclones and storms 111–12
drought **108–09**
effect of El Niño 110–11
effect of Southern Annular Mode 111
heatwaves **112**
ocean temperatures 107
rainfall 107–08
streamflows 109
projected impact
direct vs indirect 121
in the 450 and 550 cases 127–28*tab*
no-mitigation vs global mitigation comparisons 125–28, 127–28*tab*
on agriculture 125–28, 127–28*tab*, 129
on buildings in coastal settlements 137–39
on bushfires and heatwaves 117–18
on critical infrastructure 135–39
on cyclones and storms 117
on dryland cropping: wheat 131–33
on ecosystems and biodiversity 141–44
on geopolitical stability in Asia–Pacific region 145–50
on human health 139–41
on irrigated agriculture 129–31
on natural resource–based tourism 133–35
on rainfall 117
on resourced-based industries and communities 128–35
on temperatures 117
on terms of trade 145
state and territory impacts 126*fig*

climate change (global)
observed impact
air and surface temperatures 75, 76, **77**
determining human impact on 82–83
ice caps and ice sheets 81–82
ocean acidity 80
rainfall 81
regional patterns **76–77**
sea level 79–80
severe weather changes 82
projected impact
carbon cycle **95–96**
centrality of temperature changes 84, 87
confidence in climate models 84–85
heatwaves **95**
ice-sheet melting 94, 120
impact of current emissions 85
impact of past vs current actions 85
ocean acidification **96**
rainfall 92–93
risk and uncertainty in 96–101
sea level 93–94
snow and ice **95**
storms **96**
temperature 89–92, 247*tab*
tropical cyclones **96**
use of multiple models to assess 87
climate change mitigation
calculating costs and benefits xxiii–xxv, 1–22, 264–68
case for government intervention 303
conceptual aspect 302–5
costs of delay 2–3
global goals 42
global vs single country action xxvii, 25, 285–87
market vs non-market benefits 9–12, 262
methods and pathways 42–48
preferred approach for Australia 308–13
see also decision-making methodology; emissions trading scheme for Australia; emissions trading schemes; international agreement (post-Kyoto)
climate change policy *see* climate change mitigation
climate-induced migration 149–50
climate information
creation and dissemination for adaptation 365–69
improving communication of **369**
limits to quantity, quality and availability 369
Review's institutional proposals 366–69
climate models *see* modelling (climate change)
climate outcomes
defined **83**
emissions scenarios/cases 85–86
high-consequence 99–101
climate science
atmospheric changes 25–26
carbon–climate feedbacks 36–37
carbon cycle 35–37
changes in greenhouse gas concentrations 25–26
climate sensitivity 38
climate system variability 39–40
definitions of climate change 27
effect of greenhouse gases on energy balance 37–38
factors influencing global warming 29–30
geo-engineering 49
greenhouse gas accumulation in atmosphere 32–36
greenhouse gas stabilisation 43–48
greenhouse gases and temperature rise 37–38
linking emissions and climate change 30–42
natural greenhouse effect 24
natural vs human influences 24–26
need for strengthened research and modelling in Australia 366–69
radiative forcing 37
Review's recommendations 366–69
severity of climate change on Australia 125–28
uncertainty in 2, 23, 84–85, 365
climate sensitivity
defined 38, *610*
influence on temperature outcomes 89, 246, 263, 271
IPCC best-estimate 38, 89
use in Review scenarios 124*tab*
climate system
climate variability 39, 75, 83
components 27
defined *610*
energy balance 27–29, 38
inertia in 42, 46
modelling of 84
slow response of 42, 91
tipping points in 97
warming of 29–30, 78
climate variability
and agriculture 129, 131, 133, 259
and insurance markets 370–71
changes in 41*fig*, 97
defined 39
large-scale patterns 39, **39**, 111
relationship to climate change 27
severe weather events 40, 83, 111, 117

Club of Rome 5, 69
CO_2 *see* carbon dioxide
CO2CRC project **495**
coal
as dominant Australian energy source 158–59
Australia's high emissions intensity 159–60
long-term future 392, 399, 438, 493–96, **583**
near-zero emissions technologies 223, 399, 482, 489, 493, **494–95**
Coal 21 Fund 438
coal consumption
Australian electricity generation 159, 493
China 57–58
increases in 57–58
coal industry and regions, structural adjustment assistance for 396–99
coastal settlements
impact of climate change on 137–39
minimising impacts through planning 461–62
sea-wall provision **462**
commercialisation *see* demonstration and commercialisation
commitment period, defined *610*
committed warming
defined 88, *610*
IPCC estimates 88
compensatory payments
case against 315–16
for coal communities 396–99
for low-income households xxxii, 387–88, 394–96, 569
concentration goals
450 ppm 42–43, 212–13, 279
550 ppm 42–43, 212–13, 279
argument for 400 ppm 42–43
Bali numbers **280**
choice between 450 and 550: 268–70, 271–72, 595–96
concept 193–94
modelling of 193–94, 246–47, 250
reaching agreement on 212–13
relationship to Australian emissions targets 278–81
relationship to global emissions targets 280
relationship to impact goals 193
setting global goals 279
concentration profiles *see* concentration goals
concentration targets
for stabilising emissions 42–45, 193–94
modelling of 193–94, 246–47, 250
overshooting 45–46, 47–48
overshooting profiles 45–46
peaking profiles 46–47
stabilisation scenarios (*see* 450/550 mitigation scenarios)
congestion charges 457–58
conservation forests 542–43*tab*, **557**
construction **259**
Consultative Group for International Agricultural Research 226–27, **227**
'contraction and convergence'
as version of the per capita emissions approach 203–05
assumptions used in modelling **206–07**
defined *610*
coordination externalities 446, 456
'Copenhagen compromise' 282–83, 294–97
Copenhagen conference *see* international agreement (post-Kyoto)
coral reef destruction 100, 142, **143–44**, 271
cost–benefit analysis
methodological overview xxiii–xxv, 1–21
social discount rate in 20
types of costs/benefits 8–13, 253, 258–62, **259–60**
costs of unmitigated climate change 253–63, 254–57*tab*
avoidance of non-market costs under 450 and 550: 270–71
comparisons with earlier estimates 260
costs in 22nd century 262–63
methodology 3–7, 247–50
critical infrastructure
economic effects of climate change 258
impact of climate change on 135–39
impacts criteria 137*tab*
cryosphere changes 81–82
cycling 519–22
cyclones 111–12, 117

Darwin, projected hot days 117
deaths, temperature-related 128*tab*, 141–42
decision-making methodology
assessing non-market benefits 11–13
calculating benefits of mitigation 9–13
calculating costs of mitigation 3–7
confronting risk and uncertainty 7–9, 300–302
effect of adaptation on costs and benefits 13–14
framework for climate change policy 302–05
measuring mitigation benefits against climate change costs xxiii–xxiv, 14–18, 300–02
valuing future vs present welfare 18–21
see also modelling (economic effects)
defence spending **260**
deforestation emissions 235, **237**
delay, costs of xxviii, 2–3, 287–89
demonstration and commercialisation
current programs 434–37
defined 425
dengue virus 147

developed countries
need for immediate leadership from 185
need for International Low-Emissions Technology Commitment 221–23, 429
developing countries
business-as-usual emissions targets for 199
differentiated emissions targets for 185, 198–200
emissions entitlements under 450 and 550 scenarios 205–11
energy intensity 56–57
energy intensity of economic activity 56–57
growth in CO_2 emissions 56–58
land-use change and forestry emissions 235–37
need for financial and technological incentives 18
need for more active abatement effort from 185
one-sided emissions targets for 198–200
role in advancing climate change abatement 185–86
disclosure schemes 410
discount rate 19–21
distribution of income 387–89
see also distributive efficiency
distributive efficiency
assistance through social security and tax 394–95
dual rationale for government assistance 396–99, **397**
household compensation package xxxii, 387–88, 394–98, 569
role for government 393–99
domestic appliances 417–18
droughts
and agriculture 459, 533
as severe weather events 40
causes 81, **108–9**, 110
defined **108**
global projections 93
impacts 125, 133, 149, 377
in Australia 106, 108–9
water supply infrastructure and 458–59
dryland cropping: wheat 131–33

early movers
matched funding 437–41
spillovers from 433–34
early research *see under* research and development
economic growth
and use of fossil fuels 4–5
growth rates 268, 269, 570
resource limits as constraint 69–71
ecosystems
defined *610*
impact of climate change on 141–44, 377
need for adaptive measures 377–78
education and training programs 410–11
effective adaptation *see* adaptation
El Niño – Southern Oscillation
and tropical cyclones 117
as extreme climate outcome 101–2, 271
as large-scale pattern of variability **39**
changes over time 97–98
defined 97, 110, *611*
effect of climate change on 97–98, 113, 117
impact on Australia 39, 108–9, **109**, 110–12, 133
relationship to drought **109**
temporal variability 97–98
uncertainty concerning thresholds 98, **109**
electricity costs, future projections 490
electricity distribution infrastructure
feed-in tariffs 452
positive externalities of embedded generation 452
electricity emissions 160
electricity generation
dependence on coal 159, 493
dependence on water 477
effect of low-emissions coal technology 493
greenhouse gas emissions from 160
impact of transformed energy sector on 491–93
in low-emissions economy 572–74, **583**
non-coal energy options 474–77
projected technology shares 485–87, 489
retrofitting existing plants with new technologies 481, 500
electricity prices 572–73, 575
electricity supply, new technologies 448
electricity transmission infrastructure
barriers to optimal scale 449
current adequacy 447–48
effect of damage to **447**
interconnectors as public goods 446–47
market failures in network extensions 448–49
need for excess capacity 446–47
role for national transmission planner 449–51
emergency management services 381
emissions entitlements
for Australia 209–11, 571, 575
implications for Australia 210–11
national allocations using 'contraction and convergence' **206–7**
need for an internationally tradable system 196

per capita emissions proposal xxiv, 200–205
principles for allocating across countries 200–205
projected allocations 209–10
see also international trading in emissions entitlements
emissions intensity *611*
agriculture 168–70
electricity 160
energy 157–59
emissions limit, defined *611*
emissions pathways 44–47, 86, 91, 113
emissions price
distributive effects on households 387–89
effect on coal-based regions 391–93
effect on industry 390–93
effect on inflation 568–69
effect on innovation 567–68
effect on use of low-emissions products 567–68
pass-down effects 386–87
regressive effects 388–89
spatial variability of effects 388–89
emissions projections scenarios
for Garnaut Review 59–62
IPCC *Special Report on Emissions Scenarios* 58, 62–64, 85, 87–88, 113, 183, 261
post-SRES 58
emissions reduction targets (or trajectories)
approaches 194
Bali negotiatons **280**
conditional and unconditional, for Australia 278–81
Kyoto Protocol 174, 180–81
one-sided for developing countries 198–200
see also interim targets
emissions taxes *see* carbon taxes
emissions trading scheme for Australia
accounting and tax issues 334
auction vs free allocation of permits 330–31
coverage 326–27, 358*tab*, 536
deferred payment for permit purchases 333–34
domestic offsets from non-covered sectors 327, 358*tab*
emissions limit and changes to the limit 325–26, 358*tab*
emissions monitoring, reporting and verification **328**
guiding principles 323–24
impact on energy market 478
impact on household meat consumption 540, 545–48
independent carbon bank 336, 351–52, 352–53*tab*
institutional arrangements 351–53
international linking 337–41, **339–40**, 359*tab*
intertemporal use of permits 334, 335, **337**, 359*tab*
limits on international offsets 340–41
linking with offset markets 340–41
means for lowering costs of meeting targets 334–41
objective 322–23
pass-through of permit value to price of goods and services **332–33**
penalties for non-compliance 328, 359*tab*
permit issue and release 330–34, 358*tab*
point of obligation 327–28, 329–30, 358*tab*, *615*
price ceilings and floors 334–35, 359*tab*
projected impact on agriculture 536, 539–40
projected impact on forestry 540–41
projected impact on gas industry 498
projected impact on transport system 507–08
Review's preferred design 358–59*tab*
role of Mandatory Renewable Energy Target (MRET) xxxii, 353–56, 355*tab*, **356**
rules 323–24
sector-specific issues 328–30
tradability of permits 324, 358*tab*
trade-exposed, emissions-intensive industries and 341–50, **346–47**, 359*tab*
transaction cost of including agriculture in 559
transition from Greenhouse Gas Reduction Scheme to 356–57
transition period (2010 to end 2112) 350–51, 358*tab*
transitional assistance arrangements 341–50, **346–47**
use of revenue from permit sales xxxii–xxxiv, 352*tab*, 388, 395
voluntary market for emissions reductions 357
see also emissions price
emissions trading schemes
as preferred policy for Australia 310–11
as superior to carbon tax 311
avoidance of compensatory payments 315–16
baseline and credit schemes 309–10
cap and trade schemes 309
dangers of compromising with design 314–15
defined 309–10, *611*
economic impact of 311–13
effect on technological development 425–26
hybrid schemes 310
necessary rules for 311
role of public information programs 408–9

emissions trajectory *see* emissions pathways; emissions reduction targets; interim targets
Energy Efficiency Opportunities program 410–11
energy efficiency
  California 416
  defined **404**, *611*
  effect of bounded rationality on 408
  effect of information asymmetry on 407
  potential for 405
  role of government in 405–06
  role of public information programs 408–09
energy efficiency ratings 413
Energy Innovation Fund 435*tab*
energy intensity of economic activity
  Australia compared with OECD countries 158
  China 57–58
  defined *611*
  OECD compared with non-OECD countries 56–57
energy prices
  effect on CO_2 emissions 68–69
  impact on low-income households 389–90
  trends 1970–2008: 67
energy research and development
  case for international public funding 218–23
  falling levels 218
  recent initiatives 219, **219**
  see also International Low-Emissions Technology Commitment
energy sector, Australia
  character and recent developments 469–72
  fuel options 474–77
  future transformation
    economic impacts 490–99
    forces driving 472–78
    phases in 478–82
    risks facing 499–501
    technology scenarios 488–89
  impact of emissions trading scheme 478
  modelling the transformed sector 482–89
energy service contracting 410–11
Energy Technology Innovation Strategy (Victoria) 435*tab*
enforcement mechanisms 238–39
enteric fermentation emissions 163, *611*
equity
  distributive efficiency **393**
  in allocating emissions entitlements 202
  in income distribution 19
  main guarantors 385–86
  see also distributional efficiency; structural adjustment
European Union
  450 ppm CO_2-e stabilisation target 43
  global mean warming target 192
  policies to mitigate climate change 177
evapotranspiration 129, *611*
externalities
  early mover contributions to 433–34
  from early research 428
  from embedded electricity generation 452
  from new transport infrastructure 456
extreme climate outcomes
  assessing the risk of 96–99
  evaluating likelihood of 101
  tipping points 96–97

feed-in tariffs 452
fire weather 118
first-mover disadvantage 448
fluorinated gases 34–35, *611*
food production *see* agriculture
food security, Asia-Pacific region 146, **147**
forest and wooded land
  Australia vs OECD 164, 555
forest sinks 535, 541
forestry
  afforestation and reforestation 550, *608, 613*
  conservation forests 557
  impact of emissions trading scheme on 536, 539, 540–41
  impact of mitigation on 535–36
  inclusion in in emissions trading scheme 330
  plantation and production forests 550–51
  potential for reducing emissions 542–44*tab*
  projected future emissions 539
fossil fuels
  constraints on consumption 4–5, 68–69
  consumption growth 5
  growth of CO_2 emissions from 55–58
  resource availability 4–5, 70–71
  role in Australia's electricity generation 159
  role in Australia's primary energy consumption 158
France, nuclear energy research 433
free-rider problem
  in electricity network extensions 448
  solutions 288
  see also prisoner's dilemma
freight transport 522–25
fuel prices 389
fugitive emissions 329
future vs present utility 18–21

Garnaut Review projections
comparison with existing projections 62–64
Garnaut-Treasury reference case projections 59–62
impact of China **65–66**
methodology and assumptions 59–62
of CO_2-e emissions 61–64, 69–71
of country shares in global output 61
of GDP and GDP per capita 60–61
of global population 59–60
of resource availability 69–71
Platinum Age projections xxv, 59, 61–62, 63
gas transmission infrastructure 453–54
gasification **552**
Generation IV International Forum 219
geo-engineering
advantages and disadvantages 49
defined *611*
proposals 49
geopolitical stability in Asia–Pacific
climate refugees 149–50
food security 146
humanitarian disasters 148
impact of climate change on **260**
infectious diseases 147
sea-level rise 148–49
severe weather events 147–48
water availability **147**
geosequestration
and coal 392, 481–82, **495**
Australia's suitability **432**, 578
defined *611*
infrastructure needs 453–54
role of government in 454–55
technologies **495–96**, 581–82
geothermal technology 219, 273*n*2, 424, 481, 578, 581
glaciers
melting of Himalayan 99, **147**
observed changes 81–82
Global Environment Facility 218
Global Integrated Assessment Model (GIAM) 248
Global Trade and Environment Model (GTEM) 248
global warming potential
defined 35, *612*
goals *see* concentration targets; emissions reduction targets; impact targets
Great Barrier Reef xxvii, 46, 125, 126*fig*, 127*tab* , **143–44**, 262, 271
Green Car Innovation Fund 435*tab*
green credits 395–96
Green Leases 415
Greenhouse Development Rights framework 202
greenhouse effect
and the climate system 28*fig*
defined 24, *612*
greenhouse gas
accumulation in atmosphere 32–35
linking emissions and climate change 30–31
changes in concentration over time 25, 32–35, 37–38
defined *612*
effect on energy balance 28–30, 37–38
global warming potential 35, *612*
lifetime in atmosphere 32, 43–44
projected concentrations 86, 246
radiative forcing 37
sources 31–32*tab*
greenhouse gas emissions (Australia)
emissions profile 153–65, 573*figs*
energy emissions 156–59
from electricity generation 160
future growth 155–56
industry emissions profiles
agriculture 167–70
mining and manufacturing, by sector 167
sectoral shares 165–66
international comparisons 153–54, 156–65, 168–70
livestock emissions 163
other stationary energy 498–99
per capita 153–54
recent growth trends 155
transport emissions 161–62, 505
greenhouse gas emissions (global)
from fossil fuel use 55–58
major emitting countries 54–55
no-mitigation projections 59–64, 69–71
projected country shares 64, 65*tab*
projected growth, China **65–66**
SRES and post-SRES projections 58
urban transport, by mode 520*fig*
Greenhouse Gas Reduction Scheme (NSW) 356–57
greenhouse gas stabilisation *see* stabilisation (of greenhouse gas concentrations)
greenhouse gas stabilisation targets *see* emissions pathways; emissions reduction targets; interim targets
Greenland, melting of ice sheet 94, 102*tab*
Group of Eight (G8) 176
Guidelines for National Greenhouse Gas Inventories (IPCC) 163, 532, 536, 550–51

hailstorms, projections for Sydney 117
health
economic effects of climate change 254–57*tab* , 258
impact of climate change on 123*tab*, 139–41, 147
main health risks in Australia 139
non-market impact of climate change 11
heatwaves
as severe weather events 40
frequency and intensity 82, **95**, **112**
impacts on health 139
projections for Australia 117
high and extreme fire weather 118
high-consequence climate outcomes 99–101
high-speed rail 523–24
historical responsibility 202
hot rocks (geothermal) technology 219, 273*n*2, 424, 482, 578, 581
Hotelling curve 7, 20, 273
household appliances
energy efficiency rating 413
minimum performance standards 417–18
households
compensation package for xxxiii, 387–88, 394–96, 569
effect of information barriers 407–09
energy-efficiency advice and audits 411–12
houses
energy efficiency rating 413
see also coastal settlements
human health *see* health
human-induced climate change
assessing 26–27, 82–83
Australia 106
effect on energy balance of climate system 28–30
humanitarian disasters 148
hurricanes 82
hybrid policies for emissions control *see* cap and trade schemes
hybrid vehicles 513, **519**
hydroelectricity 476–77
hydrofluorocarbons 34

ice caps and ice sheets
observed changes 81–82
projected changes 94, 102*tab*
impact goals 192
India
conflicts with neighbours over water **147**
policies to mitigate climate change 179
Indian monsoon 99–100
Indigenous communities 389, 557
Indonesia
linking with Australian emissions trading scheme **340**
regional partnership with Australia **237–38**
infectious disease 147
information barriers
disclosure schemes as curb 412
education and training as curb 410–11
effect on household demand for energy 389
energy advice and audits as curb 412
in adoption of energy efficient technologies 404
information asymmetry and adverse selection 407
mandatory disclosure as curb 412–13
public information as curb 408–09
undersupply of information 406–07
infrastructure
impact of climate change 135–39, 258
impacts criteria 137*tab*
innovation
Australia as beneficiary of global 567
impact of emissions pricing 566–67
innovation chain 424–25
innovation policy 426–27
see also demonstration and commercialisation; research and development
insurance markets 370–73
insurance value of mitigation 10–11, 271–72
interim targets 282–85
modelling of 294–97
revision of 289–90
International Adaptation Assistance Commitment (proposed), 224–25, 227, 239
international agreement (post-Kyoto)
allocation of emissions entitlements across countries 200–05
arguments for interim targets 288–90
Australia's role, relevance and responsibilities xxx–xxxi, 277–81, 291–94, 307–08, 342–43
based on internationally tradable emissions entitlements 196–97, 227–30
benefits of a comprehensive agreement 285–87
conditional and unconditional interim targets 282–83
defining global goals (impact, concentration, emissions) 192–94
developed country leadership 185, 307
developing country targets 198–200
enforcement mechanisms 239

entitlements based on per capita emissions 202–05
feasibility of agreeing on 550 or 450 target 212–13
graduated levels of national emissions commitments 198–200
implications of failure 579
incorporating international aviation and shipping 234
incorporating land-use change and forestry 234–37
international public funding for adaptation 223–26
international public funding for mitigation 218–23
joint agricultural research to assist developing countries 226–27
modelling a 'Copenhagen compromise' 282–85, 294–97
moving from 550 to 450 290
overcoming free-rider problem 288–89
policy-making challenges 281–82
price-based vs quantitative national commitments 195–97
role for international offsets 341
sectoral agreements for emissions-intensive industries 230–32
trade policy aspects 232–34
international emissions permit trading 174–75
International Low-Emissions Technology Commitment (proposal) 221–23, 429
international mitigation and adaptation efforts
as a 'prisoner's dilemma' 184
funding of adaptation measures **224**
inadequate funding for mitigation 218–19
International Adaptation Assistance Commitment (proposed), 224–25, 227, 239
multilateral
Asia–Pacific Economic Cooperation (APEC) 176
Asia–Pacific Partnership on Clean Development and Climate 176
Bali Roadmap 175
Group of Eight (G8) 176
Kyoto Protocol 174–75
unilateral
Australia 177
Brazil 180
Canada 177
China 179
European countries 177
India 179
Korea 177
New Zealand 177
Papua New Guinea 180
South Africa 180
projected inadequacy for achieving stabilisation targets 183–84
role of unilateral and regional action 186–87, **238–39**
ways to accelerate progress 185–87
international tourism **259**
international trade
effect of climate change on Australia's 145
international trading in emissions entitlements
benefits and risks 227–28
options for 229
trade in offset credits 229–30
under Australian mitigation 571, 575
IPCC (Intergovernmental Panel on Climate Change)
asserts global warming is 'unequivocal' 75
Bayesian approach 9
default values 165
emissions projections (scenarios) 58, 62–64, 85, 87–88, 113, 183, 261
radiative forcing from aviation 234, 507
treatment of uncertainty 9
underestimate of growth of emissions 261
IPCC Fourth Assessment Report
changes to El Niño – Southern Oscillation 98
conclusions from 26, 75
definitions of climate change 27
estimate of climate sensitivity 38, 89, 246, 271–72
estimate of committed warming 88
estimates of global emissions 53–54
estimates of land-use change and forestry emissions 235
estimates of sea-level rise 93–94
estimates of temperature increases 261
global mitigation 405
human impact on climate 26
Review's treatment of 24, **87**, 121
stabilisation scenarios 206, 211
temperature baselines **87**
IPCC *Guidelines for National Greenhouse Gas Inventories* 163, 532, 536, 550–01
IPCC *Special Report on Emissions Scenarios* 58, 62–64, 85, 87–88, 113, 183, 261
Ireland, agricultural emissions 163–64
irrigated agriculture
impact of climate change on 127*tab*, 129–31, 377
need for adaptive changes 377

Japan
Clean Technology Fund 219
contracts for vending machines 415
nuclear energy research 433
policies to mitigate climate change 177
jatropha curcas **552**
Joint Implementation (Kyoto Protocol) 174–75

Kakadu xxvii, 126*fig*, 142, **259**
kangaroos, for meat production 540, 547–48
knowledge externalities 433–34
Korea 177
Kyoto Protocol
and global warming potential 35
as basis for new international action 185
emissions budgets for Annex I countries 201
enforcement mechanism 239
'flexibility where' mechanisms 174–75
ratification by Australia 180
non-ratification by US 178, 180, 181, 182
omission of harvested wood emissions 550
progress towards reaching targets 180–82
technology transfer 218
see also Clean Development Mechanism (CDM); international agreement (post-Kyoto)

land management
in emissions trading scheme 330
land use
inclusion in emissions trading scheme 330
land-use zoning 461
potential for reducing emissions **542–44**, 548–50
reliability of emissions measurement 558
types and areas 554*tab*
land-use change
emissions accounting rules 558–59
emissions from 235–37, 541
Latrobe Valley 398, 493
Least Developed Countries Fund **224**
Lieberman-Warner Climate Security Act (US) **178**
livestock industry
impact of climate change on 547
sheep and cattle emissions 163–64, 540, 544–45
local government planning 460
low-emissions products
minimum performance standards 415–16
principal–agent problems in uptake 413–15
low-emissions technologies
Australian early research 428–29
current demonstration and commercialisation programs 434–36
defined 442n1
need for specialist Australian research body 430
low-emissions vehicles
cost competitiveness 513, **519**
modelling transition to 515–16, 518
price distortions on 526–27
projected takeup 390, 508, 515–16, 518, 524
types and emissions from **519**
Low Emissions Technology Demonstration Fund 435*tab*
low-income households
impact of emissions price on 387–90
subsidised energy-efficiency audits 412
targeted assistance for 395–96

Major Economies Meeting on Energy Security and Climate Change 176
malaria 147
mallee eucalypts 553
Mandatory Renewable Energy Target (MRET) xxxii, 354–56
manufacturing industry
emissions 166–67, 573*fig*
impact of mitigation 266
marginal elasticity of utility of consumption 19
market-based policy mechanisms
argument against exemptions 314–17
argument against supplementary regulation 317–18
with measures to correct market failures 318
market failures
climate change risks as 299
government role in correcting 318, 426–27, **427**, 454–56
in commercialisation of innovation 433–34
in early research phase of innovation chain 426, 428–29
in provision of infrastructure 445–46, 455–56
see also agency barriers; information barriers; network infrastructure
market uptake 425
Marrakesh Accords 537, 558–59, *612*
marsupials 547–48

matched funding
  as compensation for early-mover spillovers 437–38
  criteria 439–40
  funding sources 438–39
  mechanisms for 437*tab*
  private and social rates of return 440–41
meat prices and consumption 540, 545–48
metals, availability 70–71
methane
  accumulation in atmosphere 33
  and carbon cycle 35
  and radiative forcing 30*fig*, 37
  and stratospheric water vapour 34
  as a greenhouse gas 24, 31
  carbon-climate feedbacks 37, 98
  concentration in atmosphere 24, 26*fig*, 33
  emissions from livestock 163, 540, 544–45
  global emissions of 53
  lifetime in atmosphere 32, 43, 45
  abatement options 559
  projected emissions 61
  sources of emissions 31*tab*, 163, 329
  warming influence 45, 91
Methane to Markets Partnership **219**
minerals, availability 70–71
minimum performance standards
  and principal–agent problem 415–16
  criteria 416
  for appliances 417–18
  for buildings 417–18
mining
  economic effects of climate change 258
  emissions from 167, 573*fig*
  impact of mitigation 266
mitigation
  case for government action 252, 303
  defined *612*
  net benefits of xxiii–xxv, 14–18, 264–68
  science of 42–48
  see also climate change mitigation
mitigation potential 405
modelling (climate change)
  Australian Community Climate and Earth System Simulator (ACCESS) **367**
  confidence in the models 84–85
  key role of temperature change 84, 87
  problems and uncertainties in 84–85
  use of multiple models 84
modelling (economic effects)
  limitations 305–06
  methodology
    assumptions about technology 250–51, 306, 488–89, 511, 515–16, 579, 580–82
    assumptions for 'Copenhagen compromise' and 'waiting game' scenarios **294–95**
    comparing costs of climate change and mitigation 14–18, 247–50, 252
    discount rates 269–70
    Garnaut/Treasury reference case 59–62
    inclusion of Type 2 costs of climate change **259–60**
    mitigation scenarios (450 and 550) 246
    modelling long-term costs and benefits 249–50
    modelling the mitigation scenarios 250–52, 268–72
    models used **248**
    types of costs (quantifiable and non-quantifiable) 247, 249–50
  results
    agriculture and forestry 537–41
    alternative technology scenarios 306
    carbon prices 251, 296
    comparison of 450 and 550 scenarios 268–72
    concentration goals 193, 246–47, 250
    'Copenhagen compromise' 282, 294–96
    cost of unmitigated climate change 253–63, 254–57*tab*
    costs of meeting interim targets 295–96
    costs of mitigation (gross and net) 264–68
    distributional effects of emissions price 387–89
    emissions allocations under a per capita approach 205–11
    energy sector 482–87
    'enhanced technology' scenario **583**
    expanded Mandatory Renewable Energy target **356**
    nuclear energy 488
    sectoral share of emissions reductions 571–74
    trading of emissions entitlements 571, 575
    transport sector 511–16
Monash Multi Regional Forecasting (MMRF) model **248**
monsoons 99–100
Montreal Protocol 33, 34, *613*
Murray-Darling Basin
  and Southern Annular Mode 111
  competition for water 532
  decrease in streamflow 109
  impact of climate change on 122*tab*, 125, 126, 127*tab*, 129–31, 258
  possibility of a wetter Basin **131**

National Electricity Market 446–47, 464*n*1
National Greenhouse and Energy Reporting Act 2007 **328**
National Low Emissions Coal Fund 435*tab*
natural resource–based tourism 127*tab*, 133–35
network infrastructure
  market failure in 445–46, 455–56
  see also carbon dioxide transportation; electricity distribution infrastructure; electricity transmission infrastructure; gas transmission infrastructure; transport infrastructure; urban infrastructure; water supply infrastructure
New South Wales
  Greenhouse Gas Reduction Scheme 356–57
  projected climate change impact 126*fig*
New Zealand
  agricultural emissions 163–64
  linking with Australian emissions trading scheme **339**
  mitigation policies 177
Ningaloo Reef xxvii, 11, 125, 133
nitrous oxide
  and radiative forcing 30*fig*, 37
  as a greenhouse gas 24, 31, 33
  concentration in the atmosphere 25–26, 33
  global emissions 54
  lifetime in the atmosphere 43
  mitigation and measurement 543 *tab*, 544, 550, 558
  sources of emissions 31*tab*, 544, 550
Northern Territory
  projected climate change impact 126*fig*, 141, 142
  tourism 134
  savanna fires 557
nuclear energy
  demand for 475
  in Australia 476
  modelling of 477, 488
  research 433

oceans
  acidification 80, **96**, 100
  and carbon–climate feedbacks 98
  and El Niño – Southern Oscillation 98
  and geo-engineering 49
  and precipitation 81, 92, 115
  as a carbon sink 35–37
  as part of carbon cycle 35–37, 43, 95, 98
  as part of climate system 27–28, **39**
  observed changes 79–80
  sea level (*see* sea-level rise)
  warming
    around Australia 107
    inertia in 42, 91
    projected 89
    slow rate of stabilisation 42, 91
OECD countries
  carbon intensity of energy use 56
  growth in CO_2 emissions 56–57
offset credits 229–30
oil
  impact of high prices on Australian transport system 506–10
  impact of high prices on global emissions 67–69
  prices, 1970 to 2008: 67–68
  resource supply limits 70–71
outer suburban households 390, 396
overshooting 46–48, *613*
ozone
  as a greenhouse gas 24, 31
  stratospheric ozone 27, 30*fig*
    and Montreal Protocol 33
    destruction by CFCs 34
    success in combatting depletion 33, 184
  tropospheric ozone
    accumulation in atmosphere 35
    and precursor species 35
    sources 32*tab*

Pacific countries
  impact of sea-level rise 149
Papua New Guinea
  and Australian emissions trading scheme **340**
  deforestation emissions **238**
  impact of sea-level rise 149
  policies to mitigate climate change 180
  regional partnership with Australia **237–38**
passenger transport 504*tab*, 518–22, 523–24, 525
peaking profile 47, *613*
per capita emissions entitlements
  advantages 202–05
  'contraction and conversion' approach 203–05, **206–07**
  implications for Australia 204–05, 209–11
  modelling national emissions allocations under 205–11
Perth
  TravelSmart **409**
  water supply 109, 110*fig*, 136
Pew Center Pocantico Dialogue 201
pipeline system fugitive emissions
  in emissions trading scheme 329

'Platinum Age'
defined 21*n*1
projections xxv, 59, 61–62, 63
price-based emissions control
arguments for and against 195–97 (*see also* carbon taxes)
primary energy, Australia 158–59
primary energy consumption
Australia, by sector 159
principal–agent problems
contractual relationships as solution 415
defined 404
households and rising energy prices 389–90
in adoption of energy efficiency 404
in uptake of low-emissions products 413–15
mitigated through minimum performance standards 415–16
typology 414*tab*
prisoner's dilemma xviii, 184, 288–89
defined 187*n*7
projections *see* emissions projections scenarios
public goods
electricity transmission interconnectors as 446–47
research and information 220, 365, 406–07, 428
road, bicycle and walking infrastructure 456
public-housing renters 396
public information programs 408–09
public transport 390, 396, 510, 519–21

quantity- and price-based emissions control 195–97
Queensland, projected climate change impact 126*fig*, 128*tab*
Queensland Future Growth Fund 435*tab*

radiative forcing
and climate sensitivity 38
defined 37, *613*
from aviation 507
of long-lived greenhouse gases 37–38
rail transport 505, 511–13, 521, 523–24
rainfall
and drought **108–09**
and inflow into river systems 129
effect on streamflows 109
observed changes in (Australia) **108–09**
observed changes in (global) 81
projections for Australia 115–16
temporal variations 116
rapid climate change *see* abrupt climate change
rate of pure time preference 19
reference case
defined *613*
Garnaut-Treasury reference case projections 59–62
reforestation 330
refrigerators and freezers, performance standards 418
refugees 147–50
regional industries
structural adjustment assistance for 396–99
vulnerability of 400
regional partnerships **237–38**
regulatory and legal spillovers 434
renewable energy
and electricity generation 354–56, 476–77, 485–87
Australia's opportunities 478, 481–82, 578
future demand for 69
growth in consumption 68
investment in 219, 472
Mandatory Renewable Energy Target (MRET) xxxii, 317, 472, 354–56
national targets 177, 179, 180, 354–56
projected future use 485–87, 489, 498, 581, 583
research and development **219**, 239*n*1, 435*tab*
see also geothermal; hot rocks; solar energy; tidal power; wave power; wind power
Renewable Energy and Energy Efficiency Partnership (REEEP) 219
Renewable Energy Development Initiative 435*tab*
Renewable Energy Fund 435*tab*
Renewable Energy Transmission Initiative (California) 450
rent-seeking 297, 310, 315, 322, 331
research and development
early-mover spillovers 433–34
early research
additional funding for 430
Australian funding levels 428–29
defined 425
ensuring optimal levels 428–30
funding criteria 431–33, **432**
new institutions for 430
market failures in 426–28, 433–34
technological lock-in barriers 441–42
see also demonstration and commercialisation; energy research and development
resource limits
on fossil fuel use 4–6
to growth 69–71

risk
assessment of 96–101
defined 7
extreme climate outcomes 96–101
road and bridge maintenance **259**
Ross River virus 147
Rural R&D Corporations program 436

savanna, mitigation potential 543*tab*, 557
science of climate change *see* climate science
sea-level rise
accelerated 94
historical rates of 94
ice-melt contribution to 80, 93, 94
impact on Asia–Pacific region 148–49
impact on ecosystems 142
IPCC projections 93–94
observed changes in global 79–80
thermal expansion contribution to 79–80, 93
sea walls **462**
sequestration
bringing emissions below natural level of 47–48
defined *614*
technologies 495
see also biosequestration; geosequestration; soil sequestration
severe weather events
as security challenge in Asia–Pacific 147–48
defined 40, *614*
implications for insurance markets 371, 400
observed global changes 82
sheep and cattle *see* livestock industry
shipping 234, 505, 512–13, 524–25, 526
'shocks'
defined 302
mitigation and adaptation 303*tab*
skills spillovers 434
social acceptance spillovers 434
social security and tax systems 394–95
soil carbon 548, 550
soil management, mitigation potential 543*tab*, 548–50
soil sequestration 548–49
Solar Cities (program) 435*tab*, 481
solar energy 424, 478, 481, 482, 578, 581
solar radiation 29, *614*
South Africa, national climate change policy 180
South Australia, projected climate change impact 126*fig*
South-North Dialogue 201
Southern Annular Mode
effect on Australian rainfall **109**, 111
effect on southern hemisphere climate **39**
Southern Oscillation Index *see* El Niño – Southern Oscillation
Special Report on Emissions Scenarios (IPCC) 58, 62–64, 85, 87–88, 113, 183, 261
species extinction 101, 141–42, 271
spillovers
from early movers 433–34
from early research 428
stabilisation (of greenhouse gas concentrations)
Article 2 (UNFCCC) 42
defined *614*
European Union target (*see* 450/550 mitigation scenarios)
feasibility of targets 45–47
overshooting profiles 46–48
stabilisation scenarios (*see* 450/550 mitigation scenarios)
targets 42–43
UNFCCC objective 42
stabilisation targets *see* concentration goals
Stern Review
discount rate 19
economic impact of climate change 263
effect of climate change on global GDP 260
emissions projections 58
stratosphere 27, *614*
stratospheric ozone *see* ozone
streamflows **108**, 109, 377
structural adjustment assistance
coal-generation regions 396–99, 493
equity plus efficiency rationale for 396–97
inadequate arguments for **397**
through social security and tax systems 394–95
sulphate aerosol emissions 85
support sector externalities 434
Sydney, projected hailstorms 117
'syngas' **552**

tailored information 408–10
targeted assistance *see* structural adjustment assistance
targets *see* concentration goals; emissions reduction targets
tariffs 233–34
see also feed-in tariffs
tariffs on vehicles 526
Tasmania, projected climate change impact 126*fig*
technological lock-in 441
technology
effect of carbon price on 425–26
modelling assumptions about 250–51, 306, 488–89, 511, 515–16, 579, 580–82

see also biosequestration; energy research and development; research and development; technology transfer; zero-emissions coal

technology transfer
- international initiatives 218–19
- near-zero emissions coal technologies 223
- under International Low-Emissions Technology Commitment 223
- under UNFCCC and Kyoto Protocol 218

temperature increases
- committed warming 88, *610*
- global mean temperatures, post-2030: 88–89
- observed (global) 75, 76, **77**
- positive feedback effects 263
- relationship to changes in climate system 87

temperature projections
- extreme responses to stabilisation targets 89
- for Australia 113–14
- post-2030 88–89
- post-2100 91
- short-term responses to stabilisation targets 91
- spatial variation 89

temperature reference point **87**, *614*

temperature-related deaths 141–42

terms of trade
- climate change impact on Australia's 145, 258
- climate change impact on other countries' 258
- economic effects of climate change 258
- effects of mitigation on 266, 570, 576

thermal expansion
- contribution to sea-level rise 79–80, 93
- defined *614*

thermohaline circulation
- defined *614*

thresholds *see* tipping points

Tibetan glaciers **147**

tidal power 448

tipping points
- defined 96–97, *615*
- for extreme and high-consequence outcomes 101–2

tourism 133, **134–35**, **143–44**

trade-exposed, emissions-intensive industries *615*
- adjustment assistance **295**, 297, 316–17, 344–50, **346–47**
- sectoral agreements for 230–32, 342–43

trade policy 232–34

transition economies *see* developing countries

transport
- current role and structure 504–05
- emissions 161–62, 505
- future developments
  - causes of the transformation 505–11
  - economic modelling of 511–16
  - emissions reductions 512–14, 517, 522
  - expected cost of transition 525–26
  - fuel-efficient vehicles 506–07, 513–14, 518, **519**, 523–25
  - fuel substitution 508, **519**, 523–24
  - impact of carbon price 507–08, 512
  - mode shift 509–10, 517, 518–21, 522–23
  - public transport 390, 396, 510–11, 519–22, 520*fig*
  - role of governments in 510–11, 518–22, 523–24, 526–27
  - transport demand 509, 518–22, 523–24
- in emissions trading scheme 329
- information about options **409**
- transport sector typology 504*tab*

transport energy intensity of the economy 161

transport infrastructure
- biases in expenditure on 456–57
- need for coordinated long-term planning 458
- prices for transport use 457–58
- role of governments 456, 510–11, 522, 526

TravelSmart **409**

Treasury *see under* Garnaut Review projections

tropical cyclones
- Australia 111–12
- economic effects of climate change 258–59
- projections 117

tropical deforestation 235

tropical storms
- global projections **96**
- observed changes 82

troposphere 27, *615*

tropospheric ozone *see* ozone

uncertainty
- and extreme climate outcomes 97–99
- and insurance value of mitigation 10
- and the delusion of delay 287
- climate–carbon feedbacks 98–99
- climate sensitivity 24
- defined 8–9
- future technologies 581
- in climate change modelling 84–85
- in climate science 24
- in decision making 7–9, 300–02
- in operations of mitigation regime 580

United Kingdom
- Carbon Trust 410
- Clean Technology Fund 219

United Nations Framework Convention on Climate Change (UNFCCC) 175
  Article 2 42
  capacity-based emissions entitlements 201
  definition of climate change 27
  emissions reduction goal 212
  goal for global mitigation 42
  principle of 'common but differentiated responsibility' 221
  stabilisation objective 42
United States
  Clean Technology Fund 219
  need for a credible long-term target 185
  non-ratification of Kyoto Protocol 178, 180
  policies to mitigate climate change 177, **178**
urban densities 510–11, 526
urban planning 460–63
  adaptation measures 378, 460–61
US Energy Information Administration 58
US Environmental Protection Agency **178**
utility, present vs future 18–21

vested interests xxi, 314–16, 580
  see also rent-seeking
Victoria, projected climate change impact 126*fig*

walking and cycling 518–22
waste emissions 329
water availability in Asia 145, **147**
water efficiency 404
water markets
  needed reforms 459
  Review's proposals for distortion removal 374–75
  role in adaptation 373–75
water supply infrastructure
  adequacy of current arrangements 459
  future electricity demand 477–78
  impact of climate change on 135–36
  need for adaptive measures 378
water vapour
  accumulation in atmosphere 36
  as a greenhouse gas 32
West Arnhem Land Fire Abatement project 557
Western Australia, projected climate change impact 126*fig*
wheat crops 131–33
wind power 354, **356**, 431–32, 448, 481, 585
World Trade Organization 231–33

zero-emissions coal technologies 493–96, **494–95**

Printed in the United States
By Bookmasters